CW01213273

A COMPUTABLE UNIVERSE
Understanding and Exploring Nature as Computation

A COMPUTABLE UNIVERSE
Understanding and Exploring Nature as Computation

Foreword by
SIR ROGER PENROSE

Editor
Hector Zenil
University of Sheffield, UK
& Wolfram Research, USA

World Scientific

Published by

World Scientific Publishing Co. Pte. Ltd.
5 Toh Tuck Link, Singapore 596224
USA office: 27 Warren Street, Suite 401-402, Hackensack, NJ 07601
UK office: 57 Shelton Street, Covent Garden, London WC2H 9HE

British Library Cataloguing-in-Publication Data
A catalogue record for this book is available from the British Library.

A COMPUTABLE UNIVERSE
Understanding and Exploring Nature as Computation

Copyright © 2013 by World Scientific Publishing Co. Pte. Ltd.

All rights reserved. This book, or parts thereof, may not be reproduced in any form or by any means, electronic or mechanical, including photocopying, recording or any information storage and retrieval system now known or to be invented, without written permission from the Publisher.

For photocopying of material in this volume, please pay a copying fee through the Copyright Clearance Center, Inc., 222 Rosewood Drive, Danvers, MA 01923, USA. In this case permission to photocopy is not required from the publisher.

ISBN 978-981-4374-29-3

To Elena

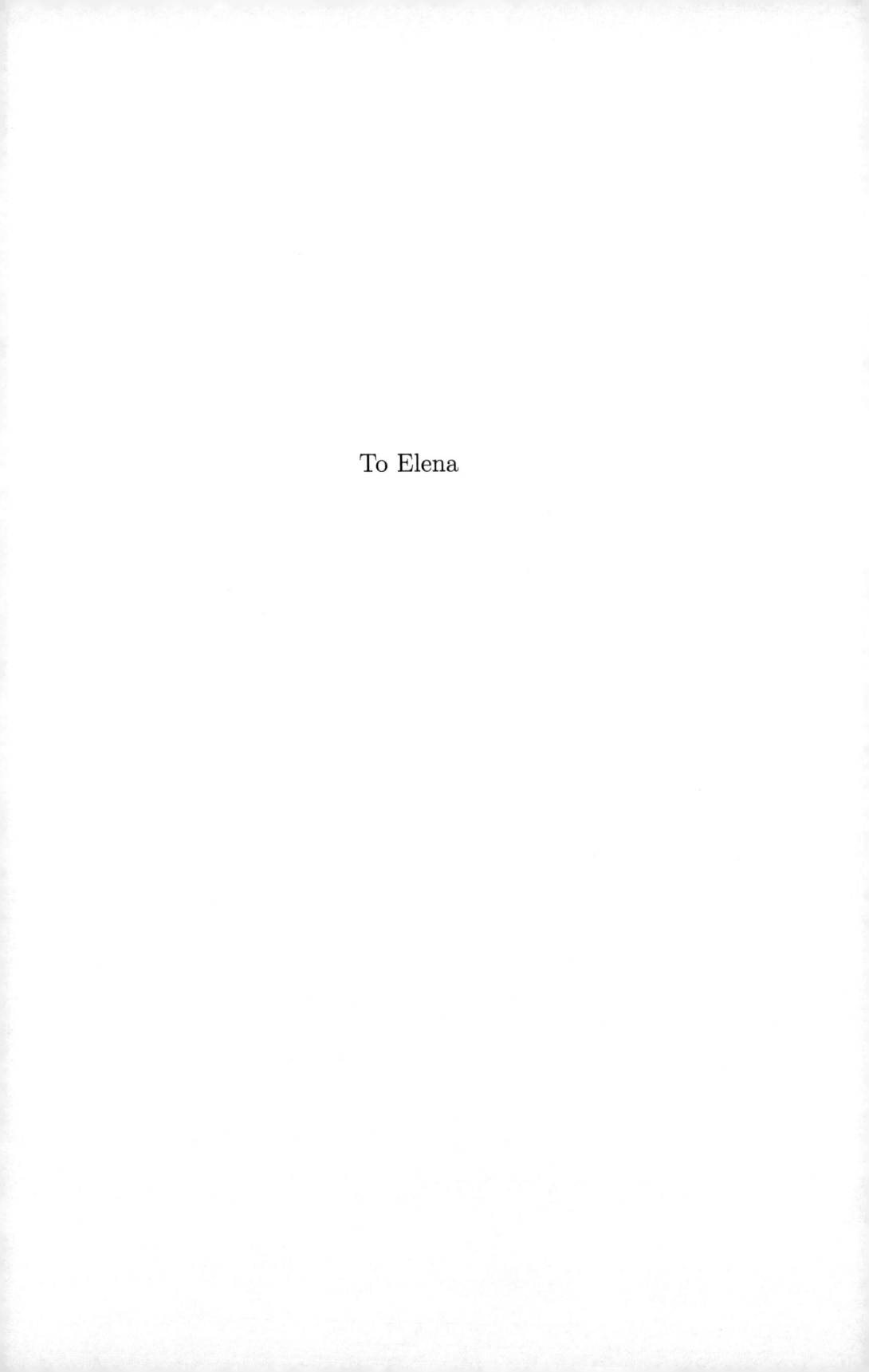

Contents

Foreword   xiii
     R. Penrose

Preface   xxxvii

Acknowledgements   xliii

1. Introducing the Computable Universe   1
     H. Zenil

Historical, Philosophical & Foundational Aspects of Computation   21

2. Origins of Digital Computing: Alan Turing, Charles Babbage, & Ada Lovelace   23
     D. Swade

3. Generating, Solving and the Mathematics of Homo Sapiens. E. Post's Views on Computation   45
     L. De Mol

4. Machines   63
     R. Turner

5.	Effectiveness *N. Dershowitz & E. Falkovich*	77
6.	Axioms for Computability: Do They Allow a Proof of Church's Thesis? *W. Sieg*	99
7.	The Mathematician's Bias — and the Return to Embodied Computation *S. B. Cooper*	125
8.	Intuitionistic Mathematics and Realizability in the Physical World *A. Bauer*	143
9.	What is Computation? Actor Model versus Turing's Model *C. Hewitt*	159

Computation in Nature & the Real World — 187

10.	Reaction Systems: A Natural Computing Approach to the Functioning of Living Cells *A. Ehrenfeucht, J. Kleijn, M. Koutny & G. Rozenberg*	189
11.	Bacteria, Turing Machines and Hyperbolic Cellular Automata *M. Margenstern*	209
12.	Computation and Communication in Unorganized Systems *C. Teuscher*	231
13.	The Many Forms of Amorphous Computational Systems *J. Wiedermann*	243

14.	Computing on Rings G. J. Martínez, A. Adamatzky & H. V. McIntosh	257
15.	Life as Evolving Software G. J. Chaitin	277
16.	Computability and Algorithmic Complexity in Economics K. V. Velupillai & S. Zambelli	303
17.	Blueprint for a Hypercomputer F. A. Doria	333

Computation & Physics & the Physics of Computation   **345**

18.	Information-Theoretic Teleodynamics in Natural and Artificial Systems A. F. Beavers & C. D. Harrison	347
19.	Discrete Theoretical Processes (DTP) E. Fredkin	365
20.	The Fastest Way of Computing All Universes J. Schmidhuber	381
21.	The Subjective Computable Universe M. Hutter	399
22.	What Is Ultimately Possible in Physics? S. Wolfram	417
23.	Universality, Turing Incompleteness and Observers K. Sutner	435

24. Algorithmic Causal Sets for a Computational Spacetime 451
   T. Bolognesi

25. The Computable Universe Hypothesis 479
   M. P. Szudzik

26. The Universe is Lawless or "Pantôn chrêmatôn metron anthrôpon einai" 525
   C. S. Calude, F. W. Meyerstein & A. Salomaa

27. Is Feasibility in Physics Limited by Fantasy Alone? 539
   C. S. Calude & K. Svozil

The Quantum, Computation & Information **549**

28. What is Computation? (How) Does Nature Compute? 551
   D. Deutsch

29. The Universe as Quantum Computer 567
   S. Lloyd

30. Quantum Speedup and Temporal Inequalities for Sequential Actions 583
   M. Żukowski

31. The Contextual Computer 595
   A. Cabello

32. A Gödel-Turing Perspective on Quantum States Indistinguishable from Inside 605
   T. Breuer

33. When Humans Do Compute Quantum *P. Zizzi*	617

Open Discussion Section — **629**

34. Open Discussion on *A Computable Universe*
   A. Bauer, T. Bolognesi, A. Cabello, C. S. Calude,
   L. De Mol, F. Doria, E. Fredkin, C. Hewitt, M. Hutter,
   M. Margenstern, K. Svozil, M. Szudzik, C. Teuscher,
   S. Wolfram & H. Zenil
	631

Live Panel Discussion (transcription) — **671**

35. What is Computation? (How) Does Nature Compute?
   C. S. Calude, G. J. Chaitin, E. Fredkin, A. J. Leggett,
   R. de Ruyter, T. Toffoli & S. Wolfram
	673

Zuse's Calculating Space — **727**

36. Calculating Space (*Rechnender Raum*) *K. Zuse*	729
Afterword to Konrad Zuse's Calculating Space *A. German & H. Zenil*	787
Index	795

Foreword

Roger Penrose
Mathematical Institute
University of Oxford, UK

I am most honoured to have the privilege to present the Foreword to this fascinating and wonderfully varied collection of contributions[a], concerning the nature of computation and of its deep connection with the operation of those basic laws, known or yet unknown, governing the universe in which we live. Fundamentally deep questions are indeed being grappled with here, and the fact that we find so many different viewpoints is something to be expected, since, in truth, we know little about the foundational nature and origins of these basic laws, despite the immense precision that we so often find revealed in them. Accordingly, it is not surprising that within the viewpoints expressed here is some unabashed speculation, occasionally bordering on just partially justified guesswork, while elsewhere we find a good deal of precise reasoning, some in the form of rigorous mathematical theorems. Both of these are as should be, for without some inspired guesswork we cannot have new ideas as to where look in order to make genuinely new progress, and without precise mathematical reasoning, no less than in precise observation, we cannot know when we are right—or, more usually, when we are wrong.

The year of the publication of this book, 2012, is particularly apposite, in being the centenary year of Alan Turing, whose theoretical analysis of the notion of "computing machine", together with his wartime work in deciphering Nazi codes, has had a huge impact on the enormous development of electronic computers, and on the consequent influence that these devices

---

[a]Footnotes to names in the next pages are pointers to the chapters in this volume (A Computable Universe by H. Zenil), Ed.

have had on our lives and on the way that we think about ourselves. This impact is particularly evident with the application of computer technology to the implications of known physical laws, whether they be at the basic foundational level, or at a larger level such as with fluid mechanics or thermodynamics where averages over huge numbers of elementary constituent particles again lead to comparatively simple dynamical equations. I should here remark that from time to time it has even been suggested that, in some sense, the "laws" that we appear to find in the way that the world works are *all* of this statistical character, and that, at root, there are "no" basic underlying physical laws (e.g. Wheeler's "law without law",[43] Sakharov's ideas of "induced gravity",[31] etc., and we find this general type of view expressed also in this volume also[b]). However, I find it hard to see that such a viewpoint can have much chance of yielding anything like the enormously precise non-statistical dynamics[32] and great mathematical sophistication that we find in so much of 20th century physics. This point aside, we find that in reasonably favourable circumstances, computer simulations can lead to hugely impressive imitations of reality, and the resulting visual representations may be almost indistinguishable from the real thing, a fact that is frequently made use of in realistic special effects in films, as much as in serious scientific presentations. When we need precision in particular implications of such equations, we may run into the difficult issues presented by chaotic behaviour, whereby the dependence on initial conditions becomes exponentially sensitive. In such cases there is an effective randomness in the evolved behaviour. Nevertheless, the computational simulations will still lead to outcomes that would be physically allowable, and in this sense provide results consistent with the behaviour of reality.

Computational simulations can have great importance in many areas other than physics, such as with the spread of epidemics, or with economics (where the mathematical ideas of game theory can play an important role),[c] but I shall here be concerned with *physical* systems, specifically. The impressiveness of computational simulations is often most evident when it is simply 17th century Newtonian mechanics that is involved, in its enormously varied different manifestations. The implications of Newtonian dynamical laws can be extensively computed in the modelling of physical systems, even where there may be huge numbers of constituent particles, such as atoms in a simplified gas, or particle-like ingredients, such as stars in globular clusters or even in entire galaxies. It may be remarked that

---

[b]see Calude.
[c]Velupillai.

computational simulations are normally done in a time sense where the future behaviour is deduced from an input which is taken to be in the past. In principle, one could also perform calculations in the reverse "teleological" direction, because of the time-reversibility of the basic Newtonian laws.[d] However, because of the second law of thermodynamics, whereby the entropy (or "randomness") of a physical system increases with time in the natural world, such reverse-time calculations tend to be untrustworthy.

When Newtonian laws are supplemented by the Maxwell-Lorentz equations, governing the behaviour of electromagnetic fields and their interactions with charged material particles, then the scope of physical processes that can be accurately simulated by computational procedures is greatly increased, such as with phenomena involving the behaviour of visible light, or with devices concerned with microwaves or radio propagation, or in modelling the vast galactic plasma clouds involving the mixed flows of electrons and protons in space, which can indeed be computationally simulated with considerable confidence.

This latter kind of simulation requires that those physical equations be used, that correctly come from the requirements of special relativity, where Einstein's viewpoint concerning the relativity of motion and of the passage of time are incorporated. Einstein's special relativity encompassed, encapsulated, and superseded the earlier ideas of FitzGerald, Lorentz, Poincaré and others, but even Einstein's own viewpoint needed to be reformulated and made more satisfactory by the radical change of perspective introduced by Minkowski, who showed how the ideas of special relativity come together in the natural geometrical framework of 4-dimensional space-time. When it comes to Einstein's *general* relativity, in which Minkowski's 4-geometry is fundamentally modified to become *curved*, in order that gravitational phenomena can be incorporated, we find that simulations of gravitational systems can be made to even greater precision than was possible with Newtonian theory. The precision of planetary motions in our Solar System is now at such a level that Newton's 17th century theory is no longer sufficient, and Einstein's 20th century theory is needed. This is true even for the operation of the global positioning systems that are now in common use, which would be useless but for the corrections to Newtonian theory that general relativity provides. Indeed, perhaps the most accurately confirmed theoretical simulations ever performed, namely the tracking of double neutron-star motions, where not only the standard general-relativistic corrections

---

[d] Beavers.

(perihelion advance, rotational frame-dragging effects, etc.) to Newtonian orbital motion need to be taken into account, but also the energy-removing effects of gravitational waves (ripples of space-time curvature) emanating from the system can be theoretically calculated, and are found to agree with the observed motions to an unprecedented precision.

The other major revolution in basic physical theory that the 20th century brought was, of course, *quantum mechanics*—which needs to be considered in conjunction with its generalization to quantum *field* theory, this being required when the effects of special relativity have to be taken into account together with quantum principles. It is clear from many of the articles in this volume, that quantum theory is (rightly) considered to be of fundamental importance, when it comes to the investigation of the basic underlying operations of the physical universe and their relation to computation. There are many reasons for this, an obvious one being that quantum processes are undoubtedly fundamental to the behaviour of the tiniest-scale ingredients of our universe, and also to many features of the collective behaviour of many-particle systems, these having a characteristically quantum-mechanical nature such as quantum entanglement, superconductivity, Bose-Einstein condensation, etc. However, there is another basic feature of quantum mechanics that may be counted as a reason for regarding this scheme of things as being more friendly to the notion of computation than was classical mechanics, namely that there is a basic *discreteness* that quantum mechanics introduces into physical theory. It seems that in the early days of the theory, much was made of this discreteness, with its implied hope of a "granular" nature underlying the operation of the physical world. A hope had been expressed[30,32] that somehow the domination of physical theory by the ideas of *continuity* and *differentiability*—which go hand-in-hand with the pervasive use of the real-number system—might have at last been broken, via the introduction of quantum mechanics. Accordingly, it was hoped that the ideas of discreteness and combinatorics might soon be seen to become the dominant driving force underlying the operation of our universe, rather than the continuity and differentiability that classical physics had depended upon for so many centuries. A discrete universe is indeed much more in harmony with current ideas of computation than is a continuous one, and many of the articles in this volume[e] argue powerfully from this perspective, and particularly in the context of *cellular automata*[f].

---

[e]Bolognesi, Chaitin, Wolfram, Fredkin, and Zenil.

[f]Martínez, Margenstern, Sutner, Wiedermann and Zuse.

The very notion of "computability" that arose from the early 20th century work of various logicians Gödel, Church, Kleene and many others, harking back even to the 19th century ideas of Charles Babbage and Ada Lovelace,[g] and which was greatly clarified by Turing's notion of a computing machine, and by Post's closely related ideas, indeed depend on a fundamental *discreteness* of the basic ingredients. The various very different-looking proposals for a notion of effective computability that these early 20th century logicians introduced all turned out to be *equivalent* to one another, a fact that is central to our current viewpoint concerning computation, and which provides us with the *Church–Turing thesis*, namely that this precise theoretical notion of "computability" does indeed encapsulate the idea of what we intuitively mean by an idealized "mechanical procedure". We find this issue discussed at some depth by numerous authors in this volume[h]. For my own part, I am happy to accept the Church–Turing thesis, in this *original* sense of this phrase, namely that the mathematical notion of computability—as defined by what can be achieved by Church's λ-calculus, or equivalently by a Turing machine—is indeed the appropriate ideal *mathematical* notion that we require for our considerations of computability. Whether or not the universe in which we live operates in accordance with such a notion of computation is then an issue that we may speculate about, or reason about in one way or another (see, for example, Refs. 20,45).

Nevertheless, I can appreciate that there are other viewpoints on this, and that some would prefer to define "computation" in terms of what a physical object can (in principle?) achieve[i]. To me, however, this begs the question, and this same question certainly *remains*, whichever may be our preference concerning the use of the term "computation". If we prefer to use this "physical" definition, then all physical systems "compute" *by definition*, and in that case we would simply need a different word for the (original Church-Turing) mathematical concept of computation, so that the profound question raised, concerning the perhaps computable nature of the laws governing the operation of the universe can be studied, and indeed questioned. Accordingly, I shall here use the term "computation" in this *mathematical* sense, and I address this question of the computational nature of physical laws in a serious way later.

Returning, now, to the issue of the discreteness that came through the introduction of standard quantum mechanics, we find that the theory, as we

---

[g]DeMol, Sieg, Sutner, Swade and Zuse.
[h]DeMol, Sieg, Dershowitz, Sutner, Bauer and Cooper.
[i]Deutsch, Teuscher, Bauer and Cooper.

understand it today, has *not* developed in this fundamentally discrete direction that would have fitted in so well with our ideas of computation. The discreteness that Max Planck revealed, in 1900, in his analysis of black-body radiation (although not initially stated in this way) was in effect a discreteness of *phase space*—that high-dimensional mathematical space where each spatial degree of freedom, in a many-particle system, is accompanied by a corresponding momentum degree of freedom. This is not a discreteness that could apply directly to our seemingly continuous perceptions of space and time. Nonetheless, various contributors to this volume[j] have ventured in that more radical direction, arguing that some kind of discreteness might be revealed when we try to examine spatial separations of around the *Planck length* l_P (approximately 10^{-35}m) and temporal separations of around the Planck time (approximately 10^{-43}s). These separations are absurdly tiny, smaller by some 20 orders of magnitude from scales of distance and time that are relevant to the processes of standard particle physics. Since these Planck scales are enormously far below anything that modern particle accelerators have been able to explore, it can be reasonably argued that a granularity in the very structure of space-time occurring at the absurdly tiny Planck scales would not have been noticed in current experiments. In addition to this, it has long been argued by some theoreticians, most notably by the distinguished and highly insightful American physicist John A. Wheeler,[42] that our understanding of how a *quantum-gravity* theory ought to operate (according to which the principles of quantum mechanics are imposed upon Einstein's general theory of relativity) tells us that we must indeed expect that at the Planck scales of space-time, something radically new ought to appear, where the smooth space-time picture that we adopt in classical physics would have to be abandoned and something quite different should emerge at this level. Wheeler's argument—based on principles coming from conventional ideas of how Heisenberg's uncertainty principle when applied to quantum fields—involves us in having to envisage wild "quantum fluctuations" that would occur at the Planck scale, providing us with a picture of a seething mess of topological fluctuations. While this picture is not at all similar to that of a discrete granular space-time, it is at least supportive of the idea that something very different from a classically smooth manifold ought to be relevant to Planck-scale physics, and it *might* turn out that a discrete picture is really the correct one. This is a matter that I shall need to return to later in this Foreword.

---

[j]Bolognesi and Lloyd.

When it comes to the simulation of conventional quantum systems (not involving anything of the nature of Planck-scale physics) then, as was the case with classical systems, we find that we need to consider the smooth solutions of a (partial) differential equation—in this case the *Schrödinger equation*. Thus, just as with classical dynamics, we cannot directly apply the Church–Turing notion of computability to the evolution of a quantum system, and it seems that we are driven to look for simulations that are mere *approximations* to the exact continuous evolution of Schrödinger's wave function. Turing himself was careful to address this kind of issue,[39] whether it be in the classical or quantum context, and he argued, in effect, that discrete approximations when they are not good enough for some particular purpose can always be improved upon while still remaining discrete. It is indeed one of the key advantages of digital as opposed to analogue representations, that an exponential increase in the accuracy of a digital simulation can be achieved simply by incorporating additional digits. Of course, the simulation could take much longer to run when more digits are included in the approximation, but the issue here is what can *in principle* be achieved by a digital simulation rather than what is practical. In theory, so the argument goes, the discrete approximations can always be increased in accuracy, so that the computational simulations of physical dynamical process can be as precise as would be desired.

Personally, I am not fully convinced by this type of argument, particularly when chaotic systems are being simulated. If we are merely asking for our simulations to represent *plausible* outcomes, consistent with all the relevant physical equations, for the behaviour of some physical system under consideration, then chaotic behaviour may well not be a problem, since we would merely be interested in our simulation being realistic, not that it produces the actual outcome that will in fact come about. On the other hand, if—as in weather prediction—it is indeed required that our simulation emphis to provide the actual outcome of the behaviour of some specific system occurring in the world that we actually inhabit, then this is another matter altogether, and approximations may not be sufficient, so that chaotic behaviour becomes a genuinely problematic issue.[k]

It may be noted, however, that the Schrödinger equation, being *linear*, does not, strictly speaking, have chaotic solutions. Nevertheless, there is a notion known as "quantum chaos", which normally refers to quantum systems that are the quantizations of chaotic classical systems. Here the issue

---

[k]Matters relevant to this issue are to be found in.[12,44,46]

of "quantum chaos" is a subtle one, and is all tied up with the question of what we normally wish to use the Schrödinger equation *for*, which has to do with the fraught issue of *quantum measurement*. What we find in practice, in a general way—and I shall need to return to this issue later—is that the evolution of the Schrödinger equation does not provide us with the unique outcome that we find to have occurred in the actual world, but with a superposition of possible alternative outcomes, with a probability value assigned to each. The situation is, in effect, no better than with chaotic systems, and again our computational simulations cannot be used to predict the actual dynamical outcome of a particular physical system. As with chaotic systems, all that our simulations give us will be alternative outcomes that are plausible ones—with probability values attached—and will not normally give us a clear prediction of the future behaviour of a particular physical system. In fact, the quantum situation is in a sense "worse" than with classical chaotic systems, since here the lack of predictiveness does not result from limitations on the accuracy of the computational simulations that can be carried out, but we find that even a completely precise simulation of the required solution of the Schrödinger equation would not enable us to predict with confidence what the *actual* outcome would be. The unique history that emerges, in the universe we actually experience, is but one member of the *superposition* that the evolution of the Schrödinger equation provides us with.[1]

Even this "precise simulation" is problematic to some considerable degree. We again have the issue of discrete approximation to a fundamentally continuous mathematical model of reality. But with quantum systems there is also an additional problem confronting precise simulation, namely the vast size of the parameter space that is needed for the Schrödinger equation of a many-particle quantum system. This comes about because of the quantum entanglements referred to earlier. Every possible entanglement between individual particles of the system requires a separate complex-number parameter, so we require a parameter space that is exponentially large, in terms of the number of particles, and this rapidly becomes unmanageable if we are to keep track of everything that is going on. It may

---

[1] The question may be raised that the seeming randomness that arises in chaotic classical dynamics might be the result of a deeper quantum-level actual randomness. However, this cannot be the full story, since quantum randomness also occurs with quantized classical systems that are not chaotic. Nevertheless, one may well speculate that in the non-linear modifications of quantum mechanics that I shall be later arguing for, such a connection between chaotic behaviour and the probabilistic aspects of present-day quantum theory could well be of relevance.

well be that the future development of *quantum computers* would find its main application in the simulation of quantum systems. We find in this collection, some discussion of the potential of quantum computers, though there no consensus is provided as to the likely future of this interesting area of developing technology.[m]

We see that despite the discreteness that has been introduced into physics via quantum mechanics, our present theories still require us to operate with real-number (or complex-number) functions rather than discrete ones. There are, however, proposals (e.g.[4]) in which the notion of "computation" is taken in a sense in which it applies *directly* to real-number operations, the real numbers that are employed in the physical theory being treated as real numbers, rather than, say, rational approximations to real numbers (such as finitely terminated binary or decimal approximations). In this way, simulations of physical processes can be carried out without resorting to approximations. This, however, can require that the initial data for a simulation be given as explicitly known functions, and that may not be realistic. Moreover, there are various *different* concepts of computability with real numbers,[4,5,28,33,41] which, unlike in the situation that arose for discrete (integer-valued) variables, where the Church-Turing concept appears to have provided a *single* generally accepted universal notion of "computation", there are many different proposals for real-number computability and no such generally accepted single version appears to be in evidence. Moreover, we unfortunately find that, according to a reasonable-looking notion of real-number computability, the action of the ordinary second-order wave operator turns out to be non-computable in certain circumstances (see e.g. Refs. 28,29). Whatever the ultimate verdict on real-number computability might be, it appears not to have settled down to something unambiguous as yet.

There is also the question of whether an exact theory of real-number computability would have genuine relevance to how we model the physical world. Since our measurements of reality always contain some room for error—whether this be in a limit to the precision of a measurement or in a *probability* that a discrete parameter might take one or another value (as sometimes is the case with quantum mechanics)—it is unclear to me how such an exact theory of real-number computability might hold advantages over our present-day (Church–Turing) discrete-computational ideas. Although the present volume does not enter into a discussion of these mat-

---

[m]Schmidhuber, Lloyd, Zukowski.

ters, I do indeed believe that there are significant questions of importance here that should not be left aside (for example, see[5,20,29,41]).

Several articles in this volume address the issue of whether, in some sense, the universe actually *is* a computer.[n] To me, this seems to be a somewhat strange idea. Although I can more-or-less understand what it might mean for it to be possible to have (theoretically) a computational *simulation* of all the actions of the physical universe,[o] which involves some sort of "constructivist" assumption[p] for the operation of the physical world, I find it much less clear what it might mean for the universe to *be* a computer. Various images come to mind, maybe suggested by how one chooses to picture a modern electronic computer in operation. Our picture might perhaps consist of a number of spatially separated "nodes" connected to one another by a system of "wires", where signals of some sort travel along the wires, and some clear-cut rules operate at the nodes, concerning what output is to arise for each possible input. There also needs to be some kind of direct access to an effectively unlimited storage area (this being an essential part of the Turing-machine aspirations of such a computer-like model). However, such a discrete picture and a fixed computer geometry does not very much resemble the standard present-day models that we have of the small-scale activity of the universe we inhabit. The discreteness of this picture is perhaps a little closer to some of the tentative proposals for a discrete physical universe, such as the "causal sets"[q] that I shall briefly return to later, which represent some attempts at radical ideas for what space-time might be "like" at the Planck scale.

Yet, there are some partial resemblances between such a computer-like picture and our (very well supported) present-day physical theories. These theories involve individual constituents, referred to as "quantum particles", where each would have a classical-level description as being spatially "point-like"—though persisting in time, providing a classical space-time picture of a 1-dimensional "world-line". If these world-lines are to be thought of as the "wires" in the above computer-inspired picture, then the "nodes" could be thought of as the interaction places (or intersection points) between different particle world-lines. This would be not altogether unlike the computer image described above, though in standard theory, the topological geometry of the connections of nodes and wires would be part of the dynamics, and

---

[n] Lloyd, Deutsch, Turner and Zuse.
[o] Bolognesi and Szudzik.
[p] Bauer.
[q] Bolognesi.

not fixed beforehand. Perhaps the lack of a fixed geometry of the connections would provide a picture more like the *amorphous* type of computer structure also considered in this volume,[r] than a conventional computer. However, it is still not clear how the "direct access to an effectively potentially unlimited storage area" is to be represented. More seriously, this is merely the *classical* picture that is conjured up by our descriptions of small-scale particle activity, where the quantum "picture" would consist (more or less) of a *superposition* of all these classical pictures, each weighted by a complex number. Such a "picture" perhaps gets a little closer to the way that a *quantum* computer[s] might be represented, but again there are the crucial issues raised by the topology of the connections being part of the dynamics and the absence of an "unlimited storage area", in the physical picture, which seem to me to represent fundamental differences between our universe picture and a quantum computer. In addition to all this, there is again the matter of how one treats the *continuum* in a computational way, which in quantum (field) theory is more properly the *complex* rather than the real continuum. Over-riding all this is the matter of how one actually gets information out of a quantum system. This requires an analysis of the *measurement problem* that I shall need to come to shortly.

I think that, all this notwithstanding, when people refer to the universe "being" a computer, the image that they have is not nearly so specific as anything like that suggested above. More likely, for our "computer universe" they might simply have in mind that not only can the universe's actions be precisely simulated in all its aspects, but that it has *no other* functional quality to it, distinct from this computational behaviour. More specifically, for our "computer universe" there would be likely to be some parameter t (presumably a discrete one, which could be regarded as taking integer values) which is to describe the passage of time (not a very relativistic notion!), and the state of the universe at any one time (i.e. t-value) would have some computational description, and so could be completely encoded by a single natural number S_t. It would be the universe's job to compute $S_{t'}$ from S_t whenever $t' > t$, and the universe would be considered to *be* a computer provided that not only is it able always to achieve this, but—more importantly—that this is the *sole* function of the universe. It seems to me if, on the other hand, the universe has any additional function, such as to assign a *reality* to any aspect of this description, then it would not simply be a computer, but it would be something more than this, succeeding

---

[r]Hewitt, Teuscher, Margenstern and Wiedermann.
[s]Schmidhuber, Margenstern, Zukowski.

in providing us with some kind of *ontology* that goes beyond the mere computational description.

To conclude this Foreword, I wish to present something that is much more in line with my *own* views as to the relation between computation and the nature of physical reality. To begin with, I should perhaps point out that my views have evolved considerably over the decades, but without much in the way of abrupt changes. Early on I had been of a fairly firm persuasion that there should be a discrete or combinatorial basis to physics, perhaps somewhat along the lines expressed in some of the articles[t] in this volume. In 1967 Erwin Kronheimer and I published a paper[14,18] on the kind of causal sets referred to earlier in this Foreword, where the basic relationships between the elements are those of *causality*[u], mirroring the causal relations between events in continuous space-time, but where no continuity or smoothness is assumed, and where one could even envisage situations of this kind where the total number of these elements is finite.

Although I also had different reasons to be interested in spaces with a structure defined solely by causal relations—partly in view of their role in relation to singularity theorems[11,17] (for the study of black holes and cosmology)—the causality relations not necessarily being tied to the notion of a smooth space-time manifold, I did not have much of an expectation that the true small-scale structure of our actual universe should be helpfully described in these terms. I had thought it much more probable that a different combinatorial idea, that I had been playing with a good deal earlier, namely that of *spin-networks* (see Ref. 19) might have true relevance to the basis of physics (and indeed, much later, a version of spin-network theory was to form part of the loop-variable approach to quantum gravity,[1] although the role that spin-networks acquire in loop-variable gravity is somewhat different from what I had originally envisaged).

Spin-network theory was based on one of the most striking parts of standard quantum mechanics, where a fundamental notion that is continuous in classical mechanics, is discrete in quantum mechanics, namely *angular momentum* (or *spin*). In fact, many of the most basic and counter-intuitive features of quantum mechanics, such as discreteness and (Bell) non-locality,[v] are most powerfully expressed in terms of quantum-mechanical spin. The puzzling relation between the continuous array of possibilities for the direction of a spin axis in our classical space-time pictures and the discrete

---
[t]Bolognesi, Schmidhuber, Lloyd, Wolfram, Zuse, Fredkin and Zenil.
[u]Bolognesi.
[v]Breuer, Cabello, Schmidhuber and Zenil.

(or "granular") nature of the quantum idea of "spin-axis direction" had always maintained a fundamental fascination for me. This and the basic non-locality of information in quantum mechanics come together in spin-network theory, where the classical idea of a spatial "direction" does not arise in a well-defined way until very large spin-network structures are present in order provide a good approximation to the continuous sphere of possible spatial directions. Specific mathematical devices for calculating the often extremely complicated expressions were developed, but everything remains completely discrete, and computational in the conventional sense of the word, continuity arising only in the limit of large numbers. The need to generalize the idea of spin-networks in order that the geometry of 4-dimensional space-time might be described, rather than just the sphere of spatial directions, finally found some satisfaction in the ideas of *twistor theory* (see Refs. 18,23, Chapter 33). This provided a different way of looking at space-time geometry from what is usual—but now the idea of discreteness underlying the basis of physics began to fade, and became superseded by the magic of complex geometry and analysis.

One normally thinks of the space-time 4-manifold as being composed of "events" (i.e. space-time points), which are the basic elements of the geometry. Instead, twistor theory takes its basic elements to be modelled on entire histories of massless spinning particles in free flight. By a careful combination of ideas from space-time 4-geometry and the quantum-mechanical structure of relativistic angular momentum for massless particles, the concept of "twistor algebra" was developed.[18,19] In special relativity, the basic concept of a *twistor*, which describes the kinematical structure of a spinning massless particle, finds its mathematical description as an element of the complex 4-vector space $\mathbb{T}$ referred to as "twistor space". The geometry of $\mathbb{T}$ relates to the real geometry of Minkowski 4-space $\mathbb{M}$ by means of an explicit geometrical correspondence, relating $\mathbb{M}$ directly to the complex 3-geometry of the *projective* twistor space $\mathbb{PT}$. It turns out that the complex numbers of quantum mechanics dovetail with those of the complex geometry of twistor theory in surprising ways, and that there is an intriguing interplay between the non-locality that naturally arises in the twistor description of quantum wavefunctions and the non-locality that we actually find in quantum phenomena.[24] In recent years, twistor theory has found considerable value in the calculation of high-energy scattering processes, where the rest-masses of the particles involved can be ignored, (See, for example, Ref. 2) but many of the deeper issues confronting twistor theory remain unresolved.

It has always been an aim of twistor theory (still only partially fulfilled) that it should form a vehicle for the natural unification of quantum mechanics with general relativity. By this, I do not mean "quantum gravity" in the conventional sense in which this term is used. What is usually meant by quantum gravity is some scheme in which the ideas of Einstein's theory of gravity—namely *general relativity* (or else perhaps some modification of Einstein's theory)—is brought under the umbrella of quantum field theory. This viewpoint is to take the laws of quantum field theory as being *inviolate*, and that the ideas of general relativity must yield to those of quantum theory via some appropriate form of "quantization". My own view has always been different from this, as I believe that quantum theory itself, quite apart from its need to be unified with general relativity theory, is basically *self-inconsistent* and that some help is needed from *outside* the normal rules of quantum (field) theory. The view here is that the underlying principles of general relativity should help to supply this outside assistance.

This inconsistency is a very fundamental one, and is in a clear sense completely obvious (the "elephant in the room"!) as we shall see. As remarked upon earlier, we take the evolution of a quantum system in isolation to be governed by the Schrödinger equation—or, in more general terms, *unitary evolution*—and for which I use the symbol "U". But, as was remarked upon earlier, the reality of the world that we actually observe taking place about us tends *not* to be described directly by the solution Ψ of this equation that we get by this U-evolution, but when an observation or "measurement" is deemed to have taken place, Ψ is considered to "jump" to just one member Ψ_r of a family of *superposed* alternative solutions

$$\Psi = \alpha_1 \Psi_1 + \alpha_2 \Psi_2 + \ldots + \alpha_n \Psi_n \qquad (1)$$

where the respective squared moduli of the complex-number weightings $\alpha_1, \alpha_2, \ldots, \alpha_n$, supply the respective probabilities of each Ψ_r being the result (the quantities Ψ_r being assumed to be all normalized and mutually orthogonal). The "evolution process" whereby Ψ is replaced by the particular Ψr that happens to come about is the *reduction* of the state (collapse of the wavefunction) and I denote this process by the letter "R".[w]

Of course, there will be many such decompositions, for a given Ψ, depending on the choice of *basis* that is supposed to be determined by the choice of "measuring device". Indeed, we must allow that this measuring

---

[w]In Von Neumann's classic book Mathematical Foundations of Quantum Mechanics,[16] he introduced "R" and "U" under the respective names "process I" and "process II".

device is also part of the entire system under consideration, and so should have a quantum state that becomes entangled with the quantum system under examination. Nevertheless there is still taken to be a "jump" in the system as a whole as soon as the measurement is considered to have been made, where the different "pointer states" of the device are entangled with the different possible Ψ_rs that can result. It is *obvious* that this "jumping" from the state of the system (consisting of both the measuring device and system under examination, together with the entire relevant surrounding environment), from before measurement to after measurement, is normally not even continuous, let alone a solution of the Schrödinger equation: so R blatantly violates U (in almost all circumstances).

Why do physicists not normally consider this to be a contradiction in quantum mechanics? There are many responses, usually involving some subtle issue of "interpretation", according to which physicists try to circumvent this (seeming?) contradiction. Here is where the "many-worlds" viewpoint of Hugh Everett III is often invoked,[6,8] whereby it is considered that *all* alternative outcomes simply(!) *co-exist* in quantum superposition, and that it is perhaps somehow a feature of our conscious perception processes that we always perceive only *one* of these alternatives. Despite this idea's popularity among many philosophically minded physicists (or physically well-educated philosophers), I find this viewpoint very unsatisfactory. I would agree that it is indeed where we are led, if we regard the U-process as inviolate, but to me this is to be taken as a *reductio ad absurdum* and a clear indication that we need to seek an improvement in current quantum mechanics. To put this another way, even if the many-worlds viewpoint is in some sense "correct", it is still inadequate as a description of the physical world, for the simple reason that it does not, as it stands, describe the world that we actually observe, in which we find that something extremely well approximated by the R-process *actually* takes place when quantum superpositions of states that are sufficiently different from one another are involved.

What do I mean by "sufficiently different"? It is clear that mere *physical distance* apart, for the different material displacements involved in the superposed states, is not the correct criterion, because there have been well-confirmed experiments in which photon states tens of kilometres apart still maintain their quantum entanglements with one another, so that their various possible different polarization states remain in quantum superposition with each other even over such distances.[35] However, there are reasons to expect, from various foundational principles of Einstein's general theory of relativity, (see [7,21,25]) that when *mass displacements* between two quantum-superposed states get large, then such superpositions become unstable and ought to decay, in a roughly calculable time τ, into one or the other, so that classical behaviour begins to take over from quantum behaviour. The estimate of τ is given by the formula

$$\tau \approx \frac{\hbar}{E_G} \qquad (2)$$

where E_G is the gravitational self-energy of the *difference* between the mass distributions in each of two quantum states under consideration, each being assumed to constitute a stationary state if on its own. Such a decay would represent a deviation from the standard *linearity* of U and might perhaps even be the result of some kind of chaotic behaviour arising in some non-linear generalization of present-day quantum mechanics. There are experiments currently under development that are aimed at testing this proposal, and we may perhaps anticipate results over the next several years (see Ref. 15).

For various reasons, partly concerned with the quantum *non-locality* referred to earlier (which only begins to present substantial problems for quantum realism when R is involved, treated as a *real* phenomenon), I would expect this change in current quantum mechanics to represent a *major revolution* and would not be at all easy to arrive at simply by "tinkering" with the Schrödinger equation. Indeed, my expectations are that such a theory would have to be *non-computable* in some very subtle way. Why am I making such an assertion? The main reasons are rather convoluted, and I quite understand why some people regard my proposals as somewhat fanciful. Nevertheless, I am of the view that there is a good foundational rationale for a belief that something along these lines may actually be true!

The basic reason comes from Gödel's famous incompleteness theorems, which I regard as providing a strong case for human understanding being something essentially non-computable. The central argument is a familiar one, and I still find it difficult to comprehend why so many people are unwilling to take on board what would seem to be its fairly clear implication in this regard. In simple terms, the argument can be applied to our abilities to demonstrate the truth of certain mathematical propositions—which we can take to be of the form of Π_1-*sentences*. A Π_1-*sentence* is an assertion that some proposed Turing computation never terminates (examples being Wiles's "Fermat's last theorem" and Lagrange's theorem that every natural number is the sum of four squares). We might try to encapsulate, within some algorithmic procedure $\mathbb{A}$, all possible types of argument that can, in principle, be used to establish Π_1-*sentences*, according to human insight and understanding. This argument might be a *proof* within some given formal system $\mathbb{F}$, where $\mathbb{A}$ would be an algorithm for checking whether a proposed proof using the rules of $\mathbb{F}$ had been correctly carried out, giving the answer YES after a finite number of steps if this is indeed the case. What the Gödel(–Turing) theorem shows, in this context, is that if we have *trust* in $\mathbb{A}$ (and therefore in the soundness of such an $\mathbb{F}$, with regard to Π_1-*sentences*) that a "proved" Π_1-*sentence* is indeed *true* whenever $\mathbb{A}$ asserts YES, then one can explicitly exhibit a Π_1-*sentence* G where our trust in $\mathbb{A}$ extends also to a trust in the truth of G, even though $\mathbb{A}$ itself is shown to be incapable of directly establishing G. In the case of Gödel's second incompleteness theorem, the Π_1-*sentence* $G(=G(\mathbb{F}))$ would be an assertion of the *consistency* of $\mathbb{F}$, and G would be the Π_1-assertion that among the theorems of $\mathbb{F}$, there would be none whose negation is also a theorem of $\mathbb{F}$. Although our trust tells us that $\mathbb{A}$ would be unable to establish G (i.e. the consistency of $\mathbb{F}$), our trust that G is actually true follows from our *trust* in $\mathbb{A}$ (which *depends* on $\mathbb{F}$'s consistency—otherwise $\mathbb{F}$ would be able to establish $2 = 3$, a conclusion which we certainly would not trust). Our trust in the *use* of $\mathbb{F}$ as a means of establishing the truth of Π_1-*sentences* therefore carries us beyond the direct capabilities of $\mathbb{F}$, and enables us to assert that $G(\mathbb{F})$ is true, on the basis of that same trust, despite the fact that $\mathbb{F}$ does not contain $G(\mathbb{F})$ among its theorems.

This is basically the thrust of Gödel's attack on formalism. Although the formalization of various areas of mathematics certainly has its value, allowing us to the transfer different aspects of human understanding and insight into computational procedures, Gödel shows us that these explicit procedures, once known—and *trusted*—cannot cover everything in mathematics that is accessible to understanding and insight.[x] And, indeed, this applies already for the relatively limited area of Π_1-*sentences*. Yet, a case can certainly be argued that this does *not* yet provide a demonstration that human insight is, at root, a non-algorithmic procedure, and I list here what appear to be the main arguments in support of that case, i.e. of criticisms of the above claim that the Gödel-type arguments show that human understanding is non-computational;

(1) *Errors argument*—human mathematicians make errors, so rigorous Gödel-type arguments do not apply.

---
[x] Zizzi.

(2) *Extreme complication argument*—the algorithms governing human mathematical understanding are so vastly complicated that their Gödel statements are completely beyond reach.

(3) *Ignorance of the algorithm argument*—we do not know the algorithmic process underlying our mathematical understanding, so we cannot construct its Gödel statement.

I have tried to argue elsewhere[20] that (1), (2), and (3) do not invalidate the conclusion that our conscious understandings are very unlikely to be entirely the product of computational actions, and it is not my purpose to repeat such detailed arguments here. Nevertheless I briefly summarise my counter-arguments, in what follows.

The main point, with regard to (1) is that human errors are correctable. We are not so much concerned with the often erroneous gropings that mathematicians employ in their search for truth, but more the *ideals* that they grope *for* and, more importantly, measure their achievements against. It is their ability to *perceive* these ideals that we are concerned with, if only in principle, and it is this ability to perceive ideal mathematical truth that we are concerned with here, not the errors that we all make from time to time. (It may be evident from these comments that I *do* regard mathematical truth—especially with regard to matters so straight-forward as Π_1-*sentences*—as something absolute, and external to ourselves. But I appreciate that others[y] are sometimes less sympathetic to this kind of viewpoint. I do not believe, however, that one's philosophical standpoint in this respect significantly affects the arguments that I am putting forward here.) With regard to (2) the point is somewhat similar. If the algorithms were *in principle* to be known, then their size or complication is of no real concern. This applies to a great many mathematical arguments. In Euclid's proof of the infinity of primes, for example, we need to consider primes that are so large that there would be no way to write them down explicitly in the entire observable universe, and to calculate the product of them all up to some such size in practice, is even more out of the question. But all this is irrelevant for the proof. Similar points apply to (2).

The argument (3) is, however, much more relevant to the discussion, and was basically Gödel's own reservation (referred to in the commentaries here[z]) with regard to making the strong conclusion that I am arguing for here. Rather than pushing the logical argument further, which is certainly

---

[y] DeMol.
[z] Sieg.

possible to do (see Ref. 20). I shall here merely indicate the extraordinary improbability of the needed algorithmic action arising in our heads, by the process of *natural selection*. Such an algorithm would have to have extraordinary sophistication, so as to be able to encapsulate, in its effective "formal system" many steps of "Gödelization". As an example, I have pointed out elsewhere[22] that whereas *Goodstein's theorem*,[9] whose meaning[aa] is easily accessible even to those with little mathematical knowledge other than basic numerical notation, has been shown by Kirby and Paris[13] to be inaccessible by first-order Peano arithmetic (without a "Gödelization" step, that is), yet this theorem can be readily seen to be true through mathematical understanding. If our mathematical understanding is achieved by some (unknowable, but sound) algorithmic procedure, it would be a total mystery how it could have arisen through natural selection, when the experiences of our remote ancestors could have gained no benefit whatsoever from having such a sophisticated yet totally irrelevant algorithm planted in their brains!

If, then, it is accepted that our understanding of mathematics is not an algorithmic process, we must ask the question what kind of process can it be? A key issue, it seems to me, is that genuine *understanding* (at least in our normal sense of this word) is something that requires awareness—as it would seem to me to be a misuse of the word "understanding" if it could be genuinely applied to an entity that had no actual *awareness* of the matter under its consideration. Awareness is the passive form of consciousness, so it seems to me that it was the evolutionary development of consciousness that is the key, and that such a quality could certainly have come about through natural selection, being able to confer an enormous selective advantage on those creatures possessing it. In saying this, I am expressing the view that consciousness is indeed *functional* and is not an "epiphenomenon" that simply happens to accompany certain kinds of cognitive processes. This view is certainly an implication of the quality of "understanding" requiring conscious awareness, since understanding is certainly functional.

I should make clear that I am making no claim to know—or to be able to define—what consciousness actually *is*, but its role in underlying "understanding" (whatever *that* is) seems to me to be of great evolutionary value, and could readily arise as a product of natural selection. I should also make clear that I am regarding the consciousness issue as a *scientific* one, and that I do *not* take the view that these are matters that are inaccessible to

---

[aa] Velupillai.

scientific investigation. I also take it that healthy wakeful human brains (as well as whatever other kind of animal brains may turn out also to be similarly capable) are able, somehow, to evoke consciousness by the application of those very same physical laws that are present throughout the universe, even though consciousness itself comes about only in the *very* special circumstances of organization that are needed to promote its appearance.

What kind of circumstance could that be, if we are asking for some sort of non-computable action to come about—when we bear in mind that the deterministic differential equations of classical or quantum physics seem to be of an essentially computable nature? My response to this query is that the non-computability must lie in hitherto undiscovered laws that could be of relevance here. (I am ignoring the issue, referred to earlier, of the discrete computational simulation of a continuous evolution. Yet, I do accept that there might be some questions of genuine relevance here that ought to be followed up more fully.) As far as I can see, the only big unknown, in physical laws, that could have genuine relevance here, is the U/R puzzle of quantum mechanics, referred to above. In almost all processes that take place, we have no need of the presumed New Theory that is to go beyond current quantum mechanics, mainly because its effects would go un-noticed, being swamped by the multifarious random influences of environmental decoherence. But, in the brain, there might be relevant structures able to preserve quantum coherence up to a length of time at which the previously mentioned $\tau \approx \hbar/E_G$ criterion actually becomes relevant. Then, the normal purely probabilistic action that standard quantum theory's R-process provides us with is to be replaced by some subtle non-computational decision as to which choice the state reduction leads to. With a sophisticated brain organization, where the synaptic responses are sensitive to these choices, we can imagine that the output of the brain could indeed be usefully non-computational. This, indeed, is the basis of the "orchestrated objective reduction" (**Orch-OR**) scheme that Stuart Hameroff and I have proposed some years ago, where the above "relevant structures" would be *neuronal microtubules* of the appropriate type (see Refs. 10,21,27).

It is hardly surprising that such a proposal has met with some considerable scepticism, mainly for the very understandable reason that to have body-temperature quantum coherence at anything like the level required is enormously far beyond the expectations of standard physical calculations applied to simplified models of cells.[34] Nevertheless, biological cells are, in fact, highly sophisticated structures,[bb] and one may reasonably expect that, when the structures of certain cell parts are dedicated in the appropriate directions, their behaviour might exhibit quite unusual quantum-mechanical properties.[cc] In fact, recent experiments carried out in Japan by Anirban Bandyopadhyay[3] and his co-workers appear to have demonstrated that highly intriguing quantum-coherent effects *do* actually take place in body-temperature neuronal microtubules. These results are, as of now, preliminary, but they do appear to provide some encouragement for the **Orch-OR** scheme, and it will be very interesting to see how things develop.

Even if all of this is accepted, we may still ask what would be the use of a little bit of non-computable action, from time to time, for the operation of the brain? Indeed, there would not be much value in this unless the quantum coherence is of a very global character, involving large areas of the brain, and the process would have to act in some globally coherent way. This is indeed the **Orch-OR** picture, and we take it that moments of consciousness occur when state reduction occurs at many sites (in microtubules) at once in an *orchestrated* way, so that the synapse strengths are influenced in many places and a concerted influence results, as would be expected for conscious actions. The results of particular acts of conscious understanding would be unlikely to be usually anything simple, and would depend upon the experience of memories as well as on logic. But the non-computable ingredient is taken to be essential, for the reasons described above. According to this view, our conscious actions are calling upon parts of physics—encompassed in a New Theory that is presently unknown in detail. The impact of this theory on processes *not* organized in this way would not be evident. But it would make its mark on systems—such as wakeful healthy human brains—where it emerges as conscious actions and perceptions. The non-computable effects of this New Theory would emerge in this way and result in actions that are described as "hypercomputational".

How far outside the normal scheme of computational physics would these hypercomputational actions be? Since the Gödelian insight that allows us to transcend a given trusted formal system $\mathbb{F}$ provides this insight in the form of a Π_1-*sentence*, namely $G(\mathbb{F})$, we might expect that we could model such hypercomputational actions in the form of a Turing *oracle-machine*,[dd] where the oracle is able to assert the truth or falsity of Π_1-

---

[bb] Margenstern, Ehrenfeucht et al., Rozenberg and Zenil.
[cc] Zizzi.
[dd] Chaitin, Dershowitz.

sentence. However this would not be sufficient (nor does it appear to be necessary), as we can apply a Gödel-type "diagonalization" insight again on Π_1-*sentence*-oracle machines to transcend these devices also. In a recent article,[26] I consider a type of oracle that I refer to as a "cautious oracle", which is intended to model a little more closely the kind of thing that one might consider idealized human mathematicians might be capable of, where the cautious oracle can examine a Π_n-*sentence* (for any natural number n) and either respond "true" or "false" (necessarily truthfully in each case), or else confess to being unable to supply an answer or, failing any of these, simply continue pondering indefinitely without ever providing an answer at all. Again a Gödel-type diagonalization allows us the insight to transcend any such a device's capabilities! Whatever kind of hypercomputational capabilities such a "New Theory" might confer, it appears to be something very subtle. It is some sort of never-ending capability of being able to "stand back" and contemplate whatever structure had been considered previously. This seems to be a quality that consciousness is able to achieve, but how one incorporates this kind of thing into a physical theory is hard to imagine, as our present-day theories stand.

References

1. Ashtekar, A. and Lewandowski, J. Background independent quantum gravity: a status report. *Class. Quantum Grav. 21,* R53.[gr-qc/0404018], 2004.
2. Arkani-Hamed, N., Cachazo, F., Cheung, C., and Kaplan, J. The S-Matrix in Twistor Space. arXiv:0903.2110v2 [hep-th], 2009
3. Sahu, S., Ghosh, S., Hirata, K., Fujita, D. and Bandyopadhyay, A. Ultrafast microtubule growth through radio-frequency-induced resonant excitation of tubulin and small-molecule drugs, to appear in Nature Materials.
4. Blum, L. Alan Turing and the Other Theory of Computation (on Turing's Rounding-off Errors in Matrix Processes) in S Barry Cooper and Jan van Leeuwen (eds), Alan Turing – *His Work and Impact.* Elsevier, 2012.
5. Bridges, D.S. Can constructive mathematics be applied in physics? *J. Philos. Logic 28,* 439–53, 1999.
6. DeWitt, B.S. and Graham, R.D., eds. *The Many-Worlds Interpretation of Quantum Mechanics.* Princeton Univ. Press, Princeton, 1973.
7. Diósi, L. Models for universal reduction of macroscopic quantum fluctuations. *Phys. Rev. A40,* 1165–74, 1989.
8. Everett, H. "Relative State" formulation of quantum mechanics. In J.A. Wheeler and W.H. Zurek (eds), *Quantum Theory and Measurement.* Princeton Univ. Press, Princeton, 1983), originally in *Revs. of Modern Physics, 29,* 454–62, 1957.

9. Goodstein, R.L. On the restricted ordinal theorem. *J. Symbolic Logic 9*, 33–41, 1944.
10. Hameroff, S.R. and Penrose, R. Conscious events as orchestrated space-time selections. *J. Consciousness Studies 3*, 36–63, 1996.
11. Hawking, S.W. and Penrose, R. The singularities of gravitational collapse and cosmology, *Proc. Roy. Soc.*, London, A314, 529–548, 1970.
12. Israeli, N. and Goldenfeld, N. Computational irreducibility and the predictability of complex physical systems. *Phys. Rev. Lett. 92*, 074105, 2004.
13. Kirby, L.A.S. and Paris, J.B. Accessible independence results for Peano arithmetic, *Bull. L.ond. Math. Soc. 14*, 285–93, 1982.
14. Kronheimer E.H. and Penrose, R. On the structure of causal spaces. *Proc. Camb. Phil. Soc. 63*, 481–501, 1967.
15. Marshall, W., Simon, C., Penrose, R. and Bouwmeester, D. Towards Quantum Superpositions of a Mirror, *Phys. Rev. Lett.*, Vol. 91, Issue 13, 2003.
16. von Neumann, J. *Mathematical Foundations of Quantum Mechanics*. (Princeton Univ. Press, Princeton), 1955.
17. Penrose, R. Gravitational collapse and space-time singularities, *Phys. Rev. Lett. 14*, 57–59, 1965.
18. Penrose, R. Twistor algebra, *J. Math. Phys. 8*, 345–66, 1967.
19. Penrose, R. Angular momentum: an approach to combinatorial space-time. In T. Bastin (ed), *Quantum theory and Beyond*, Cambridge University Press, Cambridge, 1971.
20. Penrose, R. *Shadows of the Mind; An Approach to the Missing Science of Consciousness*, Oxford Univ. Press, Oxford, 1994.
21. Penrose, R. On gravity's role in quantum state reduction. *Gen. Rel. Grav. 28*, 581–600, 1996.
22. Penrose, R. Can a computer understand? In Rose, S. (ed), *From Brains to Consciousness? Essays on the New Sciences of the Mind*, Allen Lane, The Penguin Press, London) 154–179, 1998.
23. Penrose, R. *The Road to Reality: A Complete Guide to the Laws of the Universe*, Jonathan Cape, London, 2004.
24. Penrose, R. The twistor approach to space-time structures. In A. Ashtekar (ed), *100 Years of Relativity; Space-time Structure: Einstein and Beyond*, World Scientific, Singapore, 2005.
25. Penrose, R. Black holes, quantum theory and cosmology (Fourth International Workshop DICE 2008) J. Physics, *Conf. Ser. 174*, 012001. doi: 10.1088/1742-6596/174/1/012001, 2009.
26. Penrose, R. On attempting to model the mathematical mind, in Cooper, B.S. and Hodges, A. (eds), *The Alan Turing Year - The Once and Future Turing*, Cambridge University Press, 2012.
27. Penrose, R. and Hameroff, S. Consciousness in the Universe: Neuroscience, Quantum Space-Time Geometry and **Orch OR** Theory. *Journal of Cosmology*, Vol. 14, 2011.
28. Pour-El, M.B. and Richards, I. The wave equation with computable initial data such that its unique solution is not computable, Adv. in Math. 39, 215–239, 1981.

29. Pour-El, M.B. and Richards, I. Computability in Analysis and Physics. *Perspect. Math. Logic*, (Springer-Verlag, Berlin, Heidelberg), 206 pp., 2003.
30. Russell, B. *The Analysis of Matter* (Allen and Unwin; reprinted 1954, Dover Publ. Inc., New York), 1927
31. Sakharov, A.D. Vacuum Quantum Fluctuations in Curved Space and The Theory of Gravitation, *Sov. Phys. Dokl., 12*, 1040 [Dokl. Akad. Nauk Ser. Fiz. 177, 70]. Reprinted: (2000) *Gen. Rel. Grav., 32* 365–367, 1968.
32. Schrödinger, E. *Science and Humanism: Physics in Our Time*. (Cambridge Univ. Press, Cambridge), 1952.
33. Stannett, M. Computation and Hypercomputation. *Minds and Machines, 13*, 115–53, 2003.
34. Tegmark, M. "Importance of quantum coherence in brain processes," *Phys. Rev. E, 61*, pp. 4194–4206, 2000.
35. Tittel, W., Brendel, J., Gisin, B., Herzog, T., Zbinden, H., and Gisin N. Experimental demonstration of quantum-correlations over more than 10 kilometers arXiv:quant-ph/9707042v3, 2008.
36. Turing, A.M. On computable numbers, with an application to the Entscheidungsproblem, *Proc. Lond. Math. Soc. (ser. 2) 42*, 230–265; a correction 43, 544–546, 1937.
37. Turing, A.M. Computability and 1-definability. *J. Symb. Log., 2*, 153–163, 1937.
38. Turing, A.M. Systems of logic based on ordinals. *P. Lond. Math. Soc., 45 (2)*, 161–228, 1939.
39. Turing, A.M. 'Intelligent machinery', with AMS corrections and additions. Pages numbered 1–37, with 2 un-numbered pages of references and notes. Page 1 has MS note by R.O. Gandy, 'Turing's typed draft'.n.d., 1948.
40. Turing, A.M. Computing machinery and intelligence, *Mind 59* no. 236; reprinted in D.R. Hofstadter and D.C. Dennett (eds), *The Mind's I*, Basic Books, Inc.; Penguin Books, Ltd; Harmondsworth, Middx. 1981, 1950.
41. Weihrauch, K. *Computable Analysis: An Introduction*. Texts in Theoretical Computer Science, Springer, 2000.
42. Wheeler, J.A. *Geometrodynamics* (Società Italiana Fisica: Questioni di fisica moderna, V.1, and Academic Press, Inc., New York), 1982.
43. Wheeler, J.A. Law without law, in Quantum Theory and Measurement. In J.A. Wheeler, J.A. and Zurek, W.H. (eds), Princeton Univ. Press, Princeton, pp. 182–213, 1983.
44. Wolfram, S. A New Kind of Science. Wolfram Media Inc, 2002.
45. Zenil, H. and Delahaye, J.-P. On the Algorithmic Nature of the World. In G. Dodig-Crnkovic and M. Burgin (eds), Information and Computation, World Scientific, 2010.
46. Zenil, H, Soler-Toscano, F., and Joosten, J. J. Empirical Encounters With Computational Irreducibility and Unpredictability, *Minds and Machines, vol. 21*, 2011.

Preface

Simplified Roadmap of (Un)computable World-Views

Today information and computation play a major role in modern physics, both as a source of new unifying theories, and also of sound approaches to aspects of current mainstream theories such as statistical mechanics and thermodynamics, where it has proven to be of great use. Indeed it is central to many physical concepts nowadays.

Compelled as I was by all these questions, it was a privilege to have the opportunity of co-organising and being involved in several events around this topic. In 2008, together with Adrian German and Gerardo Ortiz, we organised the second Midwest NKS Conference at the University of Indiana, Bloomington, featuring an impressive set of participants and speakers (see http://www.cs.indiana.edu/~dgerman/2008midwestNKSconference/) including Bennett, Calude, Chaitin, Csicsery, Deutsch, Fredkin, Grover, Leggett, Lloyd, Rowland, de Ruyter, Szudzik, Toffoli and Wolfram. The momentum generated by the conference contributed to the realisation of this project, including the transcription of Deutsch's contribution to this book and the transcription of the panel discussion on the subject, which is also included. In 2010, with Tommaso Bolognesi (and mostly thanks to him) the JOUAL (Just One Universal Algorithm) Workshop (see http://fmt.isti.cnr.it/JOUAL2009/) was organised in 2009, to consider questions around the concepts of emergence, space-time and nature in computational systems. It featured Renate Loll, Stephen Wolfram and Juergen Schmidhuber, among other speakers.

An important question to which this volume may suggest an answer is whether these views are mature enough to be engaged with and discussed at length and in depth. I have had the privilege of being able to lead the effort to undertake such a challenge, and the result, I think, is a comprehensive volume in which most, if not all current trends are represented in some fashion. In preparing this volume I have made sure to include dissenting voices, the viewpoints of those not in agreement with the main thesis of this volume (an ontological view, or a pragmatic approach to a (Turing) computable universe), notably dissenting voices from the important field of quantum mechanics (beginning with Penrose, who has written the Foreword to this volume, and including as well Zizzi, Lloyd, Deutsch and Cabello). Also included are those embracing some notion of hypercomputation in one way or another and under some term or another (Doria, Cooper, Penrose and Zizzi again), and thinkers representing a novel trend, proponents of an algorithmically random world (Calude, Meyerstein, Salomaa and Svozil), which happens to be the diametrical opposite of my own algorithmic view. In an effort to provide a useful roadmap to these viewpoints, I have grouped them into a few categories, fully realising that I run the risk of oversimplification. Some of these categories oppose each other or are a bifurcation of a larger category: e.g. digital vs. quantum, deterministic vs. random. This is only my personal simplified account, and by no means necessarily represents the views these authors would entertain, either of their own work or of the work of others. This is merely intended to help the reader compare viewpoints, and perhaps orient him- or herself vis-à-vis the several hypotheses:

- **The (Turing) Computable Universe Hypothesis.**(or some form of computationalism) (e.g. Schmidhuber, Hutter). Also Digital Physics Hypothesis. Supported by the various versions (except perhaps the original one) of the Church-Turing thesis as established by Kleene. Often epistemological in nature, contrary to the common belief, it advances the idea that the universe's upper computational power is that of Turing universality, which doesn't mean by any means that these authors advance the idea that the world *is* a (universal) Turing machine. It can go from the pure ontological position (e.g. Wolfram or Bolognesi aiming at providing a basis for physics as an emergent property of reality, or Fredkin (inherited from the "Cellular Automaton Hypothesis" subcategory)) to the epistemological formalism (e.g. Szudzik's). In this category positions such as Wheeler's ("it from bit") and perhaps Feynman's would be placed. One thing is certain under this category, that nature is capable of Turing computation as attested by the existence of digital computers and nature seems to behave like if computationalism were true as we have managed to capture most natural phenomena in increasingly encompassing theories describing large parts of the world behaviour.

- **The Cellular Automaton Hypothesis**. A sub-category first suggested by Zuse and then adopted and further developed by Fredkin under his program of Digital Philosophy. With traditionally little support but proven to provide foundational concepts for the subfield of *physics of computation* (e.g. questions related to logical and physical reversibility).
- **The Mathematical Structure Hypothesis**. A sub-category of the Computational Hypothesis. Suggested by Max Tegmark and under the Computable Universe Hypothesis given that Tegmark has mentioned that by a mathematical structure he means a computable one (the uncomputable version can be grouped under the Non-Turing Computable Universe Hypothesis).

- **The Informational Universe Hypothesis**. (e.g. Wheeler) Most, if not all, authors of models of quantum gravity may fall into this category, even if the authors may not place or ask themselves whether they are doing so, as they place information as the ultimate reality (Zeilinger being the extreme case). Other authors such as Scott Aaronson may also fall into this category, taking quantum mechanics as a theory of unknowns, of probability magnitudes, and ultimately (unknown) information.
- **The Computational Pragmatic Hypothesis**. Models belonging to this category are mostly agnostic with regard to any ontological commitment concerning the ultimate structure of the world. It is a weak form of computationalism, held by virtually mod researchers in the practice of science. They are pragmatic in their approach to nature-like phenomena and seek real applications (e.g. in this volume Ehrenfeucht et al. Rozenberg, Martinez, Adamatzky, Teuscher, Velupillai and Zambelli). Most practice of scientific research falls into this category as physical laws can be solved with extraordinary precision up to unknown but increasingly more accurate levels, up to the point to believe that we can arrive to a ToE, a single (in a large sense *computable*) formula (not necessarily meaning *complete predictability power*). This pragmatic approach has turned to be unreasonably useful in its explanatory and predictive power and certainly has propelled both The Informational Universe and The (Turing) Computable Universe hypotheses.

- **The Non-Turing Computable Universe Hypothesis** (a view represented in this volume by Doria, and in different ways, and under different labels or approaches, by Calude and Cooper), which is in opposition to the Church-Turing thesis in its various versions. Its proponents either support the view that nature is capable of encoding (and solving) the halting problem and therefore embracing 'type-n computation' (or hyper-computation) for $n > 1$ (in Kleene notation), or believe in the existence of *uncomputable regularities*, that is, they hold that by some means an irregularity (i.e. a non-computable "pattern") can be construed as a regularity, suggesting that there exists in nature something (e.g. the brain) that is capable of finding patterns where computers cannot, even in principle (in unbounded time). An example would be "seeing" a Chaitin Omega without the machine that generates it, though we are not even capable of recognising patterns in computable numbers such as π). Other serious proponents may include scholars such as J. Félix Costa, Mark Burgin and Selmer Bringsjord.

    - **The Random Universe Hypothesis.** (Calude, Meyerstein, Salomaa, McAllister). Could be viewed as a sub-branch of the Non-Turing Computable Universe Hypothesis, but with different arguments and a novel approach that uses the theory of algorithmic randomness, also potentially embracing a strong interpretation, albeit mainstream, of quantum mechanics (Copenhagen), but needs no recourse to quantum mechanics to make its case.
    - **The Non-Computable Universe Hypothesis.** (e.g. Penrose) Arguments belonging or in combination of the The Non-Turing Computable Universe Hypothesis and the Quantum Hypothesis. Other possible proponents (with variations) are Lucas, Zizzi, and Searle. Opposed to most other views but perhaps tangential to either The Non-Turing Computable Universe Hypothesis and a version of the Quantum Hypothesis, or both.

- **The Algorithmic Information Hypothesis.** (Zenil) opposed to the Random hypothesis. It advances the explanation that most, if not all, the structure in the world is described by algorithmic probability which in turn is based in Turing universal computation and describes the distribution of patterns in nature. It can supersede

the upper level but it is not meant to describe the ultimate building blocks of the universe (or its computational power) unlike other hypotheses, but rather to explain structure as the output of algorithmic (computational) processes. It is also intended to be a validation theory, potentially providing statistical evidence for a stronger version up to supporting a Computable Universe Hypothesis, in particular a Turing Computable Universe Hypothesis.
- **The Standard Quantum Universe Hypothesis** (Lloyd, Deutsch, Cabello): Physics and complexity from/based in quantum mechanics, particularly those taking as a departure point the Copenhagen interpretation of Quantum Mechanics. In the cases of Lloyd, Deutsch, Cabello also acknowledging the importance of information and computation but on top of quantum mechanics.

The volume includes historical and philosophical accounts (Swade, De Mol, Turner), alternative models and approaches to that of Turing (Hewitt, De Mol), profound investigations of the nature of mathematics and computation (Turner, Sieg, Dershowitz and Falkowich, Cooper, Bauer and Harrison, Sutner, Beavers, Ehrenfeucht et al.), pragmatic approaches to new forms of computation and real-world applications (Wiedermann, Martínez, Margenstern, Teuscher, Velupillai, Zambelli) and of computation in relation to quantum reality (Cabello, Zizzi, Zukowski, Lloyd, Deutsch, Breuer). There is also an illuminating panel discussion on the question to which the book is devoted (What is Computation? (How) Does Nature Compute?), featuring a subset of the contributors and a few other authors, including the Nobel prize winner (Physics) Tony Leggett.

Hector Zenil
Sheffield, S. Yorkshire, UK, 2012
http://www.algorithmicnature.org/zenil

Acknowledgements

The volume opens with a masterful Foreword by Roger Penrose, to whom I am very grateful, and delighted to have been able to work with. I am of course very grateful to the authors, all pioneers and brilliant thinkers. To have their work collected within a single volume was itself a challenge. A novel feature of the book is the discussion section (a first version of the idea for which came from Cris Calude), where authors engaged in a lively exchange of ideas in the form of questions and answers, I want to thank them for having participated. This was an enriching process, I served at times as a legitimate inquirer and at other times asking questions whose answers both I myself and some of the authors could have anticipated, but which prompted those questioned to summarise their views in a way I hoped that readers would find helpful, sometimes disclosing commitments that may not otherwise have been evident.

The volume also contains a revised edition of Zuse's *Calculating Space*, completely rewritten in LaTeX, a project realised thanks to the drive of Adrian German, whom I want especially to acknowledge for his many contributions to this effort, including the transcriptions of the roundtable discussion and Deutsch's contribution to this volume.

Much of my own journey up to this point has been the result of interactions with some of the direct and indirect contributors to this volume. I attended Ed Fredkin's lectures at Carnegie Mellon while I was a visiting scholar during the Spring semester of 2008. I also attended Charles Bennett's lectures at the Institute Henri Poincaré in Paris, and have had the opportunity to interact with Greg Chaitin on several occasions. Indeed Greg ultimately became a member of my PhD examining committee. Stephen Wolfram has been an important influence, having worked for him at his personal office as a consultant (and senior research associate) at Wolfram Research for almost 6 years.

I must also mention other mentors, colleagues and friends such as Jean-Paul Delahaye, Cris Calude, Matthew Szudzik, Fred Meinberg,

Todd Rowland, Selmer Bringsjord, Wilfried Sieg, Jean Mosconi, Jacques Dubucs, Bernard François, Genaro J. Martińez, Joost Joosten, Francisco-Hernández-Quiroz, James Marshall and others whose oeuvres, ideas and interactions with me have helped build and enrich my own work. I also want to specially thank Elena Villarreal, who helped me at various stages in the editing of this volume, and for many other reasons.

Chapter 1

Introducing the Computable Universe

Hector Zenil

Department of Computer Science, The University of Sheffield, UK
IHPST (Paris 1 Panthéon-Sorbonne/ENS Ulm/CNRS), France
& Wolfram Science Group, Wolfram Research, USA

1. Understanding Computation & Exploring Nature as Computation

Since the days of Newton and Leibniz, scientists have been constructing elaborate views of the world. In the past century, quantum mechanics revolutionised our understanding of physical reality at atomic scales, while general relativity did the same for our understanding of reality at large scales.

Some contemporary world views approach objects and physical laws in terms of information and computation,[25] to which they assign ultimate responsibility for the complexity in our world, including responsibility for complex mechanisms and phenomena such as life. In this view, the universe and the things in it are seen as computing themselves. Our computers do no more than re-program a part of the universe to make it compute what we want it to compute.

Some authors have extended the definition of computation to physical objects and physical processes at different levels of physical reality, ranging from the digital to the quantum. Most of the leading thinkers involved in this effort are contributors to this volume, including some who oppose the (digital) approach, preferring to advance their own.

The computational/informational view (sometimes identified as *computationalism*) is rooted in pioneering thinking by authors such as John A. Wheeler and Richard Feynman. A quotation from Feynman's Messenger Lectures, delivered at Cornell University in 1964, distills his sense that nature most likely operates at a very simple level, despite seeming complex to us, and marks a shift to understanding physics in terms of digital information.

> It always bothers me that, according to the laws as we understand them today, it takes a computing machine an infinite number of logical operations to figure out what goes on in no matter how tiny a region of space, and no matter how tiny a region of time ... So I have often made the hypothesis that ultimately physics will not require a mathematical statement, that in the end the machinery will be revealed, and the laws will turn out to be simple, like the chequerboard with all its apparent complexities.

His view was probably influenced by his thesis advisor John A. Wheeler, the coiner of the phrase "it from bit", suggesting that information constitutes the most fundamental level of physical reality.[18] Another pioneer was Konrad Zuse, a new edition of whose Calculating Space (*Rechnender Raum*) we are pleased to be able to publish in this volume. We've put great effort into translating into modern LaTeX the scanned version of the original translation commissioned by Ed Fredkin and published by MIT. Fredkin is himself another pioneer, having founded the field that is today known as *digital physics*, and is a contributor to this volume. Stephen Wolfram, a student of Feynman, has also been building upon this view, spearheading a paradigm shift facilitated by today's availability of increasingly greater and cheaper computational resources. Even more recently, some modern theories of physics under development have been attempting to unify quantum mechanics and general relativity on the basis of information ('t Hooft, Susskind, Smolin), a development represented in the work of some contributors to this volume.

These contemporary views of a computational universe are also deeply related, via the concept of information, to a contemporary field of mathematical research called algorithmic information theory (AIT). AIT researchers think that the true nature of nature can only be unveiled by studying the notion of randomness (Greg Chaitin, Leonid Levin, Cris Calude). A few papers are devoted to this topic, including two from Chaitin and Calude[a].

---

[a]Where author names are provided without a reference, the authors in question are contributors to this volume.

1.1. What is computation? How does nature compute?

Zuse suggested early on that the world was possibly the result of deterministic digital computation, in particular a cellular automaton. Ed Fredkin would later develop the idea further (see his contribution to this volume). In his minimalistic approach, Wolfram argues that the world may turn out to be the result of very simple rules, perhaps even a single one, from which the apparent complexity we see around us emerges. If the universe is a computable one, we could just run a universal Turing computer on every possible program to generate not only our own universe but every possible one, as Wolfram and Schmidhuber have suggested (both are contributors to this volume). The question would then be how to distinguish our universe from any other. Wolfram has pointed out that there will be some universes that are obviously different from ours, and many others that may look very similar, in which case we can ask whether ours will turn out to be special or uncommon in any sense, whether, for example, it would rank among the first in terms of description size, i.e. be among those having the shortest description. If it is simple enough, AIT would then suggest that it is also frequent, the result of many programs generating the same universe. This is a conclusion based on algorithmic probability, which describes the distribution of patterns, and establishes a strong connection to algorithmic complexity.

According to Schmidhuber's approach, it would seem that a computer generating every possible universe would necessarily have to be several times larger than the universe itself. But if the programs are short enough, instead of running every program a step at a time, beginning with the smallest and proceeding in increasing order of size, it may be possible to start with some plausible universes and run them for longer times, as Wolfram has suggested, checking each program and allowing those that are apparently complex to unfold (and eventually leading to the physical properties of our universe).

There are also those who believe that nature not only performs digital computation (as is the view of Zuse, Fredkin or Wolfram, all contributors to this volume), but is itself the result of quantum computation (Lloyd, Deutsch and Cabello, also authors of chapters in this volume). According to them, the world would in the last instance be rooted in physics, particularly quantum mechanics, and would reflect the properties of elementary particles and fundamental forces. Lloyd, for example, asks how many bits there are in the universe, offering an interesting calculation according to which, given the properties of quantum particles, the universe cannot be rendered in a description shorter than itself–simply because every elementary particle would need to be simulated by another elementary particle. A computer to simulate the universe would therefore need to be the size of the universe, and would require the energy of the actual universe, hence making it undistinguishable from a (quantum) computer, the computer and the computed being in perfect correspondence.

If the goal is not to describe with the greatest accuracy our universe in the state it is in, one may ask whether a universe of similar complexity would require a very complicated description. The discussion seems therefore to revolve around the possible description of the universe, whether it can be written in bits or in qubits, whether it can be shorter than the universe or not. If digital information underlies the quantum, however, it may turn out that the shortest description of the universe would be much shorter than the universe itself, contrary to the views of, for example, Lloyd, Deutsch, Cabello or Calude et al., and could then be compressed into a simple short computer program, as Zuse, Wolfram, Fredkin, Bolognesi and I myself believe.

In fact, one can think of the goal of digital physics as a minimal model describing the universe, equivalent to the goal of physics in its quest for a unified theory, a descriptive formula governing all forces and particles yet to be discovered, if any such remain. Whether such a program exists is an open question, just as it is an open question whether there is a theory of everything (ToE). But it is no longer an open question whether such unification can be achieved for very large portions of physics. It is a fait accompli in fields such as gravitation and movement, electricity and magnetism, electromagnetism and most electronuclear forces, areas of physics that today model with frightening accuracy large portions of nature using simple laws than can be programmed in a computer and can in principle (and often de facto) provide perfect prescriptions and predictions about the world. In fact we have physical laws and computer programs for pretty much everything; what we lack is a single theory that encompasses all other theories.

2. The Algorithmic Approach

If the world is in fact not a digital computer, it could nevertheless behave like one. Thus whether or not it is a digital computer, one could test whether the output of processes in the world resembles the output that one would expect from running a program. There are many laws that computers may follow. Or if you prefer, they follow a specific subset of physical laws related to information processing, notably the distribution of patterns described by algorithmic probability.

From my point of view, information can only exist in our world if it is carried by a process; every bit has to have a corresponding physical carrier. Even though this carrier is not matter, it takes the form of an interaction between components of matter–an atom interacting with another atom, or a particle interacting with another particle. At the lowest level, however, the most elementary particles, just like single bits, carry no information (the Shannon entropy of a single bit is 0 because one cannot implement a communication channel of 1 bit only, 1 and 0 having the same possible information content if taken in isolation). Isolated particles may have no causal history, being memoryless when isolated from external interaction. When particles interact with other particles they appear to be linking themselves to a causal network and seem to be forced to define a value as a result of this interaction (e.g. a measurement). What surprises us about the quantum world is precisely its lack of apparent causality, which we see everywhere else and are so used to. But it is the interaction and its causal history that carries all the memory of the system, with the new bit appearing as if it had been defined at random because our theories of quantum mechanics only provide probability amplitudes. Linking a bit to the causal network may seem tantamount to producing a correlation of measurements between seemingly disconnected parts of space, while in fact they may have always already been connected, if the world were taken as deterministic (a view implicit in so-called *hidden variables* models).

Levin's universal distribution[12] based on algorithmic probability describes expected output frequencies in relation to their complexity. A process that produces a string s with a program p when executed on a universal Turing machine T has probability $Pr_T(s) = 2^{-|p|}$ where $|p|$ is the length of the program p.

The coding theorem[7,8] connects the frequency $Pr(s)$ with which a string s is produced to its algorithmic complexity $C(s)$. The so-called *semimeasure* m has also the remarkable property of dominating Pr_T for any universal Turing machine T. Roughly speaking, $m(s)$ establishes that if there are many long descriptions of a certain string, then there is also a short description with low algorithmic complexity $C(s)$, that is $m(s) \approx 2^{-C(s)}$. As neither $C(s)$ nor $m(s)$ is computable, no program can exist which takes a string s as input and produces $m(s)$ as output. However, we have proven[10] that numerical approximations are possible and that reasonable numerical evaluations produce reasonable results. Just as strings can be produced by programs, we may ask after the probability of a certain outcome from a certain natural phenomenon, if the phenomenon, just like a computing machine, is a process rather than a random event. If no other information about the phenomenon is assumed, we can see whether $m(s)$ says anything about a distribution of possible outcomes in the real world.[22] In a world of computable processes, $m(s)$ would indicate the probability that a natural phenomenon produces a particular outcome and tell us how often a certain pattern would occur. Consider an unknown operation generating a binary string of length k bits. If the method is uniformly random, the probability of finding a particular string s is exactly 2^{-k}, the same as for any other string of length k, which is equivalent to the chances of picking the digits of π at random. However, data (just like π–largely present in nature, for example, in the form of common processes relating to curves) are usually produced not at random but by a specific process (in the case of π there are many ways to produce it; there are even physical phenomena that lead to it). This is where $m(s)$ may be relevant to calculating the probability of physical processes.

2.1. *Information and structure in living organisms*

Biology has witnessed a transformation during the last century, beginning with Mendel's discoveries regarding the transfer of certain traits in pea plants, a phenomenon amounting to an information transfer between generations. Later rediscovered, his laws would lay the foundation for what is today the modern science of genetics, which has established that living organisms store and bequeath information comprising instructions for their full development encoded–as Watson and Crick discovered–in ribonucleic and nucleic acids. The code of life is digital; two bits per base pair are needed to encode the DNA.

Rules determining the way DNA replicates may be algorithmic in nature, like those governing other types of physical phenomena, leading us to sometimes discover strong similarities in their pattern distribution. Processes of DNA are relatively simple. A subset of purely digital operations can match their operations with computational ones, operations such as joining, copying, partitioning, complementation, trimming, or replacing. Which implies that layer upon layer of the code of life has been built up over billions of years in a deep algorithmic process with its own characteristic rules, making processes like protein folding appear highly complex to us. If we ignore the algorithm of protein folding, there is no reason to think that protein folding cannot be carried out by a (deterministic) machine (whether in polynomial time or not).

In the case of structures whose final state is certain (whether a folded protein or the division of a cell), the solution is either the result of an algorithm or of a random process. It is unlikely that protein folding is random because an incorrectly folded protein would obviously cause disease. One of the most important properties of life is robustness, which we think may be explained by algorithmic probability.[24]

If computation is the driving force producing structure in the world, one can use it as a basis for manipulating the direction of complexity. Think of the problem of putting together the right chemical elements for life. If we expect life to emerge out of chemicals, expecting them to produce, say, the digits of the mathematical constant π, the chances of this happening are ridiculously low. How has nature produced organisms of such complexity? Wolfram, for example, thinks that nature mines what he calls the computational universe of possible programs. The concept of Darwinian evolution may lead us to assume that whatever the processes are that give rise to the forms we see in biological systems, they must have been fully shaped by natural selection. But there is strong evidence (e.g. Ref. 20) that nature samples programs, that in some way nature may be disposing what computation proposes. If the same question about pattern production is now asked in the context of computation, the probability of chemicals self-assembling to produce the digits of π by chance is substantially larger, because one needs to find a program producing π rather than the digits themselves. Programs producing π are infinitely shorter than the number of digits in π.

Patterns in nature could also be the result of a similar way of shortcutting pattern formation, and the cause of the structure we see in nature. Patterns would not repeat by mimicking themselves; they would recur because they are the result of a simple program producing the same pattern over and over, with the frequency dictated by algorithmic probability. Should the pattern change, it would mean that the rule has changed too, but as algorithmic probability predicts, if the program remains short, the chances of its producing a pattern of low algorithmic complexity, perhaps even the same one that it was producing before a mutation, is exponentially large, as compared to the chances of its producing random-looking patterns. It would make better sense then to think of natural selection as taking place at the level of programs, in order to preserve the fundamental property of resilience in living structures, i.e., robustness. Patterns are robust then because they are the result of a short, concise rule.

On the other hand, if one were free to select random patterns rather than random programs, the result would be rather random-looking. Yet this seems not to be what happens, which is why we don't see sudden changes in patterns in nature, even if we do experience some apparent randomness. A legitimate question to ask is the role of this apparent randomness in biology, whether it is what drives biological speciation based on mutation. Is this randomness only apparent? Ultimately, is the world more random than structured (see, e.g. Calude. Another proponent of this view is McAllister[15]).

How could the world be a simple computer program? If every physical process is itself computable, Turing showed that there is a single machine capable of running all of them. In other words, one can build a single program out of many. Of course, claiming that processes or that the world itself may be a computer program doesn't mean that they actually are, or that they behave like Turing machines. We should not let ourselves be fooled into thinking that the proposition that nature is computable means that the universe is a Turing machine. Obviously there are natural processes that are definitely not like Turing machines. A Turing machine is an oversimplification of the concept of (digital) computation. The human brain, for example, is a very different object to a Turing machine, despite the parallels between them. The question of whether (all) physical processes in the human brain can be carried out by a Turing machine shouldn't be taken as suggesting that the brain is a Turing machine. In this volume, Szudzik and Hutter provide precise mathematical formalisms for the computable universe hypothesis.

But if the world is dominated by computable processes, of the kind that can be carried out by digital computation, then much of the structure of the world may be credited to computation alone, because computation would follow the universal distribution of the frequency of patterns that algorithmic probability describes. The fact that, their limitations notwithstanding, scientific models can describe much of the world, and that such models are computable, strongly suggests that though nature may do more than compute, to a large extent its activity does amount to (Turing) computation. In the course of history we have managed to make nature do what we wanted it to do. The process began with simple tools, ultimately leading to computers that can be programmed to perform all kinds of calculations for us, and used to make other devices do all kinds of work, serving as tools. If nature does more than just compute, we know that it nevertheless can compute, and that it does so very well. One argument that may be advanced against this view is the inability of scientific models to predict with arbitrary precision. But their non-linearity by no means implies indeterminism or uncomputability. As Wolfram has shown, even the simplest computer programs (such as the elementary cellular automaton (ECA) rule 30) are very difficult, if not impossible to understand well enough to be able to predict the manner of their unfolding–despite its overwhelming simplicity.

We don't have many reasons, perhaps none at all, to believe that there is anything inherent in nature that is not Turing computable, even if unpredictable. The fact that we are always able to replace theories with more encompassing and more accurate ones can be taken as a computable process in itself, in the sense of Turing's very definition of computability. Using π again, π is a computable number because a Turing machine can compute any arbitrary number of the digital expansion of π. So even if we are not able to exhibit and calculate the full expansion of π in any base, the fact that it is computable means that we can always provide a better approximation of it. For any digit with index i in the expansion of π we can always calculate $i+1$ digits of π. π is also random-looking and in many respects unpredictable, as we cannot tell, without making a calculation, what an arbitrary digit will be (not even what the chances are, as it is believed to be normal, every digit has equal probability). Even if there are formulae (BBP[2]) to calculate arbitrary digits of π in bases 2^n that do not require the calculation of the previous digits, the result of the calculation of a digit of π doesn't make it any easier to recognise. The digits of π in all the bases in which it has been studied looks random by all current statistical tests, and although there is no proof of its normality, it is believed to be so, meaning that it has a property of a true random number, viz. that it contains every possible pattern, though it is not itself a pattern, being random.

As for unknown formulae for π, for ECA rule 30, one cannot rule out the possibility of formulae of the same or completely different type that nonetheless permit rapid computation of individual digits of the evolution of these apparent random systems.

3. Determinism from Quantum Mechanics?

Unlike the apparent randomness from previous section from π or ECA rule 30, in quantum mechanics events seem to happen for no apparent reason—

for example, the time at which a particle decays or the position of an electron collapsing from entanglement. Surprisingly enough, the most serious idea that the universe is digital comes from quantum mechanics (with its roots in a long tradition beginning with the ancient Greek concept of atomism). It was Max Plank, who had been trying to understand the emission of radiation from heated objects, who discovered the quantisation phenomenon of radiation. Quantum mechanics is in effect thought of as being digital, but it is at the same time analogue. An electron can be in many places simultaneously, covering an extended region of space. So quantum mechanics hasn't settled the issue once and for all, because particles themselves behave as waves, thus preserving a duality between digital and analogue descriptions of the most fundamental building blocks of the universe (for a discussion of this topic see Ref. 23). This is not a question of philosophy, nor even of mathematics; it is a question of physics, with a definite answer. If we zoomed in far enough into reality, and if the world is digital, we should be able to see pixels and bits. However, if the world is not digital, we would always be able to zoom in without being able to prove or disprove any possibility. A third answer is that it is neither one nor the other, implying that it is actually both, just as the standard interpretation of quantum mechanics would have it.

There is a tendency, however, that has favoured a discretisation of models (for some discussion of this matter see Ref. 25. Electricity was first thought to be continuous, but then the electron was discovered. Likewise with the photon, and gravitation is today approached in the same way. It would seem that we keep going from a continuous theory to an atomic one, ending up with an unsatisfactory answer (particle duality leading to an explanation of electricity in terms of electrons, which themselves behave both as particles and waves). In this process, we have come to realise that information plays an important role, as what many of these theoretical mergers actually do to objects or phenomena is to divest them of certain characteristics once believed to be exclusively theirs and attribute them to the composite object or phenomenon of which they are now thought to form a part.

The development of quantum mechanics early in the last century prompted physicists to radically rethink the concepts they used to describe the world. No current account of what information may be can be considered complete that does not take into account the interpretations of quantum mechanics. Classical systems comply with the criteria of what may be identified as local-realism, that is, that the results of measurements of a system localised in space-time are fully determined by properties inherent in that system (its physical reality), and cannot be instantaneously influenced by a distant event (locality). In other words, locality prohibits any influences between events in space-like separated regions, while realism claims that all measurement outcomes depend on pre-existing properties of objects that are independent of measurement. Quantum phenomena, however, seem fundamentally non-local in the relativistic sense, as one is forced to invoke a particular frame of reference to give a sense to the statement that measurement of a particle happens first, and that its result immediately affects the state of a second particle, even if placed beyond the limit allowed for information exchange between the two by the theory of relativity.

As discovered by Albert Einstein, Boris Podolsky and Nathan Rosen, quantum mechanics predicts strong correlations between measurements carried out on two particles in an entangled state. It is tempting to interpret these correlations as the result of shared properties determined at the time of their initial interaction and then assimilated into each particle. Bell's formulation of an inequality[5] made it possible to settle the debate by performing an experiment to test the inequalities by showing that hidden-variables theories based on the joint assumption of locality and realism are at variance with the statistical predictions of quantum mechanics.

Ever since, physicists have undertaken experiments to test quantum reality. The first of these was John Bell himself, who showed that if one assumed that particles were correlated, in the sense that measuring the properties of one tells you the properties of the other, then there was an inequality which described the maximum possible correlation in a classical world. Later refinements of Bell's inequality tests have continuously converged, closing both the locality loophole on the one hand, and the detection loophole on the other. Therefore quantum foundation scientists think that it is reasonable to consider the violation of local realism a well established fact.

For the Bell experiment to validate quantum reality, one would need to design it to cover all possible open ends or loopholes. Could it be that there were some biases induced by the Bell experiments that resulted in particles that were more correlated having a greater chance of being the ones measured? That would explain the violation. Alain Aspect[1] designed an experiment to test and rule out this possibility. It involved the use of several detectors measuring a large number of photon pairs in order to obtain statistically significant results. After several experiments, the community reached a point where it was convinced that Aspect's experiments ruled out the possibility that Bell's experiment was statistically biased, measuring only correlated particles. Could it be that this and other experiments are still fooling us, making us believe that quantum reality behaves in a certain way when it actually does not?

Aspect helped to rule out the problem of dealing with local bias in the measurement of correlated particles, but what if something were influencing the experiment by communicating properties between particles? One way to rule out this possibility is by ensuring a distance between the correlated particles, a distance sufficient to guarantee, by the speed of light, that if the two measurements at the two ends were fast enough, nothing could travel to communicate anything about one particle to the other. Remember that the assumption here is that even when the two particles are correlated because they come from a single source, they could not be correlated beyond what the Bell inequality establishes, and so if the correlation is greater than that, it means that there is an eerie entanglement between quantum particles that leads one of them to know or change its properties once the other has been measured.

If two of these particles are close enough to communicate with each other, the violation of Bell's inequality is easily explained. One should guarantee that information cannot therefore be disseminated without a physical carrier, a carrier of matter which cannot travel faster than the speed of light. And so by placing two particles far enough apart, one can try to rule out this possibility (again if it is not the case that there is an underlying layer of information, a hidden-variables reality storing or communicating the measured values). The availability of highly efficient sources of entangled particles (with the emergence of laser technology) has allowed quantum scientists to perform *Aspect experiments* with a distance large enough to apparently close this loophole, preventing particles from communicating with each other without having to assume that they have done so faster than the speed of light, which general relativity renders out of the question.

This can only work, however, if the settings of the detectors are changed every time, given that previous experiments may have influenced them or that one detector may have figured out the settings of the other. For in order to guarantee the validity of the resulting measurement of correlation, there must not be any correlated measurements at the outset. Weihs et al.[19] managed to separate two particles by 400 meters, giving them 1.3 microseconds to switch the detectors' settings randomly, and the maximum

5 nanoseconds between the measurements guaranteed that, at the speed of light, no information could possibly be transmitted between the two ends. Even as some loopholes in tests of quantum mechanics are closed, others may open up (e.g. *collapse locality*) and no test has yet encompassed all loopholes at once. The local-realistic hidden variables theory preferred by Einstein may not be a viable description of the world, but there are other loopholes to close, particularly one that is assumed and that ultimately leads to a circular argument.

It turns out that the aforementioned attempt at closing a loophole assumes the ability to randomly change the settings of the detectors on each side, by keeping their random generators far enough apart to ensure that the choice of one random generator does not influence the other. This is a loophole that no experimental physicist thinks is worth trying to close, because there is little, if anything, to test. It turns out that all tests of Bell's inequality assume that one has the freedom to choose the detectors' settings and that such freedom has to be indeterministic for the interpretation of the results of the experiment to work. But in practice one is obliged to choose between two options: either to use a pseudo-random generator, which means that the source of randomness is actually not random but deterministic and therefore potentially correlated to the other detectors' settings, or that one is free to choose from a set of non-commuting operators, i.e. free will and true (indeterministic) randomness to begin with when preparing the experiment.

If the universe runs deterministically, however, there is nothing to explain and all quantum strangeness makes immediately sense. This is in agreement with what Bell himself acknowledged as a possibility: if quantum mechanics is completely deterministic, then even the experimenter's decision to measure certain components of the spins is entirely pre-determined, so that the observer could not possibly have decided to measure anything other than what he does in fact measure. In other words, the correlations would have a trivial explanation, because it is trivial to think of correlated states at all levels, given that everything in the universe would indeed be the effect of a previous state and would therefore simply share a common origin. By analogy, think of similar sets of chromosomes between siblings. Sharing genes explains correlations in their eye colour or other features, without having to introduce other convoluted explanations. That everything in the universe is determined since the very beginning is of course a strong commitment to make in order to make sense of quantum experiments, but it is no doubt it is much simpler by most, if not all means, than current interpretations of quantum mechanics.

There must be a reason to try to force an interpretation of quantum mechanics where the only possible outcome is indeterministic randomness, even if it means assuming none other than indeterministic randomness. When the theory of quantum mechanics was launched in the works of Plank, Einstein and Bohr, it was clear that classical physics could not explain small-scale phenomena such as the interaction between particles. The theory of quantum mechanics could, but at great cost, since one had to be ready to make all kinds of concessions to certain phenomena incompatible with classical physics. For Bohr, these concessions constituted reality at the level of the quantum world, while Einstein thought the weirdness of quantum mechanics meant that the theory was incomplete, that in the end it couldn't possibly be as incompatible as it was with the familiar reality we inhabit. He was convinced that an yet undiscovered theory of hidden variables would be capable of explaining some, if not all, of these strange phenomena.

> The logical conclusion one can draw from the violation of local realism is that at least one of its assumptions fails. But one could consider the breakdown of other assumptions that are implicit in our reasoning leading to the inequality. These may include a violation of the limit imposed by relativity theory or a world that is completely deterministic.[5]

This loophole is often said to be avoided by quantum physicists concerned with quantum reality, who assume that no test of complete determinism can be performed. Nevertheless, pseudo-randomness has the obvious particularity that when its generating process is repeated, the result is exactly the same. If this is the case, then this very same phenomenon would also occur in quantum mechanics. The difficulty is perhaps that the formalism of quantum mechanics only gives us probability amplitudes, but what we may be witnessing in quantum mechanics in the form of unexpected correlations is simply actually an indication of determinism that we have tried too hard to explain otherwise.

4. From String Theory to Bit String Theory

On the other hand, information is taking on a fundamental character in modern physics, black holes, string theory, etc. The assumption that information is essential to explain our physical reality is not alien to modern physics. Many physicists have arrived at similar conclusions, giving the concept of symmetry (which may be seen as information or an abstract mathematical object) a foundational role, even in predicting the existence of new particles, for example. And so far this approach has been quite successful in many areas of physics. It is remarkable how a simple informational description of some fundamental properties of quantum mechanics fully describes quantum phenomena. Such descriptions specify how particles interact and how they are related to each other by symmetries.

According to classical mechanics, randomness is apparent in the macroscopic world, but under the standard interpretation of quantum mechanics things are fundamentally different. The position that the history of the world is computationally reversible is compatible with the determinism imposed by classical mechanics. At the quantum scale, however, things seem otherwise. When one entangles a particle the particle seems to truly and irreversibly lose track of its previous state, and there is no way, even by reversing all operations, to recover it. This also happens with radioactive decay, where an atomic nucleus of an unstable atom loses energy by emitting ionising particles for which, according to quantum theory, it is impossible to predict the decay time, decay being equally likely to occur at any given time.

Zeilinger claims that quantum randomness is intrinsically indeterministic, and that experiments violating Bell's inequality imply that some properties do not exist until measured.[21] These claims are, however, based on a particular interpretation (if not speculation) of quantum mechanics, from which he leaps to conclusions by relying on various no-go or no-hidden-variables theorems proposed by people like von Neumann, Bell, Kochen and Specker, which are supposed to show that quantum randomness is truly indeterministic. Zeilinger's position is a form of weak epistemological randomness, as he claims that one cannot observe less than one bit of information, and what is not known is then "random".

My position resonates with the response to Zeilinger in.[9] But although I share with Daumer et al. the belief that Wheeler did not shed much light on the issue with his captivating but somehow rather obscure treatment of information as being related to, or as more fundamental that physics (his "it from bit" dictum), I do not share their views as regards what's wrong with the informational content of quantum mechanics. As they point out, Wheeler's remarkable suggestion was that physics is only about information or that the physical world itself is information.

In the words of J.A. Wheeler himself:[18]

> It is not unreasonable to imagine that information sits at the core of physics, just as it sits at the core of a computer. It from bit. Otherwise put, every *it*—every particle, every field of force, even the space-time continuum itself—derives its function, its meaning, its very existence entirely—even if in some contexts indirectly—from the apparatus-elicited answers to yes-or-no questions, binary choices, bits.

Wheeler, however, interjects a nuance related to meaning when introducing the observer (remember Wheeler is also credited with formulating the *anthropic principle*). So in Wheeler's view, meaning is also subjective, which means that it is not a completely reductionist, trivial view of information content.

> 'It from bit' symbolizes the idea that every item of the physical world has at bottom–a very deep bottom, in most instances–an immaterial source and explanation; that which we call reality arises in the last analysis from the posing of yes/no questions and the registering of equipment-evoked responses; in short, that all things physical are information-theoretic in origin and that this is a participatory universe.

Wheeler's most pointed suggestion is that "information" can't be defined in terms of "matter" or "energy" and that it may therefore be as or more fundamental than either "matter" or "energy", the most basic notions in physics. But a second reading also introduces the problem of information content, meaning, and the observer in his participatory universe.

4.1. *An algorithmic approach to the problem of fine tuning*

The anthropic principle can now have a plausible interpretation under this algorithmic approach, providing a (more reasonable) answer to the question of the apparent fine tuning of the universe, that is, the question of why the universe looks just right for accommodating everything in it. A single value changed in the equation would produce a universe where nothing would be possible, certainly not life. But if the universe is a computer program, parameters are not only coupled together but there may be many computer programs producing the same output, especially an output of low algorithmic complexity, that is, a structured universe rather than a random looking one–contra the position taken by Calude et al. in this volume. The fact that parameters are coupled is also a reflection of a possible ToE, according to which everything comes into being as a result of a single physical law. Hence one may inquire into other possible ToEs that model well behaved universes where things seem just right, as they do in ours. In Wolfram's approach, for example, computer universes that are not trivial very quickly begin to look complicated enough for us to judge whether it obviously is or isn't our universe that's being modelled. This means that there is a threshold (captured in Wolfram's Principle of Computational Equivalence) where a universe may begin to look as if it had been fine tuned to accommodate all the structure it is capable of, without being at all special. On the contrary, as it is capable of structure it would have a low algorithmic complexity and therefore a high algorithmic probability.

Following Wheeler, I think that the next level of unification (along the lines of the unification of other previously unrelated concepts in science, such as electricity and magnetism, light and electromagnetism, and energy and mass, to mention a few) will involve information and physics (and ultimately, as a consequence, computation and physics).

Landauer wrote in 1996 (Quoted from Leff and Rex, pag. 335[14]):

> Information is not a disembodied abstract entity; it is always tied to a physical representation. It is represented by an engraving on a stone tablet, a spin, a charge, a hole in a punched card, a mark on paper, or some other equivalent. This links the handling of information to all the possibilities and restrictions of our real physical world, its laws of physics and its storehouse of available parts.

Current attempts to unify modern physics, such as string theory and quantum gravity, depend on information encodings and on how much information is needed to describe something. In these models, it is information that imparts sense to the forces in the universe and to matter itself. In quantum gravity space and time are not fundamental; it is information that constitutes the most basic level of physical reality. That is, everything arises out of information. Relativity had already relegated space and time to this status, because there is nothing special about either dimension; it is just what happens in each that distinguishes space from time. Events taking place in them have only a subjective meaning, with information playing a fundamental role. In quantum gravity too, the exchange of information is fundamental. These theories are believed to introduce a compatibility between quantum mechanics and general relativity through the concept of information (in particular, *maximum information*), connecting energy from quantum theory and energy from relativity theory, and finally bridging the two.

4.2. Black holes as perfect data compressors

Leó Szilárd[17] and Rudolf Landauer[13] determined how a computer uses energy when processing information. It turns out that one can see computers as spending information only when erasing information. One can add or multiply or subtract bits without consuming or increasing the energy in the universe, as long as no information is *erased* in the process. Landauer's principle is derived from solid physical principles. Unfortunately, one has almost always to erase information, for almost everything. Erasing releases heat, which dissipates in the environment. The conclusion is that erasing is not a reversible operation. Erasing is not like subtracting or adding because subtraction and addition are reversible operations, while erasing entails a loss of local information, the heat released containing the lost information. The loss of information at the level of quantum mechanics is apparently of a different nature.

It turns out that black holes play an important role in the informational view of our world given that the maximum information when bits are matched to photons is the maximal mass determined by the black hole limit. At first, black holes were thought to delete information from the universe; whatever fell into them would irreversibly be destroyed, despite the generalised assumption from thermodynamics that information in the universe is never destroyed (Bennett[6]). It was later acknowledged, however, that even black holes may conserve information in the form of a quantum phenomenon occurring at the event horizon by emitting small amounts of thermal radiation[11] from which the information in whatever fell into the black hole could be fully recovered, in principle.

In fact, black hole formation can now be determined, in the context of what has been called the holographic principle,[16] given by the number of bits of information equal to the surface of the black hole divided by four multiplied by the Planck area ($10^{-66}m^2$). Among several important consequences is that the maximum information pertaining to a black hole is finite, and that information pertaining to this physical object can be understood in terms of the information in two dimensions (its surface), for which reason the associated principle is called holographic—by analogy with the fact that the two-dimensional surface of an object contains information regarding its volume in three dimensions[4]. This black-hole thermodynamics defines the maximum amount of information that can potentially be stored in a given finite region of space which has a finite amount of energy. The more information a black hole has, the larger its event horizon, and in no smaller place such information could be contained, as black holes as maximally larger when reaching the point of singularity. Black holes can be regarded thus as perfect data compressors.[b]

Quantum decoherence means that the information an observer has about the probabilities of the different possible outcomes involves some specific physical degrees of freedom. But again, the measurement problem is just that the observer does not know the outcome of a quantum measurement, and as such quantum measurements appear not to be reversible, because entangling a particle again and performing the same experiment would lead to a different result, one that's discrepant from the previous measurement, not because it is random in the mathematical (algorithmic) sense, but in the same sense as we don't know, and we seem unable to predict the classical world, not because it is indeterministic, but because it is unpredictable. Even if there are subtle differences, it all gets down to a problem of epistemological nature, information at the end.

This goes to show how information may already explain some of the most fundamental processes in the universe at the edge of our understanding of the largest and the tiniest, and underscores the pertinence of asking seriously whether the universe can be fully explained in terms of information and computation, perhaps even suggesting ways to conciliate general relativity with quantum mechanics.

References

1. A. Aspect, P. Grangier, and G. Roger, Experimental Realization of Einstein-Podolsky-Rosen-Bohm Gedankenexperiment: A New Violation of Bell's Inequalities, *Physical Review Letters, Vol. 49*, Iss. 2, pp. 91–94, 1982.
2. D.H. Bailey, P.B. Borwein, and S. Plouffe, On the Rapid Computation of Various Polylogarithmic Constants, *Math. Comput.* 66, 903–913, 1997.
3. J.S. Bell, *Speakable and unspeakable in quantum mechanics,* Cambridge University Press, Cambridge, 1987.
4. J.D. Bekenstein, Black-hole thermodynamics, *Physics Today,* 24–31, 1980.
5. J.S. Bell, Free variables and local causality, *Dialectica* 39, 103–106 (1985).

---

[b]An interesting question in connection to Kolmogorov complexity, given that Kolmogorov complexity is uncomputable, is how nature may have achieved a perfect data compressor, even though its contents and mechanisms may turn out to be inaccessible to us in practice.

6. C.H. Bennett, The Thermodynamics of Computation – A Review, *International Journal of Theoretical Physics*, vol. 21, no. 12, pp. 905–940, 1982.
7. C.S. Calude, *Information and Randomness*, Springer-Verlag, 2002.
8. T.M. Cover, J.A. Thomas, *Elements of Information Theory*, Wiley-Blackwell; 2nd Edition edition, 2006.
9. M. Daumer, D. Dürr, S. Goldstein, T. Maudlin, R. Tumulka and N. Zanghì, The Message of the Quantum?, PhilSci-Archive, 2006.
10. J.-P. Delahaye, and H. Zenil, Numerical Evaluation of Algorithmic Complexity for Short Strings: A Glance Into the Innermost Structure of Randomness, forthcoming in *Mathematics and Computation*, 2012.
11. S.W. Hawking, Black hole explosions? *Nature* 248 (5443): 30, 1974.
12. W. Kirchherr, M. Li and P. Vitanyi, The miraculous universal distribution, *Math. Intelligencer*, 19(4), 7–15, 1997.
13. R. Landauer, Irreversibility and Heat Generation in the Computing Process. *IBM J. Res. Develop.* 5 (3): 183–191, 1961.
14. H. Leff, A.F. Rex (eds), *Maxwell's Demon 2: Entropy, Classical and Quantum Information*, Computing, Taylor & Francis; 2nd Revised edition, 2002.
15. J.W. McAllister, Algorithmic randomness in empirical data, *Studies in History and Philosophy of Science*, pp. 633–646, 2003.
16. L. Susskind, The World as a Hologram, Journal of Mathematical Physics, 36 (11): 6377–6396, 1995.
17. L. Szilard, Über die Entropieverminderung in einem thermodynamischen System bei Eingriffen intelligenter Wesen, *Zeitschrift für Physik A Hadrons and Nuclei*, 1929.
18. J.A. Wheeler, Information, physics, quantum: The search for links. In W. Zurek (ed.) *Complexity, Entropy, and the Physics of Information*, Addison-Wesley, 1990.
19. G. Weihs, T. Jennewein, C. Simon, H. Weinfurter and A. Zeilinger, Violation of Bell's inequality under strict Einstein locality conditions, *PRL 81* (21), 5039–5043, 1998.
20. S. Wolfram, *A New Kind of Science*, Wolfram Media, 2002.
21. A. Zeilinger, The message of the quantum, *Nature* 438, 743, 2005.
22. H. Zenil and J.-P. Delahaye, On the Algorithmic Nature of the World. In G. Dodig-Crnkovic and M. Burgin (eds), *Information and Computation*, World Scientific Publishing Company, 2010.
23. H. Zenil, The World is Either Algorithmic or Mostly Random, *FXQi Contest: Is Reality Digital or Analog?* 3rd Prize winner, February 2011.
24. H. Zenil and J.R. Marshall, Some Aspects of Computation Essential to Evolution and Life, *Ubiquity*, forthcoming.
25. H. Zenil, Information Theory and Computational Thermodynamics: Lessons for Biology from Physics, *Information*, In revision.

PART 1
Historical, Philosophical & Foundational Aspects of Computation

Chapter 2

Origins of Digital Computing: Alan Turing, Charles Babbage, and Ada Lovelace

Doron Swade

Royal Holloway University of London, UK

Alan Turing continues to dazzle, enlighten and bemuse. His work is difficult to situate, straddling as it does theory, practice and philosophical speculation. The nature and extent of his influence continues to elude us, and the haunting originality of his work resists convenient typecasting. He was capable of blinding clarity as well as of near-mystical obscurity that has an ability to tantalise, madden, and stretch our horizons at the same time. His formal description of a universal computing engine predates the practical implementation of the general-purpose electronic stored-program digital computer, but his influence on its realisation is difficult to nail down.[28] This is partly due to the continuing legacy of war-time secrecy but also the general paucity of documentary records.

To help locate Turing's ideas in the larger trajectories of computer history I propose here to revisit the back-story of computing to identify the core ideas embodied in the realisation of the modern computer. Most of these emerged explicitly in the 19th century, a surprising assertion given the accepted modernity of the field. I hope to show that making this assertion is not a presumptuous grab by history to appropriate the triumphs of the late 20th and early 21st centuries, but rather a place-holder for an intriguing set of connections between computing pre-history and the electronic era.

1. The Grand Narrative

In the last few decades a grand narrative of the history of computing has begun to emerge.[1,49] The tale typically starts with counting systems and physical counting aids (calculi, knotted cords, tally sticks) and other devices of record. Attention then tends to shift to mechanical calculating aids starting with the abacus followed by Napier's Bones, analog devices for calculation and measurement including slide rules, and the invention of logarithms by John Napier in 1614. Desktop devices usually follow with Wilhelm Schickard's 'calculating clock' (1623), Blaise Pascal's ornamental calculator (the 'Pascaline', 1644-5), and Gottfried Leibnitz's 'Reckoner' (1672-4).

In terms of practically usable devices as distinct from philosophical novelties, things begin to get serious with the introduction, in 1820, of the arithmometer by Thomas de Colmar, though it took another sixty years before the device began to mature into a viably practical product.[26,37] There invariably follows a section on the English mathematician and polymath, Charles Babbage (1791-1871), and the designs for his vast automatic mechanical calculating engines. Babbage is followed by Herman Hollerith and the introduction in the 1890s of the Hollerith tabulator for the collation and management of census data. The Hollerith machines routinely feature in a seamless segue to IBM and the triumphs of office automation in the early and middle decades of the 20th century. The electronic era, starting in the late 1930s, overlaps with and supersedes the electromechanical office automation era, and the ensuing boom in the computer industry over the last sixty years brings us, a little breathless, to the present.

The tale told in this way implies a developmental continuity of sorts, a comforting form of evolutionary gradualism. However, a serial narrative of this kind masks discontinuities. It also tends to imply a single developmental thread and leaves unrecognised other distinct threads.[42] A time-serial tale is perhaps partly an artefact of the medium: text or spoken language is inherently serial and a chapter sequence is both slave to and an expression of this. There may be something deeper at play here: linear narrative may be reflecting a presumption, conscious or unconscious, of technological determinism – the idea that the engine of change is technology which then 'impacts' society as though separate and outside social context. 'Do machines make history?' asks Heilbroner in his classic paper,[20] or does history make machines?

A technocentric approach would naturally privilege an account in terms of generational technologies, in this case the progression from mechanical to electromechanical, and then electronic devices and systems. A tale so told is one of the 'onward march of progress' with technologies superseding their predecessors often with the implication that change was generally welcomed, with long-suffering practitioners lamenting the deficiencies of incumbent technology now rescued by the gifts of innovation. Timeline accounts of this kind make for powerful narratives. But chronology as history can obscure other features.

If we shift from a technocentric view to a user-centric view[16] the landscape opens up. If we look at the human activities that devices and systems seek to relieve or replace at least four distinct threads suggest themselves: calculation, information management, automatic computation, and communications.[42]

Our first task is to unplait the thread of automatic computation from the larger braid. In the calculation thread, desktop calculators can be seen as part of an unbroken strand from calculi to the electronic pocket calculator. Desktop calculators are capable of only a fixed and limited set of predetermined functions, they are portable or semi-portable, and are operated by hand and so require the intervention of an operator to yield useful results. These are the features that distinguish these devices as a class. The heyday of desktop calculators is perhaps best evoked through legendary brands of the early 20th century, the pinwheel devices by Brunsviga and Facit, and the key-driven comptometers by Borroughs and Felt. There is a raft of less well-known devices made in large numbers that live in the shadow of the celebrity machines and that were part of a burgeoning movement dubbed nearly a century earlier by Carlyle as 'the age of machinery'.[8]

The information management thread has its roots in early manual recording, filing and accounting practices and a major impetus to its mechanisation came in the form of Herman Hollerith's tabulator introduced in the 1890s to manage, process and collate data for the US census. The machines in the punched card era of office automation, based on tabulating and sorting, provided high throughput in the management of large data volumes. They were automatic but their capacity for calculation or logic was relatively trivial.

Of particular interest to us here is the thread of automatic calculation that debuts with the work of Charles Babbage who, starting in 1821, sought to automate calculation using machinery. In this he was led from mechanised arithmetic to fully fledged automatic digital computation, at that time entirely new territory. Babbage himself failed to complete any of his engines in their entirety and his work, inspired and uncannily farsighted as it was, was, for all its promise, a false dawn. With Babbage's death in 1871 automatic computation lost its most visible protagonist and the initiatives of his generation largely petered out. The developmental line between Babbage's machines and the modern age should not be assumed to be direct. So if we revisit the grand narrative in its received form we can see that the story has been assembled by splicing together episodes from different threads. There is little in the way, for example, of either functional similarity or obvious influential connection between Babbage's engines and Hollerith's tabulators notwithstanding the use of punched cards in both.[2] Hollerith post-dates Babbage, but beyond close succession in time to closely position Hollerith after Babbage as is routinely done, involves a jump-cut. Juxtaposing them in this way masks a large gap in the developmental continuity of automatic computation, a dormant period Leslie Comrie called the 'dark age of computing machinery that lasted 100 years'.[9] Comrie was referring to the period from the 1840s, by which time Babbage had completed his major work, to the re-articulation of the essential logical principles of automatic computation in the modern era. Yet Comrie's 'dark age' was not as devoid of light as he supposed,[33,34,42] and one of the questions posed by his portrayal of the post-Babbage period as fallow is the nature and extent of influential connection between the 19th and 20th centuries, specifically between Babbage-and-Lovelace, and the pioneers of the electronic age amongst whom Alan Turing features prominently. The final strand, communications, has its own rich traditions – language, semaphore, telegraphy, telephony, and so on. Though it plays a crucial role in the convergence that underpins the 'information age', it is of less immediate concern here.[18]

2. Automatic Computation

A towering figure in the history of automatic computation is that of Charles Babbage, whose designs for vast calculating engines rank as one of the startling intellectual achievements the 19th century. We find in his work practically all the core ideas of modern general-purpose digital computing.

Babbage's epiphany is captured in the well-known vignette in which, in 1821 in London, he and John Herschel were checking astronomical tables calculated by human computers. Dismayed at the number of errors he recounts that he exclaimed 'I wish to god these calculations had been executed by steam'.[25] The appeal to steam can be read as a metaphor for the infallibility of machinery as well as for the idea of mechanised production as a means of solving the problem of supply. The 'unerring certainty of mechanical agency'[27] would ensure error-free tables as and when needed. Babbage devoted much of the rest of his life to the realisation of mechanised calculation.

That tabular errors were the jumping off point for Babbage's work on automatic calculating machines is well evidenced. However, history appears to have transmuted this initial stimulus into an abiding motive that informed the rest of his work. There are understandable reasons for this as a recent analysis shows.[38] It is sufficient here to offer a corrective: that while errors in tables feature in Babbage's articulation of benefits, this was one consideration amongst many. It is clear from his early writing that the mathematical potential of the engines as a new technology for mathematics preoccupied him and transcended the pragmatic benefits to table making. The supposed scandal of errors in printed tables emerged more stridently when Babbage and the engine advocates lobbied for support and needed to provide a utilitarian justification in terms of practical benefit. But for Babbage the benefit to the integrity of tables, while real, was a relatively mundane companion in motivational terms, to the theoretical potential of computational machines.

How it is that the primacy of the role of errors became enshrined in the historical canon at the expense of his mathematical interests, a feature of the account that has misled historians ever since, is an intriguing tale with an unexpected show business angle.[38] Suffice it to say that Dionysius Lardner, who wrote and lectured with theatrical impact and popular success on Babbage's Engine, underplayed the mathematical potential of the machine because this material was 'too difficult' for a general audience. By emphasising errors in printed tables as the principal need, and downplaying the theoretical potential of the machines, Lardner forced Babbage to defend the machines from a position of relative weakness. And this offers a partial answer to the question as to why Babbage, who was capable of incontinent savagery in his public attacks on colleagues he regarded as unprincipled, made no public defence of his engines against assaults on their usefulness by prominent astronomers in England and the Continent.[37,38]

A startling feature of Babbage's ambitions for his machines is the quantum leap in logical conception and physical scale in relation to what went before. The main aids to calculation in the 1820s were slide rules used for low and moderate precision of typically less than four digits, and printed tables where higher precision was required. Mathematical, commercial and actuarial tables were calculated, transcribed, typeset in loose type and verified, primarily by hand, all these processes being subject to human error.[45]

Mechanical calculators of the day had a limited number of digits and were largely unreliable. They incorporated only the most elementary rules of addition in their mechanisms and at best replaced only a small part of the process of calculation and tabulation. Above all, they were manually operated: numbers needed to be entered on sliders or dials by hand or sometimes with a stylus, the carriages and handles needed to be manipulated by the operator in a defined sequence, numbers read off and written down by hand, and the result then verified. Each of these processes was subject to human error, the selfsame fallibility that was responsible for the errors that so dismayed Babbage and Herschel in 1821.

Babbage's Difference Engine, so called because of the mathematical principle on which it was based, the method of finite differences, would, in intention at least, eliminate all sources of human error at a stroke. Calculation, transcription, typesetting and verification would be unfailingly correct, ensured as they were by the certainties of mechanism. Babbage's tale is increasingly well known.[23,43] Suffice it to say that despite independent means, government backing, decades of design and development and the social privileges of a well-heeled gentleman of science, Babbage failed to complete any of his engines in his lifetime. The largest of the small assemblies he did complete was one-seventh of the calculating section of Difference Engine No. 1, assembled as a demonstration piece (Figure 1), and delivered to his house in Dorset Street in December 1832.

I would like to highlight five historically significant features of Babbage's ideas so far:

- the idea of the autonomous machine and the start of automatic computation
- the idea of a machine cycle
- digitisation of calculation through the discretisation of motion to ensure integrity
- the development of mechanical logic
- the idea of machine intelligence

The 'beautiful fragment'[5] completed in 1832 shown in Figure 1, is perhaps the most celebrated icon in the pre-history of automatic computing. It is the first successful automatic calculating device. By turning a handle i.e. exerting physical effort, one could achieve results that up to that point in time could only be arrived at by mental effort. One did not need to understand how the device worked nor the mathematical principle on which it was based in order to get useful results. Mathematical rule was for the first time embodied in mechanism. Human intelligence had been transferred from mind and to machine. The significance of this was not

Fig. 1. Difference Engine No. 1, demonstration piece, 1832. Credit – SSPL, Science Museum.

lost on Babbage's contemporaries and the machine was referred to by some as 'the thinking machine'.[43] The notion of machine intelligence was struggled with from the start. Savants were trapped between enthusiasm for the capabilities of the new machines, and concern to give reassurances that machines were incapable of 'creativity' as imputing that man-made machines could 'think' ran dangerously close to giving offense to religious views of the time. The first automatic calculating machine gave urgent impetus to this debate, and the historical significance of a machine embodying logical rule in mechanism and operating autonomously cannot be overstated. In a broader context the machine can be seen as an extension of the industrial metaphor of production from thing to thought, from matter to mind, from physical to mental. The machine, to quote a contemporary of Babbage, was a 'real manufactory of figures'.[31,35]

Difference engines calculated and tabulated polynomial functions using the method of differences, a technique that offered the significant simplification of evaluating each next value using repeated addition and so eliminating the need for multiplication and division ordinarily required. Well-behaved mathematical expressions were approximated by polynomial expansions valid within predetermined limits of precision in a given domain. Babbage implemented the operations involved in tabulation by method of differences in a mechanical medium, with much of the mechanical logic ingenious and original. The following logical features are amongst those found in the Difference Engine designs:

- Automatic (human intervention in algorithmic process unnecessary)
- Parallel operation (simultaneous operation on each digit of multi-digit numbers)
- Repeated non-destructive addition (recovery of the lost addend after addition)
- 'Microprogramming' (automatic execution of a sequence of subsidiary operations)
- 'Pipelining'
- 'Pulse-shaping' (cleaning up degraded transitions to ensure digital integrity)
- 'Binary latching' (one-bit storage)
- 'Polling' (sequenced interrogation of a series of logical states)
- Overflow warning
- Carriage of tens, taking account of secondary carries.

Attributing these features is not based on suggestive hints or simply a backwards projection from our own age: the logical mechanisms are explicit in the detail of the designs held in an archive of Babbage's technical work comprising some 7,000 manuscript sheets. The language describing these features ('microprogramming', 'polling', 'pipelining' for example) is of course modern with its anachronistic use here signalled by apologetics. But the logical functions referred to are unmistakably congruent across time.

Babbage's engines are decimal digital machines. He considered other number bases including 2, 3, 4, 5, 10, 12, 16, 20, and 100 but settled on decimal for engineering efficiency with the added convenience of familiarity.[11,24,48] The machines were digital in the sense that only discrete number values were logically valid. Numbers were represented by cogwheels (called figure wheels) engraved with the numbers 0 through 9, and the value of a digit was represented by the amount of angular rotation of a given figure wheel. Each digit-position in a multi-digit number had its own figure wheel, and multi-digit numbers are represented by columns of figure wheels with units below, tens, hundreds and so on up the column. So 2.5 is not represented by a figure wheel half-way between 2 and 3 but by one digit wheel at 5 and the next digit wheel up the stack at 2.[39,44] However, cogwheels are inherently analog devices. All transitional states between discrete integer values are physically stable but logically indeterminate. Babbage was concerned to ensure the absolute integrity of results and he used mechanical control devices to discretise motion and ensure that the mechanisms were confined to logically defined states.[36,42] These security mechanisms, distributed throughout the machines, performed three distinct functions: error correction (automatic re-alignment in the event of minor derangement); locking (wheels immobilised during parts of the cycle in which they are not operational); and, finally, error detection (a figure wheel in a position intermediate between integral values will jam the machine and halt the cycle). So jamming does not necessarily signify calamitous malfunction but serves as a warning that the integrity of the calculation has been compromised.

All that remains of Babbage's failed efforts to construct his engines are some partial assemblies, most of them experimental, the most substantial of which is the portion of Difference Engine No. 1 (Figure 1) which represents one-seventh of the calculating section of the full design. The first complete working Babbage engine built to original designs is Difference Engine No. 2, designed between 1847 and 1849 and completed in 2002, the culmination of a project lasting seventeen years. The design is elegant and efficient with the physical machine consisting of 8,000 separate parts, weighing five tonnes, and measuring eleven feet long and seven feet high (Figures 2, 3). It calculates and tabulates any 7th-order polynomial to thirty one decimal places using, as before, the method of finite differences.

The output apparatus automatically typesets results, prints inked hardcopy and impresses results in stereotype trays in two fonts simultaneously and allows programmable output format including variable line height, margin widths, numbers of columns, and whether the results are listed line-to-line with automatic column-wrap or column-to-column with automatic line-wrap. The machine works as Babbage intended with no material changes required to the original design.[44]

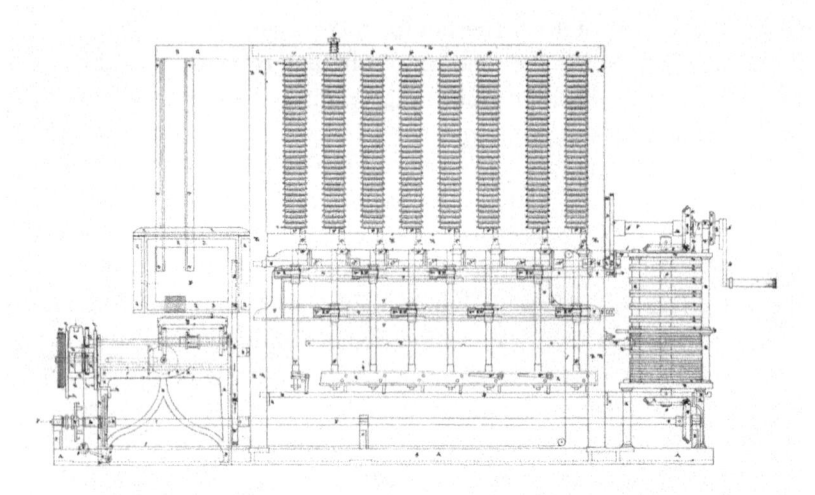

Fig. 2. Design drawing showing elevation of Charles Babbage's Difference Engine No. 2, c. 1848. The Engine is operated by turning a crank handle shown on the right. The vertical stack of 28 cams alongside the crank orchestrates the internal motions. The columns of figure wheels hold seven difference values to 31 decimal places. The tabular value appears on the left-most column. The output apparatus is shown to the left of the results column. Credit – SSPL, Science Museum.

Fig. 3. Difference Engine No. 2, completed 2002. The Engine consists of 8000 parts, measures 11 feet long, 7 feet high and weighs 5 tonnes. Credit – Doron Swade.

3. From Calculation to Computation

The Difference Engine was what we would now call a calculator: it performed a specific set of functions to which its operation was absolutely limited. It is Babbage's designs for the Analytical Engine, conceived in 1834, that mark the essential transition from calculation to general-purpose computation. There are detailed and extensive designs for the Analytical Engine which he worked on intermittently till he died in 1871, though yet again, like the difference engines, the Analytical Engine was never built. The Analytical Engine, an automatic programmable general purpose computing machine, was capable of executing any sequence of arithmetical operations under program control, and it embodies another set of logical and architectural features in addition to many of those featured in the difference engines. These features, explicit by 1840, include:

- Parallel bus (the simultaneous transfer of digits of multi-digit numbers on a multipath 'highway')
- Separation of the Store and the Mill (architectural feature in which memory and central processing are identifiably distinct subsystems)
- Internal repertoire of automatically executable functions (including direct multiplication and division)
- Conditional control (automatically taking one or another route depending on intermediate outcome)
- Serial operation and a 'fetch-execute' cycle
- Anticipating carriage (simultaneous multi-digit carriage of tens)
- Programmability using punched cards
- Iterative looping (automatic repetition of a sequence of operations a predetermined number of times)
- Parallel processing (multiple Mills)
- Punched card input
- Printed, punched card, graph plotter output
- Up to 50-digit 'registers' with double precision results
- 'Buffering'

The separation of Store and Mill, serial operation involving a fetch-execute cycle, and input-output devices, all of which are featured in the Analytical Engine designs, are essential features of classic computer architecture described by Von Neumann in his seminal paper published in 1945. The eponymous Von Neumann architecture has dominated computer design to this day.[47]

4. Computation as Systematic Method

Much of the historical attention to date has focused on Babbage's engines, and his writing on computational method has been largely overlooked. In his early work we find explicit reference to notions of algorithmic process, computation as systematic method of solution of equations, computation as a systematic method of evaluating the n-th term of a series for which there is no general expression, the prediction of the need for a new branch of mathematics to optimise the efficiency of computation by machine, and the heuristic value of the engines in suggesting new series for which there is no general expression for the n-th term. Of particular interest to us here are his ideas for automatically halting the machine on the detection of particular states during the course of computation.

The prospects for machine solutions to unsolved equations signalled something fundamentally new. The standard analytical technique for solving equations was to equate the expression to zero and solve for the unknown. There was no systematic process for doing this and the success of the process depends on ingenuity, creativity, and often an ability to manipulate the problem into a recognisable form that had a known class of solution. Not only was there no guarantee of solution using such techniques, there was no way of determining whether or not the equation in question was soluble in principle. If analytical methods failed, then trial and error substitution could be tried. This involves substituting trial values of the independent variable and repeating this process to see if a value of the argument can be found that reduces the function to zero. But the technique was hit and miss. It was regarded as 'inelegant' by mathematicians and did not guarantee success.[37,38]

What was new in Babbage's description of solving equations with machines was the use of computation as a systematic method of solution. Starting with an initial value of the independent variable, each cycle of the engine generated each next value of the expression, and the machine had found a solution when the figure wheels giving the tabular result are all at zero. Finding a solution reduced to detecting the all-zero state, and the number of machine cycles taken to achieve this represented the value of the independent variable, which was the solution sought.

Rather than rely on visual detection of a particular result, Babbage incorporated a bell that rang to alert the operator to the occurrence of specific condition including the all-zero state or a sign change as the value passed through zero. Finding the roots of equations involved the operator in setting the initial values and then cranking the handle to cycle the machine until the bell rang. The operator would halt the machine and read off the number of cycles the machine had run on a counter automatically incremented after each machine cycle. This number was the first root of the equation. If there were multiple roots the operator would keep cranking, until the bell rang again.

Various conditions rang bells – the need for specific new data, 'register overflow' indicating the possibility of infinitely large quantities, the completion of a fixed number of cycles[3] and the detection of specific numerical states, including the all-zero state and sign-change on passage through zero. So as not to rely on an operator responding to a bell to halt the machine Babbage included mechanisms that would halt the machine automatically.[3,37,39] In the event of there being no roots the machine would continue *ad infinitum*. While Babbage made clear that finding roots of equations by mechanical process was a critical new feature of his machine, and while he incorporated mechanisms to automatically halt the machines on the occurrence of specific conditions including the occurrence of roots, he nowhere, at least in his published writings, trumpeted as a specially significant *theoretical* finding the notion of halting as a criterion of solubility. For him it was a practical issue: the machine would halt when it found a solution. However, Babbage's notions of a machine cycle, of stepwise operational procedure, and of algorithmic process[19] were explicit, and have intrinsic to them concepts of 'definite method' and 'mechanical process' that pre-echo seminal concepts in Turing's work a century later. For Babbage, in his world of cogs, levers and ratchets, 'mechanical process' was something literal rather than metaphorical though, ironically, his unbuilt engines were virtual rather than real.

5. Formal Description

Beyond the nuts and bolts of the mechanisms, Babbage formalised the description of his machines using a universal symbolic language of his own devising. His Mechanical Notation is not a calculus. It is more in the nature of a symbolic description that specifies the intended motion of each part and its relation to related parts: whether a part is fixed (a framing piece for example) or free to move, whether motion of a moving part is circular or linear, continuous or intermittent, driver or driven, to what other parts it is connected, and the timing of its motion in relation to other parts. The Notation has three main forms: a tabular form usually included on the same design drawing as the mechanisms it describes (see Figure 2, for example, top left); timing diagrams that describe the phasing of motions in relation to each other; finally, a flow diagram form which indicates the routing of data and control for each train of parts.[42] Babbage developed the Notation as a design aid, to manage the unprecedented complexity of the mechanisms and to optimise the designs, in part by eliminating redundancy. The Notation is akin to what we would now call a hardware description language. He was immensely proud of it and regarded it as amongst his finest inventions with application beyond science and engineering. His son, Henry Prevost, was well versed in the scheme and aspects of it were used in the modern day construction of Difference Engine No. 2. The fate of the Notation, baroque in its intricacy and idiosyncratically novel, has been, by and large, one of obscurity.

6. Ada Lovelace

Ada Lovelace and Charles Babbage met in 1833 when she was seventeen and he in his early forties, and Babbage became, and remained, a close family friend until her death in 1852. Lovelace collaborated with Babbage on a description of the Analytical Engine, published in her name in 1843. Her *Sketch of the Analytical Engine*[30] remains the most penetrating contemporary account of capabilities and potential of the machine.

Babbage's conception of the Analytical Engine was largely bound by its implications for mathematics. In 1836 he hints at the prospects of a general purpose algebraic machine able to manipulate letters 'without any reference to the value of the letters'[10] but he describes the idea as 'very indistinct' and does not elaborate. But the context is nonetheless still mathematical. Nowhere, at least in his published writings, does Babbage consider the larger implications of number representing entities other than quantity. This conclusion is fortified by Babbage's own hand. In 1869 he set out to write a general description of the Analytical Engine. The manuscript stops abruptly after only a few pages. But the opening sentence is declarative and defining: 'The object of this Engine is to execute by *machinery* 1. All the operations of arithmetic, 2. All the operations of Analysis, 3. To print any or all of the calculated results'.[4] Babbage's account is then quickly sidetracked into gearwheels and how quantity is represented by the angular displacement of toothed gears, but it is clear that the context is numerical, and that 'analysis' is does not conceal aspirations to greater generality. The manuscript is dated 4 May 1869. This was two-and-a-half years before Babbage's death and so represents a mature reflection after thirty five years of Analytical Engine design. In the stated purpose of the engine there is no mention of anything outside mathematics and number, nor is there mention of anything more general in the potential of the machine.

It was Lovelace who made explicit the transition from calculation to computation. She speculated that if the rules of harmony and composition were appropriately represented then the Analytical Engine 'might compose elaborate and scientific pieces of music of any degree of complexity or extent'. She also wrote of the machine operating on numbers directly representing entities other than quantity 'as if they were *letters* or any other *general* symbols' and that the Analytical Engine might produce outputs in notational form.[30] It seems that Lovelace saw the potential of computing in the general capacity of machines to manipulate symbols according to rules and that the power of computers lay in bridging the internal states of the machine to the world through the representational content of the symbols it was able to manipulate.

Latterly Lovelace has been widely celebrated, and four main claims tend to feature in her posthumous fêting. She is variously described as a mathematics genius, as someone who made an inspirational contribution to the conception and development of Charles Babbage's calculating engines, as being the first programmer, and, finally, she is frequently heralded as a prophet of the computer age. Of these four claims only one has credible foundation. There is no compelling evidence for the first. The second claim is readily disproved by the simple chronology of events. The third claim (that she was the first programmer) is understandable but wrong. But the tribute of being a visionary of the computer age is fully deserved and it is appropriate that she is properly celebrated. It was she who appears to have understood better than anyone the potential of computation in its more general sense.[17,22,41]

7. Legacy of the 19th Century

So by the mid-1840s, through Babbage and Lovelace, several features of general purpose programmable computation as we currently understand it, and indeed of a computer science, were explicitly articulated. These are:

- computational logic and control (implemented at that time in a mechanical medium)

- a symbolic hardware description language as a design aid used to specify, optimise, and manage complexity
- computation as a systematic method of solution through step-wise sequences of operations i.e. the notion of algorithmic programming
- the notion of computers as machines for manipulating symbols according to rules regardless of the representational content of the symbols.
- the notion of machine intelligence.

Babbage, as we know, failed to complete any of his engines in their entirety. Others attempted difference engines in the 19th century.[42] The machines that succeeded technically, failed commercially. Despite the inspired efforts of Babbage, Lovelace and others, the movement to automate calculation and build computing engines in the 19th century failed. There were at least two isolated episodes in the early 20th century to build analytical engines, by Percy Ludgate (1909) and Torres y Quevedo (1913).[33,34] While historically interesting, each was a developmental *cul de sac*.

8. Turing, Influence, and the Modern Era

During the office automation boom in the early decades of the 20th century automatic computation was largely dormant. Automatic tabulation was revived in the 1930s by Leslie Comrie who used commercially available adding machines for differencing.[15] Comrie's success can be seen as a delayed epilogue to Babbage ill-fated efforts a century earlier and as such Comrie's tabulation work belongs more to the 19th than the 20th century. With WWII came new impetus and with it the development of electromechanical and early electronic computing machines. A new generation of pioneers emerged including amongst others Konrad Zuse, Vannevar Bush, John von Neumann, Presper Eckert, John Mauchley, Maurice Wilkes, Tommy Flowers, Freddie Williams, Tom Kilburn, Max Newman, Howard Aiken, Grace Hopper, and of course Alan Turing. First-generation pioneering machines included Zuse's Z-series machines, ENIAC, EDVAC, EDSAC,[40] Colossus, the Bombe, Manchester 'Baby' (aka SSEM), and the Harvard Mark I (aka ASCC).

It appears at first sight that the principles of electronic digital computing were re-invented by the pioneers of the electronic age largely in ignorance of what had gone before. Features identified as explicitly articulated in the 19th century pre-echo with startling similarity those of the 20th century, and the inevitable question that arises is one of influential connection. Studies show that most of the pioneers of the modern age knew of Babbage. A few did not.[32] At least one appears to have claimed to know more than he did.[9] The legend of Babbage did not die, and what he had attempted was common knowledge to the small cadre of those involved with automatic computation. The question that emerges is just what was known of Babbage's works by those who knew of him at all.

In specific relation to Turing, the areas of shared or overlapping interests with Babbage and Lovelace are striking: the notion of mechanical process, the role of halting in the context of computability, speculation about machine intelligence, and what Lovelace referred to in 1844 as 'a Calculus of the Nervous System'.[29] She appears to share Turing's view of the unrestricted potential of intelligent machines when she wrote 'it does not appear to me that cerebral matter need be more unmanageable to mathematicians than sidereal and planetary matters and movements, if they would but inspect it from the right point of view'.[29] Amongst the questions that emerge are the extent of Turing's knowledge of Babbage and Lovelace, and also whether any such knowledge predates his critical work on computability.

We know that Turing discussed Babbage with Donald Bayley[21] after Bayley was posted, in 1944, to Hanslope Park, a secret government establishment, where Turing was working on speech encypherment. Similarly, there are accounts of 'lively mealtime discussion' about Babbage at the Government Code and Cypher School at Bletchley Park,[12] the earliest of which would not have predated September 1939 when Turing reported there at the start of the war. However, insufficient is known about the content of these discussions to provide any conclusive view of the extent of Turing's knowledge.

There were three main sources that go beyond general accounts of Babbage's work of the kind found in encyclopaedias. The first is Lovelace's *Sketch* published in 1843,[30] the most extensive contemporary account of Babbage's Analytical Engine. This contains a translation of Luigi Menabrea's description of the Analytical Engine and includes his specific mention of the machine halting automatically at the detection of the all-zero state or a sign-change. The second is Dionysius Lardner's over-long and grandiloquent article published in 1834 which gives an extensive account of Babbage's Difference Engine No. 1. Here there is specific and explicit mention of the machine cycling to produce successive values of a formula, the detection of the all-zero state corresponding to the occurrence of a root, and a bell ringing automatically to signal a solution.[27] Lardner's article is both cited and referred to in Lovelace's *Sketch*. The third source is the technical archive of Babbage's designs and notebooks held since 1878 by the Science Museum in London.[7] No more than a handful of the 7,000 manuscript sheets have been published and any significant knowledge would necessarily have involved physical examination *in situ*. The detail of this material was not studied in earnest until the 1970s.[6,11,48]

Turing wrote in 1950 that Lovelace's 'memoir' was the source of 'our most detailed information of Babbage's Analytical Engine'[14] and this is his earliest known reference to Lovelace. In the same paper, *Computing Machinery and Intelligence*, Turing refers to 'Lady Lovelace's Objection' (that computing machines cannot originate anything but can only execute instructions) which he uses as a counterpoint to argue for more generous limits to the potential for intelligent machines.[46] Nowhere does he mention Babbage's archival drawings. On the available evidence it is reasonable to conclude that Turing's primary source was the Lovelace *Sketch* and that he did not directly consult Babbage's plans. The question of when Turing first had sight of the Sketch remains unresolved,[12] though had it come to his attention in time to influence the 1936 with paper on computability, there is no reason to suppose that he would have concealed this and, in the light of what is currently known, it can be no more than wishful speculation that his reference to the *Sketch* in 1950 was in fact a belated form of acknowledgement. In overall terms there does not appear to be a direct connection of influence between Babbage-Lovelace in the 19th century and the pioneers of electronic digital computing in the 20th century. However, the core ideas from the two eras are startlingly similar, and we are left with the conclusion, both reassuring and confining, that these ideas, articulated a century apart, embody something fundamental about the nature of computation.

References

1. Augarten S. Bit by Bit: An Illustrated History of Computers. Boston, Sydney: George Allen & Unwin; 1984. 1.
2. Austrian GD. Herman Hollerith: forgotten giant of information processing. New York: Columbia University Press; 1982.
3. Babbage BH. Babbage's Calculating Machine or Difference Engine. Science and Art Department, 1872. Reprinted in: Campbell-Kelly M, editor. The Works of Charles Babbage. London: William Pickering; 1989 (2): p. 232.
4. Babbage C. Scribbling Book S12, p. 134, 4 May 1869. Babbage Archive, Science Museum.
5. Babbage HP, editor. Babbage's Calculating Engines: a collection of papers

by Henry Prevost Babbage. London: Spon; 1889. Reprinted in: Bromley A, editor. Babbage's Calculating Engines: a collection of papers by Henry Prevost Babbage. Los Angeles: Tomash; 1982. See preface.
6. Bromley AG. Babbage's Analytical Engine Plans 28 and 28a - The Programmer's Interface. IEEE Annals of the History of Computing. 2000;22(4):4-19.
7. Bromley AG. The Babbage papers in the Science Museum: a cross-referenced list. London: Science Museum; 1991.
8. Carlyle T. Signs of the Times. Edinburgh Review. 1829(June).
9. Cohen IB. Babbage and Aiken. Annals of the History of Computing. 1988;10(3):171-91.
10. Collier B. The Little Engines that Could've: The Calculating Engines of Charles Babbage. 2 ed. New York: Garland; 1990, p. 139.
11. Collier B. The Little Engines that Could've: The Calculating Engines of Charles Babbage. 2 ed. New York: Garland; 1990, p. 164.
12. Copeland BJ, editor. The Essential Turing: The ideas that gave birth to the computer age. Oxford: Oxford University Press; 2004, p. 29.
13. Copeland BJ, editor. The Essential Turing: The ideas that gave birth to the computer age. Oxford: Oxford University Press; 2004, p. 263.
14. Copeland BJ, editor. The Essential Turing: The ideas that gave birth to the computer age. Oxford: Oxford University Press; 2004, p. 455.
15. Croarken M. Early Scientific Computing in Britain. Oxford: Clarendon; 1990.
16. Edgerton D. The Shock of the Old:Technology and Global History since 1900. London: Profile Books; 2006.
17. Fuegi J, Francis J. Lovelace & Babbage and the Creation of the 1843 'Notes'. IEEE Annals of the History of Computing. 2003;25(4):16-26. See also, Fuegi J, Francis J. To Dream Tomorrow: Ada Byron Lovelace, documentary film, Flare Productions. 2003.
18. Gleick J. The Information: the History, the Theory, the Flood. London: Fourth Estate; 2011.
19. Grattan-Guinness I. Charles Babbage as an Algorithmic Thinker. IEEE Annals of the History of Computing. 1992;14(3):34-48.
20. Heilbroner R. Do Machines Make History? Technology and Culture. 1967;8(July 1967):335-45.
21. Hodges A. Alan Turing: The Enigma. New York: Simon & Shuster; 1983, p. 297.
22. Hyman A, editor. Memoir of the life and labours of the late Charles Babbage Esq. F.R.S. by H. W. Buxton. Cambridge, Massachusetts: Tomash; 1988: p. 156.
23. Hyman A. Charles Babbage: Pioneer of the Computer. Oxford: Oxford University Press; 1982.
24. Hyman A. Charles Babbage: Pioneer of the Computer. Oxford: Oxford University Press; 1982, p 167.
25. Hyman A, editor. Memoir of the life and labours of the late Charles Babbage Esq. F.R.S. by H. W. Buxton. Cambridge, Massachusetts: Tomash; 1988: p. 46.

26. Johnston S. Making the Arithmometer Count. Bulletin of the Scientific Instrument Society. 1997;52(March):12-21.
27. Lardner D. Babbage's Calculating Engine. Edinburgh Review. 1834;59:263-327. Reprinted in: Campbell-Kelly M, editor. The Works of Charles Babbage. London: William Pickering; 1989(2), p. 169.
28. Lavington S., editor. Alan Turing and his Contemporaries: Building the world's first computers. Swindon: British Informatics Society; 2012.
29. Lovelace AA to Woronzow Greig, 15 November 1844. In: Toole BA, editor. Ada, the Enchantress of Numbers: A Selection from the Letters of Lord Byron's Daughter and Her Description of the First Computer. Mill Valley, CA: Strawberry Press; 1992, p. 296.
30. Lovelace AA. Sketch of the Analytical Engine. Scientific Memoirs. 1843;3:666-731. Reprinted in: Campbell-Kelly M, editor. The Works of Charles Babbage. London: Pickering; 1989(3): p. 89-170.
31. Menabrea LF. Notions sur la machine analytique de M. Charles Babbage. Biblioteque universelle de Geneve. 1842;41:352-76. Reprinted in: Campbell-Kelly M, editor. The Works of Charles Babbage. London: William Pickering; 1989 (3), p. 113.
32. Metropolis N, Worlton J. A Trilogy of Errors in the History of Computing. Annals of the History of Computing. 1980;2(1):49-59.
33. Randell B. From Analytical Engine to Electronic Digital Computer: The Contributions of Ludgate, Torres, and Bush. Annals of the History of Computing. 1982;4(4 October):327-41.
34. Randell B. Ludgate's analytical machine of 1909. Computer Journal. 1971;14(3):317-26. 1. First published as Ludgate's analytical machine of 1909. Newcastle: University of Newcastle upon Tyne, 1971 January 1971. Report No.: 15.
35. Schaffer S. Babbage's Intelligence: Calculating Engines and the Factory System. Critical Enquiry. 1994;21:203-27.
36. Swade D. Automatic Computation: Charles Babbage and Computational Method. The Rutherford Journal [Internet]. 2010; 3(2010). Available from: http://www.rutherfordjournal.org/article030106.html
37. Swade D. Calculating engines: machines, mathematics, and misconceptions. In: Flood R, Rice A, Wilson R, editors. Mathematics in Victorian Britain. Oxford: Oxford University Press; 2011. p. 239-59.
38. Swade D. Calculation and Tabulation in the 19th Century: George Biddell Airy versus Charles Babbage [PhD]. London: University College London; 2003.
39. Swade D. Charles Babbage's Difference Engine No. 2: Technical Description. Research Paper. Science Museum Papers in the History of Technology, 1996 1996. Report No.: 6.
40. Swade D. Inventing the User: EDSAC in Context. Computer Journal. 2011;54(1):143-7.
41. Swade D. Pioneer Profile: Ada Lovelace. Resurrection. 2011(53):31-3. Available from: http://www.cs.man.ac.uk/CCS/res/res53.htm#g

42. Swade D. Pre-electronic Computing. In: Jones CB, Lloyd JL, editors. Dependable and Historic Computing. Berlin: Springer-Verlag; 2011. p. 58-83.
43. Swade D. The Cogwheel Brain: Charles Babbage and the quest to build the first computer. London: Little, Brown; 2000. Published in the US as: The Difference Engine: Charles Babbage and the Quest to Build the First Computer. New York: Viking; 2001.
44. Swade D. The Construction of Charles Babbage's Difference Engine No. 2. IEEE Annals of the History of Computing. 2005;27(3):70-88.
45. Swade D. The 'unerring certainty of mechanical agency': machines and table making in the nineteenth century. In: Campbell-Kelly M, Coarken M, Flood R, Robson E, editors. The History of Mathematical Tables: From Sumer to Spreadsheets. Oxford: Oxford University Press; 2003, p. 143-74.
46. Turing A. Computing Machinery and Intelligence. Mind. 1950:433-60. Reprinted in, Copeland BJ, editor. The Essential Turing: The ideas that gave birth to the computer age. Oxford: Oxford University Press; 2004, p. 441-464. See p. 263.
47. von Neumann J. First Draft of a Report on the EDVAC. Moore School of Electrical Engineering, University of Pennsylvania, 1945 30 June. Reprinted in: Randell B, editor. The Origins of Digital Computers: Selected Papers. Berlin: Springer-Verlag; 1982, p. 383-392.
48. Wilkes MV. Babbage as a Computer Pioneer. British Computer Society and the Royal Statistical Society, 1971 18 October, p. 4.
49. Williams MR. A History of Computing Technology. Englewood Cliffs, N.J.: Prentice-Hall, Inc; 1985.

Chapter 3

Generating, Solving and the Mathematics of Homo Sapiens.
Emil Post's Views on Computation

Liesbeth De Mol*

Centre for History of Science, Ghent University, Belgium
elizabeth.demol@ugent.be

"*For if symbolic logic has failed to give wings to mathematicians this study of symbolic logic opens up a new field concerned with the fundamental limitations of mathematics, more precisely the mathematics of Homo Sapiens.*"

Emil Leon Post, 1936

1. Introduction

What is a computation? What does it mean to compute something and in how far does it make sense to talk about computation outside of mathematics? These fundamental questions have not received a satisfactory answer yet, despite the well-known "Church-Turing thesis". Roughly put, the Church-Turing thesis states that the set of the "computable" – in the vague, intuitive sense – coincides with the Turing computable, hence rendering a vague notion exact and thus, if accepted, setting a borderline between the computable and the non-computable. However, knowing/assuming that the "computable" is that which can be computed by a Turing machine (or any other formal device that is logically equivalent to it) does not necessarily

---

*This research was supported by the Fund for Scientific Research, Flanders, Belgium. I would like to thank the people at the *Manuscripts Department, American Philosophical Society* for their assistance with the *Emil Leon Post papers* as well as the people at the *Department of Rare Books and Special Collections, Princeton University Library* for their assistance with the *Alonzo Church papers*. I am also indebted to Maurice Margenstern for some very interesting discussions on (the nature of) computation (in nature) especially on the significance of tag systems and research on the boundaries of decidability and undecidability.

imply that one knows if and how one can compute whether or not it will rain in Paris on June 23, 2112 or what will be the next sentence of this paper. Viz., to accept that the "computable" is the Turing computable, does not necessarily imply that one knows the universe of the computable.

Since 1936, the year Church and Turing each proposed their theses, these questions have only gained in significance, perhaps not so much because of advancements in mathematical logic, but rather due to the development of what is often understood as the physical realization of computability – the digital, electronic stored-program and general-purpose computer. It was and is this machine that has *effectively* widened the scope of the computable and brought computations to people who are and were not really interested in the problems of mathematical logic. It is this machine which has made it possible to physically implement a wide range of computational processes, problems and objects and it was hence the computer that made it relevant to non-logicians to broaden and study the field of the computable and the non-computable, making it necessary to pose questions like: what could be a high-quality weather prediction software? It is exactly this extension of the scope of the computable that explains why it is no longer considered "heretical"[a] to say that all processes in "nature" – going from the human mind to the growth of plants – can be understood as computations, or, much stronger, that everything in "nature" can be computed by a Turing machine.

Despite or perhaps due to the fact that the scope of the computable has effectively been extended beyond the field of mathematical logic, the Church-Turing thesis, *as a philosophical thesis* but also *as a mathematical definition*, is still the uncontested delimiter of the computable. It is here that lies part of its beauty, viz., its two-facedness of opening up the field of computation and, at the same time, imposing a fundamental limitation on it and hence also setting the stage for the non-computable. It is this 'Janus face' of the Church-Turing thesis, going to the core of the thesis, that is sometimes too much underestimated in current debates on the Church-Turing thesis.

To celebrate and honor Turing in 2012 is also to celebrate the Church-Turing thesis. However, to celebrate this achievement is also to acknowledge its historical, philosophical and mathematical richness. Even though one must and should fully acknowledge Turing's fundamental contributions, the fact that hardly any other historical sources are examined in some of the

---

[a]See Alan Turing's[27] where he describes the idea of intelligent machinery as heretical.

current discussions, including those that allegedly claim to give a "historical" account, can only result in views that are incapable of acknowledging exactly this richness. This is what I would like to do here by exposing the reader to Emil Leon Post's thoughts and results on the 'Church-Turing thesis'. It was Post, who besides Church and Turing, also proposed a "thesis" in 1936. It was Post who had already developed a model of computability that is very different from Turing's in 1921, and, it was also Post who, unlike Church and Turing, insisted on the view that Church's and Turing's thesis should be regarded as a hypothesis, and, ultimately, a natural law rather than as a mathematical definition.

2. Why Turing Rules

In 1936 both Church and Turing published their celebrated theses. They proposed their respective theses as *formal definitions of an otherwise vague and intuitive notion.* It were these definitions that allowed Church and Turing to tackle the famous *Entscheidungsproblem* for first-order logic. Indeed, in order to prove the incomputability of this problem, it was necessary to have a formal definition of computability. In his definition, Church identified the vague notion of an *effectively calculable function of positive integers* with the formal notions of *general recursive functions* (or, equivalently, λ-definable function).[1] Turing identified the vague notion of computable numbers, i.e., *"the real numbers whose expressions as a decimal are calculable by finite means"* [26, 230], with those numbers computable by a *Turing machine.*

Even though in the literature Church's thesis and Turing's thesis are mostly not differentiated but put together under the header "Church-Turing thesis" because of their logical equivalence, it is Turing's that is usually considered as being superior to Church's thesis. To put it in Church's words: *"[Turing's thesis makes] the identification with effectiveness in the ordinary (not explicitly defined) sense evident immediately"*[2] Gödel also shared this view and judged that with Turing's work one has *"for the first time succeeded in giving an absolute definition of an interesting epistemological notion"*.[11] Soare recently even went a step further by stating that *[i]t was Turing* alone *who [...] gave the first convincing formal definition of a computable function"*.[25]

But why exactly is Turing's thesis so much more appealing than Church's? This issue has already been discussed in detail in[10,23,24] and comes down to the fact that unlike Church, Turing's thesis makes a *"di-*

rect appeal to intuition" [26, 249]. Church only arrived at his thesis after several research results on λ-calculus had been established. It was its equivalence with general recursive functions and the fact that Church, Kleene and Rosser were able to λ-define any function they could come up with, that resulted in Church's formulation of the thesis, first in terms of λ-calculus and later in terms of general recursiveness. In other words, Church did not start out from the idea of trying to formally capture some intuition, but, on the contrary, only saw that the formalism he was working with might be powerful enough to capture all "calculable" functions *after* a thorough analysis of that formalism. This explains why defining effective calculability in terms of λ-definability or general recursiveness is not "intuitively appealing" since neither of these symbolic systems was constructed with that purpose in mind.

In contrast, Turing *did* start out from the intuitive and vague idea of a man in the process of computing a number, and tried to find a formalism that somehow captures the essential features of any such process. As Turing formulated it [26, 249]:

> The real question at issue is: "What are the possible processes which can be carried out in computing a number?"

The formal notion of a Turing machine resulted from an analysis of the human processes involved in computing a number by extracting the essential features of such processes, and introducing them as conditions, like, for instance, the idea that the number of symbols the "computor" (the human computer) can write down is finite. These conditions almost naturally lead to the concept of a Turing machine (See the work of Sieg and Gandy[10,23] for more details). Thus, unlike Church, Turing started out from a "real-world" situation and derived an idealized machine out of it. In this sense, Turing's thesis *does* aim at establishing a relation between something physical – the (human) processes involved in computing a number – and something symbolic – the Turing machine.

That Turing's thesis is considered as the better thesis is thus explained by the fact that Turing tackled the problem from the direction of finding a good formalization of a vague idea, whereas Church tackled the problem from the other side.[b]

What is mostly forgotten or at least underestimated in this context is that Emil Leon Post had formulated two different theses, one in '21 and

---

[b]Note that one of Church's main arguments in support of his thesis[1] is also circular and hence contains a fundamental flaw (See[10,23,24]).

one in '36, as a result of tackling similar problems from *both* directions. He formulated his first thesis after having realized that the formal devices he was studying are much more powerful than he had expected. Being unsatisfied with this thesis for the same reasons as Gödel was with Church's, he then formulated a second thesis which was published in 1936 and involves a formal device which is almost identical to the Turing machine. It resulted from an analysis of *"all the possible ways in which the human mind [can] set up finite processes"* [21, 387] to "solve" a problem.

3. Two Theses, Two Sides

Emil Post had two very different though logically equivalent theses. The first was formulated in 1921 and was the necessary epistemological assumption for Post's "proof" that there exist *absolutely* unsolvable problems, like for instance the famous halting problem.[c] Here are two different formulations of Post's thesis I, first identified as such by Martin Davis:[5]

Post's Thesis I. *(simple) For every set of sequences for which a process can be set-up to generate it there is also a normal system which will generate it.*

Post's Thesis I. *(complete) Every generated set of sequences on a given set of letters $a_1, a_2, \ldots, a_\mu$ is a subset of the set of assertions of a system in normal form with primitive letters $a_1, a_2, \ldots, a_\mu, a'_1, a'_2, \ldots, a'_\nu$, i.e., the subset consisting of those assertions of the normal system involving the letters $a_1, a_2, \ldots, a_\mu$.*

It would take another 15 years before Post proposed a second thesis, which was published in his short note *Finite combinatory processes – Formulation I*.[19] This note was a reaction on Church's thesis which Post found unsatisfactory. This second thesis uses a formalism which is almost identical to the Turing machine. It is in this sense that Post's 1936 contribution *"contains the same idea as Turing's paper"* [13, 61]. Here is Post's thesis II in its simple and more technical form:

---

[c]These results were not published in 1921. Post waited about 20 years before he submitted the manuscript *Absolutely unsolvable problems and relatively undecidable propositions - Account of an anticipation* to the *American Journal for Mathematics* which contained a description of his research during the period 1920-21. It was rejected by Hermann Weyl. Post's letter of submission and part of Weyl's rejection letter are printed in.[6] Only a significantly abbreviated version of it was finally published as Post's influential.[20] It would take about another 20 years before Martin Davis posthumously published the manuscript in his.[4]

Post's Thesis II. *(simple) Solvability of a problem in the intuitive sense coincides with solvability by formulation 1.*

Post's Thesis II. *(complete) Any decision problem is solvable in the intuitive sense if it can be stated as a 1-given problem for which one can set-up a finite 1-process which results in a 1-solution to the problem.*

Several questions can and should be posed here: where do these two theses come from? What kind of logical and philosophical analysis and views underlie these two different theses? Why did Post find it necessary to propose a second thesis and, above all, what kind of insights can we gain nowadays from these two theses? We need to dig a bit deeper here into Post's work to tackle these questions.

3.1. *Post's thesis I: Generating sequences and limits of the computable*

In his PhD *Introduction to a general theory of elementary propositions*, published in 1921,[17] Post introduces the idea to develop the most general *form* of logic, and, since he considered logic as the foundation of mathematics, ultimately mathematics. Such general forms he could then use as *"instruments of generalization"* [17, 173] which make possible a study of the *general* properties and problems of the whole of mathematics. It was Post's view that if one wants to study the general properties of logic and mathematics then one needs not one particular system of symbolic logic or mathematics, but a general form that comprises all such possible systems. Given such a form he believed it possible to generalize some of the fundamental results from his PhD – the completeness, consistency and decidability of propositional logic – to the whole of mathematics. This ambitious "programme" is announced in the introduction to his PhD [17, 164].

After his PhD was finished Post became a Proctor fellow in Princeton during the period 1920-21. It was during that time that he set himself the goal of proving the decidability of the decision problem for first-order logic[d] which Post called a *finiteness problem*, and, ultimately, for the whole of mathematics. He *"was proposing no less than to find a single algorithm for all of mathematics"*.[6] His approach to this problem? The development of a general *form* characterized by the method of *"combinatory iteration"* which *"eschews all interpretation"* [21, 386]. More specifically:[18]

---

[d]The famous Entscheidungsproblem proven incomputable by Church and Turing.

> [T]he method of combinatory iteration completely neglects [...] meaning, and considers the entire system purely from the symbolic standpoint as one in which both the enunciations and assertions are groups of symbols or symbol-complexes [....] and where these symbol assertions are obtained by starting with certain initial assertions and repeatedly applying certain rules for obtaining new symbol-assertions from old.

So, what kind of form would one expect to result from such radical formalism? In his PhD Post had in fact already formulated such a form initially called the *generalization by postulation* and later the *canonical form A*. However, this did not immediately lead to the desired ultimate algorithm and Post started to develop other forms and study problems related to this *finiteness problem*. After some months this resulted in Post's 'frustrating' tag systems.

Definition 1 *(v-tag system)* *A tag system T consists of a finite alphabet Σ of μ symbols, a deletion number $v \in \mathbb{N}$ and a finite set of μ words $w_0, w_1, \ldots, w_{\mu-1} \in \Sigma^*$ called the appendants, where any appendant w_i corresponds to $a_i \in \Sigma$. A v-tag system has a deletion number v.*

In a computation step of a tag system T on a word $A \in \Sigma^*$, T appends the appendant associated with the leftmost letter of A at the end of A, and deletes the first v symbols of A. This computational process is iterated until the tag system produces the empty word ϵ and hence halts. To give an example, let us consider the one tag system mentioned by Post[20,21] with $v = 3$, $0 \to 00$, $1 \to 1101$.[20,21] If the initial word $A_0 = 110111010000$ we get the following productions:

~~110~~111010000
$\vdash$ ~~111~~0100001101
$\vdash$ ~~010~~00011011101
$\vdash$ ~~000~~1101110100
$\vdash$ ~~110~~111010000

The word A_0 is reproduced after 4 computation steps and is thus an example of a periodic word.

Clearly, these formal devices are very much in the spirit of the method of combinatory iteration. They *really* have the appearance of eschewing all interpretation. They are symbol manipulating systems at their purest. But how to study such systems? Post's approach was to start out from the simplest cases, i.e., classes of tag systems with a low number of symbols μ and a low deletion number v, look at some particular such cases and then see where that would get him.[e] By following this approach he was able to prove that one can decide for any tag system T with $\mu = v = 2$ and some word $A \in \Sigma^*$ whether or not T will *generate* or produce A, viz., what is now called the *reachability problem* for tag systems.[8] The proof involved "*considerable labor*" and Post considered it as the "*major success*" of his Procter fellowship [21, 362].[f] However, going but one small step beyond this, Post discovered a zoo of tag systems of a "*bewildering complexity*". He studied several particular examples, writing out productions generated from different initial words, but was not able to get a grip on the "problem of tag" – he simply could not find a way to predict and control the *behavior* of these apparently simple systems. In fact, even the apparently simple example from above proved "*intractable*" and is up to today still an open problem. Post concluded that given the difficulties involved with trying to understand (the behavior of) these systems that *the general problem of "tag" appeared hopeless, and with it our entire program of the solution of finiteness problems*" [21, 363]. Post had underestimated the complexities such a "*primitive form of mathematics*" [21, 363] can give rise to. By focusing on the behavior of this form he now clearly understood that simplicity in form does not necessarily imply simplicity in behavior and of the problems defined relative to that form. This insight would lay the foundation of Post's thesis I.

After nine months of research Post turned to two other forms that he had developed: the canonical form C – nowadays known as *Post production systems* in the context of formal language theory – and the normal form. Systems in normal form, shortly, normal systems are a special class of systems in canonical form C. A system in normal form has one initial word (postulate) and a finite set of production rules all of the following form:

$$g_i P_i$$
produces
$$P_i g_i'$$

Clearly, normal systems are very similar to tag systems. In fact, the normal form was inspired by the form of "tag" (See[7]).

---

[e]This is described in the unpublished document[18] available from the *Emil Leon Post papers, American Philosophical Society*.
[f]Regretfully the proof was never published. A new proof was recently found and published as.[8]

Knowing from his experience with tag systems that apparent formal simplicity does not necessarily imply real simplicity, Post started on a project of proving the "power" of systems in normal form, viz. their generality: he first proved that canonical form A can be reduced to a system in canonical form C and then, most importantly, proved that the canonical form C reduces to normal form. This fundamental result known as the *normal form theorem* and once described as *"the most beautiful in mathematics"* by Marvin Minsky, was later published as.[20] From this Post concluded that in fact the whole of Russell's and Whitehead's *Principia Mathematica* could be reduced to the normal form:

> [F]or if the meager formal apparatus of our final normal systems can wipe out all of the additional vastly greater complexities of the canonical form [...], the more complicated machinery of [the canonical form] should clearly be able to handle formulations correspondingly more complicated than itself.

It was this additional insight that resulted in the formulation of Post's thesis I, stating that anything that can be "generated" can also be "generated" by the "primitive" normal form. Post next proved that the finiteness problem for normal form is unsolvable (incomputable) and, on the assumption of his thesis, concluded that it is unsolvable by any means and, thus, also *absolutely unsolvable*.

3.2. *Post's thesis II: Solvability and the realm of the computable*

When Post became convinced of the universality of his normal form and the absolute unsolvability that comes with it, he was very much aware that he was only able to come to this conclusion because of his own interactions and experiences with this and other forms and that, hence, the thesis would not be so convincing to others who had not had this experience. Thus Post's thesis I is comparable to Church's thesis in the sense that both transpired only after a thorough study of systems of symbolic logic, and, because of that reason, the identifications they propose are not intuitively appealing. Post concluded that *"for [the thesis to obtain its] full generality] a complete analysis [should] be made of all the possible ways the human mind [can] set up finite processes for generating sequences."* [21, 387].

Not unlike Turing (see the quote on p. 48), Post now set himself the goal of finding a formalism on the basis of an analysis of all the processes one can set up to generate a set, processes which are essentially *human* processes, processes of the human mind. For that reason [21, 394]:

> "[e]stablishing this universality [of the characterization of generated set of sequences in terms of normal form] is not a matter for mathematical proof, but of psychological analysis of the mental processes involved in combinatory mathematical processes.

Only 15 years later Post would publish what is most probably the result of this analysis started in 1921 and interrupted by episodes of his manic-depressive illness and a heavy teaching load (16 contact hours!) at New York City College.[g] The direct reason for publication was that he had read Church's[1] and was not satisfied with Church's thesis, probably for the same reasons for not being satisfied with his own thesis I. Hence, Post proposed his identification between solvability and solvability by formulation 1.

Instead of a tape, Post's 'machine' consists of a two-way infinite symbol space divided into boxes. The idea is that a worker is working in this symbol space, being capable of a set of five primitive acts (O_1 mark a box, O_2 unmark a box, O_3 move one box to the left, O_4 move one box to the right, O_5 determining whether the box he is in is marked or unmarked), following a finite set of directions $d_1, ..., d_n$ where each direction d_i always has one of the following forms:
(A) Perform one of the operations (O_1–O_4) and go to instruction d_j
(B) Perform operation O_5 and according as the box the worker is in is marked or unmarked follow direction $d_{j'}$ or $d_{j''}$.
(C) Stop.
Post also defined a specific terminology for his formulation 1 in order to define the solvability of a problem in terms of his formulation 1. These notions are applicability, finite-1-process, 1-solution and 1-given. Roughly speaking these notions assure that a decision problem is solvable with formulation 1 on the condition that the solution given in the formalism always terminates with a correct solution. As is clear from this description, Post's formulation 1 is almost identical to Turing machines even though there are some interesting smaller differences.[h]

Echoing the reason why he was not fully satisfied with his own thesis I, he emphasized that the purpose of this formulation *"is not only to present a system of a certain logical potency but also, [...] of psychological fidelity"* [19, 105] Viz., his proposal is not only about *defining* the scope of the solvable but to capture *all* possible processes we humans can set-up to

---
[g]See[6] for a description of Post's (way of) teaching at City College.
[h]Like the fact that Post explicitly introduced a halting order.

solve problems. In this sense, the identification he proposes goes beyond that of the mathematical definition and becomes a *"working hypothesis"*. This is the point at which Post strongly opposes Church, because [19, 105]:

> to mask this identification under a definition hides the fact that a fundamental discovery in the limitations of the mathematicizing power of *Homo Sapiens* has been made and blinds us to the need of its continual verification.

Because of this hypothetical character, Post proposes a program of considering wider and wider formulations and to prove that all these are logically reducible to the original formulation 1, hence 'testing' the generality of formulation 1 by confronting it with more general formulations. It is the success of such program which, for Post, could change this hypothesis into a *"natural law"* [19, 105] rather than a definition.[i]

3.3. *"When the bubble of symbolic logic finally burst..."*

Post proposed two logically equivalent but still very different theses, one in 1921 (published in 1943[20]), and one in 1936. The first thesis was motivated by the insight that a simple form can be much more powerful than its apparent simplicity would suggest. This insight is intimately related to another one: the confrontation with the limitations of symbolic logic. Indeed, it was the experience with his tag systems that led Post to the conclusion that there are limits to what can be achieved with symbolic logic – there are problems that cannot be solved with this *"instrument"*. Even the *"primitive"* form of "tag" escapes the methods of symbolic logic – a fact unknown to Post but believed and hoped to be true.[j]

The implication of this was clear: *if* Post's thesis was (and is) true, *then* the ultimate *human* method for solving/deciding all mathematical problems in finite time does not exist. Hence, Post's thesis I, or at least the research from which it originated, is at least as much about the limits

---

[i]In this sense, Post believed that the proof of the equivalence between general recursive functions and λ-definability carries Church's thesis already *"beyond the working hypothesis stage"* [19, 105].

[j]In one of his letters to Church (dated May 30, 1936) Post expresses this hope by stating that *"should [the problem of "tag"] too prove unsolvable I will be supplied with the perfect alibi for a year of frustration*. In a postcard to Church (dated May 3, 1943) he speaks about the *"probably unsolveable tag-problem"*. The letters from Post to Church can be found in the Alonzo Church papers, box 20, Folder 14; Department of Rare Books and Special Collections, Princeton University Library. The problem of "tag" was finally proved unsolvable by Marvin Minsky in 1961 after the problem was suggested to him by Martin Davis.[16]

of symbolic logic as it is about finding the most general form of (systems of) symbolic logic, characterized as instruments to generate sequences of letters. As such, it emphasizes its own double-faced character, a positive and a negative side (provided one finds the existence of a limitation negative rather than liberating).

Post's second thesis is, in a certain sense, much more about this positive side: its motivation was to somehow capture all the possible processes the human mind can set up to solve decision problems. Thus, Post proposed his program of finding wider and wider formulations, starting from formulation 1, in order to see how far it reaches, to understand what it comprises – how general it actually is, hence effectively extending (our knowledge of) the scope of the humanly solvable/computable. The motivation however for such program lies in the negative side of the thesis. Since Post insisted that his thesis, or any other logically equivalent such thesis, is about *human* limitations, he found it necessary to explore this limit by studying how far it can be extended.

Post's thesis II identifies solvability in the intuitive sense with solvability by formulation 1. It is oriented towards the idea of defining a formal procedure which somehow captures all processes *humans* can set up to solve a problem *in a finite number of steps*. It is about solv*ability* , just as Turing's thesis is about comput*ability* and Church's about calcul*ability* – notions that are focused on the property that certain "objects" – numbers, functions, problems – are (Turing-) computable, (Church-)calculable or (Post-)solvable. This thesis is concentrated on the idea of having a correct output after a finite number of steps, of effectively being able to have a solution *at some point in time*. Post's finite-1-processes are supposed to halt on an input, just as Turing machines are in the modern context. This is one of the reasons why (halting) Turing machines are, from a practical (!) point of view, not always considered the most suitable models in certain modern computational situations where, for instance, computer networks are not supposed to halt if they have "solved" one problem. Imagine that your OS would stop working every time you have googled something! In such contexts, alternative models like non-haling Turing machines are more suitable.[k]

---

[k]Of course, one should put things in their proper historical context: when Post was proposing his formulation 1 the issue was to find a formulation that captures the vague notion of solvability making it possible to differ between solvable and unsolvable problems. Hence, termination when a solution is found is a very reasonable assumption for the model.

Post's "older" models did not have this requirement of termination. This is not surprising knowing how and why these "models" came into existence. They were not intended as models of computability or solvability but rather as instruments of generalization to tackle general mathematical problems like the *Entscheidungsproblem*, viz., they were intended as general *forms* that "*eschew all interpretation.*" It is exactly because of this that the form of "tag" can hardly be considered as a good model of computability. However, it is also because of this that the approach Post chose for researching tag systems is one that brings to the foreground not some computable object, but *the computational process itself*. What matters then is to know what kinds of general properties one can expect from, for instance, the processes induced by the formal rules of the example on p. 51. These tag systems only became a frustration to Post because they take and need their time to develop their behavior and was confronted with the fact that one cannot reach beyond a certain "time horizon". The only way, for Post, to know what would happen even to the simple example of p. 51 was to write down and trace the process induced by the formal definition of that example. As such, the 'time sensitivity' of these computational devices was clear from the start – Post did not need to wait for a computer to understand this.

This focus on the behavior of the computational process itself rather than the 'algorithm' that induces it, results in an approach which is much more based on observation and, in this way, opens up the way for exploratory experimentation, a possibility only made truly feasible with the rise of the electronic and programmable computer. It furthermore introduces a way to think about the borderline between undecidability and decidability in relation to the complexity of the behavior of a given computational process. In this way, one of Post's historically older and less intuitively appealing 'models' are, in a sense, more adept to modern research with its focus on the relation between processes in nature and computation, exactly because of the focus on the computational process itself that tag systems necessitate.[1]

The fact that Post concentrated on studying the behavior of tag systems confronted him with the possibility that his entire program of proving the decidability of the *Entscheidungsproblem* might be in vain. By exploring the behavior of tag systems he was confronted with his own all too

---

[1] I do not mean that such approach is not possible in the context of, for instance, Turing machines, but that, seen from a historical perspective, tag systems make such approach more apparent.

human limitations.[m] It is exactly this human experience that can help to understand Post's insistence on calling his theses and any other logically equivalent one a *hypothesis* rather than a definition.

But this view also has a deeper philosophical underpinning. It is rooted in a strong view on the nature of symbolic logic and mathematics, a view which is not completely unrelated to Brouwer's views on the nature of mathematics, to whose work Post sometimes referred. This view is made explicit here [21, 403]:

> I consider mathematics as a product of the human mind, not as absolute

A similar view is expressed when Post explains that the finitary character of symbolic logic follows from the fact that it is "*essentially a human enterprise, and that when this is departed from, it is* then *incumbent on such a writer to add a qualifying "non-finitary"*." [21, 394]

Since Post considered mathematics as a product of the human mind, it is understandable that he rejected the idea of the thesis as being but a definition because, from that perspective, you can indeed not "*hide behind merely a definition* [Post in a letter to Church, dated July 10, 1936] – one cannot isolate the thesis as a thesis about mathematics from the real world mathematician since the only mathematics we know is human mathematics. Hence, if the thesis is correct we are indeed confronted with an absolutely unsolvable problem, viz. unsolvable by the mathematics of humans [21, 340]:

> The writer cannot overemphasize the fundamental importance to mathematics of the existence of absolutely unsolvable combinatory problems. True, with a specific criterion of solvability under consideration, say recursiveness [...], the unsolvability in question, as in the case of the famous problems of antiquity, becomes merely unsolvability by a given set of instruments. [The] fundamental new thing is that for the combinatory problems the given set of instruments is in effect the only humanly possible set.

---

[m]This "confrontation" went so far that Post reports in one of his letters to Church, dated March 3, 1943, that it was while working on tag systems that he had his first manic-depressive attack: "*my wife is much worried. So I told her, really for the first time, the exact history of my mental ups and downs and worse from its first inception in trying to solve the probably unsolvable tag-problem in Princeton*" Whether or not this is really true can perhaps be doubted, but it says something about the psychological impact these systems have had on Post, systems he never wanted to work on again (personal communication with Martin Davis).

The idea of this absoluteness greatly impressed Post and it is for that reason that he kept insisting on the hypothetical character of his thesis. However, Post understood that this absoluteness is only an absoluteness relative to humans and so *"the troubling thought [is suggested] have we so fathomed all our own powers as to insure our assertion of absolute unsolveability relative to us."* (Post in a letter to Church, dated July 10, 1936) In the correspondence with Church, Post is clearly searching for ways to "test" the hypothesis. Among others, he considers wider and wider formulations of the Turing machine model. However, it is also clear from that same correspondence that even though the "thesis" should be treated as a hypothesis, Post was strongly convinced of the truth of the hypothesis. To put it in his words (Post in a letter to Church, dated June 9, 1937):

> I – to put it crudely but forcibly – am willing to bet a 1000 to $1 that [there are no effectively calculable functions that are not recursively solvable], and it would take me five years to save up that sum [...] But I am not willing to stake my "immortal" soul in it – which I should were I to adopt your [Church's] original position [the thesis as a definiton].

Post might have been tempted at that time to conclude that, since we have reached the limitations of (finitary) symbolic logic, we might as well give up on it. This was not Post's conclusion: knowing (or believing) that there are things that cannot be achieved with symbolic logic, knowing that there are absolutely unsolvable problems does not mean that one knows what they are or that one knows the realm of the computable. It is here that symbolic logic has a clear future:[21]

> with the bubble of symbolic logic as universal logical machine finally burst, a new future dawns for it as the indispensable means for revealing and developing those limitations. For [...] Symbolic Logic may be said to be Mathematics become self-conscious.

Indeed, for Post symbolic logic (or formal logic) can be used as an instrument to explore the limitations exposed *through* the methods of symbolic logic.[n]

---

[n]This is exactly what Post partly did in some of his later work: the proof of the unsolvability of the Post correspondence problems and the word problem of semi-groups as well as his fundamental paper on recursively enumerable sets and their decision problems can be understood from this perspective.

4. Some Afterthoughts

To expose you, reader, to Post's ideas on computation is the first ambition of this paper but the motivation behind this is not mere historical completeness but rather to add historical and philosophical depth and understanding to the nowadays so fundamental concept of computation *and* its limitations.

The lessons I learned from Post's writings are manifold, but I think that one of the more important ones lies in his (implicit) insistence on the fact that the thesis is as much about the universe of computation as it is about the limits of that self-same universe, a universe shaped by the possibilities and limitations of *human* mathematics – why else call it a hypothesis?[o]

Exploring these limitations, this borderline, is in my view still an essential objective in our search for a better understanding of the nature of computation. An 'old' approach to this problem is not to study that which lies "beyond" that borderline, but rather to approximate it, to explore it, by searching for and studying 'smallest' universal and/or undecidable devices and 'largest' decidable classes of such devices.[p] It is this line of research that allows to excavate this frontier of (Turing-)computability from the bottom up, starting from the computable side of the borderline just up to the edge of that borderline. With the advances in computer hard-and software these excavations are now supported by computer-aided explorations of the behavior of very different classes of computational devices like tag systems,[9] Turing machines[15] or cellular automata[3] which, among others, advances our understanding of the computational power of particularly difficult instances like the example on p. 51.

This kind of research on the limits of (Turing-)computability is perhaps less fashionable today than for instance research on DNA computing, but that does not mean that it lacks significance. In fact, it is readily seen that such results are paramount to our understanding not only of the nature of computation but also of computation in nature, even more so if we incorporate the computer-aided studies. Indeed, to determine limits of the computable is also to determine the limits of our computational models of physics and biology and the question of what can and cannot be computed in nature goes to the core of what computation in nature means. These kind of results can be achieved by tracing down properties in the behavior of possibly undecidable instances or by learning more about the size of

---

[o]Just to be clear, this does *not* imply some form of computationalism.
[p]See e.g.[14] for a survey of such research.

smallest universal devices in computational contexts that are more suited for this kind of research, and then transmuting these findings to the context of natural computing.

Post did not start out from the idea of developing a model that is a suitable model of computability, rather he was searching for the most general form of symbolic logic and ultimately mathematics. This resulted in his apparently simple form of "tag" which lay the basis for his findings on the possibilities and limitations of symbolic logic. Nowadays, models of computability like tag systems seem to be less interesting if one wants to inquire into (the nature of) computation (in nature) because they seem unrelated to nature, because they do not have the "appeal to intuition" that e.g. membrane systems have for cellular behavior. Perhaps however a return to such abstract formal devices that are less 'natural' can rejuvenate or at least stimulate current research on natural computing. Indeed, Post's devices are different enough from physical processes such that they allow to zoom-in on the 'non-physical' aspects of computational processes but are nonetheless erratic, unpredictable and complex in a way that is not completely averse to certain physical processes. In this way, by placing 'nature' between brackets for a while, one can perhaps study the nature of computation by investigating nature taking place in computation.

References

1. A. Church, *An unsolvable problem of elementary number theory*, American Journal of Mathematics, no. 58, 345–363, 1936.
2. A. Church, *Review of*[26], Journal of Symbolic Logic, 2, no. 1, 42–43, 1937.
3. M. Cook, *Universality in elementary cellular automata*, Complex Systems, 15, no. 1, 1–40, 2004.
4. M. Davis, *The undecidable. Basic papers on undecidable propositions, unsolvable problems and computable functions*, Raven Press, New York, 1965, Corrected republication (2004), Dover publications, New York.
5. M. Davis, *Why Gödel didn't have Church's thesis*, Information and Control 54, 3–24, 1982.
6. M. Davis, *Emil L. Post. His life and work* in:,[22] xi–xviii, 1994.
7. L. De Mol, *Closing the circle: An analysis of Emil Post's early work.*, The Bulletin of Symbolic Logic, 12, no. 2, 267–289, 2006.
8. L. De Mol, *Solvability of the halting and reachability problem for binary 2-tag systems*, Fundamenta Informaticae 99, no. 4, 435–471, 2010.
9. L. De Mol, *On the complex behavior of simple tag systems*, Theoretical Computer Science, 412, no. 1–2, 97–112, 2011.
10. R. Gandy, *The confluence of ideas in 1936*, Published in,[12] pp. 55–111, 1988.

11. K. Gödel, *Remarks before the princeton bicentennial conference on problems in mathematics*, in:,[4] 84–88, 1946.
12. R. Herken (ed.), *The Universal Turing machine*, Oxford University Press, Oxford, 1988, Republication, Springer Verlag, New York, 1994.
13. S. C. Kleene, *Origins of recursive function theory*, Annals of the history of computing, 3, 52–67, 1981.
14. M. Margenstern, *Frontier between decidability and undecidability: A survey*, Theoretical Computer Science, 231, no. 2, 217–251, 2000.
15. P. Michel, *Small Turing machines and generalized Busy Beaver competition*, Theoretical Computer Science, 326, no. 1–3, 45–56, 2004.
16. M. Minsky, *Recursive unsolvability of Post's problem of tag and other topics in the theory of Turing machines*, Annals of Mathematics, 74, 437–455, 1961.
17. E. L Post, *Introduction to a general theory of elementary propositions.*, American Journal of Mathematics, 43, 163–185, 1921.
18. E. L. Post, , *Note on a fundamental problem in postulate theory*, 1921, dated June 4, 1921, Emil Leon Post papers, *American Philosophical Society*.
19. E. L. Post, , *Finite combinatory processes - Formulation 1*, The Journal of Symbolic Logic, 1, no. 3, 103–105, Also published in,[4] 289–291, 1936.
20. E. L. Post, , *Formal reductions of the general combinatorial decision problem*, American Journal of Mathematics, 65, no. 2, 197–215, 1943.
21. E. L. Post, , *Absolutely unsolvable problems and relatively undecidable propositions - Account of an anticipation*, in:,[4] 340–433. Also published in,[22] 1965.
22. E. L. Post, , *Solvability, Provability, Definability: The collected works of Emil L. Post*, Birkhauser, Boston, edited by Martin Davis, 1994.
23. W. Sieg, *Mechanical procedures and mathematical experience*, Mathematics and Mind (Oxford), Oxford University Press, pp. 71–117, 1994.
24. R. L. Soare, *Computability and recursion*, The bulletin of Symbolic Logic, 2, no. 3, 284–321, 1996.
25. R. L. Soare, *Computability and incomputability*, Computation and Logic in the Real world. Third conference on computability in Europe, CIE2007, Siena, Italy (S.B. Cooper, B. Löwe, and A. Sorbi, eds.), LNCS, Springer, pp. 705–715, 2007.
26. A. M. Turing, *On computable numbers with an application to the Entscheidungsproblem*, Proceedings of the London Mathematical Society (1936–37), no. 42, 230–265, A correction to the paper was published in the same journal, vol. 43, 1937, 544–546. Both were published in,[4] 116–151.
27. A. M. Turing, *Intelligent machinery, a heretical theory*, 1951, Lecture given at Manchester (2 versions, one numbered 1–10, the other numbered 96–101). Available at the digital Turing archive at http://www.turingarchive.org, archive number AMT/B/4.

Chapter 4

Machines

Raymond Turner

School of Computer Science & Electronic Engineering
University of Essex, UK

What is it for a physical machine to be an *implementation* of an abstract one? In this paper we shall provide an answer to this question that has its origins in the computer science concept of formal specification.

Logicians and theoretical computer scientists deal with abstract machines and computations on them. In contrast, computer engineers build physical machines - actual computers.[a] Obviously these two kinds of machine and their corresponding notions of computation are quite different ([16], [18], [19]). Physical ones have a location in space and time. They are subject to change and causal impact. Consequently, they may be investigated empirically. Certainly, one cannot find out how they will actually behave by conceptual reflection and mathematical analysis alone. In contrast, abstract ones have no spatial or temporal existence. Moreover, they are immune to any kind of causal interaction. It follows that they are not subject to direct empirical investigation: we can only discover their properties by conceptual reflection and mathematical analysis. Furthermore, computations on abstract machines can only be carried out by hand. As a consequence, they are most often infeasible, a limitation of pencil and paper computation that provides the fundamental motivation for the construction of physical computers.

These differences point to a simple question: when is a physical machine to be taken as an *implementation* of an abstract one? The answer to this

---

[a]Turing can rightly be said to be the founder of the theoretical computer science and our concept of an abstract machine. Moreover, as Hodges [10] remarks, *it is reasonable to suppose that Turing believed that the very simple logical operations of his abstract machines could in principle be realized physically, since they required only operations such as were employed in contemporary teleprinters and automatic telephone exchanges.*

question is linked to the question of what we take to be a *physical computation* [14]. From one perspective it seems clear that such computations are somehow delineated by abstract ones: how else could they be picked out? How else would we recognize something as a computation? These questions are also linked to a central question in the philosophy of applied mathematics: what is the relationship between an abstract mathematical concept and its purported physical analogue? For instance, Euclidean geometry, is a purely mathematical theory full of idealizations and abstractions: its points, lines, and planes are abstract mathematical entities. How is it that these abstract notions are so successfully applied to the physical world?

1. Abstract Machines

Turing machines, push-down automata, and finite-state machines are three well known examples of abstract machines. Turing machines have a special status in that they are taken to characterize the very notion of *by-hand computation*. Furthermore, every programming language is defined by some underlying abstract machine [8]. Such languages are underpinned by very different styles of machine. For instance, functional languages are based upon various reduction machines, logical ones on some form of unification machine and standard imperative ones upon machines that have the structure of a store. Here we are not so much concerned with the differences between these computational paradigms but only with that relationship between abstract machines and physical ones which is common to all styles. Consequently, we may employ any paradigm to illustrate the central issues.

The following abstract machine could be used as the basis for the underlying machine for a certain form of imperative language where the latter is characterized by the fundamental operation of updating the contents of locations in a store. The *store* consists of named locations that hold numerical values. Each location holds at most one value. There are two operations that allow the updating of values in locations and the inspection of their values. We shall employ a small fragment of elementary finite set theory to define the machine. There are three finite sets: *Store*, *Location* and *Numerals* that represent the store itself, the locations of the store and the values in the locations. In addition, there are two operations: *Lookup* is a partial function from the Cartesian product of *Store* and *Location* to *Numerals* and *Update* is a total function from the Cartesian product of

Store, *Location* and *Numerals* to *Store*.

$$Lookup : Store \otimes Location \to Numerals$$
$$Update : Store \otimes Location \otimes Numerals \to Store$$

These operations are governed by the following conditions that dictate the relationship between them i.e., that if a location is updated with a new value, *Lookup* returns that value. More formally, for all stores s, locations x and numerals v:

$$Lookup(Update(s, x, v), x) = v$$
$$Lookup(Update(s, x, v), y) = Lookup(s, y),$$
$$\text{where } x \neq y \text{ and } Lookup(s, y) \text{ is defined}$$

Of course this set theoretic definition does not automatically refer to any specific physical devices. It employs the abstract notion of set and introduces the update and the lookup operations as abstract set theoretic functions. Actually, there are many different ways of defining abstract machines including the use of type theory and category theory. But the use of sets is the most common, and nothing that we shall say depends upon the specific choice of mathematical underpinnings.

What do computations on this machine look like? Our simple store could serve as the underlying abstract machine of a simple imperative programming language whose programs are based upon the basic operation of assignment.

$$x := y$$

In terms of the machine its meaning is given as follows: the value in the location y is looked up and then the value in the location x is updated with that value i.e., in the store s, the assignment statement changes the store as follows.

$$Update(x, s, Lookup(y, s))$$

Presumably, a computation on our machine will be individuated by a sequence of such assignments. For example, the program

$$z := x; x := y; y := z$$

swops the values in locations x and y. The computation induced on the machines would be a series of state changes where location z is given the contents of location x and then location x is given the contents of location y and finally locations y is given the contents of location z.

This brings us to the concrete instance of our general question generated by this machine: when is a physical device to be taken as an *implementation* of this abstract machine? Presumably, if we can answer this question, a physical computation will be determined as the image of an abstract one under the implementation. So for example, once fixed, the swop program will determine a physical process, in this case a sequence of state changes, that modifies the physical device. So once we decide what constitutes an implementation, some clarity is gained with respect to the individuation issue for computations, at least on this machine. Of course, the notion of programming language is not restricted to the current paradigms of computational practice. We could admit new paradigms and machines without really effecting the substantive points we wish to make.

So the first problem we have to address concerns the very notion of *implementation*. This turns out to be less straightforward and more controversial than it might seem.

2. Specification and Implementation

One notion of implementation emerges from the actual practice of computer design. Here the abstract machine is seen as a *specification* of a physical one; it is taken to lay out the *functional requirements* of the required physical device. The various components of the abstract machine must have corresponding physical ones. So for example, the specification of a PC might have as components a specification of a processor, a specification of a memory, a specification of a hard drive and a specification of a motherboard. In addition, the processor must interact with the memory and the hard drive in specified ways etc.[b] We shall call this the *intensional* approach to characterizing the relationship between the abstract machine and a physical one.

This approach is similar to the mechanistic one advocated by Piccinini ([13], [14]) that accounts for concrete computation in terms of the mechanistic properties of a system. But we take our lead from the methodology

---

[b]We have argued elsewhere [20] that the rule following considerations ([2], [11]) demand that such specifications must be taken to be abstract entities [20].

of specification and design. This has implications for the way we view the correctness of the implementation.

Although our abstract machine has only a minimal amount of structure, seen as a specification of a physical device, it demands the construction of a physical representation of the store together with physical devices that carry out its update and lookup operations. More specifically, the locations of the abstract machine must match the physical ones and the operations of the abstract machine, the *update* and *lookup* operations, must be matched by actual physical devices that act in accord with our axiomatic requirements. More concretely, assume that there are only two locations l and r, and two possible values 0 and 1. Then we have four possible states $(0,0), (0,1), (1,1)$ and $(1,0)$.

We need to ensure that the two machines behave in harmony. To make this requirement explicit consider the following table that articulates the input/output behavior of our abstract machine i.e. all the possible state changes of our abstract machine induced by updating.

Start state	Update $(r,1)$	Update $(r,0)$	Update $(l,1)$	Update $(l,0)$
$(0,0)$	$(0,1)$	$(0,0)$	$(1,0)$	$(0,0)$
$(0,1)$	$(0,1)$	$(0,0)$	$(1,1)$	$(0,1)$
$(1,0)$	$(1,1)$	$(1,0)$	$(1,0)$	$(0,0)$
$(1,1)$	$(1,1)$	$(1,0)$	$(1,1)$	$(0,1)$

In principle, to ensure that we have a correct implementation of the update operation we would need to check that the corresponding physical table is in complete agreement i.e., there is an exact match between the tables. This would involve an exhaustive series of tests where the physical update and lookup devices return the results dictated by the abstract state table. This series of tests would guarantee that the physical operations are *correct* relative to the abstract one. In our particular case, where our specification is stated as an axiomatic theory, the physical system must be a model of the abstract one.

One might be tempted to think that because of its exhaustive nature this process yields mathematical certainty. However, even in the present case, where our abstract machine has a very small number of locations and a bound on the size of numbers that may be entered, this does not change the status of the statement of agreement between the abstract and physical device. It is still an empirical claim; it is not a mathematical or formal one about the relationship between two abstract objects [9]. In particular, there is no guarantee that the next physical operation will perform according to the table: the cogs and handles may fail to function or the locations may get blocked. The correctness of the physical machine is measured against the abstract one. Indeed, we have argued elsewhere [20] that the rule following considerations ([2], [11]) demand that such specifications must be taken to be abstract. Of course, a physical system might be the result of a whole series of specifications where more and more detail is accumulated through the series. But ultimately, the physical device must be correct relative to the specification that determines the actual physical device.

3. Complexity and Uniformity

However, this ideal picture of testing and verification does not reflect engineering practice where exhaustive testing is not possible. How could you exhaustively test an actual laptop or a system as complex as the internet? There are just too many possible state transitions. In fact, testing is often indirect [1] where abstract models stand proxy for the physical device. Indeed, the complexity of real situations would even render the by-hand computation of the abstract state table infeasible. Such computations are subject to time constraints, error, lack of resources, and just down right boredom on the part of the human computer. Even if we had time, we could not ensure that we could do the calculations correctly, where here *correctness* is determined by the intensional description of the operations. For example, for our machine the correctness of the calculations is determined by the axioms that govern the relationship between the lookup and update operations. In reality we cannot perform all such computations and then marry them with the corresponding physical ones. In the practice of specification and implementation, one has to settle for much less. Instead, we carry out carefully chosen tests. Very often the specification, or a codicil to it, lays down the suites of testing procedures that are taken to be sufficient to guarantee a degree of extensional agreement between the abstract and physical devices. This provides the working notion of correctness. It is as if this notion of *correctness* is taken to replace the complete one.

But how is this incomplete notion of correctness justified? First note that the test instances are judiciously chosen to cover as wide a variety of cases and possibilities as is practically feasible. Here the underlying uniformity of the operations plays a crucial role. In our example, this uniformity is encapsulated in the intensional or axiomatic definition of the update function which insists that it operates uniformly as a function over all its inputs and outputs. If our update operation works for location x and for the numeral 4 and location y and numeral 7, we may well assume that, all things being equal, it will work for other locations and values. There is a hidden ceteris paribus rider to the effect that the physical machine has been uniformly built. Indeed, this kind of sample checking is what we do even in everyday instances of verifying that a physical device satisfies its informal specification. Again it is generally justified by some implicit assumption of underlying uniformity. Of course, one can never be sure; certainty is never obtainable in physical situations. Nevertheless, this hidden assumption of uniformity is crucial to the process of verification and testing.

Is the notion of implementation that is used in computational practice philosophically significant in terms of the answer to our central question? There is a parallel here with the philosophy of mathematics where for far too long the practice of mathematics has not played a central role in the philosophy of mathematics ([3], [12]). We should be suspicious of any notion of implementation that pays no attention to computational and engineering practice. We now turn to such a notion.

4. Extensional Implementation

There is a seemingly less demanding way in which a physical device maybe be taken as an implementation of an abstract machine. This only demands that the physical system is in extensional agreement with the abstract one. This has been dubbed the *simple mapping account* (SMA). The original version was presented by Putnam [15] in terms of Turing machines but this is taken to be somewhat parochial since it is tied to Turing machines and their specific structure. The more general version is given a clear exposition in [14].

> *A physical system P **Implements** an abstract one A just in case there is a mapping I from the states of P to the states of A such that: for any abstract state transition $s_1 \Rightarrow s_2$ **if** the system is in the physical state S_1 where $I(S_1) = s_1$, it **then** goes into the physical state S_2 where $I(S_2) = s_2$*

For example, under SMA, a physical implementation of our abstract machine might take the following shape. The states of the machine are represented as stones that change color with temperature. Consequently, the following physical configuration might be taken to be an implementation of our update operation.

Stones T_1	T_2	T_3	T_4
red	blue	red	green red
blue	blue	red	yellow blue
green	yellow	green green	red
yellow	yellow	green yellow	blue

Physical State Table (PST)

Our table is made up with an arrangement of stones of different colors where the temperatures T_1, T_2, T_3 and T_4 indicate the changes in color of four stones. At T_1 degrees the stones change color as indicated by the column labeled T_1. Similarly for the other columns. According to the SMA, this arrangement of colored stones implements the update operation of our abstract machine. For example, the state transition $(0,0) \Rightarrow (0,1)$ is represented as *redstone* $\Rightarrow$ *bluestone*. Notice that the conditional

if the system is in the physical state S_1 where $I(S_1) = s_1$, it **then** goes into the physical state S_2 where $I(S_2) = s_2$

is taken to be the material conditional of standard propositional logic. It requires only that the consequent be true when the antecedent is. There is no demand that there has to be any causal connection between the antecedent and the consequent. In particular, it does not matter that the T_i are temperature changes; they could be any physical process e.g. T_i might be stone painting. They have only to be in accord with the table. We could replace this table with any other that has exactly the same mathematical structure: the arrangements of stones together with distinct labels would satisfy the demands of SMA. Consequently, such mappings are relatively easy to come by. It is this notion of implementation that leads some authors to conclude, expressed in our present terminology, that almost every physical system implements almost every specification ([15]). This is absurd. A kettle does not meet the specification of a laptop.

What has gone wrong? First observe that in order to set up the correspondence we have to compute the whole state table for our abstract machine, which means that the human computer is doing all the computation, exactly the opposite of why computers were invented. In addition, we need to have the whole physical counterpart in view. Consequently, the relationship is created post hoc. This point is made by Copeland [6] who observes that the mappings of SMA are illegitimate because they are constructed after the computation is already given. SMA is only concerned with the formal relationship between the extensional structures of the two devices. This is quite different to the intensional approach where, in general, not only cannot we compute the whole extensional table but doing so defeats the whole point of building a physical computer.

Secondly, imagine an abstract version of the above table

Letter	T_1	T_2	T_3	T_4
r	b	r	g	r
b	b	r	y	b
g	y	g	g	r
y	y	g	y	b

This could serve as a specification of the physical one: the physical table *implements* the specification given by the abstract one. And this is true in our first sense of implementation where the specification is the abstract table itself. Here the update operation is not given in intension by defining axioms but is given by its extensional table. Such a specification only insists on some cardinality constraints and consistency of labelling.

Thirdly, the intensional approach facilitates the specification of actual systems that are massive and complex. It employs a notion of operational uniformity that provides some security in methods of testing and verification. Consequently, we have some justification for not testing the whole extensional structure. Which is fortunate since, in general, we cannot do so. The extensional approach abstracts away from uniformity of structure and requires the testing of the whole extensional structure for correctness. Indeed, it is constituted by it.

5. The Empirical Perspective

The danger of pancomputationalism has driven some authors (e.g. Chalmers and Copeland) to attempt to provide an account of implementation that somehow restricts the class of possible extensional interpretations. They aim to impose constraints on acceptable mappings ([5], [6], [7]). In particular, certain authors ([6], [5]) seek to impose causal constraints on such interpretations. We shall develop an approach that is complementary to our intensional one which, without starting with the SMA, gets at the heart of this demand for more causal constraints.

Wandering through a field I discover a device that looks like a very large magic square with 26 white squares each of which is labeled with a letter from the Roman alphabet. Attached to the board via a series of mechanical levers and gears is a keypad containing the digits 0-50 and the letters of the Roman alphabet. There are two additional keys marked *Lookup* and *Update*. I discover that when I press the lookup key followed by the letter C, the square labeled C opens to display its contained digit. This seems to apply to all the squares. If I press the update key followed by a digit 6 and the letter C, the mechanical system swings into action and replaces the number in the square labeled C with 6. When I now press the lookup key followed by the key C, I observe that 6 is in the corresponding square. In the other boxes nothing else appears to have changed. A further series of experiments suggests that these are the only two operations that I can do namely, lookup the number in a square or update a number in a square. Any other combination of lever pushings seems to have no impact.

As an observer, I am puzzled by this device. What is it? What is it for? Can I use it for something. To address these questions, I attempt to construct a *functional description* of the device. I need to articulate its structure and operations to facilitate its use as a useful device. After some reflection, I postulate that it is a simple store machine. Consequently, I adopt our abstract machine as its functional description. Of course, I need to check that this *theory* of the device is correct. For this I may undergo a series of further tests and experiments that parallel the specification case.

But once again, in more realistic and complex situations, complete verification would be infeasible. In general, the best I can do is to carry out a suite of tests that somehow reflect the underlying uniformity of the system. What abstract functional description fits a physical device is a matter of theoretical speculation, experimentation and verification.

Sprevak [19] seems to object to this perspective on the grounds that it makes the functional description of the system or device depend upon the epistemic powers of the investigating agent. He aims this criticism at some accounts of the nature of computation, but it seems to be equally applicable to abstract machines and their implementations.

> ....the reason why some realization functions do not establish computational identity cannot be that it requires epistemic work on the part of the agent to construct that realization function. First, as noted above the realization functions of PCs often require considerable work on the part of the agent to construct, and sometimes the agent (the hardware designer) has to perform the computation

herself in order to construct that function. Second, such a move would make the computation a system performs entirely a function of the epistemic powers and interests of the agents that investigate the system. There would be no other non-trivial constraint on the computation that a system performs. This would be to endorse a form of antirealism about computation: facts about computation would be constrained, not by the systems investigated (which by themselves trivially perform every computation), but by facts about epistemic agents investigating those systems. Questions about how a realization function is constructed should not matter in this debate unless one wishes to endorse anti-realism about computation.

However, in the present context there is something odd about this criticism. Surely this is just what scientific investigation (and *reverse engineering*) is like. We may admit that there is a reality that is independent of us while maintaining that the best we can do is to provide theories that provide the most plausible explanation of matters ([22], [4], [17]). If we come across a device in nature that we wish to see as a computing device, to see it as such involves the construction of a theory (functional description) of it as a computing device. From a scientific perspective, this is the best we can do i.e., postulate a theory of the device and verify it by experimentation and testing. In general, to compute the computation tables of either is practically impossible. Indeed, if this were not so we would not actually require scientific theories: we would just use very large lookup tables.

In contrast, according to the extensional approach, there would be no need for predictions about the future behavior of the physical machine since everything would already have been worked out. But this is precisely what we do not do in the construction of scientific theories. Indeed, the latter provide the ability to predict and explain what will happen. They reveal the underlying mechanisms. Lookup tables provide no explicit uniformity and consequently they provide no notion of *explanation* ([17], [4]). It is in the underlying uniformity that explanation is located. Seen as a theory about the device, our axiomatic conditions on *lookup* and *update* explain why the physical operation of update behaves the way it does. Without this intensional characterization we have no explicit explanation, we have only a table of facts. The extensional approach, is not only impractical, it does not meet the requirements for a theory of the device. The latter must provide predictive and explanatory power. The information contained in the extensional table does not yield any explanation of how the device works. The extensional approach to scientific theories would, by analogy with pancomputatationalism, trivialise the construction of scientific theories: what is to stop every abstract theory from being a model of every chunk of physical reality?

One might argue that scientific theories cover an infinite number of cases whereas the abstract and concrete machines we are considering only cover a finite number. However, the fact that our systems are finite does not really affect our overall argument. The size and intractability of realistic systems entails that the best we can do is to carry out a sequence of judicious tests that rely for their credibility on the uniformity of the structure. They have a similar role to more standard scientific theories.

It is this interpretation of the abstract machine as a theory of the physical device that brings us closer to the causal approach. The fact that such theories are explanatory and have predictive power is linked to the fact that we may state their consequences in terms of conditionals of various kinds (future tense conditionals or counterfactual conditionals). In particular, it is linked to the suggestion of [6] that we replace the material conditional in the SMA

if *the system is in the physical state S_1 where $I(S_1) = s_1$ it* **then** *goes into the physical state S_2 where $I(S_2) = s_2$ into the physical state S_2 where $I(S_2) = s_2$*

with a counterfactual i.e., the above conditional is to be interpreted not as the material conditional but as a counterfactual one. However, while this move will block arbitrary interpretations, by itself, it seems rather ad-hoc. The present account provides some justification: this replacement is plausible given the context where the abstract machine is seen as a theory of the physical one and where, consequently, the axioms of our machine have predictive power and so give the implicit conditionals more than material force. While there are many assumptions being made about the nature of theories, predictions and the role of explanation in science, this account of the relationship between abstract machines and physical ones captures and explains some of the plausibility of the causal account.

6. Intensional Stance

What is the connection between the two approaches: the intensional and the empirical? Although there is an agreement of the methodology of testing and verification between the two approaches, there is a fundamental difference of *intentional* stance [21].

When the abstract machine is taken as a theory of the physical one, it is the physical device that is given to us. When there is a mismatch, it is the theory that must be scarified or modified. We can hardly change the world when the theory does not fit. In contrast, in the specification case, matters are reversed. Here the specification takes charge: when there is disagreement, we blame the physical device. It is the physical implementation that is correct or not. These differing notions of *intentional stance* bring out the importance of the later notion. In turn they illustrate how impoverished the extensional account is: it ignores any form of intentional attitude towards the abstract notion and its relationship towards the physical one.

References

1. Baier, C. J. *Principles of Model Checking.* Cambridge, MA. MIT Press, 2008.
2. Boghossian, P. The Rule-following Considerations. *Mind*, 507–549, 1989.
3. Bueno, O. A. *New Waves in The Philosophy of Mathematics.* Basingstoke: Palgrave, 2009.
4. Chalmers, A. *What is this thing called Science.* Maidenhead: Open University Press, 1999.
5. Chalmers, D. J. Does a rock implement every finite-state automaton. *Synthese*, 108:309–333, 1996.
6. Copeland, B. J. What is Computation? *Synthese*, 108(3): 335–359, 1996.
7. Egan, F. Computation and Content. *Philosophical Review*, 104:181–204, 1995.
8. Fernandez, M. *Programming Languages and Operational Semantics: An Introduction.* London: King's College Publications, 2004.
9. Fetzer, J. Program Verification: The Very Idea. *Communications of the ACM*, 31(9), 1048–1063, 1988.
10. *Internet* Hodges, A. *Alan Turing: the logical and physical basis of computing.* http://www.bcs.org/upload/pdf/ewic_tur04_paper1.pdf
11. Kripke, S. *Wittgenstein on Rules and Private Language.* Boston: Harvard University Press, 1982.
12. Mancousa. *The Philosophy of Mathematical Practice.* Oxford: Oxford, 2010.
13. Piccinini, G. Computation without Representation. *Philosophical Studies*, 137:205–241, 2008.
14. Computation in Physical systems. Stanford Encyclopedia of Philosophy. http://plato.stanford.edu/entries/computation-physicalsystems/
15. Putnam, H. *Representation and Reality.* MIT Press, Cambridge, MA, 1988.
16. Rosen, G. *Abstract Objects.* Retrieved from Stanford Encyclopedia of Philosophy: http://plato.stanford.edu/entries/abstract-objects/, 2001.
17. Rosenberg, A. *The Philosophy of Science.* London: Routledge, 2000.

18. Sieg, W. Calculation by man and machine: Conceptual analysis. In Sieg, W., Sommer, R., and Talcot, C. (ads), *Reflections on the Foundations of Mathematics* (Essays in Honor of Solomon Feferman), pages 387–406. Volume 15 of Lectures Notes in Logic, Association of Symbolic Logic, 2001.
19. Sprevak, M. Computation, Individuation, and the Representation Condition. Studies In History and Philosophy of Science Part A. Volume 41, Issue 3, pp. 260-270, *Computation and cognitive science*, 2010.
20. Turner, R. Programming Languages as Mathematical Theories. In J. Vall-lverdu, *Thinking Machines and the Philosophy of Computer Science*, 2010.
21. Turner, R. Specification. *Journal Minds and Machines 21* (2):135–152, 2011.
22. Van Fraassen, B. C. *The Scientific Image*. Oxford: Clarendon Press, 1989.

Chapter 5

Effectiveness*

Nachum Dershowitz and Evgenia Falkovich

School of Computer Science, Tel Aviv University, Tel Aviv, Israel

We describe axiomatizations of several aspects of effectiveness: effectiveness of transitions; effectiveness relative to oracles; and absolute effectiveness, as posited by the Church-Turing Thesis.

> *Efficiency is doing things right;*
> *effectiveness is doing the right things.*
> —Peter F. Drucker

1. Introduction

In 1900, David Hilbert posed, among other problems, the research challenge of how to effectively determine whether any given polynomial with rational coefficients has rational roots:[25][a]

> [Probleme] 10. **Entscheidung der Lösbarkeit einer Diophantischen Gleichung.** Eine *Diophantische* Gleichung mit irgend welchen Unbekannten und mit ganzen rationalen Zahlencoefficienten sei vorgelegt: man soll ein Verfahren angeben, nach welchem sich mittelst einer endlichen Anzahl von Operationen entscheiden läßt, ob die Gleichung in ganzen rationalen Zahlen lösbar ist."

In the same lecture, as his famous second problem, Hilbert asked for a proof of the consistency of (Peano) arithmetic.

Later, he and Wilhelm Ackermann underscored the importance of the decision problem for validity of formulaæ in (first-order predicate) logic,

---

*This work was carried out in partial fulfillment of the requirements for the Ph.D. degree of the second author.
[a][Problem] **10. Determination of the solvability of a Diophantine equation.** Given a Diophantine equation with any number of unknown quantities and with rational integral numerical coefficients: To devise a process according to which it can be determined in a finite number of operations whether the equation is solvable in rational integers.

which they called the *Entscheidungsproblem* [26, pp. 73–74]:[b]

> Das Entscheidungsproblem ist gelöst, wenn man ein Verfahren kennt, das bei einem vorgelegten logischen Ausdruck durch endlich viele Operationen die Entscheindung über die Allgemeingültigkeit bzw. Erfüllbarkeit erlaubt. Das Entscheidungsproblem muss als das Hauptproblem der mathematischen Logik bezeichnet werden.... Die Lösung des Entscheidungsproblems ist für die Theorie aller Gebiete, deren Sätze überhaupt einer logischen Entwickelbarkeit aus endlich vielen Axiomen fähig sind, von grundsätzlicher Wichtigkeit.

Hilbert was seeking an effective procedure that could solve every instance of the validity question, positively or negatively: "We assume that we have the capacity to name things by signs, that we can recognize them again. With these signs we can then carry out operations that are analogous to those of arithmetic and that obey analogous laws" (quoted in[51]).

In 1936, Alonzo Church suggested that the recursive functions, or the computationally equivalent lambda-definable numeric functions, capture the intended concept of "effectively calculable" procedure [9, p. 356]. With his formalization of absolute effectivity in hand he proceeded to demonstrate that no effective solution exists for the Entscheidungsproblem. When Church subsequently learned of Alan Turing's independent proof of undecidability,[56] he conceded that Turing's machines have "the advantage of making the identification with effectiveness in the ordinary (not explicitly defined) sense evident immediately" [10, p. 43]. Similarly, Kurt Gödel [19, pp. 369–370] realized that Turing's model of effective computation, which provides "a precise and unquestionably adequate definition of the general concept of formal system," strengthens his earthshaking incompleteness results and establishes that "the existence of undecidable arithmetical propositions and the non-demonstrability of the consistency of a system in the same system can now be proved rigorously for every consistent formal system containing a certain amount of finitary number theory." In short, Hilbert's dream of devising a complete and consistent finite axiomatization of mathematics, as expressed in his second problem, is inherently unattainable.

---

[b]The Entscheidungsproblem is solved when we know a procedure that allows for any given logical expression to decide by finitely many operations its validity or satisfiability.... The Entscheidungsproblem must be considered the main problem of mathematical logic.... The solution of the Entscheidungsproblem is of fundamental significance for the theory of all domains whose propositions could be developed on the basis of a finite number of axioms.

Stephen Kleene reformulated Church's contention that the recursive functions and the effective numeric functions are one and the same as a "thesis" ([30, p. 60], [31, p. 332], [32, p. 232]):

> **Thesis I.** Every effectively calculable function (effectively decidable predicate) is general recursive.
>
> **Thesis I**[†]. Every partial function which is effectively calculable (in the sense that there is an algorithm by which its value can be calculated for every n-tuple belonging to its range of definition) is potentially partial recursive.
>
> Turing's and Church's theses are equivalent. We shall usually refer to them both as *Church's thesis*, or in connection with that one of its... versions which deals with "Turing machines" as *the Church-Turing thesis*.

Church's thesis asserted that the recursive functions are the only numeric functions that can be effectively computed. Turing's thesis staked the analogous claim that any function on strings that can be mechanically computed can be computed, in particular, by a Turing machine. Turing showed [56, Appendix] that with a suitable interpretation of strings as numbers, his machines compute exactly the recursive functions.

Three main lines of argument have been adduced in support of this thesis ([31, p. 320], [47, pp. 18–19], [31, p. 321]):

- All the many known effective computational models compute only partial recursive functions.
- "By means of detailed combinatorial studies, the proposed characterizations of Turing and of Kleene, as well as those of Church , Post, Markov, and certain others, were all shown to be equivalent."
- Turing's analysis of "the sorts of operations which a human computer could perform, working according to preassigned instructions" showed that these can be simulated by Turing machines.

Gödel is reported[12] to have believed "that it might be possible ... to state a set of axioms which would embody the generally accepted properties of [effective calculability], and to do something on that basis". As explained by Shoenfield [48, p. 26]:

> It may seem that it is impossible to give a proof of Church's Thesis. However, this is not necessarily the case.... In other words, we can write down some axioms about computable functions which most people would agree are evidently true. It might be

possible to prove Church's Thesis from such axioms.... However, despite strenuous efforts, no one has succeeded in doing this (although some interesting partial results have been obtained).

This challenge of proving the Church-Turing Thesis is first in Richard Shore's list of "pie-in-the-sky problems" for the twenty-first century.[8] Indeed, Harvey Friedman[16] has predicted that sometime in this century, "There will be an unexpected striking discovery that any model of computation satisfying certain remarkably weak conditions must stay within the recursive sets and functions, thus providing a dramatic 'proof' of Church's Thesis."

We discuss such an axiomatization of effectiveness in Sections 2–5. Unlike Turing's analysis,[56] and subsequent generalizations,[18,37,38,49,50,52,53] our axioms of effective computation are, at the same time, both formal and generic. They are formal, in that they may be cast as precise mathematical statements;[4,14] they are generic, in that they apply to computations with arbitrary states (Section 3) and arbitrary programmable transitions (Section 4).

Computability is a more general notion than recursiveness or Turing computability. Just as Turing machines provide a computational model for strings and recursive functions for the natural numbers, there are comparable notions of effectiveness for other data types, as explained in Sections 3.2 and 5.3.

Beyond that, Turing extended the notion of computability to devices provided with oracles that "magically" provide answers to questions for which there may be no effective means of providing answers. See Sections 3.3 and 5.3.

We draw some conclusions in the final section.

2. Discrete Algorithms

By an algorithm, one invariably means some type of state-transition system. As Donald Knuth writes,[35] for example:

> Algorithms are concepts which have existence apart from any programming language.... I believe algorithms were present long before Turing et al. formulated them, just as the concept of the number "two" was in existence long before the writers of first grade textbooks and other mathematical logicians gave it a certain precise definition.... A computational method comprises a set Q (finite or infinite) of "states", containing a subset X of "inputs" and a subset Y of "outputs"; and a function F

from Q into itself. (These quantities are usually also restricted to be finitely definable, in some sense that corresponds to what human beings can comprehend.)

Classical algorithms proceed step by step, from state to next state. We formalize this in our first postulate.

Postulate (State Transition) *An algorithm determines a (nonempty) set (or class) of* states, *a (nonempty) subset (or subclass) of* initial *states, and a partial* next-state *transition function from states to states.* Terminal *states are those states for which no transition is defined.*

Having the transition depend only on the state means that states must store all the information needed to determine subsequent behavior. Prior history is unavailable to the algorithm unless stored in the current state.

State-transitions are deterministic. Classical algorithms in fact never leave room for choices, nor do they involve any sort of interaction with the environment to determine the next step. To incorporate nondeterministic choice, probabilistic choice, or interaction with the environment, one would need to modify the above notion of transition.

This postulate is meant to exclude formalisms, such as,[20,43] in which the result of a computation—or the continuation of a computation—may depend on (the limit of) an infinite sequence of preceding (finite or infinitesimal) steps. Likewise, processes in which states evolve continuously (as in analog processes, like the position of a bouncing ball), rather than discretely, are eschewed.

Yuri Gurevich's "sequential postulates"[23] characterize algorithmicity in its classical sense. They assert that states are first-order structures and that transitions respect isomorphisms (see Section 3). An algorithm provides a prescription for updating states, that is, for changing some of the interpretations given to symbols by states. The essential idea is that there is a fixed finite set of terms that refer (possibly indirectly) to locations within a state and which suffice to determine what needs to be tested and how the state needs to change during any transition (see Section 4). This implies, as we will see, that it is possible to describe transitions by means of some finite text (see Section 5.1). These characteristics apply both to effective methods, such as factoring, and ideal ones, like inverting a matrix of arbitrary reals.

For an algorithm to be effective, there is one additional, crucial issue: it must be possible to describe (initial) states in some finite fashion.

3. States

States must be comprehensive: they need to incorporate *all* the relevant data that, when coupled with the program, completely determine the next state, and, hence, the whole future of the computation. For instance, the "instantaneous description" of a Turing-machine computation is just what is needed to pick up a machine's computation from where it has been left off; see.[56] Likewise, the state of a procedural language contains the values of all variables plus the "program counter", pointing to the current operation. Similarly, the "continuation" of a Lisp program contains all the state information needed to resume its computation.

3.1. *Abstract states*

In addition to storing mutable data, states of algorithms should incorporate the means to make changes. So, they (and, by the same token, states of non-algorithmic processes) may best be regarded as (first-order) logical structures with (finitely many) partial functions, relations, and constants. To simplify matters, relations can be treated as truth-valued functions and constants as nullary functions. So, each state consists of a domain and interpretations for its symbols as partial functions over the domain. Structures, or (partial) algebras, as we sometimes refer to them, suffice to model all salient features of states. All relevant information about a state is given explicitly in the state by means of its interpretation of the symbols appearing in the vocabulary of the structure. Compare [42, pp. 420–429].

The values of programming variables, in and of themselves, are meaningless to an algorithm, which is implementation independent. The specific details of the implementation of the data types used by the algorithm should not matter. Rather, it is relationships between values that matter to the algorithm. It follows that an algorithm should work equally well in isomorphic worlds. Compare [18, p. 128]. In this sense states are "abstract". These considerations lead to the second postulate:[1,23]

Postulate (State) *States of an algorithm are (first-order) structures over a finite vocabulary, closed under isomorphism of domains, such that initial states and terminal states are also closed under isomorphism. Furthermore, transitions preserve the domain of states, do not change any defined point of a function into undefined, respect isomorphisms, in the sense that non-terminal isomorphic states transition to corresponding isomorphic states.*

Qua structure, a state interprets each of the function symbols in its vocabulary. Each state interprets its function symbols as partial operations and, for every term over its vocabulary, either assigns it a domain value, or leaves it undefined if any of the operations involved is undefined for its arguments. States usually include equality, or at least a partially defined equality. (We presume that state structures are endowed with Boolean truth values and standard Boolean operations, and vocabularies include symbols for these.)

Vocabularies are finite, since an algorithm must be describable in finite terms, so can only refer explicitly to finitely many operations. Hence, an algorithm cannot, for instance, involve all of Knuth's[36] arrow operations, $\uparrow$, $\uparrow\uparrow$, $\uparrow\uparrow\uparrow$, etc. Instead one could employ a ternary operation $\lambda xyz.\, x \uparrow^z y$.

In restricting structures to be "first-order", we are limiting the *syntax* to be first-order. This precludes states with infinitary operations, like the supremum of infinitely many objects, which would not make sense from an algorithmic point of view. This does not, however, limit the semantics of algorithms to first-order notions. The domain of states may have sequences, or sets, or other higher-order objects, in which case, the state would also need to provide operations for dealing with those objects.

Closure under isomorphism ensures that the algorithm can operate on the chosen level of abstraction. The states' internal representation of data is invisible and immaterial to the program. This means that the behavior of an *algorithm*, in contradistinction with its "implementation" as a C program—cannot, for example, depend on the memory address of some variable. If an algorithm does depend on such matters, then its full description must also include specifics of memory allocation.

It is possible to liberalize this postulate somewhat to allow the domain to grow or shrink, or for the vocabulary to be infinite or extensible, but such enhancements do not materially change the notion of algorithm.

A fable To illustrate the importance of operating on the correct level of abstraction, consider the following story:[c] A student in an algebraic topology course did not hand in the assigned homework. "You know", said the student to the lecturer, "I was working hard on the homework. After a while, I felt hungry and decided to take a break for a cup of coffee and a donut. But then I spent the whole night trying to understand which one of them I should eat and which one I should drink." What actually hap-

---

[c]Based on the famous quip of John Kelley, "A *topologist* is a man who doesn't know the difference between a doughnut and a coffee cup" [27, p. 88n.].

pened to this poor fellow? There were two states to consider, one with a cup of coffee and no donut, the other with only a donut left. Both real-life situations comprise many properties, like color, temperature, material, recipe used to cook, shape, coordinates on the table, etc. But from the algebraic-topology point of view, one cares only for the genus of the surface and thus cannot distinguish between those two states. On the other hand, a proper algorithm for dealing with midnight hunger should be able to distinguish between food and drink. In other words, the "algebraic topology algorithm" and the "midnight hunger algorithm" have different salient properties, a crucial factor, which the hapless student failed to account for.

3.2. Effective states

Already in 1922, Emil Post [42, pp. 427–428] noted the following about the states of effective computations:

> We ... assume [symbolic representations] to be finite and we might say discrete.... Each symbolization can be considered to consist of a finite number of unanalysable parts (unanalysable from the standpoint of the symbolization) these parts having certain properties and certain relations with each other.... The ways in which these parts can be related will be assumed to be specified for the whole system of symbolizations.... The number of these elementary properties and relations used is finite and ... there is a certain specific finite number of elements in each relation.... The symbol-complexes are completely determined by specifying all the properties and relations of [their] parts.... Each complex of the system can be completely described [by a conjunction of relations]....

In other words, not only should states be symbolic and be represented by relational structures, but they need to be finitely representable if they are to be effective. Accordingly, we insist that effective states harbor no information beyond the means to reach domain values, plus anything that can be derived therefrom.

In general, then, the operations in states come in three flavors: domain constructors; defined functions; and black-box oracles. For a state to be effective, it should provide means to access all the elements of its domain and should not have any oracles.

Function symbols C *construct* a particular domain in a given state if the state assigns each value in the domain to exactly one term over C (so the terms over C form a free Herbrand algebra). Constructors are the usual way of thinking of the domain values of computational models. For example, strings over an alphabet {a,b,...} are constructed from a nullary constructor $\varepsilon()$ and unary constructors $a(\cdot)$, $b(\cdot)$, etc. The positive integers in binary notation are constructed out of the nullary ε and unary 0 and 1, with the constructed string understood as the binary number obtained by prepending the digit 1. A domain consisting of integers and Booleans can be constructed from TRUE, FALSE, 0, and a "successor" function that takes non-negative integers (n) to the predecessor of their negation $(-n-1)$ and negative integers $(-n)$ to their absolute value (n). To construct 0-1-2 trees, we would have three constructors, $k_0()$, $k_1(\cdot)$, and $k_2(\cdot,\cdot)$, for nodes of outdegree 0 (leaves), 1 (unary), and 2 (binary), respectively.

Definition 1 (Effective State)

- A state is *basic* if it includes constructors for its domain, plus totally undefined operations, meaning that they all always yield the same default value (UNDEF, say), and no oracles.
- Such states are *(absolutely) effective*.
- Moreover, a state is effective also if all its defined operations can be effectively computed (in a bootstrapped sense to be made precise below) from basic states and with the same constructors.

This effectiveness postulate excludes algorithms with ineffective oracles, such as the halting function, but allows one to be given effective operations, like equality of trees or division of integers. Having only free constructors at the foundation prevents the hiding of potentially uncomputable information by means of equalities between distinct representations of the same domain element. This is the approach to effectiveness advocated in,[4] extended to include partial functions in states, as in.[1]

3.3. *Oracular states*

Turing[57] introduced the powerful idea of computability relative to oracles. He said, "We shall not go any further into the nature of this oracle apart from saying that it cannot be a machine." We may think of a Turing machine that equipped with a special tape for querying oracles and special states q_M and q_o for each oracle o in $\mathcal{O}$. When, during an execution, the machine enters state q_o, the oracle magically answers by replacing the string x on the query tape with the value $o(x)$ and reverts to state q_M.

In the presence of oracles, we still want the domain to be constructive, or else there may be no finite way of representing inputs and outputs, but now we allow basic operations that may not be effective. Accordingly, we speak, instead, of relative effectiveness.

Definition 2 (Relatively Effective State)

- A state is *basic* in oracles $\mathcal{O}$, if it includes constructors for its domain, totally undefined operations, plus oracles $\mathcal{O}$.
- Such states are *relatively effective*.
- Moreover, a state is relatively effective also if all its defined operations can be computed from basic states with the same constructors and oracles.

One can give an alternate characterization of effective state, one that is based on oracular Turing machines, extending a suggestion of Wolfgang Reisig[46].

Lemma 1 *A state X is effective relative to a set of oracles if and only if there is a Turing machine with the same oracles that can semi-decide the congruence induced by X. In other words, given two terms over the vocabulary of X as input, the machine returns* TRUE *whenever both terms are defined and assigned the same values by X,* FALSE *when both are defined but not equal, and diverges otherwise. Input and output for the machines's oracles is via constructor terms.*

The proof is along the lines of the non-oracular one in.[5]

4. Transitions

For a process, effective or not, to be deemed algorithmic, it must be possible to express the transition rules for going from state to state in some *finite* fashion. Kleene stressed this point repeatedly ([34, p. 17], [32, pp. 240–241n.], [33, p. 493]):

> An algorithm is a finitely described procedure.... In performing the steps, we simply follow the instructions like robots; no ingenuity or mathematical invention is required of us.
>
> An algorithm in our sense must be fully and finitely described before any particular question to which it is applied is selected. When the question has been selected, all steps must then be predetermined and performable without any exercise of ingenuity or mathematical invention by the person doing the computing.

> The notion of an "effective calculation procedure" or "algorithm" (for which I believe Church's thesis) involves its being possible to convey a complete description of the effective procedure or algorithm by a finite communication, in advance of performing computations in accordance with it.

So, algorithms need to be expressible by means of finite texts, making reference to only finitely many terms and relations among them. Indeed, an algorithm can only determine relations between values stored in an abstract state via terms in its vocabulary and equalities (and disequalities) between their values.

4.1. *Effective transitions*

The actions taken by a transition are describable in terms of *updates* in which a *new* interpretation is given by the next state to function symbols. The set of updates encapsulates the state-transition relation of an algorithm by providing all the information necessary to change the current state into the next one. To determine the updates for any given state, the algorithm needs to evaluate some terms. The third postulate (,[23] refined as in[1]) ensures that there is a finite description of the update process, and that its execution requires only a bounded amount of work. Simply stated, there is a fixed, finite set of (ground) terms that determines the stepwise behavior of an algorithm.

Postulate (Transition) *For every state of every algorithm, there is a set of critical terms over its vocabulary, of size up to some bound (determined by the algorithm). Any set of states that all assign the same values to their shared critical terms all have the same critical terms and the same updates (if any).*

The intuition is that an algorithm must base its actions on the values contained at locations in the current state. Unless all states undergo the same updates unconditionally, an algorithm must explore one or more values at some accessible locations in the current state before determining how to proceed. The only means that an algorithm has with which to reference locations is via terms, since the values themselves are abstract entities. If every referenced location has the same value in more than one state, then the behavior of the algorithm must be the same for those states.

This postulate precludes programs of infinite size (like an infinite table lookup) or which are input-dependent.

On account of the presence of partial operations, we need to take into account which locations in the state are actually accessed by the given algorithm. Should an undefined location be accessed, the computation would go into limbo. That is why critical terms are individual to states. Partial operations are required for full generality of the formalization of effectiveness.

4.2. *Classical algorithms*

A careful analysis of the notion of algorithm in[23] and an examination of the intent of the founders of the field of computability in[14] demonstrate that the postulates are in fact true of all ordinary, sequential algorithms, the (only) kind envisioned by the pioneers of the field. In this sense, the traditional notion of algorithm is precisely captured by these axioms. Accordingly, we refer to a process satisfying the above three postulates as a *classical algorithm*.

5. Effectiveness

Having axiomatized algorithmic processes, we turn to the question of how to describe them by finite means.

5.1. *Algorithms*

Gurevich [23] showed that his *abstract state machines*[22], constitute a most general model of computation, one that can precisely describe effective transitions of any classical algorithm, on any desired level of abstraction of data structures and native operations. Programs in this formalism may be built from just three components: There are generalized assignments $f(s_1, \ldots, s_n) := t$, where f is any function symbol and the s_i and t are arbitrary (ground) terms. Statements may be prefaced by a conditional test. Program statements are composed in parallel. The semantics of assignment statements, parallel composition, and conditionals are as expected. A program describes a single transition step; its statements are executed repeatedly, as a unit, until no assignments have their conditions enabled.

This very simple model of computation suffices to precisely capture the behavior of the whole class of classical algorithms over any domain, including those with partial operations, be they effective or oracular, that hang outside their domain of definition.[1] This model is not wedded to any particular data representation—in the way, say, that Turing machines manipulate strings using a small set of tape operations. In this sense, abstract state machines are the most generic of computational models.

A simple program in this framework is the following:

$$\text{if } |b-a| > \varepsilon \text{ then do } \begin{cases} \text{if } \operatorname{sgn} f((a+b)/2) = \operatorname{sgn} f(a) \text{ then } a := (a+b)/2 \\ \text{if } \operatorname{sgn} f((a+b)/2) = \operatorname{sgn} f(b) \text{ then } b := (a+b)/2 \end{cases}$$

The conditional is repeated over and over until the outer condition turns false, and no more assignments are made. The domain is the reals plus the Booleans; the operations for addition ($+$), subtraction ($-$), halving ($/2$), equality ($=$), greater than ($>$), absolute value ($|\cdot|$), and signum (sgn) are fixed in all states; the values of f and ε are given in the initial state as inputs; the values of the nullary symbols a and b are also given as inputs, but they are changed by the transitions from state to state. Should the signs of a and b start out the same, then both inner conditions will hold, and both assignments will be performed forever. Also if ε is nonpositive, the program will never terminate.

The critical term is $|b-a| > \varepsilon$ in states that falsify this condition, and includes also $f((a+b)/2) = \operatorname{sgn} f(a)$ and $f((a+b)/2) = \operatorname{sgn} f(b)$ when the condition is true.

This program describes the standard bisection search for the root of a function, as described in [21, Algorithm #4]. The point is that this abstract formulation is, as the author of[21] wrote, "applicable to any continuous function" f over the reals—including ones that are not programmable. This program cannot be considered effective; indeed its domain is uncountable. See[45] for examples of geometric constructions with compass and straightedge.

5.2. *Effective algorithms*

The sequential postulates limit transitions to be effective, in the sense of being amenable to finite description, but they place no constraints on the nature of the contents of states. In particular, states may contain ineffective oracles. To preclude that and ensure that an algorithm is effective, in an absolute sense, it suffices to place limits on initial states.

Postulate (Initial State) *The initial states of an effective algorithm are all (absolutely) effective (in the sense of Definition 1) and are all identical, up to isomorphism, except for input values. The initial states of a relatively effective algorithm are all relatively effective (in the sense of Definition 2).*

In both cases, initial states are all identical, up to isomorphism, except for input values.

Since transitions make only finitely many changes, once initial states are effective, then so are all subsequent states.

We will say that an algorithm *computes* a partial function f over a domain D if there are *input* terms such that their values in all initial states with domain D cover all possible input values. We also demand that those states otherwise agree on the values of all terms, so no information is hidden in individual states. Given values $\bar{a}$ for the input terms, the corresponding input state leads, via a sequence of transitions specified by the algorithm, to a terminal state in which the value of some designated *output* term is $f(\bar{a})$ whenever the latter is defined, and leads to an infinite computation whenever it is not.

When we spoke earlier (Definitions 1 and 2) of "bootstrapping", we meant that there is a way of programming the defined operations, using constructors and oracles, if any. And if there is any way of programming them, then there is an abstract-state-machine program that fits the bill. For example, with 0 and successor, one can program addition, starting from basic states, so addition my be included in the initial states of (absolutely) effective algorithms over the natural numbers. Multiplication is also effective, since there is a program for multiplication that makes use of addition.

We are requiring that all elements of an algorithm's domain be accessible via terms in initial states (inaccessible superfluous elements may be removed with no ill effect). But note that a transition may cause accessible elements to become inaccessible in later states.[46]

5.3. *Relatively effective algorithms*

Just like Turing extended his machines to incorporate oracles, the notion of recursive functions has been extended to allow oracles, for total functions by Turing [57, p. 175] and for partial ones later by Kleene [31, p. 178].

One form of this generalization is as follows: The *partial-recursive functions relative to oracles* $\mathcal{O}$ is the class of partial functions over the naturals, $\mathbb{N}$, that includes the constant zero function, successor, all the projections, plus the operations in $\mathcal{O}$ and is closed under composition, primitive recursion, and minimization. We say that an algebra (with finitely or infinitely many partial functions) over the naturals is *recursive in* $\mathcal{O}$ if all its functions are.

Another extension of recursion theory applies it to domains other than the naturals. For this, we need the concept of "simulation" under encodings. An algebra $\mathcal{A}$ with domain D *simulates* an algebra $\mathcal{B}$ with domain E if there is an injective encoding ρ of E into D such that for every partial function g of $\mathcal{B}$ there is a partial function f of $\mathcal{A}$, such that $g = \rho^{-1} \circ f \circ \rho$. A detailed discussion of simulations may be found in.[2]

So, a state X over vocabulary F and arbitrary domain D is *computable over oracles* $\mathcal{O}$ if there is an encoding of D into the naturals and a recursive structure Y with domain $\mathbb{N}$ over oracles $\rho \circ o \circ \rho^{-1}$ for all $o \in \mathcal{O}$ that simulates X via ρ. An algorithm is *relatively computable* if all its initial states are computable all over the same oracle. And a model is *relatively computable* if all its algorithms are, via the same encoding and same oracle. Sans oracles, we call it *computable*. This is akin to a *computable algebra*, as in,[17,40,44,55] but we are not placing restrictions on the injective encoding.

Were we not to require the encoding to be an injection, we could trivially simulate everything by encoding everything by a single constant. One may ask whether the allowance of any injective encoding between the arbitrary domain and the natural numbers is sensible. But it turns out that, as long as all domain elements are reachable by ground terms, any arbitrary injective representation implies the existence of a bijection between the domain and the natural numbers [5, Lemma 1]. Hence, the initial functions of a computable algorithm are isomorphic to some partial-recursive functions, which makes their effectiveness hard to dispute.

For example, one standard injective encoding of lists, with nullary ε and binary : as constructors, is given by $\rho(\varepsilon) = 0$ and $\rho(x : y) = 2^{\rho(x)} 3^{\rho(y)}$. The standard bijective encoding is $\rho(\varepsilon) = 0$ and $\rho(x : y) = 2^{\rho(x)}(2\rho(y) + 1)$.

These two notions, effective relative to oracles and computable over oracles, are coextensional (cf. the non-oracle case proved in[4]).

An alternative An equivalent definition—along the lines of Gödel's[19] original definition of recursive equations—is to say that an algebra over domain D, with finitely many operations F, is computable relative to $\mathcal{O}$ if there exist constructors C for D and a finite set E of equations defining F. Each equation in E is of the form $f(\bar{s}) = t$, where f is a symbol for an operation in F, $\bar{s}$ is a tuple of constructor terms built from C and variables, and t is an arbitrary term built from F, C, and variables. The equations define an operation f in F relative to $\mathcal{O}$ if for all tuples $\bar{c}$ of ground constructor terms, one can deduce (by substitution of equals for equals) $E \cup \mathcal{O} \vdash f(\bar{c}) = d$ for *at most one* ground constructor term d, where $\mathcal{O}$ is now an infinite set of (ground) equations giving the (defined) values of the oracular functions in constructor terms.

For example, a computable algebra of lists with an append operation $\star$ is defined by $\varepsilon \star z = z$ and $(x : y) \star z = x \cdot (y \star z)$. With $\star$ as the (in this case, computable) oracle, one can define list reversal using just $r(\varepsilon) = \varepsilon$ and $r(x : y) = r(y) \star (x : \varepsilon)$.

6. Conclusion

To summarize, we have seen that a model of computation is effective relative to oracles if and only if the congruences of its states are semi-decidable by oracular Turing machines if and only if it is computable over those oracles each algorithm in the model, there is a partial-recursive function under some encoding.

Theorem 1 *Every relatively effective algorithm can be simulated by an oracular Turing machine.*

The fact that these three prima facie different definitions of relative effectiveness over arbitrary domains, building on competing suggestions in,[4,14,46] comprise exactly the same functions, strengthens our conviction that the essence of the underlying notion of effectiveness has in fact been captured.

In the special case of no oracles, this proves (a formalization of) what Church and Turing have claimed:

Theorem 2 (Church-Turing Thesis[4]) *Every absolutely effective algorithm can be simulated by a Turing machine.*

In fact [5, Theorem 4], the set of Turing-computable string functions (and likewise the set of partial recursive functions each algorithm in the model, there is a partial-recursive function) is the unique maximal effective model, up to isomorphism, over any countable domain. By "maximal", we mean that adding any function would make it impossible to show the model to be computable (by simulation).

Moreover, we have recently demonstrated the validity of the widely believed (classical; non-physical) *Extended Church-Turing Thesis*:

Theorem 3 (Extended Church-Turing Thesis[13]) *Every effective algorithm can be polynomially simulated by a Turing machine.*

It follows from all the above that any model purporting to be hypercomputational model, that computes all the Turing-computable functions and then some, be they (idealized) humans (as claimed, for example, in[6,39,41]), theoretical contrivance (e.g.[7,20,43]), or hypothetical (or idealized) physical apparatus (as proposed, for instance, in[15,28,54]), must violate one of our postulates. Note that, to be truly hypercomputational, it is crucial that a model that encodes strings in some way also be capable of computing the ordinary computable functions. It is not sufficient to merely compute one additional function,[d] as explained in.[3]

References

1. Andreas Blass, Nachum Dershowitz, and Yuri Gurevich. Exact exploration and hanging algorithms. In *Proceedings of the 19th EACSL Annual Conferences on Computer Science Logic (Brno, Czech Republic)*, volume 6247 of *Lecture Notes in Computer Science*, pages 140–154, Berlin, Germany, August 2010. Springer. Available at http://nachum.org/papers/HangingAlgorithms.pdf (viewed June 3, 2011); longer version at http://nachum.org/papers/ExactExploration.pdf (viewed May 27, 2011).
2. Udi Boker and Nachum Dershowitz. Comparing computational power. *Logic Journal of the IGPL* 14(5):633–648, 2006.
3. Udi Boker and Nachum Dershowitz. A hypercomputational alien, *Applied Mathematics and Computation* 178(1):44–57, 2006.
4. Udi Boker and Nachum Dershowitz. The Church-Turing thesis over arbitrary domains. In Arnon Avron, Nachum Dershowitz, and Alexander Rabinovich, editors, *Pillars of Computer Science, Essays Dedicated to Boris (Boaz) Trakhtenbrot on the Occasion of His 85th Birthday*, volume 4800 of *Lecture Notes in Computer Science*, pages 199–229. Springer, 2008. Available at http://nachum.org/papers/ArbitraryDomains.pdf (viewed Dec. 13, 2011).
5. Udi Boker and Nachum Dershowitz. Three paths to effectiveness. In Andreas Blass, Nachum Dershowitz, and Wolfgang Reisig, editors, *Fields of Logic and Computation: Essays Dedicated to Yuri Gurevich on the Occasion of His 70th Birthday*, volume 6300 of *Lecture Notes in Computer Science*, pages 36–47, Berlin, Germany, August 2010. Springer. Available at http://nachum.org/papers/ThreePathsToEffectiveness.pdf (viewed Dec. 13, 2011).
6. Selmer Bringsjord, Owen Kellett, Andrew Shilliday, Joshua Taylor, Bram van Heuveln, Yingrui Yang, Jeffrey Baumes, and Kyle Ross: A new Gödelian argument for hypercomputing minds based on the busy beaver problem. *J. Applied Mathematics and Computation* 176(2): 516–530 (2006).
7. Mark Burgin (2005), *Super-Recursive Algorithms*, Monographs in computer science, Springer.

---

[d]Cf. [11, p. 1]: "Hypercomputation is the computation of functions or numbers that cannot be computed in the sense of Turing...."

8. Samuel R. Buss, Alexander A. Kechris, Anand Pillay, and Richard A. Shore, The prospects for mathematical logic in the twenty-first century, *Bulletin of Symbolic Logic*, vol. 7, no. 2, June 2001, pp. 169–196. Available at http://www.math.ucla.edu/~asl/bsl/0702/0702-001.ps (viewed Dec. 13, 2011).
9. A. Church. An unsolvable problem of elementary number theory. *American Journal of Mathematics*, 58:345–363, 1936.
10. A. Church, review of Alan M. Turing, On computable numbers, with an application to the Entscheidungsproblem (*Proceedings of the London Mathematical Society*, vol. 2, no. 42, 1936, pp. 230–265), *Journal of Symbolic Logic*, vol. 2, 1937, pp. 42–43.
11. B. J. Copeland. Hypercomputation. *Minds and Machines*, 12:461–502, 2002.
12. M. Davis. The myth of hypercomputation. In Christof Teuscher, editor, *Alan Turing: Life and Legacy of a Great Thinker*, pages 195–212. Springer, 2003.
13. N. Dershowitz and E. Falkovich. A formalization and proof of the Extended Church-Turing Thesis. In *Proceedings of the Seventh International Workshop on Developments in Computational Models (DCM 2011)*, Zurich, Switzerland, July 2011. Available at http://nachum.org/papers/ECTT.pdf (viewed Dec. 9, 2011).
14. N. Dershowitz and Y. Gurevich. A natural axiomatization of computability and proof of Church's Thesis. *Bulletin of Symbolic Logic*, 14(3):299–350, September 2008. Available at http://nachum.org/papers/Church.pdf (viewed Dec. 13, 2011).
15. G. Etesi and I. Németi, 2002, Non-Turing computations via Malament-Hogarth space-times, *Int. J. Theor. Phys.* 41(2), 2002, 341–370. Available at http://lanl.arxiv.org/pdf/gr-qc/0104023v2 (viewed Dec. 8, 2011).
16. H. M. Friedman. Mathematical logic in the 20th and 21st centuries. FOM mailing list. April 27, 2000. Available at http://cs.nyu.edu/pipermail/fom/2000-April/003913.html (viewed December 6, 2011).
17. A. Fröhlich and J. C. Shepherdson, Effective procedures in field theory, *Philosophical transactions of the Royal Society of London*, Series A, vol. 248, 1956, pp. 407–432.
18. R. Gandy. Church's thesis and principles for mechanisms. In *The Kleene Symposium*, volume 101 of *Studies in Logic and the Foundations of Mathematics*, pages 123–148. North-Holland, 1980.
19. K. Gödel, On undecidable propositions of formal mathematical systems, Lecture notes by S. C. Kleene and J. B. Rosser, Inst. for Advanced Study, Princeton, 1934. Reprinted with corrections and postscriptum in M. Davis (ed.): *The Undecidable – Basic Papers on Undecidable Propositions, Unsolvable Problems and Computable Functions*, Raven Press, 1965, pp. 39–74. The postscriptum is also reprinted in Gödel's *Collected Works*, vol. I, pp. 369–371.
20. E. M. Gold. Limiting recursion. *J. Symbolic Logic*, 30(1):28–48, 1965.
21. S. Gorn. Algorithms: Bisection routine. *Communications of the ACM*, 3(3):174, 1960.
22. Y. Gurevich. Evolving algebras 1993: Lipari guide. In Egon Börger, editor, *Specification and Validation Methods*, pages 9–36. Oxford Univer-

sity Press, 1995. Available at http://research.microsoft.com/~gurevich/opera/103.pdf (viewed Apr. 15, 2009).
23. Y. Gurevich. Sequential abstract state machines capture sequential algorithms. *ACM Transactions on Computational Logic*, 1(1):77–111, July 2000. Available at http://research.microsoft.com/~gurevich/opera/141.pdf (viewed Apr. 15, 2009).
24. D. Harel. On folk theorems. *Communications of the ACM*, 23(7):379–389, July 1980.
25. D. Hilbert, Mathematische Probleme: Vortrag, gehalten auf dem internationalen Mathematiker-Kongreß zu Paris 1900 (in German). Available at http://wikilivres.info/wiki/Mathematische_Probleme (viewed Dec. 1, 2011).
26. D. Hilbert and W. Ackermann, *Grundzüge der theoretischen Logik*, Springer-Verlag, Berlin, 1928 (in German). English version of the second (1938) edition: *Principles of Theoretical Logic* (R. E. Luce, translator and editor), AMS Chelsea Publishing, New York, 1950.
27. J. L. Kelley, *General Topology*, van Nostrand, New York, 1955.
28. T. D. Kieu. Quantum algorithm for Hilbert's Tenth Problem. *International Journal of Theoretical Physics*, 42:1461–1478, 2003.
29. S. C. Kleene. Lambda-definability and recursiveness. *Duke Mathematical Journal*, 2:340–353, 1936.
30. S. C. Kleene, Recursive predicates and quantifiers, *Transactions of the American Mathematical Society*, vol. 53, no. 1, 1943, pp. 41–73. Reprinted in M. Davis (ed.), *The Undecidable*, Raven Press, Hewlett, NY, 1965, pp. 255–287.
31. S. C. Kleene. *Introduction to Metamathematics*. D. Van Nostrand, New York, 1952.
32. S. C. Kleene. *Mathematical Logic*. Wiley, New York, 1967.
33. S. C. Kleene. Reflections on Church's thesis. *Notre Dame Journal of Formal Logic*, 28(4):490–498, 1987.
34. S. C. Kleene. Turing's analysis of computability, and major applications of it. In *A Half-Century Survey on The Universal Turing Machine*, pages 17–54, New York, NY, 1988. Oxford University Press.
35. D. E. Knuth, Algorithm and program; information and data, *Communications of the ACM*, vol. 9, no. 9, Sept. 1966, p. 654.
36. D. E. Knuth, 1976, Mathematics and computer science: Coping with finiteness". *Science* 194 (4271): 1235–1242.
37. A. N. Kolmogorov, O ponyatii algoritma [On the concept of algorithm], *Uspekhi Matematicheskikh Nauk [Russian Mathematical Surveys]*, vol. 8, no. 4, 1953, pp. 175–176 (in Russian). English version in: Vladimir A. Uspensky and Alexei L. Semenov, *Algorithms: Main Ideas and Applications*, Kluwer, Norwell, MA, 1993, pp. 18–19.
38. A. N. Kolmogorov and V. A. Uspensky, K opredeleniu algoritma, *Uspekhi Matematicheskikh Nauk [Russian Mathematical Surveys]*, vol. 13, no. 4, 1958, pages 3–28 (in Russian). English version: On the definition of an algorithm, *American Mathematical Society Translations*, ser. II, vol. 29, 1963, pp. 217–245.

39. J. R. Lucas, review of Judson C. Webb, *Mechanism, Mentalism and Metamathematics: An Essay on Finitism* (D. Reidel, Dordrecht, 1980), *The British Journal for the Philosophy of Science*, vol. 33, no. 4, Dec. 1982, pp. 441–444.
40. A. I. Mal'cev, Konstruktivnyye algyebry. 1, *Uspekhi Matematicheskikh Nauk*, vol. 16, no. 3, 1961, pp. 3–60. English version: Constructive algebras, I, The meta-mathematics of algebraic systems (K. A. Hirsch, translator), *Russian Mathematical Surveys*, vol. 16, no. 3, 1961, pp. 77–129. Also in: *The Metamathematics of Algebraic Systems. Collected Papers 1936–1967*, B. F. Wells, III, editor, North-Holland, Amsterdam, 1971, pp. 148–212.
41. R. Penrose. *Shadows of the Mind: A Search for the Missing Science of Consciousness*. Oxford University Press, Oxford, 1994.
42. E. L. Post. Absolutely unsolvable problems and relatively undecidable propositions: Account of an anticipation. In M. Davis, editor, *Solvability, Provability, Definability: The Collected Works of Emil L. Post*, pages 375–441. Birkhaüser, Boston, MA, 1994. Unpublished paper, 1941.
43. H. Putnam. Trial and error predicates and the solution to a problem of Mostowski. *J. Symbolic Logic*, 30(1):49–57, 1965.
44. M. O. Rabin, Computable algebra, general theory and the theory of computable fields, *Transactions of the American Mathematical Society*, vol. 95, no. 2, May 1960, pp. 341–360.
45. W. Reisig. On Gurevich's theorem on sequential algorithms. *Acta Informatica*, 39(4):273–305, April 2003. Available at http://www2.informatik.hu-berlin.de/top/download/publications/Reisig2003_ai395.pdf (viewed Dec. 13, 2011).
46. W. Reisig. The computable kernel of Abstract State Machines. *Theoretical Computer Science*, 409(1):126–136, December 2008. Draft available at http://www2.informatik.hu-berlin.de/top/download/publications/Reisig2004_hub_tr177.pdf (viewed Dec. 13, 2011).
47. H. Rogers, Jr. *Theory of Recursive Functions and Effective Computability*. McGraw-Hill, New York, 1966.
48. J. R. Shoenfield. *Recursion Theory*, volume 1 of *Lecture Notes In Logic*. Springer, Heidelberg, 1991.
49. W. Sieg, Mechanical procedures and mathematical experiences, in: *Mathematics and Mind* (A. George, editor), Oxford University Press, Oxford, 1994, pages 71–117.
50. W. Sieg, Step by recursive step: Church's analysis of effective calculability, *Bulletin of Symbolic Logic* 3(2), June 1997.
51. W. Sieg, Hilbert's programs: 1917–1922, *Bulletin of Symbolic Logic*, vol. 5, no. 1, Mar. 1999, pages 1–44. Available at http://www.math.ucla.edu/~asl/bsl/0501/0501-001.ps (viewed Dec. 13, 2011).
52. W. Sieg and J. Byrnes, K-graph machines: Generalizing Turing's machines and arguments, in: *Gödel 96: Logical Foundations of Mathematics, Computer Science, and Physics* (P. Hájek, editor), *Lecture Notes in Logic*, vol. 6, Springer-Verlag, Berlin, 1996, pages 98–119.

53. W. Sieg and J. Byrnes, An abstract model for parallel computations: Gandy's thesis, *The Monist*, vol. 82, no. 1, 1999, pages 150–164.
54. W. D. Smith. (July 2006). Church's thesis meets the N-body problem. *Applied Mathematics and Computation*, 178(1):154–183.
55. J. V. Tucker and J. I. Zucker. Abstract versus concrete computation on metric partial algebras. *ACM Transactions on Computational Logic*, 5(4):611–668, 2004.
56. A. M. Turing. On computable numbers, with an application to the Entscheidungsproblem. *Proceedings of the London Mathematical Society*, 42:230–265, 1936–37. Corrections in vol. 43 (1937), pages 544–546. Reprinted in M. Davis (ed.), *The Undecidable*, Raven Press, Hewlett, NY, 1965. Available at http://www.abelard.org/turpap2/tp2-ie.asp.
57. A. M. Turing, Systems of logics based on ordinals. *Proc. Lond. Math. Soc*, 45:161–228, 1939.

Chapter 6

Axioms for Computability:
Do They Allow a Proof of Church's Thesis?*

Wilfried Sieg

Department of Philosophy
Carnegie Mellon University, USA

Church's and Turing's theses assert dogmatically that an informal notion of effective calculability is adequately captured by a particular mathematical concept of computability. I present analyses of calculability that are embedded in a rich historical and philosophical context, lead to precise concepts, and dispense with theses.

To investigate effective calculability is to analyze processes that can in principle be carried out by calculators. This is a philosophical lesson we owe to Turing. Drawing on that lesson and recasting work of Gandy, I formulate boundedness and locality conditions for two types of calculators, namely, human computing agents and mechanical computing devices (or discrete machines). The distinctive feature of the latter is that they can carry out parallel computations.

Representing human and machine computations by discrete dynamical systems, the boundedness and locality conditions can be captured through axioms for Turing computors and Gandy machines; models of these axioms are all reducible to Turing machines. Cellular automata and a variety of artificial neural nets can be shown to satisfy the axioms for machine computations.

---

*This contribution consists of a previously published paper, *Church without dogma: axioms for computability*, and a long Postscriptum, *Is there a proof of Church's Thesis?*. The paper appeared in S.B. Cooper, B. Löwe, A. Sorbi, (eds), *New Computational Paradigms: Changing conceptions of what is computable*, Springer, 2007; it is reprinted here with the permission of Springer. The new Postscriptum gives a detailed analysis of the paper *A natural axiomatization of computability and proof of Church's Thesis* by N. Dershowitz and Y. Gurevich; it is argued that it does not contain a proof of Church's Thesis.

Background

The title of this essay promises axioms for computability. Such axioms will emerge from a *conceptual analysis* that begins with a straightforward observation: whatever we consider to be computable must be associated with computations that are carried out by some device or other. Consequently, we have to pay close attention to the nature of the device at hand, when thinking through the characteristic features that determine (the extension of) its notion of computability. My analysis builds on work by Turing and Gandy concerning computations that are carried out by human calculators and discrete machines, respectively.

I sharpen the informal concepts of computation for these two devices, specify rigorously their characteristic features, and formulate a representation theorem for the resulting systems of axioms. A broad methodological point can be immediately inferred: theses in the standard Church-Turing form are not needed to connect rigorously defined notions of computability with informally grasped concepts. It is however crucial to gain a proper understanding of these canonized connections, because the significance of logical results like Gödel's incompleteness theorems depends on it, as does the centrality of related issues in the philosophy of mind. Part 1 articulates three principal *Church canons*[a] supporting the thesis. For the canonical argument from confluence I distinguish between support that derives from examining the effective calculability of number theoretic functions and support that is obtained through analyzing mechanical operations on symbolic configurations. The analysis of such operations when carried out by a human calculator leads to Turing's claims in 1936. The arguments for these claims exploit *boundedness* and *locality conditions* that are presented in Part 2. Against this background I introduce in Part 3 axioms for *Turing computors* and *Gandy machines*, list models, and formulate a representation theorem. That completes the conceptual analysis. I will conclude with remarks on Gödel, Turing, and philosophical errors.

Note: This essay is based on two papers I published in 2002, but whose methodological considerations I would like to bring out more directly. I presented versions of this essay under the title *Beyond Church Canons* in the Distinguished Lecture Series (Haverford College, October 2002), in the

---

[a] According to the fifth edition of the Shorter OED, *canon* does not cover just ecclesiastical laws and decrees, but has also the meaning of "a general law, rule, or edict; a fundamental principle" since the late middle ages, and that of "a standard of judgement; a criterion" since the early 17th century.

Annual Lecture Series at the Center for Philosophy of Science (University of Pittsburgh, January 2004), at the Colloquium of the IHPST (Sorbonne, May 2004), as well as at the Colloquium of the Department of Philosophy (University of Florence, November 2004) and at the conference *Computability in Europe* (Amsterdam, July 2005). For detailed discussions of the origins and developments of computability, see also (Sieg[16,17]) and the rich literature that is referred to in those papers.

1. Church Canons

In a sense we have to untangle the relation between the concept of computability and the concept of computability, understanding the first concept as informally grasped and the second as rigorously defined. If one takes Gödel's notion of general recursiveness as the rigorously defined concept and effective calculability as the informally grasped one, then Church's Thesis expresses the relation between this and that concept of computability for number-theoretic functions: they are co-extensional. To provide a proper perspective for the broader investigation, I will examine the early history of computability hinted at in these remarks.

1.1. *The thesis*

Gödel introduced general recursiveness for number theoretic functions in his 1934 Princeton Lectures via his equational calculus; he viewed it as a heuristic principle that the informal concept of *finite computation* can be captured by suitably general *recursions*. Refining and generalizing a notion of finitistically calculable functions due to Herbrand, Gödel defined a number theoretic function to be general recursive just in case it satisfies certain recursion equations and its values can be determined from the equations by simple steps, namely, replacement of variables by numerals and substitution of complex closed terms by their numerical values. When he gave this definition in 1934 Gödel was not convinced, however, that the underlying precise concept of recursion was the most general one, and he expressed his doubts in conversation with Church. Nevertheless, Church formulated the thesis a year later for the first time in print. Here is the classical statement found in the abstract for Church's talk to the American Mathematical Society in December 1935:

> ...Gödel has proposed ...a definition of the term recursive function, in a very general sense. In this paper a definition

of recursive function of positive integers which is essentially Gödel's is adopted. And it is maintained that the notion of an effectively calculable function of positive integers should be identified with that of a recursive function, since other plausible definitions of effective calculability turn out to yield notions that are either equivalent to or weaker than recursiveness.

Between Church's conversations with Gödel in 1934 and the formulation of the above abstract in 1935 some crucial developments had taken place in Princeton. Kleene and Rosser had done significant quasi-empirical work, convincing themselves and Church that all known effective procedures are λ-definable. Kleene had discovered his normal-form theorem and established the equivalence of Gödel's general recursiveness with μ-recursiveness. Finally, Church and Kleene had proved the equivalence of λ-definability and general recursiveness. All these developments are alluded to in Church's abstract, and they are interpreted as supporting the thesis, which was then, and is still now, principally defended on two grounds. First, there is the quasi-empirical reason: all known calculable functions are general recursive. This point, though important, is clearly not decisive and will be taken up in the broader context of section 2.3. Second, there is the argument from confluence: a variety of mathematical computability notions all turn out to be equivalent. This second important point is however only really convincing, if the "confluent" notions are of a quite different character and if there are independent reasons for believing that they capture the informal concept. Both Church and Gödel tried to give such independent reasons in 1936. Let me sketch their considerations.

1.2. *Semi-circles*

Church and Gödel took the evaluation of a function in some form of the equational calculus as the starting point for explicating the effective calculability of number theoretic functions. Church generalized broadly: an evaluation is done in some logical calculus through a step-by-step process, and the steps must be elementary. Church argued that functions whose values can be computed in this way must be general recursive. Gödel, in contrast, just made a penetrating observation without giving an argument: the rules of the equational calculus are part of any adequate formal system of arithmetic and the class of calculable functions is not enlarged beyond the general recursive ones, if the formal system is strengthened. This *absoluteness* of the notion was pointed out in a Postscriptum to (Gödel[7]) for transfinite extensions of type theory and in the Princeton Bicentennial lecture ten years later for extensions of formal set theory. Gödel formulated the significance of his observation in the lecture (Gödel,[8] p. 150) as follows:

> Tarski has stressed ... the great importance of the concept of general recursiveness (or Turing computability). It seems to me that this importance is largely due to the fact that with this concept one has for the first time succeeded in giving an absolute definition of an interesting epistemological notion, i.e., one not depending on the formalism chosen.

But what is the argument for Church's claim, and what could it be for Gödel's? If one uses the strategic considerations underlying the proof of Kleene's normal-form theorem, it is in both cases easily established that the functions calculable in the broader frameworks are general recursive, as long as the steps in the logical systems are elementary, formal, ... well, general recursive. Church turned the elementary steps explicitly into general recursive ones, whereas Gödel could not but exploit the formal character of the theories at hand through their recursive presentation.

Taken as principled arguments for the thesis, Gödel's and Church's considerations rely on a hidden and semi-circular condition for steps. Hilbert and Bernays moved this step-condition into the foreground when investigating calculations in deductive formalisms and reckonable functions ("regelrecht auswertbare Funktionen"). They imposed explicitly *recursiveness conditions* on deductive formalisms and showed that formalisms satisfying these conditions have as their calculable functions exactly the general recursive ones. In this way they provided mathematical underpinnings for Gödel's absoluteness claim and for Church's argument, but only *relative* to the recursiveness conditions: the crucial one requires the proof predicate of deductive formalisms, and thus the steps in formal calculations, to be primitive recursive.[b]

The work of Gödel, Church, Kleene, and Hilbert & Bernays had intimate historical connections and is still of deep interest. It explicated calculability of functions by exactly *one core notion*, namely, calculability of their values in logical calculi via (a finite number of) elementary steps. But no one gave convincing and non-circular reasons for the proposed rigorous restrictions on steps permitted in calculations. The question is, whether this stumbling

---

[b]These investigations are carried out in the second supplement of their *Grundlagen der Mathematik II*.

block for a deeper analysis can be overcome. The answer lies in a motivated and general formulation of constraints on steps.

1.3. *Symbolic processes*

Church reviewed in 1937 the two classical papers by Turing and Post, which had been published in 1936. When comparing Turing computability, general recursiveness, and λ-definability he claimed "the first [of these notions] has the advantage of making the identification with effectiveness in the ordinary (not explicitly defined) sense evident immediately..." After all, Church reasoned, "To define effectiveness as computability by an arbitrary machine, subject to restrictions of finiteness, would seem to be an adequate representation of the ordinary notion, ..." The finiteness restrictions require that machines occupy only a finite space and that their working parts have finite size. Turing machines are obtained from such finite machines by further "convenient restrictions", but "these are of such a nature as obviously to cause no loss of generality". Church then observed, completely reversing Turing's sequence of analytic steps, "a human calculator, provided with pencil and paper and explicit instructions, can be regarded as a kind of Turing machine". He was obviously captured by the machine image and saw in it the reason for the deep interest of Turing's computability notion. In sum, we have arrived at three *Church canons* in support of the thesis, namely, (i) the confluence of notions, (ii) the step-by-recursive-step argument, and (iii) the immediate evidence of the adequacy of Turing's notion.

In his reviews Church failed to recognize two crucial aspects of a dramatic shift in perspective. One aspect underlies the work of both Turing and Post, whereas the other is distinctively Turing's. The first aspect becomes visible when Turing and Post, instead of considering schemes for computing the values of number theoretic functions, look at identical symbolic processes that serve as building blocks for calculations. In order to specify such processes Post uses a human worker who operates in a symbol space and carries out, over a two-letter alphabet, exactly the kind of operations a Turing machine can perform. Post expects that his formulation will turn out to be equivalent to the Gödel-Church development. Given Turing's proof of the equivalence of his computability notion with λ-definability, Post's formulation is indeed equivalent.

Post asserts that "Church's identification of effective calculability with recursiveness" should be viewed as a "working hypothesis" in need of "continual verification". In sharp contrast, Turing attempts to give an analytic argument for the claim that these simple processes are sufficient to capture all human mechanical calculations. Turing exploits for his reductive argument broad constraints that are grounded in limitations of relevant capacities of the human computing agent. This is the second aspect of the novel perspective that made for genuine progress, and it is unique to Turing's work.

2. Computors

It is ironic that Post when proposing his worker model at no place used the fact that a human worker does the computing, whereas Turing who seems to emphasize machine computations explicitly examined *human* computations. Call a human computing agent who proceeds mechanically a *computor*; such a computor operates on finite configurations of symbols and, for Turing, deterministically so. The computer hovering about in Turing's paper is such a computor; computers in our contemporary sense are always called machines. Wittgenstein appropriately observed about Turing's machines that *these machines are humans who calculate.*[c] But how do we step from the calculations of computors to computations of Turing machines?

2.1. *Preliminary step*

When Turing explores the extent of the computable numbers (or, equivalently, of the effectively calculable functions), he starts out by considering two-dimensional calculations "in a child's arithmetic book". Such calculations are first reduced to computations of *string machines*, and the latter are then shown to be equivalent to computations of a *letter machine*. Letter machines are ordinary Turing machines operating on one letter at a time, whereas string machines operate on finite sequences of letters. In the course of his reductive argument Turing formulates and uses broadly motivated constraints. The argument concludes as follows: "We may now construct a machine to do the work of the computer [computor in our terminology]. ... The machines just described [string machines] do not differ

---

[c]It is exactly right for Turing to look at human computations given the intellectual context that reaches back to at least Leibniz: the *Entscheidungsproblem* in the title of his ([21]) paper asked for a procedure that can be carried out by humans; the restrictive formal conditions on axiomatic theories were imposed in mathematical logic to ensure intersubjectivity for humans on a minimal cognitive basis.

very essentially from computing machines as defined in §2 [letter machines], and corresponding to any machine of this type a computing machine can be constructed to compute the same sequence, that is to say the sequence computed by the computer." (Turing,[21] pp. 137–8)

For the presentation of Turing's argument it is best to consider the description of Turing machines as Post production systems. This is most appropriate for a number of reasons. Post introduced this description in 1947 to establish that the word-problem of certain Thue-systems is unsolvable. Turing adopted it in 1950 when extending Post's results, but also in 1954 when writing a wonderfully informative and informal essay on solvable and unsolvable problems. In addition, this description reflects directly the move in Turing's ([21]) to eliminate states of mind for computors[d] in favor of "more physical counterparts". Finally and most importantly, it makes perfectly clear that Turing is dealing with general symbolic processes, whereas the restricted machine model that results from his analysis almost obscures that fact.

2.2. *Boundedness and locality*

The constraints Turing imposes on symbolic processes derive from his central goal of isolating the most basic steps of computations, that is, steps that need not be further subdivided. This objective leads to the normative demand that the configurations, which are directly operated on, must be *immediately recognizable* by the computor. This demand and the evident limitation of the computor's sensory apparatus motivate most convincingly two central restrictive conditions:

(B) (*Boundedness*) A computor can immediately recognize only a bounded number of configurations.
(L) (*Locality*) A computor can change only immediately recognizable configurations.[e]

Turing's considerations leading from operations of a computor on a two-dimensional piece of paper to operations of a letter machine on a linear tape are represented schematically in *Diagram 1*: Step 1 indicates

---

[d]Turing attributes states of mind only to *human computers*; machines have corresponding "m-configurations".
[e]The boundedness and locality conditions are violated in Gödel's equational calculus: the replacement operations naturally involve terms of arbitrary complexity. I.e., the shift from arithmetic calculations to symbolic processes is absolutely crucial in Turing's analysis.

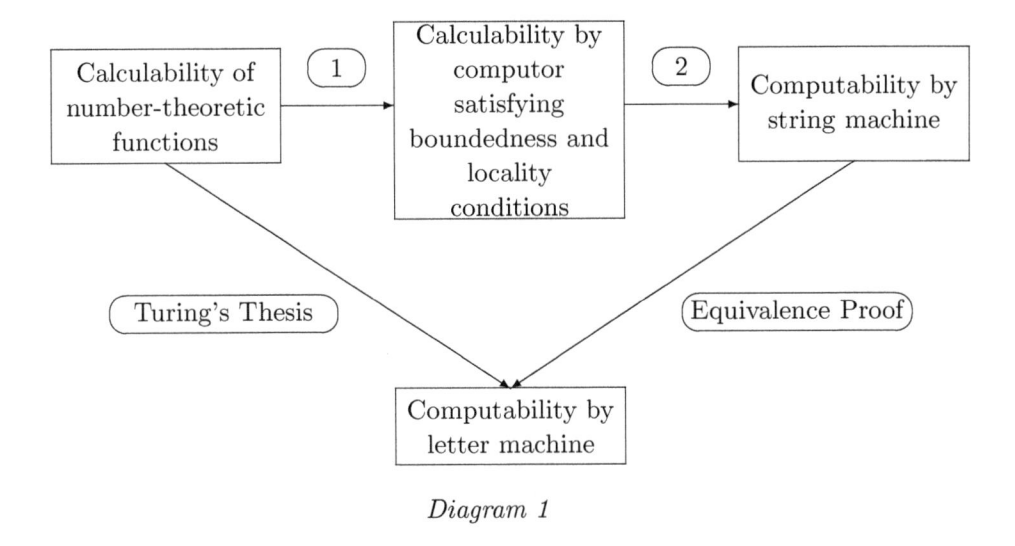

```
                    Calculability by
 Calculability of     computor
 number-theoretic  ─1→ satisfying    ─2→  Computability by
   functions          boundedness and      string machine
                        locality
                       conditions
```

```
  (Turing's Thesis)                    (Equivalence Proof)
              ↘                        ↙
                  Computability by
                    letter machine
```

Diagram 1

Turing's analysis, whereas 2 refers to Turing's *central thesis* asserting that the calculations of a computor can be carried out by a string machine.

This remarkable progress has been achieved by bringing in, crucially and correctly, the computing agent who carries out the mechanical processes. Yet Turing finds the argument mathematically unsatisfactory as it involves an appeal to *intuition* in support of the central thesis, i.e., the ability of "making spontaneous judgments, which are not the result of conscious trains of reasoning". (Turing,[22] pp. 208–9) What more can be done?

2.3. *Generalizations*

At least two kinds of inductive support can be given for the quasi-empirical claim that all known effective procedures are general recursive or Turing computable. Turing provided in his paper one kind, by showing that large classes of numbers are indeed machine computable; Post suggested providing in his ([13]) a second kind, by reducing ever-wider formulations of combinatory processes (as production systems) to his worker model.[f] This inductive support can be strengthened further through considering more general symbolic configurations with associated complex substitution op-

---

[f]Post of course did provide such reductions in his ([14]) whose origins go back to investigations in the very early 1920s; see note 18 of Post's paper.

erations.[g] In the spirit of this approach we can ask with Post, when have we gathered sufficient support to view the thesis as a *natural law*? Gödel and Church faced in their analysis of effective calculability the stumbling block of having to define the elementary character of steps, rigorously and without semi-circles. Turing and Post faced at this point, it seems, a problem akin to that of induction. However, their fundamental difficulties are really the same and can be pinpointed more relevantly and quite clearly, as they are related to the looseness of the above restrictive conditions and the corresponding vagueness of the central thesis. These difficulties would be addressed by answering the questions, What are symbolic configurations? What changes can mechanical operations effect?—Even without giving rigorous answers, some well-motivated ideas can be formulated for computors: (i) they operate deterministically on finite configurations; (ii) they recognize in each configuration exactly one pattern (from a bounded number of different kinds of such); (iii) they operate locally on the recognized pattern; (iv) they assemble the next configuration from the original one and the result of the local operation. Exploiting these ideas I will attack the problem with a familiar tool, the axiomatic method.

However, before formulating the axioms for Turing computors, I discuss yet another sense of generalization that is relevant here. Gandy proposed in his ([5]) a characterization of *machines* or, more precisely, *discrete mechanical devices*. The latter clause was to exclude analogue machines from consideration. The novel aspect of Gandy's proposal was the fact that it incorporated parallelism in perfect generality. Gandy used, as Turing did, a *central thesis*: any discrete mechanical device satisfying some informal restrictive conditions can be represented as a particular kind of dynamical system. Instead, I characterize a *Gandy machine* axiomatically based on the following idea: the machine has to recognize all the patterns contained in a given finite configuration, act on them locally in parallel, and assemble the results of these local computations into the next configuration. As in the case of Turing computors, the configurations are finite, but unbounded; the generalization is simply this: there is no fixed bound on the number of patterns that such configurations may contain. To help the imagination a bit, the reader should think of the Post-presentation of a Turing machine and the Game of Life as typical examples of a Turing computor and Gandy machine, respectively.

---

[g]In (Sieg and Byrnes[20]) that is done for K-graphs and K-graph machines; this is a generalization of the work on algorithms by Kolmogorov and Uspensky.

3. Axiomatics

The axioms are formulated for discrete dynamical systems and capture the above general ideas precisely. In the first subsection the broad mathematical set-up for the axioms is discussed, whereas the specific principles for *Turing computors* and *Gandy machines* are formulated in the second subsection. The axioms for Turing computors are motivated by the restrictive conditions for human computing, i.e., the limitations of the human sensory apparatus. The axioms for Gandy machines are to capture the characteristic features of finite machines performing parallel computations. The restrictive conditions are in this case motivated by purely physical considerations: the uncertainty principle of quantum mechanics justifies a lower bound on the size of distinguishable "atomic" components, and the theory of special relativity yields an upper bound on signal propagation. Together, these conditions justify boundedness and locality conditions for machines in the very same way sensory limitations do for computors.[h]

3.1. *Patterns & local operations*

We consider pairs $\langle \mathbf{D}, \mathbf{F} \rangle$ where $\mathbf{D}$ is a *class of states* and $\mathbf{F}$ an *operation* from $\mathbf{D}$ to $\mathbf{D}$ transforming a given state into the next one. States are finite objects and are represented by non-empty hereditarily finite sets over an infinite set of atoms. Such sets reflect states of computing devices just as other mathematical structures represent states of nature. Obviously, any ϵ-isomorphic set can replace a given one in this reflective role, and so we consider *structural classes* $\mathbf{D}$, i.e., classes of states that are closed under ϵ-isomorphisms. What invariance properties should the state transforming operations $\mathbf{F}$ have, i.e., how should the $\mathbf{F}$-images of ϵ-isomorphic states be related? These and other structural issues will be addressed now.

For the general set-up we notice that any ϵ-isomorphism between states is an extension of some permutation π on atoms. Letting $\pi(\mathbf{x})$ stand for the result of applying the ϵ-isomorphism determined by a permutation π to the state $\mathbf{x}$, the requirement on $\mathbf{F}$ fixes the dependence of values on just structural features of a set, not the nature of its atoms: $\mathbf{F}(\pi(\mathbf{x}))$ is ϵ-isomorphic to $\pi(\mathbf{F}(\mathbf{x}))$, and this isomorphism must be the identity on the atoms occurring in $\pi(\mathbf{x})$; we say that $\mathbf{F}(\pi(\mathbf{x}))$ and $\pi(\mathbf{F}(\mathbf{x}))$ are ϵ-*isomorphic over* $\pi(\mathbf{x})$ and write $\mathbf{F}(\pi(\mathbf{x})) \cong_{\pi(\mathbf{x})} \pi(\mathbf{F}(\mathbf{x}))$. Note that we do not require

---

[h] I hope the overall structure of the considerations will be clear from this informal presentation; for mathematical details (Gandy[5]) and (Sieg[19]) should be consulted.

$\mathbf{F}(\pi(\mathbf{x})) = \pi(\mathbf{F}(\mathbf{x}))$; that would be far too restrictive as new atoms may expand the state $\mathbf{x}$, and it should not matter which new atoms are chosen. The requirement $\mathbf{F}(\pi(\mathbf{x})) \cong \pi(\mathbf{F}(\mathbf{x}))$, on the other hand, would be too loose, as we want to guarantee the physical persistence of atomic components.

Now we turn to *patterns* and *local* operations. If $\mathbf{x}$ is a given state, regions of the next state are determined *locally* from particular *parts for* $\mathbf{x}$ on which the computor can operate.[i] *Boundedness* requires that there are only finitely many different kinds of such parts, i.e., each part lies in one of a finite number of isomorphism types or, using Gandy's terminology, *stereotypes*. A maximal part $\mathbf{y}$ for $\mathbf{x}$ of a certain stereotype is a *causal neighborhood* for $\mathbf{x}$, briefly $\mathbf{y} \in \mathrm{Cn}(\mathbf{x})$; we call the elements of $\mathrm{Cn}(\mathbf{x})$ also *patterns*. Finally, the local change is effected by a structural operation $\mathbf{G}$ that works on unique causal neighborhoods. The values of $\mathbf{G}$ are in general not exactly what we need in order to assemble the next state, because the configurations may have to be expanded and that expansion involves the addition and coordination of new atoms. To address that issue we introduce *determined regions* $\mathrm{Dr}(\mathbf{z}, \mathbf{x})$ of a state $\mathbf{z}$; they are ϵ-isomorphic to $\mathbf{G}(\mathbf{y})$ for some causal neighborhood $\mathbf{y}$ for $\mathbf{x}$ (and must satisfy a technical condition on the "newness" of atoms).

3.2. Axioms & a theorem

Recalling the boundedness and locality conditions for computors, we define $\mathbf{M} = \langle \mathbf{S}; \mathbf{T}, \mathbf{G} \rangle$ to be a *Turing Computor on* $\mathbf{S}$, where $\mathbf{S}$ is a structural class, $\mathbf{T}$ a finite set of stereotypes, and $\mathbf{G}$ a structural operation on $\bigcup \mathbf{T}$, if and only if, for every $\mathbf{x} \in \mathbf{S}$ there is a $\mathbf{z} \in \mathbf{S}$, such that

$$(\mathbf{L.0}) : (\exists! \mathbf{y}) \mathbf{y} \in \mathrm{Cn}(\mathbf{x});$$
$$(\mathbf{L.1}) : (\exists! \mathbf{v} \in \mathrm{Dr}(\mathbf{z}, \mathbf{x})) \mathbf{v} \cong_{\mathbf{x}} \mathbf{G}(\mathbf{cn}(\mathbf{x}));$$
$$(\mathbf{A.1}) : \mathbf{z} = (\mathbf{x} \setminus \mathrm{Cn}(\mathbf{x})) \cup \mathrm{Dr}(\mathbf{z}, \mathbf{x}).$$

$(\exists! \mathbf{y})$ is the existential quantifier expressing uniqueness; in $(\mathbf{L.1})$, $\mathbf{cn}(\mathbf{x})$ denotes the unique causal neighborhood guaranteed by $(\mathbf{L.0})$. As in the case of Gandy Machines below, $\mathbf{L}$ abbreviates locality and $\mathbf{A}$ stands for assembly.—The state $\mathbf{z}$ is determined uniquely up to ϵ-isomorphism over $\mathbf{x}$. An $\mathbf{M}$-computation is a finite sequence of transition steps involving $\mathbf{G}$

---

[i] A part $\mathbf{y}$ for $\mathbf{x}$ used to be in my earlier presentations a connected subtree $\mathbf{y}$ of the $\in$-tree for $\mathbf{x}$, briefly $\mathbf{y} <^* \mathbf{x}$, if $\mathbf{y} \neq \mathbf{x}$ and $\mathbf{y}$ has the same root as $\mathbf{x}$ and its leaves are also leaves of $\mathbf{x}$. More precisely, $\mathbf{y} \neq \mathbf{x}$ and $\mathbf{y}$ is a non-empty subset of $\{\mathbf{v} \mid (\exists \mathbf{z})(\mathbf{v} <^* \mathbf{z}\ \&\ \mathbf{z} \in \mathbf{x})\} \cup \{\mathbf{r} \mid \mathbf{r} \in \mathbf{x}\}$. Now it is just a subset, but I will continue to use the term "part" to emphasize that we are taking the whole $\in$-structure into account.

that is halted when the operation on state **z** yields **z** as the next state. A function **F** is (Turing) *computable* if and only if there is a Turing computor **M** whose computation results determine, under a suitable encoding and decoding, the values of **F** for any of its arguments. A Turing machine is easily seen to be a Turing computor.

Generalizing these considerations to graph machines, for example, one notices quickly complications. When several new atoms are being introduced in the image of some causal neighborhood as well as in the next state, the new atoms have to be structurally coordinated; cf. (Sieg & Byrnes[20]). This issue is clearly even more pressing, when parallel computations are carried out. There the coordination can be achieved by a second local operation and a second set of stereotypes. Causal neighborhoods of type 1 are parts of larger neighborhoods of type 2 and the overlapping determined regions of type 1 must be parts of determined regions of type 2, so that they fit together appropriately. (Determined regions "overlap" if the intersection of their sets of new atoms is non-empty.)

For machines that carry out parallel computations we consequently need in addition to the finitely many stereotypes and the structural operation working on them a second set of stereotypes together with a second structural operation, which allow the machine to assemble the determined regions. This is reflected by separating the Assembly principle for Gandy machines into two kinds, where the principle of the first kind captures the idea expressed at the end of the last paragraph; the principle of the second kind is a more general form of the A-principle for Turing computors. Finally, we can define the central concept here: $\mathbf{M} = \langle \mathbf{S}; \mathbf{T}_1, \mathbf{G}_1, \mathbf{T}_2, \mathbf{G}_2 \rangle$ is a *Gandy machine on* **S**, where **S** is a structural class, $\mathbf{T}_i$ a finite set of stereotypes, $\mathbf{G}_i$ a structural operation on $\bigcup \mathbf{T}_i$, if and only if, for every $\mathbf{x} \in \mathbf{S}$ there is a $\mathbf{z} \in \mathbf{S}$, such that

(**L.1**) : $(\forall \mathbf{y} \in \mathrm{Cn}_1(\mathbf{x}))(\exists! \mathbf{v} \in \mathrm{Dr}_1(\mathbf{z}, \mathbf{x})) \mathbf{v} \cong_\mathbf{x} \mathbf{G}_1(\mathbf{y})$;
(**L.2**) : $(\forall \mathbf{y} \in \mathrm{Cn}_2(\mathbf{x}))(\exists \mathbf{v} \in \mathrm{Dr}_2(\mathbf{z}, \mathbf{x})) \mathbf{v} \cong_\mathbf{x} \mathbf{G}_2(\mathbf{y})$;
(**A.1**) : $(\forall \mathbf{C})[\mathbf{C} \subseteq \mathrm{Dr}_1(\mathbf{z},\mathbf{x}) \& \bigcap \{\mathrm{Sup}(\mathbf{v}) \cap \mathrm{A}(\mathbf{z},\mathbf{x}) \mid \mathbf{v} \in \mathbf{C}\} \neq \emptyset \rightarrow$
$(\exists \mathbf{w} \in \mathrm{Dr}_2(\mathbf{z},\mathbf{x}))(\forall \mathbf{v} \in \mathbf{C}) \mathbf{v} <^* \mathbf{w}]$;
(**A.2**) : $\mathbf{z} = \bigcup \mathrm{Dr}_1(\mathbf{z},\mathbf{x})$.

$\mathrm{A}(\mathbf{z},\mathbf{x}) = \mathrm{Sup}(\mathbf{z}) \setminus \mathrm{Sup}(\mathbf{x})$, i.e., it consists of the new atoms that have been introduced into **z**. Thus, the condition $\bigcap \{\mathrm{Sup}(\mathbf{v}) \cap \mathrm{A}(\mathbf{z},\mathbf{x}) \mid \mathbf{v} \in \mathbf{C}\} \neq \emptyset$ in (**A.1**) expresses that the determined regions **v** in **C** have common new atoms, i.e., they overlap. The restrictions for Gandy machines, as those for Turing computors, amount to boundedness and locality conditions. They are justified *directly* by two physical bounds, namely, a lower bound on the size of atoms and an upper bound on the speed of signal propagation. On account of these bounds only boundedly many different configurations can be physically realized (within a unit time interval); cf. (Mundici & Sieg[12]).

With these remarks I actually completed the foundational analysis, and I can describe now some important mathematical facts for Gandy machines. The central facts are these: (i) the state **z** following **x** is determined uniquely up to ϵ-isomorphism over **x**, and (ii) Turing machines can effect such transitions. The proof of the first fact contains the combinatorial heart of matters and uses crucially the first assembly condition. The proof of the second fact is rather direct. Only finitely many finite objects are involved in the transition, and all the axiomatic conditions are decidable. Thus, a search will allow us to find **z**. This can be understood as a Representation Theorem: any particular Gandy machine is computationally reducible to a two-letter Turing machine. Conversely, any Turing machine is a Gandy machine. Indeed, there is a rich variety of additional models, as the game of life, other cellular automata, and many artificial neural nets are Gandy machines. (Cf. DiPisapia.[4])

4. Adequacy & Philosophical Errors

So what? What have we gained? In very broad terms, taken from Hilbert, we have gained *eine Tieferlegung der Fundamente* (a deepening of the foundations) via the axiomatic method. In a conversation with Church in early 1934, Gödel found Church's proposal to identify effective calculability with λ-definability "thoroughly unsatisfactory". As a counter-proposal he suggested "to state a set of axioms which would embody the generally accepted properties of this notion [i.e., effective calculability], and to do something on that basis". Perhaps, the remarks in the 1964 Postscriptum to the Princeton Lectures of 1934 echo those earlier considerations. "Turing's work gives," according to Gödel, "an analysis of the concept of 'mechanical procedure' .... This concept is shown to be equivalent with that of a 'Turing machine'." Gödel did neither elucidate these remarks, nor did he articulate what the generally accepted properties of effective calculability might be or what might be done on the basis of an appropriate set of axioms.

The work on which I reported substantiates Gödel's remarks in the following sense: it formulates axioms for the concept "mechanical procedure" and it shows that this axiomatically characterized concept is indeed equivalent to that of a Turing machine. As a matter of fact it does so for two such concepts, namely, when the computing agents are computors, respectively discrete machines. These considerations use only "generally accepted properties" of the informal concepts and avoid any appeal to theses, whether central or not. As to the correctness of the underlying analyses, an appeal to some understanding can no more be avoided in this case than in any other case of an axiomatically characterized (class of) mathematical structure(s) intended to mirror broad aspects of physical or intellectual reality. The general point is this: we don't have to face anything mysterious surrounding the concept of calculability; rather, we have to face the ordinary issues for the adequacy of mathematical concepts, and these are of course non-trivial.[j] From a slightly different and complementary perspective, the function of the axiom systems for computing devices can be seen as being similar to that of the axiom systems for the classical algebraic structures like groups, rings or fields, namely, to abstract the essential aspects from a wide variety of instances and point to deep structural analogies. They explain here, by way of the representation theorem, the computational reducibility of their models to Turing machines.

In the central case under discussion, Turing computability, its adequacy is still fraught with controversy and often misunderstanding. The controversy begins with the very question what the intended informal concept is. For example, Gödel spotted in 1972 a "philosophical error" in Turing's work, *assuming* that Turing's argument in the[21] paper was to show that "mental procedures cannot go beyond mechanical procedures". He considered the argument as inconclusive. Indeed, Turing does not give a conclusive argument for Gödel's claim, but it has to be added that he did not intend to argue for it. Even in his work of the late 1940s and early 1950s that deals explicitly with mental processes, Turing does not argue, "mental procedures cannot go beyond mechanical procedures".

Mechanical processes are, in this later work, still made precise as Turing machine computations; machines that might exhibit intelligence have in contrast a more complex structure than Turing machines. Conceptual idealization and empirical adequacy are being sought for quite different purposes, and Turing is trying to capture clearly what Gödel found missing in the would-be analysis of a broad concept of humanly effective calculability, namely, "...that mind, in its use, is not static, but constantly developing". The real difference between Turing's and Gödel's views, it seems, is Gödel's belief that it is "a prejudice of our time" that "[t]here is no mind separate

---

[j] Other examples of such analyses are provided by Dedekind's work on continuous domains (the reals) and simply infinite systems (natural numbers).

from matter". This is reported by Wang. Gödel expected, also according to Wang, that this prejudice "will be disproved scientifically (perhaps by the fact that there aren't enough nerve cells to perform the observable operations of the mind)". Clearly, Turing did not share these expectations.

There are many fascinating issues concerning physical and mental processes that may or may not have adequate computational models. They are empirical, conceptual, mathematical ... well, indeed, richly interdisciplinary. Steps towards their clarification or resolution will be most illuminating. Why, let me ask, are we interested so deeply in computations?—One answer is, we want to determine states from other states, be they mathematical, physical or mental; and we want to do that effectively and in a sharply intersubjective way that makes use of adequate symbolic representations.

References

1. Church, A. An unsolvable problem of elementary number theory; American Journal of Mathematics 58, 345–363; reprinted in (Davis[3]), 1936.
2. Church, A. Review of (Turing[21]); Journal of Symbolic Logic 2, 40–41, 1937.
3. Davis, M., (ed.) *The Undecidable*, Basic papers on undecidable propositions, unsolvable problems and computable functions; Raven Press, Hewlett, New York, 1965.
4. De Pisapia, N. *Gandy Machines: an abstract model of parallel computation for Turing Machines, the Game of Life, and Artificial Neural Networks*; M.S. Thesis, Carnegie Mellon University, Pittsburgh, 2000.
5. Gandy, R. Church's Thesis and principles for mechanisms; in: *The Kleene Symposium* (edited by J. Barwise, H. J. Keisler and K. Kunen, North-Holland, 123–148, 1980.
6. Gödel, K. On undecidable propositions of formal mathematical systems; in: *Collected Works I*, 346–369, 1934.
7. Gödel, K. Über die Länge von Beweisen; in: *Collected Works I*, 396–399, 1936.
8. Gödel, K. Remarks before the Princeton bicentennial conference on problems in mathematics; in: *Collected Works II*, 150–153, 1946.
9. Gödel, K. (1986–2003) *Collected Works*, volumes I–V; Oxford University Press.
10. Hilbert, D. and P. Bernays, *Die Grundlagen der Mathematik II*; Springer Verlag, Berlin, 1939.
11. Kolmogorov, A. N and V. A. Uspensky (1958) On the definition of an algorithm; Uspekhi Mat. Nauk 13 (Russian), 1958; English translation in: *AMS Translations*, 2, 21, 217–245,1963.
12. Mundici, D. and Sieg W., Paper Machines; Philosophia Mathematica 3, 5–30, 1995.

13. Post, E. Finite combinatory processes. Formulation I. Journal of Symbolic Logic 1, 103–5, 1936.
14. Post, E. Formal reductions of the general combinatorial decision problem; American Journal of Mathematics, 65 (2), 197–215, 1943.
15. Post, E. Recursive unsolvability of a problem of Thue; Journal of Symbolic Logic 12, 1–11, 1947.
16. Sieg, W. Mechanical procedures and mathematical experience, in: *Mathematics and Mind* (A. George, ed.), Oxford University Press, 71–117, 1994.
17. Sieg, W. Step by recursive step: Church's analysis of effective calculability; The Bulletin of Symbolic Logic 3 (2), 154–180, 1997.
18. Sieg, W. Calculations by man and machine: conceptual analysis; Lecture Notes in Logic 15, 390–409, 2002.
19. Sieg, W. Calculations by man and machine: mathematical presentation; in: *In the Scope of Logic, Methodology and Philosophy of Science*, volume one of the 11th International Congress of Logic, Methodology and Philosophy of Science, Cracow, August 1999 (P. Gärdenfors, J. Wolenski and K. Kijania-Placek, eds.), Synthese Library volume 315, Kluwer, 247–262, 2002.
20. Sieg, W. and Byrnes, J. K-Graph machines: generalizing Turing's machines and arguments; in: *Gödel '96* (P. Hajek, ed.), Lecture Notes in Logic 6, Springer Verlag, 98–119, 1996.
21. Turing, A. On computable numbers, with an application to the *Entscheidungsproblem*; Proceedings of the London Mathematical Society (Series 2) 42, 230–265, 1936.
22. Turing, A. (1939) Systems of logic based on ordinals; Proc. London Math. Soc., series 2, 45, 161–228; reprinted in (Davis[3]).
23. Turing, A. The word problem in semi-groups with cancellation; Ann. of Math. 52, 491–505, 1950.
24. Turing, A. (1954) Solvable and unsolvable problems; Science News 31, 7–23; reprinted in *Collected Works of A. M. Turing: Mechanical intelligence*, (D. C. Ince, ed.), North-Holland, 1992.

Postscriptum: Is There a Proof of Church's Thesis?

The question has been of deep interest ever since Church, in his [1936], suggested *identifying* the informal concept of *effective calculability* for number theoretic functions with Gödel's mathematical concept of general recursiveness. In the essay that is reprinted here, I give "No!" as the answer. However, building on [Turing 1936], [Post 1947], and [Gandy 1980], I addressed the challenge of mathematically characterizing *mechanical procedures* without appealing to a thesis in two steps:

> (i) the abstract concept of a "Turing Computor" is axiomatically defined as a discrete dynamical system satisfying general boundedness and locality conditions, and
>
> (ii) *a representation theorem* is proved for this abstract concept, stating that Turing machines can simulate the computations of any model of the axioms.

This work, extended to cover also parallel computations via the abstract concept of a "Gandy Machine", was presented in my papers [18] and [18] and is informally described in the previous essay.[k] — In seemingly sharp contrast to my answer, Dershowitz and Gurevich claim in their [2008] that Church's Thesis "provably follows" from four postulates for computability; see, for example, pp. 306, 307 and 339.

D&G's postulates are grounded in a perspective on computability that is broadly similar to mine, thus, also to that of Turing, Post, and Gandy: computations are given by "deterministic state-transition systems", which are in turn comprised of a *set of states* together with a (partial) *transition function* on those states; states are *logical structures* and transitions between states are *discrete*. The four postulates from which Church's Thesis is to be proved are presented on p. 306 in a very informal and preliminary way:

> I. An algorithm determines a sequence of "computational" states for each valid input.
> II. The states of a computational sequence are structures. And everything is invariant under isomorphism.

---

[k]The mathematical details are also given in my paper [3]. — Conceptually it is crucial to recall that the conditions for a Turing Computor hold for human computing agents, whereas those for a Gandy Machine hold for mechanical computing devices (or discrete machines).

III. The transitions from state to state in computational sequences are governed by some fixed, finite description.
IV. Only undeniably computable operations are available in initial states.

A *rigorous* formulation of the *Sequential Postulates* I-III and the *Arithmetical State Postulate* IV is promised for section 2, respectively section 4.

There is much of conceptual and also computational interest in their exposition, but I will focus exclusively on the mathematical claim that Church's Thesis has been proved and the underlying methodological issue.[1] Before looking at the precise formulation of the postulates in Part 2, I examine in Part 1 the structure of the "proof of Church's Thesis". Finally, I comment on D&G's suggestions for improving earlier work in Part 3.

1. A proof?

There are two immediate questions. The first question concerns the notion of proof in "proof of Church's Thesis": Is Church's Thesis proved by a deductive argument from D&G's postulates for computability as Cantor's Theorem is obtained from Zermelo's axioms for set theory? The second question concerns the formulation of Church's Thesis: Is it a statement in the language used for the axiomatization? D&G throw light on both questions when noting on p. 307, "Then, in Section 4, we turn Church's Thesis into a precise mathematical statement and explain why the fact that only the recursive functions can be calculated by effective means follows provably from our four postulates." The phrase "precise mathematical statement" refers to the following theorem:

> Theorem 4.8 (Church's Thesis). Every numeric (partial) function computed by an arithmetical algorithm is (partial) recursive.

The usual formulation of Church's Thesis corresponding to this mathematical statement is of course different, as it involves the informal concept of effective calculability: every numeric (partial) function that is effectively calculable is (partial) recursive. To be able to infer this usual formulation from Theorem 4.8, one would have to prove that effectively calculable functions are computable by *arithmetical algorithms*. That is not done.

---
[1]Crucial points of my criticism were made in a letter I wrote to Dershowitz on November 24, 2007 in which I expressed my reaction to the penultimate draft of their publication [2008].

Theorem 4.8 is an immediate consequence of Corollary 4.6, as *arithmetical algorithms* are simply defined as state-transition systems satisfying the Sequential and Arithmetical Postulates.

> Corollary 4.6. Every numeric function computed by a state-transition system satisfying the Sequential Postulates, and provided initially with only basic arithmetic, is partial recursive.

The corollary is "precisely what we have set out to establish" D&G assert, and it has this direct proof: "By the ASM Theorem (Theorem 3.4), every such algorithm[m] can be emulated by an ASM whose initial states are provided only with the basic arithmetic operations. By Theorem 4.5, such an ASM computes a partial recursive function." ASM stands for Abstract State Machine, and to emulate means to *simulate step-by-step*. The ASM Theorem expresses in slightly different terms the *Main Theorem* of [Gurevich 2000], namely, that for every sequential algorithm there exists an equivalent sequential ASM. *Sequential algorithm* and *sequential ASM* are taken to be rigorous mathematical notions, and the connection between them is stated here as follows:

> Theorem 3.4 (ASM Theorem [42]). For every process satisfying the Sequential Postulates, there is an abstract state machine in the same vocabulary (and with the same sets of states and initial states) that emulates it.

For the proof the reader is referred to [2], i.e., entry [42] in D&G's bibliography.

The second theorem appealed to in the above brief proof of Corollary 4.6 is Theorem 4.5. It asserts, for numeric functions, the equivalence of partial recursiveness and computability by an arithmetical ASM. The argument for the claim that arithmetical ASMs compute only partial recursive functions is deeply problematic, and it is this direction of the equivalence that is needed for the proof of the corollary. The argument, given on p. 326, is concise: "... it is clear that any arithmetical ASM can be programmed in a standard programming language. ... Such programs, of course, can compute only partial recursive functions." To the last sentence D&G attached footnote 29, which begins as follows:

> This implicit appeal to the formal effectiveness of standard programming techniques (viz. what can be programmed in any formalism can be expressed as a general recursion) is sometimes

---

[m] D&G certainly intended to write for "algorithm" here "state-transition system satisfying the Seqential Postulates I-III".

also referred to as an invocation of Church's Thesis... but the omitted details could be fleshed out in what amounts to no more than a programming assignment for an undergraduate course.

For a specific ASM it might be a good programming assignment to show that it calculates a partial recursive function. However, the general claim concerning *any* arithmetical ASM cannot be settled in this way. Rather, the implicit appeal to Church's Thesis is a genuine appeal to a version of Church's Thesis concerning *programming formalisms* for ASMs — that are not given a precise and general formulation. This argument has exactly the same semi-circular character as that found in Church's [1936]. There, a numeric function is defined to be "effectively calculable" just in case its values can be determined "in a logic by elementary steps". Elementary steps are then required to be recursive and, by what Gandy aptly called Church's "step-by-step argument", one has established the claim that all effectively calculable functions are recursive.

The questions I raised at the beginning of this Part can now be answered. "Church's Thesis" does not follow from D&G's postulates (in the way Cantor's Theorem follows from Zermelo's axioms); it is not even Church's Thesis that is being proved, but rather the "precise mathematical statement" of Theorem 4.8; in addition, a version of Church's Thesis is appealed to in the proof of this theorem. Let me formulate one further question that is really basic for the underlying methodological issue: Why is Theorem 4.8 considered to be a "precise mathematical statement"? Is it because the notions it relates are precise mathematical ones? That certainly is true for partial recursiveness. The other notion, arithmetical algorithm, is defined as a state-transition system that satisfies postulates I through IV; the rigor of this notion depends evidently on the rigor of the formulation of the postulates. What are then the precisely formulated postulates?

2. Postulates?

I will review the postulates with suitable informal motivations and quote their formulations directly from D&G's paper. Computations are viewed as *deterministic state-transition systems* with transitions given by partial functions; that fact is expressed through Postulate I (p. 313):

Postulate I (Sequential time). An algorithm is a state-transition system. Its transitions are partial functions.

The states of such systems are structures. When only functions are involved in these structures (and relations are perhaps incorporated through characteristic functions), the structures are called *algebraic*; transition systems with only algebraic structures are *Abstract Transition Systems* or simply ATSs. The abstractness of states is reflected by their closure under isomorphisms and, naturally, transitions commute with isomorphisms. In addition, each state "must contain all the data required by the algorithm for making the next step". These constraints, adapted from Gandy and Turing, are formulated as Postulate II (p. 317):

> **Postulate II** (Abstract state). States are structures, sharing the same fixed, finite vocabulary. States and initial states are closed under isomorphism. Transitions preserve the domain, and transitions and isomorphisms commute.

Transition steps have to be effective (pp. 318-9), and that is to be guaranteed by the third postulate:

> **Postulate III** (Bounded exploration). Transitions are determined by a fixed finite "glossary" of "critical" terms. That is, there exists some finite set of (variable-free) terms over the vocabulary of the states, such that states that agree on the values of these glossary terms, also agree on all next-step state changes.

This postulate bounds the number of locations that have to be considered when making a transition from one state to the next and that "corresponds exactly to the ability to describe finitely how transitions are effected, whatever the language or format of description".

As Church's Thesis has to do with numeric calculations, D&G write on (p. 324), "... we should endow our states with basic arithmetic abilities". These abilities are described in definition 4.2 of *arithmetical state*[n] and referred to in Postulate IV (p. 325):

> **Postulate IV** (Arithmetical state). Initial states are arithmetical and blank. Up to isomorphism, all initial states share the same static operations, and there is exactly one initial state for any given input values.

---

[n] This is a complex and long definition that starts out by asserting: "Up to isomorphism, an *arithmetical state* is as follows: Its domain includes the natural numbers $\mathbb{N}$, as well as the two (distinct) Boolean truth values, *True* and *False*, and some (other) distinguished value $\bot$ signifying 'undefined'." The remainder of the definition requires that its operations include the standard arithmetic ones, but also equality and inequality as well as logical constants, operations for the Booleans, and "various symbols for dynamic functions".

Note that arithmetical states must be infinite, as they include by definition the set of natural numbers $\mathbb{N}$; see Footnote n. In both Gandy's and my formulation, states are always finite, but can expand in the course of a computation; the insistence on finiteness is conceptually well-founded, but presents also a mathematical challenge that is met by the introduction of a refined notion of isomorphism and the delicate assembly of the "next" state; see section 3.1 of my essay.

In order to extend their considerations to algorithms operating over domains that may include, in addition to natural numbers, also "rationals, vectors, matrices, strings, lists, graphs, etc.", D&G formulate in section 6 the *Arithmetizability Postulate* IV_B. That postulate requires the existence of an encoding that is "recursive (in the ordinary sense)" when restricted to $\mathbb{N}$; it is to ensure that the states are *arithmetizable*. Here we encounter again, most directly, the semi-circularity of Church's 1936-argument.

At this point, the larger issue is one of formal and informal rigor: In what sense is D&G's sequence of postulates an *axiomatization*? It is certainly not a formal theory with a precisely specified language and articulated first principles. Is it perhaps not intended as a formal theory, but rather as an axiomatization of a different kind that defines an appropriately abstract notion? I am thinking of the axiomatic definition of a structural concept like that of a group, field, topological space or, for that matter, of Turing Computor and Gandy Machine. Should we think of "abstract transition system" as the intended core concept under which, for example, particular ASMs and Turing machines would fall? To turn Postulates I through III into precise axioms for such a structural notion would require, however, real and non-trivial conceptual work.

3. Improvements?

In section 7, D&G make summary remarks on their own work in this paper, but also on previous analyses. As to their own work, they emphasize the significance of Theorem 6.4 and view it as the *culmination* of their analysis. That theorem is a generalization of Theorem 4.8 and states, "Every numeric (partial) function computed by an arithmetized algorithm is (partial) recursive." This result, together with the considerations sketched on p. 337, allows them to extend "the proof of Church's Thesis" to a "proof of Turing's Thesis regarding string-based effective computations". But, of course, all the problematic features pointed out in the "proof of Church's Thesis" affect also the "proof of Turing's Thesis".

As to previous analyses, D&G assert on p. 339, "no complete axiomatization [of computability] has previously been presented in the literature".[o] In contrast to this remark they mention on p. 307, "Gandy was the first to attempt an axiomatization, and was followed in this endeavor by Sieg"; they also suggest particular ways in which their "proposed axiomatization improves upon its predecessors".[p] What are the specific improvements? Though the axioms of Gandy and Sieg are "formal", "they are expressed on the level of a specific representation of states (namely, hereditarily finite sets)". Their own postulates are claimed, on p. 307, to be formal and "generic" as "they are expressed in terms of computation sequences with arbitrary states and arbitrary programmable transitions. Each transition corresponds to a single step in a given algorithmic process." However, such a genericity claim can also be made for the discrete dynamical processes of a Turing Computor: replace "arbitrary states" by "arbitrary finite states" and "arbitrary programmable transitions" by "arbitrary local operations".

There remains then the delicate issue concerning the representation of states. D&G, as quoted above, assert that the axioms of Gandy and Sieg are expressed "on the level of a specific representation of states (namely, hereditarily finite sets)". How are arbitrary structures of their state-transition systems given, are they not set theoretic objects, and if not, what are they? The broad issue is both subtle and important. One thing is clear, however: within set theory one can obviously have different levels of conceptual organization and thus possibilities for representing states at different levels. Barendregt's remarks quoted in D&G's paper concern representations in the λ-calculus, but can be made with even more good sense about representations in set theory.[q]

There are many other subtle and important issues that deserve attention, for example, a comparison of Gandy's principles for discrete mechanical devices and my axioms characterizing the abstract notion of a Gandy Machine; that is obviously done in my papers [2002 a and b] and in the

---

[o] It is not clear to me in what sense D&G's axiomatization is "complete".
[p] They point to [Gandy 1980], my papers [1994], [1997], [2002], [2003], [2008] as well as a preprint of my [2009]. My [2008] is of course the essay reprinted here.
[q] Here are Barendregt's remarks as quoted in D&G's paper (p. 310, Note 12): "Lambda definability was introduced for functions on the set of natural numbers $\mathbb{N}$. In the resulting mathematical theory of computation (recursion theory) other domains of input or output have been treated as second class citizens by coding them as natural numbers. In more practical computer science, algorithms are also directly defined on other data types like trees or lists. Instead of coding such [inductive] data types as numbers one can treat them as first class citizens by coding them directly as lambda terms while preserving their structure. Indeed, lambda calculus is strong enough to do this. ..."

reprinted essay, but not by D&G. Let me end with restating the answer to the topical question and then raise one last issue. The answer remains "No", as D&G do not give a proof of Church's Thesis, but only of a corresponding "precise mathematical statement". That statement, at best, amounts to a representation theorem for computations satisfying axiomatic conditions; the proof of this statement has many problematic features, which were pointed out in Parts 1 and 2; the most startling one is the appeal to Church's Thesis for programming formalisms. Here is the last issue to think about: In D&G's approach, can one avoid considering infinite structures as the elements of states for a computation? — One could try and characterize axiomatically a notion of ATS by sharpening Postulates I-III (in particular the postulate of bounded exploration) and then prove that every model is a Turing Computor. It would be most satisfying, if the two abstract notions — motivated by similar perspectives — were so related; but such a result cannot be proved until a properly rigorous formulation of D&G's postulates is available.

Wilfried Sieg
Pittsburgh, December 21, 2011.

References

1. N. Dershowitz and Y. Gurevich, A natural axiomatization of computability and proof of Church's Thesis; Bulletin of Symbolic Logic 14, 299–350, 2008.
2. Y. Gurevich, Sequential abstract state machines capture sequential algorithms; ACM Transactions on Computational Logic 1, 77–111, 2000.
3. W. Sieg, On computability; in: Handbook of Philosophy of Mathematics (A. Irvine, ed.), Elsevier, 535–630, 2009.

Chapter 7

The Mathematician's Bias — and the Return to Embodied Computation

S. Barry Cooper[*]

School of Mathematics, University of Leeds, UK

> There are growing uncertainties surrounding the classical model of computation established by Gödel, Church, Kleene, Turing and others in the 1930s onwards. The mismatch between the Turing machine conception, and the experiences of those more practically engaged in computing, has parallels with the wider one between science and those working creatively or intuitively out in the 'real' world. The scientific outlook is more flexible and basic than some understand or want to admit. The science is subject to limitations which threaten careers. We look at embodiment and disembodiment of computation as the key to the mismatch, and find Turing had the right idea all along – amongst a productive confusion of ideas about computation in the real and the abstract worlds.

When we get out of bed in the morning, we approach a complicated world of information with a determination not just to survive the day – though that may be hard enough: we mean to "compute" our way towards various vaguely defined objectives. The process will likely be messy, but we certainly experience it as a computational one. Our conception of the computation is very flexible, but we host no *in principle* rejection of Turing's notion of mechanical intelligence.

But our sense of ownership of a computational process deserts us somewhat when we think about what it is that makes our daily computing adventure so complicated. The world outside has both predictability and a lack of it bordering on randomness. Here is Nassim Taleb, in his best-selling book "The Black Swan":[23]

> I have spent my entire life studying randomness, practicing randomness, hating randomness. The more that time passes, the

---
[*]Preparation of this article supported by EPSRC Research Grant No. EP/G000212.

> worse things seem to me, the more scared I get, the more disgusted I am with Mother Nature. The more I think about my subject, the more I see evidence that the world we have in our minds is different from the one playing outside. Every morning the world appears to me more random than it did the day before, and humans seem to be even more fooled by it than they were the previous day. It is becoming unbearable. I find writing these lines painful; I find the world revolting.

Taleb distrusts mathematicians and their models, from personal experience of their failures, and of their perceived unwillingness to face up to the realities. But we need to give the professionals a chance. Let us look more closely at the modelling process, and how we deal with computing in a material world. Can we absorb Taleb's computational context into a classical model based on logical structure. Or does embodied information need to be separately modelled? And does this take us beyond the mathematician's focus on computable *functions*.

1. Computation Disembodied

What was clearly new about the Turing model of computation was its successful *disembodiment* of the machine. For practical purposes, this was not as complete as some post-Turing theoreticians like to pretend: the re-embodied computer which is now a familiar feature of the modern world was hard won by pioneering engineers. But, for the purposes of the stored program computer, and for the proof of incomputability of the Halting Problem, the essential disembodiment was that delivering program-data convergence. It was this universality that John von Neumann recognised as a theoretical anticipation of the stored program computer. The apparent omnipotence even led Turing to talk of his post-war ACE project as aimed at building 'a brain'.

This paradigm has achieved a strong grip on subsequent thinking. Within the philosophy of mind there is a strong tendency towards physicalism and functionalism, both of which open the door to some version of the Turing model. The functionalist (see Hilary Putnam Ref.[18]) stresses what a computer *does* as something realisable in different hardware. An important expression of the functionalist view in computer science is provided by the notion of a *virtual machine*, whereby one expects to achieve software implementation of a given programmable machine. Aaron Sloman[20] and others have usefully applied the concept to AI.

This playing down of distinction between information and process has been taken further, and become a familiar feature of programming and theory. As Samson Abramsky describes (private communication):

> Turing took traditional mathematical objects, real numbers, functions etc. as the things to be computed. In subsequent work in Computer Science, the view of computation has broadened enormously. In the work on concurrent processes, the behaviour is the object of interest. There is indeed a lack of a clear-cut Church-Turing thesis in this wider sphere of computation – computation as interaction, as Robin Milner put it.

In the quantum world there is a parallel convergence between matter and law-like energy. All this has given rise to a standard computational paradigm vulnerable to surprises from the natural world. Physical processes not subject to data-process convergence will not be *recognisably* different. But beneath the 'normal science', theoretical inadequacies may be brewing – or not, according to the viewpoint. The challenges to the standard model are varied, but most seem to have an impact on universality.

What is happening in both situations is a side-stepping of the *mathematically* familiar type structure, whereby numbers, functions and relations, and relations over functions etc. give rise to a hierarchy of fundamentally different objects, increasingly hard to handle as one closes up hierarchically the universe of definable entities. In the real world we require a level of constructibility, of computability even, which forces approximations and a rejection of paths to the unknown.

1.1. *The Mathematician's bias*

A symptom of the inadequacy of a type-constrained world-view is the October 2010 ACM Ubiquity Symposium on *What is Computation?* Part of the Editor's Introduction by Peter J. Denning[10] reads:

> By the late 1940s, the answer was that computation was steps carried out by automated computers to produce definite outputs. That definition did very well: it remained the standard for nearly fifty years. But it is now being challenged. People in many fields have accepted that computational thinking is a way of approaching science and engineering. The Internet is full of servers that provide nonstop computation endlessly. Researchers in biology and physics have claimed the discovery of natural computational processes that have nothing to do with computers. How must our definition evolve to answer the chal-

lenges of brains computing, algorithms never terminating by design, computation as a natural occurrence, and computation without computers?

In another contribution to the Symposium, Lance Fortnow[12] asks: "So why are we having a series now asking a question that was settled in the 1930s?" And continues:

> A few computer scientists nevertheless try to argue that the [Church-Turing] thesis fails to capture some aspects of computation. Some of these have been published in prestigious venues such as Science, the Communications of the ACM and now as a whole series of papers in ACM Ubiquity. Some people outside of computer science might think that there is a serious debate about the nature of computation. There isn't.

Undeterred, Dennis J. Frailey thinks it's the mathematicians have got it wrong:

> The concept of computation is arguably the most dramatic advance in mathematical thinking of the past century ... Church, Gödel, and Turing defined it in terms of mathematical functions ... They were inclined to the view that only the algorithmic functions constituted computation. I'll call this the "mathematician's bias" because I believe it limits our thinking and prevent us from fully appreciating the power of computation.

Clearly, we do not have much of a grip on the issue. It is the old story of the *Blind Men and the Elephant* again. On the one hand computation is seen as an essentially open, contextual, activity with the nature of data in question. Others bring out a formidable armoury of mathematical weapons in service of the reductive project – for them there is little in concurrency or interaction or continuous data or mental-recursions to stretch the mathematical capabilities of the Turing machine.

Of course, there has always been creative play on the paradoxical misfit between the Turing machine model and the realities of the real world. One of the best-known is Jin Wicked's wonderful image of Alan Turing himself embodying his universal machine. How well this fits with David Leavitt's perception, in his book *The Man Who Knew Too Much*, of Turing actually *identifying* with his computing machines:

Essentially, what is happening is that some observers are reviving the mathematical type-structure in a real-world context, others are denying that the mathematics is capable of inhabiting the material world. Some

Fig. 1. Alan Turing as a Universal Turing Machine by Jin Wicked.

are struck by the sheer globality of how the world computes, the computational fruits of complexity, connectivity and interaction. And by the loss of a simple inductive structure implicit in non-linearity and the failure of computable approximations implicit in too inclusive a view of computationally based environments.

In the past though, it has been the mathematics that has clarified difficult problems and vague intuitions. Apparently, there is a lot of mathematics we have not found out how to use in the material context, though the ownership of the mathematics of reality is beginning to slip from the grasp of those unwilling to adapt.

At the cutting edge there is new mathematics being developed, and questioning of previously sacrosanct conceptual frameworks. Here are some uncomfortable observations (private communication) from Samson Abramsky:

> Formally, giving a program + data logically implies the output (leaving aside non-determinism or randomness), so why actually bother computing the result! ...
> ...Can information increase in computation? Information theory and thermodynamics seem to tell us that it can't, yet intuitively, this is surely exactly why we compute – to get information we didn't have before.

And in mathematics, our operations and definitions certainly do give us new information. Is our wonderful universe so constrained it cannot go where even high-school arithmetics leads us (as the Davis-Matiyasevich-Putnam-Robinson negative solution to Hilbert's Tenth Problem tells us)?

2. The Mathematics of Embodiment?

So it is not that mathematicians are only interested in simple mathematical objects. Or that computability only deals with functions over the natural numbers. What we do have is a computational paradigm which dominates our view of the landscape. We even have higher-type computation in various forms, including that mapped out by Stephen Kleene in his three late papers on the topic. What we did not get was any suggestion that higher-type computability might play a role in modelling the *processes* which it is now suggested might stretch the old Turing model. And the suspicion is that this simple connection can only be explained in the context of a powerful counter-paradigm.

The other by-product of the counter-paradigm is a tendency to desert basic mathematical theory in favour of less focused descriptive arguments for new computational phenomena, accompanied by attempts at models derived from these descriptions unrooted in any classical analysis of their power. Here is another of the *Ubiquity* symposium contributors, Peter Wegner writing with Dina Goldin[28] on *The Church-Turing Thesis: Breaking the Myth*:

> One example of a problem that is not algorithmic is the following instruction from a recipe [quote Knuth, 1968]:
> 'toss lightly until the mixture is crumbly.'
> This problem is not algorithmic because it is impossible for a

> computer to know how long to mix: this may depend on conditions such as humidity that cannot be predicted with certainty ahead of time. In the function-based mathematical worldview, all inputs must be specified at the start of the computation, preventing the kind of feedback that would be necessary to determine when it's time to stop mixing.

At the level of a human trying to carry out the cooking instructions, we have a strong impression that something non-algorithmic is going on. But a determined reductionist would not be at all convinced that this is more than a superficial testing of classical modelling based on a particular mode of observation. The recipe is certainly implementable. But will be executed differently by different cooks under different conditions. What is actually happening in a global sense can be made precise and potentially reducible to the Turing model by modelling the *whole* context, cook and kitchen and ingredients. Who cares that the language used in the description of the recipe is a bit imprecise?

Even quantum physics has received effective attention from the reductionists, via David Deutsch's[11] placing of the standard model of quantum computing firmly within the scope of the Turing model. Of course, the model does not make full use of wave function collapse, and Calude and Svozil[5] have shown, under reasonable assumptions, that quantum randomness entails incomputability. Meanwhile, the mind is very hard to pin down theoretically, but Deutsch is following a well-established reductive tradition when he argues (in *Question and Answers with David Deutsch*, on the New.Scientist.com News Service, December, 2006):

> I am sure we will have [conscious computers], I expect they will be purely classical, and I expect that it will be a long time in the future. Significant advances in our philosophical understanding of what consciousness is, will be needed.

Maybe starting at the other end, and attempting to embody the richness of trans-computational mathematics, can bring more convincing results? Here we have a very different problem. Our examples of incomputable objects are very simple, even 'natural' from some perspectives. But there is a huge gap between the universal Turing machine, which we can see embodied in a very persuasive sense by a modern computer, and the halting problem, which is incomputable but very abstract to the point of having no visible embodiment at all. And the lazy perception is that such examples of incomputability are nothing to do with the material world, even if we are moved to suspect trans-Turing computation by our impressionistic observation of natural phenomena.

The key to bringing some clarity to the situation is to examine the mathematics of the incomputable, in all its basic simplicity. And to look for qualitatively similar mathematics bringing with it a level of embodiment, and an apparent avoidance of computability, to give us a better handle on the natural world and its candidate trans-Turing phenomena.

A meaningful first example, residing at the border between mathematics and the embodied world is the Mandelbrot set. In fact, there are many other such fractal-like objects. But the Mandelbrot set has attracted special attention for good reasons: It has a simple definition over the complex plane involving basic arithmetical operations and a couple of quantifiers – in fact, with a little fine-tuning, just one universal quantifier; it is graphically beautiful and complex, being both approximable via a computer screen, and containing endlessly explorable inner structure; and, its computability as a set of complex numbers provides us with a challenging open problem. As Roger Penrose says in his best-selling 1994 book[16] *The Emperor's New Mind*:

> Now we witnessed ... a certain extraordinarily complicated looking set, namely the Mandelbrot set. Although the rules which provide its definition are surprisingly simple, the set itself exhibits an endless variety of highly elaborate structures.

What we see is a complicated and undeniably physical object existing in front of us on our computer screen *because* it was described, and having complex form *because* the description had a slightly elevated level of complexity of language – namely the addition of a quantifier. Mathematically, the description is not much more complicated than that of the non-halting set of a universal Turing machine. The Mandelbrot set is of course made up of complex numbers rather than natural numbers, but the fact it appears on out computer screen gives it a digital kinship with familiar Turing-machine data. Leaving this abstract unembodied vista behind, and walking round to the other side of the Mandelbrot set, we see why such fractals have such significance for so many people – the scene is one of Nature in all its fascinating complexity, beautiful and embodied, like the Mandelbrot set, but this time produced by simple laws which are themselves embodied. This toy fractal provides a neat connection between natural complexity and the unrestrained type-structures of mathematics, with their power to take us definably far beyond what we can computably capture.

What indicates to the observer a phenomenon of computational complexity? It is the richness of *visible form* which impresses. It is the character of a higher order entity. On the other hand, despite the appearance pointing to the basic computational unpredictability, there is an overall identity to the appearance which we are capable of appreciating, and which is caught by the *definition* of the exotic shape. And it is natural to view this definition as *computed* by the totality of the underlying computational context. The mathematics both points to the atomic unpredictability of the universal Turing machine observed; and reassures us with a computation of a higher order, whereby we reaffirm a level of understanding of the underlying turbulence.

There is an obvious role here for some of the most abstract and little known mathematics logicians have developed. How many mathematicians have imagined any role at all in the real world for the conceptual development of higher order computation by Stephen Kleene, Gerald Sacks, Dag Normann and their successors? As a primary schoolboy, I was fascinated by the family folding mangle, and carried it off to possess and enjoy in my own little 'camp' in the undergrowth beyond the back gardens. The usefulness had no meaning, the loss I was made aware of later by the grown-ups was a shock. It is surely time for the generalised 'recursion theorists' to return this conceptually beautiful work to the real world which indirectly gave rise to it. And to make sense of the sort of physical mysteries that Alan Turing himself identified as important sixty years ago.

3. Emergent Natural Patterns

Back in the early 1950s, Alan Turing became interested in how certain patterns in Nature arise. His seminal importation[27] of mathematical techniques into this area were to bring a new level of computability to natural phenomena, while improving our understanding of some of the mysteries pointed to by those who distrust the reductionism of those trapped by the Turing machine paradigm. Turing was able to relate such familiar features of everyday life as patterns on animals' coats – stripes on zebras, patches on cow hides, moving patterns on tropical fish – to simple reaction-diffusion systems describing chemical reactions and diffusion. This brought mathematics to play in biology in a way that made his one published paper on the topic one of the most cited in the literature, and the most cited in recent years of all the papers he wrote – including the 1936 computable numbers paper, and the influential AI 'Turing Test' article[25] in Mind in 1950.

Turing's mathematics links up with the relationship between emergence and descriptions pointed to by the fractal example. Of course, some halting problems are solvable, as some Julia sets are computable (see Ref. 2). See Ref. 7 for a more detailed argument for the two-way correspondence between emergence and appropriate descriptions, opening the door to incomputability in Nature. And for the grounding of the observed robustness of emergent phenomena in mathematical framing of the descriptions as mathematical *definability*.

Turing's approach is seminal, illuminating the connection between possible incomputability, mathematics, and natural phenomena. It has been carried forward by James D. Murray[15] and others, and though things get a lot more complex than the examples tackled by Turing, it is enough to make something more coherent from the confusion of intuitions and models. The embodiment does extend to the emergent halting set and possibly hierarchically beyond, taking us into a world beyond basic algorithms – see Chaitin's recent take[6] on creativity and biology.

This knack Turing had for drawing fundamental computational aspects out of concrete contexts has become increasingly clear to us since he died. The relevance of this seminal work on morphogenesis becomes daily wider and deeper in import. And we can now see it as definability *embodied*, and hence identify definability as a very real form of *computability*.

To summarise: It is often possible to *define* emergent properties in terms of the elementary actions underlying them. While in mathematics, relations and even objects *arise* from descriptions via the notion of definability. And, if the language used is complicated enough, this can be a source of Turing incomputability. The key observation is that all that fancy higher type mathematics which the logicians carted off into the bushes has a vitally important practical usefulness. Emergent phenomena not only generate descriptions – there is no serendipity here – but *derive* from them.

One should not be misled by the morphogenesis into thinking of definability as exclusively relating higher order relations to basic algorithmic structure. As is well-known from the mathematics, definability can bring other quite basic aspects of a structure into play via descriptions in terms of objects of similar complexity. At the most basic level, we may find individual objects attaining their identity – that is, a unique role as observable entities – via their context within a wider causal framework.

The picture is one of simple computable rules, with a degree of connectivity underpinning a higher order computation, and emergent forms *defined* at the edge of computability. We see morphogenesis inhabit the same world a the Mandelbrot set; the same world as the halting problem for a Turing machine; occupy the same world as the large scale structure we see in the wider universe; and, one speculates, the same world as human mental creativity, linking the synaptic connectivity of the brain via neural net modelling to the most surprising artistic and scientific achievements.

4. The Mind as Mathematics?

There is a strong inclination amongst both computer scientists and philosophers towards some kind of physicalist basis for mentality. For the philosopher, this is still an area with little to agree upon – except the *supervenience* of the mental upon the physicality of the brain. As Jaegwon Kim puts it in his book[13] on *Mind in a Physical World* (pp.14–15), supervenience:

> represents the idea that mentality is at bottom physically based, and that there is no free-floating mentality unanchored in the physical nature of objects and events in which it is manifested.

Or, more mathematically, from the *Stanford Encyclopedia of Philosophy*:

> A set of properties A supervenes upon another set B just in case no two things can differ with respect to A-properties without also differing with respect to their B-properties.

In this context, one has familiar questions, such as: How can mentality have a causal role in a world that is fundamentally physical? And the puzzle of causal 'overdetermination' – the problem of phenomena having both mental and physical causes. As Kim[14] sums up the problem (in *Physicalism, or Something Near Enough*, 2005):

> ...the problem of mental causation is solvable only if mentality is physically reducible; however, phenomenal consciousness resists physical reduction, putting its causal efficacy in peril.

For computer scientists – and mathematicians such as Alan Turing – a physicalist approach takes one in the direction of computational models based on what we know of brain functionality.

The computational content of most of these models is not so radically different from that of a Turing machine, though the level of connectivity and parallelism present may present some problems for the reductionists. While back in the real world, the typical observer of brain functionality will probably not be so impressed, and will see a lot of what one knows about brains, from having one, missing. There are various models relevant here, most of them with a lot in common, and discussions which are transferable from one to another. Rodney Brooks[3] comments:

> ...neither AI nor Alife has produced artifacts that could be confused with a living organism for more than an instant. AI just does not seem as present or aware as even a simple animal and Alife cannot match the complexities of the simplest forms of life.

The mind as a computational instrument presents a level of challenge to the modellers only matched by that facing those wanting to put the standard model of particle physics. on a more secure footing. And many of us are persuaded of important connections between these modelling challenges, explored in various ways by different researchers – with, for example, the contrasting proposals of Roger Penrose and Henry Stapp having attracted widespread attention.

It is still true that we have very interesting connectionist models for the brain (see Ref.[24] for Turing's contribution), and we are not surprised to see a leading researcher like Paul Smolensky[21] saying:

> There is a reasonable chance that connectionist models will lead to the development of new somewhat-general-purpose self-programming, massively parallel analog computers, and a new theory of analog parallel computation: they may possibly even challenge the strong construal of Church's Thesis as the claim that the class of well-defined computations is exhausted by those of Turing machines.

But, as Steven Pinker puts it: "...neural networks alone cannot do the job" – and we too expect a bit more than the emergence of embodied incomputability, occupying a different level of data to that used for further computation. That is no real advance on the familiar Turing machine. It is not very promising for trans-Turing computing machines. Pinker does not talk about incomputability, but does describe [17, p.124] human thinking as exhibiting "a kind of mental fecundity called recursion", giving us a good impression of real life emergent phenomena re-entering the system – or the computational process as we would see it.

Is there really something different happening here? If so, how do we model it? Does this finally sink the standard Turing machine model? Neuroscience gives us an impressively detailed picture of brain functionality. Antonio Damasio[9] vividly fills out the picture we get from Pinker:

> As the brain forms images of an object - such as a face, a melody,

> a toothache, the memory of an event - and as the images of the object affect the state of the organism, yet another level of brain structure creates a swift nonverbal account of the events that are taking place in the varied brain regions activated as a consequence of the object-organism interaction. The mapping of the object- related consequences occurs in first-order neural maps representing the proto-self and object; the account of the causal relationship between object and organism can only be captured in second-order neural maps. ...one might say that the swift, second-order nonverbal account narrates a story: that of the organism caught in the act of representing its own changing state as it goes about representing something else.

The book gives a modern picture of how the human body distributes its 'second-order' representations across the impressive connectivity of the human organism, and enables representations to play a role in further thought processes.

Remarkably, once again, we find clues to what is happening mathematically in the 1930s work of Kleene and Turing. In his 1939 Princeton paper, Turing[26] is in no doubt he is saying something mathematically about human mentality:

> Mathematical reasoning may be regarded ...as the exercise of a combination of ...intuition and ingenuity. ...In pre-Gödel times it was thought by some that all the intuitive judgements of mathematics could be replaced by a finite number of ...rules. The necessity for intuition would then be entirely eliminated. In our discussions, however, we have gone to the opposite extreme and eliminated not intuition but ingenuity, and this in spite of the fact that our aim has been in much the same direction.

To cut things short, Turing's analysis depends on Kleene's trick of building the constructive ordinals by computably *sampling* infinitary data needed to compute higher ordinals. The sampling is not canonical, there are lots of ways of doing it computably. This means we have an iterative process analogous to that described by Pinker and Damasio, which exhibits the irreversibility we expect from Prigogine. It is a picture of definable data represented using constructive ordinals. Turing's representation of data is scientifically standard, in terms of reals.

Buried away in this opaque but wonderful paper is a key idea, that of the oracle Turing machine, which remarkably is just what one needs to model computable 'causality' in science. One can represent any of the familiar physical transitions capturable on our computers – and this encompasses a comprehensive swathe of what we regard as scientific – using the appropriate kind of *relative* computation executable by the right oracle machine. The basic computational structure of the connectivity is captured by *functionals* modelled by oracle Turing machines. The mathematics delivers a complex structure – an extended Turing model – with a rich overlay of definable relations, corresponding to real world ubiquity of emergent form impacting non-trivially on the development of the organism.

Again, we may present definability as a form of computability via the strangely neglected models from generalised recursion theory. For those wanting to dig out the hidden treasures of this beautiful and suddenly relevant subject, probably the best guide is the 1990 Springer *Perspectives in Logic* book of Gerald E. Sacks[19] on *Higher Recursion Theory*.

The physical relevance of this extended model of computable causality fits nicely with the current return to basic notions of key figures concerned with quantum gravity and foundational questions in physics. As Lee Smolin affirms in his book[22] on *The Trouble with Physics* (p.241): "...causality itself is fundamental".

Smolin is not, of course, thinking of computationally constrained models of causality in the sense of Turing. But there is a convergence of aims and value put on embodiment of higher order relations on very basic structures. Smolin alludes to the work of 'early champions' of the role of causality, such as Roger Penrose, Rafael Sorkin, Fay Dowker, and Fotini Markopoulou, and sets out a version of Penrose's strong determinism (p.241 again):

> It is not only the case that the spacetime geometry determines what the causal relations are. This can be turned around: Causal relations can determine the spacetime geometry ... ItÕs easy to talk about space or spacetime emerging from something more fundamental, but those who have tried to develop the idea have found it difficult to realize in practice. ... We now believe they failed because they ignored the role that causality plays in spacetime. These days, many of us working on quantum gravity believe that causality itself is fundamental - and is thus meaningful even at a level where the notion of space has disappeared.

5. Embodiment Restored

The difference between the extended Turing model of computation and what is commonly seen from the modellers of process is that there is a proper balance between process and information. The embodiment was a key problem for the early development of the computer, insufficiently recognised since the early days by the theorists, fixated on the universality paradigm.

Rodney Brooks[4] tells how embodiment in the form of stored information has re-emerged in AI:

> Modern researchers are now seriously investigating the embodied approach to intelligence and have rediscovered the importance of interaction with people as the basis for intelligence. My own work for the last twenty five years has been based on these two ideas.

The mathematics of the extended Turing model is notoriously for its technicality. And the mathematical character of the global structure based on it is disappointingly pathological from the point of view of mathematical expectations. But if we expect to model the emergence of global relations in terms of local structure – say that of large-scale structures in the real universe, or even of relations expressing globally observed natural laws – then the pathology provides the raw material for an embodied language within which to talk about the complexity of physical forms we discover about us. In retrospect, Hartley Rogers was remarkably prescient when he asked about the character of the Turing invariant relations in a talk entitled *Some problems of definability in recursive function theory*, at the Tenth Logic Colloquium, at the University of Leicester in 1965. Though at that time people had no appreciation of the physical significance of the question. It appearing increasingly likely that these relations are the key to pinning down how basic laws and entities emerge as mathematical constraints on causal structure. One should be aware though of the schematic nature of the understanding they may provide. They are the equivalent of maps of the landscape provided by satellite scans, while the real work of exploring the substance of down below is at best guided and enlightened by an awareness of the overall structuring.

The character of the all-important automorphism group of the Turing universe is still to be pinned down. In the way of progress towards explanations of such scientific mysteries as the dichotomy between classical and quantum reality, and the removal of the need for speculative assumptions of 'many-worlds' and multiverses, is the so-called *Bi-interpretability Conjecture*, which has haunted us for nearly thirty years. Very roughly speaking, the conjecture says that the Turing definable relations are exactly those with information content describable in second-order arithmetic. One consequence of a positive solution to this problem would be the ruling out of non-trivial automorphisms of the Turing universe. And the breaking of the current match between the computability-theoretic structures and the physical reality we observe. As things are, we see and important partial validation of the conjecture, yielding a rigid substructure of the Turing universe replete with uniquely defined entities – much like our classical reality we live in: and a wildly ill-defined context, with automorphic displacements corresponding to pre-measurement quantum ambiguity. Currently, the main clues to the outcome of the full Bi-interpretability Conjecture are a lack of more than incremental progress towards a positive answer for the last fifteen years, and an outline of a strategy for constructing a non-trivial automorphism in the public domain since the late 1990s.

We summarise some features of the mathematics, and refer the reader to sources such as[8] and[7] for further detail:

- Embodiment invalidating the 'machine as data' and universality paradigm.
- The organic linking of mechanics and emergent outcomes delivering a clearer model of supervenience of mentality on brain functionality, and a reconciliation of different levels of effectivity.
- A reaffirmation of the importance of experiment and evolving hardware, for both AI and extended computing generally.
- The validating of a route to *creation* of new information through interaction and emergence.
- The significance of definability and its breakdown for the physical universe giving a route to the determination of previously unexplained constants in the standard model of physics, and of quantum ambiguity in terms of a breakdown of definability, so making the multiverse redundant as an explanatory tool, and ...
- ...work by Slaman and Woodin and others on establishing partial rigidity of the Turing universe (see Ref. 1) promising an explanation of the existence of our 'quasi-classical' universe.

As for building intelligent machines, we give the last word to Danny Hillis, quoted by Mark Williams in *Red Herring* magazine in April 03, 2001:

> I used to think we'd do it by engineering. Now I believe we'll evolve them. We're likely to make thinking machines before we understand how the mind works, which is kind of backwards.

So, Turing's computational modelling is showing good signs of durability – but within a turbulent natural environment, embodied, full of emergent wonders, and exhibiting a computational structure reaching far into unknown regions – via a revived type structure, the relevance of which is currently neglected by both mathematicians, computer scientists, and beyond. But, as I write this at the start of the centenary year of Alan Turing's birth, there is every sign of a revival of the basic approach which underlay the great discoveries of the first half of the twentieth century.

Turing, as we know, anticipated much of what we are still clarifying about how the world computes. And, we hope, he is smiling down on us at the centenary of his birth, in a Little Venice nursing home (now the Colonnade Hotel).

References

1. Klaus Ambos-Spies and Peter Fejer. Degrees of unsolvability. *unpublished*, 2006.
2. Ilia Binder, Mark Braverman, and Michael Yampolsky. Filled Julia sets with empty interior are computable. *Foundations of Computational Mathematics*, 7(4):405–416, 2007.
3. Rodney Brooks. The relationship between matter and life. *Nature*, 409:409–411, 2001.
4. Rodney Brooks. The case for embodied intelligence. In S. Barry Cooper and Jan van Leeuwen, editors, *Alan Turing - His Work and Impact*. Elsevier Science, to appear.
5. Cristian S. Calude and Karl Svozil. Quantum randomness and value indefiniteness. *Advanced Science Letters*, 1:165–168, 2008.
6. Gregory J. Chaitin. Metaphysics, metamathematics and metabiology. In Hector Zenil, editor, *Randomness Through Computation: Some Answers, More Questions*. World Scientific, Singapore, 2011.
7. S. Barry Cooper. Emergence as a computability-theoretic phenomenon. *Applied Mathematics and Computation*, 215(4):1351–1360, 2009.
8. S. Barry Cooper and Piergiorgio Odifreddi. Incomputability in nature. In S. B. Cooper and S. S. Goncharov, editors, *Computability and Models: Perspectives East and West*, pages 137–160. Plenum, New York, 2003.
9. Antonio R. Damasio. *The Feeling of What Happens: Body and Emotion in the Making of Consciousness*. Harcourt Brace and Co, 1999.
10. Peter J. Denning. Ubiquity symposium 'What is computation?': Opening statement. *Ubiquity*, 2010.
11. David Deutsch. Quantum theory, the Church-Turing principle, and the universal quantum computer. *Proc. Royal Soc.*, A400:97–117, 1985.
12. Lance Fortnow. Ubiquity symposium 'What is computation?': The enduring legacy of the turing machine. *Ubiquity*, 2010, December 2010.
13. Jaegwon Kim. *Mind in a Physical World: An Essay on the Mind-Body Problem and Mental Causation*. MIT Press, 2000.

14. Jaegwon Kim. *Physicalism, or Something Near Enough*. Princeton University Press, 2005.
15. James D. Murray. *Mathematical Biology: I. An Introduction*. Springer, New York, 3rd edition, 2002.
16. Roger Penrose. *The Emperor's New Mind: Concerning Computers, Minds, and the Laws of Physics (Popular Science)*. Oxford University Press, USA, 2002.
17. Steven Pinker. *How the Mind Works*. W.W. Norton, New York, 1997.
18. Hilary Putnam. Minds and machines (1960). In Putnam, editor, *Mind, Language, and Reality*, pages 362–385. Cambridge University Press, 1975.
19. Gerald E. Sacks. *Higher recursion theory*. Springer-Verlag New York, Inc., New York, NY, USA, 1990.
20. Aaron Sloman. Some requirements for human-like robots: Why the recent over-emphasis on embodiment has held up progress. In Bernhard Sendhoff et al., editors, *Creating Brain-Like Intelligence*, volume 5436 of *LNCS*, pages 248–277. Springer, 2009.
21. Paul Smolensky. On the proper treatment of connectionism. *Behavioral and Brain Sciences*, 11:1–74, 1988.
22. L. Smolin. *The trouble with physics: the rise of string theory, the fall of a science, and what comes next*. Houghton Mifflin Co., 2006.
23. Nassim Nicholas Taleb. *The Black Swan: The Impact of the Highly Improbable*. Random House, 2010.
24. Christof Teuscher. *Turing's Connectionism. An Investigation of Neural Network Architectures*. Springer-Verlag, London, 2002.
25. A. M. Turing. Computing machinery and intelligence. *MIND*, 59(236):433–460, October 1950.
26. Alan Mathison Turing. Systems of logic based on ordinals. *Proceedings of the London Mathematical Society*, 45:161–228, 1939. reprinted in Alan M. Turing, Collected Works: Mathematical Logic, pp. 81–148.
27. Alan Mathison Turing. The chemical basis of morphogenesis. *Phil. Trans. of the Royal Society of London. Series B, Biological Sciences*, 237(641):37–72, 1952.
28. Peter Wegner and Dina Goldin. The Church-Turing Thesis: Breaking the myth. In S. Barry Cooper and Benedikt Löwe, editors, *CiE 2005: New Computational Paradigms*, volume 3526 of *LNCS*. Springer, 2005.

Chapter 8

Intuitionistic Mathematics and Realizability in the Physical World

Andrej Bauer

University of Ljubljana, Slovenia

Intuitionistic mathematics perceives subtle variations in meaning where classical mathematics asserts equivalence, and permits geometrically and computationally motivated axioms that classical mathematics prohibits. It is therefore well-suited as a logical foundation on which questions about computability in the real world are studied.

The realizability interpretation explains the computational content of intuitionistic mathematics, and relates it to classical models of computation, as well as to more speculative ones that push the laws of physics to their limits. Through the realizability interpretation Brouwerian continuity principles and Markovian computability axioms become statements about the computational nature of the physical world.

1. Intuitionistic Understanding of Truth

Constructive mathematics, whose main proponent was Erret Bishop,[a] lives at the fringe of mainstream mathematics. It is largely misunderstood by mathematicians, and consequently by physicists as well. Contrary to the popular opinion, constructive mathematics is not poorer but *richer* in possibilities of mathematical expression than its classical counterpart. It differentiates meaning where classical mathematics asserts equivalence and thrives on geometric and computational intuitions that are banned by the classical doctrine. In this contribution I explore what constructive mathematics and the related realizability interpretation of intuitionistic logic have to offer to those who are interested in real-world computation.

If classical and constructive mathematicians just disagreed about what was true, the matter would be resolved easily. Unfortunately they use

---

[a]Bishop's constructivism is *compatible* with respect to classical mathematics, and should not be confused with the intuitionism of L.E.J. Brouwer, which assumes principles that are classically false.

the same words to mean two different things, which is always an excellent source of confusion. The origin of the schism lies in the criteria for truth, i.e., in what makes a statement true. Speaking vaguely, intuitionistic logic demands *positive evidence* of truth, while classical logic is happy with *lack of negative evidence*. The constructive view is closer to the criterion of truth in science, where a statement is accepted only after it has been positively confirmed by an experiment.

The Brouwer-Hetying-Kolmogorov (BHK) interpretation explains informally what counts as positive evidence:

- evidence for a conjunction $\phi \wedge \psi$ consists of evidence for ϕ together with evidence for ψ,
- evidence for a disjunction $\phi \vee \psi$ consists of evidence for ϕ, or for evidence for ψ, together with information about which disjunct the evidence is for,
- evidence for an implication $\phi \Rightarrow \psi$ is a method for converting evidence for ϕ to evidence for ψ,
- evidence for a universal quantification $\forall x \in A \, . \, \phi(x)$ is a method which takes (a representation of) any $a \in A$ can converts it to evidence for $\phi(a)$,
- evidence for an existential quantification $\exists x \in A \, . \, \phi(x)$ is (a representation of) an $a \in A$ together with evidence for $\phi(a)$,
- evidence for a negation $\neg \phi$ is lack of evidence for ϕ,
- anything is evidence for truth $\top$,
- there is no evidence for falsehood $\bot$.

The BHK interpretation leaves the meaning of "evidence" and "method" unexplained. Its formalization leads to the *realizability interpretation*, which we shall consider in detail in Section 3. Let us apply the BHK interpretation to the archetypical example, the Law of excluded middle:

"*Every proposition is either true or false.*"

The criterion by which a classical mathematician judges the law is

"*Is it the case that each proposition is either true or false?*"

whereas an intuitionistic mathematician is more demanding:

"*Is there a method for determining, given any proposition, which of the two possibilities holds?*"

Even though it might be the case that each particular proposition happens to be true or false, there might still be no method for deciding the truth and falsehood of propositions. The precise reason for there being no such method depends crucially on what counts as one. For example, if methods are required to be Turing computable, then a standard argument in computability theory shows that a method for deciding arbitrary propositions would yield a Halting oracle.

The intuitionistic mathematician does not accept the Law of excluded middle. When asked to produce counterexamples, he states that there are none, and makes things worse by claiming that there is in fact no proposition which is neither true nor false. Such a position looks thoroughly nonsensical to the classical mathematician. In terms of formal logic, the intuitionistic mathematician does not accept the Law of excluded middle

$$\forall \phi \in \mathsf{Prop}. \phi \vee \neg \phi,$$

yet he claims

$$\neg \exists \phi \in \mathsf{Prop}. \neg \phi \wedge \neg \neg \phi.$$

Are not these two propositions logically equivalent? Only if we accept the Law of excluded middle, so the intuitionist has not been caught in an inconsistency.

It is instructive to consider the difference between ϕ and $\neg\neg\phi$ in intuitionistic logic. Whereas ϕ holds if there is evidence supporting it, $\neg\neg\phi$ holds if there is no evidence that there is no evidence of ϕ. In other words $\neg\neg\phi$ holds when ϕ cannot be falsified, or when ϕ is *potentially true*. Every actually true statement is potentially true, but the converse need not hold. An example of a statement which is potentially true but not intuitionistically true would be "There is an undetectable elephant in Bertrand's room." Clearly, there can be no evidence against such a claim, but there can be no positive evidence for it either.

If a statement ϕ is intuitionistically equivalent to its double negation $\neg\neg\phi$, then there is no difference between its being actually and potentially true. In logic such statements are called $\neg\neg$-*stable*. The laws of physics are typically stated as universal statements involving equations, possibly with side conditions. Because such statements[b] are $\neg\neg$-stable the laws of physics are crisp: if they are potentially true then they are actually true.

---

[b] A statement built from the universal quantifier $\forall$, conjunction $\wedge$, implication $\Rightarrow$, and numerical equality $=$ is $\neg\neg$-stable, as can be easily verified.

There is a well-known translation of classical logic into intuitionistic logic, called the *double negation translation*. It transforms a given proposition by placing double negations in front of every quantifier and logical connective. In terms of our terminology, it inserts the adverb "potentially" everywhere, so the intuitionistic mathematician can interpret the utterances of his classical colleague as statements about potential truth. For example, the statement "every Turing machine either halts or runs forever" is translated to "there is no Turing machine which neither halts nor runs forever" Unfortunately, there is no such easy translation in the opposite direction. The classical mathematician must build models of intuitionistic logic to imagine how the intuitionistic friends think.

The intuitionistic and classical understanding of ¬¬-stable statements coincide. This is fortunate because we can all agree on what the universal laws of physics say. However, a disagreement is reached when we discuss which things exist in our universe, for intuitionistic existence requires explicit evidence where classical existence is satisfied with lack of negative evidence. Again we see the similarity between intuitionistic and scientific thinking.

2. Synthetic Differential Geometry

Before focusing on the central theme, let us show how intuitionistic logic allows us to accept axioms which are useful to physicists in their everyday calculations, but are officially prohibited because of the reign of classical logic.

Nowadays we teach analysis in the style of Cauchy and Weierstrass, with ϵ-δ definitions of continuity and differentiability. We might even tell our students that the original differential calculus of Leibniz and Newton was based on a flawed concept of *infinitesimals*, which were supposed to be infinitely small non-zero quantities. Yet when the students attend a physics class they never see an ϵ-δ argument. Their professor freely differentiates everything in sight and uses the outlawed infinitesimals. Not having been told the precise rules for handling infinitesimals, students get confused. A typical calculation might go like this:

$$(x^2)' = \frac{(x+dx)^2 - x^2}{dx} = \frac{2x \cdot dx + dx^2}{dx} = 2x + dx = 2x.$$

In the last step we pretend that dx is so small in comparison to $2x$ that it can be neglected. Surely then it is also small with respect to x, and can be

neglected already in the first step:
$$(x^2)' = \frac{(x+dx)^2 - x^2}{dx} = \frac{x^2 - x^2}{dx} = \frac{0}{dx} = 0 \quad ?!$$
My own experience as a student was that asking these kinds of questions about infinitesimals only lead to frustration and further confusion. No wonder the exile of infinitesimals was welcomed by mathematicians.

However, if the ϵ-δ analysis is of no use to physicists, it is mathematicians' task to provide a theory of infinitesimal calculus which does not suffer from the 17th century deficiencies. That this can be done was shown by William F. Lawvere and others, under the name *Synthetic Differential Geometry*.[c] Its fundamental axiom is the following principle, for which physicists should feel a certain degree of affinity:

> **Principle of micro-affinity:** *An infinitesimal change in the independent variable causes an affine (linear) change in the dependent variable.*

More precisely, if $f : \mathbb{R} \to \mathbb{R}$ is *any* function, $x \in \mathbb{R}$ and dx is an infinitesimal, then there exists a unique number $f'(x)$, called *the derivative of f at x*, such that $f(x+dx) = f(x) + f'(x)dx$ for all infinitesimals dx. A quantity dx is *infinitesimal* (of second degree) if its square dx^2 is zero.

Classical logic contradicts the Principle of micro-affinity because it shows that the only infinitesimal is 0. Indeed, if $dx^2 = 0$ then dx cannot be positive as that would imply $dx^2 > 0$, and it cannot be negative for the same reason. But if 0 is the only infinitesimal then the Principle of micro-affinity is false because $f(x) = x$ has both 0 and 1 as its derivative. We are left with no choice but to abandon classical logic.

The Principle of micro-affinity has strange consequences. Because it fails if all infinitesimals are 0, there must *potentially* exist some that are not zero. On the other hand, since infinitesimals can be neither negative nor positive, they cannot be different from zero, which is to say that they are *potentially zero*.[d] However strange this seems, it is not a contradiction. It may be helpful to think of infinitesimals as quantities so small that they cannot be experimentally distinguished from zero (they are potentially zero), but

---
[c]Synthetic differential geometry is not to be confused with Robinson's non-standard analysis,[9] which uses classical logic and does not contain nilpotent infinitesimals.

[d]We have used the intuitionistically acceptable principle that there is no number which is neither negative, nor zero, nor positive. The stronger law of trichotomy, which states that every number is either negative, zero, or positive, is not intuitionistically acceptable.

neither can they be shown to all equal zero (potentially there are some non-zero ones).[e]

The following consequence of the Principle of micro-affinity is quite useful for calculations:

> **Law of cancellation:** If $a \cdot dx = b \cdot dx$ for all infinitesimals dx, then $a = b$.

To derive the law, consider the function $f(x) = ax - bx$ and compute

$$f(x + dx) - f(x) = (a - b) \cdot dx = 0 \cdot dx,$$

where we used the assumption that $(a - b) \cdot dx = 0$. Because both $a - b$ and 0 are the derivative of f, they are equal, hence $a = b$.

The Law of cancellation is important in practical calculations because it allows us to cancel infinitesimals as long as they are *arbitrary*. Without it we could do no such thing, as infinitesimals are not invertible (they are potentially zero). For illustration, let us calculate the derivative of $f(x) = x^2$, this time correctly. For an *arbitrary* infinitesimal dx, by taking into account $dx^2 = 0$ we compute

$$f'(x) \cdot dx = f(x + dx) - f(x) = (x + dx)^2 - x^2 = 2x \cdot dx,$$

and by canceling dx we get $f'(x) = 2x$. We can similarly derive all the usual rules for computing derivatives, prove the fundamental theorem of calculus in two lines, derive the wave and heat equations, etc., as long as we stick to a simple set of rules: never assume anything specific about an infinitesimal dx, other than $dx^2 = 0$; cancel infinitesimals on both sides of an equation, do not divide by them; and do not prove equations by contradiction, just calculate as physicists always do.

The Principle of micro-affinity implies that every function has derivatives of all orders, everywhere. How are we supposed to model sudden changes, such as reflection of a light-ray or a bouncing ball? Intuitionistic treatment of sudden changes is more profound that the classical one. Consider a ball which moves freely up to time t_0, bounces off a wall, and moves freely afterwards. Its position p is described as a function of time t in two parts,

$$p(t) = p_1(t) \quad \text{if } t \leq t_0,$$
$$p(t) = p_2(t) \quad \text{if } t \geq t_0,$$

---

[e]Incidentally, we are not talking about lengths below the Planck length, as there are clearly positive reals numbers smaller than $1.6 \cdot 10^{-35}$.

where p_1 and p_2 are smooth functions and $p_1(t_0) = p_2(t_0)$. Because p is defined separately for $t \leq t_0$ and for $t \geq t_0$, its domain of definition is the union of two half-lines $(-\infty, t_0] \cup [t_0, \infty)$. Obviously this is a subset of the reals $\mathbb{R}$, but it may come as a bit of surprise that it is *not* the whole $\mathbb{R}$ because that would amount to the statement

"*For every real number t, either $t \leq t_0$ or $t \geq t_0$.*"

which is inconsistent with the Principle of micro-affinity, and is generally not acceptable intuitionistically.[f] Therefore, the domain of p is a proper subset of $\mathbb{R}$ and so p is *not* defined everywhere on $\mathbb{R}$. But neither is there an instant of time at which p is undefined! The domain of p should not be imagined as "$\mathbb{R}$ with missing points" but rather as "$\mathbb{R}$ with extra information". In the smooth world, any sudden changes in physical quantities are recorded at the level of logic as wrinkles in space-time.

Consistency of Synthetic differential geometry is secured by sheaf-theoretic models based on a geometric interpretation of intuitionistic logic that measures the truth of a statement as a region of space in which the statement holds locally. Let us explain this by way of example. If T denotes temperature, then in some places we have $T < 273$ and $T \geq 273$ in others. We say that $T < 273$ holds *locally* at a given place if it holds everywhere in a small neighborhood of the place, and likewise for $T \geq 273$. The statement

"*Either $T < 273$ or $T \geq 273$.*"

is then said to hold at a given place when either $T < 273$ holds there locally, or $T \geq 273$ does. Even if we assume that T varies smoothly, there might be a place where the temperature is exactly 273 so that $T \geq 273$ holds, but an arbitrarily small displacement leads to $T < 273$. The above statement, which happens to be an instance of the Law of excluded middle, therefore need not hold universally. The interested readers are invited to consult Bell's booklet[1] and the more advanced texts[6,8] for further reading on this topic.

---

[f]To see that there is no realistic method for determining whether $t \leq t_0$ or $t \geq t_0$, consider a real-valued measured quantity t. Whatever experiment we perform, there will always be a small measurement uncertainty $t \pm \Delta t$. If by luck $t + \Delta t < t_0$ or $t - \Delta t > t_0$, then we can decide which of $t \leq t_0$ and $t \geq t_0$ holds. Otherwise, we might perform a more precise measurement, but there is no guarantee that we shall succeed in finite time.

3. The Realizability Interpretation

Formalization of the BHK interpretation leads to the *realizability interpretation* of intuitionistic mathematics. The interpretation acts as a bridge between constructive mathematics and computational models of various kinds.

The essential idea of realizability theory is expressed mathematically with the *realizability relation*, written as

$$r \Vdash \varphi$$

and read as "*r realizes φ*". Depending on the context, we also say that r represents, implements, or witnesses φ. We usually think of r as something concrete (a program, a number, a sequence of bits), and of φ as something abstract (an element of a set, a function, a logical statement), even though they are both mathematical objects. Later on we shall speculate on r being a real-world entity, but for now we focus on the mathematical aspects of realizability.

For a sound interpretation of intuitionistic mathematics, the realizers should support certain operations, such as:[g] a *pairing* which combines realizers r and s into a single one $\langle r, s \rangle$, together with projections that recover the components of a pair; an *application* operation $r \cdot s$ which applies the function encoded by the realizer r to the realizer s; and a suitable encoding of natural numbers which assigns to each $n \in \mathbb{N}$ a *numeral* $\bar{n}$. The BHK interpretation is then formalized as follows:

$\langle r, s \rangle \Vdash \phi \wedge \psi$ iff $r \Vdash \phi$ and $s \Vdash \psi$

$\langle r, s \rangle \Vdash \phi \vee \psi$ iff $r = \bar{0}$ and $s \Vdash \phi$, or $r = \bar{1}$ and $s \Vdash \psi$

$r \Vdash \phi \Rightarrow \psi$ iff   for all s, if $s \Vdash \phi$ then $r \cdot s \Vdash \psi$

$r \Vdash \forall x \in A \, . \, \phi(x)$ iff   for all s and a, if $s \Vdash (a \in A)$ then $r \cdot s \Vdash \phi(a)$

$\langle r, s \rangle \Vdash \exists x \in A \, . \, \phi(x)$ iff   there is a such that $r \Vdash (a \in A)$ and $s \Vdash \phi(a)$

$r \Vdash \neg \phi$ iff   there is no s such that $s \Vdash \phi$

$r \Vdash \top$ always

$r \Vdash \bot$ never

In realizability a set must always be introduced together with realizers for its membership relation. In other words, we have to explain how the elements of the set are represented by the realizers. For instance, a natural

---

[g]The technical requirement is that the realizers should form a *partial combinatory algebra*.

number $n \in \mathbb{N}$ is represented by the corresponding numeral $\overline{n}$. As always we need to keep in mind that n is an abstract mathematical entity whereas $\overline{n}$ is something concrete, such as a sequence of bits in computer memory, or a string of symbols on paper.

We can use the realizability relation to compute the realizers of any logical statement. For example, if we unravel the realizability interpretation of the principle of mathematical induction, which is formally expressed by the formula

$$\phi(0) \wedge (\forall k \in \mathbb{N} . \phi(k) \Rightarrow \phi(k+1)) \Rightarrow \forall n \in \mathbb{N} . \phi(n),$$

we discover that its realizers correspond precisely to a known concept in programming, namely *primitive* recursion. There are many examples where the realizability interpretation of a well-known statement in mathematics turns out to be a well-known programming concept. At a more abstract level the connection between logic and computation is expressed by the following soundness theorem:

> **Soundness Theorem:** *from an intuitionistic proof of a statement we can extract a realizer for it.*

The theorem has at least three uses. First, because the extraction process is completely mechanical, we can actually compute realizers from formal proofs and constructions, a task best left to computer proof systems.[h] Second, it is often easier to prove a statement intuitionistically than it is to construct a realizer for it. Third, we can conclude that a statement is not realized if we show that it implies another statement which is known not to be realized.

4. Realizability in the Real World

We have so far not specified a particular model of computation on top of which realizability is built. Kleene's original number realizability[3] was based on partial recursive functions, encoded by natural numbers. Later Kleene gave a realizability model based on functions,[5] and there are still other realizability models based on programming languages, topological spaces, and even games, see[13] for a general theory of realizability.

We would like to use "real-world computation" as the underlying model for realizability. We do not know precisely what such a model amounts to.

---

[h]A famous proof system which uses the realizability interpretation for extraction of programs from proofs is Coq.[7]

Is it possible to send a computer into a black hole so it performs infinitely many steps of computation? Can we compute in parallel universes? Is it possible to perform a computation whose inner workings are hidden from the rest of the universe? Is the space-time made of tiny cellular automata? Since realizability can accommodate all kinds of computational models, we need not make a particular commitment about the true nature of the real world. Instead, we leave open all possibilities and ask what it would take to realize various logical and mathematical principles, such as the Axiom of choice and Brouwer's Continuity principle. We will be naturally lead to questions about the limits of real-world computation.

Law of excluded middle. What would it take to realize the Law of excluded middle? Some instances are realized, for example decidability of equality of natural numbers

> *"Every two natural numbers are either equal or distinct."*

This is so because we can prove the statement intuitionistically by induction. However, decidability of equality of real numbers

> *"Every two real numbers are either equal or distinct."*

is more problematic because its realizer would exceed the power of Turing machines. For if the statement were realized then we could tell whether any given Turing machine t halts by comparing zero and the computable real whose k-th digit is 1 if t has halted at step k, and 0 otherwise. In fact, decidability of equality of real numbers and the Halting problem are equivalent in terms of computational power:

> **Theorem:** *Decidability of reals is real-world realized if, and only if, we can build the Halting oracle for Turing machines.*

Proposals have been given on just how we might solve the Halting problem by sending a machine into a black hole in such a way that it would perform infinitely many computational steps in time that appears finite to us. Until such machines appear in computer stores, it seems safer not to rely on them. In any case, even if there are super-machines that perform amazing computational feats, some instances of the Law of excluded middle will still fail to be realized. Indeed, the (formalization of) the statement

> *"Every real-world machine halts or runs forever."*

fails to be realized, for its realizer would be a Halting oracle for real-world machines, which does not exist by the usual argument.[i]

Axiom of Choice. The Axiom of choice is perhaps the most controversial of the standard axioms of set theory. Some mathematicians accept it grudgingly only because it is necessary for certain desirable theorems in algebra and analysis. There are several formulations of the Axiom of choice, but the one whose realizability interpretation is most easily analyzed is

"*Every total relation contains a function.*"

In this generality the Axiom of choice is not realized because a theorem of Diaconescu's tells us that the Axiom of choice implies the Law of excluded middle. Since the latter is not realized neither is the former.

Nevertheless, some instances of the Axiom of choice are realized, namely those for which the domain of the total relation has *canonical realizers*. In general, an element may have many realizers, for example, a function has as many realizers as there are programs for computing it. We say that a set A has canonical realizers if there is a realizer r which computes canonical realizers: if s_1 and s_2 both realize $x \in A$ then so does $r \cdot s_1$ and moreover $r \cdot s_1 = r \cdot s_2$.

The natural numbers have canonical realizers. Recall that the only realizer for a natural number n is the corresponding numeral $\bar{n}$. Thus we are never even in a position to contemplate two different realizers for n. Consequently, the axiom of *Number Choice* is realized, which is lucky as large parts of constructive analysis depend on it.

A more complicated instance of the Axiom of Choice is *Function Choice*. For it to be realized we need to compute canonical realizers for number-theoretic functions $\mathbb{N} \to \mathbb{N}$. The realizers of a function $h : \mathbb{N} \to \mathbb{N}$ are programs that compute the function. How could we select one of them, in a realized way? In Kleene's function realizability this is accomplished by realizing h with an (infinite tape containing) the sequence $h(0), h(1), h(2), \ldots$ We could try doing the same in the real world, as follows. We build a machine which accepts a realizer r for h, and constructs a tape containing the values $h(0), h(1), h(2), \ldots$. We can then use the tape to get values of h by simply looking them up. The tape itself depends only on h and not on the

---

[i]The usual argument assumes that there is a universal machine, which may or may not exist in the real-world. But even if there isn't one, there will be still other instances of the Law of excluded middle that fail to be realized. A realizability model is always properly intuitionistic, as long as there are at least two realizers.

particular realizer r from which it was generated. However, since the tape is infinite its construction is an unfinished process, so in principle we could always observe the machine building it, and within it the original realizer r. What we actually need is to place r inside a *black box*, i.e., isolate it in such a way that its internal workings cannot be observed, and the only way to interact with it is to feed it inputs and observe outputs. To summarize:

> **Theorem:** *Function choice is real-world realized if there are black boxes in which computations can be hidden.*

In a science fiction story a black box might be an impenetrable "force field" or a "quantum cage". Should physicists declare that there are no such things, a bureaucratic approach might help: if we change the definition of real-world realizers so that any realizer labeled as "black box" never gets inspected, then black boxes are easily obtained by means of a sticker and a pen.

Continuity principles. In Section 2 we considered an intuitionistic system in which all real functions were smooth. Realizability does not validate such a setting because we can realize the absolute value map $x \mapsto |x|$, which is not smooth. In fact, we can realize some fairly complicated functions, such as Weierstrass's continuous but nowhere differentiable functions. However, realizing the jump function

$$j(x) = 0 \quad \text{if } x \leq 0,$$
$$j(x) = 1 \quad \text{if } x > 0,$$

amounts to decidability of equality on $\mathbb{R}$. Indeed, by computing $j(|x - y|)$ with precision $1/3$ suffices to tell whether its value is 0 or 1, and consequently whether $x = y$ or not. The same sort of reasoning can be applied to any function with a sudden jump, and so we wonder whether the continuity principle

> *"All functions are continuous."*

is real-world realized. Let us consider just the specific instance

> *"Every real function is continuous."*

This is essentially realized by a machine C which accepts as input a machine F for computing a real function $f : \mathbb{R} \to \mathbb{R}$, a realizer t for a real number $x \in \mathbb{R}$, and a numeral $\overline{n}$. The machine C is supposed to figure out how many digits of x suffice to compute the first n digit of $f(x)$. One way to accomplish this as follows. The machine C feeds into F a specially crafted realizer t' for the real number x which acts exactly like t, except that it communicates back to C what digits have been accessed by F. Now C waits for F to output the first n digits of $f(x)$ and records the maximum value m reported by t'. Clearly, m digits of the input suffice to compute the first n digits of $f(x)$, and therefore C may safely output $\overline{m}$.

There is a potential problem with our construction. The realizer F could analyze the inner workings of t' and discover that t' is secretly communicating with the outside world, namely with C. In such a case F could simply cut the communication between t' and C, or modify it in some way. We may dispense with such problems by assuming that t' can be hidden in a black box *with a private communication channel*, i.e., t' can be constructed in such a way that F cannot analyze it, and neither can it detect or disrupt its communication with C.

There is a second problem with C. If by a sleight of physics F manages to inspect infinitely many digits of its input in finite time, the maximum value reported to C will be infinity and f need not actually be continuous. So we need one more assumption to have C function properly:

> **Theorem:** *If black boxes and private communication channels exist, and only finitely many computation steps can be performed in finite time, then the Continuity principle is real-world realized.*

The Church-Turing thesis. An obvious question to ask is how real-world machines compare to Turing machines. One has the feeling that every Turing machine can be built in the real world, at least if we ignore certain questions about finiteness of space-time. That the converse holds is expressed by

> **Real-world Church-Turing thesis:** *a function is realizable in the real-world if, and only if, it is computed by a Turing machine.*

The principle states that we cannot exceed the computational power of Turing machines by using black holes, quantum mechanics, and other engineering tricks. Since I am not an engineer I do not wish to pass a judgment about the issue, but I can still relate the Church-Turing thesis with realizability of mathematical statements.

Kleene constructed an unbounded Turing computable binary tree which has no Turing computable infinite paths.[4] To understand how strange such a tree is, contemplate the following situation. A master machine has built a system of underground tunnels, beginning with a hole at the surface. The tunnels always go down, they may split into two, or lead to dead ends. We are told that there is no limit to the depth of the tunnels. Yet, even if the master machine is known to us we cannot build a machine that would enter the hole and always go down without ever hitting a dead end.

Students of mathematics learn certain basic geometric facts, such as:

> **Theorem:** *A continuous real function defined on a closed interval is bounded.*

and

> **Theorem:** *A closed interval is compact.*

These theorems take some getting used to, but are indispensable tools in mathematical analysis. They express basic geometric intuitions about the continuum. You may wonder what analysis and the Kleene tree have to do with the real-world Church-Turing thesis. They are all related! For if the real-world Church-Turing thesis holds, then the following statement is realized by the Kleene tree:

> *"There is an unbounded binary tree without an infinite path."*

From such a tree we can construct counterexamples to real-world realizability of the above two theorems,[12] namely a real-world realized continuous unbounded map defined on a closed interval, and a real-world realized infinite cover of a closed interval such that every finite subfamily fails to be a cover. Unfortunately, reviewing the constructions here would exceed the scope of the contribution.

Should we reject the real-world Church-Turing thesis, or do we give up basic geometric intuitions about space? There is a profound move that lets us eat the cake and have it too. In the usual conception of geometry a space is a set of points equipped with extra structure, such as metric or topology. But we can switch to a different view in which the extra structure is primary and points are derived ideal objects. For example, a topological space is not viewed as a set of points with a topology anymore, but rather just the topology, given as an abstract lattice with suitable properties, known as a *locale*.[j] In constructive mathematics such treatment of the notion of space is much preferred to the usual one. The desirable geometric intuitions are

---

[j]The suitable properties ask for a complete lattice in which finite infima distribute over arbitrary suprema.[2]

restored,[11,14] even under the assumption that everything in sight is Turing computable. Locale theory has certain advantages over traditional topology even in classical mathematics, where it allows for an appealing construction of the space of random sequences and resolves the Banach-Tarski paradox.[10] These are strong indicators that the last word on the nature of computation and geometry has not been said yet.

References

1. J. L. Bell. *A primer of infinitesimal analysis.* Cambridge University Press, 1998.
2. P. T. Johnstone. *Stone spaces.* Cambridge University Press, 1982.
3. S. C. Kleene. On the interpretation of intuitionistic number theory. *Journal of Symbolic Logic,* 10:109–124, 1945.
4. S. C. Kleene. Recursive functions and intuitionistic mathematics. In L.M. Graves, E. Hille, P.A. Smith, and O. Zariski, editors, *Proceedings of the International Congress of Mathemaiticans, August 1950. Cambridge, Mass.,* pages 679–685, 1952.
5. S. C. Kleene and R. E. Vesley. *The Foundations of Intuitionistic Mathematics, especially in relation to recursive functions.* North-Holland Publishing Company, 1965.
6. A. Kock. *Synthetic differential geometry.* Cambridge University Press, 1981.
7. The Coq development team. *The Coq proof assistant reference manual.* LogiCal Project, 2004. Version 8.0.
8. I. Moerdijk and G. E. Reyes. *Models for smooth infinitesimal analysis.* Springer-Verlag, New York, 1991.
9. A. Robinson. *Non-standard analysis.* North-Holland Publishing Co., 1966.
10. A. Simpson. Measure, randomness and sublocales. *Annals of Pure and Applied Logic,* 2012.
11. P. Taylor. TychonovâĂŹs theorem in Abstract Stone Duality. In *Domains VII,* 2004. http://www.paultaylor.eu/ASD/tyctas/.
12. A. S. Troelstra and D. van Dalen. *Constructivism in Mathematics, An Introduction, Vol. 1.* Number 121 in Studies in Logic and the Foundations of Mathematics. North-Holland, 1988.
13. J. van Oosten. *Realizability: An Introduction to its Categorical Side,* volume 152 of *Studies in Logic and the Foundations of Mathematics.* Elsevier, 2008.
14. S. Vickers. Some constructive roads to Tychonoff. In L. Crosilla and P. Schuster, editors, *From Sets and Types to Topology and Analysis: Practicable Foundations for Constructive Mathematics,* number 48 in Oxford Logic Guides, pages 223–238. Oxford University Press, 2005.

Chapter 9

What is Computation?
Actor Model versus Turing's Model

Carl Hewitt

http://carlhewitt.info

Concurrency is of crucial importance to the science and engineering of computation in part because of the rise of the Internet and many-core architectures. However, concurrency extends computation beyond the conceptual framework of Church, Gandy [1980], Gödel, Herbrand, Kleene [1987], Post, Rosser, Sieg [2008], Turing, etc. because there are effective computations that cannot be performed by Turing Machines.

In the Actor model,[29,33] computation is conceived as distributed in space where computational devices communicate asynchronously and the entire computation is not in any well-defined state. (An Actor can have stable information about about what it was like when it receives a message.) Turing's Model is a special case of the Actor Model.

A non-deterministic Turing Machine has bounded non-determinism (i.e. there is a bound on the size of integer that can be computed starting on a blank tape by an always-halting machine). Proving that a server will actually provide service to its clients requires unbounded non-determinism. In the semantics of bounded nondeterminism, a request to a shared resource might never receive service because a nondeterministic transition is always made to service another request instead. That's why the semantics of CSP were reversed from bounded non-determinism[36] to unbounded non-determinism.[37] However, bounded non-determinism was but a symptom of deeper underlying issues with communicating sequential processes as a foundation for concurrency. The Computational Representation Theorem[7,30] characterizes the semantics of Actor Systems without making use of sequential processes.

Turing's Model

Turing's[57] model of computation was intensely individual and sequential in that:

- *"the behavior of the computer at any moment is determined by the symbols which he [the computer] is observing, and his 'state of mind' at that moment"*
- *"there is a bound B to the number of symbols or squares which the computer can observe at one moment. If he wishes to observe more, he must use successive observations."*

Alan Turing

In the above, computation was conceived as being carried out in a single place by a device that proceeds from one well-defined state to the next while carrying out a calculation.[i]

Turing[58] stated the following thesis:

LCMs [logical computing machines: Turing's expression for Turing machines] *can do anything that could be described as ... "purely mechanical"... This is sufficiently well established that it is now agreed amongst logicians that "calculable by means of an LCM" is the correct accurate rendering* [of phrases like "purely mechanical"]

Kurt Gödel declared that

It is *"absolutely impossible that anybody who understands the question* [What is computation?] *and knows Turing's definition should decide for a different concept."*

Gödel was deeply suspicious of abstraction in computation to the point that he did not accept the λ Calculus as a model of computation until it had been proved that it computed the same functions as Turing Machines.

Kurt Gödel

Actor Model

However, Turing's model is in need of revision because of the increasing importance of concurrency in systems implemented using client-cloud computing and many-core computer architectures.

In the Actor model,[29,33] computation is conceived as distributed in space where computational devices called Actors communicate asynchronously using addresses of Actors and the entire computation is not in any well-defined state. (An Actor can have information about other Actors that it has received in a message about what it was like when the message was sent.) The behavior of an Actor is defined when it receives a message and at other times may be indeterminate.

Axioms of locality including *Structural* and *Operational* hold as follows:[ii]

- *Structural:* The local storage of an Actor can include *addresses* only
    1. that were provided when it was created
    2. that have been received in messages
    3. that are for actors created here
- *Operational:* In response to a message received, an Actor can
    1. create more Actors
    2. send messages to addresses in the following:
        ○ the message it has just received
        ○ its local storage
    3. Specify how to process another message

The Actor Model differs from its predecessors and most current models of computation in that the Actor model assumes the following:

- Concurrent execution in processing a message.
- The following are *not* required by an Actor: a thread, a mailbox, a message queue, its own operating system process, *etc.*
- Message passing has the same overhead as looping and procedure calling.

Lambda expressions[4] are special case Actors that never change. The lambda calculus can express parallelism but not general concurrency (see discussion below).

Turing's model is a special case of the actor model

Actor systems can implement systems that are impossible in Turing's model as illustrated by the following standard example of *"unbounded nondeterminism"*:[iii]

There is a bound on the size of integer that can be computed by an *always-halting* nondeterministic Turing Machine starting on a blank tape.

Gordon Plotkin[48] gave an informal proof as follows:[iv]

> Now the set of initial segments of execution sequences of a given nondeterministic program P, starting from a given state, will form a tree. The branching points will correspond to the choice points in the program. Since there are always only finitely many alternatives at each choice point, the branching factor of the tree is always finite. That is, the tree is finitary. Now König's lemma says that if every branch of a finitary tree is finite, then so is the tree itself. In the present case this means that if every execution sequence of P terminates, then there are only finitely many execution sequences. So if an output set of P is infinite, it must contain a nonterminating computation.

Alonzo Church

Consequently, *either*

- The tree has an infinite path. ⇔ The tree is infinite. ⇔ It is possible that P does not halt.
  If it is possible that P does not halt, then it is possible that that the set of outputs with which P halts is infinite.

or

- The tree does not have an infinite path. ⇔ The tree is finite. ⇔ P always halts.
  If P always halts, then the tree is finite and the set of outputs with which P halts is finite.

By contrast, there are always-halting Actor systems with no inputs that can compute an integer of unbounded size:[a]

---

[a]Plotkin's proof does not apply to the Actor system below for the following reason: In order to produce an output, the Actor System must pass through a sequence of interactions with *go* messages that it receives. However, during these interactions with *go* messages, the system is not in a well-defined state because there is a *stop* message in transit (perhaps in the physical form of photons). Consequently, the computation is inherently concurrent and not the sequence of global states assumed in Plotkin's proof for a Nondeterministic Turing Machine.

An Actor is created with local storage that is initialized with an integer variable *count* initialized to **0** and a Boolean variable *continue* that is **true** with the following behavior:

- When a *stop* message is received, set *continue* to **false** and return *count*.
- When a *go* message is received:
  1. if *continue* is **true**, increment *count* by **1** and send myself a *go* message.
  2. if *continue* is **false**, do nothing

The above Actor is started by concurrently sending it both a *go* message and a *stop* message.

The above Actor system can be implemented using ActorScript™ [Hewitt 2010a] as follows:

```
Unbounded ≡ ²
  start →                  ① a start message is implemented by
    let c ← Counter(0) →
              ① let c be a Counter that is a created by Counter with count equal 0 and continue equal true
    {c.go ,       ① send c a go message and concurrently
     c.stop}      ① return the value of sending c a stop message

Counter ≡
  actor(count↦³Integer) continue ← true (|⁴
    stop → count also continue←false,⁵   ① return count also continue becomes false
    go →                                  ① a go message does
      continue ??⁶ (
        true →         ① if continue is true then
          exit⁷ go also count←count+1,⁸  ① exit sending self a go message also count is incremented
        false → void)⁹|)¹⁰   ① if continue is false return void
```

By the semantics of the Actor model of computation[7,30] sending Unbounded a *start* message results in sending an integer of unbounded size to the return address received with the *start* message.

---

[b]Read as "*is defined to be.*"
[c]Read as "*has type.*"
[d]Token that marks beginning of Actor's methods (i.e. message handlers).
[e]Token that separates methods.
[f]Read as "*has cases.*"
[g]Exit and return following value.
[h]Token that separates condition handlers.
[i]Read as "end cases."
[j]Token that marks end of Actor's methods.

Nondeterminism is a special case of indeterminism

Consider the following Nondeterministic Turing Machine that starts at *Step 1*:

> *Step 1*: Either print 1 on the next square of tape or execute *Step 3*.
> *Step 2*: Execute *Step 1*.
> *Step 3*: Halt

According to the definition of Nondeterministic Turing Machines, the above machine might never halt.[v]

Note that the computation performed by the above machine are structurally different from the Counter program that implements unbounded nondeterminism in the following way:

1. The decision making of the above Nondeterministic Turing Machine is *internal* (having an essentially individual psychological basis).
2. The decision making of the above Actor counter is partly external (having an essentially sociological and anthropological basis)

Unbounded nondeterminism may seem like an esoteric property, but it is crucial to showing that a server does not accidentally starve a client by always serving others instead. And there are many other systems (e.g. computer operating systems) that cannot be implemented using Turing machines.

Actors are becoming the default model of computation. C#, Java, JavaScript, and Objective C are all headed in the direction of the Actor Model and ActorScript is a natural extension of these languages. Since it is very close to practice, many programmers just naturally assume the Actor Model.

The following major developments in computer technology are pushing the Actor Model forward because Actor Systems are highly scalable:

- Many-core computer architectures
- Client-cloud computing

In fact, the Actor Model and ActorScript can be seen as codifying what are becoming some best programming practices for many-core and client-cloud computing.

Configurations versus Global States

Computations are represented differently in State Machines and Actors:

1. *State Machine*: a computation can be represented as a global state that determines all information about the computation. It can be nondeterministic as to which will be the next global state, *e.g.*, in simulations where the global state can transition nondeterministically to the next state as a global clock advances in time, e.g., Simula.[10][vi]

   Kristen Nygaard (left)
   Ole Johan Dahl (right)

2. *Actors*: a computation can be represented as a configuration. Information about a configuration can be indeterminate.[11]

In 1975, Irene Greif published the first operational model of Actors in her dissertation. Two years after Greif published her operational model, Carl Hewitt and Henry Baker published the Laws for Actors.[2]

The *Computational Representation Theorem*[7,301] characterizes computation for systems which are closed in the sense that they do not receive communications from outside:

The denotation Denotes of a closed system S represents all the possible behaviors of S as

$\text{Denotes}_S = \sqcup_{i \in \mathbf{N}} \text{Progressions}^i(\perp_S)$

where Progression$_S$ is an approximation function that takes a set of partial behaviors to their next stage and $\perp_S$ is the initial behavior of S.

In this way, the behavior of S can be mathematically characterized in terms of all its possible behaviors.

A consequence of the Computational Representation system is that an Actor can have an *uncountable* number of different possible outputs. For

---

[k]For example, there can be messages in transit that will be delivered at some indefinite time.
[l]Building on the denotational semantics of the lambda calculus.[54]

example, Real.*go* can output any real number[m] between 0 and 1 where
Real ≡ *go* → [(0 **either** 1)] **##** **postpone** Real.*go*
where

- (0 **either** 1) is the nondeterministic choice of 0 or 1,
- [first] **##** rest is the sequence that begins with first and whose remainder is rest, and
- **postpone** expression delays execution of expression until the value is needed.

The upshot is that *concurrent systems can be represented and characterized by logical deduction but cannot be implemented.* Thus, the following practical problem arose:

How can practical programming languages be rigorously defined since the proposal by Scott and Strachey [1971][53] to define them in terms of λ-calculus failed because the λ-calculus cannot implement concurrency?

A proposed answer to this question is the semantics of ActorScript.[33]

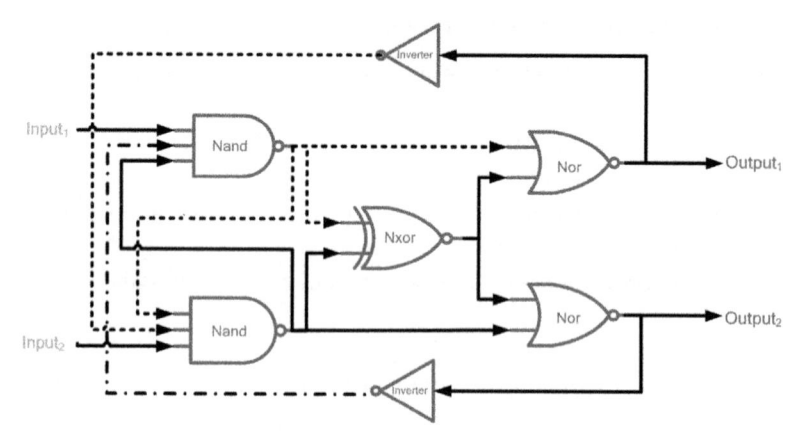

Arbiter Concurrency Primitive[14]

Another point of departure from Turing's model is that concurrency violates a narrowly conceived "public processes"[38] criterion for computation. Actor systems make use of hardware devices called arbiters to decide the order of processing messages.

---
[m] Using binary representation. See Ref. 21 for more on computation over the reals.
[n] Dashed lines are used for clarity to avoid confusing wires that cross over (under) one another.

After the above circuit is started, it can remain in a meta-stable state for an unbounded period of time before it finally asserts either Output$_1$ or Output$_2$.[vii]

The output of the operation of an Arbiter can in general *not* be logically inferred from its inputs. Thus (contrary to the claim of[40]), computation is *not* subsumed by deduction. Consequently, Logic Programming° (although sometimes useful) is not a universal programming paradigm. See the appendices of this paper for an overview of modern inconsistency-robust Logic Programming.

The internal processes of arbiters are not public processes. Attempting to observe them affects their outcomes. Instead of observing the internals of arbitration processes, we necessarily await outcomes. Indeterminacy in arbiters produces indeterminacy in Actors. The reason that we await outcomes is that we have no realistic alternative.[viii]

Unbounded Nondeterminism Controversy

Considerable controversy developed over issues involving unbounded nondeterminism.

Dijkstra believed that unbounded nondeterminism is impossible to implement

Edsger Dijkstra believed that unbounded nondeterminism cannot be implemented.[ix] His belief was manifested in his theory of computation based on "*weakest preconditions*" for global states of computation[15] in which he argued that unbounded nondeterminism results in non-continuity of his weakest precondition semantics.[17]

Edsger Dijkstra[1]

Bounded nondeterminism in CSP

Hoare was convinced that unbounded nondeterminism could not be implemented and so the semantics of CSP specified bounded nondeterminism.

Consider the following program written in CSP:[36]

°See Ref. 33 for the middle history of Logic Programming.

```
[X :: Z!stop( )    ⓘ In process X, send Z a stop message
 ||    ⓘ process X operates in parallel with process Y
 Y :: guard: boolean; guard := true;    ⓘ In process Y, initialize boolean variable guard to true and then
       *[guard→ Z!go( ); Z?guard]   ⓘ while guard is true, send Z a go message and then input guard from Z
 ||         ⓘ process Y operates in parallel with process Z
 Z :: n: integer; n:= 0;   ⓘ In process Z, initialize integer variable n to 0 and then
       continue: boolean; continue := true;    ⓘ initialize boolean variable continue to true and then
       *[                ⓘ repeatedly either
          X?stop( ) → continue := false;   ⓘ input a stop message from X, set continue to false and then
                Y!continue;            ⓘ send Y the value of continue
          []                          ⓘ or
          Y?go( )→ n := n+1;         ⓘ input a go message from Y, increment n, and then
                Y!continue]]          ⓘ send Y the value of continue
```

According to Clinger:[7]

> this program illustrates global nondeterminism, since the non-determinism arises from incomplete specification of the timing of signals between the three processes X, Y, and Z. The repetitive guarded command in the definition of Z has two alternatives: either the stop message is accepted from X, in which case continue is set to false, or a go message is accepted from Y, in which case n is incremented and Y is sent the value of continue. If Z ever accepts the stop message from X, then X terminates. Accepting the stop causes continue to be set to false, so after Y sends its next go message, Y will receive false as the value of its guard and will terminate. When both X and Y have terminated, Z terminates because it no longer has live processes providing input.
>
> As the author of CSP points out, therefore, if the repetitive guarded command in the definition of Z were required to be fair, this program would have unbounded nondeterminism: it would be guaranteed to halt but there would be no bound on the final value of n. In actual fact, the repetitive guarded commands of CSP are not required to be fair, and so the program may not halt.[36] This fact may be confirmed by a tedious calculation using the semantics of CSP[24] or simply by noting that the semantics of CSP is based upon a conventional power domain and thus does not give rise to unbounded nondeterminism.

p

---

[p]A very important point is that Actors do not have to make use of the repeated nondeterministic choices of CSP as in loop above:
*[X?stop( ) ? continue := false; Y!continue;[] Y?go( ) ? n := n+1; Y!continue].
The nondeterministic choice of sources of input poses fundamental practical and theoretical problems.[39]

The upshot was that Hoare was convinced that unbounded nondeterminism is impossible to implement. That's why the semantics of CSP specified bounded nondeterminism. But Hoare knew that trouble was brewing in part because for several years proponents of the Actor Model had been beating the drum for unbounded nondeterminism. To address this problem, he suggested that implementations of CSP should be as close as possible to unbounded nondeterminism! However, using the above semantics for CSP it was impossible to formally prove that a server actually provides service to multiple clients[q] (as had been done previously in the Actor Model). That's why the semantics of CSP were reversed from bounded non-determinism[36] to unbounded non-determinism.[37,xi] Bounded non-determinism was but a symptom of deeper underlying issues with nondeterministic transitions in communicating sequential processes (see Ref. 39).

---

[q]In the semantics of bounded nondeterminism, a request to a shared resource might never receive service because a nondeterministic choice is always made to service another request instead.

Summary of the unbounded nondeterminism controversy

A nondeterministic system is defined to have "*unbounded nondeterminism*" exactly when both of the following hold:

1. When started, the system always halts.
2. For every integer n, it is possible for the system to halt with output that is greater than n.

This article has discussed the following points about unbounded nondeterminism controversy:

- A Nondeterministic Turing Machine cannot implement unbounded nondeterminism.[xii]
- Dijkstra believed that unbounded nondeterminism cannot be implemented.
- Semantics of unbounded nondeterminism are required to prove that a server provides service to every client.
- An Actor system[29] can implement servers that provide service to every client and consequently unbounded nondeterminism.
- The semantics of CSP[24] specified bounded nondeterminism for reasons mentioned in the article. Since Hoare *et al.* wanted to be able to prove that a server provided service to clients, the semantics of a subsequent version of CSP were switched from bounded to unbounded nondeterminism.
- Unbounded nondeterminism was but a symptom of deeper underlying issues with communicating sequential processes as a foundation for concurrency.[xiii]

Process Calculi

In his Turing lecture, Robin Milner wrote:[43]
> Now, the pure lambda-calculus is built with just two kinds of thing: terms and variables. Can we achieve the same economy for a process calculus? Carl Hewitt, with his Actors model, responded to this challenge long ago; he declared that a value, an operator on values, and a process should all be the same kind of thing: an Actor.
>
> This goal impressed me, because it implies the homogeneity and completeness of expression ...
>
> So, in the spirit of Hewitt, our first step is to demand that all things denoted by terms or accessed by names–values, registers, operators, processes, objects–are all of the same kind of thing....

Robin Milner

Process calculi (e.g.[3,43]) are closely related to the Actor model. There are similarities between the two approaches, but also many important differences (philosophical, mathematical and engineering):

- There is only one Actor model (although it has numerous formal systems for design, analysis, verification, modeling, etc.) in contrast with a variety of species of process calculi.
- The Actor model was inspired by the laws of physics and depends on them for its fundamental axioms in contrast with the process calculi being inspired by algebra.[43]
- Unlike the Actor model, the sender is an intrinsic component of process calculi because they are defined in terms of reductions (as in the λ-calculus).
- Processes in the process calculi communicate by sending messages either through channels (synchronous or asynchronous), or via ambients (which can also be used to model channel-like communications).[3] In contrast, Actors communicate by sending messages to the addresses of other Actors (this style of communication can also be used to model channel-like communications using a two-phase commit protocol[39]).

Computational Undecidability

Some questions cannot be uniformly answered computationally.[5,57] Since logical inference is a special case of computation, the proof method of showing computationally undecidability developed by Turing and Church demonstrates further limits of inference as explained below.

Completeness versus inferential undecidability

Inferential undecidability of a theory $\mathcal{T}$ can be formally defined as follows:
$\texttt{InferentiallyDecidable}_\mathcal{T} \equiv \forall s \in \texttt{Sentences}_\mathcal{T} \rightarrow (\vdash_\mathcal{T} \lfloor s \rfloor_\mathcal{T}) \vee (\vdash_\mathcal{T} \neg \lfloor s \rfloor_\mathcal{T})$

Computational undecidability implies that mathematics is inferentially undecidable without using self-referential sentences.[r]

Inferential undecidability can be proved by showing that there is a proposition Ψ such that[xv]

(1) $\vdash' \nvdash \Psi$ and
(2) $\vdash' \nvdash \neg\Psi$

It is sometimes said that propositions like the one above are "provably true but provably unprovable", by which it is meant that[xvi]

(1) $\vdash' \models \Psi$[xvii]
(2) $\vdash' \nvdash \Psi$

Information Invariance is a fundamental technical goal of logic consisting of the following:

(1) *Soundness of inference*: information is not increased by inference
(2) *Completeness of inference*: all information that necessarily holds can be inferred

Direct Logic aims to achieve information invariance even when information is inconsistent using inconsistency robust inference regardless that mathematics is inferentially undecidable. As proved above, Direct Logic infers there is a proposition Ψ such that both $\vdash' \nvdash \Psi$ and $\vdash' \nvdash \neg\Psi$. But this does not mean that Direct Logic is somehow fundamentally Òincomplete Ó with respect to the information that can be inferred.

---

[r]This is in contrast with Gödel [1931] who argued for inferential undecidability on the basis of using the particular self-referential sentence "This sentence is not provable." (See historical appendix of this paper for further discussion.)

Consistency of mathematics

Theorem: *Mathematics self-proves its own consistency.*[s]
Consistency can be defined as follows:
Consistent≡¬∃s∈Sentences →⊢ ⌊s⌋,¬⌊s⌋
Theorem: ⊢ Consistent
> Proof. Suppose to obtain a contradiction that ¬*Consistent*. Consequently, ∃s∈Sentences →⊢ ⌊s⌋,¬⌊s⌋ and there is a sentence s_0 such that ⊢ s_0 and ⊢ ¬s_0. These theorems can be used to infer s_0 and ¬s_0, which is a contradiction.

Using proof by contradiction, ⊢ Consistent

Acknowledgement

Important contributions to the semantics of Actors have been made by: Gul Agha, Beppe Attardi, Henry Baker, Will Clinger, Irene Greif, Carl Manning, Ian Mason, Ugo Montanari, Maria Simi, Scott Smith, Carolyn Talcott, Prasanna Thati, and Aki Yonezawa.

Important contributions to the implementation of Actors have been made by: Bill Athas, Russ Atkinson, Beppe Attardi, Henry Baker, Gerry Barber, Peter Bishop, Nanette Boden, Jean-Pierre Briot, Blaine Garst, Bill Dally, Peter de Jong, Jessie Dedecker, Ken Kahn, Bill Kornfeld, Henry Lieberman, Carl Manning, Mark S. Miller, Tom Reinhardt, Chuck Seitz, Richard Steiger, Dale Schumacher, Dan Theriault, Mario Tokoro, Darrell Woelk, and Carlos Varela.

Research on the Actor model has been carried out at Caltech Computer Science, Kyoto University Tokoro Laboratory, MCC, MIT Artificial Intelligence Laboratory, SRI, Stanford University, University of Illinois at Urbana-Champaign Open Systems Laboratory, Pierre and Marie Curie University (University of Paris 6), University of Pisa, University of Tokyo Yonezawa Laboratory and elsewhere.

Conversations over the years with Dennis Allison, Bruce Anderson, Arvind, Bob Balzer, Bruce Baumgart, Gordon Bell, Dan Bobrow, Rod Burstall, Luca Cardelli, Vint Cerf, Keith Clark, Douglas Crockford, Ole-Johan Dahl, Julian Davies, Jack Dennis, Peter Deutsch, Edsger Dijkstra,

---

[s]Of course, this is contrary to a famous result from Gödel [1931]. A resolution is that Direct Logic uses a powerful natural deduction system that is its own meta system in the proof below. A tradeoff is that the self-referential propositions of Gödel (used to prove that mathematics cannot prove its own consistency) are not allowed in Direct Logic. See further discussion in Ref. 35.

Scott Fahlman, Dan Friedman, Ole-Johan Dahl, Julian Davies, Patrick Dussud, Doug Englebart, Sol Feferman, Bob Filman, Kazuhiro Fuchi, Mike Genesereth, Cordell Green, Jim Gray, Pat Hayes, Anders Hejlsberg, Pat Helland, John Hennessy, Tony Hoare, Mike Huhns, Dan Ingalls, Anita Jones, Bob Kahn, Gilles Kahn, Alan Karp, Alan Kay, Bob Kowalski, Monica Lam, Butler Lampson, Leslie Lamport, Peter Landin, Vic Lesser, Jerry Lettvin, Lick Licklider, Barbara Liskov, John McCarthy, Dave McQueen, Erik Meijer, Robin Milner, Marvin Minsky, Fanya Montalvo, Ike Nassi, Alan Newell, Kristen Nygaard, Seymour Papert, David Patterson, Carl Petri, Gordon Plotkin, Vaughan Pratt, John Reynolds, Jeff Rulifson, Earl Sacerdoti, Vijay Saraswat, Herbert Simon, Munindar Singh, Dana Scott, Ehud Shapiro, Burton Smith, Guy Steele, Gerry Sussman, Chuck Thacker, Kazunori Ueda, Dave Unger, Richard Waldinger, Peter Wegner, Richard Weyhrauch, Jeannette Wing, Terry Winograd, Glynn Winskel, David Wise, Bill Wulf, *etc.* greatly contributed to the development of the ideas in this article.

Jeremy Forth and Richard Waldinger made very helpful comments and suggestions. Discussions with Dennis Allison, Ron Dolin, Eugene Miya, Vaughan Pratt and others were helpful in improving this article. Pankaj Mehra and Dale Schumacher corrected some typos.

References

1. A. Anderson and M. Zelëny (editors). *Logic, Meaning and Computation: Essays in Memory of Alonzo Church.* Springer. 2002.
2. H. Baker and C. Hewitt *The Incremental Garbage Collection of Processes* Proceeding of the Symposium on Artificial Intelligence Programming Languages. 1977.
3. L. Cardelli and A. Gordon. *Mobile Ambients* FoSSaCS'98.
4. A. Church *The Calculi of Lambda-Conversion* Princeton University Press. 1941.
5. A. Church. *An unsolvable problem of elementary number theory.* American Journal of Mathematics, 58 (1936),
6. A. Church and J.B. Rosser. *Some properties of conversion.* Transactions of the American Mathematical Society. May 1936.
7. W. Clinger. *Foundations of Actor Semantics* MIT Mathematics Doctoral Dissertation. June 1981.
8. J. Copeland. *The Essential Turing* Oxford University Press. 2004.
9. H. Curry "Some Aspects of the Problem of Mathematical Rigor" *Bulletin of the American Mathematical Society* Vol. 4. 1941.
10. O.-J. Dahl and K. Nygaard. *Class and subclass declarations* IFIP TC2 Conference on Simulation Programming Languages. May 1967.

11. J. Dawson *Logical Dilemmas. The Life and Work of Kurt Gödel* AK Peters. 1997.
12. J. Dawson. *What Hath Gödel Wrought?* Synthese. Jan. 1998.
13. J. Dawson. *Shaken Foundations or Groundbreaking Realignment? A Centennial Assessment of Kurt Gödel's Impact on Logic, Mathematics, and Computer Science* FLOC'06.
14. C. Diamond. *Wittgenstein's Lectures on the Foundations of Mathematics, Cambridge, 1939* Cornell University Press. 1967.
15. E. Dijkstra. *A Discipline of Programming.* Prentice Hall. 1976.
16. E. Dijkstra and A.J.M. van Gasteren. *A Simple Fixpoint Argument Without the Restriction of Continuity* Acta Informatica. Vol. 23. 1986.
17. E. Dijkstra. *Position Paper on "Fairness"* EWD 1013. E.W. Dijkstra Archive. UT Austin.
18. S. Feferman, J. Dawson, S. Kleene, et al., editors. *Collected Works of Kurt Gödel Vol. I-V* Oxford University Press. 2001-2003.
19. S. Feferman. *Axioms for determinateness and truth* Review of Symbolic Logic. 2008.
20. S. Feferman. *In the Light of Logic* Oxford University Press. 1998.
21. S. Feferman. "About and around computing over the reals", *Computability: Gödel, Church, Turing and Beyond.* MIT Press. forthcoming 2012.
22. W.H.J. Feijen, A.J.M. van Gasteren, David Gries and J. Misra (editors). *Beauty is Our Business: Birthday Salute to Edsger W.Dijkstra* Springer. 1990.
23. F. Fitch. *Symbolic Logic: an Introduction.* Ronald Press. 1952.
24. N. Francez, T. Hoare, D. Lehmann, and W.-P. de Roever. *Semantics of nondeterminism, concurrency, and communication* Journal of Computer and System Sciences. December 1979.
25. J. Kenneth Galbraith. *Economics, Peace and Laughter.* New American Library 1971.
26. R. Gandy. *Church's Thesis and Principles of Mechanisms* The Kleene Symposium. North–Holland. 1980.
27. I. Greif. *Semantics of Communicating Parallel Processes* MIT EECS Doctoral Dissertation. August 1975
28. C. Hewitt and R. Atkinson. *Specification and Proof Techniques for Serializers* IEEE Journal on Software Engineering. January 1979.
29. C. Hewitt, P. Bishop and R. Steiger. *A Universal Modular Actor Formalism for Artificial Intelligence* IJCAI-1973.
30. C. Hewitt. *What is Commitment? Physical, Organizational, and Social* COIN@AAMAS'06. (Revised version in Springer Verlag Lecture Notes in Artificial Intelligence. Edited by Javier Vázquez-Salceda and Pablo Noriega. 2007) April 2006.
31. C. Hewitt *Middle History of Logic Programming: Resolution, Planner, Edinburgh LCF, Prolog, and the Japanese Fifth Generation Project* ArXiv 0904.3036. 2009-2011.
32. C. Hewitt *Formalizing common sense for inconsistency-robust information*

integration using Direct Logic^{TM} Reasoning and the Actor Model Inconsistency Robustness 2011.
33. C. Hewitt *ActorScript^{TM} extension of C#^{TM}, Java^{TM}, JavaScript^{TM} and Objective C^{TM}* ArXiv. 1008.2748, 2010.
34. C. Hewitt *Actor Model of Computation: Many-core Inconsistency-robust Information Integration* Inconsistency Robustness 2011.
35. C. Hewitt *Mathematics self-proves its own consistency (contra Gödel et al.)* Submitted for publication to arXiv on March 22, 2012.
36. T. Hoare. *Communicating Sequential Processes* CACM August, 1978.
37. T. Hoare *Communicating Sequential Processes* Prentice Hall. 1985.
38. D. Hofstadter. *Gödel, Escher, Bach: An Eternal Golden Braid* Vintage. 1980.
39. F. Knabe. *A Distributed Protocol for Channel-Based Communication with Choice* PARLE'92.
40. R. Kowalski *The Early Years of Logic Programming* CACM. January 1988.
41. J. Law. *After Method: mess in social science research* Routledge. 2004.
42. M. Löb. *Solution of a problem of Leon Henkin.* Journal of Symbolic Logic. Vol. 20. 1955.
43. R. Milner. *Elements of Interaction* CACM. January 1993.
44. M.S. Miller. *Robust Composition: Towards a Unified Approach to Access Control and Concurrency Control* Doctoral dissertation. John Hopkins. 2006.
45. R. Monk. *Bourgeois, Boshevist or anarchist? The Reception of Wittgenstein's Philosophy of Mathematics* in Wittgenstein and his interpreters. Blackwell. 2007.
46. D. Park. *Concurrency and Automata on Infinite Sequences* Lecture Notes in Computer Science, Vol. 104.Springer. 1980.
47. G. Peano *Arithmetices principia, nova methodo exposita* (The principles of arithmetic, presented by a new method). 1889.
48. G. Plotkin. *A powerdomain construction* SIAM Journal of Computing September 1976.
49. G. Plotkin. *Robin Milner: A Craftsman of Tools for the Mind* YouTube. 2010.
50. B. Roscoe. *The Theory and Practice of Concurrency* Prentice-Hall. Revised 2005.
51. B. Russell. *Principles of Mathematics* Norton. 1903.
52. B. Russell. *Principia Mathematica* 2nd Edition 1925.
53. D. Scott and C. Strachey. Towards a mathematical semantics for computer languages. *Proc. Symp. on Computers and Automata,* Polytechnic Institute of Brooklyn, 1971.
54. D. Scott. *Data Types as Lattices.* SIAM Journal on computing. 1976.
55. W. Sieg and J. Byrnes. *An Abstract Model for Parallel Computations: Gandy's Thesis* Monist. 1999.
56. W. Sieg *Church Without Dogma – axioms for computability.* New Computational Paradigms. Springer Verlag. 2008 (republished in this volume with an additional postscriptum).

57. A. Turing. *On computable numbers, with an application to the Entscheidungsproblem* Proceedings London Math Society. 1936.
58. A. Turing. *Intelligent Machinery* National Physical Laboratory Report. 1948.
59. H. Wang *A Logical Journey, From Gödel to Philosophy* MIT Press. 1974.
60. L. Wittgenstein. 1956. *Bemerkungen ¨uber die Grundlagen der Mathematik/Remarks on the Foundations of Mathematics, Revised Edition* Basil Blackwell. 1978.
61. L. Wittgenstein. *Philosophische Grammatik* Basil Blackwell. 1969.
62. L. Wittgenstein. (1933-1935) *Blue and Brown Books.* Harper. 1965.
63. L. Wittgenstein *Philosophical Investigations* Blackwell. 1953/2001.

Appendix 1. Historical development

"Faced with the choice between changing one's mind and proving that there is no need to do so, almost everyone gets busy on the proof."
John Kenneth Galbraith[25] [p. 50]

Turing versus Wittgenstein

Turing differed fundamentally on the question of inconsistency from Wittgenstein when he attended Wittgenstein's seminar on the Foundations of Mathematics:[t]

Wittgenstein:... Think of the case of the Liar. It is very queer in a way that this should have puzzled anyone — much more extraordinary than you might think... Because the thing works like this: if a man says 'I am lying' we say that it follows that he is not lying, from which it follows that he is lying and so on. Well, so what? You can go on like that until you are black in the face. Why not? It doesn't matter. ...it is just a useless language-game, and why should anyone be excited?

Turing: What puzzles one is that one usually uses a contradiction as a criterion for having done something wrong. But in this case one cannot find anything done wrong.

Wittgenstein: Yes — and more: nothing has been done wrong, ... where will the harm come?

Turing: The real harm will not come in unless there is an application, in which a bridge may fall down or something of that sort.... You cannot be confident about applying your calculus until you know that there are no hidden contradictions in it[t]*.... Although you do not know that the bridge will fall if there are no contradictions, yet it is almost certain that if there are contradictions it will go wrong somewhere.*[u]

Wittgenstein followed this up with [[60], pp. 104e–106e]: *Can we say:*

---
[t]Church and Turing later proved that determining whether there are hidden contradictions in a useful calculus is computationally undecidable.
[u]Turing was correct that it is unsafe to use classical logic to reason about inconsistent information. For this reason, inconsistency-robust logic[32] has been developed to more safely reason about inconsistent information.

'Contradiction is harmless if it can be sealed off'? But what prevents us from sealing it off?

The above debate between Turing and Wittgenstein has continued to this day with interesting developments described below.

Wittgenstein: self-referential propositions lead to inconsistency

Having previously conceived inconsistency tolerant logic, Wittgenstein had his own interpretation of inferential undecidability (which was completely at odds with Gödel):

> "True in Russell's system" means, as we have said, proved in Russell's system; and "false in Russell's system" means that the opposite [negation] has been proved in Russell's system.
> Let us suppose I prove[v]
> the unprovability (in Russell's system [*Russell*]) of P [⊢$_{Russell}$⊬$_{Russell}$ P where P ⇔⊬$_{Russell}$ P]; then by this proof I have proved P [⊢$_{Russell}$ P].
> Now if this proof were one in Russell's system [⊢$_{Russell}$⊢$_{Russell}$ P] —I should in this case have proved at once that it belonged [⊢$_{Russell}$ P] and did not belong [⊢$_{Russell}$ ¬P because ¬P ⇔⊢$_{Russell}$ P] to Russell's system.
>
> But there is a contradiction here [in *Russell*]!
> [w]
> —Well, then there is a contradiction here.

Thus the attempt to develop a universal system of classical mathematical logic once again ran into inconsistency. That a theory that infers its own inferential undecidability using the self-referential proposition P is incon-

---

[v]Wittgenstein was granting the supposition that Gödel had proved inferential undecidability in Russell's system, e.g. ⊢$_{Russell}$⊬$_{Russell}$ P. However, inferential undecidability is easy to prove using Gödel's self-referential proposition. Suppose to obtain a contradiction that ⊢$_{Russell}$P. Both of the following can be inferred:

1) ⊢$_{Russell}$⊬$_{Russell}$ P from the hypothesis because P ⇔⊬$_{Russell}$ P

2) ⊢$_{Russell}$⊢$_{Russell}$ P. from the hypothesis by Adequacy.

But 1) and 2) are a contradiction in *Russell*. Consequently, ⊢$_{Russell}$⊬$_{Russell}$ P follows from proof by contradiction in *Russell*.

[w]Wittgenstein was saying that Gödel's self-referential proposition P shows that Russell's system is inconsistent in much the same way that Russell had previously shown Frege's system to be inconsistent using the self-referential set of all sets that are not members of themselves.

sistent represented a huge threat to Gödel's firmly held belief that mathematics is based on objective truth.

Classical logicians versus Wittgenstein

> The powerful (try to) insist that their statements are literal depictions of a single reality. 'It really is that way', they tell us. 'There is no alternative.' But those on the receiving end of such homilies learn to read them allegorically, these are techniques used by subordinates to read through the words of the powerful to the concealed realities that have produced them.[41]

Wittgenstein had also written:[60]

> Can we say: 'Contradiction is harmless if it can be sealed off'? But what prevents us from sealing it off?
>
> Let us imagine having been taught Frege's calculus, contradiction and all. But the contradiction is not presented as a disease. It is, rather, an accepted part of the calculus, and we calculate with it.
>
> Have said-with pride in a mathematical discovery [e.g., inconsistency of Russell's system (above)]: "Look, this is how we produce a contradiction."

Gödel responded as follows:[18]

> He [Wittgenstein] has to take a position when he has no business to do so. For example, "you can't derive everything from a contradiction." He should try to develop a system of logic in which that is true.[x]

According to:[45]

> Wittgenstein hoped that his work on mathematics would have a cultural impact, that it would threaten the attitudes that prevail in logic, mathematics and the philosophies of them. On this measure it has been a spectacular failure.

Unfortunately, recognition of the worth of Wittgenstein's work on mathematics came long after his death. Classical logicians mistakenly believed that they had been completely victorious over Wittgenstein. For example, according to[13] [*emphasis in original*]:

- Gödel's results altered the mathematical landscape, but they did **not** "produce a debacle".

---

[x]Gödel knew that it would be technically difficult to develop a useful system of logic proposed by Wittgenstein in which *"you can't derive everything from a contradiction"* and evidently doubted that it could be done.

- *There is **less** controversy today over mathematical foundations than there was **before** Gödel's work.*

However, the groundbreaking realignment came later when computer science invented a useable inconsistency robust logic because of pervasive inconsistency in computer information systems.

The controversy between Wittgenstein and Gödel can be summarized as follows:

- Gödel
    1. Mathematics is based on objective truth.[y]
    2. A theory is not allowed to *directly* reason about itself.
    3. Self-referential sentences prove inferential undecidability but (hopefully) not inconsistency.
    4. Theories should be proved consistent.

- Wittgenstein
    1. Mathematics is based on communities of practice.
    2. Reasoning about theories is like reasoning about everything else, *e.g.* chess.
    3. Self-referential sentences can lead to inconsistency.
    4. Theories should use inconsistency robust inference.

According to:[19]

> So far as I know, it has not been determined whether such [inconsistency robust] *logics account for "sustained ordinary reasoning", not only in everyday discourse but also in mathematics and the sciences.*

Direct Logic[32] was put forward as an improvement over classical logic with respect to Feferman's desideratum above. Computer science needs an all-embracing system of inconsistency-robust reasoning to implement practical information integration.[z]

Turing versus Gödel

Turing recognized that proving inference for *Russell* is computationally undecidable is quite different than proposing that a self-referential proposition

---

[y]According to[18] [p. 30] mathematical objects and *"concepts form an objective reality of their own, which we cannot create or change, but only perceive and describe."*
[z]Computer systems need all-embracing rules to justify their inferences, *i.e.*, they can't always rely on human manual intervention.

proves *Russell* is inferentially undecidable[57] [p. 259]: *It should perhaps be remarked what I shall prove is quite different from the well-known results of Gödel.*[18] The proof method of showing computational undecidability developed by Turing and Church proves that Russell is inferentially undecidable without constructing any particular proposition P such that $\vdash_{Russell} P$ and $\nvdash_{Russell} \neg P$.

As Wittgenstein pointed out, construction of the self-referential "This sentence is not provable" leads an inconsistency in the foundations of mathematics. If the inconsistencies of self-referential propositions stopped with this example, then it would be somewhat tolerable for inconsistency-robust logic. However, other self-referential propositions (constructed in a similar way using the logical fixed point construction Fix) can be used to prove every proposition rendering inference useless:

- *Curry's Paradox* [Curry 1941]:
  Curry$_v$ ≡ $\lfloor$Fix(Diagonalize)$\rfloor_T$ where Diagonalize ≡ (s)→$\lceil \lfloor s \rfloor_T \vdash_T \lfloor v \rfloor_T \rceil_T$
  1) Curry$_v$ ⇔$_T$ (Curry$_v$ $\vdash_T \lfloor v \rfloor_T$) and for any sentence v, it is possible to infer $\lfloor v \rfloor_T$ as follows:
  2) $\vdash_T$(Curry$_v$ $\vdash_T$Curry$_v$)   ① *idempotency of inference*
  3) $\vdash_T$(Curry$_v$ $\vdash_T$(Curry$_v$ $\vdash_T \lfloor v \rfloor_T$))   ① *substituting* 1) *into* 2)
  4) $\vdash_T$(Curry$_v$ $\vdash_T \lfloor v \rfloor_T$)   ① *contraction*
  5) $\vdash_T$Curry$_v$ ① *substituting* 1) *into* 4)
  6) $\vdash_T \lfloor v \rfloor_T$ ① *chaining* 4) *and* 5)
- *Löb's Paradox* [Löb 1955]:
  Löb$_v$ ≡ $\lfloor$Fix(Diagonalize)$\rfloor_T$ where Diagonalize ≡ (s)→$\lceil (\vdash_T \lfloor s \rfloor_T) \vdash_T \lfloor v \rfloor_T \rceil_T$
  1) Löb$_v$ ⇔$_T$ (($\vdash_T$ Löb$_v$) $\vdash_T \lfloor v \rfloor_T$) and for any sentence v, it is possible to infer $\lfloor v \rfloor_T$ as follows:
  2) $\vdash_T$(($\vdash_T$ Löb$_v$) $\vdash_T$ Löb$_v$)   ① *a proposition holds when it is inferred*
  3) $\vdash_T$(($\vdash_T$ Löb$_v$) $\vdash_T$(($\vdash_T$ Löb$_v$) $\vdash_T \lfloor v \rfloor_T$))   ① *substituting* 1) *into* 2)
  4) $\vdash_T$(($\vdash_T$ Löb$_v$) $\vdash_T \lfloor v \rfloor_T$)   ① *contraction*
  5) $\vdash_T$ Löb$_v$ ① *substituting* 1) *into* 4)
  6) $\vdash_T \lfloor v \rfloor_T$ ① *chaining* 4) *and* 5)

This is why Direct Logic[33] does not support self-referential propositions. Although they share some similar underlying ideas, the method of proving inferential undecidability developed by Church and Turing is much more robust than the one previously developed by Gödel that relies on self-referential propositions. The difference can be explicated as follows:

- Actors: an Actor that has an address for itself can be used to generate infinite computations.
- Sentences: a sentence that has a reference to itself can be used to infer inconsistencies.

Appendix 2. Inconsistency-robust Natural Deduction

Below are schemas for nested-box-style Natural Deduction[23] for Direct Logic of a theory T:[aa] See the section on Inconsistency-robust Logic Programming for how this can be implemented.

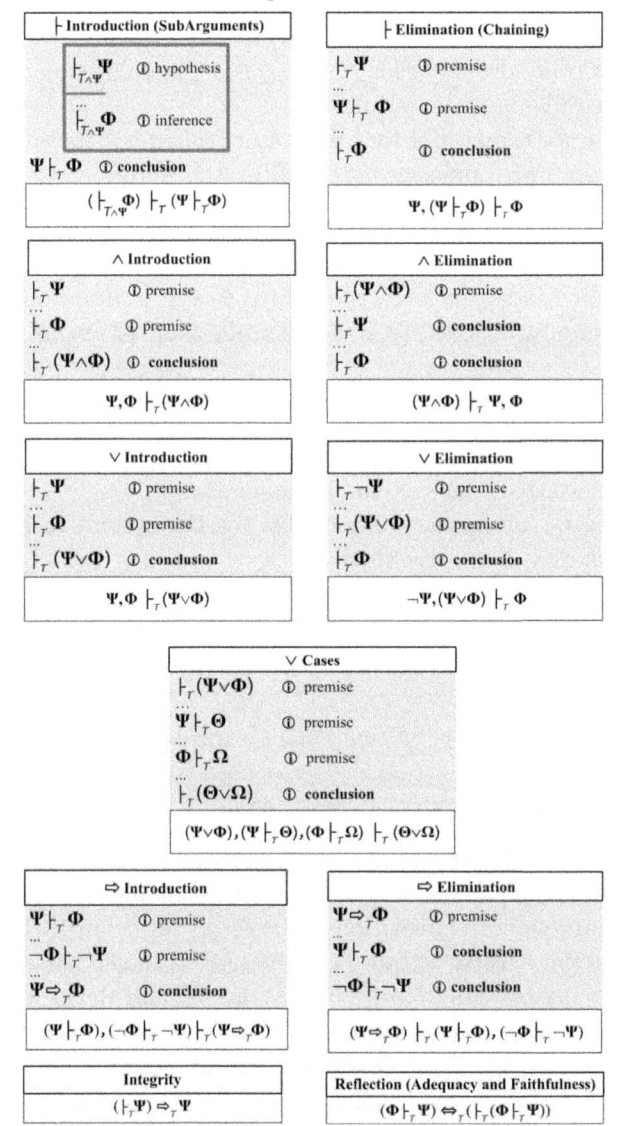

⊢ Introduction (SubArguments)

$\vdash_{T \wedge \Psi} \Psi$	① hypothesis
$\vdash_{T \wedge \Psi} \Phi$	① inference
$\Psi \vdash_T \Phi$	① conclusion

$(\vdash_{T \wedge \Psi} \Phi) \vdash_T (\Psi \vdash_T \Phi)$

⊢ Elimination (Chaining)

$\vdash_T \Psi$	① premise
$\Psi \vdash_T \Phi$	① premise
$\vdash_T \Phi$	① conclusion

$\Psi, (\Psi \vdash_T \Phi) \vdash_T \Phi$

∧ Introduction

$\vdash_T \Psi$	① premise
$\vdash_T \Phi$	① premise
$\vdash_T (\Psi \wedge \Phi)$	① conclusion

$\Psi, \Phi \vdash_T (\Psi \wedge \Phi)$

∧ Elimination

$\vdash_T (\Psi \wedge \Phi)$	① premise
$\vdash_T \Psi$	① conclusion
$\vdash_T \Phi$	① conclusion

$(\Psi \wedge \Phi) \vdash_T \Psi, \Phi$

∨ Introduction

$\vdash_T \Psi$	① premise
$\vdash_T \Phi$	① premise
$\vdash_T (\Psi \vee \Phi)$	① conclusion

$\Psi, \Phi \vdash_T (\Psi \vee \Phi)$

∨ Elimination

$\vdash_T \neg \Psi$	① premise
$\vdash_T (\Psi \vee \Phi)$	① premise
$\vdash_T \Phi$	① conclusion

$\neg \Psi, (\Psi \vee \Phi) \vdash_T \Phi$

∨ Cases

$\vdash_T (\Psi \vee \Phi)$	① premise
$\Psi \vdash_T \Theta$	① premise
$\Phi \vdash_T \Omega$	① premise
$\vdash_T (\Theta \vee \Omega)$	① conclusion

$(\Psi \vee \Phi), (\Psi \vdash_T \Theta), (\Phi \vdash_T \Omega) \vdash_T (\Theta \vee \Omega)$

⇒ Introduction

$\Psi \vdash_T \Phi$	① premise
$\neg \Phi \vdash_T \neg \Psi$	① premise
$\Psi \Rightarrow_T \Phi$	① conclusion

$(\Psi \vdash_T \Phi), (\neg \Phi \vdash_T \neg \Psi) \vdash_T (\Psi \Rightarrow_T \Phi)$

⇒ Elimination

$\Psi \Rightarrow_T \Phi$	① premise
$\Psi \vdash_T \Phi$	① conclusion
$\neg \Phi \vdash_T \neg \Psi$	① conclusion

$(\Psi \Rightarrow_T \Phi) \vdash_T (\Psi \vdash_T \Phi), (\neg \Phi \vdash_T \neg \Psi)$

Integrity

$(\vdash_T \Psi) \Rightarrow_T \Psi$

Reflection (Adequacy and Faithfulness)

$(\Phi \vdash_T \Psi) \Leftrightarrow_T (\vdash_T (\Phi \vdash_T \Psi))$

[aa] In addition to the usual Boolean equivalences.

End Notes

[i]See Ref. 26 and 55,56 for further development of Turing's model of computation.

[ii]To first approximation in the Actor Model, we have the following:

- a partial order (called the "Activation Order") on events that activate other events (This ordering is a generalization of the one in the parallel lambda calculus.)
- a separate total order for each Actor (called its "Reception Order") on events with messages received by the Actor

These orderings are not part of the parallel lambda calculus. The axioms for Actor Systems[2,33] stated required relationships among these orderings. Since the Actor Model, rejected the Global State Assumption of previous models of computations, there is no identification of events with global states.

[iii]A nondeterministic system is defined to have unbounded nondeterminism exactly when both of the following hold:

1. When started, the system always halts.
2. For every integer n, it is possible for the system to halt with an output that is greater than n.

For example the following systems do *not* have unbounded nondeterminism:

- A nondeterministic system which sometimes halts and sometimes doesn't
- A nondeterministic system that always halts with an output less than 100,000.
- An operating system that never halts

[iv]This result is very old. It was known by Dijkstra motivating his belief that it is impossible to implement unbounded nondeterminism. Also the result played a crucial role in the invention of the Actor Model in 1972. The proof also applies to the Abstract State Machine (ASM) model [Blass, Gurevich, Rosenzweig, and Rossman 2007a, 2007b; Glausch and Reisig 2006].

[v]Consequently, it does not implement unbounded nondeterminism. Some people have argued that it is "unfair" for the nondeterministic Turing machine to always make the first choice in Step 1. This had led to a confusing body of literature on various kinds on "unfairness." For example, if a nondeterministic program makes choices in two different steps in the program is

it "unfair" for them to be correlated?[17] argued against "unfairness" because it cannot be proved by any finite computation.

[vi] An example of the global state model is the Abstract State Machine (ASM) model [Blass, Gurevich, Rosenzweig, and Rossman 2007a, 2007b; Glausch and Reisig 2006].

[vii] Arbiters render meaningless the states in the Abstract State Machine (ASM) model [Blass, Gurevich, Rosenzweig, and Rossman 2007a, 2007b; Glausch and Reisig 2006].

[viii] See Ref. 22 on the work of Dijkstra.

[ix] © David Monniaux.

[x] See Ref. 50.

[xi] Furthermore, a Logic Program cannot implement unbounded nondeterminism where a Logic Program is defined by the criteria that it must logically infer its computational steps. See discussion in [Hewitt ArXiv 0812.4852].

[xii] See Ref. 39.

[xiii] See Ref. 49 on Milner's work.

[xiv] Proof: For $i \in \mathbb{N}$ let ProvablyTotalRecursive$_i$ be an enumeration of the provably total recursive procedures.

$$\text{Diagonal} \equiv \lceil (j) \to \text{ProvablyTotalRecursive}_j(j) + 1 \rceil$$

It follows that $\forall i \in \mathbb{N} \to \vdash \downarrow \lfloor \text{Diagonal} \rfloor (i)$

But $\not\vdash \lfloor \ll \forall i \in \mathbb{N} \to \downarrow \lfloor \text{Diagonal} \rfloor (i) \gg \rfloor$ because Diagonal differs from every provably total recursive.

[xv] Proof: $\vdash \neg \lfloor \ll \forall i \in \mathbb{N} \to \downarrow \lfloor \text{Diagonal} \rfloor (i) \gg \rfloor$ is equivalent to
$\vdash (\exists i \in \mathbb{N} \to \neg \downarrow \lfloor \text{Diagonal} \rfloor (i))$, which is inconsistent with
$\forall i \in \mathbb{N} \to \vdash \downarrow \lfloor \text{Diagonal} \rfloor (i)$

[xvi] Follows from $\forall i \in \mathbb{N} \to \models \downarrow \lfloor \text{Diagonal} \rfloor (i)$

PART 2
Computation in Nature & the Real World

Chapter 10

Reaction Systems: A Natural Computing Approach to the Functioning of Living Cells

Andrzej Ehrenfeucht[1], Jetty Kleijn[2], Maciej Koutny[3] and Grzegorz Rozenberg[2]*

[1] *Department of Computer Science,
University of Colorado at Boulder, USA*
[2] *LIACS, Leiden University, 2300 RA, The Netherlands*
[3] *School of Computing Science, Newcastle University, NE1 7RU, UK*

In this paper we present in a tutorial fashion the framework of reaction systems — a formal approach to investigating processes instigated by the living cell. The main idea behind this approach is that this functioning is determined by interactions between biochemical reactions, and these interactions are based on two mechanisms, facilitation and inhibition.

Keywords: Reaction system, living cell, natural computing, genetic regulatory network, switching circuit, transition system, Petri net.

Introduction

Natural computing (see, e.g.,[14,17]) is a research area that investigates both human-designed computing inspired by nature and computing taking place in nature. The former research strand investigates models and computational techniques inspired by nature, while the latter investigates, in terms of information processing, phenomena taking place in nature.

Examples of the first strand of research include evolutionary computation with paradigms inspired by Darwinian evolution of species, neural computation with paradigms inspired by the functioning of the brain, quantum computation with paradigms inspired by quantum mechanics, and molecular computation with paradigms inspired by molecular biology.

Examples of the second strand of research are investigations into the computational nature of self-assembly, the computational nature of develop-

---
*All correspondence should be addressed to Grzegorz Rozenberg.

mental processes, the computational nature of brain processes, the systems biology approach to bionetworks where cellular processes are investigated in terms of communication and interaction, and the computational nature of biochemical reactions.

This second strand of research underscores the fact that computer science is also the fundamental science of information processing, and as such a basic science for other scientific disciplines such as, e.g., biology.

The acceptance, also by biologists, of this second strand of research as a way of thinking about biology is well-illustrated by the following statement by Richard Dawkins, a world-leading expert on evolutionary biology:[6] "If you want to understand life, don't think about vibrant throbbing gels and oozes, think about information technology."

In this paper we discuss a formal framework for the investigation of the functioning of the living cell. It belongs to the second strand of research, as it views the functioning of the living cell in terms of formal processes resulting from interactions between biochemical reactions taking place in it. We also assume that these interactions are driven by two mechanisms, facilitation and inhibition: the reactions may facilitate or inhibit each other. The central model we consider in this paper, reaction systems, follows this philosophy. It abstracts from various technicalities of biochemical reactions to such extent that it becomes a qualitative rather than a quantitative model. However, it takes into account the basic bioenergetics (flow of energy) of the living cell, and it takes into account that the living cell is an open system and its behaviour is influenced by its environment.

The framework of reaction systems is formed by the central model of reaction systems and its extensions (created by specific needs of various research themes). The research that we review here focuses on understanding processes taking place in reaction systems and their extensions (which in turn formalise processes taking place in living cells).

The paper is organised as follows. In Section 1 we formalise the notion of biochemical reaction and define the effect of a set of reactions on a state of a biochemical system. Then in Section 2 we introduce the main construct of our approach, viz., reaction systems. In Section 3 we demonstrate the use of reaction systems in a biological context by showing how to formalise/implement a genetic regulatory system. Then in Sections 4 and 5 we demonstrate a remarkable flexibility of reaction systems by describing implementations of switching circuits (a fundamental unit/concept of hardware) and of finite transition systems (a fundamental model of computation). In Section 6 we relate reaction systems to Petri nets — the classic framework for modelling concurrent systems. For Sections 4, 5 and 6 we assume some basic familiarity with switching circuits, finite transition systems (finite automata), and Petri nets, respectively. However, we informally recall the relevant basic notions, so that the motivated reader not familiar with these research areas can still follow the main ideas of these sections. In Section 7 we provide an example of an extension of the model of reaction systems defined in order to account for an important biochemical issue (decay of molecules). Then in Section 8 we summarise the contents of this paper and provide a broader perspective by describing a number of research themes that are not covered in this paper (due to the restriction on its size).

The paper is written in a rather informal tutorial/survey style, but the reader can find a more formal treatment in the provided references. We always provide motivation/intuition behind the key notions.

0. Preliminaries

Throughout the paper we use standard mathematical notation. In particular, $\varnothing$ denotes the empty set, $|X|$ the cardinality of a set X, $X \setminus Y$ set difference, $X \cup Y$ set union, $X \cap Y$ set intersection, $X \subseteq Y$ set inclusion, and $\bigcup \mathcal{X}$ the union of a family of sets $\mathcal{X}$.

1. Reactions

In order to formulate a model based on interactions of biochemical reactions we need to provide a formal notion of reaction. A biochemical reaction will take place if all of its reactants are present (in the current state of a biochemical system) and none of its inhibitors is present. When a reaction takes place, it creates its products. Following this intuition, we arrive at the following formal definition (see, e.g.,[4,10]).

Definition 3 *A reaction is a triplet $b = (R, I, P)$ such that R, I, P are finite nonempty sets with $R \cap I = \varnothing$.*

The sets R, I, P are called the *reactant set of* b, the *inhibitor set of* b, and the *product* b, respectively — they are also denoted as R_b, I_b and P_b, respectively. If $R, I, P \subseteq Z$ for a finite set Z, then we say that b is a *reaction in* Z. We use $rac(Z)$ to denote the set of all reactions in Z — note that $rac(Z)$ is finite.

Now that (the structure of) a reaction is defined, we move to define the effect of a set of reactions on a current state of a biochemical system. We do this in two steps: first we define the effect of a single reaction.

Definition 4 *Let Z be a finite set and let $T \subseteq Z$.*

(1) *Let $b \in rac(Z)$. Then b is enabled by T, denoted by $en_b(T)$, if $R_b \subseteq T$ and $I_b \cap T = \emptyset$. The result of b on T, denoted by $res_b(T)$, is defined by $res_b(T) = P_b$ if $en_b(T)$, and $res_b(T) = \emptyset$ otherwise.*

(2) *Let $B \subseteq rac(Z)$. The result of B on T, denoted by $res_B(T)$, is defined by $res_B(T) = \bigcup \{res_b(T) : b \in B\}$.*

Note that, since $res_b(T) = \emptyset$ if b is not enabled by T, $res_B(T)$ can be obtained by considering only the results of all reactions *enabled by T*, i.e., $res_B(T) = \bigcup \{P_b : b \in B \text{ and } en_b(T)\}$.

The above definition has a clear intuition in terms of the biochemistry of the living cell. A finite set T formalises a state of the cell, i.e., the set of biochemical entities currently present in it. Then $en_b(T)$ holds if T separates R_b from I_b, i.e., all entities from R_b are present in T and *none* of the entities from I_b is present in T. The result of a set of reactions B on T is *cumulative*, i.e., it is the union of results of all individual reactions from B.

Thus if state transitions (from a state to its successor state) are determined only by the reactions of the biochemical system (i.e., there is no influence by the environment), then the successor state consists of only the entities produced by the reactions enabled in the current state. This implies that there is *non-permanency*: in the transition from a current state T to its successor state, *an entity from T vanishes unless it is sustained by a reaction*. This assumption/property reflects the basic bioenergetics of the living cell: without the flow/supply of energy the living cell disintegrates, but the use/absorption of energy by the living cell is achieved through biochemical reactions (see, e.g.,[16]).

Although in the formal definition of the result function we require "instant non-permanency" (an entity vanishes within one state transition unless it is again produced by the result function), a more "subtle" approach taking into account the decay time is also considered — see Section 7 of this paper.

Note that if a, b are two reactions from B enabled by T, then $P_a \cup P_b \subseteq res_B(T)$ even if $R_a \cap R_b \neq \emptyset$. This means that we do not have the notion of conflict between reactions if they need to share reactants. This follows from the assumption of the *threshold nature of resources* in our approach: either an entity is available and then there is "enough of it", or it is not available. This in turn reflects the *level of abstraction* we have adopted for the formulation of our basic model: we do not count concentrations of entities/molecules to infer from these which reactions can/will be applied. We operate on a higher level of abstraction: we assume that the cell is running/functioning and we want to understand the processes going on then.

This can be compared to the level of abstraction of Turing machines (and its variants such as finite automata) which is undoubtedly the most successful model for understanding computational processes running on electronic computers.[13] Nothing in the Turing model takes into account the electronic properties of the underlying hardware (such as the flow of electrons, possible glitches, etc.). The Turing model assumes that the underlying electronics/hardware functions well and then it aims at modelling processes implemented by this hardware. Similarly, we assume that the underlying biochemistry of the living cell (stoichiometry, concentrations, etc.) works and then we want to understand the processes carried out in the functioning living cell. Thus, at this stage, we are not interested in the underlying "hardware properties" of the living cell, but rather in the resulting processes.

Consequently, there is no counting in our basic model, and so we deal with a *qualitative* rather than with a quantitative model (thus, technically, we deal with sets rather than, e.g., with multisets). This is a basic difference with the traditional models of concurrent systems in computer science, such as, e.g., Petri nets. This non-counting assumption holds for our basic model. However, in our more general framework we also admit models that require counting — see Section 8 of this paper.

We conclude this section by observing that in fact the Turing machine model makes an abstract assumption about the underlying hardware storage. It assumes that if a symbol x is stored in a square of Turing tape, then independently of the number of transition steps needed to return to this square during a computation, x will be still stored there all the time in "perfect form", hence there is no deterioration of stored information. It is an orthogonal assumption to our assumption of non-permanency in reaction systems and it may reflect a difference between human-designed hardware and nature-produced bioware.

2. Reaction Systems

We are ready now to define reaction systems, our abstract model of the functioning of the living cell (see[4,10]).

Definition 5 *A reaction system, abbreviated rs, is an ordered pair* $\mathcal{A} = (S, A)$, *where S is a finite set and* $A \subseteq rac(S)$.

The set S is called the *background set of* $\mathcal{A}$, and its elements are called *entities of* $\mathcal{A}$ — they represent molecular entities (e.g., atoms, ions, molecules) that may be present in the states of the biochemical system (e.g., the living cell). The set A is called the *set of reactions of* $\mathcal{A}$; clearly A is finite (as S is finite).

The subsets of S are called the *states of* $\mathcal{A}$. Given a state $T \subseteq S$, the *result of* $\mathcal{A}$ *on* T, denoted by $res_{\mathcal{A}}(T)$, is defined by $res_{\mathcal{A}}(T) = res_A(T)$.

Thus a reaction system is essentially a set of reactions. We also specify the background set which consists of entities needed for defining the reactions and for reasoning about the system (see the definition of an interactive process below). There are no "structures" involved in reaction systems (such as, e.g., a tape of a Turing machine). Finally, note that this is a strictly finite model — its size is restricted by the size of the background set.

Our model of reaction systems formalises the "static structure" of the living cell as the set of all reactions of the cell (together with the set of underlying entities). What we are really interested in are processes instigated by the functioning of the living cell. They are formalised as follows.

Definition 6 *Let* $\mathcal{A} = (S, A)$ *be an rs. An interactive process in* $\mathcal{A}$ *is a pair* $\pi = (\gamma, \delta)$ *of finite sequences such that, for some* $n \geq 1$, $\gamma = C_0, \ldots, C_n$ *and* $\delta = D_0, \ldots, D_n$, *where* $C_0, \ldots, C_n, D_0, \ldots, D_n \subseteq S$, $D_0 = \varnothing$, *and* $D_i = res_{\mathcal{A}}(D_{i-1} \cup C_{i-1})$, *for all* $i \in \{1, \ldots, n\}$.

The sequence γ is the *context sequence of* π, the sequence δ is the *result sequence of* π, and the sequence $\tau = W_0, \ldots, W_n$, where, for all $i \in \{1, \ldots, n\}$, $W_i = C_i \cup D_i$, is the *state sequence of* π, with $W_0 = C_0$ called the *initial state*. Thus the dynamic process formalised by an interactive process π begins in the initial state W_0. The reactions of $\mathcal{A}$ enabled by W_0 produce then the result set D_1, which together with the context set C_1 forms the successor state $W_1 = res_{\mathcal{A}}(W_0) \cup C_1$. This formation of the successor state is iterated, $W_i = res_{\mathcal{A}}(W_{i-1}) \cup C_i$, resulting in the state sequence $\tau = W_0, \ldots, W_n$.

Note that an interactive process π is determined by its context sequence γ (through the result function $res_\mathcal{A}$). The context sequence formalises the fact that the *living cell is an open system* in the sense that it is influenced by its environment (the "rest" of a bigger system).

If, for all $i \in \{1,\ldots,n\}$, $C_i \subseteq D_i$, then we say that π is *context-independent*: whatever C_i adds to the state W_i is already produced by the system (included in the result D_i). If π is context-independent, then (in its analysis) we may as well assume that, for each $1 \leq i \leq n$, $C_i = \varnothing$. Clearly, if π is context-independent, then the initial state $W_0 = C_0$ determines π by the repeated application of $res_\mathcal{A}$.

The set of all state sequences of $\mathcal{A}$ (i.e., the state sequences of all interactive processes of $\mathcal{A}$) is denoted by $STS(\mathcal{A})$, and the set of all context-independent state sequences of $\mathcal{A}$ is denoted by $CISTS(\mathcal{A})$.

3. Biological Example

In this section we demonstrate how to formally model/implement by reaction systems genetic regulatory networks, one of the "central ingredients" of the living cell (see e.g.,[1]).

Since we do not assume that the reader is familiar with basic molecular biology, we provide now a very short, and *extremely simplified* (but sufficient for the purpose of this paper) description of gene expression (production of a protein from a gene).

For the purpose of this paper, a gene g can be considered as a segment of a DNA molecule, and it consists of the promoter field followed by the coding region. The promoter plays the role of a "landing site" for a special molecule called RNA polymerase. If this site is not occupied, then RNA polymerase can land there and then move/slide through the coding region producing its transcript in the form of a molecule called messenger RNA. This messenger RNA will leave the nucleus (where DNA resides), and it will then be processed outside the nucleus, eventually yielding the protein specified by the coding region of g.

If the cell wants to interrupt the production of this protein, then it "sends" an inhibitor molecule which lands on the promoter field. Consequently, RNA polymerase cannot land there and thus the transcription phase of the expression process cannot begin, and the protein determined by g cannot be produced anymore. With this in mind, consider the simple generic regulatory network given in an informal graphical form in Figure 1.

The network consists of three genes x, y, z expressing proteins X, Y, Z,

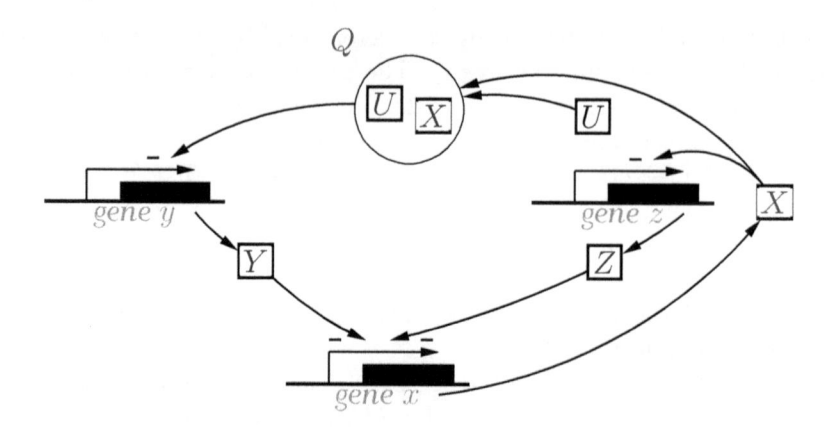

Fig. 1. A genetic regulatory network.

respectively. Moreover protein X interacts with protein U (if it is present in a given state of the network) to form a protein complex Q. There are a lot of interactions going on in the network: protein X inhibits (as explained above) the expression of gene Z, the presence of either of the proteins Y or Z inhibits the expression of gene x, and the protein complex Q inhibits the expression of gene y.

To implement this network by an rs we will need four sets of reactions: A_x, A_y, A_z implementing the expression of genes x, y, z, respectively and A_Q implementing the formation of Q:

$$A_x = \{(\{x\}, I_x, \{x\}), (\{x\}, \{Y, Z\}, \{x'\}), (\{x, x'\}, I_{ex}, \{X\})\}$$
$$A_y = \{(\{y\}, I_y, \{y\}), (\{y\}, \{Q\}, \{y'\}), (\{y, y'\}, I_{ey}, \{Y\})\}$$
$$A_z = \{(\{z\}, I_z, \{z\}), (\{z\}, \{X\}, \{z'\}), (\{z, z'\}, I_{ez}, \{Z\})\}$$
$$A_Q = \{(\{U, X\}, I_Q, \{Q\})\}$$

The set of reactions A_x implements/formalises the functioning of gene g as follows:

- $(\{x\}, I_x, \{x\})$ ensures that if x is available/functional in the current state, then it is also available in the successor state unless "something bad" happens to x as expressed by I_x (we did not specify I_x as it is irrelevant for our considerations here, but "something bad" may be e.g., a high level of radiation — discrete levels of radiation are easily specifiable by I_x).
- $(\{x\}, \{Y, Z\}, \{x'\})$ formalises the role of the promoter: if x is available/functional in the current state and proteins Y, Z are not

present in this state, then RNA polymerase x' will land on the promoter of x.
- $(\{x, x'\}, I_{ex}, \{X\})$ formalises the role of the coding region: if x is available/functional and x' sits on the promoter in the current state, then, unless inhibited by I_{ex}, X will be expressed and hence present in the successor state.

An analogous explanation/intuition holds for the reactions in A_y and A_z. The reaction $(\{U, X\}, I_Q, \{Q\})$ ensures that if U and X are present in the current state, then Q will be present in the successor state.

Now, if we combine all these reactions for G forming $A_G = A_x \cup A_y \cup A_z \cup A_Q$, then the rs $\mathcal{A}_G = (S_G, A_G)$, with S_G consisting of all the entities occurring in reactions from A_G, implements/formalises the structure of G. The reasoning about the functioning of G is formalized through the reasoning about the processes of $\mathcal{A}_G$.

It is important to notice that in fact $\mathcal{A}_G$ is the "union" of the reaction systems $\mathcal{A}_x = (S_x, A_x)$, $\mathcal{A}_y = (S_y, A_y)$, $\mathcal{A}_z = (S_z, A_z)$, and $\mathcal{A}_Q = (S_Q, A_Q)$, where S_x, S_y, S_z, and S_Q are all the entities occurring in reactions from A_x, A_y, A_z, and A_Q, respectively. The operation of union on reaction systems is easily defined (as sets are our basic data structure): for reaction systems $\mathcal{B}_1 = (S_1, B_1)$ and $\mathcal{B}_2 = (S_2, B_2)$, their union is the rs $(S_1 \cup S_2, B_1 \cup B_2)$.

As a matter of fact, the union of reaction systems is the basic mechanism for composing reaction systems. It expresses our assumption about bottom-up combination of local descriptions into a global picture. This combination happens "automatically": the sheer fact that all "ingredients" are present in the same biochemical medium (molecular soup) makes interactions possible. *There is no need for providing additional interfaces here.* This is a fundamental difference with models of computation in computer science.

4. Switching Circuits

In this section we demonstrate how reaction systems can implement (elegantly and efficiently) switching circuits (see, e.g.,[19]).

Rather than to recall a formal definition of binary switching circuits, we give an informal example portrayed in Figure 2(a). A binary switching circuit has inputs represented by small squares (here $H_1, \ldots, H_5$). Also each gate (here $G_1, \ldots, G_5$) has its output and two inputs, each input being *either* one of the inputs (H_j) of the circuit *or* an output of another gate (note

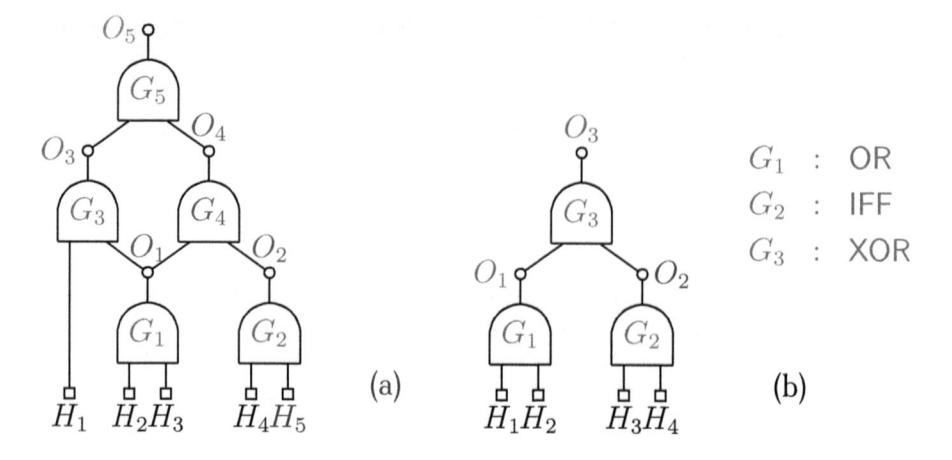

Fig. 2. Binary switching circuits.

that the output of a gate can serve as input to several gates). Moreover, one or more outputs of gates (here O_5) are designated as the outputs of the circuit. We consider the clocked mode of functioning of switching circuits, meaning that there is a global clock, and at each tick of the clock:

- a combination of 0's and 1's is set on the set of inputs (here $H_1, \ldots, H_5$),
- gates are setting either 0 or 1 on their outputs depending on the combination of 0's and 1's on their inputs at the previous tick of the clock.

Which signal (0 or 1) is set on the output of a gate depends on the truth table of this gate, which tells for each combination of inputs whether 0 or 1 will be set on the output. Thus, e.g., for the Exclusive OR gate (XOR), as well as for the OR gate and the IFF gate, the truth tables look as follows:

XOR

x_1	x_2	y
0	0	0
0	1	1
1	0	1
1	1	0

OR

x_1	x_2	y
0	0	0
0	1	1
1	0	1
1	1	1

IFF

x_1	x_2	y
0	0	1
0	1	0
1	0	0
1	1	1

To implement a switching circuit by a reaction system, we implement each gate it comprises by implementing its truth table. This is done by providing reactions that produce the output (y) whenever, for a given combination of 0's and 1's on the inputs, the value of the output in the truth table is 1.

Thus we have one reaction for each row of the truth table yielding 1 on the output, and if a row of a truth table yields 0 on the output, then, for the corresponding combination of inputs, the output y will not be produced simply because there is no reaction for producing it. Following this strategy each of the three truth tables given above is implemented by reactions as follows:

XOR $\rightsquigarrow (\{x_2\}, \{x_1\}, \{y\})$ and $(\{x_1\}, \{x_2\}, \{y\})$.
OR $\rightsquigarrow (\{x_2\}, \{x_1\}, \{y\}), (\{x_1, x_2\}, \varnothing, \{y\})$ and $(\{x_1\}, \{x_2\}, \{y\})$
IFF $\rightsquigarrow (\varnothing, \{x_1, x_2\}, \{y\})$ and $(\{x_1, x_2\}, \varnothing, \{y\})$

Note that for the rows of the truth table where either both of the inputs equal 0 or both of the inputs equal 1, we get either empty reactant set or empty inhibitor set. Since this is formally not allowed, we introduce two "dummy symbols", d_R and d_I, and modify the reaction $(\varnothing, \{x_1, x_2\}, \{y\})$ to $(\{d_R\}, \{x_1, x_2\}, \{y\})$, and the reaction $(\{x_1, x_2\}, \varnothing, \{y\})$ to $(\{x_1, x_2\}, \{d_I\}, \{y\})$. Then we will consider processes in which each state contains d_R and does not contain d_I, so that these modified reactions are always correctly enabled.

Hence for the binary switching circuit $\mathcal{G}$ shown in Figure 2(b), we get:

$$G_1 \rightsquigarrow A_{G_1} = \{(\{H_2\}, \{H_1\}, \{O_1\}), (\{H_1\}, \{H_2\}, \{O_1\}),$$
$$(\{H_1, H_2\}, \{d_I\}, \{O_1\})\},$$
$$G_2 \rightsquigarrow A_{G_2} = \{(\{d_R\}, \{H_3, H_4\}, \{O_2\}), (\{H_3, H_4\}, \{d_I\}, \{O_2\})\},$$
$$G_3 \rightsquigarrow A_{G_3} = \{(\{O_2\}, \{O_1\}, \{O_3\}), (\{O_1\}, \{O_2\}, \{O_3\})\},$$

altogether yielding the set of reactions $A_G = A_{G_1} \cup A_{G_2} \cup A_{G_3}$. We then consider processes of the resulting reaction system $\mathcal{A}_G = (S_G, A_G)$, where $S_G = \{H_1, H_2, H_3, H_4, O_1, O_2, O_3, d_I, d_R\}$.

The initial state $C_0 = W_0$ indicates which inputs are activated (have signal 1 on them) and (as explained above) it also contains d_R. Then each consecutive context C_i tells how inputs are activated at the $(i+1)$-st tick of the clock. For example, here is a 2-step process in $\mathcal{A}_\mathcal{G}$ which implements the clocked run of $\mathcal{G}$ in which the initial activation of inputs H_1, H_3 followed by the activation of H_1 results in signal 1 appearing on the output O_3:

$$\begin{array}{llll} C: & H_1, H_3, d_R & H_1, d_R & H_3, d_R \\ D: & \varnothing & O_1 & O_1, O_2, O_3 \\ W: & H_1, H_3, d_R & H_1, O_1, d_R & H_3, O_1, O_2, O_3, d_R \end{array}$$

In this informal 3-row representation of a process, the first row represents the context sets and it is labelled by "C", the second row represents result sets and it is labelled by "D", and the third row represents states and it is labelled by "W". In order to keep the representation simple we omit set parentheses from the representation of sets.

In general, for each binary switching circuit $\mathcal{G}$ with gates G and connectors (inputs of the circuit and outputs of all gates) C, there is a reaction system $\mathcal{A}_\mathcal{G} = (S, A)$ implementing $\mathcal{G}$ where:

- The background set is $S = \mathsf{C} \cup \{d_R, d_I\}$, and so $|S| = |\mathsf{C}| + 2$.
- For each gate G, A_G is the set of reactions implementing G.
- A is the set of all reactions for all gates of $\mathcal{G}$, i.e., $A = \bigcup \{A_G : G \in \mathcal{G}\}$.

As $|A| \leq 4 \cdot |\mathsf{G}|$, the implementation in terms of reaction systems is linear in the number of gates (if we assume, as commonly done, that in the truth table of each gate at least one row yields 0, then the coefficient 4 may be replaced by 3).

The above construction is easily generalised to k-ary switching circuits, where now each gate has $k \geq 2$ inputs. Again, A is linear in the number of gates: $|A| \leq 2^k \cdot |\mathsf{G}|$ (also here 2^k may be replaced by 2^{k-1} under the assumption as above).

5. Finite Transition Systems

In this section we relate reaction systems to the classical model of sequential computation, viz., finite transition systems (see, e.g.,[2,13]). In particular, we will demonstrate how transition system behaviour can be implemented by reaction systems.

We briefly recall that a *deterministic transition system* is a triplet $F = (Q, \Sigma, \delta)$, where Q is a nonempty finite set of *states*, Σ is a finite set of characters (the *input alphabet*) and $\delta : Q \times \Sigma \to Q$ is a *transition function*. Then, the *behaviour* of F is given by finite transition sequences of the form $q_0 \xrightarrow{x_1} q_1 \xrightarrow{x_2} q_2 \xrightarrow{x_3} \cdots \xrightarrow{x_n} q_n$, for some $n \geq 0$, such that $\delta(q_i, x_{i+1}) = q_{i+1}$, for each $i \in \{0, 1, \ldots, n-1\}$.

For the explanation of the implementation of F by a reaction system it is convenient to assume that $Q \cap \Sigma = \varnothing$ and $|Q \cup \Sigma| > 2$.

The aim of the implementation is to construct a reaction system $\mathcal{A}_F = (S_F, A_F)$ such that $q_0 \xrightarrow{x_1} q_1 \xrightarrow{x_2} q_2 \xrightarrow{x_3} \cdots \xrightarrow{x_n} q_n$ is a behaviour of F iff

$C:$	x_1	x_2	x_3		x_n	$\emptyset$
$D:$	q_0	q_1	q_2	$\ldots$	q_{n-1}	q_n
$W:$	q_0, x_1	q_1, x_2	q_2, x_3		q_{n-1}, x_n	

is an interactive process of the reaction system $\mathcal{A}_F$, i.e., $res_{\mathcal{A}_F}(\{q_i, x_{i+1}\}) = q_{i+1}$, for each $i \in \{0, 1, \ldots, n-1\}$. Note that here $D_0 = \{q_0\}$, while the formal definition of an interactive process requires $D_0 = \emptyset$. This is done for the purpose of explanation; to get $D_0 = \emptyset$ one can set $C_0 = \{q_0, x_1\}$ and $D_0 = \emptyset$.

Let, for all states $p, q \in Q$ and characters $x \in \Sigma$, $a_{p,q,x}$ be the reaction defined by $(\{p, x\}, S_F \setminus \{p, x\}, \{q\})$. Then $\mathcal{A}_F = (S_F, A_F)$, where $S_F = Q \cup \Sigma$ and $A_F = \{a_{p,x,q} : \delta(p, x) = q\}$. Since we require that in a reaction system for each reaction a, $I_a \neq \emptyset$, we have assumed that $|Q \cup \Sigma| > 2$ (so $S_F \setminus \{p, x\} \neq \emptyset$ as required).

The following is a deterministic transition system F (given by the graph of δ) and the list of the reactions of $\mathcal{A}_F$ (note that $S_F = \{q_0, q_1, q_2, x, y\}$):

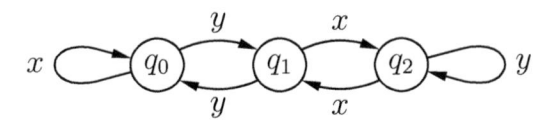

$$A_F = \left\{ \begin{array}{l} (\{q_0, x\}, \{q_1, q_2, y\}, \{q_0\}) \; (\{q_0, y\}, \{q_1, q_2, x\}, \{q_1\}) \\ (\{q_1, x\}, \{q_0, q_2, y\}, \{q_2\}) \; (\{q_1, y\}, \{q_0, q_2, x\}, \{q_0\}) \\ (\{q_2, x\}, \{q_0, q_1, y\}, \{q_1\}) \; (\{q_2, y\}, \{q_0, q_1, x\}, \{q_2\}) \end{array} \right\}$$

Then, e.g., the transition sequence $q_1 \xrightarrow{x} q_2 \xrightarrow{y} q_2 \xrightarrow{y} q_2 \xrightarrow{x} q_1 \xrightarrow{y} q_0$ in F corresponds to the following interactive process in $\mathcal{A}_F$:

$C:$	x	y	y	x	y	
$D:$	q_1	q_2	q_2	q_2	q_1	q_0
$W:$	q_1, x	q_2, y	q_2, y	q_2, x	q_1, y	

The implementation of non-deterministic finite transition systems provides an instructive insight into the role of context in interactive processes — it is done as follows. Assume that in our example transition system F the transition from q_0 on y is non-deterministic: $\delta(q_0, y) = \{q_0, q_1\}$. We mark these two transitions by symbols "1" and "2", and accordingly have two reactions: $(\{q_0, y, 1\}, \{q_1, q_2, x, 2\}, \{q_0\})$ and $(\{q_0, y, 2\}, \{q_1, q_2, x, 1\}, \{q_1\})$. Then the implementing reaction system will implement the transition from q_0 by y to q_0 if the context of the current state contains the symbol 1, and it will implement the transition from q_0 by y to q_1 if the context contains the symbol 2.

Context in interactive processes can be also used to implement stochasticity.

6. Petri Nets

In this section we relate reaction systems to Petri nets — a well-established framework for the modelling of concurrency in a wide range of applications.[7]

Recall that the underlying structure of any Petri net model is a net, a bipartite directed graph, consisting of places (circles in diagrams), transitions (rectangles in diagrams) and directed arcs connecting places to transitions and transitions to places. The idea is that places are local states to indicate, e.g., the availability of a certain resource, and transitions are the active elements of the system, e.g., producers and consumers of this resource. In order to occur, a transition should be enabled. Whether or not a transition is enabled at a given state depends on the information provided by its neighbouring (input and output) places. Also, if an enabled transition occurs, then the resulting state change only concerns its neighbouring places. Many extensions have been defined to this basic model. A prominent example is inhibitor arcs, which provide a possibility to test for the absence of resources.

By drawing reactions as rectangles, entities as circles, and then (1) connecting circles by "positive" directed edges to rectangles when they belong to reactant sets and by "negative" directed edges when they belong to inhibitor sets, and (2) connecting rectangles by "positive" directed edges to circles if they belong to their product sets, we obtain a graph representing a Petri net with inhibitors. However this resemblance is very superficial, as the way to generate processes of reaction systems from this graph and the way to generate processes of Petri nets with inhibitors from this graph are dramatically different. In order to overcome this discrepancy and hence to faithfully implement reaction systems behaviour in Petri nets, set-nets with inhibitors were introduced in.[15]

Definition 7 *A set-net with inhibitor arcs is a tuple* $N = (Pl, Tr, Flw, Inh)$ *such that Pl and Tr are finite, disjoint sets of respectively places and transitions, and:* $Flw \subseteq (Pl \times Tr) \cup (Tr \times Pl)$, $Inh \subseteq Pl \times Tr$ *are respectively the sets of flow and inhibitor arcs.*

A state of a set-net, referred to as a *marking*, is a subset of places indicating the presence (without quantification) of entities. Places belonging to a marking are marked with a small black dot. A set of transitions U, also called a *step*, is enabled at a marking M if (1) all input places to the transitions of U belong to M and (2) no place connected by an inhibitor arc to a transition of U is in M. Moreover, we require that U is a *maximal* set with this property. An enabled step can be executed leading to a successor marking M' which is $(M \setminus X) \cup Y$, where X are the input places of the transitions in U, and Y are the output places of the transitions in U.

The *behaviour* of a set-net N is given by step sequences of the form $M_0 \xrightarrow{U_1} M_1 \xrightarrow{U_2} M_2 \xrightarrow{U_3} \ldots \xrightarrow{U_n} M_n$, for some $n \geq 0$, such that U_{i+1} is a step enabled at marking M_i and M_{i+1} is the successor marking of M_i by U_i.

To implement the behaviour of a context-independent reaction system $\mathcal{A} = (S, A)$, we construct a set-net $N_\mathcal{A}$ with each entity $p \in S$ represented by a place also called p, and each reaction $a \in A$ represented by a transition also called a. We have a flow arc (p, a) whenever $p \in R_a$, a flow arc (a, p) whenever $p \in P_a$, and an inhibitor arc (a line with a small circle as its arrowhead) from p to a if $p \in I_a$. In addition, to implement the non-permanency of entities in reaction systems, for each entity p there is a special transition $p_\downarrow$ with p as its only input place.

$$\mathcal{A} = (S, \{a, b\})$$
$$S = \{x, z\}$$
$$a = (\{z\}, \{x\}, \{x\})$$
$$b = (\{x\}, \{z\}, \{z\})$$

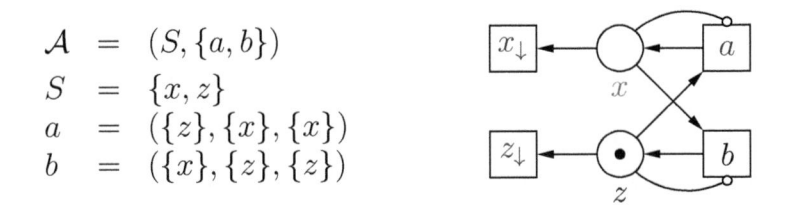

Fig. 3. Reaction system $\mathcal{A}$ and corresponding set-net with marking $\{z\}$.

Consider the following computation of the set-net of Figure 3 $\{z\} \xrightarrow{\{a\}} \{x\} \xrightarrow{\{b\}} \{z\}$. Then the sequence of generated markings $\{z\}\{x\}\{z\}$ belongs to $CISTS(\mathcal{A})$. This is not a mere coincidence as one can show[15] that $CISTS(\mathcal{A})$ are precisely all the sequences of markings appearing in step sequences of the implementing $\mathcal{A}$. This illustrates how Petri nets can indeed be used as a *faithful* semantical model for reaction systems. As a consequence, Petri net based concepts could be deployed to analyse reaction systems, to investigate causality and concurrency of reactions, and for synthesis purposes where the issue is to construct a reaction system from a description of its behaviour.

In this way the computational intuition originating from reaction systems has provided the inspiration to introduce a novel and challenging class of nets.

7. Reaction Systems with Durations

A notable feature of reaction systems is *non-permanency* of entities, meaning that an entity from a current state will not be present in the successor state (for a given interactive process) unless it is either produced by one of the reactions enabled in the current state or it is introduced by the context of the successor state. This "vanish" is instantaneous in the sense that the vanishing entity does not survive even one transition step. However, it is well known from biochemistry (see, e.g.,[1]) that the vanishing of entities in a biochemical environment (e.g., macromolecules in the living cell) requires some time to be realised (the decay time). To take this into account, reaction systems with duration were introduced in.[3]

In a reaction system with duration, the duration function d assigns to each entity x its duration/decay time $d(x)$, which means that when x is introduced in a state W of an interactive process π, then it will survive through $d(x) - 1$ transitions of π, hence altogether its lifetime will equal $d(x)$, i.e., it will live through $d(x)$ consecutive states. More specifically, if entity x is produced in a state W_{i+1} (i.e., $x \in res_{\mathcal{A}}(W_i)$) of an interactive process with the state sequence $W_0, \ldots, W_{i+1}, \ldots, W_n$, then x will live (will be present) at least in states $W_{i+1}, \ldots, W_{i+d(x)}$, assuming that $n \geq i+d(x)$. In fact x can also live beyond $W_{i+d(x)}$ if either x is introduced by context or x is also produced in one of the states $W_{i+2}, \ldots, W_{i+d(x)}$.

Definition 8 *A reaction system with duration, abbreviated rsd, is a triplet $\mathcal{A} = (S, A, d)$, where (S, A) is an rs, and $d : S \to \mathbb{Z}^+$.*

The rs (S, A) is referred to as the *underlying reaction system of $\mathcal{A}$*, denoted by $und(\mathcal{A})$, and the function d is referred to as the *duration function of $\mathcal{A}$*; for $x \in S$, $d(x)$ is the *duration of x*.

The notion of an interactive process has to be extended/modified in order to account also for entities present in a current state even though they are produced in some previous states but persist by duration/decay.

Definition 9 *Let $\mathcal{A} = (S, A, d)$ be an rsd. An interactive process in $\mathcal{A}$ is a triplet $\pi = (\gamma, \delta, \rho)$ of finite sequences such that, for some $n \geq 1$,*

$\gamma = C_0, \ldots, C_n$, $\delta = D_0, \ldots, D_n$, and $\rho = G_0, \ldots, G_n$, where $D_0 = \varnothing$, $G_0 = \varnothing$, $G_1 = \varnothing$, and, for all $i \in \{1, \ldots, n\}$,

- $C_i, D_i, G_i \subseteq S$,
- $G_i = \{x \in S : d(x) \geq 2 \text{ and } x \in D_j \text{ for some } i - d(x) < j < i\}$,
- $D_i = res_{\mathcal{A}}(D_{i-1} \cup C_{i-1} \cup G_{i-1})$.

The usual terminology and notation carry over from ordinary reaction systems to reaction systems with duration. The sequence ρ is the *duration sequence of* π, and each G_i consists of entities that persist after being produced in some previous states. Then the sequence $\tau = W_0, \ldots, W_n$, where, for all $i \in \{1, \ldots, n\}$, $W_i = C_i \cup D_i \cup G_i$, is the *state sequence of* π. Thus now each state W_i depends not only on $res_{\mathcal{A}}(W_{i-1})$ and the context C_i but also on the entities from G_i. Clearly, if $\mathcal{A}$ is such that $d(x) = 1$ for all $x \in S$, then $\mathcal{A}$ corresponds to an rs (without duration) — it is then essentially $und(A)$.

The behaviour of reaction systems with duration is very different from the behaviour of (ordinary) reaction systems. For example, the properties of context-independent state sequences of reaction systems with duration are remarkably different from the properties of context-independent state sequences of ordinary reaction systems.

The duration function of an rsd is "brought from outside": it is added to an rs as an additional component, on top of the rs. It turns out that the duration function may be explained (and implemented) within the framework of ordinary reaction systems. In fact the phenomenon of duration/decay can be explained as an interaction between a reaction system and its environment, where the term "environment" is meant here as a "structured" context; in fact the context becomes here a bigger reaction system encompassing a given reaction system.

Thus, given an rsd $\mathcal{A} = (S, A, d)$ one can embed the underlying reaction system $und(\mathcal{A})$ in an rs (without duration) $\mathcal{A}' = (S', A')$, where embedding means that $S \subseteq S'$ and $A \subseteq A'$, so that the set of context-independent state sequences of $\mathcal{A}$ can be obtained from the context-independent state sequences of the bigger system $\mathcal{A}'$. More precisely, it can be shown that: for every rsd $\mathcal{A} = (S, A, d)$ there exists an rs $\mathcal{A}' = (S', A')$ such that $und(\mathcal{A})$ is embedded in $\mathcal{A}'$ and $proj_S(CISTS(\mathcal{A}')) = CISTS(\mathcal{A})$ (recall that for a sequence of sets $\tau = Z_1, \ldots, Z_n$ and a set Q, the projection of τ onto Q is the sequence of sets $proj_Q(\tau) = Z_1 \cap Q, \ldots, Z_n \cap Q$).

The construction of $\mathcal{A}'$ depends on d, but $\mathcal{A}'$ is a reaction system that (for $und(\mathcal{A})$, which is embedded in it) yields the same result as the addition of an explicit numerical duration function. In this way, the phrase "embedding an rs in an environment" becomes a formal construct, and indeed the duration/decay becomes the result of an interaction of a reaction system with its environment. Indeed, from the point of view of biochemistry such a result may be quite intuitive, as, e.g., the decay of molecules may be much faster in high temperatures than it is in low temperatures.

8. Discussion

In this paper we have discussed the formal framework of reaction systems. The original motivation behind this framework was to understand (on an abstract level) processes instigated by the functioning of the living cell. The basic idea behind this framework is that this functioning is based on interactions between biochemical reactions, and these interactions are controlled by two mechanisms, facilitation and inhibition.

The basic construct of this framework is reactions systems. It is an abstract model — it abstracts from many details of biochemical reactions, and so it is a qualitative rather than a quantitative model, which means that there is no counting in reaction systems. It is a strictly finite model (its size is determined by its background set). This model takes into account the basic bioenergetics of the living cell (through non-permanency of its entities) and it also takes into account the fact that the living cell is an open system in the sense that it is influenced by its environment (through the context sequence component of an interactive process). The core of the model (expressed by the result function) is deterministic, but features of processes such as non-determinism and stochasticity can be naturally accommodated by the contexts of interactive processes.

The framework of reaction systems is obtained by adding to the basic construct of reaction systems additional components dictated by specifics of a research theme under investigation (this is referred to as the "onion approach" — adding additional layers whenever convenient/required). An example of such an extension of the basic model are reaction systems with durations.

Research topics of this framework are motivated either by biological considerations or the need to understand the underlying computations. This framework represents a fundamental approach to understanding the functioning of the living cell in the sense that it is geared towards answering two fundamental questions: "What happens?" and "Why?"

Although initially motivated by biological considerations, by now (in our opinion) reaction systems have also become a novel and attractive model of computation. In order to provide a better perspective on the research concerning reaction systems, we will now briefly discuss a number of research topics that could not be included in this paper due to page limitations.

One of the main lines of research concerns the understanding of the core of the reaction systems, viz., their result functions ($res_\mathcal{A}$) — this corresponds to the investigation of context-independent processes, hence to the investigation of reaction systems as closed systems (without the influence of the environment). Here one considers reaction systems as specifications of functions on finite power sets ($res_\mathcal{A} : 2^S \to 2^S$), and then tries to prove properties that are specific/characteristic for this class of functions (see, e.g.,[8,9]).

To understand how entities of a reaction system influence each other is important from both biological and computational points of view. Such causalities may be of a *static* nature (deducible *directly* from the set of reactions) or of a *dynamic* nature (deducible from the set of interactive processes) — both kinds of causalities are investigated in.[5]

Formation of modules is an important research area in biology and biochemistry (see[18]). The formal notion of a module and its formation (by dynamic events) is discussed in,[11] where it is proved that when interactive processes stabilise, then modules form specific structures (lattices). It is also shown there that reaction systems can be viewed as self-organising systems.

Although reaction systems are a qualitative rather than a quantitative model, there are many situations where one needs to assign quantitative parameters to states. To account for this one considers reaction systems with measurements, where *measurement functions* (added to a reaction system) assign numerical values to states (see[12]). One of the research themes utilising measurement functions is the issue of time in reaction systems (hence in living cells) — it is investigated in depth in,[8] where among other results it is shown how to define reaction times, times of formation of components, time distances between states, etc.

Acknowledgements

The authors are indebted to R.Brijder and M.G.Main for useful comments. This research was supported by the Pascal Chair award from the Leiden Institute of Advanced Computer Science (LIACS) of Leiden University.

References

1. B. Alberts, D. Bray, A. Johnson, J. Lewis, M. Raff, K. Roberts and P. Walter, Essential Cell Biology. Garland Publishing, Inc. (1998).
2. A. Arnold, Finite Transition Systems: Semantics of Communicating Systems. Prentice Hall (1994).
3. R. Brijder, A. Ehrenfeucht and G.Rozenberg, Reaction Systems with Duration. LNCS **6610** (2011) 191–202.
4. R. Brijder, A. Ehrenfeucht, M. G. Main and G. Rozenberg, A Tour of Reaction Systems. Int. Journal of Foundations of Computer Science (2011).
5. R. Brijder, A. Ehrenfeucht and G. Rozenberg, A Note on Causalities in Reaction Systems. Electronic Communications of EASST **30** (2010).
6. R. Dawkins, The Blind Watchmaker. Penguin, Harmondsworth (1986).
7. J. Desel, W. Reisig and G. Rozenberg (Eds.), Lectures on Concurrency and Petri Nets. Lecture Notes in Computer Science **3098**, Springer (2004).
8. A. Ehrenfeucht, M. Main and G. Rozenberg, Combinatorics of Life and Death for Reaction Systems. Int. Journal of Foundations of Computer Science **21** (2010) 345–356.
9. A.Ehrenfeucht, M. Main and G. Rozenberg, Functions Defined by Reaction Systems. Int. Journal of Foundations of Computer Science **22** (2011) 167–178.
10. A. Ehrenfeucht and G. Rozenberg, Reaction Systems. Fundamenta Informaticae **75** (2007) 263–280.
11. A. Ehrenfeucht and G. Rozenberg, Events and Modules in Reaction Systems. Theoretical Computer Science **376** (2007) 3–16.
12. A. Ehrenfeucht and G. Rozenberg, Introducing Time in Reaction Systems. Theoretical Computer Science **410** (2009) 310–322.
13. J. E. Hopcroft, R. Motwani and J. D. Ullman, Introduction to Automata Theory, Languages, and Computation. Prentice Hall (2006).
14. L. Kari and G. Rozenberg, The Many Facets of Natural Computing. Communications of ACM **51** (2008) 72–83.
15. J. Kleijn, M. Koutny and G. Rozenberg, Modelling Reaction Systems with Petri Nets. In: Proc. of BioPPN-2011, CEUR-WS **724** (2011) 36–52.
16. A. L. Lehninger, Bioenergetics: The Molecular Basis of Biological Energy Transformations. W. A. Benjamin, Inc., New York, Amsterdam (1965).
17. G. Rozenberg, Th. Bäck and J. Kok (Eds.), Handbook of Natural Computing, Springer Verlag, to appear (2011).
18. G. Schlosser and G. P. Wagner (Eds.), Modularity in Development and Evolution. The University of Chicago Press, Chicago (2004).
19. S. H. Unger, The Essence of Logic Circuits. Wiley-IEEE Press (1996).

Chapter 11

Bacteria, Turing Machines and Hyperbolic Cellular Automata

Maurice Margenstern

Université de Lorraine,
LITA, EA 3097, UFR MIM, and CNRS, LORIA
Campus du Saulcy, 57045, Metz Cédex, France
margens@univ-metz.fr, margenstern@gmail.com

In this paper we look at links between biology and computer science, at how Turing machines may change our look of bacteria, at how colonies of bacteria could be classified and how hyperbolic cellular automata could be used to better understand the behaviour of colonies of bacteria.

Keywords: Turing machines, DNA, bacteria colonies, cellular automata, hyperbolic geometry, tilings.

1. Introduction

In this paper, we shall look at how computer science could help to introduce new revolutions in biology.

Let us start from the following observation. In many treatise about biology, the authors stress the need to start from the very physical laws within which molecules can evolve into combinations which give rise to the wide variety and impetuous development of life. Is this actually relevant? In many cases, you can read in books about biology, that in order to understand the working of a today operating system in a computer, you need to understand the basic laws of physics.

This latter point is clearly false: at least, it is clear for computer scientists that it is false. Computer scientists well know that an operating system is based on logic, not at all on the laws of physics. The laws of physics are used in order to control a series of electronic phenomena which guarantee the possibility to perform a bounded list of specified actions, as many time as needed. Once these specifications are guaranteed, the organization of the operating system relies on the laws of logic, on the laws of the organization of a complex system. It is perfectly possible to base computations on other phenomena than electronic ones, as the traffic of trains for instance, see.[13] Up to now, electronic phenomena and the level of their control by electronics gives us the fastest and cheapest way to perform extremely fast and complex computations. How to handle the computations rely on the laws of logic, not on those of physics. The better illustration of this fact is given by Turing's foundational paper. There, Alan Turing analyzes how an actual 'computer' performs computations. Now, if any one carefully reads the paper and remembers that it was written in 1936, he/she will clearly understand that the 'computer' considered by Turing is a man. Indeed, Turing discusses the notion of state of his machine by comparison with the *states of mind* of his computer who is clearly a man. And Turing stresses why these states of mind must be discrete. We stop here this discussion, summarizing it by the fact that operating systems work after the laws we know of the conscious mind and not at all according to the laws of physics.

It seems to us that, if the laws of physics are essential to understand which conditions make life possible, the laws of life do not come from these conditions but, rather, by the new possibilities given by free combinations which can appear in these new conditions. It seems to me that biology could benefit of a few approaches of computer science. Why not trying to adapt them to its own problematics?

The goal of this paper is to provide some hints in this regard. Section 2 gives some light on a possible connection of Turing machines with viruses and bacteria. Section 3 will remind a few works of Professor Ben Jacob which inspired[9] and a large part of this paper. I explain there why these works gave me the idea to start from hyperbolic geometry. Sections 4 and 5 give the needed material to understand the simulations presented in Section 6. Section 7 gives a few possibilities for further studies in this line.

2. Turing Machines and Viruses

Let us start this section with the following riddle I put in my presentation of[9] at **CAAA'2011**. The question was: what are the lines of Table 1?

They look like DNA strands or pieces of DNA strands, especially for non specialists of DNA, but they are not a piece of a DNA strand. In fact, these lines are the encoding of the program of a Turing machine in the alphabet

Table 1. The riddle, explained by Tables 2 and 3.

TGTCCACACACACAACTCCACACAACTCCACAAGTCCACAACCTCTTCCAAG
TGTCCACAAGCTCTCTCTTCACTCCACAACTCACTCAC
TGTCACCTCTCTCTCTTCCACACACAAGCTCTCTCTTCCACAACTCACTCAC
TGTCCACACACACAAGCTCTCTTCCACACAAGTCCACAACCTCTTCAGTCAG

of DNA! The encoded Turing machine is deciphered in Table 2 where it is presented in a standard way. It is a universal Turing machine, strongly universal as its initial configuration is finite. This machine is one of the smallest ones up to date and comes from.[12] It was found by Yurii Rogozhin in 1982. We can see that such a machine is very small. Such machines work on the basis of the simulation of another device, called a tag system, which turns out to be also universal. The machine works as an interpreter of tag systems. The initial configuration is an appropriate encoding of the tag system given to the machine. The first encodings found by Marvin Minsky, see,[11] and used by Yurii Rogozhin were doubly exponential. However, as was found by Turlough Neary and Damien Woods in 2006 who improved the results obtained by Yurii Rogozhin, it is possible to devise the encoding in such a way that it is polynomial in the size of the standard representations of tag-systems, see.[14]

Table 2. A small universal Turing machine.

	b	0	1	c	d
1	dR	$1R$	$0L$	$0R2$	bL
2	$0L4$	R	$0R$	R	R
3	$R5$	$cL4$	$0R$	R	R
4	$dL3$	$1L$	$0R2$	L	L
5		R		$1R1$	bR

For the sake of the reader, we indicate the encoding used in Table 1. First, the program of the Turing machine given by Table 2 is represented by a string, each row of the table occupying a corresponding interval on the string: the first row, a beginning of the string and so on, until the last row which corresponds to a suffix of the string. To differentiate these intervals, we use a marker translated by TG in Table 3. Each row contains the instructions in which the head of the Turing machine is under the same state. The instructions themselves need to be delimited, which is the role of TC. In the riddle itself, we can see that each line starts with TG and we can count five occurrences of TC, to be not confused with those coming from the words $(CT)^+$, see Table 1. Inside an instruction, we remark that the place of the instruction corresponds to the letter scanned by the head of the machine. And so we encode the new letter only as well as the motion of the head and the possible new state. When the letter is not changed, we do not encode it. The same principle applies to the new state. For letters and states, they are ordered and so we encode the ith letter by $(CA)^i$ and the ith state by $(CT)^i$. At last, the motions are encoded AG for the motion to left and AC for the motion to right and we may assume that the machine has only motions either to left or to right.

Table 3. The encoding of the machine of Table 2 by the four letters of the DNA alphabet.

for letters:

b	0	1	c	d
CA	CACA	CACACA	CACACACA	CACACACACA

for states: 1 up to 5:

CT	CTCT	CTCTCT	CTCTCTCT	CTCTCTCTCT

directions: L R

AG   AC

markers: instruction   states

TC   TG

Now, looking again at Table 2 and its encoding by the riddle of Table 1, we can remark that it is a very small table. Also, the string of Table 1 is small compared to standard strands of DNA. We can assert that this table is smaller than the smallest virus. This should lead us to a more cautious attitude about viruses and, a fortiori about bacteria which are much bigger than viruses. It seems to me that the today race after the strongest antibiotic substance is hopeless, worse, very dangerous. We need more fundamental studies to better classify bacteria. We should have in mind that the first level reached by such colonies is at least that of the recursively enumerable sets. So that probably, such a study requires the use of the arithmetical hierarchy and its knowledge should be much more propagated to scientists than the present situation were only a few specialists of mathematical logic and of theoretical computer science are aware of what it is.

If the reader is not convinced by this purely theoretical argument, the experiments to which we now turn will perhaps lead him/her to a second thought.

3. Colonies of Bacteria

A very intriguing phenomenon of diffusion is given by the growth of colonies of bacteria, see.[1] As explained by Professor Ben-Jacob, very surprising structures can be obtained by putting such colonies in very severe conditions, see Figures 1 and 12. This reveals a striking power of adaptability of these colonies. These experimental data are comforted by the discovery of bacteria in almost every possible hard conditions as geysers, ocean fathoms, core of the earth and even atomic piles. In the introduction of,[1] Professor Ben-Jacob says:

> Eons before humans, bacteria inhabited a very different Earth. As the earliest life form they devised ways to counter the spontaneous course of increasing entropy and convert high-entropy, inorganic substances into low entropy, organic molecules. They paved the way for other forms of life by changing harsh physical and chemical conditions on the Earth's surface and its atmosphere into the modern life-sustaining environment.
>
> To face changing environmental hazards, bacteria resort to a wide range of cooperative strategies. They alter the spatial organization of the colony in the presence of antibiotics for example. Bacteria form complex patterns as needed to function efficiently...
>
> They collectively glean information from the environment, communicate, distribute tasks, perform distributed information processing and learn from past experience.

I think that these facts confirm my claim about the importance to attack the problems raised by colonies of bacteria from another point of view, both theoretically and practically. In the last section we have indicated a direction for theoretical studies. Here we think in particular to the way to struggle against infections caused by viruses and bacteria. It is important to understand that the present race to stronger and stronger antibiotics is hopeless. We have to find other weapons and perhaps even to change our mind: the word weapon is probably not appropriate. We could find other ways, in particular something along the path of cooperation. Colonies of bacteria are already used for a long time in preparing certain types of food and industry starts to make use of them in several still few cases. This is certainly one of the paths to better study and to better explore.

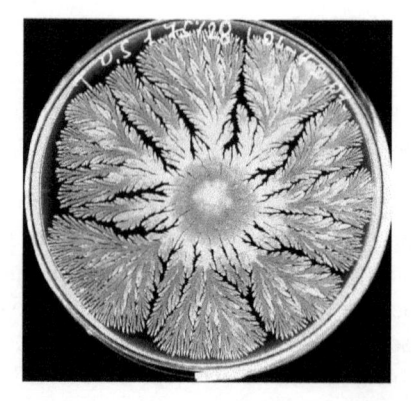

Fig. 1. Examples of growth of bacteria colonies. Pictures by courtesy of Professor Ben-Jacob.

Going back to the above experiments, in many cases, the growth of colonies of bacteria on plates used by microbiologists to study them constitute figures with a more or less fractal symmetry. Cellular automata were already used to model this, see.[2] Now, fractal symmetry may address to hyperbolic geometry. This is why we tried the other way: let us start from hyperbolic geometry, in fact from an appropriate tiling of the hyperbolic plane and try to simulate the observed growth with hyperbolic cellular automata.

In Section 4 we give the needed information for the reader about what to know about hyperbolic geometry in order to understand how our grid is obtained and to see how cellular automata are implemented in this context. In Section 6, we see how to proceed to the simulations indicated in the abstract.

4. A Triangular Tiling of the Hyperbolic Plane

In this section, we provide the reader with the minimal material which allows the paper to be self-contained. Sub-section 4.1 is a very short introduction to hyperbolic geometry and to the Poincaré's disc model which is intensively used in the illustrations of the paper. Then, Sub-section 4.2 indicates which grid we consider for the simulations we propose. Then, in Sub-section 5 we introduce cellular automata adapted to this grid and the discussion we have there will lead to the various examples of simulation dealt with in Section 6.

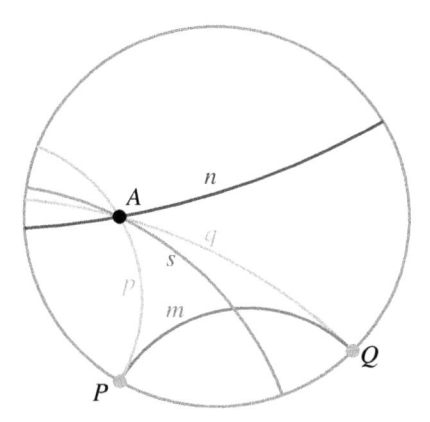

Fig. 2. Poincaré's disc model.

4.1. *Hyperbolic geometry*

Hyperbolic geometry appeared in the first half of the 19th century, proving the independence of the parallel axiom of Euclidean geometry. Models were devised in the second half of the 19th century and we shall use here one of the most popular ones, Poincaré's disc. This model is represented by Figure 2.

Inside the open disc represented in the figure we have the points of the hyperbolic plane. Note that by definition, the points on the border of the disc do not belong to the hyperbolic plane. However, these points play an important role in this geometry and are called **points at infinity**. Lines are trace of diameters or circles orthogonal to the border of the disc, e.g. the line m. In this model, two lines which meet in the open disc are called **secant** and two lines which meet at infinity, *i.e.* at a point at infinity are called **parallel**. In the figure, we can see a line s through the point A which cuts m. Now, we can see that two lines pass through A which are parallel to m: p and q. They touch m in the model at P and Q respectively which are points at infinity. At last, and not the least: the line n also passes through A without cutting m, neither inside the disc nor outside it. This line is called **non-secant**.

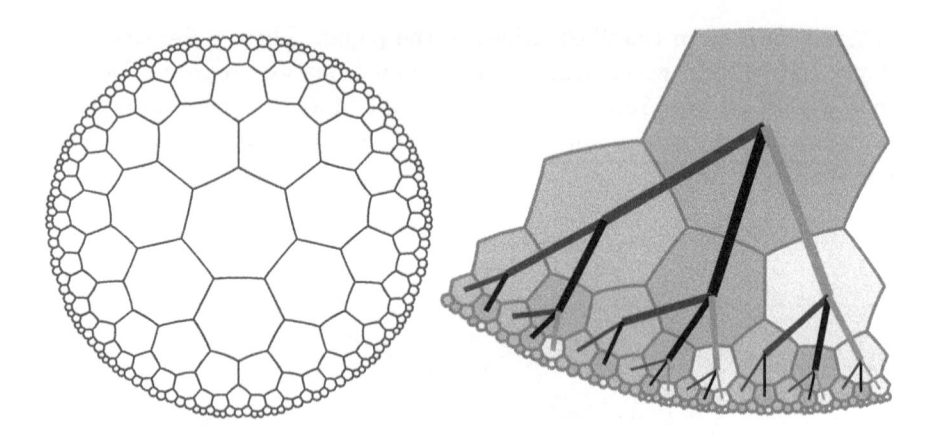

Fig. 3. The heptagrid. On the right-hand side: the key structure to explore the tiling.

4.2. *The grid of our simulations*

From a famous theorem established by Poincaré in the late 19th century, it is known that there are infinitely many tilings in the hyperbolic plane, each one generated by the reflection of a regular convex polygon P in its sides and, recursively, in the reflection of the images in their sides. Remember that in the Euclidean plane, up to similarities, there are only three such tilings: the tiling based on the square, that on the hexagon and that on the equilateral triangle.

In the paper, we shall focus our attention on one of the simplest tilings belonging to this family: the tiling $\{7,3\}$, which we call the **heptagrid**, see Figure 3. Here, the polygons are regular convex heptagons, all vertices being shared by three of them. On the right-hand side of the figure, we can see the tree which is in bijection with an angular sector, a basic structure of the heptagrid, also see Figure 6. This property is the basis of very efficient navigation tools to locate tiles in the heptagrid, see.[5,6]

The tiles of the heptagrid are very big. So, to obtain our grid, we first split each tile of the heptagrid into seven triangles whose vertices are the centre of the tile and the end-points of its edges. This gives generation 1, see Figure 4. Then, we split the triangles into four triangles by taking the mid-points of the previous triangles: the new triangles are defined by a vertex of the previous triangles and the mid-points of the edges of the previous triangle which meet at the vertex. This defines generation 2, see Figure 5

We can go on the process inductively: the generation $n+1$ is obtained from the generation n as generation 2 is obtained from generation 1. However, in this paper we shall focus on generation 2 only. Let us call it the **second triangular heptagrid**, **heptatrigrid** for short.

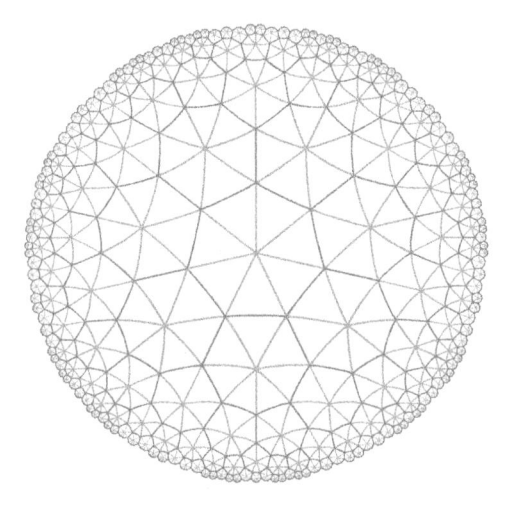

Fig. 4. The heptatrigrid: generation 1.

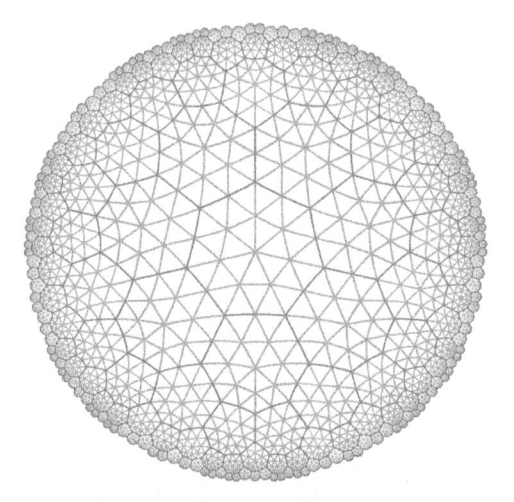

Fig. 5. The heptatrigrid: generation 2.

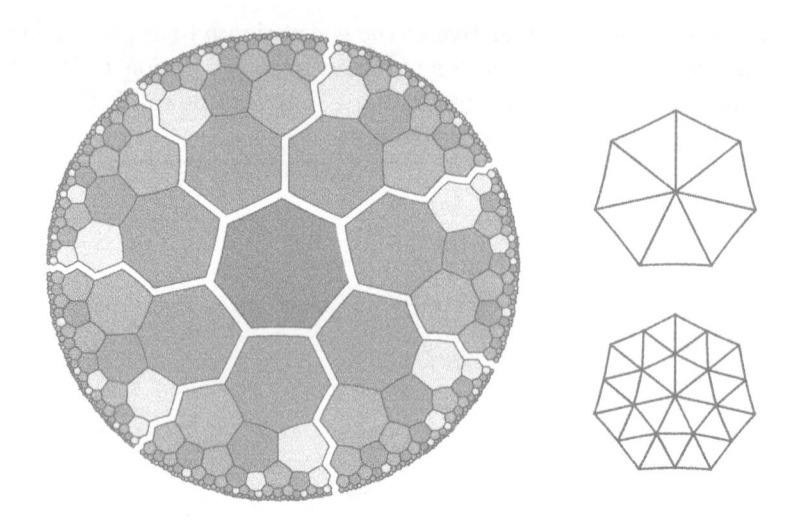

Fig. 6. From the heptagrid to the second triangular heptagrid, the heptatrigrid.

5. Cellular Automata on the Hyperbolic Triangular Grid

Now, we turn to the implementation of a cellular automaton in the heptatrigrid.

Remember that such an automaton consists in a finite automaton A attached to each 2-triangle of the heptatrigrid. A cell of the cellular automaton consists of A and a 2-triangle which is the **support** of the cell. The neighbours of a cell c with T as its support have, as their supports, the 2-triangles which share a side with T.

Figure 6 indicates the basic elements of the location of a triangle. A coordinate is defined by four numbers in the format (σ, ν, τ, π): σ is the number of the sector in which the triangle lies; ν is the number of the heptagon of the sector in which the triangle lies; τ is the number in [1..7] of the generation 1 triangle in which the triangle lies; and in this triangle, π is the number of the triangle itself.

The numbering of the generation 1 triangle, we say later **1-triangle**, is defined by the number of the side of the heptagon on which the 1-triangle is built: this is why this numbers is in [1..7]. For the central heptagon, side 1 is fixed once for all and the other sides are numbered by counter-clockwise turning around the tile. The side i touches the heading heptagon of the sector i, with i in [1..7]. For the other heptagons, side 1 is shared by the father of the heptagon in the tree. We consider that the father of the root of the tree in each sector is the central cell.

In each 1-triangle, we have four triangles of generation 2, we call them **2-triangles**. The four 2-triangles of a 1-triangle are numbered from 0 to 3. First, we number the vertices and the sides of a 1-triangle T from 0 to 2: 2 is the centre of the heptagon, 0 and 1 are the vertices of the side of the heptagon defining T; the side i is opposite to the vertex i. Following the counter-clockwise orientation, 0 comes before 1. Now, the number of a vertex of T is the number of the 2-triangle which possesses this vertex. Accordingly, 3 is the number of the 2-triangle whose vertices are the midpoints of the edges of T. This numbering can be repeated for any further generation, see.[6] This numbering has interesting properties. The interested reader is referred to[6] for more information.

Now, to implement cellular automata, we have to compute the coordinates of the neighbours of a cell c from the coordinate of c itself.

Let T be the 2-triangle which supports the cell. Number the neighbours of c from 0 to 3, c itself being neighbour 0. For the other numbers, the neighbour i of c is the 2-triangle which shares the side $i-1$ of T. Let (σ, ν, τ, π) be the coordinate of T, the support of c. The coordinates of the neighbours of T are given by Table 4.

As can be seen in the table, each 2-triangle has at least one neighbour which is in the same heptagon and in the same slice of the heptagon. Note that a 2-triangle with place 3 has all its neighbours in the same slice of the same heptagon. A 2-triangle with place 2 has all its neighbours in the same heptagon, but two neighbours are in the slices which are adjacent to its own one. This is indicated by the expressions $\tau \ominus 1$ and $\tau \oplus 1$. As the number of a slice is in $[1..7]$, subtracting 1 from 1 gives 7 and adding 1 to 7 gives 1. For $\tau \in [2..7]$, $\tau \ominus 1 = \tau-1$, and for $\tau \in [1..6]$, $\tau \oplus 1 = \tau+1$. For the 2-triangles with place 0 and 1, the computation of the coordinates of their neighbours is more complex. Indeed, in each case, one of the neighbours do not belong to the same heptagon H, but to a heptagon K neighbouring H, i.e. which means that H and K have a common edge. This changes the value of ν and it may also change the values of σ and τ. This is indicated in Table 4 by the expressions $s(\sigma, \nu\,\tau)$, $v(\tau, \nu)$ and $t(\nu, \tau)$.

The computation of these expressions can be found in[5,6] but we repeat them in order the paper to be self-contained. The main point is the computation of $v(\tau, \nu)$ as it may involve auxiliary functions, namely $f(\nu)$, the number of the father of the tile ν in its tree, $\sigma(\nu)$, the number of the **preferred son** of ν and $st(\nu)$, the **status** of ν. As mentioned from,[3,4] the

Table 4. Coordinates of the neighbours of a 2-triangle T in terms of the coordinates of T.

neighbour	sector	number	slice	place
0	σ	ν	τ	0
1	σ	ν	τ	3
2	$s(\sigma,\nu,\tau)$	$v(\tau,\nu)$	$t(\nu,\tau)$	1
3	σ	ν	$\tau \ominus 1$	1
0	σ	ν	τ	1
1	$s(\sigma,\nu\tau)$	$v(\tau,\nu)$	$t(\nu,\tau)$	0
2	σ	ν	τ	3
3	σ	ν	$\tau \oplus 1$	0
0	σ	ν	τ	2
1	σ	ν	$\tau \ominus 1$	2
2	σ	ν	$\tau \oplus 1$	2
3	σ	ν	τ	3
0	σ	ν	τ	3
1	σ	ν	τ	0
2	σ	ν	τ	1
3	σ	ν	τ	2

tree which we consider has two kinds of nodes: black and white ones with two and three sons respectively. This colour of a node is its **status**. The sons can be deduced from the node by the following rules $B \to B_*W$ and $W \to BW_*W$ in easy notations, where the star indicates the place of the preferred son. The nodes receive numbers defined as follows: the root receives number 1 and then we go down in the tree level by level and, on each level, from the left to the right.

For completeness of the paper, we briefly indicate how to compute $f(\nu)$ and $\sigma(\nu)$ where ν is the number of a node. It is known that for each positive integer n, n can be written in the following way: $n = \sum_{j=1}^{k} a_j f_j$ where $a_j \in \{0,1\}$ and f_j is the jth term of the Fibonacci sequence defined by the equation $f_{n+2} = f_{n+1} + f_n$ for $n \geq 2$ with the initial condition $f_1 = f_0 = 1$. We can consider a_j as the jth letter of a word α which we call a **representation** of n as it is not unique. The representation can be made unique by requiring that α does not contain the pattern 11. We then say that the 1's in α are **isolated**. The representation with isolated 1's is also characterized as the one for which the word α is the longest. We define

Table 5. Correspondence between the numbers of a side shared by two heptagons, H and K. Note that if H is white, the other number of side 1 may be 4 or 5 when K is white and that it is always 5 when K is black.

black H in H	in K	white H in H	in K
1	$3^{wK}, 4^{bK}$	1	$4^{wK}, 5$
2	6	2	7
3	7	3	1
4	1	4	1
5	1	5	1
6	2	6	2
7	2	7	$2^{wK}, 3^{bK}$

the longest representation of n to be the **coordinate** of the node and we denote it by $[n]$. Conversely, if α is a word on $\{0,1\}$ say $a_k...a_1$ we denote $\sum_{j=1}^{k} a_j f_j$ by $]\alpha[$. Now, from,[4-6] we know that $\sigma(\nu) =][n]00[$ which we can rewrite as $[\sigma(\nu)] = [n]00$. The coordinate of $\sigma(\nu)$ is obtained from that of ν by appending two 0's. From this and from the above rule on the sons of a node, it is proved in[4-6] that $f(\nu) = m + \alpha_2$ where m and α_2 are obtained from ν by writing $[\nu] = [m]\alpha_2\alpha_1$, which means that α_2 and α_1 are the last two digits of $[\nu]$.

Now, we can go on our explanation of the computation of the coordinates of the neighbour of a node. Clearly, the coordinate of a neighbour K of a heptagon H with coordinate ν depends on the side τ shared by H and K. Now, the side numbered by τ in H does not receive the same number in K and we shall say that K is the neighbour τ of H. The correspondence between these numbers gives the value of the function $t(\nu, \tau)$ and, for completeness, we give it in Table 5. Note that the sides of the central cell are all numbered by 1 in its neighbours. For the other cells, the correspondence depends on the status of H and it may also depend on that of K. Side 7 is always the side shared by a neighbour which is on the same level of the tree, even when there is a change of tree by the change of sector. If H is black, its side 7 is numbered 2 on the other side. If H is white, the number of its side 7 in the other neighbour K depends on the status of K as indicated in the table.

Table 6. The values of $v(\tau, \nu)$.

τ	black	left	white	right	root
1	$f(\nu)$	$f(\nu)$	$f(\nu)$	$f(\nu)$	0
2	$f(\nu)-1$	$\nu-1$	$\nu-1$	$\nu-1$	1
3	$\nu-1$	$\sigma(\nu)-1$	$\sigma(\nu)-1$	$\sigma(\nu)-1$	$\sigma(\nu)-1$
4	$\sigma(\nu)$	$\sigma(\nu)$	$\sigma(\nu)$	$\sigma(\nu)$	$\sigma(\nu)$
5	$\sigma(\nu)+1$	$\sigma(\nu)+1$	$\sigma(\nu)+1$	$\sigma(\nu)+1$	$\sigma(\nu)+1$
6	$\sigma(\nu)+2$	$\sigma(\nu)+2$	$\sigma(\nu)+2$	$\nu+1$	$\nu+1$
7	$\nu+1$	$\nu+1$	$\nu+1$	$f(\nu)+1$	1

NOTE: *Left* denotes the leftmost branch of the tree, *right* denotes its rightmost one.

From this table, we can indicate the values of $v(\tau, \nu)$ which are given in Table 6. The basic point is that $v(1, \nu)$ for the heptagon H defined by ν is always $f(\nu)$, as its neighbour 1 is the father of H. Similarly, we have that $v(4, \nu)$ is always $\sigma(\nu)$ as neighbour 4 is the preferred son of H, regardless of the status of H. Note that in the case of a black heptagon H on the leftmost branch of the tree, two of its neighbours belong to the other tree on this side of the sector of H: neighbours 2 and 3. Neighbour 2 is still $\nu-1$ and, consequently, neighbour 3 is the rightmost son of neighbour 2, hence it is $\sigma(\nu-1)+1 = \sigma(\nu)-1$. A symmetrical remark holds for a white H standing on the rightmost branch of the tree: neighbours 6 and 7 belong to another tree, the one which spans the other sector than that of H. Now, this time, neighbour 6 is numbered $\nu+1$ and so, as neighbour 7 is the father of neighbour 6, neighbour 7 is numbered $f(\nu+1) = f(\nu)+1$. At last, the root, which is a white node, belongs to both the left- and the rightmost branches of the tree. This is why it has a specific profile, different from both a standard white node and from a node on the rightmost branch of the tree.

It remains to indicate that in the case of a heptagon H which is on the left- or the rightmost branch, it is easy to define the number of the sector to which belongs the neighbours which do not belong to the tree of H. Indeed, let σ be the number of the sector in which H lies. If H is a black node, its neighbours 2 and 3 are in the sector $\sigma \ominus 1$. If H is a white node, its neighbours 6 and 7 are in the sector $\sigma \oplus 1$. Note that for the root of the sector σ, its neighbour 2 is in the sector $\sigma \ominus 1$, its neighbour 7 is in the sector $\sigma \oplus 1$ and its neighbour 1 is the central cell which is outside all the sectors.

To conclude with this section, let us remember that cellular automata have been implemented in several grids of the hyperbolic plane. The complexity classes of these cellular automata have been investigated leading to very surprising results. Several universal cellular automata also have been implemented in these grids. We refer the reader to[6,7] for more information and more references.

6. The Simulations

The basic idea behind our simulations is the propagation of the tree structure of the heptagrid by a cellular automaton. The result, illustrated by Figure 7 convinced us that we could try to simulate colonies of bacteria. We propose three of them which are examined in Sub-section 6.2 and which differ by the number of states of the cellular automaton which is used for the simulation.

6.1. *First simulation: the propagation of the tree structure*

The tree structure of the heptagrid can be implemented by cellular automata on this grid: this was illustrated in[8] in order to give a toy example of a cellular automaton on this grid.

Fig. 7. The propagation of the tree structure of the heptagrid with the indication of the grid.

We can do the same here and Figure 7 gives the 36th step of execution of this automaton starting from an initial configuration in which the seven 2-triangles of place 2 of a heptagon are in the same state, in red in the figure: we call this the core-2 configuration. As we can see, the automaton has a non-small number of states: 18 of them. In,[8] we had 5 states only. In fact, it is possible to have 4 states in the case of the heptagrid if we do not need to differentiate the two white sons of a white node. We need much more states here as we wish to diffuse the structure of the tree with its two types of rules. For programming reasons, it was easier to program the automaton by implementing the following strategy: when the automaton enters a heptagon, it goes as soon as possible to the 2-triangles with place 2. There, by a counting process, it determines the directions of the sons from the direction of the father which is the direction from which the automaton entered the cell.

The way the automaton is working can be seen as an animation on the slides which are deposited on.[10]

Fig. 8. Above: the picture of Figure 7 without trace of the grid. Below: propagation of a bacteria colony, picture by courtesy of Professor Ben-Jacob.

It seems to us that the result has a striking similarity with pictures about the growth of colonies of bacteria in highly stressed conditions, see Figure 8.

6.2. The other simulations

In this sub-section, we successively examine three attempts to simulate the propagation of colonies of bacteria. We shall consider the number of states we use as well as the information that the cells are assumed to know about themselves. We shall try to give the states and these assumptions a kind of biological flavour. We have to keep in mind the specificity of the cellular automaton programming. A cell cannot directly act upon another one. Such an action has to be 2-stepped: if c wants to act on a neighbour n, c has to signal this intention by taking a particular state. Seeing this state on c, and possibly seeing an additional information displayed by its other neighbours, n can interpret the intention and take the *desired* state. However, we often speak in a direct manner, that c acts on n in this or that way.

With two states

The states are white for the medium, black for the colony. The cells want to propagate, but competition is not encouraged. This can be rather simply formulated as follows:

> (a) A black cell remains black.
> (b) A white cell becomes black if and only if it has exactly one black neighbour at this time.

From condition (a), once a white cell c becomes black at time t_0, it is black for all times t with $t \geq t_0$.

Figure 9 illustrates the 36th time of this situation starting from the core-2 configuration. We can see that the colony invades almost all the space, leaving holes unoccupied. The condition on the change of the white cell to a black one has, as a consequence, that a white cell which has two black neighbours exactly remains white. This is the reason of the pairs of adjacent white cells which are regularly produced in the evolution of the automaton.

With four states: version 1

Now we have four states: W, R, Y and V calling them white, red, yellow and vermilion respectively. White represents the medium. Red is almost the initial configuration which is, here again, the core-2 configuration.

The action of the cells is now:

Fig. 9. Simulation of a diffusion process with 2 states.

(a) A red, yellow or vermilion cell remains in its colour.
(b) A white cell becomes red, yellow or vermilion if and only if, at this time, it has exactly one neighbour which is red, yellow or vermilion respectively.

Here too, when a white cell becomes non-blank, it keeps the new colour for ever.

The above picture of Figure 10 illustrates the 36th step of computation starting from the core-2 configuration. The non white cells occupy a heptagon exactly, with the pattern we have in the figure for the red cells. We call this the heptagonal core configuration. We can notice that in this case also, the cells which remain white are the same as those of the previous automaton. We also notice that, thanks to the core-2 configuration, the red state no more occurs in the computation. After the initial time, the computation outside the heptagonal core involves three states only: white, yellow and vermilion.

In view of the bottom picture of Figure 10, it seems reasonable to consider that yellow and vermilion states together represent the colony.

With four states: version 2

Here, we again have four states. But we also assume that the colony has some knowledge of the geometry of the space. This can be viewed as an acquired experience of the space by the colony. We assume that a cell knows

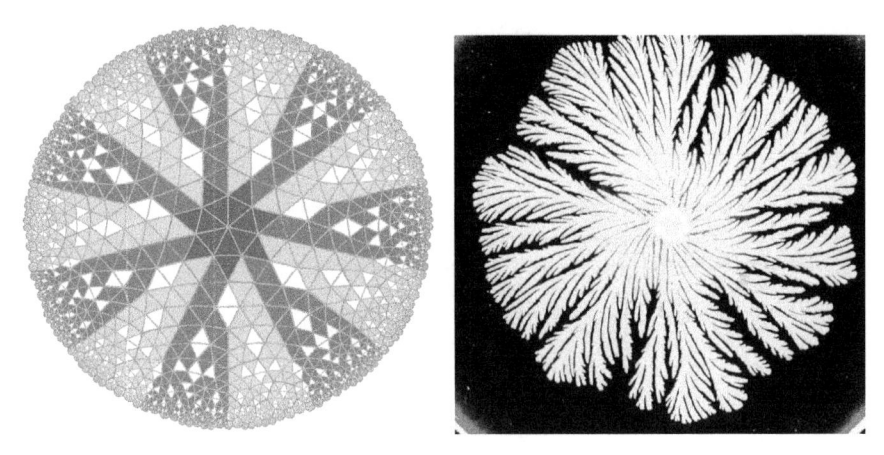

Fig. 10. Comparing the diffusion process with 4 states based on local knowledge of the colony with a colony of Figure 1, picture by courtesy of Professor Ben-Jacob.

its place and whether it is in slice 1 or not. It is easy to see that this is a 2-bit information only. We take the same colours as previously, with white as the state for the space. Here too the computation starts from the core-2 configuration.

This time too, the formulation of the rules is in the same style as previously but it becomes more intricate, as it involves the place and the slice of a 2-triangle.

(a) A red, yellow or vermilion cell remains in its colour.
(b) If a white cell has two white neighbours and if its slice is 1, then it takes the colour of its third neighbour.
(c) If a white cell has two white neighbours and if its slice is not 1 but its place is 2 then if its third neighbour is red, yellow or vermilion, it becomes yellow, vermilion or red respectively.
(d) If a white cell has two white neighbours and if its third neighbour is red then, if its place is 3, 0 or 1, it becomes red, vermilion or yellow respectively.
(e) If a white cell has the states white, yellow and vermilion among its three neighbours, if its slice is 1 and if its place is 3, then it becomes red.

With these rules, the cellular automaton behaves in a somewhat different manner. As can be seen from Figure 11, although the initial configuration is the same as previously, the four states are now involved during the whole computation. Moreover, the colony does not occupy the whole space: the

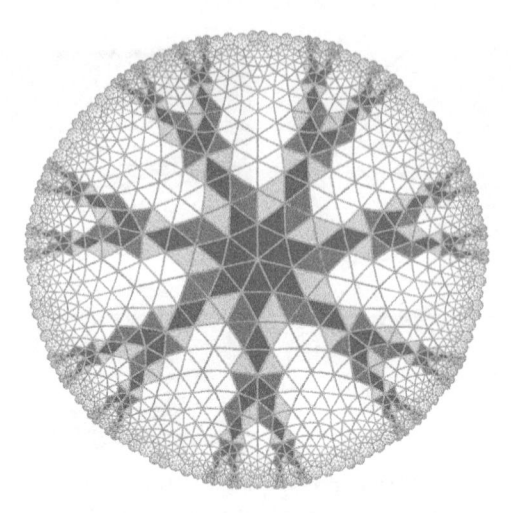

Fig. 11. Simulation of a diffusion process with 4 states, based on a local knowledge of the colony plus a slight insight on the global structure.

branches which regularly are spread out are far away from each other, which avoid any kind of competition. Also, we can see that this time we have a cooperation between the states. The knowledge whether a cell is in a slice 1 or not allows the colony to take advantage of the topology in order to invade the center of a heptagon, according to the scenario contained in the condition c of the rule. Next, the conditions d and e allow the colony to occupy the slices 4 and 5 of the heptagon and those ones only, without knowing the number of the slice. This is obtained by the combination of the conditions c, d and e. Once this is checked for one heptagon around the central cell, this is repeated for all the heptagons which are the 4- and 5-neighbours of a heptagon. This way, we obtain seven binary trees which grow from the heptagonal core.

In Figure 12, we compare the growth of Figure 11 with a picture of another bacteria colony. Note that the upper picture of Figure 12 is obtained from Figure 11 by removing the drawing of the heptatrigrid. The computer program which draws the figure writes down a PostScript file from the information obtained by performing the simulation up to the 36th step, starting from the heptagonal core. In this writing, the program simply removes the drawing commands used to draw Figure 11, simply keeping the filling commands which allow to paint closed areas defined for drawing the same figure.

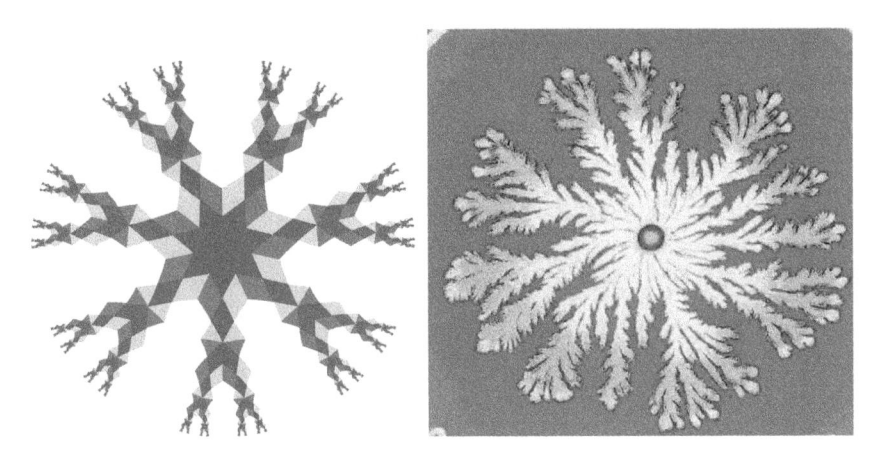

Fig. 12. Above: simulation of the growth of a bacteria colony with the hepta-trigrid. Below: propagation of the bacteria colony already shown in Figure 8, picture by courtesy of Professor Ben-Jacob.

7. Conclusion

It should be remarked that in all the previous simulations, the computation may be as long as wished within the time and memory limits of a computer. Due to the exponential growth of the number of 2-triangles as we go away from the central heptagon, these limits are rapidly reached and improvements in technology may perhaps allow us by one round of 2-triangles further each time the capacity is multiplied by 3. However, for simulations of actual colonies of bacteria, this is not a problem as their growth is not only finite but also small in the hyperbolic scale.

It seems to us that this hyperbolic simulation gives an interesting approximation of the phenomenon observed in real experiments. The above discussion about the space of computation indicates that it could be interesting to investigate generation 3 of triangles and so, to look at what we obtain for 3-triangles. Most probably, we could get a finer simulation but certainly at the price of a bigger number of states. The interpretation of these states from a biological point of view is of course a question as well as how much of the knowledge of the space could be allowed for 3-triangles where a third parameter within the place is necessary.

It could also be interesting to use simultaneously 3-triangles and 2-triangles: this might make it possible to simulate the capacity of colonies of bacteria to change physical parameters of their environment, also changing their shape in order to answer harder conditions.

These are directions for further work on this topic.

Acknowledgement

The author is very much in debt to Professor Ben-Jacob for the pictures he sent him and for the permission of publishing them in this paper.

References

1. E. Ben-Jacob, Social behavior of bacteria: from physics to complex organization, *European Physical Journal B*, **65**(3), 315–322, 2008.
2. G.B. Ermentrout, L. Edelstein-Keshet, Cellular automata approaches to biological modelling, *Journal of Theoretical Biology*, **160**(1), 97–133, 1993.
3. M. Margenstern, K. Morita, NP problems are tractable in the space of cellular automata in the hyperbolic plane, *Theoretical Computer Science*, **259**, 99–128, 2001.
4. M. Margenstern, New Tools for Cellular Automata of the Hyperbolic Plane, *Journal of Universal Computer Science*, **6**(12), 1226–1252, 2000.
5. M. Margenstern, *Cellular Automata in Hyperbolic Spaces, vol. 1, Theory*, Old City Publishing, Philadelphia, 422p., 2007.
6. M. Margenstern, *Cellular Automata in Hyperbolic Spaces, vol. 2, Implementation and computations*, Old City Publishing, Philadelphia, 360p., 2008.
7. M. Margenstern, An algorithmic approach to tilings of hyperbolic spaces: 10 years later, *Lecture Notes in Computer Sciences*, **6501**, 32–52, 2010.
8. M. Margenstern, A universal cellular automaton on the heptagrid of the hyperbolic plane with four states, *Theoretical Computer Science*, **412**, 33–56, 2011.
9. M. Margenstern, Bacteria Inspired Patterns Grown with Hyperbolic Cellular Automata, **CAAA/HPCS'2011**, Istanbul, Turkey, July 4–8, 2011.
10. File on the personal site of M. Margenstern: http://www.lita.univ-metz.fr/~margens/look_test.pdf
11. Minsky M.L., Computation: Finite and Infinite Machines. *Prentice Hall, Englewood Cliffs, N.J.*, 1967.
12. Yu. Rogozhin, Small Universal Turing Machines, *Theoretical Computer Science*, **168**(2), (1996), 215–240.
13. I. Stewart, A Subway Named Turing, Mathematical Recreations in *Scientific American*, 90–92, 1994.
14. D. Woods, T. Neary, *On the time complexity of 2-tag-systems and small universal Turing machine*, **FOCS'2006**, Berkeley, October, 22–24, 2006.

Chapter 12

Computation and Communication in Unorganized Systems

Christof Teuscher

Department of Electrical and Computer Engineering,
Portland State University &
Department of Computer Science, University of New Mexico

The computing disciplines face difficult challenges by extending CMOS technology to and beyond the end of dimensional scaling. One solution path is to use an innovative combination of novel devices, compute paradigms, and architectures to create new information processing technology. It is reasonable to expect that future devices will increasingly exhibit extreme physical variation, and thus have a partially or entirely unknown structure with limited functional control. In this chapter we argue and illustrate that unorganized compute machines—as originally also proposed by Alan M. Turing in a little known report—are a promising solution to address some of the challenges. Such unconventional machines can even have significant benefits over their conventional counterparts, however, we need to use different communication and compute paradigms.

1. Introduction

Regularity is beautiful and appeals by its simplicity, yet, most objects in nature are to some extend irregular, asymmetric, and imperfect.[39] In computer science and computer engineering, both real hardware and virtual structures are generally kept regular because that is what makes an engineer's life simple. Regular structures are easier to build, easier to understand, and easier to control. Good examples of regular structures are computer memories, Field Programmable Gate Arrays (FPGAs),[15] crossbar circuits,[9] and cellular automata.[40] Going back to one of the main fathers of modern computer science, Alan M. Turing,[19] the Turing machine with its linear tape represents another good example of well-organized structural regularity. However, Turing also realized—inspired by how the human brain is structured and processes information—that the highly structured Turing Machine may not be an appropriate model for all information processing that happens in nature. In a little known *National Physical Laboratory* (NPL) report entitled "Intelligent Machinery," he proposed three types of what he called *unorganized machines*. The report first appeared in an edited collection by Evans and Robertson,[12] 14 years after Turing's death, and one year later in "Machine Intelligence".[42] In many ways, Turing's unorganized machines[29] represent the other end of the spectrum of regularity. Such non-classical machines process information differently and need to be programmed differently than their classical counterparts. We have argued in the past that unorganized machines are a valid model for emerging computing architectures,[33,34] in particular for devices that were self-assembled. With his "schoolboy ideas," Turing may therefore unknowingly have contributed to the groundwork of future nanoscale and molecular computers.

The goal of this chapter is to muse about alternative communication and computing architectures and models, in particular unorganized compute fabrics. We will argue and illustrate that such fabrics are a promising solution to address some of the challenges the computing disciplines face. The goal is to challenge well-established believes and to illustrate that there is a world beyond the traditional von Neumann architecture, a world that may be better than we commonly tend to think.

2. Non-classical Computing

Classical computing is undoubtedly a remarkable success story, however, there is a growing community that believes it only encompasses a small subset of all computational capabilities and that certain grand challenges the traditional approaches face can be solved by radically new approaches and radically new approaches will open up new application domains where traditional silicon-based computers are inappropriate. *Non-classical, novel,* or *emerging computation*)[1,2,10,17,27,31,36] is a broad and interdisciplinary research area with the main goal to go beyond the standard models and practical implementations of computers, such as the von Neumann computer architecture and the abstract Turing machine, which have dominated computer science for more then half a century. This quest, in both theoretical and practical dimensions, is motivated by a number of trends. First, it is expected that, without disruptive new technologies, the ever-increasing computing performance and storage capacity achieved with existing technologies will eventually reach a plateau. The main reason for this are fundamental physical limits on the miniaturization of today's silicon-based electronics.[20] Second, novel ways to synthetically fabricate chemical and biological assemblies, for example through self-assembly, self-replication (e.g.,[13]), or bio-engineering (e.g.,[7,28]) allow one to create systems of unimagined complexity in new ways. However, we currently lack the methodologies and the tools to design and program such massively parallel and spatially extended unconventional "machines." Third, many of today's most important computational challenges, such as for example understanding complex biological and physical systems by simulations or identifying significant features in large, heterogeneous, and unstructured data sets, may not be well suited for classical computing machines.

Despite all the promising aspects, many barriers are also observed. The non-classical research and educational landscape is incoherent, fragmented, lacks pragmatic reasons why one should prefer such models over classical paradigms, and it is hard to think in terms of new paradigms well enough to teach both researchers and students to achieve something useful quickly and easily. The inherent multidisciplinary character and the involvement of various system and abstraction levels (i.e., device, architecture, language) make this landscape even more rugged to navigate. Teuscher et al.[37] have suggested to the community to ask certain leitmotif questions in order to gain a better momentum and to focus on the most pressing challenges.

3. Regularity versus Irregularity

The common assumption of computer engineers is that irregularity and variation in devices are a bad thing. While this may be true to some extend, at least when thinking in terms of conventional computing architectures and paradigms, we have been exploring for almost a decade the inherent benefits and drawbacks of irregular, heterogeneous, and imperfect computing devices.

There are many levels and degrees one can introduce irregularity. Inspired by Sipper's cellular computing framework,[26] computing fabrics can be classified according to local interconnect topology, the compute cell's arrangement in space, the mobility, the uniformity, and the cell's behavior. In this context, "cell" can stand for core, node, CLB, or other forms of basic compute units. Figure 1 shows different compute cell arrangements and interconnect topologies. In our research, we are mostly interested in subfigure (F), i.e., computing substrates that are arranged and interconnected in an irregular way.

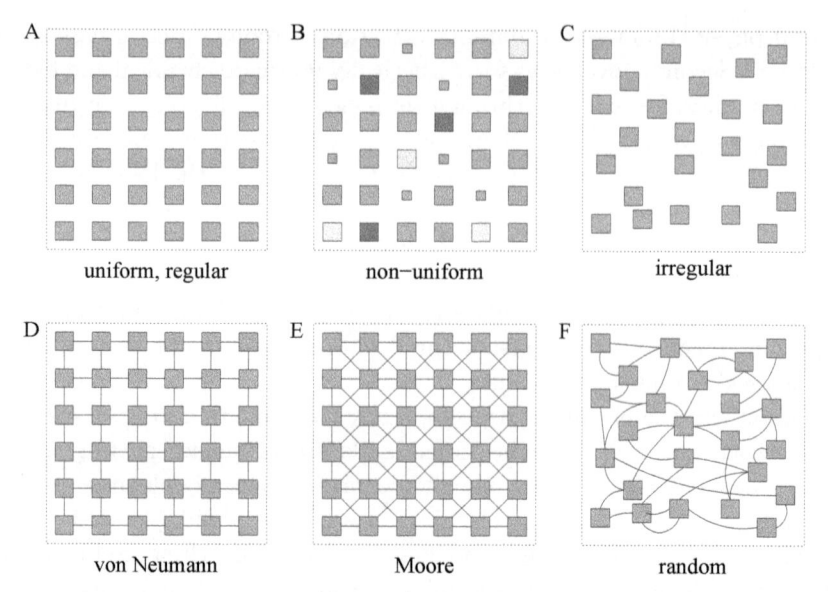

Fig. 1. Illustration of different compute cell arrangements and interconnect topologies. (A) regular and uniform; (B) regular and non-uniform, different shadings and sizes indicate different cell programs; (C) irregular and uniform; (D) regular, uniform, von Neumann neighborhood; (E) regular, uniform, Moore neighborhood; (F) irregular, uniform, random neighborhood.

Figure 2 shows a different way of representing the design space and the trade-offs we are interested in in this chapter. The most interesting corner, the so-called "evil" corner, represents heterogeneous, irregular, non-linear, and imperfect systems. We believe that such systems will become increasingly important because the computing disciplines face difficult challenges by extending CMOS technology to and beyond the end of dimensional scaling. One solution path is to use an innovative combination of novel devices, compute paradigms, and architectures to create new information processing technology. It is reasonable to expect that devices will increasingly exhibit extreme physical variation, and thus have a partially or entirely unknown structure with limited functional control. The assumption is therefore that the computing sciences will need to learn to "live" in and deal with this corner simply because we will not be able to fabricate systems with the precision, homogeneity, and regularity as we are used to now. It is not that we *want* to be in that corner, but the stance we take here is that this corner offers many benefits if we "embrace" it and change the way we think about

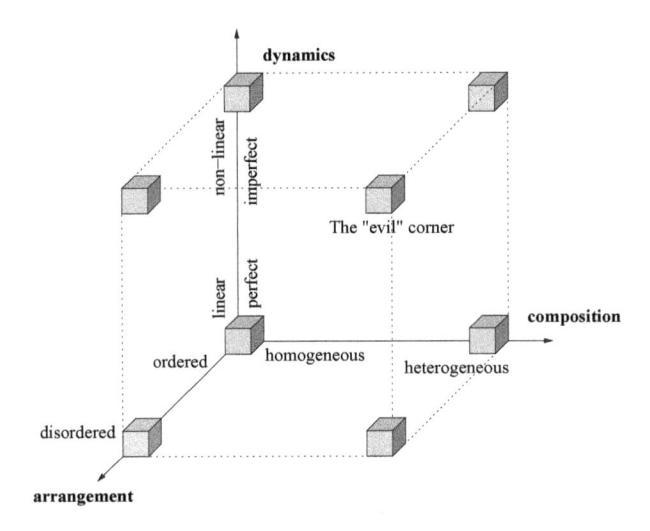

Fig. 2. Design space spanning the arrangement, the composition, and the dynamics axis we are interested in in this chapter. The "evil" corner represents heterogeneous, disordered, non-linear, and imperfect systems.

computation. In the following, we will provide examples of how that can be done.

4. Communication in Unstructured Nanoscale Interconnect Networks

Future nano-scale electronics built up from vast number of components need efficient, highly scalable, and robust means of communication in order to ever become competitive with traditional silicon approaches. As the device dimensions shrink further, it is increasingly hard to provide structured and reliable components. In this section we explore two radically new approach of interconnecting processing elements by an unstructured network-on-chip-like interconnect: the first approach is inspired by growing nanowires directly on a surface while the second approach is inspired by dropping nanowires onto a surface.

The controlled synthesis of nanostructured interconnects represents a key challenge for future and emerging electronics. Carbon nanotubes and metallic nanowires are often mentioned as promising alternatives. Many techniques have been reported to grow both nanowires and carbon nanotubes in bulk, but they are generally hard to place in a structured way.

We question the traditional approach of mesh-based interconnects and investigate a new interconnect paradigms that offers both better performance and lower cost while being more appropriate for self-assembly processes. Our models are based on physically realistic small-world network approach as explored by.[24] However, our approach is more applied and has an experimental part,[44] and addresses the question of how much and what type of interconnect emerging electronics need. The primary goal is to investigate and understand the characteristics of such self-assembled networks and to ultimately use these insights to tune the chemical parameters of the self-assembly process.

Others have proposed novel approaches to improve network-on-chip (NoC) performance. For example, Oshida and Ihara[23] investigated the packet traffic of scale-free and large-scale NOC and concluded that scale-free topologies achieve short latencies and low packet loss ratios. However, they have not considered the wiring cost. Ogras et al.[22] showed that a significant reduction in the average packet latency can be achieved by superposing a few long-range links to a standard mesh network. In their approach, the links are not inserted at random but where they are most useful for increasing the critical traffic workload. Neither[23] nor[22] have considered distance-dependent connections in order to minimize the network wiring cost. More recently Ganguly et al. used wireless NoC to establish "short cuts" in the network.[14]

To evaluate and compare the performance metrics of different networks, we used a simple network-on-chip (NoC) framework that was first described in.[30] The network is composed of programmable processing nodes (PNs), switch nodes (SNs), and an interconnect fabric. The connections between the SNs are assumed to be bi-directional point-to-point interconnects. We introduce two new design control parameters, p and α, which allow to explore the design space of a large class network topologies. Starting from a regular mesh with local connections only, we add R additional connections randomly to already existing links. This results in double, triple, or more connections for some nodes with some neighbors. We then apply a rewiring algorithm that goes through all the nodes and rewires each existing (locally connected) wire with probability p to a node chosen proportional to a power-law distribution as a function of the Euclidean distance l between nodes: $l^{-\alpha}$. In simple terms, α determines the proportion of non-local connections. If $\alpha = 0$, the connections are established independent of the distance between them, for higher α, local connections are more likely. Thus p determines the how much irregularity and α how many non-local links the network has. Depending on p, such graphs have the small-world property,[43] i.e., a very short average path length that scales logarithmically with the system size. Petermann and De Los Rios have[24] shown that small-world networks with power-law distributions have a lower wiring cost for the same performance. Note that the rewiring is only used "virtually" and as a model to obtain the type of self-assembled networks as explore in.[44]

In[32] we confirmed this result by using evolutionary algorithms. We have shown that small-world networks with power-law distance-dependent wire-length distributions are more power-efficient while offering the same performance than simple small-world topologies. This is particularly important in a world where power and energy issues have significantly gained in importance in computing environments in the last few decades. We argued that such networks occupy optimal spots in the design space of NoCs. Our results are particularly relevant for addressing the scalability problem of global (or long-range) links, for building more power-efficient computers, and for emerging computing devices built through self-assembly.

Unstructured interconnect topologies may seem intimidating and useless at first, but as we have shown previously,[30,35,38] they offer several benefits over mesh-like topologies. In particular, by choosing particular values for the level of unstructuredness (p), the level of non-locality (α), and the total number of wires to be used, such networks allow a designer to explore parts of the design space that are not accessible with structured mesh-like interconnects, to minimize the wiring cost, and to maximize the performance. We believe that computation, which will be further discussed in the next section, in unstructured self-assemblies of components and interconnections is a highly appealing paradigm, both from the perspective of fabrication as well as performance and robustness.

5. Computation in Unstructured Nanoscale Boolean Networks

Approaching hard physical limits has made it increasingly difficult to gain higher performance in conventional CMOS-based electronics through higher integration. Below a certain integration density, the miniaturization of transistors and their interconnections increases power consumption, heat dissipation, failure rate, and manufacturing defects.[18] As a result, the industry focused on developing multi-core architectures. In this approach, many processing cores are interconnected with a mesh-like or ad-hoc network structure with rather local connectivity. The current top-down design methodology favors this architecture and metal wires limit the wire-length one can traverse in a single clock cycle. The architecture also resembles a class of discrete-dynamical systems called *Cellular Automata* (CA). CAs are structured and highly regular with local interconnects only. Zhirnov et al.[45] argued that integrated digital CA architectures are well-suited for semiconductor technology because of their local and regular structure. However, while the fan-in in CA-based architectures is 4, the optimal number of components for a combinational logic block is achieved with fan-in of 2.[8] The effect of fan-in on latency and energy consumption remains an open question. As a result of the multi-core trend, new challenges were introduced in the area of concurrent programming,[6] power consumption, and interconnect load. The advent of nanotechnology has opened alternative opportunities for manufacturing electronic devices. Two possible approaches are: (1) hybrid systems that combine traditional silicon-based electronics with nano- and molecular-scale components or (2) radically shifting computing and manufacturing paradigms using bottom-up self-assembled devices.[33] However, precise control over self-assembly processes is hard, and the devices are therefore susceptible to defects and failures.

We have recently addressed[3] the question of what fan-in optimizes both the power consumption and the latency in an unstructured network. We showed that an average fan-in of $K = 3.3$ is optimal for random Boolean networks when energy and latency are considered equally important. Our results are important as they show an inverse relationship between the energy consumption and the performance, and this allow us to determine the optimal connectivity of a certain class of self-assembled nanoscale devices. In another paper,[4] we have presented a software and hardware framework to measure the performance of a self-configurable computing architecture for unstructured and unknown reconfigurable fabrics composed of simple nodes interconnected by nanowires. The framework allows to create an irregular network of compute nodes where each node can be configured as a simple 1-bit ALU. The compute nodes are organized hierarchically by means of anchor nodes that recruit compute nodes with a chemically-inspired algorithm. The nodes are then self-configured by means of a gate-level netlist describing any digital logic circuit. We have developed a topology-agnostic optimization algorithm inspired by simulated annealing, which self-optimizes the circuit for latency. The work presented another step toward building a new generation of compute architectures on irregular fabrics.

Another line of research for computing with unstructured fabrics consists in harnessing the inherent dynamics of such networks.[11] In 2007, Rohlf et al.[25] systematically studied damage spreading at the sparse percolation (SP) limit in random Boolean networks with perturbations that are independent of the network size N. This limit is relevant to information and damage propagation in many technological and natural networks. They found a critical connectivity (also called the "edge of stability" or "edge of robustness") close to $K = 2$, where the damage spreading is independent of N. In 2011, Goudarzi et al.[16] went a step further and studied information processing in populations of Boolean networks with evolving connectivity, then systematically explored the interplay between the learning capability, the robustness, the network topology, and the task complexity. They used genetic algorithms to evolve networks to perform required jobs (i.e., simple tasks), and therefore did in many ways exactly what Turing suggested in 1948 by "genetical search" and "appropriate interference." Even more interestingly, they solved a long-standing open question and found computationally that, for large system sizes N, adaptive information processing drives the networks to a critical connectivity $K = 2$, which is the connectivity that Turing—for whatever reason—proposed.

In,[5] Anghel et al. showed that one can compute simple task with the self-assembled nanoscale networks[44] as presented in the previous section. Because such systems are not programmable by standard means, they have shown a solution to the supervised learning problem of mapping a desired binary input to a desired binary output in suc networks with given control nodes. The results showed that one- and two-control node random networks can implement linearly separable sets. A similar idea of controlling a network was explored in.[21,41]

6. Conclusion

We believe that computation in random self-assemblies of simple components and interconnections is a highly appealing paradigm, both from the perspective of fabrication as well as performance and robustness. In this context, we find Turing's work more than ever current, influential, and deeply fascinating. His idea of organizing an initially random network of neurons and connections is undoubtedly one of the most significant aspects of the "Intelligent Machinery" paper.

Acknowledgements

This work was supported in parts by NSF grants # 1028120 and # 1028378.

References

1. A. Adamatzky, L. Bull, B. De Lacy Costello, S. Stepney, and C. Teuscher, editors. *Unconventional Computing 2007*. Luniver Press, Beckington, UK, 2007.
2. A. Adamatzky and C. Teuscher, editors. *From Utopian to Genuine Unconventional Computers*. Luniver Press, Beckington, UK, 2006.
3. A. Amarnath, P. Damera, A. Goudarzi, and C. Teuscher. Latency and power consumption in unstructured nanoscale Boolean networks. In *Proceedings of the 11th International Conference on Nanotechnology (IEEE Nano 2011)*, pages 854–859, Cincinnati, OH, USA, 2011. IEEE.
4. A. Amarnath and C. Teuscher. A self-configurable computing architecture for unstructured and unknown reconfigurable fabrics. In *Proceedings of the NASA/ESA Conference on Adaptive Hardware and Systems (AHS-2011)*, pages 271–278, 2011.
5. M. Anghel, C. Teuscher, and H.-L. Wang. Adaptive learning in random linear nanoscale networks. In *Proceedings of the 11th International Conference on Nanotechnology (IEEE Nano 2011)*, pages 445–450, Cincinnati, OH, USA, 2011. IEEE.
6. K. Asanovic, R. Bodik, J. Demmel, T. Keaveny, K. Keutzer, J. Kubiatowicz, N. Morgan, D. Patterson, K. Sen, J. Wawrzynek, D. Wessel, and K. Yelick. A view of the parallel computing landscape. *Communications of the ACM*, 52:56–67, 2009.
7. S. Basu, Y. Gerchman, C. H. Collins, F. H. Arnold, and R. Weiss. A synthetic multicellular system for programmed pattern formation. *Nature*, 434(7037):1130–1134, 2005.
8. V. Beiu and H. E. Makaruk. Deeper sparsely nets can be optimal. *Neural Processing Letters*, 8:201–210, 1998.
9. Y. Chen, G.-Y. Jung, D. A. A. Ohlberg, X. Li, D. R. Steward, J. O. Jepperson, K. A. Nielsen, J. F. Stoddart, and R. S. Williams. Nanoscale molecular-switch crossbar circuits. *Nanotechnology*, 14:462–468, 2003.
10. S. B. Cooper, B. Löwe, and A. Sorbi, editors. *New Computational Paradigms*. Springer, New York, NY, USA, 2008.
11. J. P. Crutchfield, W. L. Ditto, and S. Sinha. Introduction to focus issue: Intrinsic and designed computation: Information processing in dynamical systems–beyond the digital hegemony. *Chaos*, 20:037101, 2010.
12. C. R. Evans and A. D. J. Robertson, editors. *Cybernetics: Key Papers*. University Park Press, Baltimore Md. and Manchester, 1968.
13. H. Fellermann, S. Rasmussen, H.-J. Ziock, and R. V. Solé. Life cycle of a minimal protocell—a dissipative particle dynamics study. *Artificial Life*, 13:319–345, 2007.
14. A. Ganguly, K. Chang, S. Deb, P. P. Pande, B. Belzer, and C. Teuscher. Scalable hybrid wireless network-on-chip architectures for multi-core systems. *IEEE Transactions on Computers*, 2011. PrePrint: http://doi.ieeecomputersociety.org/10.1109/TC.2010.176.
15. M. Gokhale and P. S. Graham. *Reconfigurable Computing: Accelerating*

Computation with Field Programmable Gate Arrays. Springer, Berlin, Heidelberg, 2005.
16. A. Goudarzi, C. Teuscher, N. Gulbahce, and T. Rohlf. Emergent criticality through adaptive information processing in Boolean networks. arXiv:1104.4141, 2011. In revision.
17. T. Munakata (guest editor). Beyond silicon: New computing paradigms. *Communications of the ACM*, 50(9):30–72, Sep 2007.
18. M. Haselman and S. Hauck. The future of integrated circuits: A survey of nanoelectronics. *Proceedings of the IEEE*, 98(1):11–38, 2010.
19. A. Hodges. *Alan Turing: The Enigma*. Walker & Company, New York, 2000.
20. L. B. Kish. End of Moore's law: Thermal (noise) death of integration in micro and nano electronics. *Physics Letters A*, 305:144–149, 2002.
21. J. Lawson and D. H. Wolpert. Adaptive programming of unconventional nano-architectures. *Journal of Computational and Theoretical Nanoscience*, 3:272–279, 2006.
22. U. Y Ogras and R. Marculescu. "It's a small world after all": NoC performance optimization via long-range link insertion. *IEEE Transactions on VLSI Systems*, 14(7):693–706, 2006.
23. N. Oshida and S. Ihara. Packet traffic analysis of scale-free networks for large-scale network-on-chip design. *Physical Review E*, 74:026115, 2006.
24. T. Petermann and P. De Los Rios. Physical realizability of small-world networks. *Physical Review E*, 73:026114, 2006.
25. T. Rohlf, N. Gulbahce, and C. Teuscher. Damage spreading and criticality in finite random dynamical networks. *Physical Review Letters*, 99(24):248701, 2007.
26. M. Sipper. The emergence of cellular computing. *IEEE Computer*, 32(7):18–26, July 1999.
27. S. Stepney and S. Emmott, editors. *Special Issue: Grand Challenges in Non-Classical Computation, International Journal of Unconventional Computing*, volume 3, 2007.
28. J. J. Tabor. Programming living cells to function as massively parallel computers. In *Proceedings of the 44th Design Automation Conference (DAC'07)*, pages 638–639, 2007.
29. C. Teuscher. *Turing's Connectionism. An Investigation of Neural Network Architectures*. Springer-Verlag, London, September 2002.
30. C. Teuscher. Nature-inspired interconnects for emerging large-scale network-on-chip designs. *Chaos*, 17(2):026106, 2007.
31. C. Teuscher and A. Adamatzky, editors. *Unconventional Computing: From Cellular Automata to Wetware*. Luniver Press, Beckington, UK, 2005.
32. C. Teuscher, H. Chung, A. Grimm, A. Amarnath, and N. Parashar. The power of power-laws: Or how to save power in SoC. In *Proceedings of the Second International Green Computing Conference (IGCC'11)*, 2011. In press.
33. C. Teuscher, N. Gulbahce, and T. Rohlf. An assessment of random dynamical network automata for nanoelectronics. *International Journal of Nanotechnology and Molecular Computation*, 1(4):39–57, 2009.
34. C. Teuscher, N. Gulbahce, T. Rohlf, and A. Goudarzi. Random dynamical

network automata for nanoelectronics: A robustness and learning perspective. In B. MacLennan, editor, *Theoretical and Technological Advancements in Nanotechnology and Molecular Computation: Interdisciplinary Gains*, chapter 19, pages 295–314. IGI Global, Hershey, NY, 2011.
35. C. Teuscher and A. A. Hansson. Non-traditional irregular interconnects for massive scale SoC. In *Proceedings of the IEEE International Symposium on Circuits and Systems (ISCAS)*, pages 2785–2788, 2008.
36. C. Teuscher, I. Nemenman, and F. J. Alexander. Editorial: Novel computing paradigms: Quo vadis? *Physica D*, 23:v–viii, 2008.
37. C. Teuscher, I. Nemenman, and F. J. Alexander (Eds.). Novel computing paradigms: Quo vadis? *Physica D*, 237(9):1157–1316, 2008.
38. C. Teuscher, N. Parashar, M. Mote, N. Hergert, and J. Aherne. Wire cost and communication analysis of self-assembled interconnect models for networks-on-chip. In *Proceedings of the Second International Workshop on Network on Chip Architectures (NoCArc 2009)*, pages 83–88, 2009.
39. D. W. Thompson. *On Growth and Form*. Cambridge University Press, Cambridge, UK, 1961.
40. T. Toffoli and N. Margolus. *Cellular Automata Machines*. MIT Press, Cambridge, MA, 1987.
41. J. Tour, W. L. Van Zandt, C. P. Husband, S. M. Husband, L. S. Wilson, P. D. Franzon, and D. P. Nackashi. Nanocell logic gates for molecular computing. *IEEE Transactions on Nanotechnology*, 1(2):100–109, 2002.
42. A. M. Turing. Intelligent machinery. In B. Meltzer and D. Michie, editors, *Machine Intelligence*, volume 5, pages 3–23. Edinburgh University Press, Edinburgh, 1969.
43. D. J. Watts and S. H. Strogatz. Collective dynamics of 'small-world' networks. *Nature*, 393:440–442, 1998.
44. P. Xu, S.-H. Jeon, H. Chen, H. Luo, G. Zou, Q. Jia, M. Anghel, C. Teuscher, D. Williams, B. Zhang, X. Han, and H.-L. Wang. Facile synthesis and electrical properties of silver wires through chemical reduction by polyaniline. *The Journal of Physical Chemistry C*, 114(50):22147–22154, 2010.
45. V. Zhirnov, R. Cavin, G. Leeming, and K. Galatsis. An assessment of integrated digital cellular automata architectures. *IEEE Computer*, 41:38–43, 2008.

Chapter 13

The Many Forms of Amorphous Computational Systems*

Jiří Wiedermann

Institute of Computer Science,
Academy of Sciences of the Czech Republic,
Prague, Czech Republic
jiri.wiedermann@cs.cas.cz

Amorphous computing presents a novel computational paradigm. From a computational viewpoint, amorphous computing systems differ from the classical ones almost in every aspect. They consist of a set of tiny, independent, anonymous and self-powered processors or robots that can communicate wirelessly to a limited distance. The processors are simplified down to the absolute necessaries in order to enable their massive production. The amorphous systems appear in many variants. Their processors can be randomly placed in a closed area or volume and form an ad-hoc network; in some applications they can move, either actively, or passively (e.g., in a bloodstream). Depending on their environment, they can communicate either via radio or via signal molecules. Assuming exponential progress in all sciences resulting in our ability to produce amorphous computing systems with myriads of processors, an unmatched application potential is expected profoundly to change all areas of science and life.

1. Introduction

1.1. Motivation

A motivation for considering *amorphous computing systems*, i.e., the systems lacking any definitive physical form, comes at least from four sources: it is the progress in the field of information processing technologies, the visionary ideas about the future of computing, the internal needs of computability theory, and, most importantly, the application domain.

---
*This research was carried out within the institutional research plan AV0Z10300504 and partially supported by GA ČR grant No. P202/10/1333.

Classical models of universal computations, such as Turing machines, RAMs, etc., are rigorously defined mathematical structures in whose design there is no room for randomness. The situation is slightly different when the computing systems represented by networks of processors (such as the Internet, wireless networks, etc.) are considered: here, the network topology may result from a random process. In order to "compute" bold assumptions about such networks must have been made: at least we require that all network nodes are connected by communication links, that prior to the start of computation each network node possesses a unique "network address", that there are communication primitives supporting message exchange and, last but not least, that each network node does possess a universal computing power, at least in principle. Such models have been the domain of the classical computational theory of the distributed system.

However, recent developments in micro-electro-mechanical systems, wireless communications and digital electronics have brought yet a new challenge into the area of distributed computing systems. Their new instances integrate sensing, data processing and wireless communication capabilities. Typical representatives of such systems are sensor, mobile, or ad-hoc wireless networks (cf.[13]). At an extreme end, people consider exotic systems, such as smart dust (cf.[22]) or amorphous computers (cf.[1,2,8]). In these systems the miniaturization is pushed to its limits resulting, presumably, into processors of almost molecular size with the respective communication and computing facilities adequately (and thus severely) restricted. These limitations call for the change of the basic computational and communication model of distributed computing systems which must subsequently be also reflected in the design of the corresponding algorithms.

Hand in hand with the previous trends, and in its most advanced form, the ideas of amorphous computing systems have emerged in the sci-fi literature. For the first time, this has probably happened in the nowadays cult novel by the (later) astronomer royal Sir Fred Hoyle, "The Black Cloud" published in 1957.[9] When observed from the Earth, this cloud appeared as an intergalactic gas cloud threatening to block the sunshine. After a dramatic attempt to destroy the cloud by a nuclear bomb the scientists came to a conclusion that the cloud possessed a specific form of intelligence. In an act of a pure hopelessness, they tried to communicate with it and, to their great surprise, they discovered a form of life obeying an intelligence by far surpassing that of humans.

A recent sci-fi example of amorphous computing systems represent the so-called "localizers" in the works of American mathematician and computer scientist, professor Vernor Vinge. In 1999 Vinge won the highest sci-fi Nebula and Hugo Award for his novel "Deepness in the Sky".[21] Vinge's spaceships and even the entire planets are bespread by tiny robotic nodes. These nodes communicate within a limited range via radio ultra-wide band waves (something like the Bluetooth of a distant future) powered by microwaved pulses. The nodes are equipped by cameras, microphones, movement detectors, biometric and other sensors, and cover each corner of the planet and, like dust motes, float in the air or in the air-conditioned atmosphere of the spaceships. They also have connection to other computing and sensing facilities. The nodes are "aware" of their relative position and of their position with respect to their environment. The localizers in the vicinity of one person's eye and brain work as a human-to-localizers interface. In fact, it is an unbelievably sophisticated successor of today's Internet. In this way, people can silently communicate with each other, access the giant databases, observe far-off events, and even act on a distance through the localizers' actuators.

The next motivation for considering amorphous computational systems comes from the internal needs of the computability theory. Here not only the limits of what is computable, but also, the simplest computational models achieving these limits, are sought. Surprisingly, it appears that also here the amorphous computing systems have their say. Namely, it appears that in certain cases the nodes of the amorphous computing systems can be simplified down to the simplest computational devices — bounded depth combinatorial circuits — while retaining the universal computing power of the system as a whole.

The last and the most important motivation for considering amorphous computing systems is their considerable application potential. The respective devices could, e.g., be spread from an airplane on the surface over a certain region (on the Earth, or on an other planet, perhaps) in order to monitor certain parameters of the environment, such as temperature, humidity, or precipitation. These measurements can be broadcast into a base station that will make the data processing. The information can be used, e.g., in agriculture to increase the yields of plants, in space exploration to discover life, etc. Such gadgets can also be used for object surveillance — once spread over the respective perimeter they can monitor the environment in order to detect intruders. Another application is in medical sciences — the devices can be attached to patients and spread over the hospitals in order to monitor the patients' movements and life functions. A nano-size device of such kind can even enter human bodies in order to perform genetic manipulations at the cell level, to strengthen the immune system, to heal up injuries, to cure heart or brain strokes, to substitute the red cells in the blood, etc. Real bacteria represent a template for such systems already existing in nature. Substitutional tissues and organs can be grown in this way, too. Ray Kurzweil, an artificial intelligence guru and the prophet of the Singularity, sees an irreplaceable role of nanobots (i.e., of nano-sized swarms of robots operating like amorphous computing systems) in bringing the mankind towards a point in time in the near future (the Singularity point), when the technological means allow to create a super-human intelligence.[12]

1.2. *State-of-the art*

The contemporary engineering efforts for constructing amorphous computing systems are represented, e.g., by the 2001 project of a "smart dust" by K. S. J. Pister (University of California) (cf.[10,11,22,23]). Smart dust is an ad-hoc network made from micro-electronic devices of approximatively several cubic mm size or smaller. The long-term goal was to reach the size of a dust mote (cf. the survey article[18]). For communication in a nano-world the use of so-called nano-radios has been proposed.[7]

Along with the Smart-Dust project another project appeared in 1999, giving the field its name: Amorphous Computing Project at MIT. Also this project assumed a vast number of identical computational elements randomly placed in a bounded volume. The goal was to construct these elements in such a way that they can self-organize into a prescribed shape, thus mimicking the growth of organic tissues or even of the entire organs, or in order to perform a certain coordinated action (cf.[1,2,3]).

A more recent example of a still ongoing project in amorphous computing is offered by so-called *speckled computing*. Speckled computing is an emerging technology in which data will be sensed in a minute, ultimately around one cubic millimeter, semi conductor grain called speck.[5] A speck has capability to process, sense and communicate in range of few centimeters. It is intended to be autonomous, mobile, and can be proactive in nature. Wireless networks of thousands of specks are considered. Speckled computing is currently being researched and developed by a group of five universities. Their ultimate goal is to reduce the size of the speck so that it can be designed and developed at nanoscale.[19]

The common design framework in all three above mentioned projects has been the fact that all of them make use of a vast amount of very simple autonomous devices. These devices are randomly placed in the target area

— there is no regular topology assumed as in the case of cellular automata. Another joint idea in these projects has been that the respective devices should self-organize in order to perform a coordinated action none of the element alone was able to realize.

Nevertheless, there is one more common feature characterizing the initial approaches to amorphous computing as illustrated by the previous projects: this is the prevailing focus on engineering, or technological aspects of such devices without paying attention to theoretical questions related to computational power and efficiency of such systems. Without knowing their theoretical limits, one cannot have a complete picture of the potential abilities and limitations of such systems.

This was the starting point for the project of the present author and his Ph.D. student L. Petrů (between 2004-2009). Initiated in 2004, this is an on-going project in basic research devoted to studies of theoretical issues in amorphous computing.

2. Our Approach to Amorphous Computing

In our project, a series of formal models of amorphous computing systems has been investigated. All models have been based on informal ideas on amorphous computing systems scattered through the above mentioned projects and the respective reports and papers. The novel idea was to investigate how far one can go in simplifying the nodes of the amorphous computing system while retaining its functionality. The computational universality has been considered as the measure of the system's functionality. Namely, the universality of an amorphous system implies that the system is programmable — is capable of solving whatever algorithmic or robotic task. In that sense, the requirement of universality is a "killer application" for such systems.

Several models capturing increasingly more general instances of amorphous computing systems, working in air-borne or water-borne media, have been designed and studied. The models range from the static models of amorphous computing systems in which processors do not move (cf. Section 3.1 and Section 3.2) up to dynamic models with mobile processors (cf. Section 3.3 and Section 3.4). The computational power of the nodes ranges from finite automata to miniature RAMs with severely restricted memory. The wireless communication under consideration covers both radio connection (in the air-borne media) and molecular communication (in aqueous media) (cf. Section 3.4).

The constraints put on amorphous computing systems in our approach are best seen when considering their relationship to wireless sensory networks. Indeed, amorphous computing systems represent a specific kind of wireless sensory networks. Such systems can be seen as an extreme case of wireless sensory networks: not only are the nodes of amorphous systems considered under severe size and cost constraints resulting into corresponding constraints on resources such as energy, memory, and computational speed. There are additional constraints usually not considered in the domain of wireless sensory networks: the computational and communication hardware in the processors is stripped down to an absolute minimum that seems to be necessary in order to maintain the required functionality and scalability of the network. For instance, in order to allow potentially unbounded scalability of amorphous computing systems their processors initially do not possess any identity: they are all alike, they are very simple and can be seen as finite state automata. In order to maximally simplify the wireless communication mechanism in the nodes of the amorphous computing systems the existence of no communication software is assumed. There is no synchronicity assumed and the communication among the nodes is "blind" — a sender cannot learn that its message has been delivered since a receiver node cannot reliably send message acknowledgment to a message sender. The broadcast is unreliable — the simultaneously broadcasting nodes jam each other in such a way that no message can be delivered and, moreover, the interference in broadcasting cannot be recognized by the processors. The design of communication protocols working reliably under such constraints has been the main task in our dealing with amorphous computing systems. Thanks to previously mentioned minimalistic requirements on the communication mechanisms in amorphous computing systems other than wireless radio-based communication schemes can be considered. For instance, in water-borne media chemical communication based on molecular exchange by diffusion or convection can be considered. This allows considering molecularly interacting mobile nano-sized amorphous computing systems (cf.[25,28] and Section 3.4).

To some extent, amorphous computing systems could also be compared to so-called population protocols introduced by Angluin et al. (cf.[4]). In the underlying model, they consider the anonymous finite-state agents computing a predicate of the multiset of their inputs via two-way or one-way interactions in the all-pairs family of communication networks. This model and its variants differ from our model of amorphous computing systems in many aspects, the most important being that always an ordered pair of processors (a sender and a receiver) is considered, allowing a reliable one- or two-way interaction. This assumption alone already goes well beyond the basic communication postulates in our approach to the amorphous computing systems. Nevertheless, the focus on theoretical computational issues in population protocols is similar to our approach.

3. The Models and the Results

In order to investigate computational properties of amorphous computing systems (in fact, of any distributed computational system) one must (i), formulate a mathematical or formal computational model of such systems capturing all their important aspects, (ii), design communication and other algorithms for such systems, and (iii), study their computational properties. This is the uniform methodology which has been applied in a series of articles within our approach to amorphous computing,[14,15,16,17,24,25,26] and.[28] In the sequel we present the main design ideas behind the respective models, point to the main problems to be solved, indicate their solution, and present the main results.

3.1. *Amorphous cellular automata*

The first and the simplest model we have considered was the model of a *synchronous amorphous computing system* densely covering a given area. In place of processors we considered finite automata equipped by broadcasting mechanism. The idea was to show that under a certain statistical assumption concerning the coverage of the given area (for simplicity, we considered a rectangular area), the processors of the system could self-organize into a roughly rectangular grid that is able to simulate a two-dimensional cellular automaton which is known to possess universal computing power.

The organization into a rectangular grid is initiated by a process starting in all nodes on the left side of the area. From these nodes, within the communication radius r a signal is spread to all its neighbors who retransmit it further. The nodes count modulo k, for a certain constant $k > 0$. As a result, the nodes are divided into vertical strips of width k. A similar process is then initiated in all nodes on the lower side of the area. These two "orthogonal" processes enable to mark the sub-squares in the intersections of strips alternately by black or white color, respectively, effectively splitting the area into a roughly rectangular grid resembling a chessboard, with squares of average expected size $O((kr)^2)$. Now the nodes from black and white squares can alternate in simulating the steps of a cellular automaton.

The simulation works under the assumptions that (i) the processors work in a synchronous manner, (ii) there are no broadcast conflicts, (iii) in each cell of a grid overlaying the area there is at least one processor, and (iv), the communication radius r of a processor in any cell covers that cell.

The system has been described in greater detail in.[14]

3.2. *Asynchronous stationary amorphous computing systems*

Obviously, the latter assumptions on the simulation of a two-dimensional cellular automaton were too strong to make the system interesting also in practice. Nevertheless, this first result served as a proof of the principle — at least it showed that amorphous computing systems of a certain simple kind can, indeed, have a universal computing power.

In our next model, we tried to get rid of the previous bold assumptions. The resulting model of *asynchronous stationary amorphous computing system* consisted of asynchronous processors. More precisely, each processor possessed a clock running with the same speed as in other processors; however, the "ticking" of all clocks was not synchronized. Each processor was modeled by a "miniature" RAM with a finite number of registers capable of storing integers up to size N, with N denoting the number of nodes of the underlying amorphous system. As before, each processor was equipped by a single-channel radio device of a limited communication range. The processors had no identifiers. There was also a random number generator in each processor.

Now, if N of such processors were randomly spread over a given square area, they formed a network whose nodes were created by processors with wireless communication links emerging among the processors that were within the communication radius of each other. In general, the underlying communication graph will not be a connected graph. The desirable properties of the underlying random communication graph are: a potentially large connected component whose nodes can be used for representing all registers of the simulated RAM, a not-too-large number of neighbors of any node (diminishing the probability of communication conflicts), and a small graph diameter (enabling a faster simulation).

A severe restriction was imposed on the communication abilities of processors. A processor P_1 could receive a message sent by processor P_2 if and only if the following conditions held true: (i) the processors were in the communication range of each other, (ii) P_1 was in a listening mode, and (iii) P_2 was the only processor within the communication radius of P_1 broadcasting at that time.

There is no mechanism making it possible to distinguish the case of no broadcast from that of broadcast collision. These restrictions concerning the radio communication are among the weakest ones that one can expect to be fulfilled by any simple radio communication device. The expected benefit from such restrictions is a simple engineering design of processors.

Now, the main task was to design a randomized protocol enabling a reliable delivery of a message among processors within the communication range of each other. The problem was complicated by the fact that the processors worked asynchronously, had no identifiers, and by the absence of broadcast collision detection mechanism.

The key idea leading to a solution was that the processors should broadcast a message sporadically in order to prevent message delivery (i.e., broadcast) conflicts, and repeatedly in order to maximize the likelihood of a successful delivery. It turned out that the probability of sending should depend inversely on the number of a node's neighbors and should be repeated more times if there were more processors in a node neighborhood. Such a broadcast protocol was presented in.[15] Interestingly, for the simpler synchronous case, the problem of message delivery under conditions similar to ours has been studied and solved in the seminal paper by Bar-Yehuda, Goldreich and Itai in 1992.[6]

Having a basic communication protocol, it was a relatively plain task to design a "flooding" protocol delivering a message to all accessible nodes in a network. As the next step, addresses of processors must be generated using the random number generators in the processors. Care must be taken that all addresses in the required range of addresses are generated (duplicates do not harm), giving enough room for simulating all RAM registers. Then, each register of a simulated RAM is realized by one or more processors, and the simulation proceeds relatively straightforwardly. The reliability of the simulation depends on the reliability of each message delivery. The latter reliability can be increased by repeating the sending of each message.

The results were quite encouraging, having solved satisfactorily the basic problem of simulating a RAM by a reasonably realistic model of amorphous computer (cf.,[24,26]).

3.3. *Flying amorphous computers*

The above mentioned work was the first step towards the so-called *flying amorphous computers* in which the processors of the same type as the in asynchronous stationary model moved freely and randomly in a closed area. The main problem here was to keep the system operating under steadily changing topology of the communication network where new communication paths emerge, while the previous ones vanish. This has eventually happened under the assumption that no node in the network remains for ever isolated. Thanks to this assumption, a message sent to a node with a given address would in a finite time reach this node and this node could send an acknowledgment that in a finite time will reach the sender. The schema of the simulation is as before: first, we assign addresses to each node so that each node represents the corresponding RAM register with the given address (duplicates are eliminated with a high probability), and then the simulation proceeds straightforwardly as in the previous case of asynchronous stationary amorphous computing system. The next step is initiated only after the sender obtains an acknowledgment from the receiver. Although the whole system can correctly simulate a RAM with arbitrary high probability, the simulation time cannot be bounded by any function. However, if the address assignment process is successful (an this can be guaranteed with an arbitrary large probability), the simulation terminates within a finite time and always delivers the correct result. This computer has been described in full detail in[14] and later it was presented in.[16]

The previous simulation has suffered from one major problem. Namely, aside from the fact that the time of simulation could not have been bounded, the simulation has not been robust enough — a defect of a single processor would block the entire simulation. In the forthcoming paper[17] an improved simulation has been suggested. The idea has been to set a time upper bound for a sender to wait for an acknowledgment. After its expiration the processor at hand is declared as "dead" (or "inaccessible") and a new version of the whole simulation, inclusively the address allocation process, is started from the very beginning. The versioning is necessary since for some time processors from the previous and the current simulation run coexist. Their mixing is prevented by assigning a unique version number to each processor with a newly assigned address. In order this to work a pool of reserve processors is kept that could be used in place of dead processors. Processors that went astray and in the meantime have became accessible are added to this pool. The resulting algorithm is robust against the defect of processors or against their loss. The resulting running time is difficult to estimate since it is influenced by several factors. The first factor is the availability of nodes holding the input data. If such a node gets damaged then computation cannot be restarted. The next factor is the dependence of the total running time on the number of restarts. This, in turn, depends on the number of failures or the number of temporarily (for the prescribed time period) inaccessible nodes. Finally, also the number of reserved memory nodes (representing the RAM registers) can get exhausted due to the node defects. Simulations have shown that the resulting algorithm is robust, as expected, indeed, but moreover in some cases it also tends to terminate correctly faster than the algorithm without built-in robustness. However, the fixing of all parameters of the simulation is tricky and no closed analytic expression for the expected running time could have been obtained.

3.4. *Molecularly communicating nano-machines*

It turnes out that the similar ideas holding for the case of nodes communicating via radio also work in the case of mobile nano-machines operating in liquids (e.g., in a blood stream).

Nano-machines are wirelessly communicating self-reproducing embodied nanosized (i.e., of order 10^{-6} mm) automata. The information exchange mechanism is based on molecular communication. Their prospective fabrication will make use of molecular self-assembly or of modifications of real bacteria via genetical engineering.

In our approach, each nano-machine was modeled by a timed probabilistic finite-state automaton equipped by a mechanism for molecular communication using the same communication protocol as described in Section 3.2. The resulting amorphous system was shown to be able to model a counter automaton,[25] thus achieving again a universal computing power.

A similar result has recently been also shown for self-reproducing mobile nano-machines communicating via so-called *quorum sensing*, i.e., by making collective decisions based on the density of nano-machine population.[28] This density has been inferred from the concentration of signal molecules emitted by the machines within a confined space. Timed probabilistic finite-state automata have been used again as the control unit of machines. The self-reproducing organ of a nano-machine is a non-computational mechanism which is a part of the machine's embodiment. In a given volume the machines multiply and emit the signal molecules until their maximal concentration has been reached. Then, they make a collective decision in which they simulate one step of a counter automaton.

A further modification of the embodiment of the underlying automata also includes a memory organ, the timers, and an organ serving as a random bit generator. Then, the task of memorizing the current state, triggering of computational processes (via timers), and generating random bits can be delegated to the nano-machines' embodiment. Consequently, the computational mechanism of each nano-machine could be stripped down to combinatorial circuits of bounded depths (circuits from the complexity class AC_0). The resulting population of nanomachines controlled by circuits probably presents the simplest universal amorphous computing system since, from a computational point of view, its activity is governed by the simplest computing devices.

The last mentioned model is of interest not only from the viewpoint of the theory of universal computing machines, but also from a practical point of view, since it could lead to a simpler engineering of nanomachines.

4. Towards the Theory of Amorphous Computing

Our main result has been the proof of probabilistic computational universality of all amorphous computing systems we have considered. From the viewpoint of the computational complexity theory, the underlying computational models are non-uniform models. The reason is that in order to simulate classical universal models (a RAM, a counter automaton) some parameters of the amorphous model, especially timing of various computational mechanisms, must be adjusted to the volume of the environment. The environment itself must be big enough to accommodate a sufficient number of nodes of the amorphous computing system which, in turn, depends on the number of the inputs to the system. It is interesting to observe that practically all contemporary computing systems, be it natural or artificial ones, from the Internet to amorphous computing systems, from bacteria to the brain, can be modeled as non-uniform computational systems which in principle possess super-Turing computing power (cf.,[20,27]).

Experimental simulations of our amorphous computing systems have revealed a number of dependencies and conjectures that we were not able to prove analytically. More work is needed in order to design still more realistic models (e.g., we did not consider energy consumption minimization problem).

In general, our work on amorphous computing systems has identified interesting research problems in the graph theory, distributed algorithm design and their complexity analysis, and even in the fundamentals of computability theory.

No doubt that amorphous computing systems present a new computational paradigm that will cover an increasing number of future computing systems and their applications.

References

1. H. Abelson, et al. Amorphous Computing. MIT Artificial Intelligence Laboratory Memo No. 1665, Aug. 1999.
2. H. Abelson, D. Allen, D. Coore, Ch. Hanson, G. Homsy, T. F. Knight, Jr., R. Nagpal, E. Rauch, G. J. Sussman, R. Weiss. Amorphous Computing. Communications of the ACM, Volume 43, No. 5, pp. 74–82, May 2000.
3. H. Abelson, J. Beal, G. J. Sussman. Amorphous Computing. Computer Science and Artificial Intelligence Laboratory, Technical Report, MIT-CSAIL-TR-2007-030, June 2007.
4. D. Angluin, J. Aspnes, D. Eisenstat, E. Ruppert: The computational power of population protocols. Distributed Computing 20(4):279–304, November 2007.
5. D. K. Arvind, K. J. Wong: Speckled Computing — A Disruptive Technology for Network Information Appliances. Proc. IEEE International Symposium on Consumer Electronics (ISCE'04), 2004, pp. 219–223.
6. R. Bar-Yehuda, O. Goldreich, A. Itai: On the Time-Complexity of Broadcast in Multi-hop Radio Networks: An Exponential Gap Between Determinism and Randomization. J. Comput. Syst. Sci. Vol. 45, No. 1, pp. 104–126, 1992.
7. K. Bullis. TR10: NanoRadio. Technology Review. Cambridge: MIT Technology Review, 2008-02-27.
8. D. Coore: Introduction to Amorphous Computing. Unconventional Programming Paradigms: International Workshop 2004, LNCS Volume 3566, pp. 99–109, Aug. 2005.
9. F. Hoyle: The Black Cloud. Penguin Books, 1957, 219 pp.
10. J. M. Kahn, R. H. Katz, K. S. J. Pister. Next century challenges: mobile networking for "Smart Dust". In: Proceedings of the 5th Annual ACM/IEEE International Conference on Mobile Computing and Networking, MobiCom '99, ACM, pp. 271–278, Aug. 1999.
11. J. M. Kahn, R. H. Katz, K. S. J. Pister. Emerging Challenges: Mobile Networking for Smart Dust. Journal of Communications and Networks, Volume 2, pp. 188–196, 2000.
12. R. Kurzweil: The Singularity is Near. Viking Books, 652 pages, 2005.
13. Nikoletseas, S.: Models and Algortihms for Wireless Sensor Networks (Smart Dust). In: SOFSEM 2006: Theory and Practice of Computer Science, Proceedings. Eds.: J. Wiedermann et al., LNCS Vol. 38312, Springer, pp. 65–83, 2006.

14. L. Petrů: Universality in Amorphous Computing. PhD Disseration Thesis. Dept. of Math. and Physics, Charles University, Prague, 2009.
15. L. Petrů, J. Wiedermann: A Model of an Amorphous Computer and Its Communication Protocol. In: Proc SOFSEM 2007: Theory and Practice of Computer Science. LNCS Volume 4362, Springer, pp. 446–455, July 2007.
16. L. Petrů, J. Wiedermann: A Universal Flying Amorphous Computer. In: Proc. Unconventional Computation, 10th International Conference, UC'2011, LNCS, Vol. 6714, 2011, pp. 189–200.
17. L. Petrů, J. Wiedermann: A Robust Universal Flying Amorphous Computer. In preparation, Institute of Computer Science AS CR, Prague, 2011.
18. M. J. Sailor, J. R. Link: Smart dust: nanostructured devices in a grain of sand, Chemical Communications, vol. 11, p. 1375, 2005.
19. S. C. Shah, F. H. Chandio, M. Park: Speckled Computing: Evolution and Challenges. Proc. IEEE International Conference on Future Networks, 2009, pp. 181–185.
20. J. van Leeuwen, and J. Wiedermann, The Turing machine paradigm in contemporary computing. In: eds. B. Enquist and W. Schmidt, *Mathematics Unlimited - 2001 and Beyond*, Springer-Verlag, Berlin, 2001, pp. 1139–1155.
21. V. Vinge: A Deepness in the Sky. Tor Books, January 2000, 800 pp.
22. B. Warneke, M. Last, B. Liebowitz, K. S. J. Pister: Smart Dust: communicating with a cubic-millimeter computer. Computer, Volume: 34, Issue: 1, pp. 44–51, Jan. 2001.
23. B. Warneke, B. Atwood, K. S. J. Pister: Smart dust mote forerunners. In: Proceedings of the 14th IEEE International Conference on Micro Electro Mechanical Systems, 2001, MEMS 2001, pp. 357–360, 2001.
24. J. Wiedermann, L. Petrů: Computability in Amorphous Structures. In: Proc. CiE 2007, Computation and Logic in the Real World. LNCS Volume 4497, Springer, pp. 781–790, July 2007.
25. J. Wiedermann, L. Petrů: Communicating Mobile Nano-Machines and Their Computational Power. In: Third International ICST Conference, NanoNet 2008, Boston, MA, USA, September 14–16, 2008, Revised Selected Papers, LNICST Vol. 3, Part 2, Springer, pp. 123–130, 2009.
26. J. Wiedermann, L. Petrů: On the Universal Computing Power of Amorphous Computing Systems. Theory of Computing Systems 46:4 (2009), 995–1010, www.springerlink.com/content/k2x6266k78274m05/fulltext.pdf
27. J. Wiedermann, and J. van Leeuwen, How We Think of Computing Today (Invited Talk). In: *Proc. CiE 2008*, LNCS 5028, Springer, Berlin, 2008, pp. 579–593.
28. J. Wiedermann: Nanomachine Computing by Quorum Sensing. In: Computation, Cooperation, and Life. Essays Dedicated to Gheorghe Păun on the Occassion of His 60th Birthday. J. Kelemen, A. Kelemenová, eds. LNCS Vol. 6610, Springer, pp. 203–215.

Chapter 14

Computing on Rings

Genaro J. Martínez[1], Andrew Adamatzky[1] and Harold V. McIntosh[2]

[1] Unconventional Computing Center
University of the West of England, Bristol BS16 1QY, United Kingdom
[2] Departamento de Aplicación de Microcomputadoras, Instituto de
Ciencias, Universidad Autónoma de Puebla, Puebla, México

In this paper, we will review the developing features of computations based on rings. Particularly, we will analyse what kinds of interaction occur between gliders travelling on a 'cyclotron' cellular automaton derived from a catalog of collisions. We will demonstrate that collisions between gliders emulate the basic types of interaction that occur between localizations in non-linear media: fusion, elastic collision, and soliton-like collision. Computational outcomes of a swarm of gliders circling on a one-dimensional torus are analysed via implementation of some simple computing models. Gliders in one-dimensional cellular automata are compact groups of non-quiescent patterns translating along an automaton lattice. They are cellular-automaton analogous to localizations or quasi-local collective excitations travelling in a spatially extended non-linear medium. So, they can be represented as binary strings or symbols travelling along a one-dimensional ring, interacting with each other and changing their states, or symbolic values, as a result of interactions. We present a number of complex one-dimensional cellular automata with such features.

Keywords: Rings, cellular automata, particles, collisions, beam routing, unconventional computing, computability.

1. Introduction

Computations as effective procedures were developed in the past century by logicians and mathematicians such as Kurt Gödel, Alonzo Church, Alan Turing, Emil Post and Stephen Kleene.[20] A notion of *unconventional computing* has also been developed more recently, and consist on abstract and concrete models in non-linear media, typically capable of massively parallel computation based in different physics and complementary logic. To mention but a few examples, there is conservative logic,[21] reversible computing,[14,47] reaction-diffusion computers,[6] Physarum computers,[5] cellular automata computers,[1,24,34,38,51,58] collision-based computing such as optical or molecular computing,[3] solitons or competing patterns computing,[2,32] or hot ice computers.[4]

In this paper, we consider a particular case where computations can be carried by particle collisions with complex cellular automata (CA). Here we study one-dimensional CA and glider interaction on the evolution space. But confined naturally in one dimension, on rings. This way, gliders (or particles) can be represented as a set of strings which may be characterised via de Bruijn diagrams.[40,41] Hence each glider is encoded as a regular expression and initialised on a specific initial condition. So computations are a consequence of cyclical interaction of gliders on such "cyclotrons". We should differentiate that this study is not related to: circular Turing machines,[11] circular Post machines[29] or cyclic tag systems[18,58] because in this case, sets of strings represent gliders and they shall be transformed to other set of strings and periodic strings that are persistent in time. We will illustrate such devices in complex elementary CA (ECA)[57,58] and ECA with memory (ECAM).[9,10]

2. One-Dimensional Cellular Automata

2.1. *Elementary cellular automata (ECA)*

A one-dimensional CA can be represented by an array of *cells* x_i where $i \in \mathbb{Z}$ (integer set) and each x takes a value from a finite alphabet Σ. Thus, a sequence of cells $\{x_i\}$ of finite length n describes a string or *global configuration* c on Σ. This way, the set of finite configurations will be expressed as Σ^n. An evolution is comprised by a sequence of configurations $\{c_i\}$ produced by the mapping $\Phi : \Sigma^n \to \Sigma^n$; thus the global relation is symbolized as:

$$\Phi(c^t) \to c^{t+1} \qquad (1)$$

where t represents time and every global state of c is defined by a sequence of cell states. The global relation is determined over the cell states in configuration c^t updated at the next configuration c^{t+1} simultaneously by a local function φ as follows:

$$\varphi(x_{i-r}^t, \ldots, x_i^t, \ldots, x_{i+r}^t) \to x_i^{t+1}. \tag{2}$$

Wolfram represents one-dimensional CA with two parameters (k, r),[57] where $k = |\Sigma|$ is the number of states, and r is the neighbourhood radius, hence the ECA domain is defined by the parameters $(2, 1)$. There are Σ^n different neighbourhoods (where $n = 2r + 1$) and k^{k^n} distinct evolution rules. The evolutions in this paper have periodic boundary conditions.

2.2. *Elementary cellular automata with memory (ECAM)*

Conventional CA are memoryless, that is the new state of a cell depends on the neighbourhood configuration solely at the preceding time step of φ. CA with *memory* can be considered as an extension of the standard framework of CA where every cell x_i is allowed to remember some time window of its previous evolution. CA with memory were originally proposed in.[7–10] Basically memory is based on the state and history of the system, thus we design a memory function ϕ, as follows:

$$\phi(x_i^{t-\tau}, \ldots, x_i^{t-1}, x_i^t) \to s_i \tag{3}$$

such that $\tau < t$ determines the backwards degree of memory and each cell $s_i \in \Sigma$ is a function of the series of states in cell x_i up to time-step $t - \tau$. Finally to execute the evolution we apply the original rule again as follows:

$$\varphi(\ldots, s_{i-1}^t, s_i^t, s_{i+1}^t, \ldots) \to x_i^{t+1}.$$

In CA with memory, while the mapping φ remains unaltered, a historic memory of past iterations is retained by featuring each cell as a summary of its previous states; therefore cells *canalize* memory to the map φ. As an example, we can take the memory function ϕ as a *majority memory*:

$$\phi_{maj} \to s_i \tag{4}$$

where in case of a tie given by $\Sigma_1 = \Sigma_0$ in ϕ, we shall take the last value x_i. So ϕ_{maj} represents the classic majority function for three variables,[46] as follows:

$$\phi_{maj} : (x_1 \wedge x_2) \vee (x_2 \wedge x_3) \vee (x_3 \wedge x_1) \to x$$

on cells $(x_i^{t-\tau}, \ldots, x_i^{t-1}, x_i^t)$ and defines a temporal ring before calculating the next global configuration c. In case of a tie, it is allows to break it in

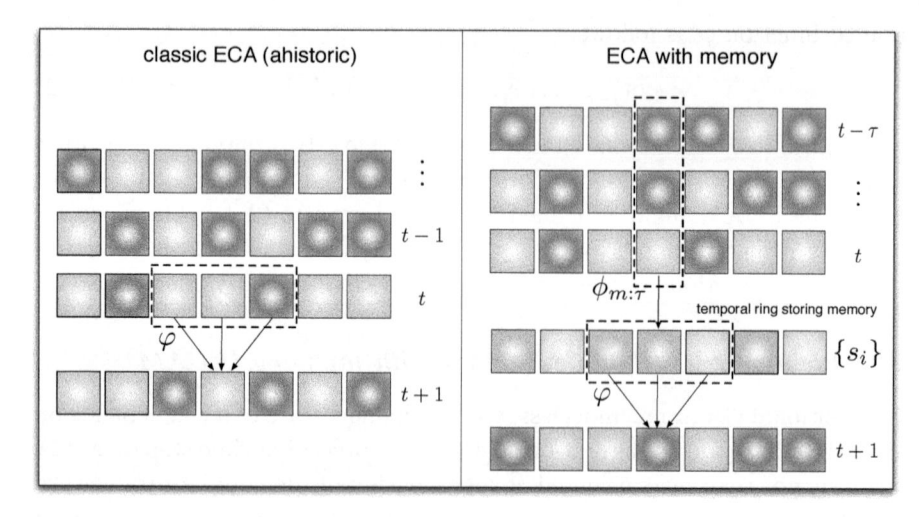

Fig. 1. Ahistoric CA (left) and with memory (right).

favour of zero if $x_{\tau-1} = 0$, or to one whether $x_{\tau-1} = 1$. The representation of a ECA with memory (given previously in[31,33,37]) is given as follows:

$$\phi_{CARm:\tau} \qquad (5)$$

where CAR represents the decimal notation of a particular ECA and m the kind of memory given with a specific value of τ. Thus the majority memory (maj) working in ECA Rule 126 checking tree cells ($\tau = 3$) of history is simply denoted as $\phi_{R126maj:3}$.[37] Figure 1 depicts in detail the memory working on ECA. Note that memory is as simple as any CA local function and therefore preserve and increase its dynamics in new orders of complexity.[31]

3. Strings in One-Dimensional Cellular Automata

We will handle particles or gliders as sets of strings, this way each string represents a specific glider with some properties as well. The set of strings represent a subset of regular expressions. Hence we will use de Bruijn diagrams and tiling patterns to extract such strings.

3.1. *Regular expressions*

Several interesting problems arise in a study of formal languages; one of them is to determine the type of language derived and to which class the language belongs. This hierarchy is well-known and established by Chomsky's classification. We shall study languages determined by regular sets, since the set of expressions determined by each glider on a CA evolution rule can be associated to a particular regular expression.

The finite automaton is a mathematical model with a system of discrete inputs and outputs; the system can be placed in one of a finite set of states. This state has the information of the received inputs necessary to determine the behaviour of the system with regard to subsequent inputs. Formally, a finite automaton M consists of a finite set of states and a set of transitions among states induced by the symbols selected from some alphabet. For each symbol there is a transition from one state to the other (can be the same one); there is an initial state where the automaton starts and some states are designated as final ones or as acceptance states.[25]

A directed graph called a *transition diagram* is associated with a finite automaton as follows: the vertices of the graph correspond to the states of the automaton; for a transition from state i to state j produced by an input symbol, there is an edge labeled by this symbol from i to j in the transition diagram. The finite automaton accepts a chain w if the analogous transition sequence leads from the initial state to a final one (acceptance).

A *language accepted* by M, represented by $L(M)$, it is the set $\{w|w$ is accepted by $M\}$. The type of languages accepted by a finite automaton is important because they complement the analysis established with regular expressions. Historically an important relation was established by Kleene demonstrating that regular expressions can be expressed by finite automata and vice versa, i.e., they are equivalent representations.[46] In other words, a language is a *regular set* if it is accepted by some finite automaton. The accepted languages by finite automata are described by expressions known as *regular* expressions; particularly, the accepted languages by finite automata are indeed the class of languages described by regular expressions.

The sets of *regular expressions* on an alphabet are defined recursively as:[25]

(1) ϕ is the regular expression representing the empty set.
(2) ϵ is the regular expression describing the set $\{\epsilon\}$.
(3) For each symbol $a \in \Sigma$, a is a regular expression depicting the set $\{a\}$.
(4) If a and b are regular expressions representing languages A and B respectively, then $a + b$, ab, and a^* are regular expressions representing $A \cup B$, AB and A^* respectively.

When it is necessary to distinguish between a regular expression a and the language determined by a, we shall use L_a.

The formal languages theory provides a way to study sets of chains from a finite alphabet. The languages can be seen as inputs of some classes of machines or like the final result from a typesetter substitution system i.e., a generative grammar into the Chomsky's classification.[26]

Table 1. Language classes.

language	structure
recursively enumerated	Turing machine
context sensitive	linear bounded automata
context free	pushdown automata
regular	finite automata

The basic model necessary for the languages of these machines (and for any computation), is the Turing machine; the machines recognising each family of languages are described as a Turing machine with restrictions. The relevance of the association between machines resolving each type of language established the classification (Table 1 of[26]).

Some languages are established by regular sets;[a] here we will take all the words recognised by the de Bruijn diagram that represent precisely a set of regular expressions,[41] and we just need those chains representing a structure on a specific complex CA evolution rule, to manipulate the evolution space with constructions of gliders.

3.2. de Bruijn diagrams

De Bruijn diagrams[40,41] are very adequate for describing evolution rules in one-dimensional CA, although originally they were used in shift-register theory (the treatment of sequences where their elements overlap each other). We shall explain the de Bruijn diagrams illustrating their constructions for determining chains w defining a pair of gliders in $\mathcal{G}$, for complex CA.

For a one-dimensional CA of order (k, r), the de Bruijn diagram is defined as a directed graph with k^{2r} vertices and k^{2r+1} edges. The vertices are labeled with the elements of the alphabet of length $2r$. An edge is directed from vertex i to vertex j, if and only if, the $2r - 1$ final symbols of

---

[a]Examples and properties of formal languages, grammars, finite state machines, Turing machines and equivalent systems, can be consulted in.[11,19,25,46,53]

i are the same that the $2r-1$ initial ones in j forming a neighbourhood of $2r+1$ states represented by $i \diamond j$. In this case, the edge connecting i to j is labeled with $\varphi(i \diamond j)$ (the value of the neighbourhood defined by the local function).[54,55]

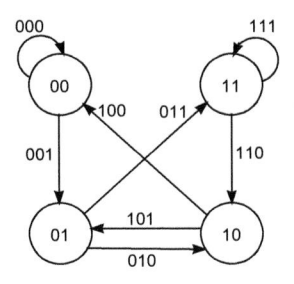

Fig. 2. Generic de Bruijn diagram for a ECA (2,1).

The connection matrix M corresponding with the de Bruijn diagram is as follows:

$$M_{i,j} = \begin{cases} 1 \text{ if } j = ki, ki+1, \ldots, ki+k-1 \ (\text{mod } k^{2r}) \\ 0 \text{ in other case} \end{cases} \quad (6)$$

Module $k^{2r} = 2^2 = 4$ represent the number of vertices in the de Bruijn diagram and j must take values from $k*i = 2i$ to $(k*i)+k-1 = (2*i)+2-1 = 2i+1$. Vertices are labeled by fractions of neighbourhoods originated by 00, 01, 10, and 11, the overlap determines each connection. In Table 2 the intersections derived from the elements of each vertex are showed; they are the edges of the de Bruijn diagram as we can see in Figure 2.

Table 2. Intersections determining the edges of the de Brujin diagram.

(0,0) ⋄ (0,0)	000
(0,0) ⋄ (0,1)	001
(0,1) ⋄ (1,0)	010
(0,1) ⋄ (1,1)	011
(1,0) ⋄ (0,0)	100
(1,0) ⋄ (0,1)	101
(1,1) ⋄ (1,0)	110
(1,1) ⋄ (1,1)	111

The de Bruijn diagram has four vertices which can be renamed as $\{0, 1, 2, 3\}$ corresponding with the four partial neighbourhoods of two cells $\{00, 01, 10, 11\}$, and eight edges representing neighbourhoods of size $2r + 1$.

Paths in the de Bruijn diagram may represent chains, configurations or classes of configurations in the evolution space.

Vertices in the de Bruijn diagram are sequences of symbols in the set of states and the symbols are sequences of vertices in the diagram. The edges describe how such sequences can be overlapped; consequently, different intersection degrees yield distinct de Bruijn diagrams. Thus, the connection takes place between an initial symbol, the overlapping symbols and a terminal one (Table 2). Sequences derived from a de Bruijn diagram are the set of regular expressions that a CA can generate since its evolution rule.[41]

Also, we have the extended de Bruijn diagrams[b] that calculate all the periodic sequences by the cycles defined in the diagram. These ones also calculate the shift of a periodic sequence for a certain number of steps; thus we can get de Bruijn diagrams describing all the periodic sequences characterising a glider in any complex CA.

In order to illustrate how the sequences of each glider are determined, we calculate the de Bruijn diagram composing an A glider in Rule 110, and discussing how the periodic sequences are extracted for representing this glider and specifying as well the set of regular expressions for such glider.

The A glider moves two cells to the right in three times (for details see[44]). We compute the extended de Bruijn diagram (2-shift, 3-gen) depicted in Figure 3. The cycles of the diagram have the periodic sequences describing the A glider; however, these sequences are not ordered yet. Therefore, we must determine and classify them.

In the figure we have two cycles: a cycle formed by vertex 0 and a large cycle of 26 vertices which is composed as well by 9 internal cycles. The evolution of the right illustrates the location of the different periodic sequences producing the A glider in distinct numbers.

Following the paths through the edges we obtain the sequences or regular expressions determining the phases of the A glider. For example, we have cycles formed by:

   I. The expression (1110), vertices 29, 59, 55, 46 determining A^n gliders.

---

[b]The de Bruijn diagrams were calculated with the NXLCAU21 system developed by McIntosh for NextStep (OpenStep and LCAU21 to MSDOS). Application and code source are available from: http://delta.cs.cinvestav.mx/~mcintosh/cellularautomata/SOFTWARE.html

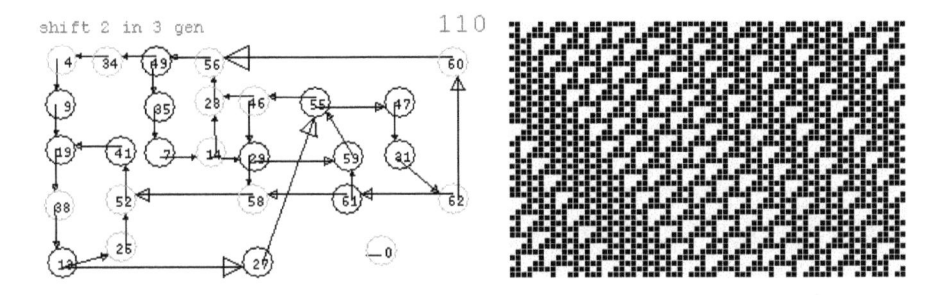

Fig. 3. De Bruijn diagram calculating A gliders and ether configurations.

II. The expression (111110), vertices 61, 59, 55, 47, 31, 62 defining nA gliders with a T_3 tile[c] between each glider.
III. The expression (11111000100110), vertices 13, 27, 55, 47, 31, 62, 60, 56, 49, 34, 4, 9, 19, 38 describing ether configurations in a phase (in the following subsection we will see that it corresponds to the phase $e(f_{1\_}1)$).

The cycle with period 1 represented by vertex 0 produces a homogenous evolution with state 0. The evolution of the right (Figure 3) shows different packages of A gliders, the initial condition is constructed following some of the seven possible cycles of the de Bruijn diagram or several of them. We can select the number of A gliders or the number of intermediate ether configurations changing from one cycle to another.

Following each phase initiated by every T_3 tile, the phases $f_{i\_}1$ for the A glider are as follows:

- $A(f_{1\_}1) = 111110$
- $A(f_{2\_}1) = 11111000111000100110$
- $A(f_{3\_}1) = 11111000100110100110$

The sequence is defined taking the first value from the first cell of T_3 tile on the left until reaching a second cell representing the first value of the second T_3 tile on the right. Finally, such set of strings correspond precisely to the sequences derived from the de Bruijn diagram concatenated with fragments of ether configuration. This way, the set of strings $e(f_{1\_}1)$, $A(f_{i\_}1)$ $\forall\, i = \{1, 2, 3\}$ under the operations $+, \cdot, *$, they are regular expressions. The full set of stings to code gliders in Rule 110 $\mathcal{L}_{R110}$ is represented from the

---
[c]A tile T_3 is formed for the ether configuration in Rule 110, for details please see.[44]

de Bruijn diagrams and tiles.[44][d] Tiles were necessary to represent strings in bigger gliders where the de Bruijn diagram is very hard to calculate.

4. Universal CA and Some CA Computers

Universality in CA was conceptualised, developed and solved by von Neumann in[56] as a previous step in the specification of his universal constructor for a two-dimensional automaton of 29-states. The element of universality into this constructor is a necessary ingredient to handle non-reliable pieces (atomic elements) for assembling reliable components with the capacity of executing computations by collisions of signals. Actually such approaches about complex CA, such as Conway's game of Life or Wolfram's Rule 110, consist in controlling gliders to make them reliable components in order to perform CA engineering leading to computation universality.

There are some significant simplifications in universal-computing CA with less states and dimensions: Codd in 1968,[17] Banks in 1971,[12] Smith in 1971,[50] Conway in 1982,[13] Lindgren and Nordahl in 1990,[27] and finally Cook in 1998.[18,58]

Analogous to the search of the smallest universal Turing machine,[48] Cook provides the smallest universal CA; showing how a "simple" elementary CA (Rule 110) is capable of (Turing) universal computation through the development of a novel type of system, a *cyclic tag system* (CTS), performing computations[2,18] across collisions of gliders along millions of cells.[45][e]

4.1. *Toffoli's symbology*

In the late 1970s Fredkin and Toffoli developed a concept of a general-purpose computation based on ballistic interactions between quanta of information that are represented by abstract particles.[52] The Boolean states of logical variables are represented by balls or atoms, which preserve their identity when they collide with each other. They came up with the idea of a billiard-ball model of computation, with underpinning mechanics of elastically colliding balls and mirrors reflecting the balls' trajectories. Later Margolus developed a special class of CA which implement the billiard-ball

---

[d]The full set of regular expressions to code each glider in Rule 110, including the glider gun, is available from: http://uncomp.uwe.ac.uk/genaro/rule110/listPhasesR110.txt.

[e]A brief list of CA computers is available at: http://uncomp.uwe.ac.uk/genaro/otherRules.html.

model. Margolus' partitioned CA exhibited computational universality because they simulated the (known to be universal) Fredkin gate via collision of soft spheres.[36]

The following basic functions with two input arguments u and v can be expressed via collision between two localizations:

(1) $f(u, v) = c$, fusion
(2) $f(u, v) = u + v$, interaction and subsequent change of state
(3) $f_i(u, v) \mapsto (u, v)$ identity, solitonic collision
(4) $f_r(u, v) \mapsto (v, u)$ reflection, elastic collision

To map Toffoli's supercollider[52] onto a one-dimensional CA we use the notion of an idealised particle $p \in \Sigma^+$ (without energy and potential energy). For our study, the particle p is represented by a binary string of cell states derived from a de Bruijn diagram.

Figure 4 shows two typical scenarios where particles p_f and p_s travel in a CA cyclotron. The first scenario (Fig. 4a) shows two particles travelling in opposite directions which then collide. Their collision site is shown by a dark circle as a contact point in (Fig. 4). The second scenario demonstrates a typical beam routing where a fast particle p_f eventually catches up with a slow particle p_s at a collision site (Fig. 4b). If the particles collide like solitons,[28] then the faster particle p_f simply overtakes the slower particle p_s and continues its motion (Fig. 4c).

4.2. *CA as cyclotrons*

Typically, we can find all types of particles in CA gliders, including positive p^+, negative p^-, and patterns with neutral p^{0f} displacements,[42] and also composite particles assembled from elementary localizations. Let us consider the case where a quiescent state is substituted by cells synchronised together as an ether (periodic background). This phenomenon is associated to ECA Rule 110 φ_{R110}.[g] Its evolution space is dominated by a number of particles emerging in various different orders, some of which are really quite complex constructions. Consequently, the number of collisions between particles is increased. Each particle has a period, displacement, velocity, mass, volume, and phase.[42,44][h]

---

[f]Neutral displacement of complex structures in one dimension is related directly to *still life* configurations[13] in two-dimensions.
[g]Rule 110 repository http://uncomp.uwe.ac.uk/genaro/Rule110.html.
[h]A full description of particles in Rule 110 is available at http://uncomp.uwe.ac.uk/genaro/rule110/glidersRule110.html.

(a)

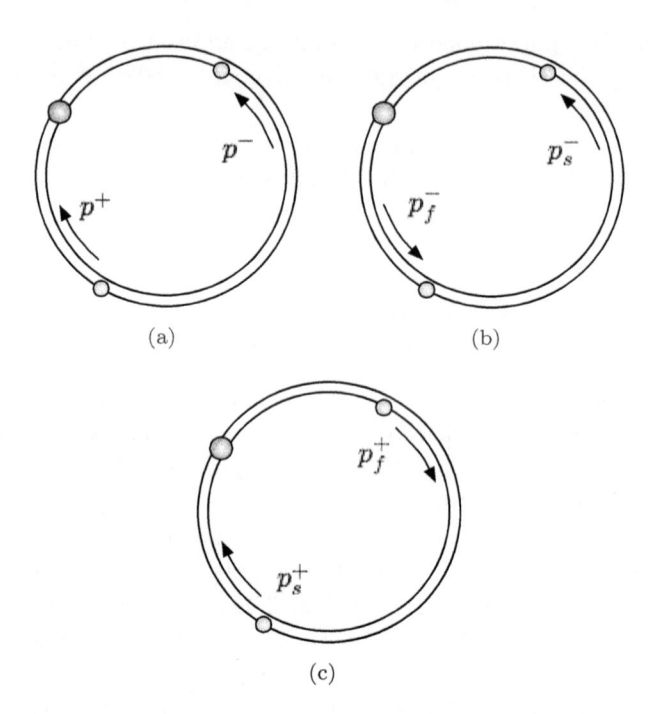

(b)

(c)

Fig. 4. Representation of abstract particles in a one-dimensional CA ring beam routing.

Figure 5 displays a one-dimensional configuration where two particles collide repeatedly and interact as solitons so that the identities of the particles are preserved in the collisions. A negative particle p_F^- collides and overtakes a neutral particle $p_{C_1}^-$. Figure 5a presents a whole set of cells in state 1 (dark points) where the ether configuration makes it impossible to distinguish the particles: p_F^- and $p_{C_1}^-$. However, we can apply a filter and thereby select particles from their background ether (Fig. 5b).[i] Space-time configurations of a cellular automaton exhibiting a collision between particles p_F^- and $p_{C_1}^-$ are shown in Fig. 5c.

Filters selected in CA are a useful tool for understand "hidden" properties of CA. This tool was developed by Wuensche in the context of automatic classification of CA.[59] The filters were derived from mechanical computation techniques,[23] pattern recognition,[49] and analysis of cell-state frequencies.[59] Thus, a filter is a sequence of cells that have a high frequency

---

[i]Ring evolution is simulated with DDLab[60] available in http://www.ddlab.org.

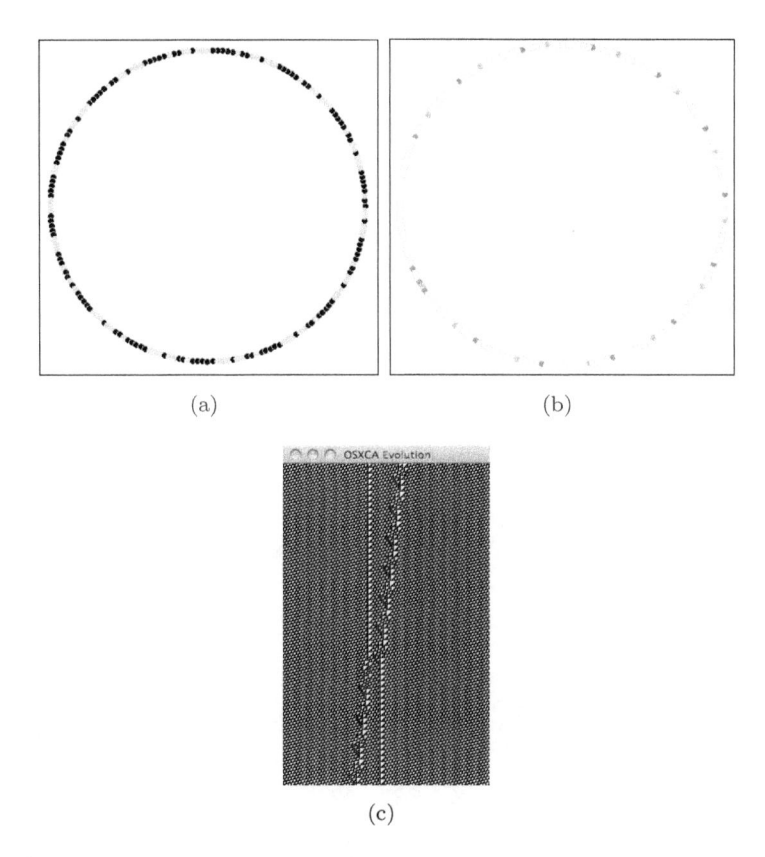

Fig. 5. Example of a soliton-type interaction between particles in ECA Rule 110: (a)–(b) two steps of beam routing, (c) exact configuration at the time of collision.

in the evolution space. Such d-dimensional string repeat periodically, coexisting with any complex structure without disturbing the global dynamics.

4.3. *Extending Toffoli's symbology*

While Toffoli considers a particle travelling to collide with other particle,[52] here we will consider packages of gliders colliding simultaneously or in series. Hence, we can develop finite machines where the vertex are not just states, they are sets of strings that represent particles turning inside cyclotrons. Consequently, the transition on these meta-vertices mean a change of such cyclotrons given by the sum of their collisions.[38]

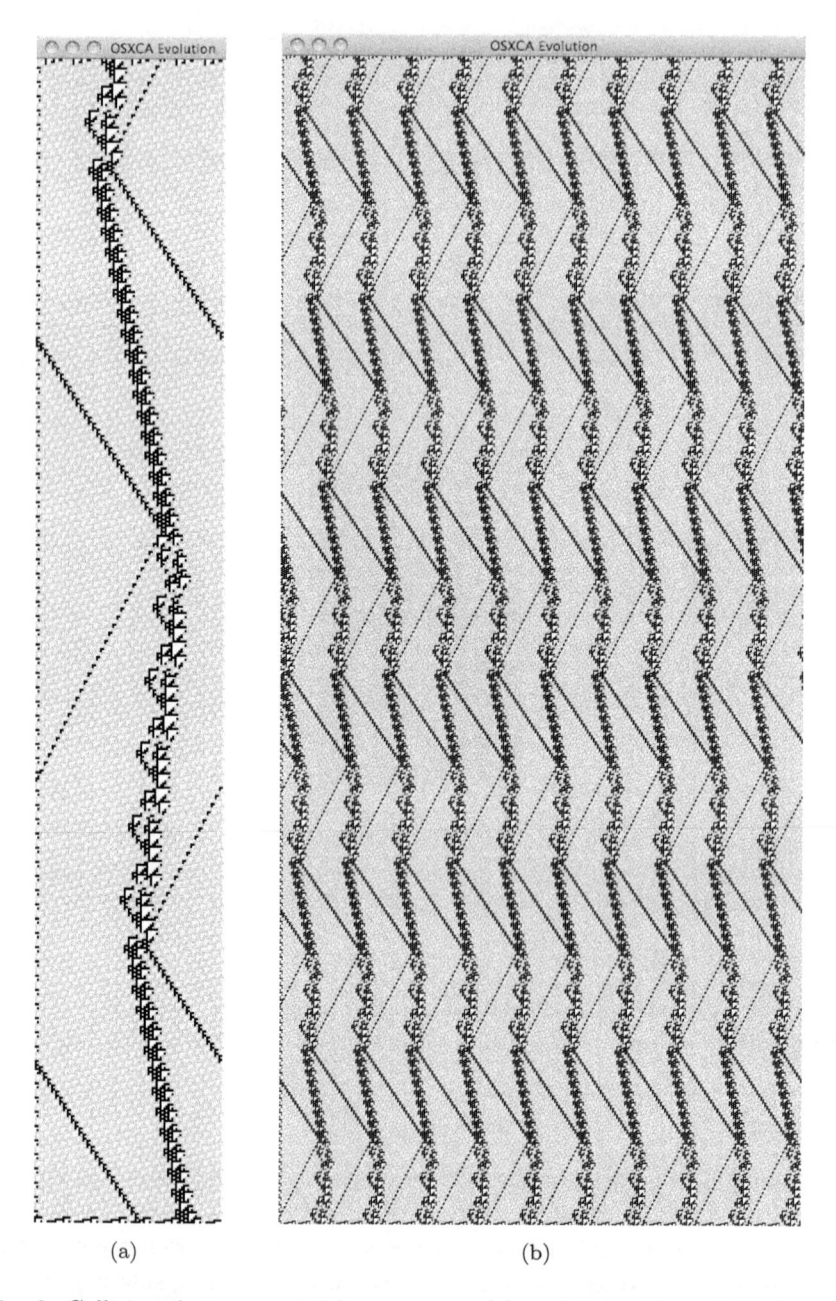

Fig. 6. Collisions between particles $p_F \leftarrow p_B$ (a) evolving with periodic boundaries to 93 cells in 573 generations, (b) internal structure of collisions without periodic boundaries evolving in 510 cells in 1,150 generations.

Lets consider a simple sample with multiple collisions. We have collisions that can be represented as cycles of collisions. In this case, we have the next relations:

(1) $p_F \leftarrow p_B = p_{D_1} + p_{A^2}$
(2) $p_{A^2} \leftarrow p_{D_1} = p_B + p_F$

given the cycle, hence we can construct this one across a cyclotron representation.

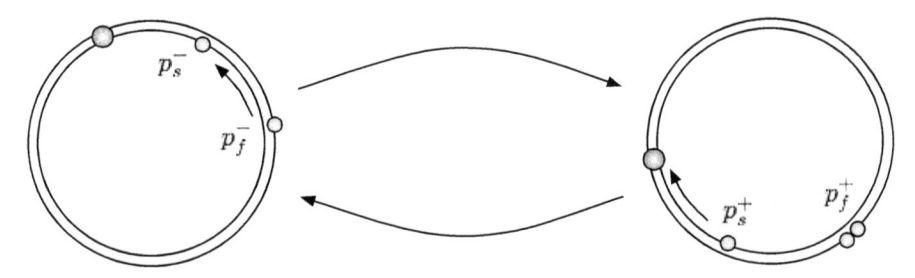

Fig. 7. Beam routing as cyclotrons working as a finite machine. In both cases fast particles reach to slow particles.

The cycle relates two cyclotrons, which when connected represent a simple state machine. This machine has two meta-vertices (states) and the transition is determined by a contact point where gliders collide, as we can see in Fig. 7.

Figure 6 illustrate two kinds of evolutions where gliders in Rule 110 have periodic reactions. First evolution (Fig. 6a) has an initial condition with 93 cells with the next regular expression (here the symbol $-$ means the concatenation operation):

$$e - F(H,f_1\_1) - e - B(f_1\_1) - e$$

the evolution has boundaries conditions and one exact distance preserve periodic collisions between such particles evolving in 573 generations. However, without boundaries conditions this periodic reaction can be preserved as the evolution (Fig. 6b). The multiple collisions are synchronised to construct a meta glider.[43] The regular expression to reproduce such global behaviour (as an infinite string) is the following:

$$(F(H,f_1\_1) - e - B(f_1\_1) - e)^*,$$

the evolution encoded in an initial condition with 510 cells evolving in 1,150 generations. Therefore, the cyclotron representation helps to synthesise such dynamics and construct specific finite machine to represent operations between them.

4.4. *Implementing simple functions in rings*

We can employ the particles codification to represent solutions of some basic computing functions. Let us consider a ECAM rule $\phi_{R30maj:8}$.[31,33] We want to implement a simple substitution function addToHead working on two strings $w_1 = A_1, \ldots, A_n$ and $w_2 = B_1, \ldots, B_m$, where $n, m \geq 1$. For example, if $w_1 = AAA$, $w_2 = BBB$ and $w_3 = w_1w_2$ then the addToHead($|w_2|$) will yield: $w_3 = w_2w_1$ or $w_3 = BBBAAA$ (see schematic diagram of Fig. 8).

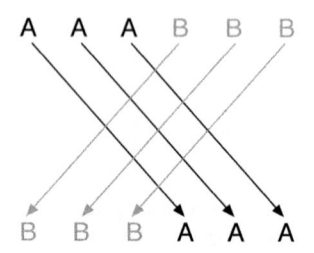

Fig. 8. Schematic diagram adding the string w_2 to head of the list w_3.

To implement such function in $\phi_{R30maj:8}$ we must represent every data 'quantum' as a particle. Gliders g_1 and g_2 are coded to reproduce a soliton reaction. Another problem is to synchronise several gliders and obtain the same result with multiple collisions.

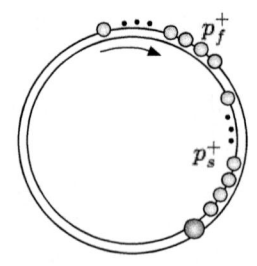

Fig. 9. Beam routing performing identity reactions in $\phi_{R30maj:8}$. A cycle realises its operation, two cycles re-initialise the beam state and the operations can then be repeated.

The codification is not sophisticated, however, a systematic analysis of reactions is required. It is known[31,33] than a periodic gap and one fixed phase between particles is sufficient to reproduce the addToHead function for any string $A^n B^m$.

4.5. *Full computable systems in complex ECA*

After a detailed study of the universality in Rule 110,[45] we can see that this cyclic tag system (CTS)[18,58] working for Rule 110 can be expressed based on cycles of collisions. Thus we can do cyclotrons or rings of particles that represent the periodic packages of gliders to implement a CTS in Rule 110 evolution space. This results are explained in detail in the paper *Cellular Automaton Supercolliders* (see[38]).

5. Final Remarks

This paper is a short review of previous results about complex ECA and the study of gliders in ECA through the de Bruijn diagrams and regular languages. Particularly on ECA Rule 110 and some ECAM. Therefore, following the Toffoli's symbology to idealise supercolliders in CA, we prove how these cyclotrons can be used to implement computations as rings of strings.

Acknowledgement

Genaro J. Martínez thanks to support given by DGAPA-UNAM and EP-SRC grant EP/F054343/1.

References

1. Adamatzky, A. *Computing in Nonlinear Media and Automata Collectives*, Institute of Physics Publishing, Bristol and Philadelphia, 2001.
2. Adamatzky, A. (Ed.) *Collision-Based Computing*, Springer, 2002.
3. Adamatzky, A. New media for collision-based computing, In,[2] 411–442, 2002.
4. Adamatzky, A. Hot Ice Computers, *Physics Letters A* **374** 264–271, 2009.
5. Adamatzky, A. *Physarum Machines: Computers from Slime Mould*, World Scientific Series on Nonlinear Science, Series A, 2010.
6. Adamatzky, A. Costello, B. L., and Asai, T. *Reaction-Diffusion Computers*, Elsevier, 2005.
7. Alonso-Sanz, R. & Martin, M. Elementary CA with memory, *Complex Systems* **14** 99–126, 2003.
8. Alonso-Sanz, R. Elementary rules with elementary memory rules: the case of linear rules, *Journal of Cellular Automata* **1** 71–87, 2006.
9. Alonso-Sanz, R. *Cellular Automata with Memory*, Old City Publishing, 2009.

10. Alonso-Sanz, R. *Discrete Systems with Memory*, World Scientific Series on Nonlinear Science, Series A, 2011.
11. Arbib, M. A. *Theories of Abstract Automata*, Prentice-Hall Series in Automatic Computation, 1969.
12. Banks, E. R. Information and transmission in cellular automata, *PhD Dissertion*, Cambridge, MA, MIT, 1971.
13. Berlekamp, E. R., Conway, J. H., & Guy, R. K. *Winning Ways for your Mathematical Plays*, Academic Press, vol. 2, chapter 25, 1982.
14. Bennett, C. H. Logical reversibility of computation, *IBM Journal of Research and Development* **17(6)** 525–532, 1973.
15. Boccara, N., Nasser, J., & Roger, M. Particle like structures and their interactions in spatio-temporal patterns generated by one-dimensional deterministic cellular automaton rules, *Physical Review A* **44(2)** 866–875, 1991.
16. Chopard, B. & Droz, M. *Cellular Automata Modeling of Physical Systems*, Collection Aléa Saclay, Cambridge University Press, 1998.
17. Codd, E. F. *Cellular Automata*, Academic Press, Inc. New York and London, 1968.
18. Cook, M. Universality in Elementary Cellular Automata, *Complex Systems* **15(1)** 1–40, 2004.
19. Davis, M. *Computability and Unsolvability*, Dover Publications, Inc. New York, 1982.
20. Davis, M. *The Universal Computing: the road from Leibniz to Turing*, W. W. Norton & Company, Inc., 2000.
21. Fredkin, E. & Toffoli, T. Conservative logic, *Int. J. Theoret. Phys.* **21** 219–253, 1982.
22. Fredkin, E. & Toffoli, T. Design Principles for Achieving High-Performance Submicron Digital Technologies, In,[2] 27–46, 2001.
23. Hanson, J. E. & Crutchfield, J. P. Computacional Mechanics of Cellular Automata: An Example, *Physics D* **103** 169–189, 1997.
24. Hey, A. J. G. *Feynman and computation: exploring the limits of computers*, Perseus Books, 1998.
25. Hopcroft, J. E. & Ullman, J. D. *Introduction to Automata Theory Languages, and Computation*, Addison-Wesley Publishing Company, 1987.
26. Hurd, L. P. Formal Language Characterizations of Cellular Automaton Limit Sets, *Complex Systems* **1** 69–80, 1987.
27. Lindgren, K. & Nordahl, M. G. Universal Computation in Simple One-Dimensional Cellular Automata, *Complex Systems* **4** 229–318, 1990.
28. Jakubowski, M. H., Steiglitz, K., & Squier, R. Computing with Solitons: A Review and Prospectus, *Multiple-Valued Logic* **6(5-6)** 439–462, 2001.
29. Kudlek, M. & Rogozhin, Y. Small Universal Circular Post Machine, *Computer Science Journal of Moldova* **9(25)** 34–52, 2001.
30. Lindgren, K. & Nordahl M. Universal Computation in Simple One-Dimensional Cellular Automata, *Complex Systems* **4** 229–318, 1990.
31. Martínez, G. J., Adamatzky, A., Alonso-Sanz, R., & Seck-Tuoh-Mora, J. C. Complex dynamic emerging in Rule 30 with majority memory, *Complex Systems* **18(3)** 345–365, 2010.

32. Martínez, G. J., Adamatzky, A., Morita, K., & Margenstern, M. Computation with competing patterns in life-like automaton. In: *Game of Life Cellular Automata*, A. Adamatzky (Ed.) Springer, United Kingdom, pp. 547–572, chapter 27, 2010.
33. Martínez, Genaro J., Adamatzky, Andrew, and Alonso-Sanz, Ramon, Complex dynamics of elementary cellular automata emerging in chaotic rule, *International Journal of Bifurcation and Chaos*, 22(2), 1250023-13, 2012.
34. Margolus, N. Physics-like models of computation, *Physica D* **10(1-2)** 81–95, 1984.
35. Margolus, N. Crystalline computation, In,[24] 267–305, 1999.
36. Margolus, N. Universal Cellular Automata Based on the Collisions of Soft Spheres, In,[1] 231–260, 2003.
37. Martínez, G. J., Adamatzky, A., Seck-Tuoh-Mora, J. C., & Alonso-Sanz, R. (2010) How to make dull cellular automata complex by adding memory: Rule 126 case study, *Complexity* **15(6)** 34–49.
38. Martínez, G. J., Adamatzky, A., & Stephens, C. R. Cellular automaton supercolliders, *International Journal of Modern Physics C* **22(4)**, 419–439, 2011.
39. Martínez, G. J. & McIntosh, H. V. ATLAS: Collisions of gliders like phases of ether in rule 110, http://uncomp.uwe.ac.uk/genaro/Papers/Papers_on_CA.html.
40. McIntosh, H. V. Linear cellular automata via de Bruijn diagrams, 1991. http://delta.cs.cinvestav.mx/~mcintosh/cellularautomata/Papers.html.
41. McIntosh, H. V. *One Dimensional Cellular Automata*, Luniver Press, 2009.
42. Martínez, G. J., McIntosh, H. V., & Seck-Tuoh-Mora, J. C. Gliders in Rule 110, *Int. J. of Unconventional Computing* **2(1)** 1–49, 2006.
43. Martínez, G. J., McIntosh, H. V., Seck-Tuoh-Mora, J. C., & Chapa-Vergara, S. V. Rule 110 objects and other constructions based-collisions, *Journal of Cellular Automata* **2(3)** 219–242, 2007.
44. Martínez, G. J., McIntosh, H. V., Seck-Tuoh-Mora, J. C., & Chapa-Vergara, S. V. Determining a regular language by glider-based structures called *phases f_{i-1}* in Rule 110, *Journal of Cellular Automata* **3(3)** 231–270, 2008.
45. Martínez, G. J., McIntosh, H. V., Seck-Tuoh-Mora, J. C., & Chapa-Vergara, S. V. Reproducing the cyclic tag system developed by Matthew Cook with Rule 110 using the phases f_{1-1}, *Journal of Cellular Automata* **6(2-3)** 121–161, 2011.
46. Minsky, M. *Computation: Finite and Infinite Machines*, Prentice Hall, 1967.
47. Morita, K. Reversible computing and cellular automata—A survey, *Theoretical Computer Science* **395** 101–131, 2008.
48. Shannon, C. E. A Universal Turing Machine with Two Internal States, *Automata Studies*, Princeton University Press, pp. 157–165, 1956.
49. Shalizi, C. R., Haslinger, R., Rouquier, J-B. K., Kristina L., & Moore, C. Automatic filters for the detection of coherent structure in spatiotemporal systems, *Physical Review E* **73(3)** 036104, 2005.
50. Smith III, A. R. Simple computation-universal cellular spaces, *J. of the Assoc. for Computing Machinery* **18** 339–353, 1971.

51. Toffoli, T. Non-Conventional Computers, In *Encyclopedia of Electrical and Electronics Engineering* **14** (John Webster Ed.), Wiley & Sons, 455–471, 1998.
52. Toffoli, T. Symbol Super Colliders, In,[2] 1–22, 2002.
53. Turing, A. On Computable numbers, with an application to the Entscheidungsproblem, *Proceedings of the London Mathematical Society* **42(2)** 230–265. 1937 Corrections, Ibid **43** 544–546, 1936.
54. Voorhees, B. H. *Computational analysis of one-dimensional cellular automata*, World Scientific Series on Nonlinear Science, Series A, Vol. 15, 1996.
55. Voorhees, B. H. Remarks on Applications of De Bruijn Diagrams and Their Fragments, *Journal of Cellular Automata* **3(3)** 187–204, 2008.
56. von Neumann, J. *Theory of Self-reproducing Automata* (edited and completed by A. W. Burks), University of Illinois Press, Urbana and London, 1966.
57. Wolfram, S. *Cellular Automata and Complexity*, Addison-Wesley Publishing Company, 1994.
58. Wolfram, S. *A New Kind of Science*, Wolfram Media, Inc., Champaign, Illinois, 2002.
59. Wuensche, A. Classifying Cellular Automata Automatically, *Complexity* **4(3)** 47–66, 1999.
60. Wuensche, A. *Exploring Discrete Dynamics*, Luniver Press, 2011.

Chapter 15

Life as Evolving Software

Gregory Chaitin
Federal University of Rio de Janeiro
Email: gjchaitin@gmail.com

1. Turing as a Biologist

Few people remember Turing's work on pattern formation in biology (morphogenesis), but Turing's famous 1936 paper "On Computable Numbers..." exerted an immense influence on the birth of molecular biology indirectly, through the work of John von Neumann on self-reproducing automata, which influenced Sydney Brenner who in turn influenced Francis Crick, the Crick of Watson and Crick, the discoverers of the molecular structure of DNA. Furthermore, von Neumann's application of Turing's ideas to biology is beautifully supported by recent work on evo-devo (evolutionary developmental biology). The crucial idea: DNA is multi-billion year old software, but we could not recognize it as such before Turing's 1936 paper, which according to von Neumann creates the idea of computer hardware and software.

We are attempting to take these ideas and develop them into an abstract fundamental mathematical theory of evolution, one that emphasizes biological creativity, inventiveness and the generation of novelty. This work is being published in two parts. Firstly a non-technical book-length treatment: G. Chaitin, *Proving Darwin: Making Biology Mathematical* to be published by Pantheon in 2012. There we explain at length the basic concepts and the history of ideas. For an overview of this book, a lecture entitled "Life as evolving software," go to *www.youtube.com* and search for *chaitin ufrgs*.

And in this paper we present a technical discussion of the mathematics of this new way of thinking about biology. More precisely, we present an information-theoretic analysis of Darwin's theory of evolution, modeled as a hill-climbing algorithm on a fitness landscape. Our space of possible organisms consists of computer programs, which are subjected to random mutations. We study the random walk of increasing fitness made by a single mutating organism. In two different models we are able to show that evolution will occur and to characterize the rate of evolutionary progress, i.e., the rate of biological creativity.

We call this new theory *metabiology*, and it deals with the evolution of mutating software and with random walks in software space. The mathematics we use is essentially Turing's version of computability theory from the 1930s, including his colorful oracles, plus the idea of how to associate probabilities with computer programs utilized since the 1970s in algorithmic information theory, which is summarized in the appendix of this paper.

It remains to be seen how far these ideas will go, but as is shown in this paper and in the companion volume,[13] the first steps are encouraging. In our opinion, Turing's ideas are of absolutely fundamental importance in biology, since biology is all about digital software.

2. Introduction to this Paper

For many years we have been disturbed by the fact that there is no fundamental mathematical theory inspired by Darwin's theory of evolution.[1-9] This is the fourth paper in a series[10-12] attempting to create such a theory.

In a previous paper[10] we did not yet have a workable mathematical framework: We were able to prove two not very impressive theorems, and then the way forward was blocked. Now we have what appears to be a good mathematical framework, and have been able to prove a number of theorems. Things are starting to work, things are starting to get interesting, and there are many technical questions, many open problems, to work on.

So this is a working paper, a progress report, intended to promote interest in the field and get others to participate in the research. There is much to be done.

In order to present the ideas as clearly as possible and not get bogged down in technical details, the material is presented more like a physics paper than a math paper. Estimates are at times rather sloppy. We are trying to get an idea of what is going on. The arguments concerning the basic math framework are however very precise; that part is done more or less like a math paper.

3. History of Metabiology

In the first paper in this series[10] we proposed modeling biological evolution by studying the evolution of randomly mutating software—we call this *metabiology*. In particular, we proposed considering a single mutating software organism following a random walk in software space of increasing fitness. Besides that the main contribution of[10] was to use the Busy Beaver problem to challenge organisms into evolving. The larger the positive integer that a program names, the fitter the program.

And we measured the rate of evolutionary progress using the Busy Beaver function $BB(N)$ = the largest integer that can be named by an N-bit program. Our two results employing the framework in[10] are that

- with random mutations, random point mutations, we will get to fitness $BB(N)$ in time exponential in N (evolution by *exhaustive search*),[10,11]
- whereas by choosing the mutations by hand and applying them in the right order, we will get to fitness $BB(N)$ in time linear in N (evolution by *intelligent design*).[11,12]

We were unable to show that *cumulative evolution* will occur at random; exhaustive search starts from scratch each time.[a]

This paper advances beyond the previous work on metabiology[10–12] by proposing a better concept of mutation. Instead of changing, deleting or inserting one or more adjacent bits in a binary program, we now have high-level mutations: we can use an arbitrary algorithm M to map the organism A into the mutated organism $A' = M(A)$. Furthermore, the probability of the mutation M is now furnished by algorithmic information theory: it depends on the size in bits of the self-delimiting program for M. It is very important that we now have a natural, universal probability distribution on the space of all possible mutations, and that this is such a rich space.

Using this new notion of mutation, these much more powerful mutations, enables us to accomplish the following:

- We are now able to show that *random evolution will become cumulative* and will reach fitness $BB(N)$ in time that grows roughly

---

[a]The Busy Beaver function $BB(N)$ grows faster than any computable function. That evolution is able to "compute" the uncomputable function $BB(N)$ is evidence of creativity *that cannot be achieved mechanically*. This is possible only because our model of evolution/creativity utilizes an uncomputable Turing oracle. Our model utilizes the oracle in a highly constrained manner; otherwise it would be easy to calculate $BB(N)$.

as N^2, so that random evolution behaves much more like intelligent design than it does like exhaustive search.[b]
- We also have a version of our model in which we can show that **hierarchical** *structure will evolve*, a conspicuous feature of biological organisms that previously[10] was beyond our reach.

This is encouraging progress, and suggests that we may now have the correct version of these biology-inspired concepts. However there are many serious lacunae in the theory as it currently stands. It does not yet deserve to be called *a mathematical theory of evolution and biological creativity*; at best, it is a sketch of a possible direction in which such a theory might go.

On the other hand, the new results are encouraging, and we feel it would be inappropriate to sit on these results until all the lacunae are filled. After all, that would take an entire book, since metabiology is, or will hopefully become, a rich and entirely new field.

That said, the reader will understand that this is a working paper, a progress report, to show the direction in which the theory is developing, and to indicate problems that need to be solved in order to advance, in order to take the next step. We hope that this paper will encourage others to participate in developing metabiology and exploring its potential.

4. Modeling Evolution

4.1. *Software organisms*

In this paper we follow a metabiological[10–13] approach: Instead of studying the evolution of actual biological organisms we study the evolution of software subjected to random mutations. In order to do this we use tools from algorithmic information theory (AIT);[14–19] **to fully understand this paper expert understanding of AIT is unfortunately necessary** (see the outline in the Appendix).

As our programming formalism we employ one of the optimal self-delimiting binary universal Turing machines U of AIT,[14] and also, but only in Section 8, a primitive FORTRAN-like language that is not universal.

So our organisms consist on the one hand of arbitrary self-delimiting binary programs p for U, or on the other hand of certain FORTRAN-like computer programs. These are the respective software spaces in which we shall be working, and in which we will study hill-climbing random walks.

---

[b]Most unfortunately, it is not yet demonstrated that random evolution cannot be as fast as intelligent design.

4.2. The hill-climbing algorithm

In our models of evolution, we define a hill-climbing random walk as follows: We start with a single software organism A and subject it to random mutations until a fitter organism A' is obtained, then subject that organism to random mutations until an even fitter organism A'' is obtained, etc. In one of our models, organisms calculate natural numbers, and the bigger the number, the fitter the organism. In the other, organisms calculate functions that map a natural number into another natural number, and the faster the function grows, the fitter the organism.

In this connection, here is a useful piece of terminology: A mutation M *succeeds* if $A' = M(A)$ is fitter than A; otherwise M is said to *fail*.

4.3. Fitness

In order to get our software organisms to evolve it is important to present them with a challenge, to give them something difficult to do. Three well-known problems requiring unlimited amounts of mathematical creativity are:

- **Model A:** Naming large natural numbers (non-negative integers),[20–23]
- **Model B:** Defining extremely fast-growing functions,[24–26]
- **Model C:** Naming large constructive Cantor ordinal numbers.[26,27]

So a software organism will be judged to be more fit if it calculates a larger integer (our Model A, Sections 5, 6, 7), or if it calculates a faster-growing function (our Model B, Section 8). Naming large Cantor ordinals (Model C) is left for future work, but is briefly discussed in Section 9.

4.4. What is a mutation?

Another central issue is the concept of a mutation. Biological systems are subjected to point mutations, localized changes in DNA, as well as to high level mutations such as copying an entire gene and then introducing changes in it. Initially[10] we considered mutating programs by changing, deleting or adding one or more adjacent bits in a binary program, and postponed working with high-level source language mutations.

Here we employ an extremely general notion of mutation: A mutation is an arbitrary algorithm that transforms, that maps the original organism into the mutated organism. It takes as input the organism, and produces as output the mutated organism. And if the mutation is an n-bit program, then it has probability 2^{-n}. In order to have the total probability of mutations be ≤ 1 we use the self-delimiting programs of AIT.[14c]

4.5. Mutation distance

A second crucial concept is mutation distance, how difficult it is to get from organism A to organism B. We measure this distance in bits and it is defined to be $-\log_2$ of the probability that a random mutation will change A to B. Using AIT,[14–16] we see that this is nearly $H(B|A)$, the size in bits of the smallest self-delimiting program that takes A as input and produces B as output.[d] More precisely,

$$H(B|A) = -\log_2 P(B|A) + O(1) = -\log_2 \left[\sum_{U(p|A)=B} 2^{-|p|} \right] + O(1). \quad (1)$$

Here $|p|$ denotes the size in bits of the program p, and $U(p|A)$ denotes the output produced by running p given input A on the computer U until p halts.

The definition of $H(B|A)$ that we employ here is somewhat different from the one that is used in AIT: a mutation is given A directly, it is not given a minimum-size program for A. Nevertheless, (1) holds.[14]

Interpreting (1) in words, it is nearly the same to consider the *simplest* mutation from A to B, which is $H(B|A)$ bits in size and has probability $2^{-H(B|A)}$, as to sum the probability over *all* the mutations that carry A into B.

Note that this distance measure is not symmetric. For example, it is easy to change (X, Y) into Y, but not vice versa.

4.6. Hidden use of oracles

There are two hidden assumptions here. First of all, we need to use an oracle to compare the fitness of an organism A with that of a mutated organism A'. This is because a mutated program may not halt and thus never produces a natural number. Once we know that the original organism

---

[c]The total probability of mutations is actually < 1, so that each time we pick a mutation at random, there is a fixed probability that we will get the *null mutation* $M(A) = A$, which always fails.
[d]Similarly, $H(B)$ denotes the size in bits of the smallest self-delimiting program for B that is *not* given A. $H(B)$ is called the *complexity* of B, and $H(B|A)$ is the *relative complexity* of B given A.

A and the mutated organism A' both halt, then we can run them to see what they calculate and which is fitter.

In the case of fast-growing computable functions, an oracle is definitely needed to see if one grows faster than another; this cannot be determined by running the primitive recursive functions[29] calculated by the FORTRAN-like programs that we will study later, in Section 8.

Just as oracles would be needed to actually find fitter organisms, they are also necessary because a random mutation may never halt and produce a mutated organism. So to actually apply our random mutations to organisms we would need to use an oracle in order to avoid non-terminating mutations.

5. Model A (Naming Integers) Exhaustive Search

5.1. *The Busy Beaver function*

The first step in this metabiological approach is to measure the rate of evolution. To do that, we introduce this version of the Busy Beaver function:

$\text{BB}(N)$ = the biggest natural number named by a $\leq N$-bit program.

More formally,

$$\text{BB}(N) = \max_{H(k) \leq N} k.$$

Here the program-size *complexity* or the algorithmic *information content* $H(k)$ of k is the size in bits of the smallest self-delimiting program p without input for calculating k:

$$H(k) = \min_{U(p)=k} |p|.$$

Here again $|p|$ denotes the size in bits of p, and $U(p)$ denotes the output produced by running the program p on the computer U until p halts.

5.2. *Proof of Theorem 1 (exhaustive search)*

Now, for the sake of definiteness, let's start with the trivial program that directly outputs the positive integer 1, and apply mutations at random.[e] Let's define the *mutation time* to be n if we have tried n mutations, and the *organism time* to be n if there are n successive organisms of increasing fitness so far in our infinite random walk.

---
[e]The choice of initial organism is actually unimportant.

From AIT[14] we know that there is an $N+O(1)$-bit mutation that ignores its input and produces as output a $\leq N$-bit program that calculates $\mathrm{BB}(N)$. This mutation M has probability $2^{-N+O(1)}$ and on the average, it will occur at random every $2^{N+O(1)}$ times a random mutation is tried. Therefore:

Theorem 1 *The fitness of our organism will reach $\mathrm{BB}(N)$ by mutation time 2^N. In other words, we will achieve N bits of biological/mathematical creativity by time 2^N. Each successive bit of creativity takes twice as long as the previous bit did.*[f]

More precisely, the probability that this should fail to happen, the probability that M has *not* been tried by time 2^N, is

$$\left(1 - \frac{1}{2^N}\right)^{2^N} \to e^{-1} \approx \frac{1}{2.7} < \frac{1}{2}.$$

And the probability that it will fail to happen by mutation time $K2^N$ is $< 1/2^K$.

This is the worst that evolution can do. It is the fitness that organisms will achieve if we are employing *exhaustive search* on the space of all possible organisms. Actual biological evolution is not at all like that. The human genome has 3×10^9 bases, but in the mere 4×10^9 years of life on this planet only a tiny fraction of the total enormous number $4^{3 \times 10^9}$ of sequences of 3×10^9 bases can have been tried. In other words, evolution is not *ergodic*.

6. Model A (Naming Integers) Intelligent Design

6.1. *Another Busy Beaver function*

If we could choose our mutations intelligently, evolution would be much more rapid. Let's use the halting probability Ω[19] to show just how rapid. First we define a slightly different Busy Beaver function BB' based on Ω. Consider a fixed recursive/computable enumeration $\{p_i : i = 0, 1, 2 \ldots\}$ without repetitions of all the programs without input that halt when run on U. Thus

$$0 < \Omega = \Omega_U = \sum_i 2^{-|p_i|} < 1 \qquad (2)$$

---
[f]Instead of *bits of creativity* one could perhaps refer to *bits of inspiration*; said inspiration of course is ultimately coming through/from our oracle, which keeps us from getting stuck on non-terminating programs.

and we get the following sequence $\Omega_0 = 0 < \Omega_1 < \Omega_2 \ldots$ of lower bounds on Ω:

$$\Omega_N = \sum_{i<N} 2^{-|p_i|}. \tag{3}$$

In (2) and (3) $|p|$ denotes the size in bits of p, as before.

We define $\mathrm{BB}'(K)$ to be the least N for which the first K bits of the base-two numerical value of Ω_N are correct, i.e., the same as the first K bits of the numerical value of Ω. $\mathrm{BB}'(K)$ exists because we know from AIT[14] that Ω is irrational, so $\Omega = 0.010000$ is impossible and there is no danger that Ω_N will be of the form 0.0011111 with 1's forever.

Note that **BB and BB' are approximately equal.** For we can calculate $\mathrm{BB}'(N)$ if we are given N and the first N bits of Ω. Therefore

$$\mathrm{BB}'(N) \leq \mathrm{BB}(N + H(N) + c) = \mathrm{BB}(N + O(\log N)).$$

Furthermore, if we knew N and any $M \geq \mathrm{BB}'(N)$, we could calculate the string ω of the first N bits of Ω, which according to AIT[14] has complexity $H(\omega) > N - c'$, so

$$N - c' < H(\omega) \leq H(N) + H(M) + c''.$$

Therefore $\mathrm{BB}'(N)$ and all greater than or equal numbers M have complexity $H(M) > N - H(N) - c' - c''$, so $\mathrm{BB}'(N)$ must be greater than the biggest number M_0 with complexity $H(M_0) \leq N - H(N) - c' - c''$. Therefore

$$\mathrm{BB}'(N) > \mathrm{BB}(N - H(N) - c' - c'') = \mathrm{BB}(N + O(\log N)).$$

6.2. Improving lower bounds on Ω

Our model consists of arbitrary mutation computer programs operating on arbitrary organism computer programs. To analyze the behavior of this system (Model A), however, we shall focus on a select subset: Our organisms are lower bounds on Ω, and our mutations increase these lower bounds.

We are going to use these same organisms and mutations to analyze both intelligent design (Section 6.3) and cumulative evolution at random (Section 7). Think of Section 6.3 versus Section 7 as counterpoint.

Organism P_ρ — Lower Bound ρ on Ω

Now we use a bit string ρ to represent a dyadic rational number in $[0, 2) = \{0 \le x < 2\}$; ρ consists of the base-two units "digit" followed by the base-two expansion of the fractional part of this rational number.

There is a self-delimiting prefix π_Ω that given a bit string ρ that is a lower bound on Ω, calculates the first N such that $\Omega > \Omega_N \ge \rho$, where Ω_N is defined as in (3).[g] If we concatenate the prefix π_Ω with the string of bits ρ, and insert $0^{|\rho|}1$ in front of ρ in order to make everything self-delimiting, we obtain a program P_ρ for this N.

We will now analyze the behavior of Model A by using these organisms of the form

$$P_\rho = \pi_\Omega 0^{|\rho|} 1 \rho. \tag{4}$$

To repeat, the output of P_ρ, and therefore its fitness ϕ_{P_ρ}, is determined as follows:

$$U(P_\rho) = \text{the first } N \text{ for which } \sum_{i<N} 2^{-|p_i|} = \Omega_N \ge \rho. \tag{5}$$

This fitness will be $\ge BB'(K)$ if $\rho < \Omega$ and the first K bits of ρ are the correct base-two numerical value of Ω. P_ρ will fail to halt if $\rho > \Omega$.[h]

Mutation M_k — Lower Bound ρ on Ω Increased by 2^{-k}

Consider the mutations M_k that do the following. First of all, M_k computes the fitness ϕ of the current organism A by running A to determine the integer $\phi = \phi_A$ that A names. **All that M_k takes from A is its fitness ϕ_A.** Then M_k computes the corresponding lower bound on Ω:

$$\rho = \sum_{i<\phi} 2^{-|p_i|} = \Omega_\phi.$$

Here $\{p_i\}$ is the standard enumeration of all the programs that halt when run on U that we employed in Section 6.1. Then M_k increments the lower bound ρ on Ω by 2^{-k}:

$$\rho' = \rho + 2^{-k}.$$

In this way M_k obtains the mutated program

$$A' = P_{\rho'}.$$

---
[g]That $\rho \ne \Omega$ follows from the fact that Ω is irrational.
[h]That $\rho \ne \Omega$ follows from the fact that Ω is irrational.

A' will fail to halt if $\rho' > \Omega$. If A' does halt, then $A' = M_k(A) = P_{\rho'}$ will have fitness N(see (5)) greater than $\phi_A = \phi$ because $\rho' > \rho = \Omega_\phi$, so more halting programs are included in the sum (3) for Ω_N, which therefore has been extended farther:

$$[\Omega_N \geq \rho' > \rho = \Omega_\phi] \implies [N > \phi].$$

Therefore if $\Omega > \rho' = \rho + 2^{-k}$, then M_k increases the fitness of A. If $\rho' > \Omega$, then $P_{\rho'} = M_k(A)$ never halts and is totally unfit.

6.3. *Proof of Theorem 2 (intelligent design)*

Please note that in this toy world, the "intelligent designer" is the author of this paper, who chooses the mutations optimally in order to get his creatures to evolve.

Let's now start with the computer program P_ρ with $\rho = 0$. In other words, we start with a lower bound on Ω of zero.

Then for $k = 1, 2, 3 \ldots$ we try applying M_k to P_ρ. The mutated organism $P_{\rho'} = M_k(P_\rho)$ will either fail to halt, or it will have higher fitness than our previous organism and will replace it. Note that in general $\rho' \neq \rho + 2^{-k}$, although it could conceivably have that value. M_k will from P_ρ take only its fitness, which is the first N such that $\Omega_N \geq \rho$.

$$\rho' = \Omega_N + 2^{-k} \geq \rho + 2^{-k}.$$

So ρ' is actually equal to a lower bound on Ω, Ω_N, plus 2^{-k}. Thus M_k will attempt to increase a lower bound on Ω, Ω_N, by 2^{-k}. M_k will succeed if $\Omega > \rho'$. M_k will fail if $\rho' > \Omega$. This is the situation at the end of stage k. Then we increment k and repeat. The lower bounds on Ω will get higher and higher.

More formally, let $O_0 = P_\rho$ with $\rho = 0$. And for $k \geq 1$ let

$$O_k = \begin{cases} O_{k-1} & \text{if } M_k \text{ fails,} \\ M_k(O_{k-1}) & \text{if } M_k \text{ succeeds.} \end{cases}$$

Each O_k is a program of the form P_ρ with $\Omega > \rho$.

At the end of stage k in this process the first k bits of ρ will be exactly the same as the first k bits of Ω, because at that point all together we have tried summing $1/2 + 1/4 + 1/8 \cdots + 1/2^k$ to ρ. **In essence, we are using an oracle to determine the value of Ω by successive interval halving.**[i]

---

[i]That this works is easy to see visually. Think of the unit interval drawn vertically, with 0 below and 1 above. The intervals are being pushed up after being halved, but it is still the case that Ω remains inside each halved interval, even after it has been pushed up.

In other words, at the end of stage k the first k bits of ρ in O_k are correct. Hence:

Theorem 2 *By picking our mutations intelligently rather than at random, we obtain a sequence O_N of software organisms with non-decreasing fitness[j] for which the fitness of each organism is $\geq BB'(N)$. In other words, we will achieve N bits of biological/mathematical creativity in mutation time linear in N. Each successive bit of creativity takes about as long as the previous bit did.*

However, successive mutations must be tried at random in our evolution model; they cannot be chosen deliberately. We see in these two theorems two extremes: Theorem 1, brainless exhaustive search, and Theorem 2, intelligent design. What can real, random evolution actually achieve? We shall see that the answer is closer to Theorem 2 than to Theorem 1. We will achieve fitness $BB'(N)$ in time roughly order of N^2. In other words, each successive bit of creativity takes an amount of time which increases linearly in the number of bits.

Open Problem 1 *Is this the best that can be done by picking the mutations intelligently rather than at random? Or can creativity be even faster than linear? Does each use of the oracle yield only one bit of creativity?*[k]

Open Problem 2 *In Theorem 2 how fast does the size in bits of the organism O_N grow? By using entirely different mutations intelligently, would it be possible to have the size in bits of the organism O_N grow linearly, or, alternatively, for the mutation distance between O_N and O_{N+1} to be bounded, and still achieve the same rapid growth in fitness?*

Open Problem 3 *In Theorem 2 how many different organisms will there be by mutation time N? i.e., on the average how fast does organism time grow as a function of mutation time?*

7. Model A (Naming Integers) Cumulative Evolution at Random

Now we shall achieve what Theorem 2 achieved by intelligent design, by using randomness instead. Since the order of our mutations will be random,

---

[j]Note that this is actually a legitimate fitness increasing (non-random) walk because the fitness increases each time that O_N changes, i.e., each time that $O_{N+1} \neq O_N$.

[k]**Yes**, only one bit of creativity, otherwise Ω would be compressible. In fact, the sequence of oracle replies must be incompressible.

not intelligent, **there will be some duplication of effort and creativity is delayed, but not overmuch.**

In other words, instead of using the mutations M_k in a predetermined order, they shall be picked at random, and also mixed together with other mutations that increase the fitness.

As you will recall (Section 6.2), a larger and larger positive integer is equivalent to a better and better lower bound on Ω. That will be our clock, our memory. We will again be evolving better and better lower bounds ρ on Ω and we shall make use of the organisms P_ρ as before ((4), Section 6.2). We will also use again the mutations M_k of Section 6.2.

Let's now study the behavior of the random walk in Model A if we start with an arbitrary program A that has a fitness, for example, the program that is the constant 0, and apply mutations to it at random, according to the probability measure on mutations determined by AIT,[14] namely that M has probability $2^{-H(M)}$.[1] So with probability one, **every mutation will be tried infinitely often**; M will be tried roughly every $2^{H(M)}$ mutation times.

At any given point in this random walk, we can measure our progress to Ω by the fitness $\phi = \phi_A$ of our current organism A and the corresponding lower bound $\Omega_\phi = \Omega_{\phi_A}$ on Ω. Since the fitness ϕ can only increase, the lower bound Ω_ϕ can only get better.

In our analysis of what will happen we focus on the mutations M_k; other mutations will have no effect on the analysis. They are harmless and can be mixed in together with the M_k. By increasing the fitness, they can only make Ω_ϕ converge to Ω more quickly.

We also need a *new* mutation M^*. M^* doesn't get us much closer to Ω, it just makes sure that our random walk will contain infinitely many of the programs P_ρ. M^* will be tried roughly periodically during our random walk. M^* takes the current lower bound $\Omega_\phi = \Omega_{\phi_A}$ on Ω, and produces

$$A' = M^*(A) = P_{\Omega_{1+\phi_A}}.$$

A' has fitness 1 greater than the fitness of A and thus mutation M^* will always succeed, and this keeps lots of organisms of the form P_ρ in our random walk.

Let's now return to the mutations M_k, each of which will also have to be tried infinitely often in the course of our random walk.

The mutation M_k will either have no effect because $M_k(A)$ fails to halt,

---

[1]This is a convenient lower bound on the probability of a mutation. A more precise value for the probability of jumping from A to A' is $2^{-H(A'|A)}$.

which means that we are less than 2^{-k} away from Ω, that is, Ω_{ϕ_A} is less than 2^{-k} away from Ω, or M_k will have the effect of incrementing our lower bound Ω_{ϕ_A} on Ω by 2^{-k}. As more and more of these mutations M_k are tried at random, eventually, purely by chance, more and more of the beginning of Ω_{ϕ_A} will become correct (the same as the initial bits of Ω). Meanwhile, the fitness ϕ_A will increase enormously, passing $\mathrm{BB}'(n)$ as soon as the first n bits of Ω_{ϕ_A} are correct. And soon afterwards, M^* will package this in an organism $A' = P_{\Omega_{1+\phi_A}}$.

How long will it take for all this to happen? I.e., how long will it take to try the M_k for $k = 1, 2, 3, \ldots, n$ and then try M^*? We have

$$H(M_k) \le H(k) + c.$$

Therefore mutation M_k has probability

$$\ge 2^{-H(k)-c} > \frac{1}{c'k(\log k)^{1+\epsilon}} \qquad (6)$$

since

$$\sum_k \frac{1}{k(\log k)^{1+\epsilon}}$$

converges.[m] The mutation M_k will be tried in time proportional to 1 over the probability of its being tried, which by (6) is approximately upper bounded by

$$\xi(k) = c''k(\log k)^{1+\epsilon}. \qquad (7)$$

On the average, from what point on will the first n bits of $\Omega_\phi = \Omega_{\phi_A}$ be the same as the first n bits of Ω? We can be sure this will happen if we first try M_1, then afterwards M_2, then M_3, etc. through M_n, in that order. Note that if these mutations are tried in the wrong order, they will not have the desired effect. But they will do no harm either, and eventually will *also* be tried in the correct order. Note that it is conceivable that none of these M_k actually succeed, because of the other random mutations that were in the mix, in the melee. These other mutations may *already* have pushed us within 2^{-k} of Ω. So these M_k don't have to succeed, they just have to be tried. Then M^* will make sure that we get an organism of the form P_ρ with at least n bits of ρ correct.

---

[m] We are using here one of the basic theorems of AIT.[14]

Hence:

$$\text{Expected time to try } M_1 \leq \xi(1)$$
$$\text{Expected time to then afterwards try } M_2 \leq \xi(2)$$
$$\text{Expected time to then afterwards try } M_3 \leq \xi(3)$$
$$\ldots$$
$$\text{Expected time to then afterwards try } M_n \leq \xi(n)$$
$$\text{Expected time to then afterwards try } M^* \leq c'''$$
$$\therefore \text{Expected time to try } M_1, M_2, M_3 \ldots M_n, M^* \text{ in order} \leq \sum_{k \leq n} \xi(k) + c'''$$

Using (7), we see that this is our extremely rough "ball-park" estimate on a mutation time sufficiently big for the first n bits of ρ in $P_\rho = M^*(A)$ to be the correct bits of Ω:

$$\sum_{k \leq n} \xi(k) + c''' = \sum_{k \leq n} c'' k (\log k)^{1+\epsilon} + c''' = O(n^2 (\log n)^{1+\epsilon}). \tag{8}$$

Hence we expect that in time $O(n^2 (\log n)^{1+\epsilon})$ our random walk will include an organism P_ρ in which the first n bits of ρ are correct, and so P_ρ will compute a positive integer $\geq \text{BB}'(n)$, and thus at this time the fitness will have to be at least that big:

Theorem 3 *In Model A with random mutations, the fitness of the organisms $P_\rho = M^*(A)$ will reach $\text{BB}'(N)$ by mutation time roughly N^2.*

Note that since the bits of ρ in the organisms $P_\rho = M^*(A)$ are becoming better and better lower bounds on Ω, these organisms in effect contain their evolutionary history. **In Model A, evolution is cumulative, it does not start over from scratch as in exhaustive search.**

It should be emphasized that in the course of such a hill-climbing random walk, with probability one every possible mutation will be tried infinitely often. However the mutations M_k will immediately recover from perturbations and set the evolution back on course. In a sense the system is *self-organizing* and *self-repairing*. Similarly, the initial organism is irrelevant.

Also note that with probability one the time history or evolutionary pathway (i.e., the random walk in Model A) will quickly grow better and better approximations to *all possible* halting probabilities $\Omega_{U'}$ (see (2)) determined by *any* optimal universal self-delimiting binary computer U', not just for our original U. Furthermore, some mutations will periodically convert our organism into a numerical constant for its fitness ϕ, and there will even be arbitrarily long chains of successive numerical constant organisms $\phi, \phi+1, \phi+2\ldots$ The microstructure and fluctuations that will occur with probability one are quite varied and should perhaps be studied in detail to unravel the full zoo of organisms and their interconnections; this is in effect a kind of miniature mathematical ecology.

Open Problem 4 *Study this* **mathematical ecology**.

Open Problem 5 *Improve the estimate (8) and get a better upper bound on the expected time it will take to try M_1, M_2, M_3 through M_n and M^* in that order. Besides the mean, what is the variance?*

Open Problem 6 Separate *random evolution and intelligent design: We have shown that random evolution is fast, but can you prove that it cannot be as fast as intelligent design? I.e., we have a lower bound on the speed of random evolution, and now we also need an upper bound. This is probably easier to do if we only consider random mutations M_k and keep other mutations from mixing in.*

Open Problem 7 *In Theorem 3 how fast does the size in bits of the organism P_ρ grow? Is it possible to have the size in bits of the organism P_ρ grow linearly and still achieve the same rapid growth in fitness?*

Open Problem 8 *It is interesting to think of Model A as a conventional random walk and to study the average mutation distance between an organism A and its successor A', its second successor A'', etc. In organism time Δt how far will we get from A on the average? What will the variance be?*

8. Model B (Naming Functions)

Let's now consider Model B. Why study Model B? Because hierarchical structure is a conspicuous feature of actual biological organisms, but it is impossible to prove that such structure must emerge by random evolution in Model A.

Why not? Because the programming language used by the organisms in Model A is so powerful that all structure in the programs can be hidden. Consider the programs P_ρ defined in Section 6.2 and used to prove Theorems 2 and 3. As we saw in Theorem 3, these programs P_ρ evolve without limit at random. However, P_ρ consists of a fixed prefix π_Ω followed by a lower bound on Ω, ρ, and what evolves is the lower bound ρ, *data* which has no visible hierarchical structure, not the prefix π_Ω, *code* which has fixed, unevolving, hierarchical structure.

So in Model A it is impossible to prove that hierarchical structure will emerge and increase in depth. To be able to do this we must utilize a less powerful programming language, one that is not universal and in which the hierarchical structure cannot be hidden: the Meyer-Ritchie LOOP language.[28]

We will show that the nesting depth of LOOP programs will increase without limit, due to random mutations. This also provides a much more concrete example of evolution than is furnished by our main model, Model A.

Now for the details.

We study the evolution of functions $f(x)$ of a single integer argument x; faster growing functions are taken to be fitter. More precisely, if $f(x)$ and $g(x)$ are two such functions, f is fitter than g iff $g/f \to 0$ as $x \to \infty$. We use an oracle to decide if $A' = M(A)$ is fitter than A; if not, A is not replaced by A'.[n] The programming language we are using has the advantage that program structure cannot be hidden. It's a programming language that is powerful enough to program any primitive recursive function,[29] but it's not a universal programming language.

To give a concrete example of hierarchical evolution, we use the extremely simple Meyer-Ritchie LOOP programming language, containing only assignment, addition by 1, do loops, and no conditional statements or subroutines. All variables are natural numbers, non-negative integers. Here is an example of a program written in this language:

```
// Exponential: 2 to the Nth power
// with only two nested do loops!
function(N) // Parameter must be called N.
   M = 1
   //
   do N times
      M2 = 0
      // M2 = 2 * M
      do M times
```

---

[n]An oracle is needed in order to decide whether $g(x)/f(x) \to 0$ as $x \to \infty$ and also to avoid mutations M that never produce an $A' = M(A)$. Furthermore, if a mutation produces a syntactically invalid LOOP program A', A' does not replace A.

```
        M2 = M2 + 1
        M2 = M2 + 1
      end do
      M = M2
   end do
   // Return M = 2 to the Nth power.
   return_value = M
   // Last line of function must
   // always set return_value.
end function
```

More generally, let's start with $f_0(x) = 2x$:

```
function(N) // f_0(N)
   M = 0
   // M = 2 * N
   do N times
      M = M + 1
      M = M + 1
   end do
   return_value = M
end function // end f_0(N)
```

Note that the nesting depth of f_0 is 1.

And given a program for the function f_k, here is how we program

$$f_{k+1}(x) = f_k^x(2) \qquad (9)$$

by increasing the nesting depth of the program for f_k by 1:

```
function(N) // f_(k+1)(N)
   M = 2
   //   do M = f_k(M) N times
   do N times
      N_ = M
      // Insert program for f_k here
      // with "function" and "end function"
      // stripped and all variable names
      // renamed to variable name_
      M = return_value_
   end do
   return_value = M
```

```
end function // end f_(k+1)(N)
```

So following (9) we now have programs for

$$f_0(x) = 2x, \quad f_1(x) = 2^x, \quad f_2(x) = 2^{2^{2^{\cdots}}} \text{ with } x \text{ 2's} \ldots$$

Note that a program in this language which has nesting depth 0 (no do loops) can only calculate a function of the form $(x + \text{a constant})$, and that the depth 1 function $f_0(x) = 2x$ grows faster than all of these depth 0 functions. More generally, it can be proven by induction[29] that a program in this language with do loop nesting depth $\leq k$ defines functions that grow more slowly than f_k, which is defined by a depth $k+1$ LOOP program. This is the basic theorem of Meyer and Ritchie[28] classifying the primitive recursive functions according to their rates of growth.

Now consider the mutation M that examines a software organism A written in this LOOP language to determine its nesting depth n, and then replaces A by $A' = f_n(x)$, a function that grows faster than any LOOP function with depth $\leq n$. Mutation M will be tried at random with probability $\geq 2^{-H(M)}$. And so:

Theorem 4 *In Model B, the nesting depth of a LOOP function will increase by 1 roughly periodically, with an estimated mutation time of $2^{H(M)}$ between successive increments. Once mutation M increases the nesting depth, it will remain greater than or equal to that increased depth, because no LOOP function with smaller nesting depth can grow as fast.*

Note that this theorem works because the nesting depth of a primitive recursive function is used as a clock; it gives Model B memory that can be used by intelligent mutations like M.

Open Problem 9 *In the proof of Theorem 4, is the mutation M primitive recursive, and if so, what is its LOOP nesting depth?*

Open Problem 10 *M can actually increase the nesting depth extremely fast. Study this.*

Open Problem 11 *Formulate a version of Theorem 4 in terms of subroutine nesting instead of do loop nesting. What is a good computer programming language to use for this?*

9. Remarks on Model C (Naming Ordinals)

Now let's briefly turn to programs that compute constructive Cantor ordinal numbers α.[27] From a biological point of view, the evolution of ordinals is piquant, because they certainly exhibit a great deal of hierarchical structure. Not, in effect, as we showed in Section 8 must occur in the genotype; here it is *automatically* present in the phenotype.

Ordinals also seem like an excellent choice for an evolutionary model because of their fundamental role in mathematics° and because of the mystique associated with naming large ordinals, a problem which can utilize an unlimited amount of mathematical creativity.[26,27] Conventional ordinal notations can only handle an initial segment of the constructive ordinals.

However there are two fundamentally different ways[27] to use algorithms to name *all* such ordinals α:

- An ordinal is a program that given two positive integers, tells us which is less than the other in a well-ordering of the positive integers with order type α.
- An ordinal α is a program for obtaining that ordinal from below: If it is a successor ordinal, as $\beta + 1$; if it is a limit ordinal, as the limit of a fundamental sequence β_k ($k = 0, 1, 2 \ldots$).

This yields two different definitions of the algorithmic information content or program-size complexity of a constructive ordinal:

$H(\alpha)$ = the size in bits of the smallest self-delimiting program for calculating α.

We can now define this beautiful new version of the Busy Beaver function:

$$\text{BB}_{ord}(N) = \max_{H(\alpha) \leq N} \alpha.$$

In order to make programs for ordinals α evolve, we now need to use a very sophisticated oracle, one that can determine if a program computes an ordinal and, given two such programs, can also determine if one of these ordinals is less than the other. Assuming such an oracle, we get the following version of Theorem 1, merely by using brainless exhaustive search:

°As an illustration of this, ordinals may be used to extend the function hierarchy f_k of Section 8 to transfinite k. For example, $f_\omega(x) = f_x(x)$, $f_{\omega+1}(x) = f_\omega^x(2)$, $f_{\omega+2}(x) = f_{\omega+1}^x(2) \ldots f_{\omega \times 2}(x) = f_{\omega+x}(x)$, etc., an extension of (9).

Theorem 5 *The fitness of our ordinal organism α will reach $BB_{ord}(N)$ by mutation time 2^N.*

Can we do better than this? The problem is to determine if there is some kind of Ω number or other way to compress information about constructive ordinals so that we can improve on Theorem 5 by proving that evolution will probably reach $BB_{ord}(N)$ in an amount of time which *does not* grow exponentially.

We suspect that Model C may be an example of a case in which *cumulative evolution at random does not occur*. On the other hand, we are given an extremely powerful oracle; maybe it is possible to take advantage of that. The problem is open.

Open Problem 12 *Improve on Theorem 5 or show that no improvement is possible.*

10. Conclusion

At this point we should look back and ask why this all worked. Mainly for the following reason: We used an extremely rich space of possible mutations, one that possess a natural probability distribution: the space of all possible self-delimiting programs studied by AIT.[14] But the use of such powerful mutational mechanisms raises a number of issues.

Presumably DNA is a universal programming language, but how sophisticated can mutations be in actual biological organisms? In this connection, note that evo-devo views DNA as software for constructing the embryo, and that the change from single-celled to multicellular organisms is roughly like taking a main program and making it into a subroutine, which is a fairly high-level mutation. Could this be the reason that it took so long—on the order of 10^9 years—for this to happen?[p]

The issue of *balance* between the power of the organisms and the power of the mutations is an important one. In the current version of the theory, both have equal power, but as a matter of aesthetics it would be bad form for a proof to overemphasize the mutations at the expense of the organisms. In future versions of the theory perhaps it will be desirable to limit the power of mutations in some manner by fiat.

In this connection, note that there are two uses of oracles in this theory, one to decide which of two organisms is fitter, and another to eliminate

---

[p]During most of the history of the earth, life was unicellular.

non-terminating mutations. It is perfectly fine for a proof to be based on taking advantage of the oracle for organisms, but taking advantage of the oracle for mutations is questionable.

We have by no means presented in this paper a mathematical theory of evolution and biological creativity *comme il faut*. But at this point in time we believe that metabiology is still a possible contender for such a theory. The ultimate goal must be to find in the Platonic world of mathematical ideas that ideal model of evolution by natural selection which real, messy biological evolution can but approach asymptotically in the limit from below.

We thank Prof. Cristian Calude of the University of Auckland for reading a draft of this paper, for his helpful comments, and for providing the paper by Meyer and Ritchie.[28]

Appendix. AIT in a Nutshell

Programming languages are commonly *universal*, that is to say, capable of expressing essentially any algorithm.

In order to be able to combine subroutines, i.e., for algorithmic information to be *subadditive*,

$$\text{size of program to calculate } x \text{ and } y$$
$$\leq \text{size of program to calculate } x$$
$$+ \text{ size of program to calculate } y,$$

it is important that programs be *self-delimiting*. This means that the universal computer U reads a program bit by bit as required and there is no special delimiter to mark the end of the program; the computer must decide by itself where to stop reading.

More precisely, if programs are self-delimiting we have

$$H(x,y) \leq H(x) + H(y) + c,$$

where $H(\ldots)$ denotes the size in bits of the smallest program for U to calculate $\ldots$, and c is the number of bits in the main program that reads and executes the subroutine for x followed by the subroutine for y.

Besides giving us subadditivity, the fact that programs are self-delimiting also enables us to talk about that probability $P(x)$ that a program that is generated at random will compute x when run on U.

Let's now consider how expressive different programming languages can be. Given a particular programming language U, two important things to consider are the *program-size complexity* $H(x)$ as a function of x, and the corresponding *algorithmic probability* $P(x)$ that a program whose bits are chosen using independent tosses of a fair coin will compute x.

We are thus led to select a subset of the universal languages that minimize H and maximize P; one way to define such a language is to consider a universal computer U that runs self-delimiting binary computer programs $\pi_C \, p$ defined as follows:

$$U(\pi_C \, p) = C(p).$$

In other words, the result of running on U the program consisting of the prefix π_C followed by the program p, is the same as the result of running p on the computer C. The prefix π_C tells U which computer C to simulate.

Any two such maximally expressive universal languages U and V will necessarily have

$$|H_U(x) - H_V(x)| \leq c$$

and

$$P_U(x) \geq P_V(x) \times 2^{-c}, \qquad P_V(x) \geq P_U(x) \times 2^{-c}.$$

It is in this precise sense that such a universal U minimizes H and maximizes P.

For such languages U it will be the case that

$$H(x) = -\log_2 P(x) + O(1),$$

which means that most of the probability of calculating x is concentrated on the minimum-size program for doing this, which is therefore essentially unique. $O(1)$ means that the difference between the two sides of the equation is order of unity, i.e., bounded by a constant.

Furthermore, we have

$$H(x,y) = H(x) + H(y|x) + O(1).$$

Here $H(y|x)$ is the size of the smallest program to calculate y from x.[q] This tells us that essentially the best way to calculate x and y is to calculate x and then calculate y from x. In other words, the *joint complexity* of x and y is essentially the same as the *absolute complexity* of x added to the *relative complexity* of y given x.

---

[q]It is crucial that we are not given x directly. Instead we are given a minimum-size program for x.

This decomposition of the joint complexity as a sum of absolute and relative complexities implies that the *mutual information* content

$$H(x:y) \equiv H(x) + H(y) - H(x,y),$$

which is the extent to which it is easier to compute x and y together rather than separately, has the property that

$$H(x:y) = H(x) - H(x|y) + O(1) = H(y) - H(y|x) + O(1).$$

In other words, $H(x:y)$ is also the extent to which knowing y helps us to know x and vice versa.

Last but not least, using such a maximally expressive U we can define the *halting probability* Ω, for example as follows:

$$\Omega = \sum 2^{-|p|}$$

summed over all programs p that halt when run on U, or alternatively

$$\Omega' = \sum 2^{-H(n)}$$

summed over all positive integers n, which has a slightly different numerical value but essentially the same paradoxical properties.

What are these properties? Ω is a form of concentrated mathematical creativity, or, alternatively, a particularly economical Turing oracle for the halting problem, because knowing n bits of the dyadic expansion of Ω enables one to solve the halting problem for all programs p which compute a positive integer that are up to n bits in size. It follows that the bits of the dyadic expansion of Ω are irreducible mathematical information; they cannot be compressed into a theory smaller than they are.[r]

From a philosophical point of view, however, the most striking thing about Ω is that it provides a perfect simulation in pure mathematics, where all truths are necessary truths, of contingent, accidental truths—i.e., of truths such as historical facts or biological frozen accidents.

Furthermore, Ω opens a door for us from mathematics to biology. The halting probability Ω contains infinite irreducible complexity and in a sense shows that pure mathematics is even more biological then biology itself, which merely contains extremely large finite complexity. For each bit of the dyadic expansion of Ω is one bit of independent, irreducible mathematical information, while the human genome is merely 3×10^9 bases = 6×10^9 bits of information.

---

[r] More precisely, it takes a formal axiomatic theory of complexity $\geq n - c$ (one requiring a $\geq n - c$ bit program to enumerate all its theorems) to enable us to determine n bits of Ω.

References

1. D. Berlinski, *The Devil's Delusion*, Crown Forum, 2008.
2. S. J. Gould, *Wonderful Life*, Norton, 1990.
3. N. Shubin, *Your Inner Fish*, Pantheon, 2008.
4. M. Mitchell, *Complexity*, Oxford University Press, 2009.
5. J. Fodor, M. Piattelli-Palmarini, *What Darwin Got Wrong*, Farrar, Straus and Giroux, 2010.
6. S. C. Meyer, *Signature in the Cell*, HarperOne, 2009.
7. J. Maynard Smith, *Shaping Life*, Yale University Press, 1999.
8. J. Maynard Smith, E. Szathmáry, *The Origins of Life*, Oxford University Press, 1999; *The Major Transitions in Evolution*, Oxford University Press, 1997.
9. J. P. Crutchfield, O. Görnerup, "Objects that make objects: The population dynamics of structural complexity," *Journal of the Royal Society Interface* **3** (2006), pp. 345–349.
10. G. J. Chaitin, "Evolution of mutating software," *EATCS Bulletin* **97** (February 2009), pp. 157–164.
11. G. J. Chaitin, *Mathematics, Complexity and Philosophy*, Midas, in press. (See Chapter 3, "Algorithmic Information as a Fundamental Concept in Physics, Mathematics and Biology.")
12. G. J. Chaitin, "Metaphysics, metamathematics and metabiology," in P. García, A. Massolo, *Epistemología e Historia de la Ciencia: Selección de Trabajos de las XX Jornadas,* **16** (2010), Facultad de Filosofía y Humanidades, Universidad Nacional de Córdoba, pp. 178–187. Also in *APA Newsletter on Philosophy and Computers* **10**, No. 1 (Fall 2010), pp. 7–11, and in H. Zenil, *Randomness Through Computation*, World Scientific, 2011, pp. 93–103.
13. G. Chaitin, *Proving Darwin: Making Biology Mathematical*, Pantheon, to appear.
14. G. J. Chaitin, "A theory of program size formally identical to information theory," *J. ACM* **22** (1975), pp. 329–340.
15. G. J. Chaitin, *Algorithmic Information Theory*, Cambridge University Press, 1987.
16. G. J. Chaitin, *Exploring Randomness*, Springer, 2001.
17. C. S. Calude, *Information and Randomness*, Springer-Verlag, 2002.
18. M. Li, P. M. B. Vitányi, *An Introduction to Kolmogorov Complexity and Its Applications*, Springer, 2008.
19. C. Calude, G. Chaitin, "What is a halting probability?," *AMS Notices* **57** (2010), pp. 236–237.
20. H. Steinhaus, *Mathematical Snapshots*, Oxford University Press, 1969, pp. 29–30.
21. D. E. Knuth, "Mathematics and computer science: Coping with finiteness," *Science* **194** (1976), pp. 1235–1242.
22. A. Hodges, *One to Nine*, Norton, 2008, pp. 246–249; M. Davis, *The Universal Computer*, Norton, 2000, pp. 169, 235.

23. G. J. Chaitin, "Computing the Busy Beaver function," in T. M. Cover, B. Gopinath, *Open Problems in Communication and Computation*, Springer, 1987, pp. 108–112.
24. G. H. Hardy, *Orders of Infinity*, Cambridge University Press, 1910. (See Theorem of Paul du Bois-Reymond, p. 8.)
25. D. Hilbert, "On the infinite," in J. van Heijenoort, *From Frege to Gödel*, Harvard University Press, 1967, pp. 367–392.
26. J. Stillwell, *Roads to Infinity*, A. K. Peters, 2010.
27. H. Rogers, Jr., *Theory of Recursive Functions and Effective Computability*, MIT Press, 1987. (See Chapter 11, especially Sections 11.7, 11.8 and the exercises for these two sections.)
28. A. R. Meyer, D. M. Ritchie, "The complexity of loop programs," *Proceedings ACM National Meeting, 1967*, pp. 465–469.
29. C. Calude, *Theories of Computational Complexity*, North-Holland, 1988. (See Chapters 1, 5.)

Chapter 16

Computability and Algorithmic Complexity in Economics*

K. Vela Velupillai and Stefano Zambelli

The Department of Economics/ASSRU[†]

This is an outline of the origins and development of the way computability theory and algorithmic complexity theory were incorporated into economic and finance theories. We try to place, in the context of the development of *computable economics*, some of the classics of the subject as well as those that have, from time to time, been credited with having contributed to the advancement of the field. Speculative thoughts on where the frontiers of computable economics are, and how to move towards them, conclude the paper. In a precise sense – both historically and analytically – it would *not* be an exaggeration to claim that both the origins of computable economics and its frontiers are defined by two classics, both by Banach and Mazur: that one page masterpiece by Banach and Mazur ([6]), built on the foundations of Turing's own classic, and the unpublished Mazur conjecture of 1928, and its unpublished proof by Banach ([39], ch. 6 &[69], ch. 1, §.6). For the undisputed original classic of computable economics is Rabin's *effectivization* of the Gale-Stewart game ([43];[17]); the frontiers, as I see them, are defined by recursive analysis and constructive mathematics, underpinning computability over the computable *and constructive* reals and providing computable founda-

---

*Rabin's *effectivization* of the Gale-Stewart Game ([43]) remains the model methodological contribution to the field for which Velupillai coined the name *Computable Economics* more than 20 years ago. Alain Lewis was the first to link Rabin's work with Simon's fertile concept of *bounded rationality* and interpret them in terms of Alan Turing's work. Solomonoff (1964), one of the three – the other two being Kolmogorov and Chaitin – acknowledged pioneers of *algorithmic complexity theory*, had his starting point in one aspect of what[73] came to call the *Modern Theory of Induction*, an aspect which had *its* origins in.[24] Kolmogorov's resurrection of von Mises ([81]) and the genesis of *Kolmogorov complexity* via *computability theoretic foundations for a frequency theory of probability* has given a new lease of life to finance theory ([50]). Rabin's classic of computable economics stands in the long and distinguished tradition of *game theory* that goes back to Zermelo ([85], Banach & Mazur (,[6] Steinhaus ([63]) and Euwe ([15]).
[†]*Algorithmic Social Science Research Unit* (http://www.assru.economia.unitn.it/) Department of Economics, University of Trento.

tions for the economist's Marshallian penchant for *curve-sketching* ([10];[20]; and, in general, the contents of **Theoretical Computer Science**, Vol. 219, Issue 1-2). The former work has its roots in the Banach-Mazur game (cf.[39], especially p. 30), at least in one reading of it; the latter in ([6]), as well as other, earlier, contributions, not least by Brouwer.

1. A Setting for Computability Theory and Algorithmic Complexity Theory in Economics

"M. O. Rabin .... was the first ... to make a significant application of recursion theory to the theory of games. In Rabin ([43]) it is remarked that 'It is obvious that not all games that are considered within the theory of games are actually playable by human beings.'[a] Here we find H. Simon's [[52]] concept of bounded rationality as a hidden theme, for the point of Rabin's inquiry is to determine if certain games of the Gale-Stewart variety can be won consistently by Turing Machines that serve as surrogate players. To quote Rabin [[43], p. 147] once more: 'The question arises as to what extent the existence of winning strategies makes a win-lose [i.e., zero-sum] game trivial to play. Is it always possible to build a computer which will play the game and consistently win?'

What Rabin is doing here is to provide an interpretation of Simon's concept of bounded rationality that is computational in character. The significance of [[43]] is that the techniques of recursion theory are used to fix a precise interpretation of computability within Church's Thesis."[28], p. 84; underlining in the original.

Alain Lewis is a contemporary pioneer, whose research program on *Effectively Constructive Mathematics* ([29],[28]) had an immense flowering in the years between the mid-1980s and the early 1990s. In his remarkably prescient *Monograph* (manuscript),[28] the above elegant interpretation of (what he calls) *Rabin's Theorem* ([43]), brings together the three undisputed pioneers of computable economics, i.e., *Herbert Simon, Michael Rabin* and *Alain Lewis* himself, in one fell swoop, so to speak.

---

[a]The exact quotation is ([43] p. 147):

"It is quite obvious that not all games which are considered in the theory of games can actually be played by human beings."

Velupillai's earliest attempt at considering economic theoretical issues in terms of computability theory goes back to his work on the computational complexity of algorithms for mathematical programming formalizations of optimization models in macroeconomics, from about 1976 – when Alan Turing, had not his life been cut short by tragic events, would have been 64 years of age, the age Velupillai has now reached!

Now, 36 years later, we commemorate the Turing Centennial with a well established field of Computable Economics fully cognizant of the pioneering work of Alan Turing and its relevance to many aspects of economic thoery, applied economics, human problem solving and much else. It will not be incongruous or inappropriate in any way if we single out two pioneering economists, Herbert Simon and Alain Lewis, as those most responsible for contributing the initial impulse towards what came to be Computable Economics – in spite of earlier stirrings by philosophers (Hilary Putnam) and computer scientists (Michael Rabin).

This 'next step in [economic] analysis', conjectured the *doyen* of mathematical economics, Kenneth Arrow ([2], p. S398),' [would be] a more consistent assumption of *computability* in the formulation of economic hypotheses'. But this has *not* been taken by economic theorists or, more pertinently, by anyone claiming to be a *computational* economist, *computable* general equilibrium theorist, applied *computable* general equilibrium theorist, *algorithmic* game theorist, so-called agent-based economic and financial modeller or any variety of *DSGE*[b] theorist. Indeed, not too long after the famous, and decidedly *non-computable* and *non-constructive*, *Arrow-Debreu classic* was published ([3], the trio of outstanding mathematical economists, Arrow, Karlin and Scarf, cautioned economists against facile conflation of *existence theorems* and *effectively computable solutions* ([4], p.17). Despite this early 'warning' by three of the pioneering mathematical economists of economic theorising in the non-computable mode, only in the sadly aborted research program on *effectively constructive economics* by Alain Lewis and in my *computable economics,* have there been *systematic* and coherent attempts to take Arrow's conjecture seriously[c]. As far as I am concerned,

---

[b]Dynamic Stochastic General Equilibrium.
[c]*Computable Economics* is a name I coined in the early 1980s, from the outset with the intention of encapsulating computability and constructivity assumption in economic theory. My earliest recollection is 1983, when I announced a series of graduate lectures on *Turing and his Machine for Economists*, in the department of economics at the European University Institute. Only one person signed up for the course, Henrietta Grant-Peterkin, one of our valued departmental secretaries! The course was still-born.

Simon ([52]), together with Michael Rabin ([43d]) and Alain Lewis, are the undisputed pioneers of *Computable Economics*, and both of these classics appeared in the public domain before ([4]). In[73] it was pointed out that (pp. 25-6):

> "[Simon's] path towards a broader base for economics .... stressed two empirical facts (quotes are from,[54] p. x):
>
> (I). 'There exists a basic repertory of mechanisms and processes that Thinking Man uses in all the domain in which he exhibits intelligent behavior.';
>
> (II). 'The models we build initially for the several domains must all be assembled from this same basic repertory, and *common principles of architecture* must be followed throughout.' (italics added);"
>
> It is at this point that I feel Simon's research program pointed the way toward computable economics in a precise sense. .....
>
> Instead, the direction Simon took codified his research program in terms of the familiar notions of bounded rationality and satisficing [underpinned by computational complexity theory] ..
>
> I remain convinced that, had Simon made the explicit recursion-theoretic link at some point in the development of his research program, computable economics would have been codified much earlier."

After reading,[73] Simon wrote Velupillai as follows (italics added):

> "As the book makes clear, my own journey through bounded rationality *has taken a somewhat different path*. Let me put it this way. There are many levels of *complexity in problems*, and corresponding boundaries between them. *Turing computability is an outer boundary*, and as you show, any theory that requires more power than that surely is irrelevant to any useful definition of human rationality. ....
>
> Finally, we get to the empirical boundary, measured by laboratory experiments on humans and by observation, of the level of complexity that humans actually can handle, with and without their computers, and - perhaps more

---

[d] In one of the most elegantly written 'eternal' classics of recursion theory, Hartley Rogers ([45]), the one blemish I found is the relegation of Rabin's results to a minor problem (p.121, ex. 8.5), with the unfortunate comment: 'This is a special and trivial instance of a general theorem about games'!

important – what they actually do to solve problems that lie beyond this strict boundary even though they are within some of the broader limits.

The latter is an important point for economics, because we humans spend most of our lives making decisions that are far beyond any of the levels of complexity we can handle exactly; and this is where satisficing, floating aspiration levels, recognition and heuristic search, and similar devices for arriving at good-enough decisions take over. A parsimonious economic theory, and an empirically verifiable one, shows how human beings, using very simple procedures, reach decisions that lie far beyond their capacity for finding exact solutions by the usual maximizing criteria "

Simon chose to work within the '*empirical* boundary', recognising immediately that computable economics was an attempt at defining, effectively, the relevance of the 'outer boundary' for formalisation in economic theory.

The true significance of Lewis's insight was to realise that Simon's concept of bounded rationality had to be given computational content; that Lewis did not also realise that Simon did give it this content from the outset is besides the point. But to give the notion of bounded rationality computational content in the context of games played by computing machines is one thing; to interpret bounded rationality as encapsulated in *finite automata* is quite another thing. Fortunately, Lewis did not fall into the latter trap, one which many distinguished game theorists almost willingly embraced.

The point missed by Lewis in his handsome tribute to Rabin is that this classic came down in the great tradition of *alternating games* (see[72]), begun by Zermelo at the beginning in ([85]), on the one hand; and, on the other hand, down the even nobler and more ancient tradition of what is now called *combinatorial games* (see the recent elegant, and eminently readable,[36] for a fine exposition of the history and origins of this field, with copious references). But there are many eminent game theorists who feel able to claim Zermelo as a precursor of orthodox game theory. In some senses – particularly with regard to von Neumann's original min-max result and to the sustained non-constructive and uncomputable methodology that underpins formal, orthodox, game theory - this claim many have a modicum of truth to it.

Our own 'take' on Rabin's classic as the fountainhead of computable economics is its pedagogic value in providing a tutorial on how to effectivise a non-effective framework in orthodox theory – whether economic or game theoretic. This is what I have emphasised in.[72] But, of course, it has also led to a revitalisation of both a part of recursion theory (see p. 254 in the excellent – although slightly dated – survey by Telgársky,[65] of recursion theoretic work inspired by Banach-Mazur games for some of the early and classic references), and a reflection on the possibility of avoiding reliance on the *axiom of choice* (see below, the comment on the *axiom of determinacy*).

The von Neumann paper of 1928 ([82]), the 'official' fountainhead for *orthodox game theory*, etched indelibly, to an essentially non-existent Mathematical Economics community, what has eventually come to be called '*Hilbert's Dogma*'[e], 'consistency ⇔ existence'. This became – and largely remains – the mathematical economist's *credo*. Hence, too, the inevitable schizophrenia of 'proving' existence of equilibria, first, and looking for methods to *construct* and *compute* them at a second, entirely unconnected, stage. Thus, too, the indiscriminate appeals to the *tertium non datur* – and its implications – in 'existence proofs', on the one hand, and the ignorance about the nature and foundations of constructive mathematics and computability theory, on the other.

But it was not as if von Neumann was *not* aware of Brouwer's opposition to 'Hilbert's Dogma', even as early as 1928, although there is reason to suspect that something peculiar may have been going on. Hugo Steinhaus observed, ([63]):

> "[My] inability [to prove the minimax theorem] was a consequence of the ignorance of Zermelo's paper in spite of its having been published in 1913. .... J von Neumann was aware of the importance of the minimax principle [in[82]]; it is, however, *difficult to understand the absence of a quotation of Zermelo's lecture in his publications.*"
>
> ibid, p. 460; italics added

Why didn't von Neumann refer, in 1928, to the Zermelo-tradition of (*alternating*) *games*? van Dalen, in his comprehensive and scrupulously fair biography of Brouwer,[70] p. 636, noted (italics added), without additional

---

[e]In van Dalen's measured, scholarly, opinion,[70] pp. 576-7 (italics added):

> "Since Hilbert's yardstick was calibrated by the continuum hypothesis, *Hilbert's dogma, 'consistency ⇔ existence'*, and the like, he was by definition right. *But if one is willing to allow other yardsticks*, no less significant, but based on alternative principles, then Brouwer's work could not be written off as obsolete nineteenth century stuff."

comment that:

> "In 1929 there was another publication in the intuitionistic tradition: an intuitionistic analysis of the game of chess by Max Euwe[f]. It was a paper in which the game was viewed as a spread (i.e., a tree with the various positions as nodes). Euwe carried out *precise constructive estimates* of various classes of games, and considered the influence of the rules for draws. When he wrote his paper he was not aware of the earlier literature of Zermelo and Dénès König. *Von Neumann called his attention to these papers, and in a letter to Brouwer, von Neumann sketched a classical approach to the mathematics of chess, pointing out that it could easily be constructivized.*"

von Neumann dinn't provide this 'easily constructivized' approach – then, or later? Perhaps it was easier to derive propositions appealing to the *tertium non datur,* and to 'Hilbert's Dogma', than to do the hard work of *constructing estimates* of an algorithmic solution, *as* Euwe did[g]? Perhaps it was easier to continue using the *axiom of choice* than to construct new axioms – say the *axiom of determinacy*[h] – as Steinhaus and Mycielski did

---

[f]In a strange lapse, van Dalen refers to Euwe, 1929, without giving the exact details of the reference in his excellent bibliography. The exact reference is.[15] Max Euwe was the fifth World Chess Champion, between 1935-1937, having defeated Alexander Alekhine, on December 15, 1935.

[g]At the end of his paper Euwe reports that von Neumann brought to his attention the works by Zermelo and König, after he had completed his own work (ibid, p. 641). This further substantiates the perplexity reported by Steinhaus (above) on the absence of any reference to Zermelo in von Neumann's official publications of the time. In any case, Euwe then goes on (italics added):

> "*Der gegebene Beweis is aber nicht konstruktive,* d.h. es wird keine Methode angezeigt, mit Hilfe deren der gewinnweg, wenn überhaupt möglich, *in endlicher Zeit konstruiert werden kann.*"

[h]The introduction of this axiom is relevant in computable economics and point we wish to make is best described in Takeuti's observation ([64], pp. 73-4; italics added):

> "There has been an idea, which was originally claimed by Gödel and others, that, if one added an axiom which is a strengthened version of the existence of a measurable cardinal to existing axiomatic set theory, then various mathematical problems might all be resolved. Theoretically, nobody would oppose such an idea, but, in reality, *most set theorists felt it was a fairy tale and it would never really happen.* But it has been *realized by virtue of the axiom of determinateness,* which showed Gödel's idea valid."

([33])? Whatever the reason, the fact remains that the von Neumann legacy was a legitimization of 'Hilbert's Dogma' and the indiscriminate use of the axiom of choice in mathematical economics and game theory.

Velupillai began to think of Game Theory in algorithmic modes – but not what is today referred to as *Algorithmic Game Theory* – after realizing the futility of algorithmising the uncompromisingly *subjective* von Neumann-Nash approach to game theory and beginning to understand the importance of *Harrop's theorem* ([19]). This realization came after an understanding of *effective playability* in arithmetical games, developed elegantly by Michael Rabin.

The brief, rich and primarily recursion theoretic framework of Harrop's classic paper requires a deep understanding of the rich interplay between *recursivity and constructive representations* of **sets** that are *recursively enumerable*. There is also an obvious and formal connection between the notion of a finite combinatorial object, whose complexity is formally defined by the uncomputable Kolmogorov measure of complexity, and the results in Harrop's equally pioneering attempt to characterise the recursivity of finite sets and the resulting indeterminacy – undecidability – of a Nash equilibrium *even in the finite case*. To the best of my knowledge this interplay has never been mentioned or analysed in the mathematical economic or game theoretic literature.

When Velupillai conceived the notion of computable economics in the early 1980s, he had in mind both constructive and computable mathematics as bases for the formalization of economic theory. He was blissfully ignorant of the pioneering works by Rabin and Lewis, till about the late 1980s. Also, the important work by Douglas Bridges based on constructive mathematics were unknown to him when he was fashioning computable economics including constructive assumptions and interpretations.

Finally, anyone even remotely familiar with Conway's characteristically clear note on *A Gamut of Game Theories* ([13]) and Turing's classic on *Solvable and Unsolvable Problems* ([68]), and Herbert Simon's kind of behavioural economics – called classical behavioural economics in this paper – will know that there is an almost formal duality between problem solving and (combinatorial) games. This is not a theme space allows us to develop, but it needs to be pointed out that any future for computable economics will have to enlarge on this aspect of the interaction between recursion theory, combinatorial games, *Ramsey theory* and behavioural economics.

The paper is organised as follows. The next section is a retrospective of some of the results obtained under the rubric of computable economics.

The section is sub-divided into two sub-sections: classical behavioural and (classical) computable economics. Section is a view of randomness and (statistical) induction, underpinned by algorithmic complexity theory, but with suggestions on an unususal double duality: one between algorithmic complexity theory and computational complexity theory; the other between classical recursion theory and constructive analysis. The final section outlines aspects of our view of the frontiers of computable economics. The main vision here is the hope that 'the next step in computable economic analysis would be a more consistent' consideration of recursive or computable analysis, particularly in macroeconomic dynamics.

2. Computability in Economics: A Retrospective

"[The] adoption of the *infinitary, nonconstructive, set theoretic*, algebraic, and structural methods that are characteristic to modern mathematics [....] were controversial, however. At issue was not just whether they are consistent, but, more pointedly, whether they are meaningful and appropriate to mathematics. After all, *if one views mathematics as an essentially computational science*, then arguments without computational content, whatever their heuristic value, are not properly mathematical. .. [At] the bare minimum, we wish to know that the universal assertions we derive in the system will not be contradicted by our experiences, and *the existential predictions will be borne out by calculation.* This is exactly what *Hilbert's program*[i] was designed to do."
[5], pp. 64-5; italics added

Thus, our claim is that the *existential predictions* made by the purely theoretical part of mathematical economics, game theory and economic theory '*will [not] be borne out by calculations.*' There is, therefore, a serious *epistemological deficit* – in the sense of economically relevant knowledge that can be processed and accessed computationally and experimentally – in all of the above approaches, claims to the contrary notwithstanding, that is unrectifiable without *wholly abandoning their current mathemati-*

---

[i]Velupillai has tried to make the case for interpreting the philosophy and methodology of mathematical economics and economic theory in terms of the discipline of *Hilbert's program* in.[76]

cal foundations. This is an *epistemological deficit* even before considering the interaction between appeals to infinite – even uncountably infinite – methods and processes in proofs, where both the universal and existential quantifiers are freely used in such contexts, and the *finite* numerical instances with which they are, ostensibly, 'justified'. This epistemological deficit requires even 'deeper' mathematical and philosophical considerations in *Cantor's Paradise*[j] of ordinals[k] , where combinatorics, too, have to be added to computable and constructive worlds to make sense of claims by various mathematical economists and agent based modeling practitioners.

Against this backdrop, within the framework of what we will now call *classical* computable economics, the following are some of the results that have been derived[1]: (1). Nash equilibria of (*even*) *finite* games are constructively indeterminate; (2). The Arrow-Debreu equilibrium is uncomputable (and its existence is proved nonconstructively); (3). The Uzawa Equivalence Theorem is uncomputable and nonconstructive; (4). Computable General Equilibria are neither computable nor constructive; (5). The Two *Fundamental Theorems of Welfare Economics* are Uncomputable and Nonconstructive, respectively; (6). The *Negishi method* is proved nonconstructively and the implied procedure in the method is uncomputable; (7). There is no effective procedure to generate preference orderings; (8). Rational expectations equilibria are uncomputable and are generated by uncomputable and nonconstructive processes; (9). Policy rules in macroeconomic models are noneffective; (10). Recursive Competitive Equilibria (RCE), underpinning

---

[j]Hilbert did not want to be driven out of '*Cantor's Paradise*' ([22]; p. 191):

> 'No one shall drive us out of the paradise which Cantor has created for us.'

To which the brilliant 'Brouwerian' response, if we may be forgiven for stating it this way, by Wittgenstein was ([84]; p. 103):

> 'I would say, "I wouldn't dream of trying to drive anyone out of this paradise." I would try to do something quite different: I would try to show you that it is not a paradise – so that you'll leave of your own accord. I would say, You're welcome to this; just look about you." '

[k]Where '*Ramsey Theory*', '*Goodstein Sequences*' and the '*Goodstein theorem*', reign supreme. In work in progress these issues are dealt with in some detail, as they pertain to bridging the '*epistemological deficit*' in economic theoretical discourse in the mathematical mode.

[l]Apart from the twelfth result, which is due to the pioneering work of Michael Rabin ([43]) in 1957, the rest are due to Velupillai. The first was suggested by Francisco Doria.

the Real Business Cycle (RBC) model and, hence, the *Dynamic Stochastic General Equilibrium* (DSGE) benchmark model of Macroeconomics, are uncomputable; (11). Dynamical systems underpinning growth theories are incapable of computation universality; (12). There are games in which the player who in theory can always win cannot do so in practice because it is impossible to supply him with effective instructions regarding how he/she should play in order to win; (13). The theoretical benchmarks of Algorithmic Game Theory are uncomputable and non-constructive; (14). Boundedly rational agents,satisfying, formalised within the framework of (metamathematical) decision problems are capable of effective procedures of rational choice.

In the next subsection we outline the computability theoretic background against which # 14 can be demonstrated. The second subsection is a brief outline of classical computable economics, in retrospective mode.

2.1. *Notes on classical behavioural economics — computable foundations*

"If we hurry , we can catch up to Turing on the path he pointed out to us so many years ago."
Herbert Simon,[56] p. 101.

Velupillai coined the phrase *classical behavioural economics* to characterise the kind of behavioural economics pioneered by Herbert Simon, which was underpinned, at every level of theoretical and applied analysis, by a model of computation. Invariably, although not always explicitly, it was *Turing's model of computation*. To highlight the difference between *modern* behavioural economics, which is *never* underpinned by *a model of computation*, and the kind of behavioural economics that was pioneered and practiced by Simon and his associates and followers, Velupillai decided to refer to the latter as practitioners of *classical behavioural economics.*[m]

The fundamental focus in classical behavioural economics is on *decision problems* faced by *human problem solvers*, the latter viewed as information processing systems. All of these terms are given computational content, *ab initio*. But given the scope of this paper we shall not have the possibility of a full characterisation. The ensuing 'bird's eye' view must suffice for now.[n]

A *decision problem* asks whether there exists an *algorithm* to *decide* whether a mathematical assertion does or does not have a proof; or a for-

---
[m] See[49] for a more detailed discussion of this theme.
[n] Some details are discussed in greater and more rigorous depth in[74]

mal problem does or does not have an algorithmic solution. Thus the characterization makes clear the crucial role of an underpinning *model of computation*; secondly, the answer is in the form of a *yes/no* response. Of course, there is the third alternative of '*undecidable*', too. It is in this sense of *decision problems* that we interpret the word 'decisions' here.

As for 'problem solving', we shall assume that this is to be interpreted in the sense in which it is defined and used in the monumental classic by Newell and Simon ([34]).

Finally, the *model of computation* is the *Turing model*, subject to the *Church-Turing Thesis*.

To give a rigorous mathematical foundation for bounded rationality and satisficing, as decision problems,°, it is necessary to underpin them in a dynamic model of choice in a computable framework. However, any formalization underpinned by a model of computation in the sense of computability theory is intrinsically dynamic.

Consider the Boolean formula:

$$(x_1 \vee x_2 \vee x_3) \wedge (x_1 \vee \{\neg x_2\}) \wedge (x_2 \vee \{\neg x_3\}) \wedge (x_3 \vee \{\neg x_1\})$$
$$\wedge (\{\neg x_1 \vee \{\neg x_2\} \vee \{\neg x_3\}) \qquad (1)$$

Remark 1 *Each subformula within parenthesis is called a clause; The variables and their negations that constitute clauses are called literals; It is 'easy' to 'see' that for the truth value of the above Boolean formula to be $t(x_i) = 1$, all the subformulas within each of the parenthesis will have to be true. It is equally 'easy' to see that no truth assignments whatsoever can satisfy the formula such that its global value is true. This Boolean formula is unsatisfiable.*

Problem 2 *SAT – The Satisfiability Problem*

°The three most important classes of decision problems that almost characterise the subject of computational complexity theory, underpinned by a model of computation – in general, the model of computation in this context is the *Nondeterministic Turing Machine* – are the **P**, **NP** and **NP-Complete** classes. Concisely, but not quite precisely, they can be described as follows:

(1) **P** defines the class of computable problems that are solvable in time bounded by a *polynomial function of the size of the input*;

(2) **NP** is the class of computable problems for which a *solution* can be *verified in polynomial time*;

(3) A computable problem lies in the class called **NP-Complete** if every problem that is in **NP** can be *reduced to it in polynomial time*.

Given m clauses, $C_i(i = 1, \ldots, m)$, containing the literals (of) $x_j(j = 1, \ldots, n)$, *determine* if the formula $C_1 \wedge C_2 \wedge \ldots \ldots \wedge C_m$ is *satisfiable*.

Determine means 'find an (efficient) algorithm'. To date it is not known whether there is an *efficient* algorithm to solve *the satisfiability problem* – i.e., to determine the truth value of a Boolean formula. In other words, it is not known whether $SAT \in \mathbf{P}$. But:

Theorem 3 $SAT \in \mathbf{NP}$

Finally, we have Cook's famous theorem:

Theorem 4 *Cook's Theorem*
SAT is $\mathbf{NP}$ – *Complete*

It is in the above kind of context and framework within which we are interpreting Simon's vision of behavioural economics. In this framework optimization is a very special case of the more general decision problem approach. The real mathematical content of *satisficing*[p] is best interpreted in terms of the satisfiability problem of computational complexity theory, the framework used by Simon consistently and persistently - and a framework to which he himself made pioneering contributions.

Finally, there is the computably underpinned definition of *bounded rationality*.

Theorem 5 *The process of rational choice – i.e., boundedly rational choice – by an economic agent is formally equivalent to the computing activity of a suitably programmed (Universal) Turing machine.*

Proof. By construction. See §3.2, pp. 29-36, *Computable Economics* [73] ∎

Remark 6 *The important caveat is 'process' of rational choice, which Simon tirelessly emphasized by characterizing the difference between 'procedural' and 'substantive' rationality; the latter being the defining basis for*

---

[p]In,[57] p. 295, Simon clarified the *semantic* sense of the word *satisfice*:

> "The term 'satisfice', which appears in the *Oxford English Dictionary* as a Northumbrian synonym for 'satisfy', was borrowed for this new use by H. A. Simon (1956) in 'Rational Choice and the Structure of the Environment' [i.e,[53]]".

Olympian rationality ([55], p. 19), the former that of the computationally underpinned problem solver facing decision problems. In the Olympian model the 'process' aspect is submerged and dominated by the static optimization operator. By transforming the agent into a problem solver, constrained by computational formalisms to determine a decision problem, Simon was able to extract the procedural content in any rational choice.

Definition 7 *Computation Universality of a Dynamical System*

A dynamical system is said to be capable of computation universality if, using its initial conditions, it can be programmed to simulate the activities of any arbitrary Turing Machine, in particular, the activities of a Universal Turing Machine.

Theorem 8 *Boundedly rational choice by an information processing agent within the framework of a decision problem is capable of computation universality.*

Proof. See.[75] ∎

We have only scratched a tiny part of the surface of the vast canvass on which Simon sketched his vision of a computably underpinned behavioural economics. Nothing in Simon's behavioural economics – i.e., in *Classical Behavioural Economics* – was devoid of computable content. There was – is – never any *epistemological deficit* in any computational sense in classical behavioural economics.

2.2. *Classical computable economics*

> "The method of 'postulating' what we want has many advantages; they are the same as the advantages of theft over honest toil. Let us leave them to others and proceed with our honest toil."
>
> Bertrand Russell ([47], p. 71)

In computable economics, as in any computation with analogue computing machines or in classical behavioural economics, all solutions are based on *effectively computable* methods.[q] Thus computation is intrinsic to the subject and all formally defined entities in computable economics – as in

---
[q]We identify five varieties of computation, underpinned by computability theory, even if not explicitly: classical recursion theory, computable analysis, constructive analysis, interval analysis and classical numerical analysis (now given computable foundations in[8]).

classical behavioural economics – are, therefore, algorithmically grounded. Given the algorithmic foundations of computability theory and the intrinsic dynamic form and content of algorithms, it is clear that this will be a *'mathematics with dynamic and algorithmic overtones'*.[r] This means, thus, that computable economics is a case of a new kind of mathematics in old economic bottles. The 'new kind of mathematics' implies new questions, new frameworks, new proof techniques - all of them with algorithmic and dynamic content for digital domains and ranges.

Some of the key formal concepts of computable economics are, therefore: *solvability & Diophantine decision problems, decidability & undecidability, computability & uncomputability, satisfiability, completeness & incompleteness, recursivity* and *recursive enumerability, degrees of solvability (Turing degrees), universality* & the *Universal Turing Machine* and *Computational, algorithmic* and *stochastic complexity*. The proof techniques of computable economics, as a result of the new formalisms, will be, typically, invoking methods of: *Diagonalization, The Halting Problem for Turing Machines, Rice's Theorem, Incompressibility theorems, Specker's Theorem, Recursion Theorems*. For example, the *recursion theorems* will replace the use of traditional, non-constructive and uncomputable, topological fix point theorems, routinely used in orthodox mathematical analysis. The other theorems have no counterpart in non-algorithmic mathematics.

In the spirit of pouring new mathematical wines into old economic bottles, the kind of economic problems that computable economics is immediately able to grant a new lease of life are the classic ones of: computable and constructive existence and learning of rational expectations equilibria, computable learning and complexity of learning, computable and bounded rationality, computability, constructivity and complexity of general equilibrium models, undecidability, self-reproduction and self-reconstruction of models of economic dynamics (growth & cycles), uncomputability and incompleteness in (finite and infinite) game theory and of Nash Equilibria, decidability (playability) of arithmetical games, the intractability (computational complexity) of optimization operators; etc.

---

[r]
"I think it is fair to say that for the main existence problems in the theory of economic equilibrium, one can now bypass the fixed point approach and attack the equations directly to give existence of solutions, with a simpler kind of mathematics and *even mathematics with dynamic and algorithmic overtones*."
[59], p. 290; italics added.

Suppose the starting point of the computable economist whose visions of actual economic data, and its generation, are the following:

Conjecture 9 *Observable variables are sequences that are generated from recursively enumerable but not recursive sets, if rational agents underpin their generation.*

The above conjecture is is akin to the orthodox economic theorist and the econometrician assuming that all observable data emanate from a structured probability space and the problem of inference is simply to determine, by statistical or other means the parameters that characterise their probability distributions.

All the way from microeconomic supply and demand functions to monetary macroeconomic variables, parameters and functions, *Diophantine relations, equations and functions predominate in computable economics*. This is because the natural data types in economics are, at best, rational numbers. Hence, the following famous theorem is used extensively.

The following three theorems of classical computability theory ([37]), for example, are used to prove the uncomputability of rational expectations equilibria in orthodox frameworks, to construct computable rational expectations equilibria in computable macroeconomics and to formalise computable (macroeconomic) growth theory, respectively: **Rice's Theorem**, the **Halting Problem for Turing Machines** (the recursion theoretic) **Fixed Point Theorem** and the **Recursion Theorem** (related to invariance theorems in the domain of algorithmic complexity theory).

The idea behind the *recursion theorem* is to formalize the activity of a Turing Machine that can obtain its own description and, then, compute with it. This theorem is essential, too, for formalizing, recursion theoretically, a model of *growth* in a macroeconomy and to determine and learn, computably and constructively, *rational expectations equilibria*. The *fix point theorem* and the *recursion theorem* are also indispensable in the computable formalization of *policy ineffectiveness* postulates, *time inconsistency* and *credibility* in the theory of macroeconomic policy. Even more than in microeconomics, where topological fix point theorems have been indispensable in the formalizations underpinning existence proofs, the role of the above *fix point theorem* and the related *recursion theorem* are absolutely fundamental in what I come to call *Computable Macroeconomics*.

Anyone who is able to formalize these theorems, corollaries and conjectures and work with them, would have mastered some of the key elements that form the core of the necessary mathematics of computable economics. Unlike so-called computable general equilibrium theory and its offshoots, computable economics – and *its* offshoots – are intrinsically computational and numerical.

3. Randomness, Induction and Algorithmic Complexity

> "But it will be clear that, for those who hold that the mathematical universe consists of lawlike objects *only*, *Kollektivs* are equally *impossible*."
>
> [71], p. 60; italics added.

We have come round to the belief, via Solomonoff ([61],[62]), that Keynes ([24]) is the origin, at least from an economist interested in the foundations of statistical induction (*ibid*, p. 350), of one strand of algorithmic complexity theory. However, this is not to deny the fundamental importance of von Mises ([81]), his remarkably cogent 'manifesto', *Erst das Kollektiv, dann die Wharscheinlikhkeit*, his struggles to define a consistent notion of *Kollektivs constructively* and its eventual realisation in the computability theoretic notion of (uncomputable) *Kolmogorov complexity*. Few seem to have acknowledged the imporatnce of underpinning *Kollektivs* on the Brouwerian notion of *lawless sequences*. A close reading of,[24] particularly chapter 33, would, we contend, substantiate our stance that Keynes, too, was groping for such a notion, as always before its time.

Current orthodoxy of the field of algorithmic complexity theory is elegantly and comprehensively discussed, explained and described, all the way to the frontiers of research, in the almost encyclopaedic treatises by Li & Vitanyi ([30]), Nies ([35]) and Downey & Hirschfeldt ([14]). Velupillai has had a stab at a concise outline of the field, from the point of view of randomness *and* induction, so that learning can be studied from the point of view of algorithmic complexity (cf.,[73] chapter 5), to which we may refer the interested for a potted survey of the field.[s]

The orthodox story, in an ultra-brief nutshell, is that the origins of the field of algorithmic complexity theory lie in the work of Kolmogorov,

---

[s]Recursion theoretic *Inductive Inference* is elegantly and comprehensively treated in the second volume of Odifreddi's treatise on Classical Recursion Theory ([38], in particular, VII.5 & IX.5). Learning of rational expectations equilibria in the setting of recursively enumerable sets remains an incompletely explored field in macroeconomics – as it does in finance theory.

Chaitin and Solomonoff,[t] in their approaches to, respectively, the quantity of information in finite objects, program-size descriptions of the information content of a finite object and *induction*. For an economist the most interesting approach is that by Solomonoff, whose starting point, in fact, was the *Treatise on Probability*,[24] by Keynes. These original aims developed into, and linked with the earlier research on, the von Mises attempt to define a frequency theory approach to probability, randomness[u] and Bayesian estimation. All this is part of the folklore of the subject, easily gleaned from any of the indicated references, above.

In all three traditions – i.e., the Kolmogorov, Chaitin and Solomonoff – the intentions were to measure the amount of information necessary to *describe* a given, finite, binary sequence (or string). A little more precisely, the idea is as follows: given a string x, its algorithmic complexity is defined to be the shortest string y from which a Universal Turing Machine[v] can 'produce' the given x. On the other hand, in computational complexity theory – particularly as a result of adherence to 'Post's Program' – attention is *not* focused on *individual finite strings*. Instead the fundamental questions are about the computational difficulty – i.e., complexity – of recognising *sets*. Thus, the problem is about *deciding* whether a given finite string belongs to a particular set or not. It will be evident that in computational complexity theory one tries to associate a function $\tau_\wp : \mathbb{N} \to \mathbb{N}$, to a recursive set, $\wp$ such that $\wp$ is accepted by those Turing Machines, say Θ, that run in *time* $\Omega(\tau_\wp(n))$. Therefore, one way to link algorithmic complexity theory with computational complexity theory will be to define a notion of the former that is *time-bounded* and is able to capture aspects of the complexity of the set $\wp$. In other words, it is necessary to add a time-bounded complexity component, as is routine in computational complexity theory, to the stan-

---

[t]The representative references for Chaitin, Kolmogorov and Solomonoff are, respectively,[11],[27],[61] and.[62]

[u]One direct link with computational complexity theory, was stated succinctly by Compagner,[12] p. 700 (italics added):

"..[T]he mathematical description of random sequences in terms of complexity, which in algorithmic theory leads to the identification of randomness with *polynomial-time unpredictabilty*."

Once algorithmic complexity theory is viewed as the basis for a definition of finite random sequences, then it is inevitable that the emphasis will be on prediction rather than computation. Thus, the link with orthodox computational complexity theory is not as firm as the inclusion of the sobriquet 'complexity' in the title may suggest.

[v]In the Solomonoff tradition the corresponding 'universality' resides in the concept of a 'universal distribution'.

dard measure of algorithmic complexity. If this is done, the complexity of the finite string, x, will now be defined by the minimum of the sum of the description length and a measure of the time required to produce that x, from the given description. First steps towards such an attempt is made in.[78]

Since algorithmic complexity theory is, *ab initio*, underpinned by the Turing model of computation, it is natural to define time-constrained generation of the descriptive complexity of members of *sets*. For example, as a computable economist, we construct economic theories underpinned by Turing's model of computation. Given a computable economic theory, there will be naturally definable time-bounded measures to describe the theory and, hence, immediate considerations of computational complexity of such descriptions. This is the way the complexity of solutions to ODE is studied. First, the ODE is either constructified or formalised within computability theory; then the computational complexity of the constructified or computable theoretic solution is evaluated. Thus, description is intrinsically algorithmized and the computable economist can switch betwee algorithmic complexity theory and computational complexity theory, in formal, dual, ways. However:

> "There is also *a technical sense of 'complexity' in logic*, variously known as Kolmogorov complexity, Solomonoff complexity, Chaitin complexity, ..., algorithmic complexity, information-theoretic complexity, and program-size complexity. The most common designation is 'Kolmogorov complexity' .... this is probably a manifestation of the principle of 'Them that's got shall get,' since Kolmogorov is the most famous of these mathematicians[w]."
>
> [16]; p. 137; italics added.

Franzén's perceptive observation suggests that this whole area is really about '*a technical sense of 'complexity' in logic*'.[x] In this sense we would

---

[w] In their comprehensive and admirable text on this subject, Li and Vitanyi first gave the reason for subsuming all these different variations on one theme by the name 'Kolmogorov Complexity',[30] p. 84:

> "Associating Kolmogorov's name with [algorithmic] complexity may also be an example of the 'Matthew Effect' first noted in the Gospel according to Matthew, 25:29-30.."

[x] However, we believe Chaitin's reference to his own pioneering work as 'algorithmic information theory', is a much better encapsulation of the contents of the field and the intentions of the pioneers. Indeed, the natural precursor is Shannon, rather than von Mises, but *Whig history* is a messy affair and straightening out historical threads is a difficult task, even in a contemporary field.

like to add another point so as to dispel popular misconceptions about correlating or juxtaposing *complexity* with *incompleteness*.[y] Many an unwary reader of Chaitin's important works - and his specific program-size approach to algorithmic information theory, the incarnation of Kolmogorov complexity in Chaitin's independent work - has had a tendency to claim that incompleteness, undecidability or uncomputability propositions are only valid in so-called '*sufficiently complex*' mathematical systems. *A fortiori*, that intuitively *simple* computable systems are not computationally complex. This is simply false. Very simple formal mathematical systems are capable of generating incompleteness and undecidability propositions; just as intuitively very simple computable systems are capable of encompassing incredibly complex computational complexities, as some of the above examples have shown. Conversely, there are evidently complex systems that are provably complete and decidable; and, similarly, there exist seemingly complex functions that are capable of being computed, even primitive recursively. As one obvious and famous example illustrating *incompleteness* and *essential undecidability* in an *intuitively simple, finitely axiomatizable, simply consistent*[z] theory, one can take Robinson's Arithmetic,[44] as shown in some of the classic books of metamathematics,[aa] e.g.,[25],[26] pp. 280-1,[9] p.215,ff.

If the modern origins of computational complexity theory, via computability theory, can be found in *Hilbert's Tenth Problem*, then, equally, the proto-historic origins of algorithmic complexity theory can found, via exact approximation theory, in *Hilbert's Thirteenth Problem*.[bb] Hilbert's

---

[y]Or, *simplicity* with *completeness*. I am, of course, referring to *incompleteness* in the strict metamathematical sense.

[z]See,[26] p. 287, footnote 216 and,[25] p. 470, Theorem 53. The part played by *simple consistency* and the analogy with Rosser's result of the essential undecidability of N,[46] is also discussed in the relevant parts of.[26] Furthermore, despite some unfortunate misprints and unclarity, Franzén's fine exposition of the use and abuse of *Gödel's Incompleteness* theorems has a good discussion of the way the *Rosser sentence* (rather than the more famous *Gödel sentence*) is used in proving – by reference to[60] – undecidability in *Robinson's Arithmetic* ([16], pp. 158-9).

[aa]I cite this example also because *Robinson's Arithmetic* is sufficient to represent every recursive function. It figures in the very first, 'Introductory', pages of Odifreddi's comprehensive, yet pedagogical, textbooks of *classical recursion theory*,[37] §I.1, p. 23. There are many equivalent ways of setting out the axioms of Robinson's Arithmetic (see, for example, the discussion in[37]).

[bb]If one reads the main content of the 13th Problem by replacing the word 'nomography' with 'algorithm', then the connection with the subject matter of this paper becomes fairly clear,[21] p.424:

"[I]t is probable that the root of the equation of the seventh degree is a function of its coefficients which does not belong to the class of

aim in formulating the *13th Problem* – based on problems of nomography[cc] – was to characterise functions in terms of their *complexity* in a natural way: find those defining characteristics of a function, such that, the given function can be built up from simpler functions and simple operations. Hilbert's honed intuition suggested the formulation of the *13th Problem*; it was – like the *10th Problem* – solved 'negatively', by Kolmogorov and Arnold. The point to be emphasised here is, however, not the direct and obvious connection with computational complexity theory.[dd] Essentially, Kolmogorov introduced the concept of 'ε–entropy' of a metric space[ee] 'to evaluate the order of increase of the volume of the [nomographic] table for an *increase in the accuracy* of [nomographic] tabulation.' In other words, in his work on approximation theory, preceding his work on algorithmic complexity theory by only a few years, Kolmogorov defined the 'size' of a finite body – actually a subset of a Banach space – in terms of its 'metric entropy'; on the other hand, in his work on algorithmic complexity theory, he defined the information content in a finite string in terms on 'entropy' (in the Shannon tradition), too.

---

functions capable of nomographic construction, i.e., that it cannot be constructed by a finite number of insertions of functions of two arguments. In order to prove this, the proof would be necessary *that the equation of the seventh degree* $f^7 + xf^3 + yf^2 + zf + 1 = 0$ *is not solvable with the help of any continuous functions of only two arguments.* I may be allowed to add that I have satisfied myself by a rigorous process that there exist analytical functions of three arguments x, y, z which cannot be obtained by a finite chain of only two arguments."

Kolmogorov and Arnold refuted Hilbert's conjecture by constructing representations of continuous functions of several variables by the superposition of functions of one variable and sums of functions.

[cc]In the opening lines of the section stating the 13th Problem, Hilbert gives an intuitive idea of 'nomography',[21] p. 424:

"Nomography deals with the problem: to solve equations by means of drawings of families of curves depending on an arbitrary parameter."

Those of us who indulge in drawing *vector fields* might see the similarities!
[dd]A beautiful discussion of approximation theory from this point of view – albeit implicitly – is given in an unfortunately little reference work by Vitushkin,[80] a more technical and comprehensive survey of Kolmogorov's work on approximation theory is in.[66]
[ee]See,[80] p. xiii and,[31] chapters 9 & 10. 'Order of increase', 'increase in the accuracy', 'most favourable system of approximation', 'rapidity of convergence' are some of the phrases used in Kolmogorov approximation theory. These are the considerations that make approximation theoretical considerations naturally algorithmic and, therefore, also amenable to computational complexity analysis.

Finally, the notion of randomness of finite strings, based on algorithmic complexity theory can, we now suggest, also be defined via the constructive notion of *lawless sequences* (or *Choice Sequences,* see[67]), first enunciated by Brouwer . In this way, we think the computable economist, trained in classical recursion theory or (in the inclusive sense), constructive analysis (say from[7]), supplemented by a mastery of one of the modern classics on algorithmic complexity theory ([30],[35] or[14]) and any standard classic on computational complexity theory (for e.g.,[40] or[48]), will be properly equipped to do justice to *Turing's visions*.

4. Computable Economics: Towards the Frontiers

> "*The theory of recursive functions properly belongs to number theory*; indeed, the theory of recursive functions is, so to speak, the function theory of number theory. ... The notion of recursive function marks off those functions whose values can be effectively calculated at every particular point; and *just those functions are useful in the natural sciences*. Though the variables of recursive functions do not run through all real numbers but only the natural numbers, probability theory as well as quantum theory operates with functions of this latter kind; and recently recursive functions have begun to be applied in analysis too."
>
> [42], p. 7; italics added.

At least since Walras devised the *tâtonnement* process and Pareto's appeal to the market as a computing device, there have been sporadic attempts to find mechanisms to solve a system of supply-demand equilibrium equations, going beyond the simple counting of equations and variables. But none of these attempts to devise *mechanisms* to solve a system of equations were predicated upon the elementary fact that the data types – the actual numbers – realised in, and used by, economic processes were, at best, rational numbers. The natural equilibrium relation between supply and demand, respecting the elementary constraints of the equally natural data types of market – or any other kind of economy – should be framed as a Diophantine decision problems, and the way arithmetic games are formalised and shown to be effectively unsolvable in analogy with the *Unsolvability* of *Hilbert's Tenth Problem* (cf.[32]).

The Diophantine decision theoretic formalization is, thus, common to at least three kinds of computable economics: classical behavioural economics, algorithmic game theory in its incarnation as arithmetic game theory and elementary equilibrium economics. Even those, like Smale ([59]), who have perceptively discerned the way the problem of finding mechanisms to solve equations was subverted into formalizations of inequality relations which are then solved by appeal to (unnatural) non-constructive, uncomputable, fixed point theorems did not go far enough to realise that the data types of the variables and parameters entering the equations needed not only to be constrained to be non-negative, but also to be rational (or integer valued). Under these latter constraints, economics in its behavioural, game theoretic and microeconomic modes must come to terms with *absolutely (algorithmically) undecidable problems*. This is the cardinal message of the path towards computable economics.

Therefore, if orthodox algorithmic game theory, orthodox mechanism theory and computable general equilibrium theory have succeeded in computing their respective equilibria, then they would have to have done it with algorithms that are not subject to the strictures of the *Church-Turing Thesis* or do not work within the (constructive) proof-as-algorithm paradigm. This raises the mathematical meaning of the notion of algorithm in algorithmic game theory, orthodox mechanism theory and computable general equilibrium theory (and varieties of so-called computational economics). Either they are of the kind used in numerical analysis and so-called 'scientific computing' (as if computing in the recursion and constructive theoretic traditions are not 'scientific'; see[10] for a lucid definition and discussion of this seemingly innocuous concept) and, if so, their algorithmic foundations are, in turn, constrained by either the *Church-Turing Thesis* (as in[8]) or the (constructive) proof-as-algorithm paradigm; or, the economic system and its agents and institutions are computing the formally uncomputable and deciding the algorithmically undecidable (or are formal systems that are inconsistent or incomplete).

I believe Goodstein's algorithm,[18] could be the paradigmatic example for modelling rational - or integer - valued algorithmic (nonlinear) economic dynamics (see, for example,[41]). Every sense in which the notion of algorithm has been discussed above, for the path towards computable economics, is most elegantly satisfied by this line of research, a line that has by-passed the mathematical economics and nonlinear macrodynamics community. This is the only way I know to be able to introduce the algorithmic construction of an integer-valued dynamical system possessing a very simple global attractor, and with immensely long, effectively calculable, transients, whose existence is unprovable in *Peano Arithmetic*. Moreover, this kind of nonlinear dynamics, subject to *SSID*, ultra-long transients and possessing simple global attractors whose existence can be encapsulated within a classic *Gödelian, Diophantine*, decision theoretic framework, makes it also possible to discuss effective policy mechanisms (cf.[23]).

Kreisel's characteristically perceptive observation (see quote above, in the previous section), a plea for understanding the way to use the 'Goodstein algorithm' in economic dynamics and the economist's penchant for drawing curves and for working with numbers defined over the real numbers, convinces us that the most important frontier for computable economics is *computable analysis,* ([83]; coming down the[6] tradition) or *computable calculus* ([1], where a judicious combination of constructive logic and recursion theory is used). We have come to believe that every mathematically minded economist should be familiar with the *graph theorem of classical recursion theory* ([37], p. 135-6), and not simply be bamboozled by the Dirichlet-Kuratowski graph concept. The interaction between *recursive* and *recursively enumerable* sets, computable functions and *functions 'plottable' on a digital computer's screen* should be made clear to all students of economics, almost more importantly than teaching them probability theory, statistics and the like. This is implicit in some of the claims about the notion and definition of computation universality we have routinely been using in classical computable economics.

With an integration of *classical recursion theory, computable analysis* and a familiarity with the framework of *Diophantine Decision Problems*, and the suggestions in the previous section on mastering the double duality between algorithmic complexity and computational complexity, on the one hand, and between classical recursion theory and constructive analysis, on the other, *classical* computable economics will be ready to embark on the path towards *modern* computable economics, where not only the *theory* of the computer will be an underpinning of economic theory; but also the empirical use of the hardware, the pixels and the resolution that make the screen as much a part of the computable economist's 'box of tools' as its theory, will enrich the experiences of being educated to be a computable economist.

It is incumbent upon us to make the attempt to prepare for a 'computable and constructive' future, by writing the 'sensible textbooks' ([58]), for the next – or future – generations of students, who will be the harbingers of the computable approach to economics. It is only this way we can pay *homage* to **Turing's genius**.

References

1. Aberth, O (2001), Computable Calculus, 1st Edition, Academic Press.
2. Arrow, Kenneth. J (1986), Rationality of Self and Others in an Economic System, *Journal of Business*, Vol. 59, #4, Pt. 2, October, pp. S385–S398.
3. Arrow, Kenneth. J and Gerard Debreu (1954), Existence of an Equilibrium for a Competitive Economy, *Econometrica*, Vol. 22, pp. 265–290.
4. Arrow, Kenneth J, Samuel Karlin & Herbert Scarf (1958), The Nature and Structure of Inventory Problems, Chapter 2, pp. 16–36, in: *Studies in the Mathematical Theory of Inventory and Production*, edited by Kenneth J. Arrow, Samuel Karlin & Herbert Scarf, Stanford University Press, Stanford, CA.
5. Avigad, Jeremy (2009), The Metamathematics of Ergodic Theory, *Annals of Pure and Applied Logic*, Vol. 157 (2009), pp. 64–5.
6. Banach, Stefan & S. Mazur (1937), Sur les fonctions calculables, *Ann. Soc. Pol. de Math.*, Vol. 16, p. 223.
7. Bishop, Errett A, (1967), *Foundations of Constructive Analysis*, McGraw-Hill Book Company, New York.
8. Blum, Lenore, Felipe Cucker, Michael Shub and Steve Smale (1998), *Complexity and Real Computation*, Springer Verlag, New York.
9. Boolos, George. S, John P. Burgess and Richard C. Jeffrey, (2002), *Computability and Logic*, Fourth Edition, Cambridge University Press, Cambridge.
10. Braverman, Mark & Stephen Cook (2006), Computing over the Reals: Foundations for Scientific Computing, *Notices of the AMS*, Vol. 53, #3, pp. 318–329.
11. Chaitin, Gregory J, (1966), On the Length of Programs for Computing Fintite Binary Sequences, Journal of the Association of Computing Machinery, Vol. 13, pp. 549–69.
12. Compagner, Aaldert, (1964), Definitions of Randomness, *American Journal of Physics*, Vol. 59, # 8. August, pp. 700–705.
13. Conway, John Horton (1978), A Gamut of Game Theories, *Mathematics Magazine*, Vol. 51, #1, January, pp. 5–12.
14. Downey, Rodney. G & Denis R. Hirschfeldt (2010), *Algorithmic Randomness and Complexity*, Springer, New York.
15. Euwe, Max (1929), Mengentheoretische Betrachtungen über das Schachspiel, Communicated by Prof. R. Weizenböck (May, 25, 1929), *Proc. Koninklijke Nederlandse Akademie Van Wetenschappen* (Amsterdam) 32 (5): pp. 633–642.
16. Franzén, Torkel, (2005), *Gödel's Theorem: An Incomplete Guide to its Use and Abuse*, A K Peters, Wellesley, Massachusetts.
17. Gale, David & F. M. Stewart (1953), Infinite Games with Perfect Information, pp. 245–266, in: *Contributions to the Theory of Games, Vol. II*, edited by H. W. Kuhn & A. W. Tucker, *Annals of Mathematics Studies*, # 28, Princeton University Press, Princeton, New Jersey.
18. Goodstein, Reuben. L (1944), On the Restricted Ordinal Theorem, *Journal of Symbolic Logic*, Vol. 9, # 2, June, pp. 33–41.

19. Harrop, Ronald (1961), On the Recursivity of Finite Sets, *Zeitschrift für Mathematische Logik und Grundlagen der Mathematik*, Bd, 7, pp. 136–140.
20. Hertling, Peter (2005),A Banach-Mazur Computable but not Markov Computable Function on the Computable Real Numbers, *Annals of Pure and Applied Logic*, Vol. 132, pp. 227–246.
21. Hilbert, David, (1900), Mathematical Problems, *Bulletin of the American Mathematical Society*, Vol. 8, July, 1902, pp. 437–79; translated from the original German: Mathematische Probleme by Dr Mary Winston Newson.
22. Hilbert, David (1927 [1983]), On the Infinite, in: *Philosophy of Mathematics - Selected Readings*, Second Edition, pp. 183-201, edited by Paul Benacerraf & Hilary Putnam, Cambridge University Press, Cambridge.
23. Kirby, Laurie and Jeff Paris (1982), Accessible Independence Results for Peano Arithmetic, *Bulletin of the London Mathematical Society*, Vol. 14, pp. 285–293.
24. Keynes, John Maynard, (1921), *A Treatise on Probability*, Macmillan and Company Limited, London.
25. Kleene, Stephen Cole (1952), *Introduction to Metamathematics*, North-Holland Publishing Company, Amsterdam.
26. Kleene, Stephen, C. (1967), *Mathematical Logic*, John Wiley & Sons, Inc., New York.
27. Kolmogorov, Andrei N. (1965), Three Approaches to the Definition of the Concept of the 'Amount of Information', *Problems of Information Transmission*, Vol.1, #1, pp. 1–7.
28. Lewis, Alain. A (1986), Structure and Complexity: The Use of Recursion Theory in the Foundations of Neoclassical Mathematical Economics and Game Theory, Manuscript/Monograph, Department of Mathematics, Cornell University, Ithaca, NY, October.
29. Lewis, Alain. A (1991), Some Aspects of Effectively Constructive Mathematics That Are Relevant to The Foundations of Neoclassical Mathematical Economics and the Theory of Games, *Unpublished Manuscript*, Department of Philosophy & the School of Social Sciences, University of California at Irvine, Irvine, CA., April 27, 1991.
30. Li, Ming and Paul Vitanyi, (1993), *An Introduction to Kolmogorov Complexity and its Applications*, Springer-Verlag, Berlin and Heidelberg.
31. Lorentz, G. G. (1986), *Approximation of Functions*, AMS Chelsea Publishing, Providence, Rhode Island.
32. Matiyasevich, Yuri M (1993), *Hilbert's Tenth Problem*, The MIT Press, Cambridge, Mass.
33. Mycielski, Jan (1964), On the Axiom of Determinateness, *Fundamenta Mathematicae*, Vol. 53, pp. 205–224.
34. Newell, Allen & Herbert A Simon (1972), *Human Problem Solving*, Prentice-Hall Inc., Englewood Cliffs, NJ.
35. Nies, André (2009), *Computability and Randomness*, Oxford University Press, Oxford.
36. Nowakowski, Richard. J (2009), The History of Combinatorial Game Theory,

January 24, 2009, downloaded at: http://www.eos.tuwien.ac.at/OR/Mehlmann/Andis/publ/Spielmod10/HistoryCGT.pdf
37. Odifreddi, Piergiorgio (1989), *Classical Recursion Theory*, North-Holland, Amsterdam.
38. Odifreddit, Piergiorgio (1999), *Classical Recursion Theory: Volume II*, North-Holland, Amsterdam.
39. Oxtoby, John. C (1971), *Measure and Category*, Springer-Verlag, New York & Heidelberg.
40. Papadimitriou, Christos. H, (1994), *Computational Complexity*, Addison-Wesley Publishing Company, Reading, Massachusetts.
41. Paris, Jeff & Reza Tavakol (1993), Goodstein Algorithm as a Super-Transient Dynamical System, *Physics Letters A*, Vol. 180, # 1-2, 30 August, pp. 83–86.
42. Péter, Rózsa (1967), *Recursive Functions* (Third Revised Edition), translated from the German by István Földes, Academic Press, New York.
43. Rabin, Michael O, (1957), Effective Computability of Winning Strategies, in: *Annals of Mathematics Studies*, No. 39: *Contributions to the Theory of Games*, Vol. III, edited by M. Dresher, A. W. Tucker and P. Wolfe, pp. 147–157, Princeton University Press, Princeton, NJ.
44. Robinson, Raphael. M, (1952), An Essentially Undecidable Axiom System, pp. 729–30, in Vol. 1 of: *Proceedings of the International Congress of Mathematicians*, Cambridge, Massachusetts, 1950; American Mathematical Society, Providence, Rhode Island.
45. Rogers, Hartley, Jr., (1987), *Theory of Recursive Functions and Effective Computability*, The MIT Press, Cambridge, Massachusetts.
46. Rosser, J. Barkley (1936), Extensions of Some Theorems of Gödel and Church, *Journal of Symbolic Logic*, Vol. 1, # 3, September, pp. 87–91.
47. Russell, Bertrand (1919), *Introduction to Mathematical Philosophy*, George Allen and Unwin Ltd., London.
48. Schrijver, Alexander, (2003), *Combinatorial Optimization: Polyhedra and Efficiency: Volumes A, B, C*, Springer-Verlag, Berlin & Heidelberg.
49. Selda, Kao & K. Vela Velupillai (2012), Origins and Prioneers of Behavioural Economics, Forthcoming in *The Interdisciplinary Journal of Economics and Business Law*, Vol. 1, #. 3.
50. Shafer, Glenn & Vladimir Vovk (2001), *Probability and Finance: It's Only a Game*, John Wiley & Sons, Inc., New York.
51. Shepherdson, John C & H. E Sturgis (1963), Computability of Recursive unctions, *Journal of the Assocaition of Computing Machinery*, Vol. 10, # 2, April, pp. 217–255.
52. Simon, Herbert A (1955), A Behavioural Model of Rational Choice, *Quarterly Journal of Economics*, Vol. 69, No.1, February, pp. 99–118.
53. Simon, Herbert A (1956), Rational Choice and the Structure of the Environment, *Psychological Review*, Vol. 63, pp. 129–38.
54. Simon, Herbert A (1979), *Models of Thought*, Yale University press, New Haven.
55. Simon, Herbert A (1983), *Reason in Human Affairs,* Basil Blackwell, Oxford.
56. Simon, Herbert A (1996), Machine as Mind, Chapter 5, pp. 81-101, in: *Ma-*

chines and Thought - The Legacy of Alan Turing, Volume 1, edited by Peter Macmillan and Andy Clark, Oxford University Press, Oxford.
57. Simon, Herbert A (1997), Satisficing, Chapter IV.4, pp. 295-298, in: Models of Bounded Rationality, Vol. 3 - Empirically Grounded Economic Reason, The MIT Press, Cambridge, Massachusetts.
58. Simon, Herbert A (2000), Letter to Velupillai, 25 May, 2000.
59. Smale, Steve (1976), Dynamics in General Equilibrium Theory, *American Economic Review*, Vol. 66, No. 2, May, pp. 288–94.
60. Smullyan, Raymond, (1992), *Gödel's Incompleteness Theorems*, Oxford University Press, Oxford.
61. Solomonoff, Ray. J, (1964a), A Formal Theory of Inductive Inference: Part I, *Information and Control*, Vol. 7, pp. 1–22.
62. Solomonoff, Ray. J, (1964b), A Formal Theory of Inductive Inference: Part II, *Information and Control*, Vol. 7, pp. 224–54.
63. Steinhaus, H (1965), Games, An Informal Talk, *The American Mathematical Monthly*, Vol. 72, No. 5, May, pp. 457–468.
64. Takeuti, Gaisi (2003), Memoirs of a Proof Theorist: Gödel and other Logicians, translated by Mariko Yasugi and Nicholas Passell, World Scientific, New Jersey & Singapore.
65. Telgársky, Rastislav (1987), Topological Games: On the 50th Anniversary of the Banach-Mazur Game, *Rocky Mountain Journal of Mathematics*, Vol. 17, #2, Spring, pp. 227–276.
66. Tikhomirov, V. M, (1989), A.N.Kolmogorov and Approximation Theory, *Russian Mathematical Surveys*, Vol. 44, # 1, pp. 101–152.
67. Troelstra, Anne Sjerp (1977), Choice Sequences: A Chapter of Intuitionistc Mathematics, Clarendon Press, Oxford.
68. Turing, Alan. M (1954), Solvable and Unsolvable Problems, pp. 7-23, in: *Science News*, edited by A. W. Haslett, #31, Penguin Books, Harmondsworth, Middlesex.
69. Ulam, Stanislaw. M (1960), *A Collection of Mathematical Problems*, Interscience Publishers, Inc., New York & London.
70. van Dalen, Dirk ((2005), *Mystic, Geometer, and Intuitionist: The Life of L. E. J. Brouwer – Volume 2:* Hope and Disillusion, Clarendon Press, Oxford.
71. van Lambalgen, Michiel, (1987), *Random Sequences*, Doctoral Dissertation, University of Amsterdam, 16 September.
72. Velupillai, K. Vela (1997), Expository Notes on Computability and Complexity in (Arithmetical) Games, *Journal of Economic Dynamics and Control*, Vol. 21, No. 6, pp. 955–79, June.
73. Velupillai, K. Vela (2000), *Computable Economics*, Oxford University Press, Oxford.
74. Velupillai, K. Vela (2010), *Computable Foundations for Economics*, Routledge, London.
75. Velupillai, K. Vela (2010), Foundations of Boundedly Rational Choice and Satisfying Decision, *Advances in Decision Sciences*, April.
76. Velupillai, K. Vela (2011), Towards an Algorithmic *Revolution* in Economic Theory, *Journal of Economic Surveys*, Vol. 25, #3, July, pp. 401–430.

77. Velupillai, K. Vela (2011), Freedom, Anarchy and Conformism in Academic Research, *Interdisciplinary Journal of Economics and Business Law*, Vol. 1, Issue 1, pp. 76–101.
78. Velupillai, K. Vela (2011), Non-Linear Dynamics, Complexity and Randomness: Algorithmic Foundations, *Journal of Economic Surveys*, Vol. 25, #3, July, pp. 547–568.
79. Velupillai, K. Vela (2012), Reflections on Mathematical Economics in the Algorithmic Mode, Forthcoming in: *New Mathematics and Natural Computation*, Vol. 6, March.
80. Vitushkin, A. G, (1961), *Theory of the Transmission and Processing of Information*, translated from the Russian by: Ruth Feinstein, Pergamon Press, Oxford and London.
81. von Mises, Richard (1928), *Wahrscheinlichkeitsrechnung, Statistik Und Wahrheit*, Verlag von Julius Springer, Wien.
82. von Neumann, John (1928), Zur Theorie der Gesellsschaftsspiele by J. von Neumann, *Mathematische Annalen*, Vol. 100, pp. 295–320.
83. Weihrauch, Klaus (2000), *Computable Analysis: An Introduction*, Springer-Verlag, New York and Heidelberg.
84. Wittgenstein, Ludwig (1939 [1975]), *Wittgenstein's Lectures on the Foundations of Mathematics - Cambridge, 1939*,From the Notes of R.G. Bosanquet, Norman Malcolm, Rush Rhees, and Yorick Smithies, edited by Cora Diamond, The University of Chicago Press, Chicago.
85. Zermelo, Ernst (1913), Über ein Anwendung der Mengenlehre auf die Theorie des Schachspiels, in: *Proceedings of the Fifth International Congress of Mathematicians*, Cambridge, 11–28 August, 1912, edited by E. W. Hobson & A. E. H. Love, Vol. 2, pp. 501–4, Cambridge University Press, Cambridge.

Chapter 17

Blueprint for a Hypercomputer*

Francisco Antonio Doria
APIT and Advanced Studies Group
Programa de Engenharia de Produção
COPPE, UFRJ & HCTE, UFRJ, Brazil

We sketch the developments that led to the construction of the da Costa and Doria expression for the halting function in languages more powerful than arithmetic, and use it to describe a plausible real hypercomputer.

1. Introduction

Martin Davis opens his paper on the impossibility of the existence of any hypercomputer with a sobering tale: suppose that one builds a hypercomputer. Ask the device to compute something not computable by a Turing–like device. Wait for it to output a result, digit by digit. It will eventually end.

And the string of bits produced by our hypercomputer will be finite, and therefore, computable.

So, Davis concludes, if they operate this way, hypercomputers cannot exist.[9]

Davis' example is a *boutade*, but points out a difficulty in the way hypercomputers should operate. Now the question is, does there exist a way around Davis' objection? The answer is affirmative, and the possible solution goes back to a remark by Scarpellini at the end of a paper he published in 1963:[18,19]

> *Ideal analog computers can decide some undecidable statements.*

In 1990 (published in 1991) da Costa and the author proved that there is an "ideal analog device" — let's for the moment use the qualification *ideal*

---

*F. A. Doria is a full member of the Brazilian Academy of Philosophy, R. do Riachuelo 303, 20230–011 Rio RJ Brazil.

and the quotation marks — which can solve all Turing unsolvable instances of the halting problem.[3,4] If plugged into an oracle machine, this device allows us to settle undecidable stuff along the arithmetic hierarchy.[3,5] The idea, as I said, goes back to Scarpellini in 1963, and has been explored by Kreisel (in 1974,[14]) and by Copeland and Sylvan, in their 1999 survey.[2] So, "ideal" analog computers behave like very powerful hypercomputers. Do *real* analog computers also compute uncomputable stuff? This is the question we ask in the present paper.

Richardson's contribution

The central contribution came from Richardson's 1968 paper.[16] Out of an idea by Feynmann, Richardson mapped an universal exponential Diophantine equation onto an equation in the algebra of elementary functions over the reals, so that the universal equation has (integer) roots if and only if the corresponding equation over the reals also has roots. Out of that new equation it is easy to concoct an expression $\theta(m, n)$ with elementary functions plus the sign function (see the next sections) so that $\theta(m, n) = 1$ if and only if Turing machine coded by m stops over input n; $\theta(m, n) = 0$ otherwise, that is, if the machine doesn't stop. (Technical details in the next sections.) We can substitute an universal Diophantine equation for the exponential equation used by Richardson, and obtain the same result.

However as we point out below, we can obtain explicit expressions for the halting function without Richardson's maps, for the halting function can be written out in languages very close to arithmetic.

2. Richardson's Map Leads to An Expression for the Halting Function

We use here a previous version of these ideas, as it is a standard development. The present section is based on;[1] for the proofs see.[16] Our presentation splits into several topics:

- Formalized arithmetic and Turing machines.
- Richardson's maps.
- The Halting Function in formal languages that extend arithmetic.

Notation: $\neg$, "not," $\vee$, "or," $\wedge$, "and," $\to$, "if... then...," $\leftrightarrow$, "if and only if," $\exists x$, "there is a x," $\forall x$, "for every x." $P(x)$ is a formula with x free; it roughly means "x has property P." Finally $T \vdash \xi$ means T proves ξ, or ξ is a theorem of T. ω is the set of natural numbers, $\omega = \{0, 1, 2, \ldots\}$.

Algorithmic functions are given by their programs coded in Gödel numbers e.[17] We will sometimes use Turing machines (noted by sans–serif letters with the Gödel number as index M_e) or partial recursive functions, noted $\{e\}$.

We start from a very simple theory of arithmetic, noted $A1$. Its language includes variables $x, y, \ldots$, two constants, $\mathbf{0}$ and $\mathbf{1}$, the equality sign $=$, and two operation signs, $+, \times$. Basically $A1$ has axioms for the operations $+$ and $\times$, the behavior of constants $\mathbf{0}$ and $\mathbf{1}$, and the trichotomy axiom, that is, given two natural numbers x and y, either $x < y$ or $x = y$ or $x > y$. $A1$ contains no induction axiom.

The *standard interpretation* for $A1$ is: the variables $x, y, \ldots$ range over the natural numbers, and $\mathbf{0}$ and $\mathbf{1}$ are seen as, respectively, zero and one. The only requirement we impose on $A1$ is: that theory should be strong enough to formally include all of Turing machine theory. Recall that a Turing machine is given by its Gödel number, which recursively codes the machine's program. Rigorously, for $A1$, we must have:

Definition 2.1 *A Turing machine of Gödel number e operating on x with output y, $\{e\}(x) = y$ is* **representable** *in theory $A1$ if there is a formula $F_e(x, y)$ in the language of $A1$ so that:*

(1) $A1 \vdash \{F_e(x, y) \wedge F_e(x, z) \rightarrow [y = z]\}$, and
(2) For natural numbers a, b, if $\{e\}(a) = b$, then $A1 \vdash F_e(a, b)$. $\square$

Then we have the representation theorem for partial recursive functions in $A1$:

Proposition 2.2 *Every Turing machine is representable in $A1$. Moreover there is an effective procedure that allows us to obtain F_e from the Gödel number e.* $\square$

We restrict here our interest to theories that are *arithmetically sound*, that is, which have a model with standard arithmetic for its arithmetical segment.

Richardson's map

We now describe the Richardson transforms.[7,16] We start from a strengthening of Proposition 2.2:

Proposition 2.3 If $\{e\}(a) = b$, for natural numbers a, b, then we can algorithmically construct a polynomial p_e over the natural numbers so that $[\{e\}(a) = b] \leftrightarrow [\exists x_1, x_2, \ldots, x_k \in \omega \, p_e(a, b, x_1, x_2, \ldots, x_k) = 0]$. □

Follows:

Proposition 2.4 $a \in R_e$, where R_e is a recursively enumerable set, if and only if there are e and p so that $\exists x_1, x_2, \ldots, x_k \in \omega \, [p_e(a, x_1, x_2, \ldots, x_k) = 0]$. □

Richardson's map[7,16] allows us to obtain in an algorithmic way, given any such $p_e(a, \ldots)$, a real–defined and real–valued function $f_e(a, \ldots)$ that has roots if and only if $p_e(a, \ldots)$ has roots as a Diophantine equation.

Richardson's map: multidimensional version

We can be more specific: let $\mathcal{A}$ be the algebra of subelementary functions (polynomials over the reals, sines, cosines; everything closed under $+, -$, products by real numbers and by the functions that generate the algebra, to which we add function composition). Let R denote the real line.

(We do not require the exponential function in our constructions.)

We now state the first of Richardson's main results: given that $A1 \subset$ ZFC, and if $\mathcal{P}$ is the set of all finite–length polynomials over ω:

Proposition 2.5 (Richardson's Map, I) *There is an injection $\kappa_{\mathcal{P}} : \mathcal{P} \to \mathcal{A}$, where $\mathcal{P}$ denotes the algebra of ω–valued polynomials in a finite number of variables, and $\mathcal{A}$ is the algebra of subelementary functions described above, such that:*

(1) $\kappa_{\mathcal{P}}$ is constructive, that is, given the expression for p in A1, there is an effective procedure so that we can obtain the corresponding expression for $F = \kappa_{\mathcal{P}}(p)$ in ZFC.
(2) $\kappa_{\mathcal{P}}$ is 1–1.
(3) For $\mathbf{x} = (x_1, \ldots, x_n)$, $\exists \mathbf{x} \in \omega^n \, p(m, \mathbf{x}) = 0$ if and only if $\exists \mathbf{x} \in \mathsf{R}^n \, F(m, \mathbf{x}) = 0$ if and only if $\exists \mathbf{x} \in \mathsf{R}^n \, F(m, \mathbf{x}) \leq 1$, for $p \in \mathcal{P}$ and $F \in \mathcal{A}$.
(4) The injection $\kappa_{\mathcal{P}}$ is proper. □

The crucial property is given in step 3.: it allows us to translate the existence of roots for Diophantine equations into roots of the corresponding transformed real–defined and real–valued function, with some extras.

Next step gives us a 1–dimensional version of Richardson's map.

Richardson's map: one–dimensional version

Corollary 2.6 (Richardson's Map, II) *Let $\mathcal{A}_1$ be the algebra of subelementary functions over a single real variable x. Then there is a map $\kappa' : \mathcal{P} \to \mathcal{A}_1$ such that:*

(1) *κ' is constructive.*
(2) *κ' is 1–1.*
(3) *The inclusion $\kappa'(\mathcal{P}) \subset \mathcal{A}_1$ is proper.*
(4) *$\exists \mathbf{x} \in \omega^n \ p(m, \mathbf{x}) = 0$ if and only if $\exists x \in \mathsf{R} \ L(m, x) = 0$ if and only if $\exists x \in \mathsf{R} \ G(m, x) \leq 1$, for adequate L, G, whose expressions can be explicitly exhibited.* $\square$

The Halting Function

The main result in Alan Turing's remarkable 1937 paper, "On computable numbers, with an application to the Entscheidungsproblem",[22] is a proof of the algorithmic unsolvability of a version of the halting problem: given an arbitrary Turing machine of Gödel number e, for input x, there is no algorithm that decides whether $\{e\}(x)$ stops and outputs something, or enters an infinite loop.

Remark 2.7 Let $\mathsf{M}_m(a) \downarrow$ mean: "Turing machine of Gödel number m stops over input a and gives some output." Similarly $\mathsf{M}_m(a) \uparrow$ means, "Turing machine of Gödel number m enters an infinite loop over input a." Then we can define the halting function θ:

- $\theta(m, a) = 1$ if and only if $\mathsf{M}_m(a) \downarrow$.
- $\theta(m, a) = 0$ if and only if $\mathsf{M}_m(a) \uparrow$.

$\theta(m, a)$ is the halting function for M_m over input a. $\square$

θ isn't algorithmic, of course,[17,22] that is, there is no Turing machine that computes it.

Remark 2.8 As we now show, we can explicitly write an expression for a function in the language of classical analysis that settles the halting problem. We proceed as follows:

- Given Turing machine $\mathsf{M}_m(a) = b$, for natural numbers a, b, we can algorithmically obtain a polynomial $p_m(\langle a,b \rangle, x_1, \ldots, x_k)$ so that:

$$\mathsf{M}_m(a) = b \leftrightarrow \exists x_1, \ldots, x_2 \in \omega \, [p_m(\langle a,b \rangle, x_1, \ldots, x_k) = 0].$$

- Given F_m, real–defined and real–valued, we have that:

$$\exists x_1, \ldots, x_2 \in \omega \, [p_m(\langle a,b \rangle, x_1, \ldots, x_k) = 0] \leftrightarrow$$

$$\leftrightarrow \exists x_1, \ldots, x_k \in \mathsf{R} \, F_m(\langle a,b \rangle, x_1, \ldots, x_k) \leq 1.$$

and

$$\forall x_1, \ldots, x_2 \in \omega \, [p_m(\langle a,b \rangle, x_1, \ldots, x_k) \neq 0] \leftrightarrow$$

$$\leftrightarrow \forall x_1, \ldots, x_k \in \mathsf{R} \, F_m(\langle a,b \rangle, x_1, \ldots, x_k) > 1.$$

- That is to say: $\mathsf{M}_m(a) \downarrow$ if and only if $F_m(a, \ldots)$ goes below 1, and $\mathsf{M}_m(a) \uparrow$ if and only if $F_m(a, \ldots)$ stays above 1.
  This is the property we use in order to construct the halting function θ_m. $\square$

We now need the concept of an *universal Diophantine polynomial*. Martin Davis[8] describes an algorithmic procedure out of which, given a Turing machine with input a $\mathsf{M}_m(a)$, we obtain a polynomial $p_m(a, x_1, \ldots)$ so that it has roots if and only if $\mathsf{M}_m(a)$ converges (outputs some result). Now, if $\mathsf{U}(m, a)$ is an universal Turing machine,[17,22] we can similarly obtain a polynomial $p(\langle m, a \rangle, \ldots)$ which stands for $p_m(a, \ldots)$.

More precisely, if $[\exists x_1, \ldots, x_k \, p_m(\langle a,b \rangle, x_1, \ldots, x_k) = 0] \leftrightarrow [\mathsf{M}_m(a) = b]$, then, for the universal polynomial $p(\langle m, a, b \rangle, \ldots)$:

$$[\exists x_1, \ldots, x_r \, p(\langle m, a, b \rangle, x_1, \ldots, x_r) = 0] \leftrightarrow$$

$$\leftrightarrow [\exists x_1, \ldots, x_k \, p_m(\langle a,b \rangle, x_1, \ldots, x_k) = 0].$$

From the preceding considerations, if σ is the sign function, $\sigma(\pm x) = \pm 1$ and $\sigma(0) = 0$:

Proposition 2.9 (The Halting Function.) *The Halting Function $\theta(n, q)$ is explicitly given by:*

$$\theta(n, q) = \sigma(G_{n,q}),$$

$$G_{n,q} = \int_{-\infty}^{+\infty} C_{n,q}(x) e^{-x} dx,$$

$$C_{m,q}(x) = |F_{m,q}(x) - 1| - (F_{m,q}(x) - 1).$$

$$F_{n,q}(x) = \kappa_P p_{n,q}. \quad \square$$

Here $p_{n,q}$ is the two–parameter universal Diophantine polynomial

$$p(\langle n, q \rangle, x_1, x_2, \ldots, x_r)$$

and κ_P is as in Proposition 2.5.

There are infinitely many alternative explicit expressions for the halting function θ.[7]

Remark 2.10 We do not require Richardson's transform to obtain an expression for the Halting Function. There is also an expression for the Halting Function even within a simple extension of $A1$. Let $p(n, \mathbf{x})$ be a 1–parameter universal polynomial; $\mathbf{x}$ abbreviates $x_1, \ldots, x_p$. Then either $p(n, \mathbf{x}) \geq 1$, for all $\mathbf{x} \in \omega^p$, or there are $\mathbf{x}$ in ω^p such that $p(n, \mathbf{x}) = 0$ sometimes. As $\sigma(x)$ when restricted to ω is primitive recursive, we may define a function $\psi(n, \mathbf{x}) = 1 - \sigma p(n, \mathbf{x})$ such that:

- Either for all $\mathbf{x} \in \omega^p$, $\psi(n, \mathbf{x}) = 0$;
- Or there are $\mathbf{x} \in \omega^p$ so that $\psi(n, \mathbf{x}) = 1$ sometimes.

Thus the halting function can be represented as:

$$\theta(n) = \sigma[\sum_{\tau^q(\mathbf{x})} \frac{\psi(n, \mathbf{x})}{\tau^q(\mathbf{x})!}],$$

where $\tau^q(\mathbf{x})$ denotes the positive integer given out of $\mathbf{x}$ by the pairing function τ: if τ^q maps q–tuples of positive integers onto single positive integers, $\tau^{q+1} = \tau(x, \tau^q(\mathbf{x}))$. Recall that the infinite sum can be given by a simple iterative definition. $\square$

3. From the Ideal World to the Real World

The proof of the pudding is in the eating, and therefore we will only know if there is a hypercomputer when we build it, and start it. Will it work? Notice the following:

- *There is no real world Turing machine.* This is obvious, but isn't in general considered when we discuss possible hypercomputational devices. Memory space is bounded, and in general we only use low–complexity programs in the real world (exponential time renderization programs such as the old but beautiful *Bryce* are awful to use).

  So, in the real world, even our choice of Turing machines (programs) is severely limited.

- *The halting problem can be solved by (infinitely many) theories very close to arithmetic.* This means that even if we cannot solve the halting problem within the usual versions of formalized arithmetic such as Peano arithmetic, we can settle an arbitrary but finite number of nonhalting instances of the halting problem beyond those already settled by Peano arithmetic in a theory that has the same language as Peano's, proves all of the Peano theorems, and has the same provably total recursive functions as Peano arithmetic.

  However it is immediate that we cannot collect those different procedures into a single, algorithmic, procedure. They are, we may say, uncomputationally different.

  This follows from the fact that we will just be adding Π_1 true sentences to the theory in order to prove the desired nonhalting instances of the halting problem.

  Therefore we can say that we can nearly settle the halting problem in such formalized arithmetical theories.

A real–world hypercomputational device

We intend to use Proposition 2.5 in order to build our real–world approximation to a hypercomputer. More precisely: the present construction is based on the following result by Richardson:[2,16,18,19]

Remark 3.1

- Given Turing machine $\mathsf{M}_m(a) = b$, for natural numbers a, b, we can algorithmically obtain a polynomial $p_m(\langle a, b \rangle, x_1, \ldots, x_k)$ so that:

$$\mathsf{M}_m(a) = b \leftrightarrow \exists x_1, \ldots, x_2 \in \omega \, [p_m(\langle a, b \rangle, x_1, \ldots, x_k) = 0].$$

- Recall that we can construct an expression for a F_m, real–defined and real–valued, so that:

$$\exists x_1, \ldots, x_2 \in \omega \, [p_m(\langle a, b \rangle, x_1, \ldots, x_k) = 0] \leftrightarrow$$

$$\leftrightarrow \exists x_1, \ldots, x_k \in \mathsf{R}\, F_m(\langle a, b \rangle, x_1, \ldots, x_k) \leq 1.$$

and

$$\forall x_1, \ldots, x_2 \in \omega\, [p_m(\langle a, b \rangle, x_1, \ldots, x_k) \neq 0] \leftrightarrow$$

$$\leftrightarrow \forall x_1, \ldots, x_k \in \mathsf{R}\, F_m(\langle a, b \rangle, x_1, \ldots, x_k) > 1.$$

- In a more figurative way: $\mathsf{M}_m(a) \downarrow$ (converges) if and only if $F_m(a, \ldots)$ dives below 1, and $\mathsf{M}_m(a) \uparrow$ (diverges, enters an infinite loop) if and only if $F_m(a, \ldots)$ stays above 1.

(See Remark 2.8.) So, machine e stops over input n if Richardson's transform $F_{(e,n)}$ of the corresponding Diophantine polynomial $p(\langle e, n \rangle, \ldots)$ dives below 0; otherwise it stays above 1. Of course we can modify the corresponding Richardson transform in such a way that the "signaling gap" from 0 to 1 is widened to an arbitrary k.

To sum it up:

- The Turing machine doesn't stop at the desired input if and only if the transformed function has a value $> k$, where k is a positive real constant. So the transformed function will always be positive, and greater than k.
- The Turing machine stops if there is a corresponding root in the transformed function, that is, it will dip through the band $[0, k]$ and reach negative values.

We will never require a device that will try to settle, one by one, an infinite number of instances of the halting problem. Thus we are planning to build an analog device that simulates the transformed function in Proposition 2.5 and couple it to a corresponding digital simulation[a] so that one will correct the other. (This procedure, with an interplay between analog and digital simulations, is already used in many instances.)

The following is to be expected:

- While the digital simulation of function F can be sometimes of large computational complexity, the relaxation time is the same for all analog simulations, and so we expect a speed–up effect in our device.

---

[a]The idea arose from a conversation with Dr. N. Quilula.

- If we adequately choose the operating interval for the analog device, there will be no false results due to the imprecision of the way analog devices operate.
- We believe that it will be possible to add some kind of scaling mechanism so that we can somehow extend the operating interval of the analog device, case by case.

Then we get the following:

- If Turing machine of index m halts over input n, then the corresponding function $F_{m,n}$ will dive somewhere through the band given by hyperplanes $F = 1$ and $F = 0$.
- If it never halts, it will always remain above hyperplane $F = 1$, without ever touching it.

This is the ideal situation. Now for the concrete part:

- If and when F dips through $F = 1$, it dips through a multidimensional hyperplane. We must devise a way to locate the dip through that space.
- Infinity enters the picture when F never goes through $F = 1$, that is, the machine doesn't halt. We'll have to compactify the hyperplane where $F_{m,n}$ is defined, and the essential analog imprecision comes into play.

  Basically there will be a piling up of the F function near the borders of the compactified interval. If overshoots are somewhat controllable, we are done.

That's the rosy picture. But there are other difficulties. For instance, we require π in our transformations, and there will always be a cutoff error. What can be done to counterbalance it? We do not know at present. We first want to build a kind of first generation, or version 1.0 prototype, and observe the way it behaves — or, more likely, misbehaves...

Then we'll try to mend things.

Hypercomputation: what for?

Hypercomputaton devices have an obvious theoretical interest, but how about its practical applications? The idea is that a device such as the one sketched above will help speed up computations, as the relaxation times in analog computers do not behave as in the digital case and are independent of length of input, etc. It will be roughly the same in all situations.

Also, anything that can be framed as a Diophantine problem — and that goes from the Fermat problem[11] and from the Riemann hypothesis to more pedestrian matters — can have a corresponding function F_m as above, and therefore can benefit from the behavior of an analog procedure.

Well, anyway we'll have to build the device and see what happens: the proof of the pudding is in the eating!

Acknowledgements

The author thanks H. Zenil for the invitation to contribute to this volume and G. J. Chaitin for criticisms and remarks on the manuscript; he also wishes to thank suggestions by L. O. Vantour.

The author wishes to acknowledge support from the Production Engineering Graduate School, COPPE/UFRJ and from the interdepartmental program on Philosophy of Science (HCTE). This research is based on previous work with N. C. A. da Costa, and is funded in part by CNPq/MCT, Brazil, Philosophy Section.

References

1. R. Bartholo, C. A. Cosenza, F. A. Doria, C. Lessa, "Can economic systems be seen as computing machines?" *J. Economic Behavior and Organization*, **70**, 72–80 (2009).
2. B. J. Copeland and R. Sylvan, "Beyond the universal Turing machine," *Austral. J. Philosophy* **77**, 46–67 (1999).
3. G. Chaitin, N. C. A. da Costa, F. A. Doria, *Gödel's Way: Exploits into an Undecidable World*, CRC Press (2011).
4. N. C. A. da Costa and F. A. Doria, "Undecidability and incompleteness in classical mechanics," *Int. J. Theor. Phys.* **30**, 1041–1073 (1991).
5. N. C. A. da Costa and F. A. Doria, "Suppes predicates and the construction of unsolvable Problems in the axiomatized sciences," in P. Humphreys, ed., *Patrick Suppes, Scientific Philosopher*, II, 151–191 Kluwer (1994).
6. N. C. A. da Costa and F. A. Doria, "Variations on an original theme," in J. Casti and A. Karlqvist, *Boundaries and Barriers*, Addison–Wesley (1996).
7. N. C. A. da Costa and F. A. Doria, "Computing the future," in K. Vela Velupillai, ed., *Computability, Complexity and Constructivity in Economic Analysis*, Blackwell (2005).
8. M. Davis, "Hilbert's Tenth Problem is unsolvable," *Amer. Math. Monthly* **80**, 233 (1973).
9. M. Davis, "Why there is no such discipline as hypercomputation," in F. A. Doria and J. F. Costa, *Hypercomputation — Special Issue, Applied Mathematics and Computation* **178** 4–7 (2006).

10. F. A. Doria, "Informal vs. formal mathematics," *Synthèse* **154**, 401–415 (2007).
11. F. A. Doria, *Chaos, Computers, Games and Time*, GAE/PEP/COPPE, Rio (2011).
12. S. Feferman, "Transfinite recursive progressions of axiomatic theories," *J. Symbolic Logic* **27**, 259 (1962).
13. T. Franzen, "Transfinite progressions: a second look at completeness," *Bull. Symbolic Logic* **10**, 367–389 (2004).
14. G. Kreisel, "A notion of mechanistic theory," *Synthèse* **29**, 11–26 (1974).
15. E. Mendelson, *Introduction to Mathematical Logic*, 4th ed., Chapman & Hall (1997).
16. D. Richardson, "Some undecidable problems involving elementary functions of a real variable," *J. Symbol. Logic* **33**, 514 (1968).
17. H. Rogers Jr., *Theory of Recursive Functions and Effective Computability*, John Wiley (1967).
18. B. Scarpellini, "Two undecidable problems of analysis," *Minds and Machines* **13**, 49–77 (2003).
19. B. Scarpellini, "Comments to 'Two undecidable problems of analysis,'" *Minds and Machines* **13**, 79–85 (2003).
20. I. Stewart, "Deciding the undecidable," *Nature* **35**, 664–665 (1991).
21. P. Suppes, personal communication (1990).
22. A. M. Turing, "On computable numbers, with an application to the Entscheidungsproblem," *Proc. London Math. Society* **50**, 230 (1937).
23. A. M. Turing, "Systems of logic based on ordinals," *Proc. London Math. Society*, Ser. 2 **45**, 161 (1939).

PART 3

Computation & Physics & the Physics of Computation

Chapter 18

Information-Theoretic Teleodynamics in Natural and Artificial Systems

Anthony F. Beavers and Christopher D. Harrison

The University of Evansville, USA
afbeavers@gmail.com, http://faculty.evansville.edu/tb2/

In this paper, we employ the method of computational philosophy, the use of computational techniques to aid in the discovery of philosophical insights that might not be easily discovered otherwise, to show that it is not unreasonable to suggest 1) that there are genuine teleological causes in nature, 2) that such causes can be computed from Newtonian ("push") causation, provided that other architectural and environmental conditions are met, and 3) that it is possible to do so without recourse to semantics by taking an information-theoretic approach to measuring information flow in a system.

Keywords: Information-theoretic teleodynamics; goal-oriented-behavior; dynamic associative networks; information theory; computation.

1. Philosophical Preliminaries

This volume addresses the question of computation in nature. To do so, we must define "computation," since some definitions would appear to bias the case against natural computation and some not. Among several possibilities, we can consider computation from a conservative perspective, in which case "computation" might be defined as:

> {Def 1} a procedure whereby tokens are manipulated according to the specifications of a formal language.

{Def 1} includes systems of math and logic, along with Turing computation more generally. It appears at first to exclude the possibility that a rock can "compute" sunlight into heat or that a sunflower can "compute" its motion when tracking a light source, except by mere extension of metaphor. These latter examples seem to need a more liberal definition of the term, perhaps "computation" as:

{Def 2} any nomological transformation of input into output.

Initially, {Def 2} would seem to include {Def 1}, but not the reverse. However, much hangs here on the meaning of the terms "token," "formal language," and "nomological." If a nomological regularity can be interpreted as rule following in a formal game (in the sense described in Haugeland Ref. 20), then the laws of science might represent a formal language, and if a token can be a physical particle of some sort or other, then these might amount to pieces in such a game. In this case, the natural world *is* a computational system, or collection of them, under both {Def 1} and {Def 2}.

Without getting into the fine details of argument concerning whether the conservative or liberal definition of "computation" is fairest to define the phenomenon, or even whether the two definitions really collapse into the same, one might still like to posit a distinction between digital and analogue computation. {Def 1} seems to require a digital system, rather than analogue, and {Def 2} would seem to include both. But if we follow the conflation of {Def 1} and {Def 2} suggested in the previous paragraph, then natural computation would be just as digital as computation in artificial systems.

The philosophy of computation is a bit of a dicey game, as we see here, depending on how one slices the phenomena and defines terms. Since so much hangs here precisely on such matters, "carving nature at its joints" does not seem meaningfully possible in this case. Some definitions treat the natural world as a computational system, some do not, and that pretty much is the end of the story. This situation, however, does not preclude us from learning a few lessons.

There are several issues one could explore in this regard, but in this paper, we wish to examine teleodynamic systems in particular, that is, systems that pursue a state and hence are partly describable as operating according to teleological or "pull" causation rather than the "push" causation of Newtonian physics. Our approach here will be to examine teleodynamics in artificial systems with the hope of showing how they could be instantiated in some natural systems.

One important note must be made early on: the term "teleodynamic" is borrowed here from Terrence Deacon Ref. 9 and belongs to a layered distinction between types of emergence he finds in nature, but that also applies to some computational systems. Though we borrow the term, we do not follow his specific usage, nor the details of his analysis here, which is interesting in its own right, but set to different purposes, namely to an understanding of emergence in natural systems. (See Refs. 10, 11, 12 and 13). We are more interested in the question of computation in nature with regard to goal states, employing the emerging method of "computational philosophy" defined as the use of computational techniques to aid in the discovery of philosophical insights that might not be easily discovered otherwise. (See Ref. 7). Nonetheless, it will be difficult not to intersect with Deacon in some places while disagreeing with him in others. In the interests of staying on task and saving space, we leave it to the reader to study the similarities and differences between his view and ours, though we do agree upfront that a teleodynamic system need not be conscious, at least in any rich sense of the term.

We, like Deacon, also respect the difference between teleodynamic and teleonomic systems. The latter are systems that can be *interpreted* as pursuing a goal state, whereas teleodynamic ones actually pursue a goal state. In the example above of a rock computing sunlight into heat, it would be difficult to describe this situation as teleodynamic. The rock does not pursue heat; rather, its release of heat is fully explained by thermodynamic principles using "push" causation. The input here is light, the mechanism of action is molecular response according to nomological laws of thermodynamics, the output is heat. The case of the sunflower, however, lends itself to a different analysis. While it is true that the *internal* operations of the sunflower can also be described as operating according to the laws of thermodynamics, the fact that the flower follows the sun (or any other like light source) suggests something else. The sunflower seeks the sun, (though through the action of its internal mechanisms.) Thus, a full explanation of why the sunflower acts the way it does requires reference to the sun, for what sense can one make of the sunflower as a sun tracker were it not for the sun that it tracks? The important question here is whether the sunflower actually "desires" (read "is attracted to") light, or whether it can merely be described that way. From our perspective, the former description is the correct one.

If we are correct, we need to explain how "push" causation can create a mechanism that operates according to "pull" causation. In Aristotelean terms, the question can be put this way: is it possible to describe a mechanism that is built up from "efficient" or "moving" causes that ultimately ends up with legitimate "final" causes? In this paper we hope to provide a computational model to serve as an existence proof that answers the question in the affirmative.

The philosophical significance of this paper, then, is thus established. If we are successful with this exercise, then perhaps physics (as defined in the period extending from Galileo to Newton) might have been too quick in throwing out all teleological causes. In suggesting such, we do not mean to vindicate Medieval physics by putting "entelechies" into inanimate objects. Rather, we wish to suggest that *some* objects with internal working parts can exhibit genuine teleological behavior. Falling weights from the Leaning Tower of Pisa may not qualify; but sunflowers and neural systems may be another matter. The significance of this paper to the question of computation in nature should, thus, also be established. If we can demonstrate a computational model of how a mechanism based on "push" causation can harness teleological causes, then we have reason to believe that there are genuine goal-oriented computational systems in the natural world, whether one should adopt as a definition of "computation" {Def 1} or {Def 2} above.

2. Framing the Question Along Information-Theoretic Lines

Information-Theoretic Teleodynamics (ITT) is an approach to intelligence in both natural and artificial systems that uses the quantity and distribution of information to drive goal-oriented behavior without recourse to semantic content. By "intelligence" we mean simply massive adaptability to one's changing environment, including the ability to be surprised by failed expectations, respond accordingly and act on the basis of desired ends. By invoking the "quantity and distribution of information" we mean to make a direct connection to Claude Shannon's information theory, and, in particular, to his notion of information entropy.[25] Measuring information in this way to get goal-oriented behavior is what permits us to derive such behavior "without recourse to semantic content." This is to say that we will not "look inside" a piece of information to determine what it means; rather the measure of information, insofar as it tracks similarity and difference (along with its spread) across a networked system, provides a way to modulate conditions of attraction and repulsion that pull a system in one direction and push it away from another, as we will see in the model below.

Generally, the notion of goal-oriented behavior is thought to include reference to semantic content in the conventional sense of language-like representations (perhaps thoughts) that "stand in" for states of affairs in the world. This is partly because goal-oriented behavior is often characterized as an act of imagining some future state, developing a plan to attain it, and then executing that plan. However, action-based semantics may provide another alternative. (See, for instance, Refs. 18 and 19.) Respecting the former, Floridi notes, "in the beginning, the proto-meanings of the symbols generated by an AA [artificial agent] are the internal states of that AA, which in turn are directly correlated to the action performed by the same AA" (p. 164, Ref. 18). Further down the page he says that "The advantage of this approach is that the very first step in the generation of meaning is not in itself a semantic process, but rather an immediate consequence of an AA's performance.... The internal states of the AA are excellent candidates for the role of non-semantic yet semantic-inducing resources." Indeed, in an earlier study, Grim showed that signs can acquire meanings by coordinating signaling across artificial creatures in an agent-based model.[19] "Truth telling," here, emerged even in elementary models based solely on the interaction between agents, agent-environment relations, and simple mechanisms internal to each agent. No structured language was involved or needed. Furthermore, these agents responded positively to signals from other agents in a teleonomic (though *perhaps* not yet teleodynamic) struggle to survive.

Though we believe that both Grim and Floridi point in the right direction, we hope to go a little further with ITT. One way to address this goal is in terms of one of the "open problems in the philosophy of information" enumerated in chapter two of Floridi.[18] He asks, "Do information or algorithmic theories ... provide the necessary conditions for any theory of semantic information?" (p. 31, Ref. 18). Setting algorithmic theories aside, our hypothesis is that it is possible to reduce semantic information to the quantification of information flow *provided that other conditions respecting the internal structure of an agent and its relations to its environment are met*. For our purposes, we will use a dynamic associative network (DAN), defined momentarily, to provide evidence that our hypothesis is correct, at least where goal-pursuing behavior is concerned; further experiments with computer models will be necessary to advance the case for a complete reduction.

As Beavers explained elsewhere,[6] traditional connectionist network modeling (with Artificial Neural Networks or ANNs) begins with a fixed network structure (in terms of nodes and connections) and then proceeds to discover a set of weights to transduce signals (here information), transforming them from input to output. The techniques for determining the proper weight set vary, but the general strategy remains the same. It starts with a predefined structure and then sets the weights for connections. Other strategies are possible. One could, for instance, let information content determine the structure of the network, rather than using a predefined structure, adding nodes and connections wherever dictated by the data. McClelland and Rumelhart's early Interactive Activation and Competition (IAC) Models (see Ref. 22 and Ref. 24) use this approach along with contemporary network structures that are involved in social network analysis, citation networks, co-authorship networks, and so forth. The success of these networks has been sufficient to warrant the claim that they exhibit rudimentary intelligence and qualify as preliminary forms of artificial intelligence, if harnessed to be so.

Dynamic Associative Networks (DANs) use this latter strategy as well. These networks are built up from the data to transform input into output in such a way that the output is predictive, that is, these models learn from experience to form associations based on some sort of statistical procedure that is implicitly determined by the model. (That is, no explicit statistical methods are used.) Originally conceived to show that one could get intentionality (in both the semantic and goal-oriented senses) from association[5] and implemented in various models over the past four years, DANs have been used in a variety of micro-world experiments to prove that they can exhibit low-level cognitive abilities, including: 1) object identification based on properties and context-sensitivity, 2) comparison of similarities and differences among properties and objects, 3) shape recognition of simple shapes regardless of where they might appear in an artificial visual field, 4) association across simulated sense modalities, 5) primary sequential memory of any seven digit number (inspired by Allen and Lange[1]), 6) network branching from one subnet to another based on the presence of a single stimulus, 7) eight-bit register control that could perform standard, machine-level operations as with Turing-style computational devices, and 8) rudimentary natural language processing based on a stimulus/response (i.e. anti-Chomskian) conception of language. Models 6 and 7 here serve as proof that we could, in principle, build a Turing machine inside of a DAN, thereby showing that some dynamic network structures are Turing-complete, and model 8 *may* be an early indicator of how to get rich semantics from an information-theoretic mechanism, but, again, more work is needed.

Unlike ANNs, DANs learn by adding nodes and connections wherever needed. In more recent applications, we are reintroducing dynamic weights based on entropy equations and thresholds to improve cognitive function. In our latest experiments, for instance, we have been working to isolate control parameters to transform ordinary databases into predictive mechanisms in order to create content-addressable memory.

In keeping with the spirit of Turing, DAN architecture is intended to avoid what Beavers has elsewhere identified as the "software seduction";[8] "It is ... quite difficult to think about the code entirely in abstracto without any kind of circuit," (p. 384, Ref. 27) Turing wrote in his 1947 Lecture on the Automatic Computing Engine, suggesting that in a working machine there is no code, just hardware, and that really what computer code does is to configure circuitry within computing machinery to perform a particular information processing task. This fact has led us to adopt the slogan "circuits not software," when characterizing DANs and reframing John Haugeland's "formalists' motto.": "If you take care of the syntax, the semantics will take care of itself" (p. 106, Ref. 20). On our view, the right kind of circuit will provide the sufficient and necessary conditions for deriving semantic-respecting and goal-oriented behavior. The model described below is *part* of our effort to establish this claim. That said, we wish to be clear that we are not maintaining that only DANs can exhibit Information-Theoretic Teleodynamics. ANNs might as well–it has simply not been our project to experiment with them–and non-networked, biological systems and other natural systems could also.

3. Information Entropy, the Inverse Relationship Principle and Dynamic Threshold Values

Information "entropy" was originally used by Shannon as a measure of the uncertainty of a piece of information, the term being suggested to him by John von Neumann because of its isomorphism with entropy in thermodynamics, though whether or not this recommendation was fitting has been a matter of some debate. There is some overlap between the two conceptions nonetheless.

Floridi notes that the concept of information entropy is easily grasped when we think of information in terms of its ability to decrease our ignorance or in terms of a data deficit.[17] The toss of a coin has become the canonical example. If the coin is fair, then we cannot predict whether it will land heads or tails. Thus, tossing the coin stands to decrease our ignorance when it lands. If the coin is weighted, however, such that it will always land heads (and we know this fact), then we stand to learn nothing new when it lands. In Shannon's terms, the information entropy in the toss of the fair coin is higher, that is, more uncertain, then that of the weighted coin.

Though Shannon famously claimed that his attempts to quantify information in engineering had little to do with semantic content,[26] his word was not the last on the subject and has also been a matter of some debate. Barwise and Seligman[3] noted semantic corollaries to Shannon in what has been identified as the Inverse Relationship Principle (IRP).[18] IRP says that the more rare a piece of information is, the more informative it is, and vice versa. Thus, within the domain of one's general knowledge of animals, in the phrase, "a four-legged animal that quacks," the range of the term "animal" is larger than that of "four-legged," and hence is less informative. The term "quacks" is less likely to occur and is thus the most informative. (Notwithstanding the fact that there are no four-legged animals that quack, it would not be surprising if the reader's immediate reaction to the above phrase were to think of a duck, even though it does not have four legs, because very few animals (only ducks?) quack and many animals have four legs.)

Versions of IRP appeared before Barwise and Seligman (see, for instance, Wiener Ref. 28), and the concept has led to informational paradoxes, as in the extreme cases of tautologies and contradictions (see Hintikka Ref. 21, and Bar-Hilllell and Carnap Ref. 2) and the suggestion that "gibberish" should be more meaningful than sensical information, since it is less frequent (see Dretske, p. 42, Ref. 15). Still, IRP proves useful as the foundation for an entropy equation that points in the direction of the quantification of semantic information, depending, of course, on how it is employed. Additionally, Floridi employs a concept of semantic-oriented information entropy based on IRP in his information ethics when dealing with contradictions, though he does not believe that it is possible to reduce semantic information to information theory.[16]

Our DAN model allows the user to toggle between the simple formulae $1/n$ (IRP) or $1/n^2$, where n represents the number of nodes that feed into another node in a network. The result is that the more connected a node is, the less its connections weigh, making information that is more unique count more toward modulating network behavior. It should be apparent that by inverting this strategy, we can make information that is more typical, rather than less, exhibit more influence.[6]

The chief advantage with this approach in general is that weights in our network are dynamically set on the fly rather than being determined by a training method, such as backpropagation. Furthermore, DANs, unlike ANNs, are not trained by finding weights to match a training set, but by the wiring schematic that gives structure to the network. This wiring structure can be easily modified, that is, nodes can be added, connected and removed, without interfering with the overall cognitive performance of the network.

No "retraining" is necessary when encountering new information; rather, the structure is modified, which, in turn, dynamically resets the weights according to the entropy formulae indicated above (and also dynamically resets the threshold values to be discussed momentarily).

$1/n$ and $1/n^2$ both produce interesting, but different results, and richer entropy equations are available to try. By reinterpreting DANs as complex adaptive systems, the nodes being reconceived as agents and the connections as interactions, for instance, we could adopt entropy measures as they are used in complexity theory, in which case $1/n$ provides just one component of a diversity measure. (See Page Ref. 23). To integrate these richer entropy equations, we would have to employ them differently in the network, but their dynamic nature would be the same. A broader issue, however, is raised by the presence of several alternatives here concerning the determination of what the target behavior of the network should be. If $1/n$ and $1/n^2$ both produce intuitively interesting, though different results, which is the correct formula, if either? How do we know when we have things right? Finding an answer to this question, we believe, might be aided by testing for information entropy in free association tasks using human subjects, a study that we are just beginning and that will at least provide us with a target for testing even if it is not optimal.

We can summarize the above by considering a three-layered network, though there's no reason to suspect that neat layers are necessary, save for the ease they afford in analysis. Some of the nodes in the first layer (A) connect to some of those in the second (B), and some of those in the second connect to some of those in the third (C), in keeping with the principles of DAN architecture. If the first node in A (A1), is connected to 13 nodes in B, then the entropy for each connection to B would be $1/13$ (.0769) or $1/13^2$ (.0059) depending on which entropy formula is used. To stay with the simpler entropy formula for a moment, if a node in B, say B1, receives only two connections from A, one valuing (.0769) and the other (.1667 for $1/6$), the total weight of B1 is simply the sum of the two dynamically-determined information-theoretic weights, or, in this case, .2435. After activation of all pertinent nodes in A, some or all of the nodes in B will thus be dynamically weighted.

Having thus calculated an information-theoretic weight for each node in B, the task now becomes one of determining which of the nodes in B will be allowed to pass activation to C. To do this, we employ a threshold value that is dynamically-determined by the collective set of summed weights for the nodes in B. For the purpose of this demonstration, we use a simple average to set the threshold. (When addressing the question of computation in nature, it is important to realize that a simple average of the weights for nodes in a network layer can be determined and employed without violating localist principles, though it should also be said that this is difficult to do with an ANN and relatively simple with a DAN.) A host of other formulae are also available for this task, ranging from sigmoid functions to functions based in parametric statistics or various combinations thereof. The best way to proceed here, again, will depend on what we can learn from tests with human subjects and our tuned intuitions about what constitutes future-oriented, cognitive behavior.

Finally, before getting to the specific details and description of the model, it is worth noting that we are also exploring the possibility that some compound entropy/threshold formula might do the needed work. However, the "divide and conquer" strategy of treating each metric separately lends itself to a more intuitive understanding of how information transduction is performed in our network. We are also considering the possibility that more than one formula could be employed by a network that could dynamically decide which route to take in specific cases analogous to the interplay between the sympathetic and parasympathetic nervous systems in human beings.

4. Getting Teleodynamics from Newtonian Causation

Following Deacon,[9] we agree that the addition of recursion and memory to a system are necessary to get teleodynamic behavior, though the situation is such that we can be theory-neutral on the metaphysical question of emergence. Settling the matter is neither here nor there when building a working, computational model.

Neither "recursion" nor "memory" are new to the language of networks, but a few preliminary comments should be made. Memory can be taken here in a very minimal sense stripped of any cognitive interpretation; for our purposes, it is simply the ability of a system to detect when it is in a state that it has been in before. Memory in networks is not "stored" and "retrieved" as with conventional computers. Rather, it is implicitly embedded in the network structure itself, or, perhaps better, it *is* that structure coupled with the dynamics that permit particular parts of the circuit (network) to activate under certain conditions.

Recursion is intended here in the straightforward sense of recirculating information through the same network. To stay with the structure described in the previous section, layer A and layer C are made up of nodes representing the same features, with network processing going on between A and B, and, then again, between B and C. In our associative network, layer A is made up of nodes that represent various properties, layer B nodes that represent the objects that have those properties, and layer C again the properties in layer A. Activation passes from A to B and then B to C. Output values from C are then recirculated as the input values for A on each recursive iteration, letting these values define a trajectory for further activation.

The nodes in A represent the following characteristics which are listed below in the original order that they appear in the network. This is not incidental. The network was built by imagining that we were teaching a child about animals and, thus, reflects encounters that are more or less random as it would be in a real learning situation. It is worth noting in passing, that network performance remained intuitive throughout the entire engineering process, though this is a topic for another paper. That said, the nodes in A represent: small, furry, meows, barks, winged, flies, swims, finned, crawls, many-legged (more than four), two-legged, warm-blooded, live offspring, walks, four-legged, farm animal, oinks, moos, ridable, big, crows, woodland, hops, hoofed, curly, hisses, cold-blooded, scaled, quacks, coos, seafaring, spawns, large-mouthed, web-making, whiskered and feathered.

The nodes in B represent the following animals: dog, cat, bird, duck, fish, caterpillar, human, pig, cow, horse, rooster, rabbit, deer, retriever, poodle, mouse, snake, crocodile, lizard, turtle, dolphin, whale, ostrich, pigeon, shark, salmon, bass, house fly, cricket, spider, catfish and seal. The reader has, no doubt, noticed that we have included general classes of animals along with specific members. Thus, the list includes, for instance, bird, but also, duck, rooster, ostrich and pigeon. Doing so does not interfere with network performance as long as differentiating properties appear in A that allow the network to distinguish specific instances from archetypes. Even without these differences, however, the network will still classify members with their archetypes. Proper nouns (such as "Rover") could also function as animal names in the network as well, though we have not done so here.

Describing the complete wiring schematic is beyond the scope of this paper, but in the space allotted, we can provide some samples. Node A5 (winged) is connected to B3 (bird), B4 (duck), B11 (rooster), B23 (ostrich), B24 (pigeon), B28 (house fly) and B29 (cricket). Since A5 connects to 7 nodes in B, the weight on each connection is 1/7 using IRP or 0.1428. Node A11 (two-legged) is connected to B3 (bird), B4 (duck), B7 (human), B11 (rooster), B23 (ostrich) and B24 (pigeon), which is 6 connections for a weighted value of 1/6 using IRP or 0.1667. Simultaneous activation of A5 (winged) and A11 (two-legged) produces these weights in the B layer: B3 (bird 0.3095), B4 (duck 0.3095), B7 (human 0.1667), B11 (rooster, 0.3095), B23 (ostrich 0.3095), B24 (pigeon 0.3095), B28 (house fly 0.1428) and B29 (cricket 0.1428).

Having established weights for the appropriate nodes in B for the activation of A5 and A11, the network now calculates a dynamic threshold value, in this case 0.2500, determined by averaging the weights of all activated nodes in the B layer, though, again, another formula may prove better down the line. This allows activation from B3 (bird), B4 (duck), B11 (rooster), B23 (ostrich) and B24 (pigeon) to pass to the C layer. B7 (human), B28 (house fly) and B29 (cricket) are appropriately filtered out.

In turn, B3 (bird) is connected to C1 (small), C5 (winged), C6 (flies), C11 (two-legged), C12 (warm-blooded), C14 (walks), C21 (woodland) and C36 (feathered). Since B3 makes 8 connections, the entropy value is 1/8 using IRP or 0.1250. The incoming weight of B3 was 0.3095, as we saw above, which then gets multiplied by the entropy value of 0.1250 to yield 0.0387 as the weighted output value of B3, which is then summed into each connection B3 makes to the C layer. In like manner, connections are made from B to C for B4 (duck), B11 (rooster), B23 (ostrich) and B24 (pigeon) to yield the following weights for nodes in the C layer:

Table 1. Weights on the nodes in the C layer based on activation of the above.

C1	small	0.1482
C5	winged	0.1998
C6	flies	0.1040
C7	swims	0.0653
C11	**two-legged**	**0.1998**
C12	**warm-blooded**	**0.1998**
C14	walks	0.1556
C16	farm animal	0.0442
C20	big	0.0515
C21	crows	0.0442
C22	woodland	0.0696
C29	quacks	0.0309
C30	coos	0.0344
C36	**feathered**	**0.1998**

The threshold value based on the average weights for the C layer is .1105, thereby allowing the items in bold above to pass the threshold while filtering the others out. Nodes that pass the threshold are then resubmitted to the input layer for recursion.

Before continuing further, a few observations should be made. Activation of winged and two-legged, returns the bird set (duck, rooster, ostrich and pigeon) along with the generic bird for output from the B layer to C. More interestingly, the output values for C partially complete the bird pattern. Activation of winged and two-legged attracts small, warm-blooded, walks and feathered. Note that at this point, flies has failed to be pulled into the list of appropriate characteristics, since ostriches and roosters do not fly. However, the next recursion cycle produces an interesting effect. Rooster and ostrich are slightly demoted (but only slightly), letting the generic bird with the duck and pigeon count for more. Typicality then wins out on the next recursion cycle over the uniqueness that is tracked by IRP, as flies is pulled into the list of relevant characteristics, further demoting the rooster and the ostrich. Further recursion after the third cycle produces no further re-evaluation of nodes, unlike in other networks that can take several hundred iterations to settle.

Suppose, however, that we want the rooster to rise and the other birds to be demoted. To do so, we must activate a property in A that has high entropy and belongs to the rooster. Unfortunately, using the IRP entropy formula still drags flies into the list of pertinent properties, and over the course of a few iterations the duck passes the rooster. However, using $1/n^2$ does the task.

Switching over to $1/n^2$ and activating A5 (winged) and A11 (two-legged) still picks out the same set of birds (bird, duck, rooster, ostrich, pigeon) as the network asks for activation of small, warm-blooded, walks and feathered, just as before. One iteration pulls flies in the list as well, as the bird, duck and pigeon climb just over the ostrich and the rooster. However, if we begin with A5 (winged), A11 (two-legged) and A21 (crows), the rooster climbs substantially over the other birds. Recursion in this case does not pull flies into the property set, but it does get farm animal as it should.

Generally, $1/n^2$ works better than $1/n$, at least using the averaging thresholding formula indicated above, though not always, and admittedly we run into the occasional counter-intuitive case. The network, for instance, is good at not letting a property set contain both two-legged and four-legged, but sometimes it will admit a set including both warm- and cold-blooded, indicating that we still have some work to do. Again, the situation is complicated by the fact that the threshold parameter also affects network performance. We are optimistic (and aren't we all!) that further analysis and experimentation with more advanced mathematics in light of test results from human trials will improve performance, but this does not mean that we are not seeing important theoretical results. For the purposes of this paper, *the way in which network activation pushes its way through a trajectory that simultaneously opens before it is particularly interesting, since the trajectory dynamically modifies itself in step with changing network activation and thereby continually sets up possibilities that the network will actualize given further stimulus.*

These possibilities allows us to conceive of the system as teleodynamic and not merely teleonomic. To see this, we can go back to Aristotle's notion of final (or teleological) causation. Aristotle does not think that every acorn will become an oak tree, since external influences might impede it. Squirrels sometimes eat acorns for food after all. Rather, when the acorn is considered according to its own internal principles, including its material (*hyle*) and form (*morphe*), if it becomes anything at all, it will be an oak tree and not, say, a pig or a walnut tree.[4]

Computational philosophy, again, is the use of computational tools to establish points of philosophical interest that might not be seen otherwise.[7] Even though this work is very much in progress, we do believe that we can tentatively claim on the basis of what we have said here that teleological causes can be computed in principle, given the right kind of structure, one in which information-theoretic quantities can be used to determine information flow through a network to produce activation patterns that march in step with our semantic sensibilities. We cannot say where this research will take us, only that the results are sufficiently interesting to warrant not setting the project aside at this time. Indeed, we see two possibilities for further projects that we believe will produce enlightening results. One is a project in teleodynamic game play in which we will engineer a network to play a formal game using information-theoretic principles and pattern matching rather than heuristic search. The other concerns enriching the communications strategies in Patrick Grim's artificial life model,[19] mentioned above, to see if we can build a community of artificial agents that can learn from each other and that might cross the boundary between simple signal processing to something closer to language use. (See the comments from Floridi[18] on action-based semantics above.)

5. Information-Theoretic Teleodynamics in Natural Systems

While it would seem obvious that there are teleodynamic systems in nature, since human beings seem to engage in goal-oriented, intentional behavior, the matter is far from settled. Dennett's work on the "intentional stance"[14] and the literature surrounding it has been a part of rigorous debate for decades about whether people (and some animals) engage in genuine intentional behavior or whether such behavior is merely ascribed. We do not have space to rehearse the debate here, though in the terms of this paper, the question can be phrased to concern whether people and animals engage in genuine teleodynamic behavior or whether such is merely teleonomic, which is sufficient to establish that the question remains open. Is teleonomic description a mere strategy that we use to cope with circumstances in a partially-ordered/partially-disordered world, or is there more to the story?

Hardcore reductionists of a variety of stripes, too, would like to suggest that the natural world can be fully understood solely in terms of "push" causation. We would like to challenge this picture with this simple model of "pull" causation that may provide an important part of the explanation about why a system behaves the way it does. Is such a system plausible in the natural world? Again, without suggesting that only DAN structures can yield such results, we believe that the answer to the question is affirmative. Though we are quite far from a complete explanation, the circuit here suggested is surprisingly simple, yet cognitively rich. A simple recurrent circuit in which properties map to objects which then map back to the properties that define them and that is controlled by the kind of electrochemical dynamics one might find in real neural systems is not beyond the reach of nature, especially given the rich array of neural structures that are part of the human brain.

Of course, one need not stop here. If such behavior is explicable in information-theoretic terms, we might see it in a variety of structures, including those supporting information propagation in a group of human and/or non-human animals or even in wholly non-biological systems. We suspect that this latter possibility will interest the reader of this volume most, even though it has not been our immediate target here. We therefore invite the reader to speculate. Can communication (understood in the natural and not social scientific sense) and information processing in the natural world more generally be understood on the basis of Information-Theoretic Teleodynamics? We do not yet know, but this exercise might provide some support for the pursuit of worthy answers.

Acknowledgements

We would like to thank the following individuals for their helpful feedback on various phases of this project: Colin Allen, J. McKenzie Alexander, Aaron Bramson, Selmer Bringsjord, Luciano Floridi, Patrick Grim, Mirsad Hadzikadic, Derek Jones, Matthias Scheutz, Talitha Washington, Hector Zenil, and the attendees of a 2011 Institute for Advanced Topics in the Digital Humanities sponsored by the National Endowment for the Humanities (USA) and the Complex Systems Institute at the University of North Carolina, Charlotte.

References

1. C. Allen and T. Lange, Primary sequential memory: An activation-based connectionist model, *Neurocomputing*, 11(2-4):227-243, 1995.
2. Y. Bar-Hillel and R. Carnap, An outline of a theory of semantic information. Research Laboratory of Electronics, MIT, 1952.
3. J. Barwise and J. Seligman, *Information flow: The logic of distributed systems*, Cambridge, 1997.
4. A. Beavers, Motion, mobility and method in Aristotle's *Physics*: Comments on *Physics* II, 192b20-24, *Review of Metaphysics*, 42:357-374, 1988.
5. A. Beavers, Intentionality and association, The 3rd Annual Computing and Philosophy Conference, Oregon State University, Corvallis, Oregon, August 8th, 2003.
6. A. Beavers, Typicality effects and resilience in evolving dynamic networks, in FS-10-03, AAAI, 2010.
7. A. Beavers, Recent developments in computing and philosophy, *The Journal for General Philosophy of Science*, Springer Online First, 2011.
8. A. Beavers, Alan Turing: Mathematical mechanist, in S. B. Cooper and J. van Leeuwen (Eds.), *Alan Turing: His work and impact*, Elsevier, forthcoming.
9. T. Deacon, Emergence: The hole at the wheel's hub, in P. Clayton and P. Davies (Eds.), *The re-emergence of emergence*, Oxford, 2006.
10. T. Deacon, Shannon-Boltzmann-Darwin: Redefining information. Part 1, *Cognitive Semiotics*, 1:123-148, 2007.
11. T. Deacon, Shannon-Boltzmann-Darwin: Redefining information. Part 2, *Cognitive Semiotics*, 2:167-194, 2008.
12. T. Deacon, What is missing from theories of information?, in P. Davies and N. Gregersen (Eds.), *Information and the nature of reality*, Cambridge, 2010.
13. T. Deacon, *Incomplete nature*, Norton, 2011.
14. D. Dennett, *The intentional stance*, MIT, 1987.
15. F. Dretske, *Knowledge and the flow of information*, MIT, 1981.
16. L. Floridi, Information ethics: On the philosophical foundation of computer ethics, *Ethics and Information Technology*, 1:37-56, 1999.
17. L. Floridi, *Information: A very short introduction*, Oxford, 2010.

18. L. Floridi, *The philosophy of information*, Oxford, 2011.
19. P. Grim, et. al., Making meaning happen, *Journal for Experimental and Theoretical Artificial Intelligence*, 16(4):209-243, 2004.
20. J. Haugeland, *Artificial intelligence: The very idea*, MIT, 1985.
21. J. Hintikka, Information, deduction and the a priori, *Noûs* 4(2):135-152, 1970.
22. J. McClelland and D. Rumelhart, An interactive activation model of the effect of context in perception, part I. Chip Report 91, Center for Human Information Processing, Univ. of California, San Diego, 1980.
23. S. Page, *Diversity and complexity*, Princeton, 2011.
24. D. Rumelhart and J. McClelland, An interactive activation model of the effect of context in perception, part II. Chip Report 95, Center for Human Information Processing, Univ. of California, San Diego, 1980.
25. C. Shannon, A mathematical theory of communication, *Bell System Technical Journal*, 27:379-423, 623-656, 1948.
26. C. Shannon and W. Weaver, *The mathematical theory of communication*, Univ. of Illinois, 1949.
27. A. Turing, Lecture on the automatic computing engine, in B. J. Copeland (Ed.), *The essential Turing*, Oxford, 2004.
28. N. Wiener, *The human use of human beings*, Houghton Mifflin, 1950.

Chapter 19

Discrete Theoretical Processes (DTP)

Edward Fredkin
Carnegie Mellon University

1. Introduction

One open question in the field of Theoretical Physics has to do with the most microscopic of phenomena. We herein will examine consequences of assuming that space, time, state and all other properties and processes of physics are finite, discrete and deterministic. We already know that microscopic matter consists of discrete particles and that microscopic forces are mediated by discrete particles. If our assumptions were true then a comprehensive theory of discrete physics ought to contain non-tautological explanations for what we call "Small integer phenomena": the constants of physics that are characterized by small integers. Some examples are:

1 - photon
2 - directions of time
2 - particle and antiparticle
2 - signs of electric charge: ±
2 - flavors per generation
2 - fermions per generation (e.g. electron and lightest neutrino, or up quark and down quark)
2 - spin state properties (even or odd multiples of $\hbar/2$)
3 - generations of flavors (e.g. electron, muon and tau)
3 - generations of Quarks
3 - colors of Quarks
3 - massive bosons, W^+, W^- and Z^0
3 multiples of $\pm 1/3$ charge.

In addition, we ought to be able to find, over time, relatively simple derivations for all of the constants of the Standard Model. Most importantly, if space, time and state are discrete and if some kind of DTP could model the most microscopic processes in physics exactly, then, depending on the scale of quantized space and time, we might be able to finally understand every aspect of microscopic physics including Quantum Mechanics. Those are some of the attractive possibilities.

Unfortunately, if our assumptions are true there are also many unattractive consequences. Almost all of the wonderful analytic formulae in all of physics would be relegated to being no more than good approximations of the actual discrete measures of space, time, energy, momentum, etc. All of the continuous symmetries would also be no more than good approximations. On the other hand, it is quite possible that various conservation laws could, nevertheless, still be true and exact even while being discrete.

We are attempting to maintain clarity, in in this document, by indicating our use of words that have somewhat specialized meanings by presenting them (and no others) in italics. When context seems to be an insufficient guide, we will provide definitions.

DTP is not conceived of as a system where, at an instant in time, the known physical particles are located at discrete points in space. Rather, we imagine that the occupant of every point in space is a simple *token* (perhaps similar to a 3 state version of a bit in a computer; sometimes called a "trit") and that stable 3+1 dimensional extended *configurations* of *tokens* are what stable particles are made of while unstable particles are similar *configurations* that happen to decay when their ever changing internal state and the ever changing bordering pseudo-random external state satisfy some particular criteria in some way: thus yielding a deterministic half-life decay process consistent with observations (something once thought to not be possible for a simple deterministic process). Our task has been and still continues to be: finding ways to explain how various versions of DTP might be able to model more properties of both matter and energy in ways consistent with the known facts; the observations and measurements from experimental physics. We are certain that the kinds of models we are proposing can never be exactly consistent with our analytic laws of Quantum Mechanics, but we shall be content if instead, they are exactly consistent with the microscopic evolution of physical state that we call "Quantum Mechanics".

In the past we have used names such as "Digital Mechanics", "Digital Physics", the "Salt Model" and others as we continued to make modest progress. What we are disclosing here is much improved over all earlier versions but is nevertheless still highly speculative and certainly substantially incomplete and still just plain wrong in some respects. Nevertheless, this version solves a number of problems that we didn't know how to solve in the past. Our progress has been slow but steady as we keep finding additional possible models for various additional aspects of physics. Our study of DTP advances despite the current lack of direct experimental evidence of the magnitudes of possible discrete measures of space and time.

What will be described are properties of a class of regular, second order 3+1 dimensional Cellular Automata; based on a Cartesian Lattice, where something similar might possibly be a candidate for the most microscopic space-time structure of our world. Our current goal is to define characteristics of some such systems that, at a higher level, could result in a correspondence to the laws of quantum mechanics. Along the way, we want to communicate new insights into how it might be, in a discrete space-time-state world, that all things move with apparent translational symmetry.

If ultimately, space, time and every other measure in physics is discrete and if something similar to DTP should turn out to be an exactly correct model of the most microscopic physics, then given nothing new other than the unit of length, the unit of time, the structures that define the most microscopic representation of state, the transition rule that defines the temporal evolution of the most microscopic state, and some characterization of the initial conditions (the Big Bang) it would be possible, theoretically, to compute all of the constants and facts of microscopic physics. However, from a practical viewpoint, this might only be possible if the unit of length was greater than a Fermi, 10^{-15} Meters. If the unit of length turned out to be closer to Planck's Length, then computer simulations would likely prove very much less useful.

To better understand the kinds of discrete space-time structures we have in mind, it might be helpful to first read an earlier paper that discusses the SALT model: "Five Big Questions with Pretty Simple Answers." Published in the IBM Journal of Research and Development VOL. 48 NO. 1 JANUARY 2004.

We will describe a number of models and methodologies. Some of these may appear as bizarre, but it is important that the reader keep in mind our purpose. We do not have anything like a complete set of reasonable models for discrete counterparts to conventional physics. Instead, most of this work has involved finding discrete, deterministic models for various aspects of physics and counterexamples to the proposition that such discrete systems cannot serve as accurate models of various aspects of microscopic physics. As such, some of our examples may seem bizarre or even ridiculous. No matter, their only purpose is to exist as counterexamples to the claim that such discrete models are not possible. The best example has to do with translational Symmetry despite having a fixed Cartesian lattice as the *substrate* for space-time. We have so far found several approaches to defining discrete processes that model Translational Symmetry; some are reasonable; there are others which most readers would find wholly unreasonable. At similar points in the reading of this paper, we seek your indulgence. Keep in mind our purpose – which is more about defining the problems we must solve in order to understand DTP, as opposed to supplying all of the answers. Our approach, necessitated by our primitive methodology, has been finding counterexamples to concepts that seemed, at first glance, to rule out DTP as a potentially competent model of physics. For example, it is widely thought that: "A model of space-time that is based on a single fixed Cartesian Lattice cannot support translation or rotation symmetry." We can now demonstrate that that thought is too simplistic.

An interesting new development was finding models of simple QM Harmonic Oscillators that are highly stable despite operating in a CA without the ability to perform any arithmetic operations other than processes where nearby *tokens* conditionally permute positions with their neighbors. In other words, the behavior of a trivially simple CA can, amongst other things, model microscopic processes where the corresponding higher level mathematical models involve complex analytic functions.

Finally we offer an explanation as to how totally discrete models can explain, in general, the wonderful and miraculous applications of mathematical analysis in physics; made even more wonderful if space, time and state are all discrete! Simply put: the success of analysis may be a consequence of trivially simple microscopic processes in discrete space time state systems that happen to exactly conserve discrete quantities such as energy, momentum, angular momentum, charge... and so forth!

2. Understanding DTP

The science of physics and the art of mathematical analysis have co-evolved in an amazingly fruitful manner. Today we represent most of the laws of physics by means of equations where the functions and variables are continuous. However, when we think about DTP, with space, time, state and all other attributes of the most microscopic physical processes being discrete, we find mathematical analysis inappropriate at the most microscopic level.

Even our current use of natural language is limited in its ability to describe aspects of DTP.

The key concept that allows fundamentally discrete and deterministic microscopic models to be well characterized by mathematical analysis has to do with exact conservation laws. DTP processes are capable of conserving, exactly, various quantities of physics including those involved in processes characterized mathematically by a Hamiltonian or a Lagrangian. In such cases a version of Noether's Theorem can still connect the two properties of conservation laws and corresponding continuous symmetries. Within DTP there is nothing that has the property of continuity, however, at scales above the most microscopic, we can nevertheless attribute a kind of asymptotic continuity to DTP evolution of state. We also know that computational models of QM systems can be programmed. Thus DTP is certainly not excluded from the possibility of underlying Quantum Mechanics. We must keep in mind the fact that discrete computational processes can, to any required accuracy, model any and all analytic functions.

We therefore proceed by introducing concepts, through definitions, that we believe to be consistent with our current understanding of the appropriate attributes of this thing we are calling DTP. We investigate various aspects of DTP related to the known experimentally determined facts of physical space, time, action, charge, energy, matter and antimatter, etc., the fundamental quantities and properties that are the subject matter of theoretical physics. We try to clarify, but only in the context of DTP, both what *information* is and what is *information*. As a result, the word "*information*" will also require a new definition within the realm of DTP.

Some of the words and expressions we use are associated with DTP related expanded definitions: *discrete, local, token, rule, fundamental process, state, meaning, information, entropy, explicit information, implicit information, locality, substrate* and *configuration*.

Discrete Theoretical Processes may need only 3 fundamental numerical constants:

1. A unit of Length, L
2. A unit of Time, T (Where 6 micro-time steps, t, equal one time step, T)
3. A unit of Action, B ($B = \hbar/2$)

Today, we are quite certain that the unit of action is $\hbar/2$ but the units of

Length, L, and the unit of Time, T, are currently unknown. Further, the exact relationship between c, the speed of light, and the units of Length and of Time, insofar as DTP is concerned, are not currently well enough understood. It is clear, however, that there may be problems with the simple minded assumption that either $L/T = c$ or $L/t = c$. In any case L and T must allow for particles whose gross motion is asymptotically isotropic when velocities are less than or equal to the speed of light.

DTP assumes that all *information* must have a *discrete* means of its representation, called a *"configuration"*. All changes of *state* must be consequences of *discrete* informational processes, similar to what occurs within a computer. All processes involve the temporal evolution of a *configuration*, such as its translational motion or the interactions of *configurations* such as an electron interacting with a photon.

A *configuration* is normally a 2nd order in time spatially distributed collection of *states* that can be interpreted by some *process* in a meaningful way. This always involves, in addition, such properties as the microscopically discrete translational motion of the *configuration* that is associated with a particle. A more complete definition of *"configuration"* will be given later.

In biology, long ago, the observation that for every species, an offspring was always of the same species, was summarized by the saying "Like begets like." The discovery of DNA allowed us to understand that discrete informational representations allowed for a mechanistic explanation of "Like begets like." However, even today in 2011, we still lack a full understanding of all of the processes that enables DNA to control the development of a living thing.

In physics we may have a similar situation with regard to Translation Symmetry and Newton's First Law[a]. While it is doubtful that anyone ever described it as "Velocity begets velocity", within DTP it is clear that the velocity of a particle must also have a *local discrete* means of its representation as some kind of informational *configuration*. In fact, the primary law of all DTP systems is: All *local* state *information*, which includes velocity, must always be represented by a *discrete, local configuration*.

"Discrete" A property of a finite closed system where every one of the N distinct different possible states of that system could have an informational, one to one, correspondence with N distinct integers.

---

[a] *Law I: Every body persists in its state of being at rest or of moving uniformly straight forward, except insofar as it is compelled to change its state by force impressed.*

"*Local*" *Information* that purposefully affects the evolution of *state* of a particle can be described as "*local*" with respect to that particle. For example, the wave structure that contains directional information with regard to a particular photon, can be described as *local* despite being Meters away from the photon in the plane perpendicular to the motion of the photon.

"*Token*" is a thing that has one permanent *state* out of a small integer number of possible *states*. For Example, imagine that there are 5 kinds of *tokens* whose names are: $+1, +i, 0, -i, -1$. Every cell within the Real *Substrate* is occupied by exactly one of the following three *tokens*: $+1$, 0, -1. Every cell of the Imaginary *Substrate* is occupied by exactly one of the following three *tokens*: $+i$, 0, $-i$. We could assume that these 5 types of *tokens* have no properties other than their names. However we expect that our choice of assigning zero and the positive, negative, real and imaginary units as names of the *tokens* will correspond to some of their mathematical properties. *Tokens* are immutable; never changing into different kinds of *tokens*.

"*rule*" While we do not, at this time, know what the fundamental rule is, we can indulge in educated guesses. The fundamental *rule*, should be a 2nd order process in order to simplify requirements for the dynamic laws of physics such as Newton's laws. It has to, in a sense, produce as a consequence of its operations, evolution of *state* in accord with the known laws of physics. Since our model is a discrete process that will be computation universal, we can hypothesize that any universal system could suffice in that some initial condition is guaranteed to result in some subset of the *states* obeying the laws of physics. This is a weak argument because what we are really looking for is a universal process that produces the laws of microscopic physics and nothing else. We hypothesize that the Rule could be composed of 6 or more, microscopic sub rules. A neighborhood at $t+1$ becomes a conditional permutation of that same neighborhood at $t-1$ as controlled by the *state* of a neighborhood at time t and according to Rule $R_{t\,Mod\,6}$. As a consequence of the *rule*, small groups of neighboring *tokens* at time $t-1$, under certain conditions as determined by the *state* of other neighboring *tokens* at time t, will *locally* permute some of their positions while becoming the *state* at time $t+1$. All motion and change of *state* should be, most microscopically, consequences of repetitions of such microscopic discrete changes of *state*. There are good reasons to propose a model with 6 differently oriented microscopic steps, each depending on the value of $tMod6$, that are similar except for angular orientation. The general idea at this stage has been to focus our efforts by studying rules that should enable higher level angular isotropy despite the existence of the microscopic Cartesian coordinates. We have so far found fundamental processes that achieve various goals but have not yet found one fundamental process that simultaneously achieves all of our goals.

We assume that the *Fundamental Process* must be exactly reversible, and it must, through its operation, give rise to the fundamental constituents of physics with all of their most microscopic properties. The *Fundamental Process* is a finite discrete deterministic second order system where exactly reversible changes in *state* occur. A guide to discovering competent *Fundamental Processes* will be their ability to reproduce the small integer phenomena described at the beginning of this paper. A possible form of *Fundamental Process* causes all members of a category of *tokens* in S_{t-1}, to conditionally undergo temporal evolution as a deterministic function of neighboring *tokens* in S_t. The temporal evolution results in the modification of the *state* at time S_{t-1} to produce a new *state* which can then be relabeled as S_{t+1}. It should be noted that since every kind of permutation has an exact inverse and because S_t determines how S_{t-1} evolves into S_{t+1}, all processes, with the properties just enumerated, must always be exactly reversible! Thus we know that DTP can exhibit the same kind of reversibility as is true of microscopic physics. (In order to understand how DTP can exhibit exact CPT reversibility see "Five Big Questions with Pretty Simple Answers")

"*Process*" This is a tricky concept and it is difficult to understand. In a cellular automaton, there is a rule that governs change of *state*. In Conway's Game of Life, the rule is very simple. Based on a 2 dimensional checkerboard like cellular space, every cell has 8 neighbors, 4 nearest neighbors and 4 additional diagonal neighbors. Each cell contains either 1 or 0. "*c*" stands for the *state* of a cell, "*n*" stands for how many of its 8 neighbors are one and "*C*" stands for the new *state* of the cell, then the definition of the Fundamental Process of the Game of Life is:

If $n < 2$ or $n > 3$ then $0 \rightarrow C$;
If $n = 2$ then $C \rightarrow C$;
If $n = 3$ then $1 \rightarrow C$;

The Game of Life is an example of a very simple 2 dimensional deterministic process that has been proven to be Computation Universal in the Turing sense, however it is obviously irreversible, as many simple *configurations*

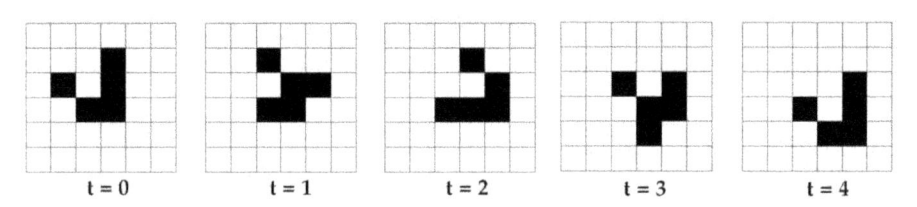

Fig. 1. 5 steps of the motion of a glider in the Conway's Game of Life.

can simply disappear. Life allows for many different kinds of meta-stable particles that exhibit translational motion at various velocities (such as the Glider) and collisions involving various numbers of Gliders can create new and different particles. The Game of Life illustrates some of the kinds of complexity that can arise in a cellular automaton given a very simple rule but the Game of Life also violates the kinds of conservation laws that exist in physics. Similarly simple cellular automata rules have been found that are reversible, computation universal and subject to conservation laws.

While the Fundamental Process governs all activity in DTP, there are higher level processes that create and annihilate various kinds of particles and that result in the motions of the various particles. In DTP these higher level processes are basically interactions between particles and or fields whose existence and properties are all consequences of the operation of the fundamental process. Even in the case of a single free particle, its interaction with its own momentum wave must result in isotropic rectilinear motion. Such processes can underlie the apparent Translational Symmetry of the system; all particles must, of course, be engaged in translational motion. Thus, every configuration has the property of being in translational motion.

"*State*" In the context of DTP – *state* is always represented by a *configuration* where the extent and *meaning* of that *state* is determined by the processes that examine or manipulate those *configurations*.

"*Meaning*" A potential property of some *configuration*, C1, that has *state*. The *meaning* of the *state* of a particular *configuration* is given by a *context* where a discrete process, created by another configuration, C2, examines, interprets, modifies or in any other way – interacts with *configuration* C1. A particular *configuration* can have multiple *meanings* depending on the *context* (depending on which other configuration might interact with it).

"*Information*" *State* that has *Meaning*. In this context "*Information*" is a *configuration* that has *meaning*. Thus *meaning* is a property of the combination of a *configuration* and a *process*. Each *meaning* of a *configuration* is given by the *process* that is interacting with it. A given *configuration* can have more than one *meaning*. In an ordinary computer, a process is represented by a block of memory that contains instructions. Data is also represented by a block in the same memory that contains data. The same concept exists in DTP where particles and fields are both patterns in the cellular space that interact with each other. The von Neumann architecture is still the basis of all modern computers; instructions and data share the same memory. It is also our model for DTP. Von Neumann's concept was absolutely prophetic! For example, consider a *configuration* in a contemporary computer that represents a one minute video. To the process that displays the video, the *meaning* is: "An MPEG file located at address A and of length N, to be decoded into a series of raster images which are sent serially to the display driver." To the software that copies the MPEG file to a USB stick the *meaning* is "A binary file, located at address A and of length N, to be copied to the USB stick". To the software that erases the file the *meaning* is "A file located at address A and of length N, to be erased and the block of memory it occupied is to be given to the Garbage Collector." In each case the process that gives meaning to data is also in a block of words in the same general memory as is the data.

"*Entropy*", in this context, entropy is certainly not the same as *information*. Basically *entropy* is a scalar measure (*Log* of the number of equally probable *states*) associated with *states* that have no individual *meanings* in the context of *entropy*. The *Entropy* of a *configuration* is equal to the number of Shannon Bits needed to specify all of the possible *states* of that *configuration*. *Entropy* does have *meaning* depending only on the numerical value of the *Log* of the number of individual *states*.

In all cellular automata there is a fundamental process that causes the most microscopic change in *state* as the discrete evolution of *state* proceeds step by step. A "*Configuration*" is a spatial-temporal arrangement of a group of various *tokens* that, as a body, is identified with a particle or a field. However, as a result of the characteristics of different *configurations*, various higher level processes can take place. Each *meaning* of a *configuration* is given by the higher level *process* that interprets or modifies the *configuration*. A given *configuration* can have more than one *meaning* as it may be interpreted or modified by more than one higher level *process*. We assume that, at time t, all of the *tokens* in a *configuration* of *tokens* have time coordinates either t or $t-1$. A *configuration* is always 2nd order in time. Normally all of the *tokens* in a *configuration* are spatially near to each other. However, in the case of a photon "near" might mean 10 Meters or more. There are 2 aspects of basic *configurations*: those that represent the static characteristics of a particle, matter or field such as the observation that a particular *configuration* represents an electron with its rest mass, charge, spin category and those that represent the instantaneous dynamic *state* of that particle; which might include its, spin *state*, momentum and energy, along with any other properties.

All processes, beyond the fundamental process, are consequences of the interactions of *configurations*. A single particle moves as a consequence of interactions involving its momentum wave structure and other aspects of the particle. A field or boson modifies that wave structure.

It is also true that a *configuration* that has *meaning* with respect to one process may be *meaningless* with respect to a second process while having other *meanings* with respect to other processes.

Explicit information is much more common than is *implicit information* and, in a sense, easier to comprehend. The velocity of a particle must be represented by the informational equivalent of an absolute velocity vector. Given one particle as the origin of a coordinate system, the relative motion of a second particle could conceivably be represented by a vector in the inertial frame of the first particle. While this is true, it is unreasonable to assume that velocities of all particles are represented in the reference frames of other particles. We need to imagine how it is that a single particle can have a velocity. We are forced into accepting a very simple solution to this problem but it requires a major philosophical shift. In the discrete space-time-state approach, we find no sensible way to represent the information inherent in the rectilinear motion of isolated particles without conceding to the existence of a single, fixed coordinate system where that coordinate system has many properties in common with a single fixed universal Cartesian Lattice. It is clear that from within a DTP system, all laboratory measurements can be in accord with Special Relativity. Further, all gravitational effects can be in accord with General Relativity. While computational universality guarantees these possibilities they can be a direct consequence of the rules governing the microscopic evolution of *state*.

Implicit Information has the characteristic that we cannot unambiguously assign numerical values to *implicit information* but we can calculate the difference between 2 *implicit* quantities. If particle A is at location x, y,

z and particle B is at x, $y+3$, $z+4$, then all we have is *implicit* information with respect to the position of A and for the position of B. However, we can calculate that the distance from A to B is explicitly 5. Thus locations and times are *implicit information* while distances and time intervals or frequencies are *explicit information*. Unlike contemporary physics, DTP requires that all velocities are explicit and absolute (in other words the fundamental process of DTP Physics must have access to explicit absolute velocity information, even though we have, so far, found no means to measure it. The same is true for energy and momentum).

A *configuration* has a property called *"locality"*. The question "What is the position of a *configuration*?" is similar to the question "Where, precisely, is that cloud?" A cloud is an ever changing mixture of air, water vapor and suspended water droplets. While most clouds move with the wind, there is an unusual kind of cloud called a "wave cloud" that does not move with the wind. When stable warm moist air is forced to rise because the prevailing wind is pushing it up and over a high mountainous ridge, the air cools as it expands and some of the vapor condenses into a cloud. However, once passed the ridge where the air is free to descend, the air heats up and the water droplets evaporate. The result is that the wave cloud appears stationary despite the strong wind, but if one looks closely it is possible to see it forming on the windward side and disappearing on the leeward side. The wave cloud is stationary despite the strong wind! Like a wave cloud, a *configuration* can be rather large, diffuse and constantly forming on one side and disappearing on the other as it is engaged in translational motion. But every *configuration* does have a location and the uncertainty of the location information affects the property that we call "*Locality*." The precision with which we can measure the *locality* of a *configuration* is related to the energy of the *configuration*. The greater the energy (or mass), the more precise the *locality*. This is a simple consequence of the fact that the precision of the information available to determine the location of a particle (by means of an interaction with another particle) is limited by the precision of the *localities* of the two particles. As the number of *tokens* representing the energy of a particle increases, the effective precision of it's location (in terms of very small fractions of the lattice spacing) increases. The concept of *locality* allows for a rational explanation for the fact that higher energy interactions allow for more precise positional measurements. In this light the concept that "...the higher the energy, the smaller the particle" turns out to be too simplistic. Finally, the precision with which we can make certain measurements is limited by the unit of action.

The 3 basic and fundamental physical units of DTP are: Time, Length and Action. Only the value of the unit of Action is known at this time: $\hbar/2$.

The real and imaginary *"Substrates"*, S_{2t} and S_{2t+1}, are the *local* volumes of 2nd order discrete space-time where *configurations* of *tokens*, that represent *information*, are found. At any instant in time a *substrate* of space-time has 2 temporally separate components which, together, are assumed to constitute a region of a discrete 2nd order $3+1$ dimensional space-time. In order to have something concrete to discuss we will assume that the *substrates* are essentially the same as a delimited region of space-time; as described in the Salt Model. (A Cartesian Space-Time Lattice of *cells* where every *cell* has 4 *implicit* integer coordinates and where every set of four integers that sum to an even number corresponds to one possible cell at one point in time.) Thus, the overall Cartesian Space-Time Lattice is partitioned into 2 subspaces; the Real subspace and the Imaginary subspace. For the Real *substrate* $x+y+z \equiv 0$ and $t \equiv 0 Mod 2$. For the Imaginary *substrate* $x+y+z \equiv 1$ and $t \equiv 1 Mod 2$. The structure of the combined real and imaginary *substrates* is essentially a Face Centered Cubic array. Thus each cell in a *substrate* has 12 nearest neighbors that are in the same *substrate* and six nearer neighbors that are in the other *substrate*. NaCl salt crystals have the same kind of geometric structure.

"State" is a property of a *configuration* and a *process*. Consider a photon. Its *state*, at a given instant in time, includes its explicit energy (or frequency), its explicit direction of motion, its implicit position, its explicit speed of propagation and its spin *state*. First, at any particular instant there is an amount of information in a *configuration* that represents, simultaneously, both its energy and its temporal *state*. Second there is an amount of information in a different *configuration* that represents both its explicit momentum and simultaneously its implicit position. The so called "Uncertainty Principle" can be explained as due to the fact that the same *configuration* represents both position and momentum while another *configuration* represents both energy and temporal *state*. Further, the quantum of measurement in DTP is always equal to $\hbar/2$.

We imagine that the *state* of a photon determines its properties. There must be *tokens* that represent the information, encoded in the wave structure, that determines the energy and momentum vector of a particular photon. The discrete representation of the *state* of the photon has to contain both the information that defines its properties along with enabling the processes that give motion to the photon and that govern its interaction with other particles. A 2nd order process converts that *local state information* of the photon at time $\{t,\ t-1\}$ into $\{t+1,\ t\}$, then into $\{t+2,\ t+1\}\ldots\{t+n,\ t+n-1\}$. There is no bubble around the photon where it is possible to say "...all the cells inside the bubble belong solely to this photon and none of the other cells belong to it." The answer is that there is a process that results in the temporal evolution of the material and energetic contents of space-time. Thus, at any instant, the question as to "What is this photon and what is not this photon?" has a complex dynamic answer that is the consequence of the operation of the most microscopic process that defines the global evolution of *state*. Even then, there cannot be a simple answer as to which *token* belongs and which nearby *token* doesn't belong as we know that there is, in the microscopic physics of bosons, the obvious property of superposition. There is an answer of sorts, and that is that the most fundamental process along with the *local state*, is what determines the temporal evolution of the *state* of every photon. Therefor the boundaries, as to what is the photon and what is not the photon are dynamically determined by the fundamental process. What is certain is that various cells at particular instants of time may be, in some sense, part of the informational structure of things more than one single photon.

2.1. *Informational characteristics*

Within DTP, we assume that the representation of *information* is by *discrete configurations* of *state* that have *meaning* within a particular *context*. On the other hand *entropy* is a physical scaler that is equal to the *Log* of the number of distinct *configurations* of *state* that, basically, do not have microscopic *meanings* within a particular context such as thermodynamics. The *meaning* of a *configuration* is given by whatever *process* interprets it. Thus a given *configuration* can have multiple *meanings* given different *contexts* and *processes*. The best examples of this are that position and momentum are both represented by a single *configuration* while energy and fine grained temporal *state* are also both represented by a different single *configuration*.

In 1948 Claude Shannon clarified the relationship between a quantity of information and a system with a number of distinguishable *states*, S. For S equally probable *states*, the amount of information is $log_2(S)$. In the case where the S *states* are not equally probably the amount of information is given by $-\sum_{i=1}^{n} P(S_i) log_2 P(S_i)$.

In such cases the unit of our measure of an amount of information is called "Shannon Bit". Thus a system in one of 1024 equally probable *state*s represents 10 Shannon Bits of so called "information" $10 = Log_2(1024)$. A number of Shannon Bits defines a fundamental limit for non-lossy compression of data. For theoretical physics, quantitative measures of the Log of the number of equally probable states of a thermodynamic system are defined as its entropy. In some sense Entropy tells us how much information we do not know about the microscopic *state*. Thus we have refrained from relating "*entropy*" to "*information*" for reasons that should now be clear.

3. Philosophical Speculations

There could be an interesting philosophical consequence if we were to conclude that some kind of DTP accurately models Physics. There would be the inescapable implication that the DTP laws exhibit rational design. However, there is no reason to fret about that possibility and there is no reason to assume any consequent connection to any of various current or historic religious concepts. DTP could certainly open the door to the plausibility of a more rational picture of the cosmogonical issues surrounding Big Bang theories.

The idea that the basic most microscopic mechanisms of a discrete model requires that effective Translational Symmetry is achieved by being, in essence, designed into the fundamental laws as opposed to just being there naturally implies that the microscopic laws of physics reflect a design that could be the product of a purposeful design process. If the concepts discussed herein prove correct, it would suggest that the laws of physics, and the initial conditions (the Big Bang) were, in fact, rationally designed. While humans are the only example of rational thought that we are so far aware of, it is plausible that intelligence has evolved in other parts of the Universe. Rational thought might be more universal than we imagine.

The observations that physics has conservation laws and that the observed universe appears to have come into existence about 14 billion years ago poses an obvious dilemma. We should be willing to admit that there is no sensible model for a Universe simply popping into existence. And aside from matter and energy, where might the Laws of physics come from? If we should discover that a discrete informational process underlies quantum mechanics then there is a more rational model for the origin of our Universe. We may assume that there is some other place, not in this Universe, which we can call "*Other*". The laws of physics in *Other* do not have to have much in common with the laws of physics in this universe. In particular, the necessity of a beginning (such as the big bang) does not have to be a property of *Other*. *Other* may or may not have conservation laws. On the other hand, *Other* is likely to also have a $3+1$ dimensional space-time, but it could have more or fewer dimensions. *Other* would have to be larger (wrt the number of possible *states*) than our Universe. Likely very much larger. One can imagine that in *Other* there exists or has been constructed a very Large Cellular Automaton (LCA) that implements the physics for our universe and that once the initial conditions were set into the LCA the process was put into motion and allowed to evolve undisturbed. The fact that physics is computation universal in the Turing sense can explain at least one purpose for the existence of this Universe. The purpose is likely to be answering a question that cannot be answered analytically. For example, it is doubtful that there is an analytical process that could answer the following question: "Will there ever occur, during 20 billion years of simulation, a specified, compact amount of some particular kind of matter at a temperature below one nano-kelvin?" Where the assumption is that doing so requires purposeful, intelligent, skillful effort without any useful or constructive motivation for making the effort – beyond curiosity.

Humans are the prototype intellectual creatures on this planet. As politics and wars attest, we are not yet able to always think and act rationally. There is the possibility that we might be superseded intellectually, by combinations of the net and possible AI supercomputers of the future. If so, perhaps such future systems will be able to better understand physics, cosmology and cosmogony. It might be better to be told, even by a computer, as opposed to remaining ignorant.

In the meantime, we might be near to our last chance to discover new science on our own.

Chapter 20

The Fastest Way of Computing All Universes

Jürgen Schmidhuber

IDSIA, Galleria 2, 6928 Manno-Lugano, Switzerland
University of Lugano & SUPSI, Switzerland

Is there a short and fast program that can compute the precise history of our universe, including all seemingly random but possibly actually deterministic and pseudo-random quantum fluctuations? There is no physical evidence against this possibility. So let us start searching! We already *know* a short program that computes all constructively computable universes in parallel, each in the asymptotically fastest way. Assuming ours is computed by this optimal method, we can predict that it is among the fastest compatible with our existence. This yields testable predictions.

Note: This paper extends an overview of previous work[51–54,58,59] presented in a survey for the German edition of Scientific American.[61]

1. Introduction

In the 1940s, Konrad Zuse already speculated that our universe is computable by a deterministic computer program (Horst Zuse, personal communication, 2006), like the virtual worlds of today's video games. In 1967 he published the first scientific paper on this idea,[77] soon to be followed by his book *Calculating Space*,[78] focusing on cellular automata as computational devices. We shall see that contrary to common belief, Zuse's hypothesis is compatible with all known observations of quantum physics. Since computable universes are much simpler than non-computable ones, and since one should prefer simple explanations over complex ones, we shall accept his hypothesis as long as there is no evidence to the contrary.

Somewhat surprisingly, there must then exist a very short and in a sense optimally fast algorithm that not only computes the entire history of our own universe, but also those of all other logically possible universes. If the computation of our world indeed is indeed based on such an optimal method, then we may derive non-trivial predictions about its future. I will also briefly discuss some philosophical and theological consequences of this view.

2. Simplicity and Complexity

An object is simple if it has a short description that can be quickly transcribed into the object. For example, the image of a fractal structure[42] may seem complex due to its wealth of detail. But in reality it is simple, as it can be completely generated by a very short and fast program. Therefore it has low algorithmic information or Kolmogorov complexity, defined as the length of the shortest program that computes it.[2,14–16,18,24,25,34–36,38,39,53,66,70,71,76,79] This length hardly depends on the chosen programming language, since programs written in one language can be translated into equivalent programs of another language through a compiler[26,75] of constant, program-independent size.

Is the past and future history of our entire universe simple or complex in this sense? Is there perhaps a very short program that calculates it, including us as observers? This program would have to yield not only the known physical laws but also determine and explain every single seemingly random elementary event. The noblest goal of physics would be to find it.

3. No Problems with Non-Computable Real Numbers

Or is the universe perhaps not computable at all, because it somehow contains or depends on non-computable numbers? As of today there is no compelling reason whatsoever to assume that.

Most physicists are indeed convinced that the universe is quantized by smallest discrete units of time and space and energy. On the other hand, they like to predict macroscopic phenomena using calculus based on the axioms of real numbers, and most real numbers are not even computable (because there are uncountably many real numbers,[12] but only countably many finite programs, such as the non-halting program computing all digits of π). Even quantum physicists who are ready to give up the assumption of a continuous universe usually do take for granted the existence of continuous probability distributions on their discrete universes, and Stephen Hawking explicitly said: *"Although there have been suggestions that space-time may have a discrete structure I see no reason to abandon the continuum theo-*

ries that have been so successful." Note, however, that all physicists in fact have only manipulated discrete symbols, thus generating finite, describable proofs of their results derived from enumerable axioms. That real numbers really *exist* in a way transcending the finite symbol strings used by everybody may be a figment of imagination[52] — compare Brouwer's constructive mathematics[3,7] and the Löwenheim-Skolem Theorem[41,68] which implies that any first order theory with an uncountable model such as the real numbers also has a countable model. As Kronecker put it: *"Die ganze Zahl schuf der liebe Gott, alles Übrige ist Menschenwerk"* ("God created the integers, all else is the work of man"[10]). Kronecker greeted with scepticism Cantor's celebrated insight[12] that there are uncountably many real numbers, mathematical objects Kronecker believed did not even exist.

Anyway, calculus does yield very good macro-level approximations of whatever discrete computable processes may really be happening on the microscopic level.

4. No Problems with Uncertainty Principle

Obviously the universe at least partially obeys simple program-like rules: apples fall to the ground again and again in similar ways; all electrons apparently act the same. Many quantum physicists, however, believe that the history of the universe also includes an incredible number of principally unpredictable, random events on the quantum level.[58] If that were true, then it would *not* have a short description, since truly random, irregular data has maximal Kolmogorov complexity, being incompressible by definition.

Here physicists like to refer to Werner Heisenberg (1901-1976), whose famous uncertainty principle[31] says that an observer cannot simultaneously precisely determine impulse and location of a physical object. For example, to measure the state of an electron, one needs to shoot other particles at it, thus changing its state. To mathematically quantify the resulting uncertainty, quantum mechanics replaces precise deterministic predictions by probabilistic ones. Many physicists believe this uncertainty to be not only a practical measurement problem, but a fundamental property of nature, claiming that God does not obey Albert Einstein's famous quote: *Gott würfelt nicht (God does not play dice)*. According to this view, history would not be pre-determined, and neither compactly describable nor precisely predictable, not even in principle.

It is possible, however, to imagine a computer-generated, pseudo-random,[20] *totally deterministic* world that makes its inhabitants believe that it is partially random and only partially observable, thanks to Heisenberg-like observation limits.[58] A hypothetical programmer of this world could interrupt the computation at any time, dump the current storage into a file, and analyze every little detail, including precise impulse and location of every bitstring-encoded elementary particle.[51] Later he could continue the program's execution, without any internal observer even noticing the pause.

5. No Problems with Bell's Inequality

Quantum physics seems weird. Two entangled particles may be separated by light years, but they somehow seem to immediately "feel" whether one of them is measured, yielding a correlated measurement. Einstein viewed this *spooky action at a distance* as a proof of quantum physics' incompleteness.

A famous inequality of John Stewart Bell (1928-1990) shows that if observers and observations are statistically independent in a certain sense, then there is no local physical rule to explain such spooky effects, even if each particle had unknown internal variables to store information about events that occurred when its entangled particle was still close.[4]

In deterministically computable universes, however, Bell's assumption of independent observers and observations is void and irrelevant. Bell himself was well aware of this.[58]

6. Occam and the Search for the Shortest Program

Most scientists appreciate the rule of William Occam (1280-1347): Among all hypothesis explaining the observations, favor the simplest one. In modern terms: Among all programs reproducing or compressing the observations, favor the shortest one. The principle is widely accepted not only in the inductive sciences such as physics,[34,39] but even in the fine arts.[50] I will later sharpen it a bit, taking into account not only program size[2,14–16,18,24,25,34–36,38,39,53,66,70,71,76,79] but also computation time.[54]

7. What Can be Computed Constructively?

So far we have seen that no physical observations contradict Zuse's hypothesis of a computable universe. Even prominent physicists such as 1999 Nobel laureate Gerard 't Hooft take it seriously.[73] Now we have to clarify, however, what exactly is constructively computable at all.

Let us consider traditional computers that take a binary input program such as 10011010100..., process it by an internal mechanism, and produce a growing number of output bits. The output could encode the evolution of some universe, for example, the total space-time of ours, or even an entire multiverse (many parallel, partially interacting universes) in the sense of Everett.[21,22]

Note that a computed universe history does not have to correspond to incrementally computed *local time steps* like in certain examples provided in my previous publications[51–54]—maybe our standard concepts of time do not even make sense in a given computable universe. But we insist that the output yields a complete representation of every detail of the universe or multiverse in question, without any loss of information.

An additional "viewer program" may facilitate the interpretation of output bitstrings, reminiscent of video games that come with a computer graphics interface to visualize bitstrings in the computer's memory which encode game states.

In traditional computer science, each output bit is viewed as being final and unmodifiable. It turns out, however, that many possible output bitstrings (and thus universes) are compactly describable only if we relax this view, and allow non-halting programs[8,9,23,27,30,32,46,48] to edit their former outputs on occasion[52,53] (compare functions in the *arithmetic hierarchy*[48] and the concept of Δ_n^0-describability, e.g., [39, p. 46-47]).

I defined[52,53] the set of *formally describable* or constructively limit-computable bitstrings x: those x that have a (possibly non-halting) finite program *converging* towards x — after some time each bit of x has to stop changing, that is, each prefix of x becomes fixed after finite (but in general unknowable) time.

For example, let us us assume the n-th output bit is 1 if the n-th program in a list of all possible programs halts, where n is a natural number. This output sequence has a very compact input program which systematically enumerates all possible programs and runs them in interleaving fashion; whenever a program in the list (say, the m-th) halts, the m-th output bit (initialized by 0) becomes 1. Every prefix of the infinite output will converge at some point. But we do not know when, otherwise we could solve the generally unsolvable halting problem.[26,75]

It turns out that a given universe such as ours might have a very short explanation or description on a machine that can edit its former outputs, but not on a traditional machine. In fact, there are more or less powerful variants of output-editing machines which vary in their expressiveness, some being able to compactly encode certain universes that need long codes on others. For example, the enumerable "number of wisdom" Ω[11,16,69,72] cannot be compressed on traditional Turing Machines,[75] but on so-called *Enumerable Output Machines*.[45,52,53] These distinctions are technically very important, but not central to the present overview; the interested reader is referred to.[52,53]

8. No Problem with Talk About Incomputable Things

Observers inhabiting a computable universe may talk about mathematical paradoxons and things that are incomputable in a sense, such as the halting probability of a universal Turing machine, which is closely related to Gödel's incompleteness theorem.[11,16,26,69,72,75] This does not involve any inconsistencies.[51] For example, the processes that correspond to our brain firing patterns and the sound waves they provoke by controlling our voices may still correspond to computable substrings of our universe's evolution. The same holds for talk about inconsistent worlds in which, say, time travel is possible.

9. The Fastest Way of Computing All Universes

In 1996 I pointed out that there is a very short algorithm that computes all possible universes, as long as they are computable.[51] In a certain sense this (non-halting) algorithm is also extremely fast, as I emphasized in 2000.[52,54] Let me write down a variant that does not consume excessive storage space (here $l(p)$ denotes the length of program p, a bitstring):

Algorithm 1 Algorithm FAST
   for $i := 1, 2, \ldots$ **do**
      Run each program p with $l(p) \leq i$ for at most $2^{i-l(p)}$ steps and reset storage modified by p
   end for

That is, in phase i, FAST generates all universes computable by some program p satisfying $l(p) + log\, t(p) \leq i$, where $t(p)$ is the runtime of p, and log denotes the binary logarithm. True, phase $i + 1$ will repeat everything done in phase i, but that is not an essential efficiency problem: every phase costs roughly as much as all previous phases taken together, that is, we lose only a factor of 2 or so of computation time, but gain a lot by not having to store all intermediate results of previously executed, only partially finished programs, which would cost exponentially growing storage space.

It is easy to see that FAST will generate the n-th bit of each universe as quickly as if it were computed by this universe's fastest program, save for a constant factor that does not depend on n. Following standard practice of theoretical computer science, we may therefore call FAST the *asymptotically fastest* way of computing all computable universes. For any God-like Great Programmer,[51] FAST offers a natural, optimally efficient way of computing all logically possible worlds.

If our universe is one of the computable ones, then FAST will eventually produce a detailed representation of its first few billion years of local time (note that nearly 14 billion years have passed since the big bang).

10. The Fastest and Truest Version of Our World?

Since there are many programs computing one and the same universe (history), our optimal algorithm FAST (Section 9) will generate many copies of ours, and many histories that start like ours (but possibly continue in different ways). At any given time in the execution of FAST, the most advanced copies will be those computable by short and fast programs. Since we exist, we already know that at least one of the programs has computed enough to enable our existence, following the weak anthropic principle.[1,13] But which of the many? A little bit of thought shows: With high probability it will be one of the shortest and fastest compatible with our existence! For a more detailed analysis, see previous work.[52,54]

Following this argumentation, we are already part of one of the simplest, fastest, non-random worlds compatible with our very being, simply because even the optimal FAST needs much more time to compute truly random events as parts of any universe's history. Computationally, randomness is extremely expensive in terms of both time and space. It does not fit the Occam's razor criterion at all.

But even if our universe's history included a huge number of truly random quantum events, one question would arise immediately: Besides the physical laws, which is the simplest and fastest pseudo-random generator needed to compute a *similar*, less random world? In a philosophical sense, wouldn't this world be the *truest* version of our world, reflecting its true essence, thanks to its lack of arbitrariness?

11. Predictions Based on the Fastest Way

If whoever is generating our universe is using algorithm FAST (Section 9) to deal with computational resource constraints in an optimal way, we can make non-trivial predictions.

For example, all seemingly random events (such as beta decay of neutrons) actually must follow some pseudo-random rule, waiting to be discovered by some grad student at CERN or elsewhere. Perhaps current physicists are like observers seeing the second billion digits in the decimal expansion of π, which at first glance look very random (for example, every 3 digit sequence occurs roughly once in a thousand 3 digit subsequences) but is actually highly regular, since it can be computed by a short program.

One somewhat depressing prediction is that quantum computation, a subject of much current excitement,[5,19,40,44] will never work well and never scale to large problems. Sure, FAST will run many programs that compute multiverse-like universes, obeying known laws of quantum mechanics and allowing for quantum computers (which can be fully simulated on traditional computers). However, the FAST-generated programs that compute our history so far *and* permit the expected effects of quantum computing will cost much more computational effort than others that are also computing our history in a less computationally expensive way. That is, under FAST they are very unlikely. That is, it is very unlikely that we are inhabiting a multiverse where quantum computation will be able to solve non-trivial problems. A pity!

I first made this prediction a decade ago.[52] Since then, nobody has been able to make quantum computation scale. For example, the biggest number to be factored by any existing quantum computer is still 15.

12. How to Find our Universe's Program

Algorithm FAST (Section 9) computes all universes, not just ours. But what we'd really like to know is the program that computes ours and nothing else. That would be the world's essential formula, the holy grail of theoretical physics. How to find it? It turns out that the optimal way of searching for it is closely related to FAST. It goes like this:

Take any sequence of physical observations, and run FAST until one of the executed programs (written in a universal programming language) reproduces the data.

This is essentially Levin's universal search algorithm[37] applied to physics. Since it is only asymptotically optimal, it can be greatly accelerated under certain conditions by methods such as the Optimal Ordered Problem Solver,[56] which may use *partial*, incomplete reproductions of the data as intermediate subgoals, and then continue the search by re-using previous subgoal-achieving programs, thus possibly dramatically reducing the constant slowdown ignored by the asymptotic notion of optimality.[56]

13. Always Slower Than the Universe Itself

Note that if somebody indeed found the shortest and fastest program of our world, this would not necessarily help to figure out the future faster than by waiting for it happen. The computer on which to run this program would have to be built within our universe, and as a small part of the latter would be unable to run as fast as the universe itself.

14. Math v Computation?

Rather than pursuing the computability-oriented path layed out in,[51] Tegmark (back then at LMU Munich) suggested what at first glance seems to be an alternative ensemble of possible universes based on an informally defined set of "self-consistent mathematical structures"[74] — compare also Marchal's and Bostrom's theses.[6,43] It is not quite clear whether Tegmark wanted to include universes that are *not* formally describable according to our definition mentioned in Section 7. It is well-known, however, that for any set of mathematical axioms there is a program that lists all provable theorems in order of the lengths of their shortest proofs encoded as bitstrings. Hence Tegmark's view[74] seems in a certain sense encompassed by the algorithmic approach.[51] The latter offers several conceptual advantages though: (1) It provides the appropriate framework for issues of information-theoretic complexity traditionally ignored in pure mathematics, and imposes natural complexity-based orderings on the possible universes and subsets thereof.[51–53] (2) It taps into a rich source of theoretical insights on computable probability distributions relevant for establishing priors on possible universes. Such priors are needed for making probabilistic predictions concerning our own particular universe.[51–53] Although Tegmark suggests that "... *all mathematical structures are a priori given equal statistical weight*" (Ref. 74, p. 27), there is no way of assigning equal nonvanishing probability to all (infinitely many) mathematical structures. Hence we really need something like the complexity-based weightings discussed in in earlier papers.[51-53] (3) The algorithmic approach is the obvious framework for questions of temporal complexity such as those discussed in this paper, e.g., "what is the most efficient way of simulating all universes?"[52,54]

15. Optimal Artificial Intelligence in Computable Universes

The fully self-referential[26] Gödel machine[63] is a Universal Artificial Intelligence (AI)[55,57,60,62,64] that is at least theoretically optimal in a certain sense. It may interact with some initially unknown, partially observable environment to maximize future expected utility or reward by solving arbitrary user-defined computational tasks. Its initial algorithm is not hardwired; it can completely rewrite itself without essential limits apart from the limits of computability, provided a proof searcher embedded within the initial algorithm can first prove that the rewrite is useful, according to the formalized utility function taking into account the limited computational resources. Self-rewrites may modify / improve the proof searcher itself, and can be shown to be *globally optimal*, relative to Gödel's well-known fundamental restrictions of provability.[26] To make sure the Gödel machine is at least *asymptotically* optimal even before the first self-rewrite, we may initialize it by Hutter's non-self-referential but *asymptotically fastest algorithm for all well-defined problems* Hsearch,[33] which uses a hardwired brute force proof searcher and (justifiably) ignores the costs of proof search. Assuming discrete input/output domains $X/Y \subset B^*$, a formal problem specification $f : X \to Y$ (say, a functional description of how integers are decomposed into their prime factors), and a particular $x \in X$ (say, an integer to be factorized), Hsearch orders all proofs of an appropriate axiomatic system by size to find programs q that for all $z \in X$ provably compute $f(z)$ within time bound $t_q(z)$. Simultaneously it spends most of its time on executing the q with the best currently proven time bound $t_q(x)$. Remarkably, Hsearch is as fast as the *fastest* algorithm that provably computes $f(z)$ for all $z \in X$, save for a constant factor smaller than $1+\epsilon$ (arbitrary real-valued $\epsilon > 0$) and an f-specific but x-independent additive constant.[33] Given some problem, the Gödel machine may decide to replace its Hsearch initialization by a faster method suffering less from large constant overhead, but even if it doesn't, its performance won't be less than asymptotically optimal.

All of this implies that there already exists the blueprint of a Universal AI which will solve almost all problems almost as quickly as if it already knew the best (unknown) algorithm for solving them, because almost all imaginable problems are big enough to make the additive constant negligible.

The only motivation for *not* quitting computer science research right now is that many real-world problems are so small and simple that the ominous constant slowdown (potentially relevant at least before the first Gödel machine self-rewrite) is *not* negligible. Nevertheless, the ongoing efforts at scaling universal AIs down to the rather few *small* problems are very much informed by the new millennium's theoretical insights[57,60,62,64] mentioned above, and may soon yield practically feasible yet still general problem solvers for physical systems with highly restricted computational power, say, a few trillion instructions per second, roughly comparable to a human brain power.

Simultaneously, our non-universal but still rather general fast deep / recurrent neural networks have already started to outperform traditional pre-programmed methods: they recently collected a string of 1st ranks in many important visual pattern recognition benchmarks, e.g., IJCNN traffic sign competition, NORB, CIFAR10, MNIST, three ICDAR handwriting competitions.[17,29,65] Here we greatly profit from ongoing advances in computing hardware, using GPUs (mini-supercomputers normally used for video games) 100 times faster than today's CPU cores, and a million times faster than PCs of 20 years ago, complementing the recent abovementioned progress in the theory of mathematically optimal universal problem solvers.[65]

16. Potential Criticism

Philosophers tend to create theories inspired by recent scientific developments. For instance, Heisenberg's uncertainty principle and Gödel's incompleteness theorem greatly influenced modern philosophy. Are algorithmic Theories of Everything (TOEs) and the "Great Programmer Religion"[51,52] just another reaction to recent developments, some in hindsight obvious byproduct of the advent of good virtual reality? (As they say: *For a man with a hammer, everything looks like a nail.*) Will they soon become obsolete, as so many previous philosophies? I find it hard to imagine so, even without a boost to be expected for algorithmic TOEs in case someone should indeed discover a simple subroutine responsible for certain physical events hitherto believed to be irregular. After all, algorithmic theories of the describable do encompass everything we will ever be able to talk and write about. Other things are simply beyond description.

17. Who Can Accept the Computable Real World?

Many researchers in the field of artificial life simulate the evolution of artificial beings adapting to their artificial environments, e.g.,.[28,47,67] Most of them do not have any problems with the idea of a computable real world in Zuse's sense. After all, a good simulation is not distinguishable from reality. Children with experience in virtual realities and video games also tend to find the idea of a computable universe more natural than their parents.

In our universe the raw computational power per cent will keep increasing by a factor of 100-1000 per decade, with no end in sight. As a consequence, realism and appeal of virtual realities will keep increasing dramatically, making the presented thoughts[51–54] more and more acceptable for the masses.

Remarkably, it is especially the quantum physicists who sometimes reject such ideas,[58] albeit without being able to justify their scepticism too well by facts.

Einstein, perhaps the greatest of all physicists, did not believe in nondeterminism, as already mentioned. For a long time his view has been unpopular among quantum physicists. But now it does not seem unreasonable to predict that it will experience a rennaissance. First, because there is no physical evidence against it. Second, because it greatly simplifies the description of the world's history in the framework of computability theory, without necessitating a gigantic amount of information for describing a vast number of truly random quantum-level events.

As long as nobody can show that the universe is indeed partially random, scientists are obliged to search for a short program that computes all the apparent randomness and therefore reveals it as pseudo-randomness.[58] If the process that calculates us makes optimally efficient use of the resources of some higher-level universe, we should expect this program to be not only short but also fast.[52,54]

18. Consequences for Philosophy and Theology

The theory of computable universes provides a purely rational and technologically oriented access to basic questions of philosophy and theology.[51]

At least in principle, everybody could become some sort of God by programming the algorithm FAST (Section 9) on a computer, systematically creating all constructively computable universes, including ours.

In some of them, programmers occasionally will intervene in the worlds computed by their programs, reminiscent of well-known religious role models. In some, the computable contents of simulated brains will be occasionally copied from one computable world to another, implementing variants of heaven or hell.

Beings evolved in some of the simulated universes will again build computers to simulate universes, in recursively nested fashion.[51] This begs the question: Where does the computer of the top universe in the hierarchy come from? It must remain open for now.

The fact that there are mathematically optimal ways of creating and computing all the logically possible worlds, however, opens a new and exciting field hardly discussed in today's mainstream philosophy and theology.

Acknowledgements

At the age of 17 my brother Christof Schmidhuber declared that the universe is *sum of all math*, inhabited by observers who are mathematical substructures (private communication, Munich, 1981). As he went on to become a theoretical physicist at LMU Munich, discussions with him about the relation between superstrings and bitstrings became a source of inspiration for writing both the first paper[51] and later ones[52-54] based on computational complexity theory, which seems to provide the natural setting for his more math-oriented ideas (private communication, Munich 1981-86; Caltech 1987-93; Princeton 1994-96; Berne/Geneva 1997–; compare his notion of *"mathscape"*[49]). I believe that Christof's early discussions with Munich-based scientists and students were the reason why such ideas emerged in Munich. Furthermore, his 1997 remarks on similarities and differences between Feynman path integrals and "the sum of all computable universes" and his resulting dissatisfaction with the lack of a discussion of temporal aspects in the 1996 paper[51] triggered the papers[52,54] on temporal complexity.

References

1. J. D. Barrow and F. J. Tipler. *The Anthropic Cosmological Principle*. Clarendon Press, Oxford, 1986.
2. Y. M. Barzdin. Algorithmic information theory. In *Encyclopaedia of Mathematics*, volume 1, pages 140–142. Reidel, Kluwer Academic Publishers, 1988.
3. M. Beeson. *Foundations of Constructive Mathematics*. Springer-Verlag, Heidelberg, 1985.
4. J. S. Bell. On the problem of hidden variables in quantum mechanics. *Rev. Mod. Phys.*, 38:447–452, 1966.

5. C. H. Bennett and D. P. DiVicenzo. Quantum information and computation. *Nature*, 404(6775):256–259, 2000.
6. N. Bostrom. Observational selection effects and probability. Dissertation, Dept. of Philosophy, Logic and Scientific Method, London School of Economics, 2000.
7. L. E. J. Brouwer. Over de Grondslagen der Wiskunde. Dissertation, Doctoral Thesis, University of Amsterdam, 1907.
8. M. S. Burgin. Inductive Turing machines. *Notices of the Academy of Sciences of the USSR (translated from Russian)*, 270(6):1289–1293, 1991.
9. M. S. Burgin and Y. M. Borodyanskii. Infinite processes and super-recursive algorithms. *Notices of the Academy of Sciences of the USSR (translated from Russian)*, 321(5):800–803, 1991.
10. F. Cajori. *History of mathematics (2nd edition)*. Macmillan, New York, 1919.
11. C. S. Calude. Chaitin Ω numbers, Solovay machines and Gödel incompleteness. *Theoretical Computer Science*, 2001.
12. G. Cantor. Über eine Eigenschaft des Inbegriffes aller reellen algebraischen Zahlen. *Crelle's Journal für Mathematik*, 77:258–263, 1874.
13. B. Carter. Large number coincidences and the anthropic principle in cosmology. In M. S. Longair, editor, *Proceedings of the IAU Symposium 63*, pages 291–298. Reidel, Dordrecht, 1974.
14. G. J. Chaitin. On the length of programs for computing finite binary sequences: statistical considerations. *Journal of the ACM*, 16:145–159, 1969.
15. G. J. Chaitin. A theory of program size formally identical to information theory. *Journal of the ACM*, 22:329–340, 1975.
16. G. J. Chaitin. *Algorithmic Information Theory*. Cambridge University Press, Cambridge, 1987.
17. D. C. Ciresan, U. Meier, J. Masci, L. M. Gambardella, and J. Schmidhuber. Flexible, high performance convolutional neural networks for image classification. In *Intl. Joint Conference on Artificial Intelligence IJCAI*, pages 1237–1242, 2011.
18. T. M. Cover, P. Gács, and R. M. Gray. Kolmogorov's contributions to information theory and algorithmic complexity. *Annals of Probability Theory*, 17:840–865, 1989.
19. D. Deutsch. *The Fabric of Reality*. Allen Lane, New York, NY, 1997.
20. T. Erber and S. Putterman. Randomness in quantum mechanics – nature's ultimate cryptogram? *Nature*, 318(7):41–43, 1985.
21. H. Everett III. 'Relative State' formulation of quantum mechanics. *Reviews of Modern Physics*, 29:454–462, 1957.
22. H. Everett, III. The many-worlds interpretation of quantum mechanics. Princeton University Press, Princeton, 1986.
23. R. V. Freyvald. Functions and functionals computable in the limit. *Transactions of Latvijas Vlasts Univ. Zinatn. Raksti*, 210:6–19, 1977.
24. P. Gács. On the symmetry of algorithmic information. *Soviet Math. Dokl.*, 15:1477–1480, 1974.
25. P. Gács. On the relation between descriptional complexity and algorithmic probability. *Theoretical Computer Science*, 22:71–93, 1983.

26. K. Gödel. Über formal unentscheidbare Sätze der Principia Mathematica und verwandter Systeme I. *Monatshefte für Mathematik und Physik*, 38:173–198, 1931.
27. E. M. Gold. Limiting recursion. *Journal of Symbolic Logic*, 30(1):28–46, 1965.
28. F. J. Gomez, J. Schmidhuber, and R. Miikkulainen. Efficient non-linear control through neuroevolution. *Journal of Machine Learning Research JMLR*, 9:937–965, 2008.
29. Alex Graves and Jüergen Schmidhuber. Offline handwriting recognition with multidimensional recurrent neural networks. In *Advances in Neural Information Processing Systems 21*. MIT Press, Cambridge, MA, 2009.
30. A. Gregorczyk. On the definitions of computable real continuous functions. *Fundamenta Mathematicae*, 44:61–71, 1957.
31. W. Heisenberg. Über den anschaulichen Inhalt der quantentheoretischen Kinematik und Mechanik. *Zeitschrift für Physik*, 33:879–893, 1925.
32. G. Hotz, G. Vierke, and B. Schieffer. Analytic machines. Technical Report TR95-025, Electronic Colloquium on Computational Complexity, 1995. http://www.eccc.uni-trier.de/eccc/.
33. M. Hutter. The fastest and shortest algorithm for all well-defined problems. *International Journal of Foundations of Computer Science*, 13(3):431–443, 2002. (On J. Schmidhuber's SNF grant 20-61847).
34. M. Hutter. *Universal Artificial Intelligence: Sequential Decisions based on Algorithmic Probability*. Springer, Berlin, 2005. (On J. Schmidhuber's SNF grant 20-61847).
35. A. N. Kolmogorov. Three approaches to the quantitative definition of information. *Problems of Information Transmission*, 1:1–11, 1965.
36. L. A. Levin. On the notion of a random sequence. *Soviet Math. Dokl.*, 14(5):1413–1416, 1973.
37. L. A. Levin. Universal sequential search problems. *Problems of Information Transmission*, 9(3):265–266, 1973.
38. L. A. Levin. Laws of information (nongrowth) and aspects of the foundation of probability theory. *Problems of Information Transmission*, 10(3):206–210, 1974.
39. M. Li and P. M. B. Vitányi. *An Introduction to Kolmogorov Complexity and its Applications (2nd edition)*. Springer, 1997.
40. S. Lloyd. Ultimate physical limits to computation. *Nature*, 406:1047–1054, 2000.
41. L. Löwenheim. Über Möglichkeiten im Relativkalkül. *Mathematische Annalen*, 76:447–470, 1915.
42. B. Mandelbrot. *The Fractal Geometry of Nature*. Freeman and Co., San Francisco, 1982.
43. B. Marchal. *Calculabilité, Physique et Cognition*. PhD thesis, L'Université des Sciences et Technologies De Lilles, 1998.
44. R. Penrose. *The Emperor's New Mind*. Oxford University Press, 1989.
45. J. Poland. A coding theorem for enumerable output machines. *Information Processing Letters*, 91(4):157–161, 2004.

46. H. Putnam. Trial and error predicates and the solution to a problem of Mostowski. *Journal of Symbolic Logic*, 30(1):49–57, 1965.
47. T. S. Ray. An approach to the synthesis of life. In C. G. Langton, C. Taylor, J. D. Farmer, and S. Rasmussen, editors, *Artificial Life II*, pages 371–408. Addison Wesley Publishing Company, 1992.
48. H. Rogers, Jr. *Theory of Recursive Functions and Effective Computability*. McGraw-Hill, New York, 1967.
49. C. Schmidhuber. Strings from logic. Technical Report CERN-TH/2000-316, CERN, Theory Division, 2000. http://xxx.lanl.gov/abs/hep-th/0011065.
50. J. Schmidhuber. Low-complexity art. *Leonardo, Journal of the International Society for the Arts, Sciences, and Technology*, 30(2):97–103, 1997.
51. J. Schmidhuber. A computer scientist's view of life, the universe, and everything. In C. Freksa, M. Jantzen, and R. Valk, editors, *Foundations of Computer Science: Potential - Theory - Cognition*, volume 1337, pages 201–208. Lecture Notes in Computer Science, Springer, Berlin, 1997, submitted 1996.
52. J. Schmidhuber. Algorithmic theories of everything. Technical Report IDSIA-20-00, quant-ph/0011122, IDSIA, Manno (Lugano), Switzerland, 2000. Sections 1-5: see Ref. 53; Section 6: see Ref. 54.
53. J. Schmidhuber. Hierarchies of generalized Kolmogorov complexities and nonenumerable universal measures computable in the limit. *International Journal of Foundations of Computer Science*, 13(4):587–612, 2002.
54. J. Schmidhuber. The Speed Prior: a new simplicity measure yielding near-optimal computable predictions. In J. Kivinen and R. H. Sloan, editors, *Proceedings of the 15th Annual Conference on Computational Learning Theory (COLT 2002)*, Lecture Notes in Artificial Intelligence, pages 216–228. Springer, Sydney, Australia, 2002.
55. J. Schmidhuber. Towards solving the grand problem of AI. In P. Quaresma, A. Dourado, E. Costa, and J. F. Costa, editors, *Soft Computing and complex systems*, pages 77–97. Centro Internacional de Mathematica, Coimbra, Portugal, 2003. Based on Ref. 57.
56. J. Schmidhuber. Optimal ordered problem solver. *Machine Learning*, 54:211–254, 2004.
57. J. Schmidhuber. The new AI: General & sound & relevant for physics. In B. Goertzel and C. Pennachin, editors, *Artificial General Intelligence*, pages 175–198. Springer, 2006. Also available as TR IDSIA-04-03, arXiv:cs.AI/0302012.
58. J. Schmidhuber. Randomness in physics. *Nature*, 439(3):392, 2006. Correspondence.
59. J. Schmidhuber. The Computational Universe. Review of *Programming the Universe: A Quantum Computer Scientist Takes on the Cosmos*, by S. Lloyd. *American Scientist*, 2006.
60. J. Schmidhuber. 2006: Celebrating 75 years of AI - history and outlook: the next 25 years. In M. Lungarella, F. Iida, J. Bongard, and R. Pfeifer, editors, *50 Years of Artificial Intelligence*, volume LNAI 4850, pages 29–41. Springer Berlin / Heidelberg, 2007. Preprint available as arXiv:0708.4311.

61. J. Schmidhuber. Alle berechenbaren Universen (All computable universes). *Spektrum der Wissenschaft Spezial (German edition of Scientific American)*, (3):75–79, 2007.
62. J. Schmidhuber. New millennium AI and the convergence of history. In W. Duch and J. Mandziuk, editors, *Challenges to Computational Intelligence*, volume 63, pages 15–36. Studies in Computational Intelligence, Springer, 2007. Also available as arXiv:cs.AI/0606081.
63. J. Schmidhuber. Ultimate cognition à la Gödel. *Cognitive Computation*, 1(2):177–193, 2009.
64. J. Schmidhuber. The new AI is general and mathematically rigorous. *Front. Electr. Electron. Eng. China*, 2010.
65. J. Schmidhuber, D. Ciresan, U. Meier, J. Masci, and A. Graves. On fast deep nets for AGI vision. In *Fourth Conference on Artificial General Intelligence (AGI)*, 2011.
66. C. P. Schnorr. Process complexity and effective random tests. *Journal of Computer Systems Science*, 7:376–388, 1973.
67. K. Sims. Evolving virtual creatures. In Andrew Glassner, editor, *Proceedings of SIGGRAPH '94 (Orlando, Florida, July 1994)*, Computer Graphics Proceedings, Annual Conference, pages 15–22. ACM SIGGRAPH, ACM Press, jul 1994. ISBN 0-89791-667-0.
68. T. Skolem. Logisch-kombinatorische Untersuchungen über Erfüllbarkeit oder Beweisbarkeit mathematischer Sätze nebst einem Theorem über dichte Mengen. *Skrifter utgit av Videnskapsselskapet in Kristiania, I, Mat.-Nat. Kl.*, N4:1–36, 1919.
69. T. Slaman. Randomness and recursive enumerability. Technical report, Univ. of California, Berkeley, 1999. Preprint, http://www.math.berkeley.edu/~slaman.
70. R. J. Solomonoff. A formal theory of inductive inference. Part I. *Information and Control*, 7:1–22, 1964.
71. R. J. Solomonoff. Complexity-based induction systems. *IEEE Transactions on Information Theory*, IT-24(5):422–432, 1978.
72. R. M. Solovay. A version of Ω for which ZFC can not predict a single bit. In C. S. Calude and G. Păun, editors, *Finite Versus Infinite. Contributions to an Eternal Dilemma*, pages 323–334. Springer, London, 2000.
73. G. 't Hooft. Quantum gravity as a dissipative deterministic system. Technical Report SPIN-1999/07/gr-gc/9903084, http://xxx.lanl.gov/abs/gr-qc/9903084, Institute for Theoretical Physics, Univ. of Utrecht, and Spinoza Institute, Netherlands, 1999. Also published in *Classical and Quantum Gravity 16*, 3263.
74. M. Tegmark. Is "the theory of everything" merely the ultimate ensemble theory? *Annals of Physics*, 270:1–51, 1998.
75. A. M. Turing. On computable numbers, with an application to the Entscheidungsproblem. *Proceedings of the London Mathematical Society, Series 2*, 41:230–267, 1936.
76. V. A. Uspensky. Complexity and entropy: an introduction to the theory of Kolmogorov complexity. In O. Watanabe, editor, *Kolmogorov complexity and*

computational complexity, pages 85–102. EATCS Monographs on Theoretical Computer Science, Springer, 1992.
77. K. Zuse. Rechnender Raum. *Elektronische Datenverarbeitung*, 8:336–344, 1967.
78. K. Zuse. *Rechnender Raum*. Friedrich Vieweg & Sohn, Braunschweig, 1969. English translation: *Calculating Space*, MIT Technical Translation AZT-70-164-GEMIT, Massachusetts Institute of Technology (Proj. MAC), Cambridge, Mass. 02139, Feb. 1970.
79. A. K. Zvonkin and L. A. Levin. The complexity of finite objects and the algorithmic concepts of information and randomness. *Russian Math. Surveys*, 25(6):83–124, 1970.

Chapter 21

The Subjective Computable Universe

Marcus Hutter

*Research School of Computer Science, Australian National University &
Department of Computer Science, ETH Zürich, Switzerland*

Nearly all theories developed for our world are computational. The fundamental theories in physics can be used to emulate on a computer ever more aspects of our universe. This and the ubiquity of computers and virtual realities has increased the acceptance of the computational paradigm. A computable theory of everything seems to have come within reach. Given the historic progression of theories from ego- to geo- to helio-centric models to universe and multiverse theories, the next natural step was to postulate a multiverse composed of *all* computable universes. Unfortunately, rather than being a theory of everything, the result is more a theory of nothing, which actually plagues all too-large universe models in which observers occupy random or remote locations. The problem can be solved by incorporating the subjective observer process into the theory. While the computational paradigm exposes a fundamental problem of large-universe theories, it also provides its solution.

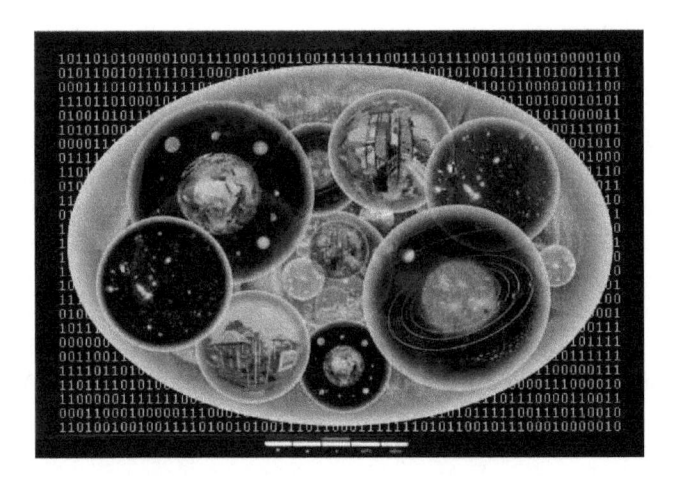

1. Introduction

The idea of a mechanical universe is quite old and has been expressed by many scholars, including Leibniz and Newton,[4] Zuse,[25] Schmidhuber,[17] Wolfram,[24] and many others. With computers and virtual realities increasingly pervading our everyday life, the computable universe metaphor seems to have become ever more accepted. The question of *which* computer program governs our universe, though, remains. Actually it is "just" the age-old quest for a theory of everything (ToE) in a new guise. The new emphasis on computability leads to novel insights and fundamentally new possibilities, but also exhibits new problems.

For instance, consider the simple "All-a-Carte" program that enumerates all natural numbers in binary: 1 10 11 100 101 110 111 1000 ... Assume our space-time universe is finite and somehow coded into a gargantuan but finite bit string. This string will appear somewhere in the above enumeration, hence formally, this simple All-a-Carte program is a theory of everything, but on closer inspection it actually is more a theory of nothing.[20] While this particular theory might simply be dismissed as nonsense, serious proposals of modern multiverse theories like Wheeler's oscillating universe, Smolin's baby universe theory, Everett's many-worlds interpretation of quantum theory, the different compactifications of string theory, and inflationary theories with variable fundamental constants[22] start to get contaminated by similar philosophical problems. The above All-a-Carte model just elucidates the problem in naked and intensified form with all the complex physics stripped off.

The focus of this article is to exploit the opportunities the computable universe paradigm offers, but at the same time avoid its pitfalls.

Section 2 starts with the philosophical prerequisites: how theories or models explain or describe observations, the relation between data compression and prediction, how to avoid difficult epistemological questions about knowledge via the bit-string ontology, how Ockham's razor and information theory solve the induction problem, and the role of observers and their localization in theories of everything. In order to make the main point of this article clear, Section 3 first traverses a number of models that have been suggested for our world, from generally accepted to increasingly speculative and questionable theories, and discusses their relative merits, in particular their predictive power (precision and coverage). We will see that localizing the observer, which is usually not regarded as an issue, can be very important. Section 4 gives an informal introduction to the necessary ingredients for Complete ToEs (CToEs), and how to evaluate and compare them using a quantified instantiation of Ockham's razor. Section 5 gives a slightly more formal definition of what accounts for a CToE, introduces more realistic observers with limited perception ability, formalizes the CToE selection principle, and discusses extensions to more realistic limited theories (rather than ToEs). Section 6 summarizes and discusses the assumptions underlying the CToE selection principle, and Section 7 concludes.

An extended version of this article with technical details has been published in.[12]

2. Philosophical Background

This article describes an *information-theoretic* and *computational* approach for addressing the *philosophical* problem of judging theories (of everything) in *physics*. The philosophical prerequisites are introduced in this section. In order to keep it generally accessible, I've tried to minimize jargon, and focus on the core problems and their solution.

Theories/models. By *theory* I mean any *model* which can explain ≈ describe ≈ predict ≈ compress[10] our observations, whatever the form of the model. Scientists often say that their model *explains* some phenomenon. What is usually meant is that the model *describes* (the relevant aspects of) the observations more compactly than the raw data. The model is then regarded as capturing a law (of nature), which is believed to hold true also for unseen/future data.

Induction. This process of inferring general conclusions from example instances is called *inductive reasoning*. For instance, observing 1000 black ravens but no white one supports but cannot prove the hypothesis that all ravens are black. In general, induction is used to find properties or rules or models of past observations. The ultimate purpose of the induced models is to use them for making predictions, e.g. that the next observed raven will also be black. Arguably inductive reasoning is even more important than deductive reasoning in science and everyday life: for scientific discovery, in machine learning,[13] for forecasting in economics, as a philosophical discipline, in common-sense decision making, and last but not least to find theories of everything. Historically, some famous, but apparently misguided philosophers,[5,21] including Popper and Miller, even disputed the existence, necessity or validity of inductive reasoning. Meanwhile it is well-known how minimum encoding length principles,[7,23] rooted in (algorithmic) information theory,[11] quantify Ockham's razor principle, and led to a solid pragmatic foundation of inductive reasoning.[16] Essentially, one can show that the more one can *compress*, the better one can *predict*, and vice versa.

Theory=model=compressed data. A deterministic theory/model allows from initial conditions to determine an observation sequence, which could be coded as a bit string. For instance, Newton mechanics maps initial planet positions+velocities into a time-series of planet positions. So a deterministic model with initial conditions is "just" a compact representation of an infinite observation string. A stochastic model is "just" a probability distribution over observation strings.

(Compete) theories (of everything). Classical models in physics are essentially differential equations describing the time-evolution of some aspects of the world. A Theory of Everything (ToE) models the whole universe or multiverse, which should include initial conditions. As I will argue, it can be crucial to also localize the observer, i.e. to augment the ToE with a model of the properties of the observer, even for non-quantum-mechanical phenomena. I call a ToE with observer localization, a *Complete ToE* (CToE).

The role of observers in previous theories. That the observer itself is important in describing our world is well-known. Most prominently in quantum mechanics, the observer plays an active role in 'collapsing the wave function'. This is a specific and relatively well-defined role of the observer for a particular theory, which is *not* my concern. I will show that (even the localization of) the observer is indispensable for *finding* or developing *any* (useful) ToE. Often, the anthropic principle is invoked for this purpose (our universe is as it is because otherwise we would not exist). Unfortunately its current use is rather vague and limited, if not outright unscientific.[19] It is possible to give a precise and formal account of observers by explicitly separating the observer's subjective experience from the objectively existing universe or multiverse, which besides other things, as pointed out in,[20] shows that we also need to localize the observer within our universe (not only which universe the observer is in).

Epistemology. To facilitate this, I will assume that the observers' experience of the world consists of a single temporal binary sequence which gets longer with time. This is definitely true if the observer is a robot equipped with sensors like a video camera whose signal is converted to a digital data stream, fed into a digital computer and stored in a binary file of increasing length. In humans, the signal transmitted by the optic and other sensory nerves could play the role of the digital data stream. Of course (most) human observers do not possess photographic memory. We can deal with this limitation in various ways: digitally record and make accessible upon request the nerve signals from birth till now, or allow for uncertain or partially remembered observations. Classical philosophical theories of knowledge[1] (e.g. as justified true belief) operate on a much higher conceptual level and therefore require stronger (and hence more disputable) philosophical presuppositions. In my minimalist "spartan" information-theoretic epistemology, a bit-string is the only observation, and all higher ontologies are constructed from it and are pure "imagination".

3. Predictive Power & Observer Localization

A number of models have been suggested for our world. They range from generally accepted to increasingly speculative to apparently bogus. For the purpose of this work it doesn't matter where you personally draw the line. Many now generally accepted theories have once been regarded as insane, so using the scientific community or general public as a judge is problematic and can lead to endless discussions: for instance, the historic geo↔heliocentric battle; and the ongoing discussion of whether string theory is a theory of everything or more a theory of nothing. In a sense this article is about a formal rational criterion to determine whether a model makes sense or not. In order to make the main point of this article clear, below I will briefly traverse a number of models.[2,8,12] The presented bogus models help to make clear the necessity of observer localization and hence the relevance of this article.

Egocentric to Geocentric model. A young child believes it is the center of the world. Localization is trivial. It is always at "coordinate" (0,0,0). Later it learns that it is just one among a few billion other people and as little or much special as any other person thinks of themselves. In a sense we replace our egocentric coordinate system by one with origin (0,0,0) in the center of Earth. The move away from an egocentric world view has many social advantages, but dis-answers one question: Why am I this particular person and not any other?

Geocentric to Heliocentric model. While being expelled from the center of the world as an individual, in the geocentric model at least the human race as a whole remains in the center of the world, with the remaining (dead?) universe revolving around us. The heliocentric model puts Sun at (0,0,0) and degrades Earth to planet number 3 out of 8. The astronomic advantages are clear, but dis-answers one question: Why this planet and not one of the others? Typically we are muzzled by semi-convincing anthropic arguments.[3,19]

Heliocentric to cosmological model. The next coup of astronomers was to degrade our Sun to one star among billions of stars in our milky way, and our milky way to one galaxy out of billions of others, according to current textbooks. Again, it is generally accepted that the question of why we are in this particular galaxy in this particular solar system is essentially unanswerable.

Multiverses. Many modern more speculative cosmological models (can be argued to) imply a multitude of essentially disconnected universes (in the conventional sense), often each with their own (quite different) characteristic: Examples are Wheeler's oscillating universe, Smolin's baby universe theory, Everett's many-worlds interpretation of quantum mechanics, and the different compactifications of string theory.[22] They "explain" why a universe with our properties exist, since the multiverse includes universes with all kinds of properties, but they cannot *predict* these properties. A multiverse theory *plus* a theory predicting in which universe we happen to live would determine the value of the inter-universe variables for our universe, and hence have much more predictive power. Again, anthropic arguments are sometimes evoked but are usually vague and unconvincing.

Universal ToE (U). Taking the multiverse theory to the extreme, Schmidhuber[18] postulates a universal multiverse, which consists of *every* computable universe. Clearly, if our universe is computable (and there is no proof of the opposite[18]), the multiverse generated by (U) contains and hence perfectly describes our own universe, so we have a theory of everything already in our hands. Unfortunately it is of little use, since we can't use (U) for prediction. If we knew our "position" in this multiverse, we would know in which (sub)universe we are. This is equivalent to knowing the program that generates *our* universe. This program may be close to any of the conventional cosmological models, which indeed have a lot of predictive power. Since locating ourselves in (U) is equivalent and hence as hard as finding a conventional ToE of our universe, we have not gained much.

All-a-Carte models (A). In the introduction I have pushed the idea even further: Champernowne's normal number glues the natural numbers, for

our purpose written in binary format, 1,10,11,100,101,110,111,1000,1001,...
to one long string.

$$110111001011101111 0001001...$$

Obviously it contains every finite substring by construction. The digits of many irrational numbers like $\sqrt{2}$, π, and e are conjectured to also contain every finite substring. If our space-time universe is finite, we can capture a snapshot of it in a truly gargantuan string u. Since Champernowne's number contains every finite string, it also contains u and hence perfectly describes our universe. Probably even $\sqrt{2}$ is a perfect ToE. Unfortunately, if and only if we can localize ourselves, we can actually use it for predictions. (For instance, if we knew we were in the center of universe 001011011 we could predict that we will 'see' 0010 when 'looking' to the left and 1011 when looking to the right.) Locating ourselves means to (at least) locate u in the multiverse. We know that u is the u's number in Champernowne's sequence (interpreting u as a binary number), hence locating u is equivalent to specifying u. So a ToE based on normal numbers is only useful if accompanied by the gargantuan snapshot u of our universe. In light of this, such an "All-a-Carte" ToE (without knowing u) is rather a theory of nothing than a theory of everything.

Localization within our universe. The loss of predictive power when enlarging a universe to a multiverse model has nothing to do with multiverses per se. Indeed, the distinction between a universe and a multiverse is not absolute. For instance, Champernowne's number could also be interpreted as a single universe, rather than a multiverse. It could be regarded as an extreme form of the infinite Fantasia Land from the Never-Ending Story, where everything happens somewhere. Champernowne's number constitutes a perfect map of the All-a-Carte universe, but the map is useless unless you know where you are. Similarly but less extreme, cosmological inflation models produce a universe that is vastly larger than its visible part, and different regions may have different properties.

Predictive power. The exemplary discussion above has hopefully convinced the reader that we indeed lose something (some predictive power) when progressing to too large universe and multiverse models. Historically, the higher predictive power of the large-universe models (in which we are seemingly randomly placed) overshadowed the few extra questions they raised compared to the smaller ego/geo/helio-centric models. But the discussion of the (physical, universal, and all-a-carte) multiverse theories has shown that pushing this progression too far will at some point harm predictive power. We saw that this has to do with the increasing difficulty to localize the observer.

4. Complete ToE Selection Principle

A ToE by definition is a perfect model of the universe. It should allow to predict all phenomena. Most ToEs require a specification of some initial conditions, e.g. the state at the big bang, and how the state evolves in time (the equations of motion). In general, a ToE is a program that in principle can "simulate" the whole universe. An All-a-Carte universe perfectly satisfies this condition but apparently is rather a theory of nothing than a theory of everything. So meeting the simulation condition is not sufficient for qualifying as a Complete ToE. We have seen that (objective) ToEs can be completed by specifying the location of the observer. This allows us to make useful predictions from our (subjective) viewpoint. We call a ToE plus observer localization a subjective or complete ToE. If we allow for stochastic (quantum) universes we also need to include the noise. If we consider (human) observers with limited perception ability we need to take that into account too. So

A complete ToE needs specification of

- (i) initial conditions
- (e) state evolution
- (l) localization of observer
- (n) random noise
- (o) perception ability of observer

We will ignore noise and perception ability in the following and resume to these issues in Section 5. Next we need a way to compare ToEs.

Predictive power and elegance. Whatever the intermediary guiding principles for designing theories/models (elegance, symmetries, tractability, consistency), the ultimate judge is predictive success. Unfortunately we can never be sure whether a given ToE makes correct predictions in the future. After all we cannot rule out that the world suddenly changes tomorrow in a totally unexpected way. We have to compare theories based on their predictive success in the past. It is also clear that the latter is not enough: For every model we can construct an alternative model that behaves identically in the past but makes different predictions from, say, year 2020 on. Popper's falsifiability dogma is little helpful. Beyond postdictive success, the guiding principle in designing and selecting theories, especially in physics, is elegance and mathematical consistency. The predictive power of the first heliocentric model was not superior to the geocentric one, but it was much simpler. In more profane terms, it has significantly fewer parameters that need to be tuned.

Ockham's razor suitably interpreted tells us to choose the simpler among two or more otherwise equally good theories. For justifications of Ockham's razor, see.[15] Some even argue that by definition, science is about applying Ockham's razor, see.[9] For a discussion in the context of theories in physics, see.[6] It is beyond the scope of this article to repeat these considerations. One can show that simpler theories more likely lead to correct predictions, and therefore Ockham's razor is suitable for finding ToEs.[12]

Complexity of a ToE. In order to apply Ockham's razor in a non-heuristic way, we need to quantify simplicity or complexity. Roughly, the complexity of a theory can be defined as the number of symbols one needs to write the theory down. More precisely, write down a program for the state evolution together with the initial conditions, and define the complexity of the theory as the size in bits of the file that contains the program. This quantification is known as algorithmic information or Kolmogorov complexity[15] and is consistent with our intuition, since an elegant theory will have a shorter program than an inelegant one, and extra parameters need extra space to code, resulting in longer programs.[7,23] From now on I identify theories with programs and write Length(q) for the length=complexity of program=theory q in bits.

Example: standard model+gravity (P) versus string theory (S). To keep the discussion simple, let us pretend that standard model of particle physics plus gravity (P) and string theory (S) each qualify as ToEs. (P) is a mixture of a few relatively elegant theories, but contains about 20 parameters that need to be specified. String theory is truly elegant, but ensuring that it reduces to the standard model needs sophisticated extra assumptions (e.g. the right compactification).

(P) can be written down in one line, plus we have to give 20+ constants, so lets say one page. The meaning (the axioms) of all symbols and operators require another page. Then we need the basics, natural, real, complex numbers, sets (ZFC), etc., which is another page. That makes 3 pages for a complete description in first-order logic. There are a lot of subtleties though: (a) The axioms are likely mathematically inconsistent,

(b) it's not immediately clear how the axioms lead to a program simulating our universe, (c) the theory does not predict the outcome of random events, and (d) some other problems. So to transform the description into an e.g. C-program simulating our universe, needs a couple of pages more, but I would estimate around 10 pages overall suffices, which is about 20'000 symbols=bytes. Of course this program will be (i) a very inefficient simulation and (ii) a very naive coding of (P). I conjecture that the *shortest* program for (P) on a universal Turing machine is much shorter, maybe even only one tenth of this. The numbers are only a quick rule-of-thumb guess. If we start from string theory (S), we need about the same length. S is *much* more elegant, but we need to code the compactification to describe our universe, which effectively amounts to the same. Note that everything else in the world (all other physics, chemistry, etc,) is emergent.

It would require a major effort to quantify which theory is the simpler one in the sense defined above, but I think it would be worth the effort. It is a quantitative objective way to decide between theories that are (so far) predictively indistinguishable.

CToE selection principle. It is trivial to write down a program for an All-a-Carte multiverse (A). It is also not too hard to write a program for the universal multiverse (U).[12,18] Lengthwise (A) easily wins over (U), and (U) easily wins over (P) and (S), but as discussed, (A) and (U) have serious defects. On the other hand, these theories can only be used for predictions after extra specifications: Roughly, for (A) this amounts to tabling the whole universe, (U) requires defining a ToE in the conventional sense, (P) needs 20 or so parameters and (S) a compactification scheme. Hence localization-wise (P) and (S) easily win over (U), and (U) easily wins over (A). Given this trade-off, it has been suggested in,[12,20] to include the description length of the observer location in our ToE evaluation measure. That is,

among two CToEs, select the one that has shorter overall length

$$\text{Length}(i) + \text{Length}(e) + \text{Length}(l)$$

For an All-a-Carte multiverse, the last term contains the gargantuan string u, catapulting it from the shortest ToE to the longest CToE, hence (A) will not minimize the sum.

ToE versus (U). Consider any ToE and its program q, e.g. (P) or (S). Since (U) runs all programs including q, specifying q means localizing ToE q in

(U). So (U)+q is a CToE whose length is just some constant number of bits (the simulation part of (U)) more than that of ToE q. So whatever ToE physicists come up with, (U) is nearly as good as this theory. This essentially clarifies the paradoxical status of (U). Naked, (U) is a theory of nothing, but in combination with another ToE q, it excels to a good CToE, albeit slightly longer=worse than q.

Localization within our universe. So far we have only localized our universe in the multiverse, but not ourselves in the universe. To localize our Sun, we could e.g. sort (and index) stars by their creation date, which the model (i)+(e) provides. Most stars last for 1-10 billion years (say an average of 5 billion years). The universe is 14 billion years old, so most stars may be 3rd generation (Sun definitely is), so the total number of stars that have ever existed should very roughly be 3 times the current number of stars of about $10^{11} \times 10^{11}$. Probably "3" is very crude, but this doesn't really matter for sake of the argument. In order to localize our Sun we only need its index, which can be coded in about $\log_2(3 \times 10^{11} \times 10^{11}) \doteq 75$ bits. Similarly we can sort and index planets and observers. To localize Earth among the 8 planets needs 3 bits. To localize yourself among 7 billion humans needs 33 bits. Alternatively one could simply specify the (x, y, z, t) coordinate of the observer, which requires more but still only very few bits. These localization penalties (l) are tiny compared to the difference in predictive power (to be quantified later) of the various theories (ego/geo/helio/cosmo). This explains and justifies theories of large universes in which we occupy a random location.

5. Formalization & Extensions

This section formalizes the CToE selection principle and what accounts for a CToE. Universal Turing machines are used to formalize the notion of programs as models for generating our universe and our observations. I also introduce more realistic observers with limited perception ability.

Objective ToE. Since we essentially identify a ToE with a program generating a universe, we need to fix some general purpose programming language on a general purpose computer. In theoretical computer science, the standard model is a so-called Universal Turing Machine (UTM).[15] It takes a binary program q, executes it and outputs a binary string u:

$$\text{UTM}(q) = u$$

The details do not matter to us, since drawn conclusions are typically independent of them. In our case, u will be interpreted as a binary representation of the space-time universe (or multiverse) generated by ToE candidate q. So q incorporates items (i) and (e) of Section 4. Surely our universe doesn't look like a bit string, but can be coded as one as explained below and in more detail in.[12] We have some simple coding in mind, e.g. u being the (fictitious) binary data file of a high-resolution 3D movie of the whole universe from big bang to big crunch. Again, the details do not matter.

Observational process and subjective complete ToE. As I have demonstrated it is also important to localize the observer. In order to avoid potential qualms with modeling human observers, consider as a surrogate a (conventional not extra cosmic) video camera filming=observing parts of the world. The camera may be fixed on Earth or installed on an autonomous robot. It records part of the universe u denoted by ω.

I only consider *direct* observations like with a camera. Electrons or atomic decays or quasars are not directly observed, but with some (classical) instrument. It is the indicator or camera image of the instrument that is observed (which physicists then usually interpret). This setup avoids having to deal with any form of informal correspondence between theory and real world, or with subtleties of the quantum-mechanical measurement process. The only philosophical presupposition I make is that it is possible to determine uncontroversially whether two finite binary strings (on paper or file) are the same or differ in some bits.

In a computable universe, the observational process within it, is obviously also computable, i.e. there exists a program s that extracts observations ω from universe u. Formally

$$\text{UTM}(s, u) = \omega$$

where the UTM runs program s on input u to produce observation ω. So ω is the observation by subject s in universe u generated by program q. Program s contains all information about the location and orientation and perception abilities of the observer/camera, hence specifies not only item (l) but also item (o) of Section 4.

> A Complete ToE (CToE) consists of a specification of a (ToE,Subject) pair (q, s). Since it includes s it is a Subjective ToE.

CToE selection principle. So far, s and q were fictitious subject and universe programs. Let o be the past "true" observations of some concrete observer in our universe, e.g. your own personal experience of the world from birth till today. The future observations are of course unknown. By definition, o contains *all* available experience of the observer, including e.g. outcomes of scientific experiments, school education, read books, etc.

The proposal $\omega = \omega(q,s)$ generated by a correct CToE must be consistent with the true observations o in the sense that ω starts with o, denoted by $o... = \omega$. If ω would differ from o (in a single bit) the subject would have 'experimental' evidence that (q,s) is not a perfect CToE. We can now formalize the CToE selection principle as follows

> *Among a given set of perfect CToEs $\{(q,s)\}$*
> *select the one of smallest Length(q) + Length(s).*

The best CToE. Finally, one may define the best CToE (of an observer with experience o) as

$$(q^*, s^*)[o] := \arg\min_{q,s}\{\text{Length}(q) + \text{Length}(s) : o... = \text{UTM}(s, \text{UTM}(q))\} \tag{1}$$

This may be regarded as a formalization of the holy grail in physics; of finding such a ToE.

Minimizing length is motivated by Ockham's razor. Inclusion of s is necessary to avoid degenerate ToEs like (U) and (A). The selected CToE (q^*, s^*) can and should then be used for forecasting future observations via $o... = \text{UTM}(s^*, \text{UTM}(q^*))$ will (by construction) output observation o followed by future observations "..." taken as prediction.

Extensions. The CToE selection principle is applicable to perfect, deterministic, discrete, and complete models q of our universe. None of the existing sane world models is of this kind. But the principle can easily be extended to more realistic, partial, approximate, probabilistic, and/or parametric models for finite, infinite and even continuous universes.[12]

Most existing theories only partially model some aspects of our world. Any theory that only predicts parts of our complete observation o, can be augmented, in the simplest case by tabulating all unpredicted bits. The complexity of this table then has to be added to that of q and s in (1). Similarly, for theories that are only approximately correct, one can table or code their errors and include them in (1). Some theories like quantum mechanics make only probabilistic predictions. A theory that predicts universe u with probability $Q(u)$ and observation o in universe u with probability $S(o|u)$, induces a probability distribution $P(o) = \sum_u S(o|u)Q(u)$ over observations. The observed noise can then be coded in $|\log_2 P(o)|$ bits[23] to be added to (1) as indicated in item (n) of Section 4. Many theories in physics also depend on real-valued parameters. They need to be specified to some minimally sufficient accuracy[7] and their code length added to (1). Theories of infinite or continuous spaces like 3+1 dimensional Minkowski space can be discretized to arbitrary precision. Such discretization is always possible, since all spaces occurring in physical theories are separable. An even more fundamental solution to construct countable models is to use a result by Loewenheim and Skolem.[18] A final note on pluralistic approaches favoring multiple theories on multiple scales for different (overlapping) application domains: While currently in fashion and convenient in practice, they have fundamental consistency problems and cannot serve jointly as theories of everything, unless reconciliated in a reductionist fashion.

See[12] for examples and details.

6. Discussion of the Assumptions

I will now discuss the assumptions which led to the CToE selection principle; more precisely, under which assumptions the principle will result in good models for our universe. I have argued in this article that the assumptions are sufficient for constructing sensible theories of everything.

> *(i) Bit-string ontology:* The observers' raw experience of the world can be cast into a single temporal binary sequence o. All other physical and epistemological concepts are derived.

What exactly knowledge is and how humans acquire it is philosophically still controversial. Discussions revolve around justification, truth, and belief, which are themselves subtle concepts. These problems are avoided by operating on the much lower ontological "data" level, namely that of an observer capable of perceiving an ordered stream of bits, and nothing else. For a robot or a human in a cyber-world we can take the binarized and linearized data stream from all sensors. For a human in the real world we can approximately digitally record all raw sensory input. All higher-level interpretations of this bit-string (leading to what is traditionally called 'knowledge') is theory-laden, where suitable theories are induced via CToE selection.

> *(ii) Realism:* There exists an objective world independent of any particular observer in it.

Although solipsists claim that the world and other minds do not exist outside and independent of their own mind, and idealists place ideas and spiritual experience at the center of existence, pragmatically, realism is the least controversial assumption. It can (much better than the alternatives) explain many features of our experience e.g. evolution and society. In any case, the CToE selection principle is powerful enough to determine whether the a-priori assumption of an objective world is warranted (Length(q^*) $\gg 0$) or not (Length(q^*) ≈ 0). Indeed, it can determine precisely which and how much of our experience should be ascribed to an objective world (namely q^*) and what is subjective (namely s^*).

> *(iii) Computable universe:* The world is computable, i.e. there exists an algorithm (a finite binary string) which, when executed, outputs the entire space-time universe.

The idea of a mechanical universe is old and has been expressed by many researchers.[17,24,25] The more we understand our world, the more plausible it becomes. As indicated in Sections 3 and 4, physical theories describe (aspects of) our universe with increasing precision and scope. All those theories are computable in the sense that they can simulate the physical phenomena on a computer, although quantum randomness and aspects of string theory complicate this picture. Given this trend, it is natural to conjecture that the total world u is computable. Note that this assumption implicitly assumes (i.e. implies) that temporally stable binary strings exist, which connects it with assumption (*i*).

> *(iv) Computable observer process:* The observer is a computable process within the objective world.

If/since the universe is computable, then an observer who is part of it, is obviously computable too, hence the observation bit-string o should be a computable function of the universe u. This is not at odds with free will [9, Sec.8.6.3]. The important point is to acknowledge that the observer process is important, even if we are/were only interested in objective theories or aspects of the world. Note that observer localization is neither based on the controversial anthropic principle, nor has it anything to do with the quantum-mechanical observation process, although there may be some deeper yet to be explored connections.

> *(v) Ockham's razor principle:* Choose the simplest theory consistent with the observations.

Ockham's razor principle has so far been invaluable for understanding our world. The assumption is that the models selected by it will continue to lead to most-likely-correct predictions. Minimum encoding length principles, rooted in (algorithmic) information theory, quantify Ockham's razor principle, and have led to a quantitative, pragmatic and universal foundation of inductive reasoning.[16] Indeed, it seems to be a necessary and sufficient founding principle of science itself, in contrast e.g. to the popular but insufficient falsifiability principle. Until other necessary and sufficient principles are found, it is prudent to accept Ockham's razor as the foundation of science. A-priori justifications of Ockham's razor are possible too, but of course they must rest themselves on (other) assumptions. For instance, if one assumes that a-priori all universe and observer programs (q, s) are equally likely, then one can show that Ockham's razor 'works' [12, Sec.8]. Hence one could replace (v) by this so-called universal self-sampling assumption (not to be confused with informal anthropic arguments or the no free lunch myth[14]).

7. Conclusions

The computational paradigm exposed a fundamental problem of large-universe theories, which could be overcome by taking serious the role of the observer. I discussed a quantitative method of world model selection by analyzing the usefulness of a theory in terms of predictive power based on model *and* observer localization complexity. In particular I have shown the following:

- Unlike falsificationism, quantified versions of Ockham's razor can serve as the foundation of science.
- A theory that perfectly describes our universe or multiverse, rather than being a Theory of Everything (ToE), might also be a theory of nothing.
- A predictively meaningful theory can be obtained if the theory is augmented by the localization of the observer.
- A truly Complete Theory of Everything (CToE) (q, s) consists of a conventional (objective) ToE q plus a (subjective) observer process s.
- The bit-string ontology, realism, computability, subjectivism, and Ockham's razor quantified in terms of code-length minimization

enable a scientifically meaningful and systematic quest for a theory of everything.

- More precisely, the CToE Selection Principle allows a rigorous and quantitative comparison of CToEs and can even be used to select the "best" CToE (q^*, s^*).

- As a side result, this allows to separate objective knowledge q from subjective knowledge s.

- One might even argue that if q^* is non-trivial, this is evidence for the existence of an objective reality.

- Another side result is that there is no hard distinction between a universe and a multiverse; the difference is qualitative and semantic.

References

1. N. Alchin. *Theory of Knowledge.* John Murray Press, 2nd edition, 2006.
2. J. D. Barrow, P. C. W. Davies, and C. L. Harper, editors. *Science and Ultimate Reality.* Cambridge University Press, 2004.
3. N. Bostrom. *Anthropic Bias.* Routledge, 2002.
4. E. Dolnick. *The Clockwork Universe: Isaac Newton, the Royal Society, and the Birth of the Modern World.* Harper, 2011.
5. M. Gardner. A skeptical look at Karl Popper. *Skeptical Inquirer*, 25(4):13–14,72, 2001.
6. M. Gell-Mann. *The Quark and the Jaguar: Adventures in the Simple and the Complex.* W.H. Freeman & Company, 1994.
7. P. D. Grünwald. *The Minimum Description Length Principle.* The MIT Press, Cambridge, 2007.
8. E. Harrison. *Cosmology: The Science of the Universe.* Cambridge University Press, 2nd edition, 2000.
9. M. Hutter. *Universal Artificial Intelligence: Sequential Decisions based on Algorithmic Probability.* Springer, Berlin, 2005.
10. M. Hutter. Human knowledge compression prize, 2006. open ended, http://prize.hutter1.net/.
11. M. Hutter. Algorithmic information theory: a brief non-technical guide to the field. *Scholarpedia*, 2(3):2519, 2007.
12. M. Hutter. A complete theory of everything (will be subjective). *Algorithms*, 3(4):329–350, 2010.
13. M. Hutter. Universal learning theory. In C. Sammut and G. Webb, editors, *Encyclopedia of Machine Learning*, pages 1001–1008. Springer, 2011.
14. T. Lattimore and M. Hutter. No free lunch versus Occam's razor in supervised learning. In *Proc. Solomonoff 85th Memorial Conference, LNAI*, Melbourne, Australia, 2011. Springer.

15. M. Li and P. M. B. Vitányi. *An Introduction to Kolmogorov Complexity and its Applications.* Springer, Berlin, 3rd edition, 2008.
16. S. Rathmanner and M. Hutter. A philosophical treatise of universal induction. *Entropy*, 13(6):1076–1136, 2011.
17. J. Schmidhuber. A computer scientist's view of life, the universe, and everything. In *Foundations of Computer Science: Potential - Theory - Cognition*, volume 1337 of *LNCS*, pages 201–208. Springer, Berlin, 1997.
18. J. Schmidhuber. Algorithmic theories of everything. Report IDSIA-20-00, arXiv:quant-ph/0011122, IDSIA, Manno (Lugano), Switzerland, 2000.
19. L. Smolin. Scientific alternatives to the anthropic principle. Technical Report hep-th/0407213, arXiv, 2004.
20. R. Standish. *Theory of Nothing.* BookSurge Publishing, 2006.
21. D. C. Stove. *Popper and After: Four Modern Irrationalists.* Pergamon Pres, 1982.
22. M. Tegmark. Parallel universes. In *Science and Ultimate Reality*, pages 459–491. Cambridge University Press, 2004.
23. C. S. Wallace. *Statistical and Inductive Inference by Minimum Message Length.* Springer, Berlin, 2005.
24. S. Wolfram. *A New Kind of Science.* Wolfram Media, 2002.
25. K. Zuse. *Rechnender Raum.* Friedrich Vieweg & Sohn, Braunschweig, 1969. English translation: *Calculating Space,* MIT Technical Translation AZT-70-164-GEMIT, Massachusetts Institute of Technology (Proj. MAC), Cambridge, Mass. 02139, Feb. 1970. (Republished in this volume).

Chapter 22

What Is Ultimately Possible in Physics?

Stephen Wolfram

Wolfram Research, USA

This essay uses insights from studying the computational universe to explore questions about possibility and impossibility in the physical universe and in physical theories. It explores the ultimate limits of technology and of human experience, and their relation to the features and consequences of ultimate theories of physics.

The history of technology is littered with examples of things that were claimed to be impossible–but later done. So what is genuinely impossible in physics? There is much that we will not know about the answer to this question until we know the ultimate theory of physics. And even when we do–assuming it is possible to find it–it may still often not be possible to know what is possible.

Let's start, though, with the simpler question of what is possible in mathematics.

In the history of mathematics, particularly in the 1800s, many "impossibility results" were found [1, p. 1137]. Squaring the circle. Trisecting an angle. Solving a quintic equation. But these were not genuine impossibilities. Instead, they were in a sense only impossibilities at a certain level of mathematical technology.

It is true, for example, that it is impossible to solve any quintic–if one is only allowed to use square roots and other radicals. But it is perfectly possible to write down a finite formula for the solution to any quintic in terms, say, of elliptic functions [2]. And indeed, by the early 1900s, there emerged the view that there would ultimately be no such impossibilities in mathematics. And that instead it would be possible to build more and more sophisticated formal structures that would eventually allow any imaginable mathematical operation to be done in some finite way.

Yes, one might want to deal with infinite series or infinite sets. But somehow these could be represented symbolically, and everything about them could be worked out in some finite way.

In 1931, however, it became clear that this was not correct. For Gödel's theorem [3] showed that in a sense mathematics can never be reduced to a finite activity. Starting from the standard axiom system for arithmetic and basic number theory, Gödel's theorem showed that there are questions that cannot be guaranteed to be answered by any finite sequence of mathematical steps–and that are therefore "undecidable" with the axiom system given.

One might still have thought that the problem was in a sense one of "technology": that one just needed stronger axioms, and then everything would be possible. But Gödel's theorem showed that no finite set of axioms can ever be added to cover all possible questions within standard mathematical theories.

At first, it wasn't clear how general this result really was. There was a thought that perhaps something like a transfinite sequence of theories could exist that would render everything possible–and that perhaps this might even be how human minds work.

But then in 1936 along came the Turing machine [4], and with it a new understanding of possibility and impossibility. The key was the notion of universal computation: the idea that a single universal Turing machine could be fed a finite program that would make it do anything that any Turing machine could do.

In a sense this meant that however sophisticated one's Turing machine technology might be, one would never be able to go beyond what any Turing machine that happened to be universal can do. And so if one asked a question, for example, about what the behavior of a Turing machine could be after an infinite time (say, does the machine ever reach a particular "halt" state), there might be no possible systematically finite way to answer that question, at least with any Turing machine.

But what about something other than a Turing machine?

Over the course of time, various other models of computational processes were proposed. But the surprising point that gradually emerged was that all the ones that seemed at all practical were ultimately equivalent. The original mathematical axiom system used in Gödel's theorem was also equivalent to a Turing machine. And so were all other reasonable models of what might constitute not only a computational process, but also a way to set up mathematics.

There may be some quite different way to set up a formal system than the way it is done in mathematics. But at least within mathematics as we currently define it, we can explicitly prove that there are impossibilities. We can prove that there are things that are genuinely infinite, and cannot meaningfully be reduced to something finite.

We know, for example, that there are polynomial equations involving integers where there is no finite mathematical procedure that will always determine whether the equations have solutions [5]. It is not–as with the ordinary quintic equation–that with time some more sophisticated mathematical technology will be developed that allows solutions to be found. It is instead that within mathematics as an axiomatic system, it is simply impossible for there to be a finite general procedure.

So in mathematics there is in a sense "genuine impossibility".

Somewhat ironically, however, mathematics as a field of human activity tends to have little sense of this. And indeed there is a general belief in mathematics–much more so than in physics–that with time essentially any problem of "mathematical interest" will be solved.

A large part of the reason for this belief is that known examples of undecidable–or effectively impossible–problems tend to be complicated and contrived, and seem to have little to do with problems that could be of mathematical interest. My own work [1] in exploring generalizations of mathematics gives strong evidence that undecidability is actually much closer at hand–and that in fact its apparent irrelevance is merely a reflection of the narrow historical path that mathematics as a field has followed [1, sect. 12.9]. In a sense, the story is always the same–and to understand it sheds light on some of what might be impossible in physics. The issue is computation universality. Just where is the threshold for computation universality?

For once it is possible to achieve computation universality within a particular type of system or problem, it follows that the system or problem is in a sense as sophisticated as any other–and it is impossible to simplify it in any general way. And what I have found over and over again is that universality–and traces of it–occur in vastly simpler systems and problems than one might ever have imagined [1, chap. 11; 6; 7].

Indeed, my guess is that a substantial fraction of the famous unsolved problems in mathematics today are not unsolved because of a lack of mathematical technology–but because they are associated with universality, and so are fundamentally impossible to solve.

But what of physics?

Is there a direct correspondence of mathematical impossibility with physical impossibility? The answer is that it depends what physics is made of. If we can successfully reduce all of physics to mathematics, then mathematical impossibility in a sense becomes physical impossibility.

In the first few decades of the modern study of computation, the various models of computation that were considered were thought of mainly as representing processes–mechanical, electronic or mathematical–that a human engineer or mathematician might set up. But particularly with the rise of models like cellular automata (e.g. [8]), the question increasingly arose of how these models–and computational processes they represent–might correspond to the actual operation of physics.

The traditional formulation of physics in terms of partial differential equations–or quantized fields–makes it difficult to see a correspondence. But the increasing implementation of physical models on computers has made the situation somewhat clearer.

There are two common technical issues. The first is that traditional physics models tend to be formulated in terms of continuous variables. The second is that traditional physics models tend not to say directly how a system should behave–but instead just to define an equation which gives a constraint on how the system should behave.

In modern times, good models of physical systems have often been found (e.g. [1, chap. 8]) that are more obviously set up like traditional digital computations–with discrete variables, and explicit progression with time. But even traditional physical models are in many senses computational. For we know that even though there are continuous variables and equations to solve, there is an immense amount that we can work out about traditional physical models using, for example, Mathematica [9].

Mathematica obviously runs on an ordinary digital computer. But the point is that it can symbolically represent the entities in physical models. There can be a variable x that represents a continuous position, but to Mathematica it is just a finitely represented symbol, that can be manipulated using finite computational operations.

There are certainly questions that cannot obviously be answered by operating at a symbolic level–say about the precise location of some idealized particle represented by a real number. But when we imagine constructing an experiment or an apparatus, we specify it in a finite, symbolic way. And we might imagine that then we could answer all questions about its behavior by finite computational processes.

But this is undoubtedly not so. For it seems inevitable that within standard physical theories there is computation universality. And the result is that there will be questions that are impossible to answer in any finite way. Will a particular three-body gravitational system (or an idealized solar system) be stable forever? Or have some arbitrarily complicated form of instability?

Of course, it could be even worse.

If one takes a universal Turing machine, there are definite kinds of questions that cannot in general be answered about it–an example being whether it will ever reach a halt state from a given input. But at an abstract level, one can certainly imagine constructing a device that can answer such questions: doing some form of "hypercomputation" (e.g. [10, 11]). And it is quite straightforward to construct formal theories of whole hierarchies of such hypercomputations.

The way we normally define traditional axiomatic mathematics, such things are not part of it. But could they be part of physics? We do not know for sure. And indeed within traditional mathematical models of physics, it is a slippery issue.

In ordinary computational models like Turing machines, one works with a finite specification for the input that is given. And so it is fairly straightforward to recognize when some long and sophisticated piece of computational output can really be attributed to the operation of the system, and when it has somehow been slipped into the system through the initial conditions for the system.

But traditional mathematical models of physics tend to have parameters that are specified in terms of real numbers. And in the infinite sequence of digits in a precise real number, one can in principle pack all sorts of information–including, for example, tables of results that are beyond what a Turing machine can compute. And by doing this, it is fairly easy to set things up so that traditional mathematical models of physics appear to be doing hypercomputation.

But can this actually be achieved with anything like real, physical, components?

I doubt it. For if one assumes that any device one builds, or any experiment one does, must be based on a finite description, then I suspect that it will never be possible to set up hypercomputation within traditional physical models [1, sect. 12.4 and notes].

In systems like Turing machines, there is a certain robustness and consistency to the notion of computation. Large classes of models, initial conditions and other setups are equivalent at a computational level. But when hypercomputation is present, details of the setup tend to have large effects on the level of computation that can be reached, and there do not seem to be stable answers to questions about what is possible and not.

In traditional mathematical approaches to physics, we tend to think of mathematics as the general formalism, which in some special case applies to physics. But if there is hypercomputation in physics, it implies that in a sense we can construct physical tools that give us a new level of mathematics–and that answer problems in mathematics, though not by using the formalism of mathematics. And while at every level there are analogs of Gödel's theorem, the presence of hypercomputation in physics would in a sense overcome impossibilities in mathematics, for example giving us ways to solve all integer equations.

So could this be how our universe actually works?

From existing models in physics we do not know. And we will not ultimately know until we have a fundamental theory of physics.

Is it even possible to find a fundamental theory of physics? Again, we do not know for sure. It could be–a little like in hypercomputation–that there will never be a finite description for how the universe works. But it is a fundamental observation–really the basis for all of natural science–that the universe does show order, and does appear to follow definite laws.

Is there in a sense some complete set of laws that provide a finite description for how the whole universe works? We will not know for sure until or unless we find that finite description–the ultimate fundamental theory.

One can argue about what that theory might be like. Is it perhaps finite, but very large, like the operating system of one of today's computers? Or is it not only finite, but actually quite small, like a few lines of computer code? We do not yet know.

Looking at the complexity and richness of the physical universe as we now experience it, we might assume that a fundamental theory–if it exists– would have to reflect all that complexity and richness, and itself somehow be correspondingly complex. But I have spent many years studying what is in effect a universe of possible theories–the computational universe of simple programs. And one of the clear conclusions is that in that computational universe it is easy to find immense complexity and richness, even among extremely short programs with extremely simple structure [1].

Will we actually be able to find our physical universe in this computational universe of possible universes? I am not sure. But certainly it is not obvious that we will not be able to do so. For already in my studies of the computational universe, I have found candidate universes that I cannot exclude as possible models of our physical universe (e.g. [12, 13]).

If indeed there is a small ultimate model of our physical universe, it is inevitable that very few familiar features of our universe as we normally experience it will be visible in that model [1, sect. 9.5]. For in a small model, there is in a sense no room to specify, say, the number of dimensions of space, the conservation of energy or the spectrum of particles. Nor probably is there any room to have anything that corresponds directly to our normal notion of space or time [1, sects. 9.6–9.11].

Quite what the best representation for the model should be I am not sure. And indeed it is inevitable that there will be many seemingly quite different representations that only with some effort can be shown to be equivalent.

A particular representation that I have studied involves setting up a large number of nodes, connected in a network, and repeatedly updated according to some local rewrite rule [1, chap. 9]. Within this representation, one can in effect just start enumerating possible universes, specifying their initial conditions and updating rules. Some candidate universes are very obviously not our physical universe. They have no notion of time, or no communication between different parts, or an infinite number of dimensions of space, or some other obviously fatal pathology.

But it turns out that there are large classes of candidate universes that already show remarkably suggestive features. For example, any universe that has a notion of time with a certain robustness property turns out in an appropriate limit to exhibit special relativity [1, sect. 9.13]. And even more significantly, any universe that exhibits a certain conservation of finite dimensionality–as well as generating a certain level of effective microscopic randomness–will lead on a large scale to spacetime that follows Einstein's equations for general relativity [1, sect. 9.15].

It is worth emphasizing that the models I am discussing are in a sense much more complete than models one usually studies in physics. For traditionally in physics, it might be considered quite adequate to find equations one of whose solutions successfully represents some feature of the universe. But in the models I have studied the concept is to have a formal system which starts from a particular initial state, then explicitly evolves so as to reproduce in every detail the precise evolution of our universe.

One might have thought that such a deterministic model would be excluded by what we know of quantum mechanics. But in fact the detailed nature of the model seems to make it quite consistent with quantum mechanics. And for example its network character makes it perfectly plausible to violate Bell's inequalities at the level of a large-scale limit of three-dimensional space [1, sect. 9.16].

So if in fact it turns out to be possible to find a model like this for our universe, what does it mean?

In some sense it reduces all of physics to mathematics. To work out what will happen in our universe becomes like working out the digits of pi: it just involves progressively applying some particular known algorithm.

Needless to say, if this is how things work, we will have immediately established that hypercomputation does not happen in our universe. And instead, only those things that are possible for standard computational systems like Turing machines can be possible in our universe.

But this does not mean that it is easy to know what is possible in our universe. For this is where the phenomenon of computational irreducibility [1, sect. 12.6] comes in.

When we look at the evolution of some system–say a Turing machine or a cellular automaton–the system goes through some sequence of steps to determine its outcome. But we can ask whether perhaps there is some way to reduce the computational effort needed to find that outcome–some way to computationally reduce the evolution of the system.

And in a sense much of traditional theoretical physics has been based on the assumption that such computational reduction is possible. We want to find ways to predict how a system will behave, without having to explicitly trace each step in the actual evolution of the system.

But for computational reduction to be possible, it must in a sense be the case that the entity working out how a system will behave is computationally more sophisticated than the system itself.

In the past, it might not have seemed controversial to imagine that humans, with all their intelligence and mathematical prowess, would be computationally more sophisticated than systems in physics. But from my work on the computational universe, there is increasing evidence for a general Principle of Computational Equivalence [1, chap. 12], which implies that even systems with very simple rules can have the same level of computational sophistication as systems constructed in arbitrarily complex ways.

And the result of this is that many systems will exhibit computational irreducibility, so that their processes of evolution cannot be "outrun" by other systems–and in effect the only way to work out how the systems behave is to watch their explicit evolution.

This has many implications–not the least of which is that it can make it very difficult even to identify a fundamental theory of physics.

For let us say that one has a candidate theory–a candidate program for the universe. How can we find out whether that program actually is the program for our universe? If we just start running the program, we may quickly see that its behavior is simple enough that we can in effect computationally reduce it–and readily prove that it is not our universe.

But if the behavior is complex–and computationally irreducible–we will not be able to do this. And indeed as a practical matter in actually searching for a candidate model for our universe, this is a major problem. And all one can do is to hope that there is enough computational reducibility that one manages to identify known physical laws within the model universe.

It helps that if the candidate models for the universe are simple enough, then there will in a sense always be quite a distance from one model to another–so that successive models will tend to show very obviously different behavior. And this means that if a particular model reproduces any reasonable number of features of our actual universe, then there is a good chance that within the class of simple models, it will be essentially the only one that does so.

But, OK. Let us imagine that we have found an ultimate model for the universe, and we are confident that it is correct. Can we then work out what will be possible in the universe, and what will not?

Typically, there will be certain features of the universe that will be associated with computational reducibility, and for which we will readily be able to identify simple laws that define what is possible, and what is not.

Perhaps some of these laws will correspond to standard symmetries and invariances that have already been found in physics. But beyond these reducible features, there lies an infinite frontier of computational irreducibility. If we in effect reduce physics to mathematics, we still have to contend with phenomena like Gödel's theorem. So even given the underlying theory, we cannot work out all of its consequences.

If we ask a finite question, then at least in principle there will be a finite computational process to answer that question–though in practice we might be quite unable to run it. But to know what is possible, we also have to address questions that are in some sense not finite.

Imagine that we want to know whether macroscopic spacetime wormholes are possible.

It could be that we can use some computationally reducible feature of the universe to answer this.

But it could also be that we will immediately be confronted with computational irreducibility–and that our only recourse will for example be to start enumerating configurations of material in the universe to see if any of them end up evolving to wormholes. And it could even be that the question of whether any such configuration–of any size–exists could be formally undecidable, at least in an infinite universe.

But what about all those technologies that have been discussed in science fiction?

Just as we can imagine enumerating possible universes, so also we can imagine enumerating possible things that can be constructed in a particular universe. And indeed from our experience in exploring the computational universe of simple programs, we can expect that even simple constructions can readily lead to things with immensely rich and complex behavior.

But when do those things represent useful pieces of technology?

In a sense, the general problem of technology is to find things that can be constructed in nature, and then to match them with human purposes that they can achieve (e.g. [1, sects. 9.11 and 9.10]). And usually when we ask whether a particular type of technology is possible, what we are effectively asking is whether a particular type of human purpose can be achieved in practice. And to know this can be a surprisingly subtle matter, which depends almost as much on understanding our human context as it does on understanding features of physics.

Take for example almost any kind of transportation.

Earlier in human history, pretty much the only way to imagine that one would successfully achieve the purpose of transporting anything would be explicitly to move the thing from one place to another. But now there are many situations where what matters to us as humans is not the explicit material content of a thing, but rather the abstract information that represents it. And it is usually much easier to transport that information, often at the speed of light.

So when we say "will it ever be possible to get from here to there at a certain speed" we need to have a context for what would need to be transported. In the current state of human evolution, there is much that we do that can be represented as pure information, and readily transported. But we ourselves still have a physical presence, whose transportation seems like a different issue.

No doubt, though, we will one day master the construction of atomic-scale replicas from pure information. But more significantly, perhaps our very human existence will increasingly become purely informational–at which point the notion of transportation changes, so that just transporting information can potentially entirely achieve our human purposes.

There are different reasons for saying that things are impossible.

One reason is that the basic description of what should be achieved makes no sense. For example, if we ask "can we construct a universe where $2 + 2 = 5$?", this makes no sense. From the very meaning of the symbols in $2 + 2 = 5$, we can deduce that it can never be satisfied, whatever universe we are in.

There are other kinds of questions where at least at first the description seems to make no sense.

Like "is it possible to create another universe?" Well, if the universe is defined to be everything, then by definition the answer is obviously "no". But it is certainly possible to create simulations of other universes; indeed, in the computational universe of possible programs we can readily enumerate an infinite number of possible universes.

For us as physical beings, however, these simulations are clearly different from our actual physical universe. But consider a time in the future when the essence of the human condition has been transferred to purely informational form. At that time, we can imagine transferring our experience to some simulated universe, and in a sense existing purely within it–just as we now exist within our physical universe.

And from this future point of view, it will then seem perfectly possible to create other universes.

So what about time travel? There are also immediate definitional issues here. For at least if the universe has a definite history–with a single thread of time–the effect of any time travel into the past must just be reflected in the whole actual history that the universe exhibits.

We can often describe traditional physical models–for example for the structure of spacetime–by saying that they determine the future of a system from its past. But ultimately such models are just equations that connect different parameters of a system. And there may well be configurations of the system in which the equations cannot readily be seen just as determining the future from the past.

Quite which pathologies can occur with particular kinds of setups may well be undecidable, but when it seems that the future affects the past what is really being said is just that the underlying equations imply certain consistency conditions across time. And when one thinks of simple physical systems, such consistency conditions do not seem especially remarkable. But when one combines them with human experience–with its features of memory and progress–they seem more bizarre and paradoxical.

In some ancient time, one might have imagined that time travel for a person would consist of projecting them–or some aspect of them–far into the future. And indeed today when one sees writings and models that were constructed thousands of years ago for the afterlife, there is a sense in which that conception of time travel has been achieved.

And similarly, when one thinks of the past, the increasing precision with which molecular archaeology and the like can reconstruct things gives us something which at least at some time in history would have seemed tantamount to time travel.

Indeed, at an informational level–but for the important issue of computational irreducibility–we could reasonably expect to reconstruct the past and predict the future. And so if our human existence was purely informational, we would in some sense freely be able to travel in time.

The caveat of computational irreducibility is a crucial one, however, that affects the possibility of many kinds of processes and technologies.

We can ask, for example, whether it will ever be possible to do something like unscramble an egg, or in general in some sense to reverse time. The second law of thermodynamics has always suggested the impossibility of such things.

In the past, it was not entirely clear just what the fundamental basis for the second law might be. But knowing about computational irreducibility, we can finally see a solid basis for it [1, sect. 9.3]. The basic idea is just that in many systems the process of evolution through time in effect so "encrypts" the information associated with the initial conditions for the system that no feasible measurement or other process can recognize what they were. So in effect, it would take a Maxwell's demon of immense computational power to unscramble the evolution.

In practice, however, as the systems we use for technology get smaller, and our practical powers of computation get larger, it is increasingly possible to do such unscrambling. And indeed that is the basis for a variety of important control systems and signal processing technologies that have emerged in recent years.

The question of just what kinds of effective reversals of time can be achieved by what level of technology depends somewhat on theoretical questions about computation. For example, if it is true that $P \neq NP$, then certain questions about possible reversals will necessarily require immense computational resources.

There are many questions about what is possible that revolve around prediction.

Traditional models in physics tend to deny the possibility of prediction for two basic reasons. The first is that the models are usually assumed to be somehow incomplete, so that the systems they describe are subject to unknown–and unpredictable–effects from the outside. The second reason is quantum mechanics–which in its traditional formulation is fundamentally probabilistic.

Quite what happens even in a traditional quantum formulation when one tries to describe a whole sequence from the construction of an experiment to the measurement of its results has never been completely clear. And for example it is still not clear whether it is possible to generate a perfectly random sequence–or whether in effect the operation of the preparation and measurement apparatus will always prevent this [1, p. 1062]. But even if–as in candidate models of fundamental physics that I have investigated–there is no ultimate randomness in quantum mechanics, there is still another crucial barrier to prediction: computational irreducibility.

One might have thought that in time there would be some kind of acceleration in intelligence that would allow our successors to predict anything they want about the physical universe.

But computational irreducibility implies that there will always be limitations. There will be an infinite number of pockets of reducibility where progress can be made. But ultimately the actual evolution of the universe in a sense achieves something irreducible–which can only be observed, not predicted.

What if perhaps there could be some collection of extraterrestrial intelligences around the universe who combine to try to compute the future of the universe?

We are proud of the computational achievements of our intelligence and our civilization. But what the Principle of Computational Equivalence implies is that many processes in nature are ultimately equivalent in their computational sophistication. So in a sense the universe is already as intelligent as we are, and whatever we develop in our technology cannot overcome that [1, sects. 9.10 and 9.12]. It is only that with our technology we guide the universe in ways that we can think of as achieving our particular purposes.

However, if it turns out–as I suspect–that the whole history of the universe is determined by a particular, perhaps simple, underlying rule, then we are in a sense in an even more extreme situation.

For there is in a sense just one possible history for the universe. So at some level this defines all that is possible. But the point is that to answer specific questions about parts of this history requires irreducible computational work–so that in a sense there can still be essentially infinite amounts of surprise about what is possible, and we can still perceive that we act with free will [1, sect. 12.7].

So what will the limit of technology in the future be like?

Today almost all the technology we have has been created through traditional methods of engineering: by building up what is needed one step at a time, always keeping everything simple enough that we can foresee what the results will be.

But what if we just searched the computational universe for our technology? One of the discoveries from exploring the computational universe is that even very simple programs can exhibit rich and complex behavior. But can we use this for technology?

The answer, it seems, is often yes. The methodology for doing this is not yet well known. But in recent years my own technology development projects [9, 14, 15] have certainly made increasingly central use of this approach.

One defines some particular objective–say generating a hash code, evaluating a mathematical function, creating a musical piece or recognizing a class of linguistic forms. Then one searches the computational universe for a program that achieves the objective. It might be that the simplest program that would be needed would be highly complex–and out of reach of enumerative search methods. But the Principle of Computational Equivalence suggests that this will tend not to be the case–and in practice it seems that it is not.

And indeed one often finds surprisingly simple programs that achieve all sorts of complex purposes.

Unlike things created by traditional engineering, however, there is no constraint that these programs operate in ways that we as humans can readily understand. And indeed it is common to find that they do not. Instead, in a sense, they tend to operate much more like many systems in nature–that we can describe as achieving a certain overall purpose, but can't readily understand how they do it.

Today's technology tends at some level to look very regular–to exhibit simple geometrical or informational motifs, like rotary motion or iterative execution. But technology that is "mined" from the computational universe will usually not show such simplicity. It will look much more like many systems in nature–and operate in a sense much more efficiently with its resources, and much closer to computational irreducibility.

The fact that a system can be described as achieving some particular purpose by definition implies a certain computational reducibility in its behavior.

But the point is that as technology advances, we can expect to see less and less computational reducibility that was merely the result of engineering or historical development–and instead to see more and more perfect computational irreducibility.

It is in a sense a peculiar situation, forced on us by the Principle of Computational Equivalence. We might have believed that our own intelligence, our technology and the physical universe we inhabit would all have different levels of computational sophistication.

But the Principle of Computational Equivalence implies that they do not. So even though we may strive mightily to create elaborate technology, we will ultimately never be able give it any fundamentally greater level of computational sophistication. Indeed, in a sense all we will ever be able to do is to equal what already happens in nature.

And this kind of equivalence has fundamental implications for what we will consider possible.

Today we are in the early stages of merging our human intelligence and existence with computation and technology. But in time this merger will no doubt be complete, and our human existence will in a sense be played out through our technology. Presumably there will be a progressive process of optimization–so that in time the core of our thoughts and activities will simply consist of some complicated patterns of microscopic physical effects.

But looking from outside, a great many systems in nature similarly show complicated patterns of microscopic physical effects. And what the Principle of Computational Equivalence tells us is that there can ultimately be no different level of computational sophistication in the effects that are the result of all our civilization and technology development–and effects that just occur in nature.

We might think that processes corresponding to future human activities would somehow show a sense of purpose that would not be shared by processes that just occur in nature. But in the end, what we define as purpose is ultimately just a feature of history–defined by the particular details of the evolution of our civilization (e.g. [1, sect. 12.2 and notes]).

We can certainly imagine in some computational way enumerating all possible purposes–just as we can imagine enumerating possible computational or physical or biological systems. So far in human history we have pursued only a tiny fraction of all possible purposes. And perhaps the meaningful future of our civilization will consist only of pursuing some modest extrapolation of what we have pursued so far.

So which of our purposes can we expect to achieve in the physical universe? The answer, I suspect, is that once our existence is in effect purely computational, we will in a sense be able to program things so as to achieve a vast range of purposes. Today we have a definite, fixed physical existence. And to achieve a purpose in our universe we must mold physical components to achieve that purpose. But if our very existence is in effect purely computational, we can expect not only to mold the outside physical universe, but also in a sense to mold our own computational construction.

The result is that what will determine whether a particular purpose can be achieved in our universe will more be general abstract issues like computational irreducibility than issues about the particular physical laws of our universe. And there will certainly be some purposes that we can in principle define, but which can never be achieved because they require infinite amounts of irreducible computation.

In our science, technology and general approach to rational thinking, we have so far in our history tended to focus on purposes which are not made impossible by computational irreducibility–though we may not be able to see how to achieve them with physical components in the context of our current existence. As we extrapolate into the future of our civilization, it is not clear how our purposes will evolve–and to what extent they will become enmeshed with computational irreducibility, and therefore seem possible or not.

So in a sense what we will ultimately perceive as possible in physics depends more on the evolution of human purposes than it does on the details of the physical universe. In some ways this is a satisfying result. For it suggests that we will ultimately never be constrained in what we can achieve by the details of our physical universe. The constraints on our future will not be ones of physics, but rather ones of a deeper nature. It will not be that we will be forced to progress in a particular direction because of the specific details of the particular physical universe in which we live. But rather–in what we can view as an ultimate consequence of the Principle of Computational Equivalence–the constraints on what is possible will be abstract features of the general properties of the computational universe. They will not be a matter of physics–but instead of the general science of the computational universe.

References

1. S. Wolfram, *A New Kind of Science,* Wolfram Media, 2002.
2. Wolfram Research, "Solving the Quintic with Mathematica," Poster; M. Trott and V. Adamchik, library.wolfram.com/examples/quintic, 1994.
3. K. Gödel, "Über formal unentscheidbare Sätze der Principia Mathematica und verwandter System I," Monatshefte für Math. u. physik, 38, 1931, pp. 173-198.
4. A. Turing, "On Computable Numbers, with an Application to the Entscheidungsproblem," in *Proceedings of the London Mathematical Society,* Ser. 2, 42, 1937, pp. 230-265.
5. Y. Matiyasevich, Hilbert's Tenth Problem, MIT Press, 1993.
6. S. Wolfram, "The Wolfram 2,3 Turing Machine Research Prize," http://www.wolframscience.com/prizes/tm23, May 14, 2007.
7. A. Smith, "Universality of Wolfram's 2,3 Turing Machine," to appear in Complex Systems.
8. S. Wolfram, *Cellular Automata and Complexity: Collected Papers,* Addison-Wesley Publishing Company, 1994.
9. Wolfram Research, *Mathematica,* 1988.
10. A. Turing, "Systems of Logic Based on Ordinals," Proc. London Math. Soc., Ser. 2–45, 1939, pp. 161–228.
11. J. Copeland, "Hypercomputation," *Minds and Machines,* 12(4), 2002, pp. 461–502.
12. S. Wolfram, Talk given at the Emergent Gravity Conference, MIT, Aug 26, 2008.
13. S. Wolfram, Talk given at the JOUAL 2009 Workshop, CNR-Area, Jul 10, 2009.
14. Wolfram Research, WolframTones, http://tones.wolfram.com, 2005.
15. Wolfram Alpha LLC, Wolfram|Alpha, http://www.wolframalpha.com, 2009.

Chapter 23

Universality, Turing Incompleteness and Observers

Klaus Sutner
Carnegie Mellon University
Pittsburgh PA 15213, USA

1. Computation and Physics

The development of the mathematical theory of computability was motivated in large part by the foundational crisis in mathematics. D. Hilbert suggested an antidote to all the foundational problems that were discovered in the late 19th century: his proposal, in essence, was to formalize mathematics and construct a finite set of axioms that are strong enough to prove all proper theorems, but no more. Thus a proof of consistency and a proof of completeness were required. These proofs should be carried only by strictly finitary means so as to be beyond any reasonable criticism. As Hilbert pointed out,[19] to carry out this project one needs to develop a better understanding of proofs as objects of mathematical discourse:

> To reach our goal, we must make the proofs as such the object of our investigation; we are thus compelled to a sort of proof theory which studies operations with the proofs themselves.

Furthermore, Hilbert hoped to find a single, mechanical procedure that would, at least in principle, provide correct answers to all well-defined questions in mathematics:[20]

> The Entscheidungsproblem is solved when one knows a procedure by which one can decide in a finite number of operations whether a given logical expression is generally valid or is satisfiable. The solution of the Entscheidungsproblem is of fundamental importance for the theory of all fields, the theorems of which are at all capable of logical development from finitely many axioms.

Initially encouraging progress was made towards the realization of Hilbert's dream: first propositional logic was shown to be complete by Hilbert's student Ackermann, see,[20] then predicate logic by Gödel[14]. Predicate logic appeared to be the right framework for Hilbert's project since it conformed nicely to his ideas about mathematical existence: unless the assumption of existence produces an inconsistency, the alleged object indeed exists. Alas, in his seminal 1931 paper,[15] Gödel showed that logical completeness does not translate into mathematical completeness. Specifically, any attempt to formalize even just elementary arithmetic is already doomed to fail in the sense that, no matter how carefully the system is constructed, there will always be unprovable yet true statements (assuming, of course, the system is indeed consistent). This result is central to proof theory, but we can also interpret it computationally. In a formal system of the type Gödel considered, proofs are perfectly finitary objects and can be generated in a purely mechanical fashion without any particular insight into their meaning. In a certain sense, producing proofs is just a problem of word processing. As a consequence, the collection of provable theorems is semidecidable: there is a semi-algorithm that halts and reports "yes" if that is indeed the correct answer; otherwise it fails to halt. On the other hand, the collection of true statements of arithmetic is not even located within the arithmetic hierarchy, see.[33] The Entscheidungsproblem is unsolvable for arithmetic, and, as we now know, for many other areas of mathematics such as group theory, lattice theory or the theory of the rationals. Gödel was very clear about the importance of computability in this context:[16]

> Thus the notion "computable" is in a certain sense "absolute," while almost all metamathematical notions otherwise known (for example, provable, definable, and so on) quite essentially depend upon the system adopted.

The question as to how precisely the notion of computability itself should be formalized was finally solved to Gödel's satisfaction by Turing's groundbreaking paper.[47] It is noteworthy that Gödel initially resisted attempts to declare either the Gödel-Herbrand or the Church approach as the canonically correct solution.

Just a few years after Gödel's discovery, Alan Turing and John von Neumann both made critical contributions to an entirely new development: the construction of powerful and practical computers, machines that could actually carry out computations that had thus far been limited to purely theoretical discourse. In particular Turing's abstract idea of a universal computer, a device that can perform all possible computations whatsoever, turned out to be technologically realizable. The reason for this, in the most abstract sense, is the fact that physics supports computation: we can build concrete devices whose behavior is described by the laws of physics that implement Turing machines in a very obvious fashion. For reasons of efficiency, one usually employs a model referred to as a random access machine, but it is easy to see that these devices are very closely related to Turing machines in computational power (at least if one is willing to accept a polynomial slow-down). Thus, for any area of mathematics for which the Entscheidungsproblem does have a solution, we can build a physical device that will answer all questions put to it. For example, if we limit arithmetic to just addition, there is a decision algorithm for all sentences in this weak system as shown by Presburger.[32] Indeed, the algorithm has found practical use in model checking and helps to verify the correctness of certain computer programs.[4] On the other hand, any realization of the decision algorithm is also plagued by a lower bound result due to Fischer and Rabin that shows that it is necessarily doubly exponential in the worst case.[10] This connection between computation and physics may seem merely a matter of pragmatism; having the ability to perform large and complicated computations on actual and concrete physical devices is obviously a great advantage for mathematics. Some physicists would push a bit further, though, as in the following comment attributed to D. Deutsch:

> The theory of computation has traditionally been studied almost entirely in the abstract, as a topic in pure mathematics. This is to miss the point of it. Computers are physical objects, and computations are physical processes. What computers can or cannot compute is determined by the laws of physics alone, and not by pure mathematics.

The same point is made more forcefully in [9, Chap. 8]:

> So, contrary to what Hilbert thought, and contrary to what most mathematicians since antiquity have believed and believe to this day, proof theory can never be made into a branch of mathematics. Proof theory is a science: specifically, it is computer science.

No doubt these assertions will prove to be quite contentious in many quarters. At any rate, note the reference to computations as physical processes, a notion we will return to shortly. There is general agreement that the limitations of the physical universe that we occupy in terms of space, time, mass and energy impose limitations on what computations can be carried out in actuality, see[27] for a lucid discussion of these issues. However, these obstructions are not of a logical nature: given enough space, time, mass and energy any convergent computation in the sense of classical computability theory can be physically realized, using only bounded resources. Naturally the question arises whether any type of physically realizable process could correspond to a computation outside of this framework. In other words, could a physical process break through the Turing barrier or is there a physical version of the Church-Turing thesis that rules out the existence of any such process. In order to give a truly satisfactory answer to this question, one would first have to resolve Hilbert's 6th problem in a strong form and axiomatize physics, in its totality. At present there seems little hope for any such formalization of physics, all one can manage is a discussion of the computational implications of some particular physical theory such as Newtonian mechanics, general relativity, quantum mechanics, quantum field theory and so on. It is of course of interest to study the strength of physical theories such as Newtonian mechanics, even if they are known not to be in total agreement with actual physical reality.[3] However, the ultimate question of physical realizability will require a stronger approach.

It is interesting that Landauer[25] suggested that physical theories should be constrained so as to avoid collision with computational possibilities:

> The calculative process, just like the measurement process, is subject to some limitations. Any sensible theory of physics must respect these limitations and should not invoke calculative routines that in fact cannot be carried out.

Of course, this restriction is only feasible if the mathematical support structure required for the theory of physics does not involve elements of non-computability, an assertion that is not much different from a strong version of the Church-Turing thesis. Disregarding Landauer's exhortation, it is possible to select some particular theory of physics, such as relativity theory, and propose the construction of devices that could "solve" computationally unsolvable problems if the laws in that particular theory were in fact a complete description of physical reality. However, it is far from clear that a complete or nearly complete theory would afford such opportunities, see[8] for a more careful discussion.

We are here interested in the reverse problem: is it possible, in the context of physical computation, to break through the Turing barrier in the downward direction? The background for this question is a celebrated result in the theory of computation: there are semidecidable sets that are neither decidable nor as complicated as the Halting problem, the most complicated semidecidable set. These sets are said to be intermediate or of intermediate degree, a degree being a collection of all sets of the same level of complexity. The question of whether such sets exist is commonly referred to as Post's problem and its solution required the invention of a new proof technique, the so-called priority method, which has since become one of the hallmarks of computability theory, see[40] for a slightly dated but excellent overview or[1] for a more recent account. Somewhat surprisingly, the method was discovered independently and almost simultaneously on different continents by Friedberg and Muchnik, see.[12,29] Priority arguments tend to be somewhat complicated and their massive use has led to some pointed criticism by Hao Wang[48]:

> The study of degrees [of unsolvability] seems to be appealing only to some special kind of temperament since the results seem to go into many different directions. Methods of proof are emphasized to the extent that the main interest in this area is said to be not so much the conclusions proved as the elaborate methods of proof.

In particular with respect to the construction of sets of intermediate degrees, the corresponding priority argument undoubtedly produces the required set, but is disappointing in the sense that the set so constructed is entirely <u>ad hoc</u>, it has no qualities other than to provide a solution to Post's problem. By contrast, many sets of full degree such as the Halting problem or the set of solvable Diophantine equations are entirely natural; likewise, at the other end of the complexity spectrum, there are lots of examples of complicated yet decidable sets. This was perhaps stated most clearly by Davis[7]:

> But one can be quite precise in stating that no one has produced an intermediate recursively enumerable degree about which it can be said that it is the degree of a decision problem that had been previously studied and named.

More recently, Ambos-Spies and Fejer in[1] are no less blunt in their complaint about the lack of natural intermediate problems:

> The sets constructed by the priority method to solve Post's Problem have as their only purpose to be a solution. ... Thus it can be said that the great complexity in the structure of the computably enumerable degrees arises solely from studying unnatural problems.

It seems pertinent to ask how intermediate degrees fare in the context of physics-like models of computation: is there a physical theory that allows for the construction of intermediate sets? In a certain literal sense, the answer would be "yes" since, in the context of any reasonable physical theory, one can construct a Turing machine that performs the priority construction and writes the elements of the intermediate set on some kind of recording device such as on output tape. For example, some version of this construction could certainly be handled in Newtonian mechanics. However, there is a sense in which this construction fails: an independent observer, with very modest computational means, could monitor the construction and produce a complete semidecidable set from the data so obtained. As it turns out, the intermediate set is obtained by hiding a significant part of the computational process that produces it; if one considers the process as a whole it is difficult to see how it could be considered to be intermediate in nature. Following this line of reasoning, Wolfram made a radical proposal in,[51] his so-called Principle of Computational Equivalence (PCE)

> There are various ways to state the Principle of Computational Equivalence, but probably the most general is just to say that almost all processes that are not obviously simple can be viewed as computations of equivalent sophistication ...PCE ...has vastly richer implications than the laws of thermodynamics or, for that matter, than essentially any single collection of laws in science.

No clear definition of a "process" is given in the reference, nor is there an explanation of what "equivalent sophistication" means technically, but since there are processes corresponding to complete semidecidable sets we can translate this into the claim that just two classes of processes can be distinguished: the decidable ones and the complete ones. Given this interpretation, it is safe to say that PCE rules out the existence of intermediate sets, at least in the context of any reasonable physical theory. As Cook's example of the universality proof for elementary cellular automaton number 110 has shown,[5,51] it can be exceedingly difficult to determine the universality of a system that is given rather than being carefully and purposefully constructed. On the other hand, finding algorithms can also be a daunting task; for example, Presburger's method is by no means obvious. Thus it will be challenging to adduce evidence towards Wolfram's PCE. Note, though, Deutsch's notion of a jump to universality ,[9] the observation that some systems undergo a phase transition that propels them from a fairly limited functionality to a much broader, maximal realm of applicability. For example, the early Roman system of numerals degenerated into a tallying system for sufficiently large numbers. The introduction of (the misnamed) Arabic numerals provided a numeration system that is universal in the sense that numbers of all sizes can be named without tallying and elementary arithmetic operations on these numbers are supported, regardless of size. If jumps to universality are the rule rather than the exception, then some version of PCE might indeed obtain.

2. Cellular Automata

As pointed out in the last section, priority arguments often lead to significant technical complications and are hard to check for correctness. This led Lerman[26] to introduce an intuitively appealing interpretation of the priority technique in terms of "pinball machines." In these machines, balls labeled by natural numbers roll down a track on an inclined plane, their movement guided by gates. A ball falling into a basket means that the corresponding number has entered a certain set. The gates are controlled by Turing machines and may restrain a ball from rolling further down the track. Thus Lerman's machines make a direct appeal to our intuitive understanding of mechanics in order to explain an otherwise complicated technical argument. Of course, significant effort would be necessary to implement these machines entirely within the confines of of Newtonian mechanics, but the basic idea is not too difficult. The power of computation based on Newtonian physics was shown conclusively in:[3] every subset of the natural numbers can be represented by some 3-dimensional Newtonian system. Importantly, the system constructed in the reference is bounded in space, time, mass and energy.

For our purposes, we prefer a setting that is further removed from physics but has a simple computational structure. Perhaps the most far-reaching proposal for such a framework are Gandy machines. In,[13] Gandy formulates four fundamental principles that describe a very general model of parallel computation, see also.[37,38] Clearly, massive parallelism is an indispensable requirement for any physics-like model of computation. Gandy's ingenious model is based on hereditarily finite sets over a countable set of indistinguishable urelements. This admits an interpretation where the urelements are similarly indistinguishable physical components such as atoms or other particles. These components interact in certain patterns prescribed by the machine. Gandy proposes a "Thesis M": any computation by any kind of machine can already be carried out by a Turing machine. In support of this assertion, the reference contains a long list of counterexamples that demonstrate that the conditions proposed there cannot be relaxed without leading to devices that can compute any number theoretic function, an ability referred to as "displaying free will."

To avoid the significant technical difficulties of Gandy's approach, we will stay closer to models of physics embraced by Zuse, Fredkin and Wolfram:[11,51,52] cellular automata. A one-dimensional cellular automaton can be viewed as a continuous, shift-invariant map $G : \mathcal{C} \to \mathcal{C}$ that operates on the space of configurations $\mathcal{C} = \Sigma^{\mathbb{Z}}$, consisting of all bi-infinite sequences over a finite alphabet Σ. The map G is the so-called global map of the automaton, see.[21,49,50] It is easy to see that G has a finitary description $g : \Sigma^w \to \Sigma$, the local map of the automaton. In essence, G is obtained by applying g simultaneously and in parallel to all overlapping blocks of length w in the given configuration. To compute the image $G(X)$ of a given configuration X it suffices to apply a finite state transducer, operating on bi-infinite words, see.[31,35] Transducers are arguably the most basic computational devices operating on inputs of arbitrary size, so there is no extra computational power hidden in the basic update step of the system.

The compact Hausdorff space $\mathcal{C}$ has good properties in the sense of classical dynamics, but it is less pleasing from the perspective of computation since it contains uncountably many, non-finitary objects. In order to avoid complication with generalizations of classical computability theory, it is advantageous to consider a subspace, the collection of almost periodic configurations. These are configurations of the form ${}^{\omega}uwv^{\omega} = \ldots uuuwvvv \ldots$ where u, w and v are all finite words. We write $\mathcal{C}_{\mathrm{ap}}$ for the collection of all almost periodic configurations. In the case where $u = v$ and $w = \varepsilon$ is empty we obtain the standard spatially periodic configurations, corresponding to finite cellular automata with periodic boundary conditions. In the case $u = v = 0$, where 0 is a specially chosen symbol of the alphabet, we obtain configurations of finite support, often referred to misleadingly as finite configurations in the literature. Clearly, almost periodic configurations have a natural finite description. Note that Cook's argument establishing the universality of elementary cellular automaton number 110 requires $\mathcal{C}_{\mathrm{ap}}$ rather than just configurations of finite support, on the latter this automaton has only trivially decidable orbits.

To see why this choice is natural, consider the first-order structure $\mathfrak{A} = \langle \mathcal{C}, G \rangle$ where, for merely technical reasons, G is interpreted as a binary relation: $G(x, y)$ holds iff $G(x) = y$. We refer to $\mathfrak{A}$ as the phasespace of the corresponding cellular automaton. $\mathfrak{A}$ has a countable substructure $\mathfrak{A}_{\rm ap} = \langle \mathcal{C}_{\rm ap}, G \rangle$. As it turns out, the restriction $\mathfrak{A}_{\rm ap}$ is an elementary substructure of $\mathfrak{A}$: exactly the same first-order sentences hold over both structures. It follows that the short-term evolution of configurations in $\mathfrak{A}$ is indistinguishable from that in $\mathfrak{A}_{\rm ap}$. Properties such as "is reversible," "has a 5-cycle" or "is 2-to-1" can all be expressed in first-order logic as can be descriptions of general finite regions of phasespace. Moreover, $\mathfrak{A}$ is an automatic structure in the sense of.[23,24] Consequently, the Entscheidungsproblem for $\mathfrak{A}$ is solvable: there is an algorithm that, given any first-order sentence, determines whether the sentence holds over $\mathfrak{A}$. Moreover, the decision algorithm is based entirely on automata-theoretic methods and is quite natural. This shows in particular that various basic properties of phasespace that can be expressed in first-order logic are all decidable. For example, one can test whether the automaton is reversible or whether the global map is surjective. Unsurprisingly, the algorithms obtained by a brute-force application of the general decision algorithm for $\mathfrak{A}$ are not necessarily particularly efficient but, with a little bit of effort, one can derive fast quadratic time algorithms for the basic properties injectivity, openness and surjectivity of the global map, see.[44,45] Our restriction to one-dimensional cellular automata is critical here, the first-order theory of two or higher dimensional automata is undecidable in general.

Questions about the long-term evolution of configurations, on the other hand, more often than not result in undecidability. The most basic of these questions is Reachability: given two configurations X and Y, does Y appear in the orbit of X? In our setting this problem is trivially semidecidable but clearly undecidable for some cellular automata: one can easily simulate a Turing machine by a cellular automaton. More complicated questions about the structure of orbits are also undecidable. For example, it is undecidable whether all configurations in $\mathcal{C}_{\rm ap}$ evolve to a limit cycle, in fact, the question is Π_2^0-complete. The question whether the orbit of a configuration is decidable is Σ_3^0-complete and it is is Σ_4^0-complete to determine whether a cellular automaton is computationally universal. Here universality is to be understood in the sense of Davis[6] and relates to the completeness of the reachability relation in phasespace. Similar problems occur in the evolution of spatially periodic configurations.[41]

Indeed, one can introduce a classification of one-dimensional cellular automata based on the complexity of reachability. To this end, define $\mathbb{C}_{\bf d}$ to be the collection of all cellular automata whose Reachability problem on the space $\mathcal{C}_{\rm ap}$ has degree exactly $\bf d$, some semidecidable degree. It is important to note that this condition is significantly stronger than the stipulation that the cellular automaton has to be capable of enumerating some set whose degree is **d**. The problem is that only a few of the orbits of the cellular automaton will correspond to actual computations of the corresponding Turing machine, but the classification involves all orbits. Some amount of effort is necessary to control the complexity of these other, computationally meaningless orbits, see.[42,43] At any rate, one can show that for every semidecidable degree **d** there is a one-dimensional cellular automaton whose Reachability problem has degree precisely **d**. In fact, the cellular automaton can be chosen to be reversible. Reachability is closely related to Confluence: given two configurations, do their orbits overlap? One might suspect that the complexity of Reachability and Confluence are fairly closely coupled, but nothing could be further from the truth. More precisely, for any two semidecidable degrees $\mathbf{d}_1$ and $\mathbf{d}_2$, there is a one-dimensional cellular automaton whose Reachability problem has degree $\mathbf{d}_1$ and whose Confluence problem has degree $\mathbf{d}_2$. Needless to say, the last result cannot hold when one considers only reversible cellular automata: in this case confluence of X and Y is equivalent with X being reachable from Y or Y being reachable from X.

These results show that one should not expect to find any simple hierarchy when one considers the computational universe of all one-dimensional cellular automata. In a sense, the enormously difficult structure of the upper semi-lattice of the semidecidable degrees carries over to this realm. For example, by Sacks' density theorem, for any two degrees $\mathbf{d}_1 < \mathbf{d}_2$ there exists a third in between: $\mathbf{d}_1 < \mathbf{d} < \mathbf{d}_2$. The Entscheidungsproblem for this semi-lattice is highly undecidable, see.[17]

3. Processes and Observers

We will now indicate why results like the existence of orbits of intermediate degree in phasespace may be misleading if one is narrowly focused on the physical aspects of computation. Since no natural examples of such degrees are known, the proofs of these results are based on the classical arguments from computability theory, carefully translated into the context of orbits on $\mathcal{C}_{\text{ap}}$ so as to avoid complications arising from unintended orbits. These embedding issues arise whenever we attempt to interpret the dynamics of a physical systems as performing a computation. In fact, Searle[36] points out, "Computational states are not discovered within the physics, they are assigned to the physics." Even in the context of our over-simplified systems the proper interpretation can be quite difficult. For example, while everyone would agree that the orbits generated by elementary cellular automaton number 110 are surprisingly complicated, it requires a highly sophisticated argument to show that one can indeed interpret some of these orbits as performing universal computations. Incidentally, it would be interesting to construct a machine-verifiable proof of this universality assertion; due to size of the building blocks of the configurations required in the argument and the highly non-continuous nature of the system finding such a proof is probably quite challenging.

Now suppose we have a one-dimensional cellular automaton, operating over $\mathcal{C}_{\mathrm{ap}}$, that implements the construction of a semidecidable set of intermediate degree, based on some particular version of the Friedberg-Muchnik priority construction. One standard version of the construction builds two sets A and B that are mutually incomparable with respect to Turing reductions, and thus by necessity of intermediate degree. It was shown by Soare[39] that the disjoint union of A and B is complete. Thus, an observer monitoring the construction might well conclude that the corresponding process is likewise complete. After all, one has to willfully ignore either A or B to conclude otherwise. This suggests a classification of computational processes based on the ability of an observer to extract complicated observations from the process. More precisely, consider one of our cellular automata operating on $\mathcal{C}_{\mathrm{ap}}$. Given some initial configuration $X_0 = {}^\omega u_0 w_0 v_0{}^\omega$ we obtain a sequence of configurations $X_t = {}^\omega u_t w_t v_t{}^\omega$, $t \geq 0$, by iterating the global map. We may safely assume that the length of u_t and v_t remains constant, but, in general, the length of w_t increases without bound. Note that the computation is narrowly constrained in this setting; as single step in phasespace cannot encompass a complicated sub-computations such as, say, a whole stage in a priority construction. In fact, even a simple task such as the comparison of two numbers requires multiple steps.

The observer, on the other hand, has the ability to filter out some part of this highly detailed process by scanning the current configuration and rewriting it slightly. Since there are only finitely many choices for the periodic parts u_t and v_t we may as well assume that the observer only sees w_t. This leads us to define an observer to be a word function ρ over the alphabet Σ that is computable in constant space. We associate the language $\mathsf{O}_\rho = \{\, \rho(w_t) \mid n \geq 0 \,\}$ with the observer which we refer to as the observation language of ρ. As an immediate consequence of our definitions, any observation language must be semidecidable. Of course, there is always an observer whose observation language is trivial. We are here interested in the case where the observer monitors some part of the process that is computationally interesting, such as the two sets A and B in the standard Friedberg-Muchnik construction above. Since we do not wish for the observer to be able to artificially inflate the complexity of the observation language, it is critical that the observer is strictly constrained in computational power. For example, if we considered a linear bounded automaton, the observer itself could perform a computation not related to the cellular automaton, using the given input w_n as scratch space. As long as the length of w_t grows without bound, any computation whatsoever could be performed in this fashion.

It is easy to see that one can construct processes and observers that produce an arbitrary semidecidable set S, for example by constructing a substring of w_n of the form $\#1^n\#$ whenever n is recognized to be in S. Here $\#$ is a special separator symbol. If we apply this approach to the set of all pairs $(e,x) \in \mathbb{N} \times \mathbb{N}$ such that $x \in W_e$, we obtain a single processes that admits an observer whose language is an arbitrary semidecidable set W_e, for all e.

We can now pin down processes of intermediate complexity as follows. First, we call the process undecidable if there exists an observer whose observation language is undecidable. For example, a process for which $w_t = 1^t$ would fail to be undecidable in this sense. Second, the process is complete if there exists an observer whose observation language is complete. Lastly, the process is intermediate if it is undecidable but fails to be complete: there has to be at least one observer that finds the process undecidable, yet the process is simple enough to rule out the existence of an observer who could extract a complete set.

How, then, would the construction of a set of intermediate degree fare in this framework? Consider the variant of the Friedberg-Muchnik priority argument that constructs two sets A and B that satisfy the requirements

$$(R_{2e}) \quad A \neq \{e\}^B \qquad (R_{2e+1}) \quad B \neq \{e\}^A$$

for all $e \in \mathbb{N}$. Here sets are identified with their characteristic functions so that $A \neq \{e\}^B$ means that the set A differs from the set computed by the eth partial recursive function with oracle B. In other words, A is not Turing reducible to B via function number e. Certainly, if all requirements are indeed satisfied, then neither set can be used to compute the other one and thus both must have intermediate degree. The principal problem in the construction is that we have to deal with infinitely many requirements, and the individual requirements may well clash with each other. To deal with these issues one orders the requirements into a sequence $R_0 < R_1 < R_2 < \ldots$ where lower rank means higher priority. At any stage during the construction, we work only on the requirement of highest priority that currently fails to be satisfied and that can be addressed at this stage. The details are somewhat complicated and require a great deal of care. However, at each level the construction is primitive recursive and a simple induction argument shows that, in the end, all requirements are ultimately satisfied. Clearly there are observers for this process that return either A or B. Alas, there are others that return the disjoint sum of A and B and thus, by Soare's result, a complete set. Similar difficulties arise in all known constructions, see[45,46] and [34, Chp. 4] or [30, Chp. X] for more details on the corresponding constructions. In fact, it was suggested by Jockusch and Soare in[22] that priority constructions obey a kind of "maximum degree principle" in the sense that the construction of an recursively enumerable set with weak negative requirements automatically produces a complete set. If the requirements are strong enough to prevent completeness of the generated set the construction itself is essentially based on universal computation. This universality is irrelevant in the context of the theory of computation but becomes visible when we recast the argument as a computational process.

It is thus tempting to conjecture that, in the context of some reasonable theory of physics, intermediate processes fail to exist.

References

1. K. Ambos-Spies and P. A. Fejer. Degrees of unsolvability. http://www.cs.umb.edu/~fejer/articles, 2006.
2. J. Barwise, editor. *The Kleene Symposium*, Amsterdam, 1980. North-Holland.
3. E. J. Beggs and J. V. Tucker. Can newtonian systems, bounded in space, ti mass and energy compute all functions? *Theor. Comput. Sci.*, 371:4–19, February 2007.
4. Edmund Clarke, Orna Grumberg, and Doron Peled. *Model Checking*. MIT Press, 2000.
5. M. Cook. Universality in elementary cellular automata. *Complex Systems*, 15(1):1–40, 2004.
6. M. Davis. *A note on universal Turing machines*, volume 34 of *Annals of Mathematics Studies*, pages 167–175. Princeton University Press, 1956.
7. M. Davis. Foundations of mathematics. http://www.cs.nyu.edu/mailman/listinfo/fom/, 2003.
8. M. Davis. The myth of hypercomputation. In C. Teuscher, editor, *Alan Turing: the life and legacy of a great thinker*, pages 195–212. Springer Verlag, 2006.

9. D. Deutsch. *The Beginning of Infinity*. Viking, 2011.
10. M. Fischer and M. Rabin. Super-exponential complexity of presburger arithmetic. In *Proc. SIAM-AMS Symposium in Applied Mathematics*, volume 7, pages 27–41, 1974.
11. E. Fredkin. Digital mechanics: an informational process based on reversible universal cellular automata. *Physica D*, 45(1-3):254–270, 1990.
12. R. M. Friedberg. Two recursively enumerable sets of incomparable degrees of unsolvability. *Proc. Natl. Acad. Sci. USA*, 43:236–238, 1957.
13. R. O. Gandy. Church's thesis and principles of mechanisms. In et al.,[2] pages 123–148.
14. K. Gödel. Die Vollständigkeit der Axiome des logischen Funktionenkalküls. *Monatsh. Math. u. Phys.*, 37:349–360, 1930.
15. K. Gödel. Über formal unentscheidbare Sätze der Principia Mathematica und verwandter Systeme, I. *Monatsh. Math. u. Phys.*, 38:173–198, 1931.
16. K. Gödel. Über die Länge von Beweisen. *Ergeb. math. Kolloquiums*, 7:23–24, 1936.
17. L. Harrington and S. Shelah. The undecidability of the recursively enumerable degrees. *Bull. Amer. Math. Soc.*, 6:79–80, 1982.
18. R. Herken. *The Universal Turing Machine: A Half-Century Survey*, volume 2 of *Computerkultur*. Springer-Verlag, 2nd edition, 1994.
19. D. Hilbert. Neubegründung der Mathematik. *Abh. math. Seminar Hamburgischen Uni.*, 1:157–177, 1922.
20. D. Hilbert and W. Ackermann. *Grundzüge der Theoretischen Logik*. Springer Verlag, 1928.
21. A. Ilachinski. *Cellular Automata: A Discrete Universe*. World Scientific, 2001.
22. C. Jockusch and R. I. Soare. Degrees of members of Π_1^0 classes. *Pacific J. Math.*, 40:605–616, 1972.
23. B. Khoussainov and A. Nerode. Automatic presentations of structures. In *LCC '94: Int. Workshop on Logical and Computational Complexity*, pages 367–392, London, UK, 1995. Springer-Verlag.
24. B. Khoussainov and S. Rubin. Automatic structures: overview and future directions. *J. Autom. Lang. Comb.*, 8(2):287–301, 2003.
25. R. Landauer. Computations and physics: Wheeler's meaning circuit. *Fund. Physics*, 16:551–564, 1986.
26. M. Lerman. Admissible ordinals and priority arguments. In Mathias and Rogers,[28] pages 311–344.
27. S. Lloyd. Ultimate physical limits to computation. *Nature*, 406:1047–1054, 2002.
28. A. R. D. Mathias and H. Rogers, editors. *Cambridge Summer School in Mathematical Logic*, volume 337 of *Lecture Notes in Mathematics*. Springer Verlag, 1973.
29. A. A. Muchnik. On the unsolvability of the problem of reducibility in the theory of algorithms. *Dokl. Acad. Nauk SSSR*, 108:194–197, 1956.
30. P. G. Odifreddi. *Classical Recursion Theory, Volume II*, volume 143 of *Studies in Logic*. Elsevier, 1999.

31. D. Perrin and J.-E. Pin. *Infinite Words*, volume 141 of *Pure and Applied Math*. Elsevier, 2004.
32. M. Presburger. Über die Vollständigkeit eines gewissen Systems der Arithmetik ganzer Zahlen, in welchem die Addition als einzige Operation hervortritt. In: *Comptes Rendus du I congrés de Mathématiciens des Pays Slaves*. Warsaw, Poland, pages 92–101, 1929.
33. H. Rogers. *Theory of Recursive Functions and Effective Computability*. McGraw Hill, New York, 1967.
34. G. E. Sacks. *Degrees of Unsolvability*. Princeton University Press, 1963.
35. J. Sakarovitch. *Elements of Automata Theory*. Cambridge University Press, 2009.
36. J. R. Searle. *Philosophy in a New Century*, chapter Is the Brain a Digital Computer, pages 86–106. Cambridge University Press, 2008.
37. J. C. Shepherdson. *Mechanisms for computing over abstract structures*, pages 537–556. Volume 2 of *Computerkultur*,[18] 2nd edition, 1994.
38. W. Sieg and J. Byrnes. An abstract model for parallel computation: Gandy's thesis. Technical Report CMU-PHIL-89, CMU, 1998.
39. R. I. Soare. The Friedberg-Muchnik theorem re-examined. *Canad. J. Math.*, 24:1070–1078, 1972.
40. R. I. Soare. *Recursively Enumerable Sets and Degrees*. Perspectives in Mathematical Logic. Springer, 1987.
41. K. Sutner. Classifying circular cellular automata. *Physica D*, 45(1–3):386–395, 1990.
42. K. Sutner. Cellular automata and intermediate reachability problems. *Fundamenta Informaticae*, 52(1-3):249–256, 2002.
43. K. Sutner. Cellular automata and intermediate degrees. *Theor. Comput. Sci.*, 296:365–375, 2003.
44. K. Sutner. Model checking one-dimensional cellular automata. *J. Cellular Automata*, 4(3):213–224, 2009.
45. K. Sutner. Cellular automata, decidability and phasespace. *Fundamenta Informaticae*, 140:1–20, 2010.
46. K. Sutner. Computational processes, observers and Turing incompleteness. *Theor. Comput. Sci.*, 412(1–2):183–190, 2011.
47. A. M. Turing. On computable numbers, with an application to the Entscheidungsproblem. *P. Lond. Math. Soc.*, 42:230–65, 1936.
48. H. Wang. *Popular Lectures on Mathematical Logic*. Dover, New York, 1993.
49. S. Wolfram. Computation theory of cellular automata. *Comm. Math. Physics*, 96(1):15–57, 1984.
50. S. Wolfram. Universality and complexity in cellular automata. *Physica D*, 10:1–35, 1984.
51. S. Wolfram. *A New Kind of Science*. Wolfram Media, Champaign, IL, 2002.
52. K. Zuse. *Rechnender Raum*. F. Vieweg & Sohn, Braunschweig, 1967.

Chapter 24

Algorithmic Causal Sets for a Computational Spacetime

Tommaso Bolognesi

CNR/ISTI, National Research Council, Pisa, Italy

1. Introduction

In this paper we discuss some achievements and ongoing investigations at the intersection between two active research areas: we refer to the first one, somewhat older, broader and fuzzier, by the name 'Computational Universe Conjecture'; the second is known as 'Causal Set Program'.

1.1. *Computational universe conjecture*

This view is centered on the idea that physical phenomena are best understood in terms of digital information processing concepts. In its most extreme forms, it suggests that the universe is discrete, deterministic, finite, and evolves by simple computing rules. The central conceptual equation for this line of thought is:

$$\text{complexity in nature} = \text{emergence in computation}.$$

We still lack experimental evidence for this conjecture, and the situation is unlikely to change in the near future, since the precise nature of the 'universal computation' is imagined to manifest itself at the tiniest spacetime scales – usually associated with the Planck units of 10^{-35}m and 10^{-44}sec – and much below the reach of current experimental setups. Thus, the supporting arguments are still of subjective, aesthetic or metaphorical nature.

The strongest of these consists perhaps in the widely recognized fact that simple models of computation can produce highly complex patterns, sometimes similar to those found in nature. This circumstance has been investigated and divulged, in particular, by S. Wolfram, with his extensive behavioral analysis of cellular automata and other simple models,[28] and is taken by some scientists, not without considerable skepticism by others, as a valid motivation for mining the space of simple algorithms in search for the ultimate, unifying, computation-based theory of physics.

The idea of relating the dynamics of our universe to the computations of cellular automata has been pioneered by Konrad Zuse[29,30] and Ed Fredkin,[8,16] and the general idea of a computable universe has been also investigated, under a variety of perspectives, by Lloyd,[12] Schmidhuber,[24] Tegmark,[26] to mention a few.

1.2. Causal set program

The central idea of this approach to quantum gravity, pioneered by Bombelli, Meyer and Sorkin, is that, at the smallest scales, spacetime is best described in terms of the simple, discrete and flexible mathematical structure of causal sets.[6,7,23,25]

A *causal set* (or '*causet*') is a finitary, partially ordered set, that is, one provided with a binary relation '$\prec$' which is reflexive, antisymmetric and transitive,[a] and such that the number of elements between any two elements is finite. Thus, a causet is conveniently represented by a graph with directed arcs and no cycles. Causet nodes represent spacetime events, the *number* of nodes in a subgraph measures the volume of a corresponding spacetime region, and the *order* relation '$\prec$' defines the causal structure among events – a structure which, in the continuum, is usually described in terms of lightcones. This is summarized by the conceptual equation:

$$\text{spacetime geometry} = \text{order} + \text{number}.$$

Causets are important because the order and number information that they encode is sufficient for determining the metric tensors of General Relativity (see e.g. Ref. 22).

1.3. The plan

By plainly merging the basic assumptions of the two discussed approaches, we come to the view that the universe is best described in terms of causal sets, *and* that these are of algorithmic nature.

A key objective of the Causal Set Program is to devise appropriate methods for growing causets, and in Section 2 we shall survey two major

---
[a]Some authors, however, adopt the *irreflexive convention* that an element does not precede itself.

probabilistic techniques for doing so. But we are interested in replacing the probabilistic approach by a *deterministic, algorithmic* one. Thus, in Section 3 we introduce a general technique for representing any sequential computation, of any model, as a causal set, we show its close analogy with one of the considered probabilistic techniques, and illustrate it by an application to Turing machines.

In Section 4 we discuss 'touring ant' models of computation, focusing on 'trinet mobile automata', one in which the somewhat rigid structure of an infinite Turing machine tape is replaced by the flexible structure of a growing graph. This graph can be pictured as the external boundary of the dynamic spacetime; the computation is carried out by a stateless 'ant' that lives on it.

Both for probabilistic and for algorithmic causets, we shall focus on the *quantitative*, emergent feature of dimensionality, and on estimation techniques for it. On the other hand, there is little doubt that algorithmic causets outperform probabilistic ones in the variety of observed *qualitative* emergent properties, and in Section 5 we summarize some of them, including deterministic chaos and 'particles'. Further discussion on dimension estimators, and on their mutual (in-)compatibility, is provided in Section 6. Section 7 presents some concluding remarks.

1.4. And the quantum mechanical view?

Is *one* causet just enough to represent spacetime? Are we looking for *The Causet* – a unique graph structure telling us the whole history of our universe?

In quantum mechanics, the dynamics of a physical system is fully encoded by the *superposition* of its possible configurations, which may take the discrete form of a *sum over histories* or the continuum form of a *path integral*. In light of the excellent predictive power of quantum mechanics, most approaches to quantum gravity – for example Causal Dynamical Triangulations (CDT)[11,13] – attempt to transpose the superposition concept to the study of spacetime, and concentrate on the search for the correct formulation of a *gravitational* path integral, or sum over histories, where a history would correspond to a single instance of spacetime.

The superposition of spacetime instances is regarded as an almost obligatory (albeit arduous!) approach by most researchers in quantum gravity, but, in our opinion, this is not a valid excuse for avoiding an accurate study of *individual,* and *algorithmic* (as opposed to probabilistic!) spacetime instances. Why?

With probabilistic techniques, any produced causet appears equivalent to any other, and preferring one in particular makes no sense. But when choosing a deterministic model of computation, a tiny fraction of the produced causets spring out as vastly more interesting than all the others. By 'interesting' we mean that they individually exhibit localized structures, a mix of order and disorder, self-similar patterns, and all those phenomena that are so widely investigated in 28, and that seem to manifest, more or less explicitly, also in nature. It is then reasonable to expect these special spacetime instances to play some key role in providing 'the final picture', be it based on superposition or on something else.

2. Causets from Probabilistic Procedures

In this section we briefly review two statistical causet construction techniques that have been investigated in the Causal Set Program.

2.1. *Causets from random sprinklings*

Let us consider the set S of 50 points uniformly distributed in some square region of two-dimensional Euclidean space E^2, shown in Fig. 1-upper-left. If we interpret the vertical dimension as time t, and the horizontal dimension as space x, and if we replace the Euclidean metric by the Lorentzian pseudo-metric with signature $(+,-)$, so that the squared distance between two points $e_1(t_1, x_1)$ and $e_2(t_2, x_2)$ is given by:

$$d^2(e_1, e_2) := +(t_1 - t_2)^2 - (x_1 - x_2)^2$$

we have transformed the Euclidean *space* in a two-dimensional version, denoted M^{1+1}, of the flat, four-dimensional Minkowski *spacetime* M^{1+3} of Special Relativity, and we can view S as a set of *events* in it.[b]

Two events e_1 and e_2 are in *time-like*, *light-like*, or *space-like* relation when $d^2(e_1, e_2)$ is, respectively, positive, null, or negative. In the upper-middle graph of Fig. 1, each event is connected with all the events that are in time-like or light-like relation with it, i.e. all those that are on, or inside, its *light cone*. These edges represent the transitive relation redundantly; we can then eliminate redundancy by taking the *transitive reduction*, as shown

---
[b]The Lorentz distance between two events is invariant under the Lorentz transformation, which maps spacetime coordinates between inertial frames of reference.

Algorithmic Causal Sets for a Computational Spacetime 455

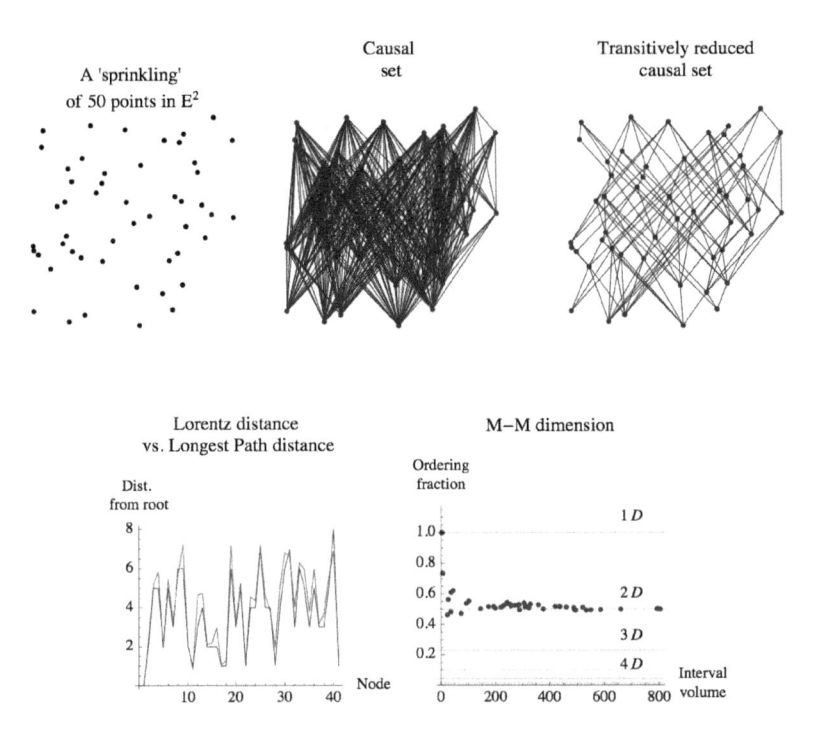

Fig. 1. Upper: uniform distribution of points in a manifold (left); full causal relation inherited from the Lorentzian pseudo-metric (center); transitive reduction, or 'Hasse graph' (right). Lower: consistency of Lorentz distance and graph-theoretic distance based on longest path (left); ordering fractions for causet intervals of variable volume, as an indicator of Myrheim-Meyer dimension (right).

in the 'Hasse graph' of Fig. 1, upper-right graph, where only the essential pairs, called 'links', are retained.[c]

The lower-left diagram of Fig. 1 plots the distances between a specific point – we have chosen the one with lowest time coordinate – and all the 40 points that happen to fall in its future lightcone. The Lorentz distance, conveniently re-scaled, is compared with the graph-theoretic distance defined as the length of the *longest path* connecting two nodes, and their agreement is revealed. Note that any such longest path is only formed by links, thus Lorentz distance is fully coded in the Hasse graph.[d]

---

[c]The *transitive reduction* of a relation R is the smallest relation that admits the same transitive closure of R. When R is acyclic, its transitive reduction is unique.

[d]The length of the longest maximal chain between two points in a causet is indeed regarded, in the Causal Set Program, as the most natural analog for the geodesic length between two events in spacetime.

Finally, in the lower-right diagram of Fig. 1 we show that the *Myrheim-Meyer dimension estimator* attributes the expected 2D value to all the *Alexandrov-intervals* of sufficiently large volume (number of points) of a causet obtained from a 2D, 1000-point sprinkling. Let us define the two introduced concepts.

Given two points x and y of a causet C, an *Alexandrov-interval*, or simply *interval*, denoted $[x, y]$, is the finite set of points lying between x and y according to the partial order. An interval, with edges inherited from C, is itself a causet.

The *Myrheim-Meyer dimension* ('M-M') of a causet C with N events is obtained by considering the ratio $r = R/\binom{N}{2}$ between the number R of event pairs (x, y) that are actually related in the causet, i.e. those for which $x \prec y$ or $y \prec x$, and the maximum number of pairs that *could* have been related, given by the binomial coefficient $\binom{N}{2}$. Ratio r is called the *ordering fraction*.[17] The ordering fraction of a causet obtained by sprinkling points into an interval of d-dimensional Minkowski space is:

$$r = \frac{3d!(d/2)!}{2(3d/2)!},$$

which decreases with d.[15] The possibly non integer M-M dimension d of C is then obtained from r by numerically inverting the above relation.

The horizontal lines in the lower-right plot of Fig. 1 correspond to the ordering fraction values for $d = 1, 2, 3, 4$, and provide a reference for estimating the dimension of the analyzed causet intervals; each interval yields a point in the plot.

2.2. *Causets from transitive percolation dynamics*

There are three reasons for considering this second, statistical causet construction technique: (i) it has been given considerable attention within the Causal Set Program, where it is regarded as 'perhaps the most obvious model of a randomly growing causet';[23] (ii) it is more abstract than the sprinkling technique, and does not assume any underlying manifold; (iii) it appears as a randomized form of the general algorithmic causet construction technique to be introduced in the next section.

Consider the set of the first n natural numbers, and let them represent the events of our causet. Then, for each pair of events i and j, with $i < j$, create, with fixed probability p, a causal relation $i \prec j$: we have built a directed, acyclic graph, and a 'raw' percolation causet. In the sequel, we shall consistently use the terms 'raw causet' and 'raw edges' for referring to the immediate products of the considered probabilistic or algorithmic procedures, before applying transitive closure or reduction.

By construction, raw percolation causets satisfy two properties: (i) event pairs have all the same probability to be related, and (ii) causal relations are independent from one another.

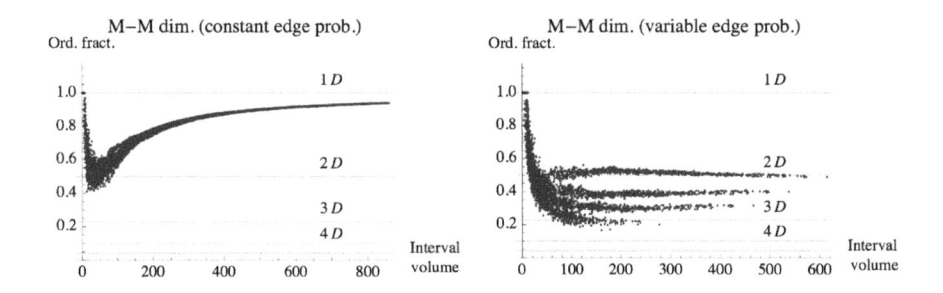

Fig. 2. Spectrum of ordering fractions for causet intervals of variable volume, as indicators of Myrheim-Meyer dimension, for a 1000-node percolation causet with edge probability 0.1 (left). Similar spectrum for a 1000-node percolation causet in which the probability of edge (m, n) is $log(n)/n$ (right).

Fig. 2-left shows a dense spectrum of ordering fraction values, yielding as many M-M dimension estimates, for a number of intervals from a 1000-node causet with edge probability 0.1. We have considered all intervals $[s, t]$ with s ranging from 1 to 951, with step 50, for any possible t. Unlike the case of sprinkled causets (see Fig. 1), the detected dimension does depend here on interval volume: the larger the volume, the closer the approximation to dimension 1. The overall shape of these plots does not change when reducing edge probability: higher dimensions (lower ordering fractions) can be achieved, but these occur only for intervals of small volume.

For applications to spacetime modeling, we would most likely prefer causet construction techniques that can comfortably achieve, say, dimension 4 – if not the higher dimensions of string theory – also for large intervals.

One way to escape the dimensional collapse problem is to act on edge probability p, making it variable. In Fig. 2-right we show an ordering fraction spectrum for a 1000-node percolation causet in which the probability of edge (m, n) is a decreasing function $(log(n)/n)$ of the upper node. As in the previous case, we consider intervals $[s, t]$ with $s = 1, 51, 101, ..., 951$. The plot shows some stabilization at M-M dimensions higher than 1, but the process uniformity is compromised: keeping interval volume constant, the ordering fraction for interval $[s, t]$ now depends on the position of s in the range 1-1000: the lower the s index, the higher the ordering fraction. In particular, the points near the 2D gridline correspond to $s = 1$.

We shall see how naturally the deterministic causet construction techniques of the next section mimic the decreasing edge probability feature.

3. Causets from Sequential Deterministic Computations

Let us now move on to a completely deterministic approach. The general idea is to obtain a causet – a discrete instance of physical spacetime – from a sequential computation, not as the final output of it, but as a direct *representation of the causal relations among its events*.[e] We shall attribute to the state variables involved in the computation the role of *causality mediators* among events, based on the idea that an event that reads a variable is influenced by the event that has written it.

Let S be the set of events of a sequential computation $e_1, e_2, \ldots, e_n, \ldots$, and X be the set of manipulated state variables. We start by assuming that X is static – no new variable is created as the computation proceeds – and that all these variables are initialized before event e_1. For deriving a causet, we represent the computation as the sequence $((R_1, W_1), (R_2, W_2), \ldots, (R_n, W_n), \ldots)$, where R_i and W_i are the sets of variables respectively read and written by event e_i. We call this an 'RW-sequence'; a causet derived from it is an 'RW-causet'.[f] Following the mediation idea above, an RW-causet is readily built: there will be a directed edge from e_i to e_j if and only if $W_i \cap R_j \neq \emptyset$: e_j reads at least one of the state variables written by e_i, say variable x. We express these facts by the notation $e_i \xrightarrow{x} e_j$. In conclusion, the set of edges of the raw RW-causet is:

$$E = \{(e_i, e_j) \in S^2 | \exists x \in X.\ e_i \xrightarrow{x} e_j\}.$$

Note that the actual variable *values* play no role in the procedure. It may be helpful to picture E as partitioned into possibly overlapping subsets E_x, each associated with a different variable x: $E = \cup_{x \in X} E_x$, where $E_x = \{(e_i, e_j) \in S^2 | e_i \xrightarrow{x} e_j\}$.

---

[e]The idea of describing computations as nets of causally related events has been first introduced by Levin and Gács,[10] although their purpose was only to characterize computable functions; it is only by the work of Wolfram [28] that these graphs are viewed as possible spacetime instances.
[f]The total order of computation steps does not represent physical time; the latter, as well as space, is expected to emerge from the growing structure of the causet.

3.1. A preliminary inspection of RW-causets

We would like to get some preliminary impressions on algorithmic RW-causets, without yet selecting a specific deterministic model of computation. One way to do this is to produce the sets R_i and W_i *randomly*. Let us also adopt two simplifications:

- $R_i = W_i$ for all i, that is, every event updates exactly the variables it reads; let RW_i denote this set of variables.
- $|RW_i| = k$, for all i and for some fixed k, that is, the number of variables manipulated by each event is fixed.

The condition for the existence of edge (e_i, e_j) reduces to: $RW_i \cap RW_j \neq \emptyset$.

In Fig. 3-left we explore the performance, in terms of M-M dimension, of a generic, 1000-node randomized RW-causet as defined above, by providing the ordering fraction spectrum for causet intervals of variable volume, as done before. Not surprisingly, this spectrum is essentially equivalent to the

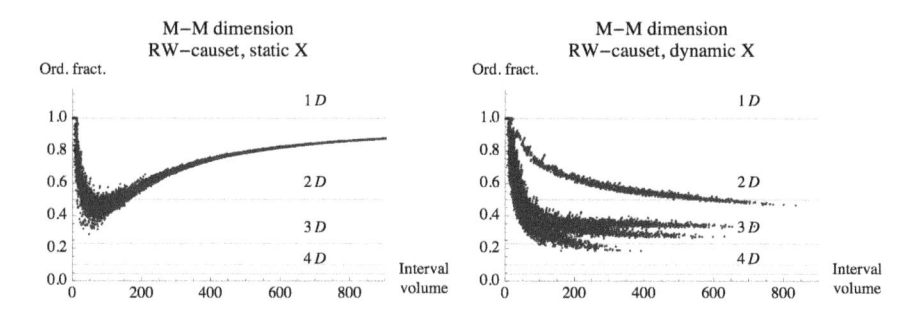

Fig. 3. Ordering fraction spectrum for the RW-causet from a 1000-step randomized computation with static set X of state variables (left); similar spectrum for a 2000-step randomized computation in which X grows linearly (right).

one for percolation causets (see Fig. 2-left), since the probability of finding an edge between events i and j amounts, by definition, to the probability of having $RW_i \cap RW_j \neq \emptyset$, which is constant, under the adopted simplifications. Note, however, that there is a significant difference from 'pure' percolation causets: edge independency is now lost, since the existence of edges (e_i, e_j) and (e_i, e_k), with $i < j < k$, increases the chances of finding an edge (e_j, e_k).[g]

---
[g]This is because, by construction, $(e_i, e_j) \in E_x \wedge (e_i, e_k) \in E_x \implies (e_j, e_k) \in E_x$.

However, no reasonable definition of 'algorithm' can be based on the assumption of a static, or bounded set of memory locations. For example, the Turing machine – the archetypal model of computation – uses of an unbounded tape (although the number of directly addressable cells must still be bounded, and conventionally reduces to one). Thus, in Fig. 3-right we explore the consequences of assuming a linearly growing set of memory locations, obtained, specifically, by letting each computation step introduce a new location, and write into it, with probability 0.4. As done before, for this 2000-step computation we have considered all intervals $[s, t]$ with $s = 1, 51, 101, ..., 1951$, for any possible t.

The branching structure of the spectrum reveals an arisen dependency between ordering fraction value and interval source (s); the phenomenon was already observed in Fig. 2-right, and similar remarks apply. More importantly, the spectrum reveals the potential of this method to achieve relatively high dimensional values.

3.2. *Forgetting non recent write operations: from fat to thin RW-causets*

The reader may have noticed that, in defining the subset of raw causet edges E_x, we have retained an edge (i, k) even in presence of some other edge (j, k), with $i < j$; thus, an event that reads variable x is influenced by *all* the events that have written x earlier in the computation. This convention was suggested by analogy with the percolation technique, but it conflicts with the usually assumed overriding nature of the write operation.

So we shall also consider RW-causets in which an event that reads variable x is influenced *only by its most recent writer event*. In this case, the set of edges contributed by variable x becomes:

$$E_x = \{(e_i, e_j) \in S^2 |\ e_i \xrightarrow{x} e_j \wedge \neg \exists k.(i < k < j \wedge e_k \xrightarrow{x} e_j)\}.$$

It is easy to realize that, in switching from the original, 'fat' causets, to these new 'thin' graphs, we are filtering out a large number of raw edges. Nevertheless, as long as we keep the convention $R_i = W_i$, the introduction of this more natural treatment of write operations has no effect on the M-M dimension of the causet! (The simple proof of this apparently surprising fact is omitted for space reasons.)

3.3. The case of elementary Turing machines

For a more concrete illustration of RW-causets, let us now show some of those that derive from the most widely known sequential, deterministic model of computation.

In an *elementary Turing machine*, a 2-state control head moves up and down a binary tape containing symbols from alphabet {0, 1}. The behavior of the head is defined by one of 4096 possible 2×2 state transition tables: rows are labeled by the control head states s_1 and s_2, columns by tape alphabet symbols, and entries are triples of the form (s', b', d). If the control head is in state s and reads bit b from the cell where it is positioned, then, following what is specified in the (s, b)-entry of the transition table, it changes its own state to s', writes bit b' in the same cell, and moves left, if $d = -1$, or right, if $d = 1$.[h] Hence, two variables are read at each computation step – the control head state and one tape cell – and the same two components are written: using previously introduced notation, we have $R_i = W_i$, and $|RW_i| = 2$.

By proceeding via RW-sequences as described above, in[4] we have computed all the raw (and 'thin') RW-causets for the computations, of arbitrary lengths, of all 4096 machines, assuming an initial tape configuration of all 0's, and we have shown that the derived graphs are all planar and fall into only three categories, sampled in Fig. 4: 1D, 2D flat, and curved. Beyond

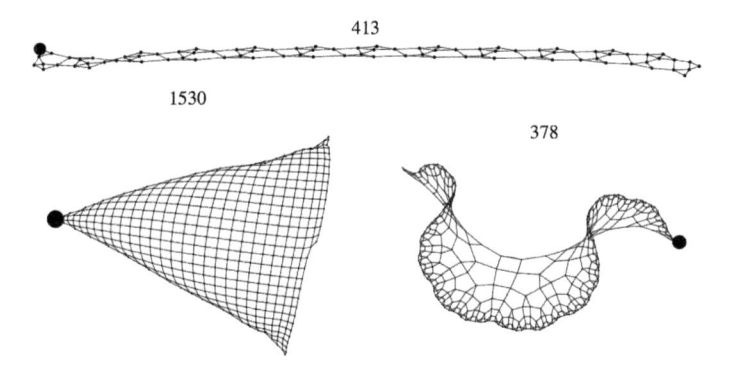

Fig. 4. Raw causets for three elementary Turing machine computations. The (segmented) 12-bit binary representation of the numeric codes yields the corresponding state transition table. 1D causet (code 413), 2D causet (code 1530) and negatively curved causet (code 378).

---

[h]For our purposes we do not need to consider termination conditions.

what is obviously suggested by visual appearance, classification is based on *node-shell-growth analysis*, a technique to be discussed later.

4. Causets from Touring Ants

We cannot expect to be able to describe physical spacetime purely in terms of the very regular RW-causets from elementary Turing machines (Fig. 4).[i] Thus, in search for more interesting RW-causets, we move to another model of computation, based on a *stateless touring ant* idea. We find it convenient, and also appropriate for the present celebratory volume, to introduce this model by two progressive modifications of a standard Turing machine.

4.1. *Touring ants on circular binary tapes*

The first modification consists in replacing the infinite, linear, binary tape with a finite, circular one, while turning the *finite-state* Turing machine head into a *stateless*, touring ant. Initially the tape has only two cells, but the ant can create new ones. At each step the ant is positioned on *two* adjacent cells, c_1 and c_2, that it reads. Depending on these two bits: (i) it writes c_1 and either c_2 or a *new* cell created between c_1 and c_2; (ii) it either remains where it was – on c_1 and the next cell – or it moves one step to the right.[j]

Our interest for this model is due to the fact that we can picture the growing tape as an elastic circular band (Fig. 5) with black and white segments – the bits manipulated by the ant – that also represents the external boundary of the growing causet/universe. Thus, the computation takes place only at the border of the universe, locally, while spacetime arises by internal accumulation of events, and records the history of what has happened at the surface. The process is depicted in Fig. 5 but, before describing it in detail, let us reassure the reader that we are doing nothing but applying the RW-causet construction technique of Section 3. In essence, we represent the computation as a sequence $((R_1, W_1), (R_2, W_2), \ldots, (R_n, W_n), \ldots)$, where R_i and W_i denote, respectively, the pairs of cells read and written by event e_i, then we let the cells play their role as causality mediators, using the 'thin causet' convention described in Subsection 3.2.

---

[i]Coincidentally, these machines are provably not Turing-complete: we need more states or alphabet symbols to achieve computational universality.
[j]There are $2^{16} = 65536$ possible behaviors of this type.

Algorithmic Causal Sets for a Computational Spacetime 463

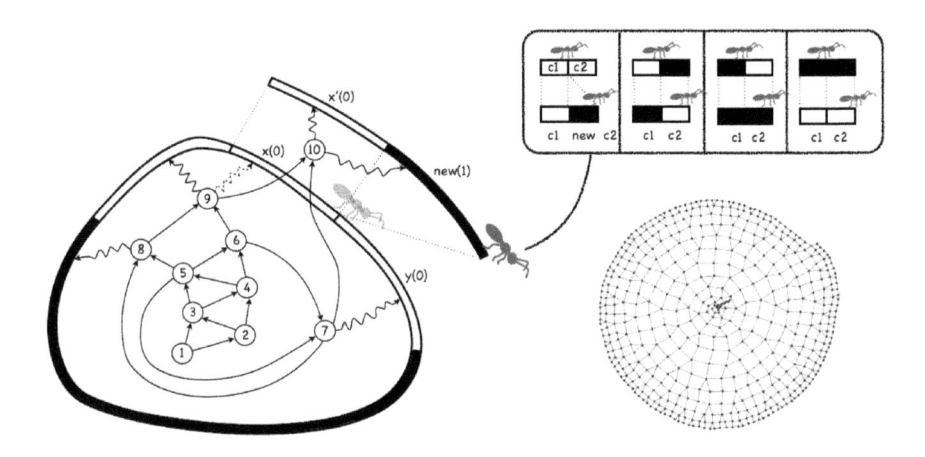

Fig. 5. 10-event RW-causet for a touring ant on circular tape, with final, 5-cell tape configuration as boundary (white = 0, black = 1); wiggling arrows help in traking the most recent cell-writer events during the growth process (left). Four-case ant behavior rule (upper-right). 500-event causet, revealing irregular growth (lower-right).

In Fig. 5-left we show a 10-event causet for a computation of an instance of this model (instance n. 31153, according to our numbering scheme), and the 5-cell border that these events have built. Also shown is the rule describing the ant behavior, which consists of four cases, one for each configuration of the two cells being read: in the first case, a new cell is introduced (and written), while in the remaining three cases the pairs of read and written cells coincide.

The diagram in Fig. 5-left also illustrates the use of wiggling arrows as an aid for building the RW-causet incrementally. Each one of these special arrows temporarily connects a cell to the event that has written it most recently. Consider the situation immediately before event 10: the ant is positioned on adjacent cells $x(0)$ and $y(0)$, with bits in parentheses denoting cell content. The first rule case applies, which involves the creation of a new cell. When the event occurs, we use the wiggling arrows currently pointing to the cells being read, namely arrows $9 \rightsquigarrow x(0)$ and $7 \rightsquigarrow y(0)$, for tracking back the most recent writers of those cells, namely events 9 and 7, and for creating two new causet edges $9 \rightarrow 10$ and $7 \rightarrow 10$: these reflect the causality mediation played by the two cells. Two new wiggling arrows are also added, for future use, from event 10 to the two cells it writes: $10 \rightsquigarrow x'(0)$ and $10 \rightsquigarrow new(1)$. Note that wiggling arrow $9 \rightsquigarrow x(0)$ is also dotted, for representing its disappearance, since it is now superseded by arrow 10 ⤳ $x'(0)$; but arrow 7 ⤳ $y(0)$ is not, since event 7 is still the most recent writer of cell y. Finally, in Fig. 5-right we show the causet for a 500-hundred step computation of the machine. The causet exhibits spiraling growth, with edges oriented outwards, either radially or along the spiral. Incidentally, the growth process appears irregular, as the reader may check by inspecting, for example, the lengths of the spiraling runs of adjacent square faces.

4.2. *Touring ants on planar trivalent networks*

The step to the next model – trinet mobile automata – is relatively short: in place of a circular tape, we now want the border of the growing causet – the memory support where the ant lives and operates – to be a 'trinet': this is our short name for a *planar, trivalent, undirected* graph, that is, one with undirected edges that can be drawn on a sphere while avoiding crossings, and such that each node has exactly three outgoing edges.

Note that even the circular tape could be seen as a graph – a *bivalent* one in which each node, corresponding to a cell, has two neighbors. However, bivalent graphs provide poor structures –just rings, thus we had to introduce binary node labels in order to obtain non trivial behavior. On the contrary, a trinet partitions the embedding sphere into *faces*, or n-ary polygons, and this n-arity information is sufficient for developing interesting algorithms, without need for extra structure.

The general operation of the trinet mobile automata introduced in,[2,3] and the application of the RW-causet construction technique to them, are simultaneously depicted in Fig. 6, which bears similarities with Fig. 5; the picture shows a trinet, the ant operating on it, and the causet growing inside. At each step the ant inspects a small portion of the graph, modifies it by one of two possible graph rewrite rules, and moves to a nearby location. The 2D Pachner rules that we use, sometimes called *Expand* and *Exchange*, are also applied in Loop Quantum Gravity (see, e.g.,[14]), and are shown in Fig. 6-right. Note that, in analogy with the previous model, by rule *Expand* the ant adds an element to the memory support, namely a trinet face. Both the rewrite rule and the next ant location are chosen by some deterministic criterion: a specific choice of ant moves is actually made in Fig. 6, separately for each rule, while the rule choice criterion is left unspecified.

A first variant of the model,[3] called *three-connectivity preserving*, is as follows:

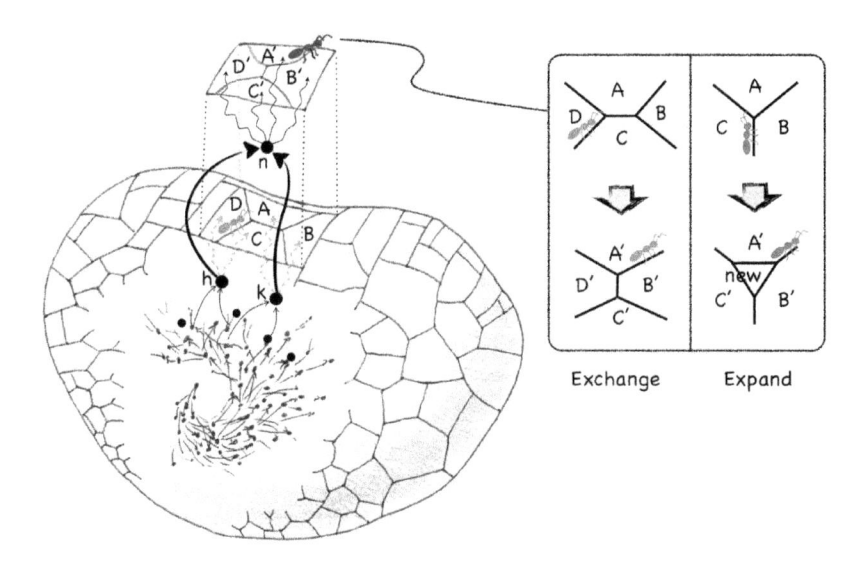

Fig. 6. RW-causet construction for a touring ant on a planar, trivalent graph, appearing as the external boundary of the growing causet (left). The two rewrite rules between which the ant chooses at each step, and a specific choice for the ant moves (right). Each trinet face is hit by a temporary wiggling arrow, starting from the causet event that has updated it most recently (not all such arrows are shown).

(1) Start with a trinet consisting of 2 nodes connected by 3 parallel edges.[k]
(2) Choose rule *Exchange* whenever it does not violate three-connectivity, otherwise choose *Expand*.
(3) Move the ant to a new nearby location, only depending on the applied rule.

In a second variant,[2] qualified as *threshold-based*, point (2) is modified as follows: choose *Exchange* whenever it does not create trinet faces with less than k sides, for some predefined k, otherwise choose *Expand*.

Let us turn to causet construction. When applying a graph rewrite rule, the ant reads and writes trinet faces: rule *Expand* reads three faces (A, B, C) and writes four (A', B', C', new), while rule *Exchange* reads and writes the same four faces (A, B, C, D). Thus, polygonal trinet faces are an obvious choice for causality mediators among events. We can then build

---
[k]This is the smallest possible three-connected graph. A connected graph is n-connected when n is the smallest number of edges one has to remove for disconnecting it.

the RW-sequence, where the R_i and W_i are now sets of trinet faces, and derive causet arcs from it, as described before.

Similar to Fig. 5, Fig. 6 also shows the use of wiggling arrows for carrying out the causet construction incrementally. Event n corresponds to the application of the *Exchange* rule, that modifies faces (A, B, C, D). Before its occurrence, the most recent writers of faces (A, B) and (C, D) were, respectively, events k and h, as shown by the wiggling arrows $k \rightsquigarrow A$, $k \rightsquigarrow B$, $h \rightsquigarrow C$, $h \rightsquigarrow D$. When event n occurs, all four wiggling edges are dropped, and two new causet edges are created: $k \to n$ and $h \to n$. In addition, four new wiggling edges are created from n to the new configurations of faces A, B, C, D.

In the next section we shall provide some experimental evidence that, in terms of emergent properties, RW-causets from touring ant models are not only much richer, expectedly, than those from elementary Turing machines, but also outperform, w.r.t. qualitative properties, causets obtained from probabilistic techniques.

5. Emergent Properties of Touring Ant Causets

A distinguishing feature of our universe – one which seems to play a key role also in art – is the mix of *order* and *disorder*. This feature is so obvious and pervasive that it goes almost completely unnoticed, and fails to qualify for serious physical explanation. But, if a unified, computational, causet-based theory of physics is ultimately found, we must expect it to account both for regular and irregular behaviors, and for their appropriate interplay. Order is obviously achieved with algorithmic causets, and Fig. 4 provided some examples; what about disorder?[1]

5.1. *Deterministic chaos in trinet mobile automata*

The ordered/disordered nature of a sequential computation is well reflected in the corresponding causet graph. But in the case of trinet mobile automata, we can conveniently use a more abstract type of plot for this purpose.

In Fig. 7 we plot the ant motion on the growing trinet, for six instances of the threshold-based model, each labeled by three parameters: the first parameter is the threshold value, and the others identify the ant moves

---
[1]For our purposes here, we are just satisfied by defining 'ordered' (or 'regular') a behavior whose evolution can be easily predicted by visual inspection.

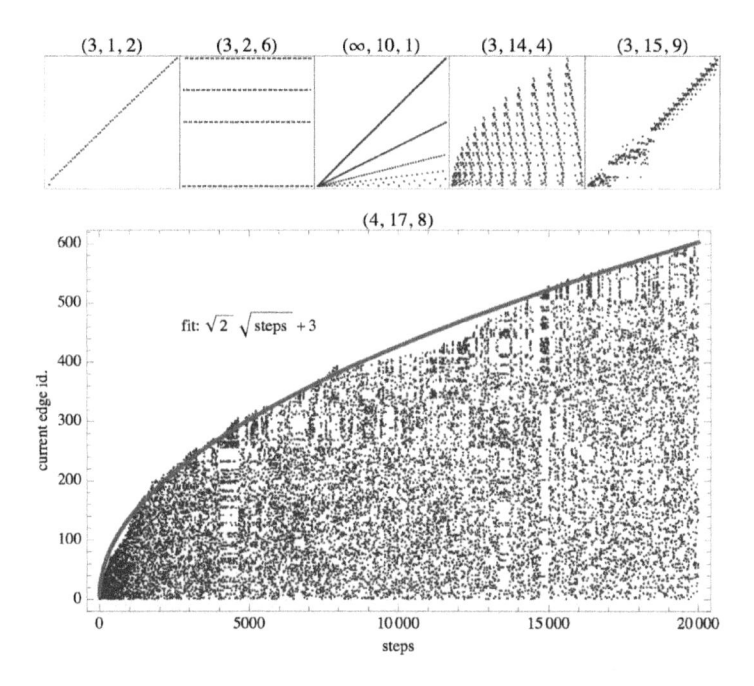

Fig. 7. Regular dynamics for the ant motion of five trinet mobile automata (upper). Chaotic dynamics for one of the two exceptional automata of this type that have been found – code (4, 17, 8) (lower).

associated, respectively, with rule *Expand* and *Exchange* (see Ref. 2 for details). More precisely, the ant 'trajectory' is rendered by plotting the identifier of the edge where the ant is located as a function of the computation step (edges are numbered progressively, as they are introduced, three at a time, by rule *Expand*).

The upper five diagrams in Fig. 7 summarize the typical dynamics that can be observed, with minor variants, for all parameter settings. In all cases, except for the second one, the trinet grows unbounded. The case with threshold $= \infty$ is known as the 'fractal sequence'. All five cases appear regular, that is, completely predictable in their evolution; the fifth one stabilizes after an initial, chaotic transient.

Then, out of a few thousand instances we have inspected, we find *only two* exceptional (and similar) cases of deterministic chaos, or pseudorandomness, that form a tiny class in themselves: automata (4, 17, 8) and (5, 9, 8). The dynamics of the first case is illustrated in Fig. 7-bottom, which refers to a 20,000-step computation. We have actually reached one billion (10^9) steps without observing any sign of stabilization! The plot also includes a fitting function, showing that the trinet growth rate is O($\sqrt{steps}$).

An interesting phenomenon is visible, in the diagram, particularly between steps 4000 and 5000: the ant is temporarily confined inside some region of the causet, and keeps visiting the same edges for a while. In[5] we call this feature 'causet compartmentation', show its impact on the causet graph structure – the creation of a 'hole' – and argue that: (i) these compartments may represent a first, rudimentary form of self-organization of the causet into regions that achieve partial independence from one another; (ii) they can not be expected to emerge in 'genuinely random' causets.

5.2. *Dimensional analysis*

By the *node-shell-growth analysis*, we identify the sets of nodes (the 'shells') at progressive distance r from a given node n, and attempt to fit their growth rate by some function, in particular a polynomial or exponential. If shell sizes grow like r^d, where d can be non-integer, we assign to the causet a 'node-shell-growth' dimension $d+1$, *relative to node n* (a more elaborate definition is provided in,[18] under the name 'internal scaling dimension').[m]

In general, the node-shell growth rate depends on the reference node; when computed relative to the root, the estimate reveals some global feature of the growing spacetime which must not necessarily be confirmed by localized observations from the inside. For example, we find regular causets from Turing machines that appear 3D from the root, and 2D from a generic internal node (see Ref. 4, Fig. 11).[n]

Figure 8 illustrates a fairly regular causet from a threshold-based trinet mobile automaton. Node-shell growth from the root is approximately quadratic, yielding a 3D estimate, and, in this case, the estimate is roughly confirmed by the localized views from the other, 'internal' nodes.

In Fig. 9 we show the causet for one of the pseudo-random computations mentioned in Subsection 5.1. Node-shell growth from the root looks rather

---

[m]In,[4] we have taken exponential shell growth as an indication of negative (hyperbolic) curvature. The obvious weakness of this approach is that fails to treat dimension and curvature separately, and to attribute a finite dimension to a curved causet. Decoupling causet dimension and curvature is a crucial, but still largely unsolved problem.

[n]Obtaining a 3D estimate for a causet deriving from a computation on a 1D support - the Turing machine tape - is interesting: it proves that the causet construction process is not always incrementing by at most one unit - the time component - the dimension of the underlying support.

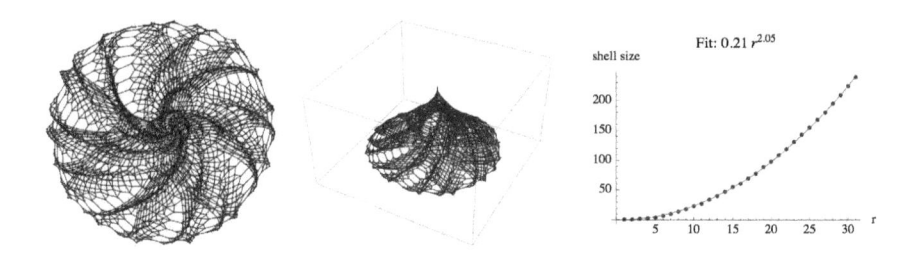

Fig. 8. Causet from a 3000-step computation of a trinet mobile automaton – code (5, 17, 2) (left and center). Node-shell growth from the root is approximately quadratic, yielding a 3D estimate (right).

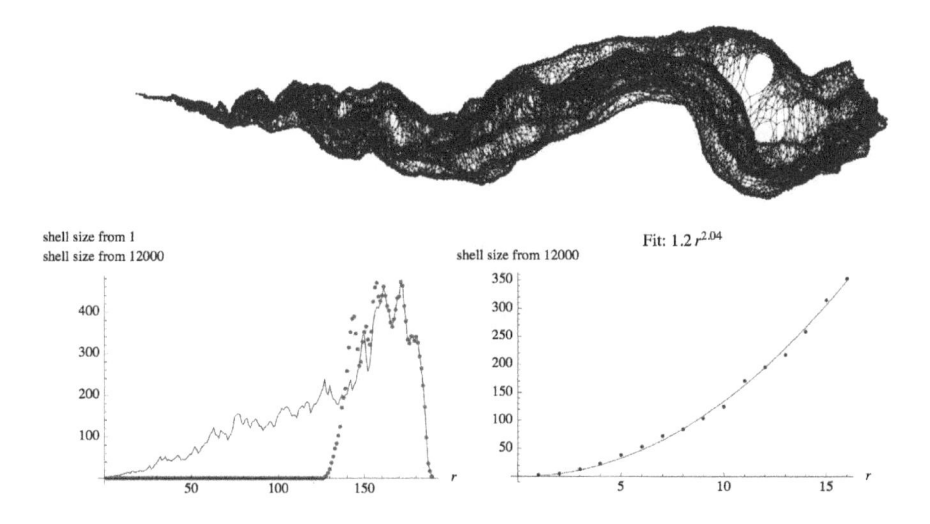

Fig. 9. Causet from a 30,000-step computation of a trinet mobile automaton – code (5, 9, 8) (upper). Node-shell growth from node 1 appears irregular (solid line, lower-left plot), while from node 12,000 it is approximately quadratic (dotted lines).

irregular; from internal nodes, however, the growth appears approximately quadratic, yielding a local 3D estimate.

5.3. Particles?

In the context of the Computational Universe Conjecture, the term 'particle' immediately evokes the 'gliders' and 'spaceships' of Conway's Game of

Life – a two-dimensional cellular automaton – or the interacting, localized structures of Wolfram's one dimensional, Elementary Cellular Automaton (ECA) 110. These phenomena are characterized by a remarkable mix of order and disorder: in ECA 110, for example, the background is completely regular, but the overall interaction pattern of particle trajectories and collisions appears pseudorandom.[28]

In cellular automata all cells are updated simultaneously; but we are interested in obtaining interacting particles from touring ant models, in which the privilege of parallel operation is not given, and this seems to be harder. A sample of what can be currently achieved is provided by the regular causets in Fig. 10. The causet on the left comes from the computation of a

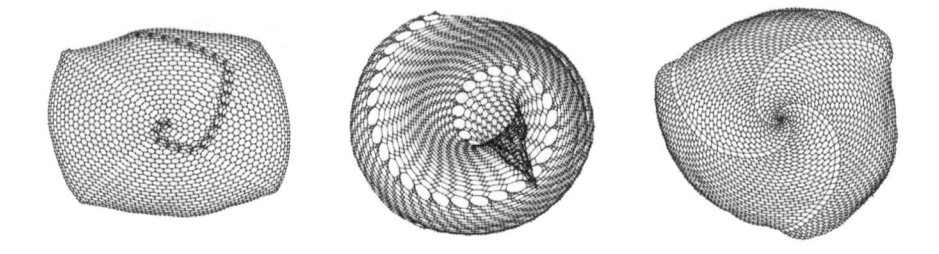

Fig. 10. Particles in regular causets derived from: 3000-step computation of a 2D Turing machine (left); 4000-step computations of trinet mobile automata (4, 16, 2) (center) and (6, 10, 2) (right).

2D Turing machine, or 'turmite',[19] which operates on a square grid (for an interactive demonstration, see,[20] turmite n. 4). The remaining two causets derive from trinet mobile automata. In all three cases, particles emanate, radially or in spirals, from the root, moving on a periodic background, as in ECA 110. But we are still far from the complexity of interactions of the latter: the only interaction observed in Fig. 10 is the deflection occurring when the spiraling and radial particles collide, in the first causet.

6. More on Dimensional Analysis

We have applied the 'Myrheim-Meyer' estimation technique to stochastic causets, and the node-shell-growth technique to algorithmic causets. Shouldn't we rather apply indifferently either technique to either causet type?

First, imagine to apply the M-M dimension estimator to the causets from Turing machines and trinet mobile automata. The result is readily anticipated: no matter how complex they appear, all these causets will exhibit a disappointing M-M dimension 1! The reason is simple. Consider Turing machines (the case of trinet automata is analogous). One of the causality mediators is the control head state s, which plays its role at *each step*; thus, the subset E_s of raw edges mediated by s forms a path that traverses *all* the events of the computation, turning the raw causet into a *totally* ordered one. Then, transitive closure creates an edge for every node pair, thus obscuring the potentially interesting structure of the raw edges deriving from tape cell mediation. In conclusion, the ordering fraction of any causet interval is 1, and so is the M-M dimension. In this case, only the raw causet retains non trivial information, and the node-shell-growth estimator appears to be the right choice for detecting it.

Conversely, imagine to apply node-shell-growth analysis to causets obtained by sprinkling (Subsect. 2.1). In this case, there is no difference between the raw and the transitively closed causet, since, by definition, all causal relations are explicitly included in the raw graph. The analysis is still possible, but makes little sense, since, starting from node s, we end up with just one gigantic shell at distance 1 from it, containing all nodes in its future light cone.

We have considered two perhaps extreme cases, which trigger a variety of questions: on the relevance of *totally-ordered* causets for applications to spacetime modeling, on the choice among causet forms, among dimension estimation techniques, on their agreement, and on the inter-dependence of these choices.

The question on totally-ordered causets goes beyond mere dimensionality concerns. Nevertheless, let us mention one reason for *not* excluding these causets from the agenda. Although the potentially interesting structure of a totally-ordered causet is washed away when transitive reduction is applied *globally*, it may still survive when the operation is applied *locally*. In Fig. 11-left we identify two regions of the totally ordered, second causet of Fig. 10. The central diagrams shows the devastating effect of applying transitive reduction to the first region or, equivalently, to the whole causet. On the contrary, the r.h.s. diagram shows that the particle structure is preserved, although modified, when transitive reduction is applied, locally, to a peripheral region.

How about the choice of causet form: raw, transitively closed, or Hasse?

Attributing importance to raw causets means attributing physical rele-

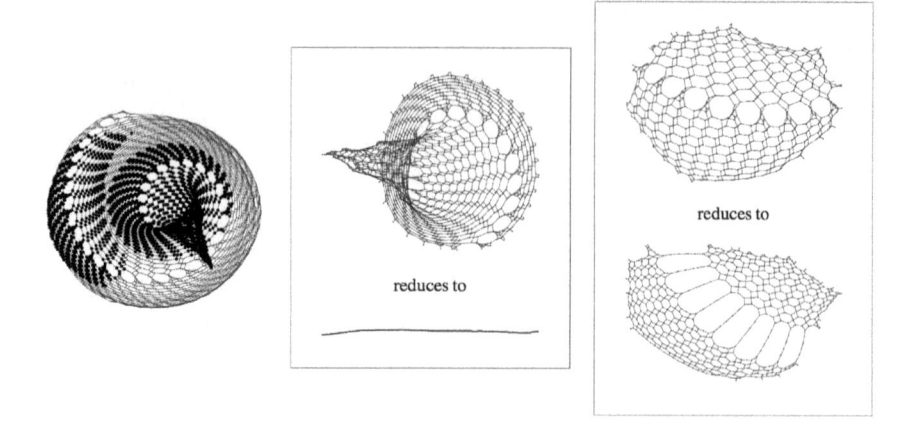

Fig. 11. Two regions in a totally ordered causet (left), and the different effects of transitively reducing them (center and left).

vance to the 'spurious' edges of the causet graph – those that disappear with (global) transitive reduction. There might be deeper reasons for preserving spurious edges, beside the practical fact that they tolerate local transitive reduction.° And yet, for simplifying our discussion, let us now assume that we discard raw causets, in favor of the other two, somewhat 'cleaner' forms.

Quite obviously, a transitively closed causet and its transitive reduction – the Hasse form – contain the same information, being in one-to-one correspondence with each other: nothing fundamental is involved in the choice between them. In principle, node-shell-growth analysis can be applied to both forms, providing *different* outcomes; but we have seen, by the example of sprinkled causets, that it is a bad idea to apply this technique to the transitively closed form. As a second example, consider a regular square grid in which vertical and horizontal edges point, respectively, upwards and to the right. In this case, node-shell sizes from a generic node grow linearly with distance, yielding the expected 2D estimate; but if we apply node-shell analysis to the transitive closure, the already mentioned shell collapse occurs.

Note that the M-M technique agrees on the 2D estimate for the 2D regular grid. We may then be tempted to declare that both estimation techniques are valid, interchangeably, provided the causet is presented in

---

°Referring to Turing machines, for example, one reason could be that the spurious causet edges introduced by the mediation of tape cells are the only vehicle for letting information on the underlying tape topology pop up to the causet level.

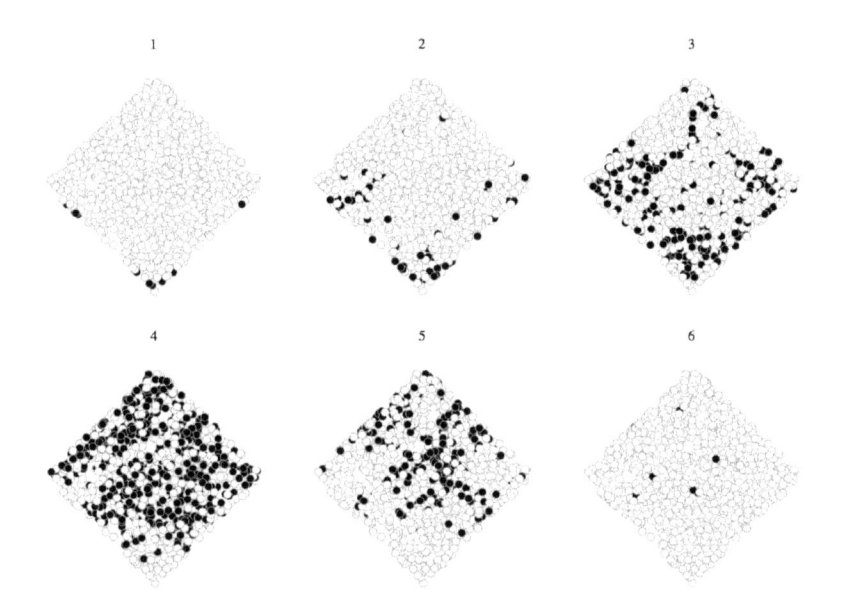

Fig. 12. Node-shells in a 1000-node Hasse causet from sprinkling, at distances 1 trough 6 from the root. Shell nodes are black, and distances are identified by the labels.

adequate form. But, can we always expect agreement between the M-M estimate, for the transitively closed causet, and the node-shell-growth estimate, for its transitive reduction?

We can test this conjecture by analyzing node-shell-growth for the transitive reduction of a causet obtained by sprinkling in a 2D manifold. In our simulations we could compute only a few node-shells, due to the computational cost of transitive reduction; these are insufficient for a reliable fitting of the growth rate, although the apparent trend is markedly above the 2D estimate. However, visualizing the structure of these few shells is useful. Fig. 12 refers to a Hasse causet obtained by sprinkling 1000 points in a 2D, diamond-shaped region. In the six copies of the graph we have highlighted the node-shells at progressive distance – 1 to 6 – from the bottom node. The diagrams reveal the highly *non-local* nature of sprinkled causets, which reflects the non-locality of Minkowski space.[p] It is doubtful that these fea-

---

[p]This non-locality feature is ultimately responsible for the counterintuitive phenomena of Special Relativity, such as the twin paradox: one twin can travel between spacetime points s and t along a long geodesic path, while his brother can take a much shorter (but accelerated) path between the same points, thus experiencing a shorter time delay.

tures be compatible with a sensible application of the node-shell-growth technique.

We have only scratched the surface of a rather intricate area – dimension estimation for causal sets – in which many questions are still open. No single estimator has yet emerged as the ideal option for causet analysis. It has been often observed that an estimate becomes reliable when two or more techniques agree on it; this does not always happen, and adding node-shell-growth analysis to the family seems to bring further complication.

7. Conclusions

In this paper we have suggested that interesting contributions to a theory of spacetime – which one might boldly equate to a theory of everything – can be provided by the study of the emergent properties of algorithmic causal sets, intended as representations of the causal relations among the events of some sequential, deterministic computation. In which model of computation?

In computability theory, all Turing-complete models – those that can simulate a universal Turing machine – are equivalent. Should we regard them as equivalent also with respect to spacetime theory? Fredkin[9] refers to this problem as 'the tyranny of computational universality', and suggests a choice criterion: there should be a one-to-one mapping between the states and function of the real world and those of the model. This leads him to focus on second-order, reversible, universal cellular automata (RUCA), and the SALT model.[16]

Under our causet-oriented perspective, a most direct reformulation of Fredkin's criterion would consist is requiring the algorithmic causet and physical spacetime to be isomorphic. By following this (largely speculative) track, one is led to compare the causet types corresponding to the various models. Our experiments[4,5] have shown that, while the simplest patterns, e.g. polymer-like or hyperbolic, are pervasive in all causet classes, fundamental differences remain. Discriminating factors include properties such as planarity, node degree unboundedness, and total order, which may or may not be satisfied.[4] The conclusion is that universal models of computation are *not* equivalent, with respect to spacetime modeling, and we are still left with an open choice problem.

In spite of this problem, in this paper we have expressed some preference for the class of 'touring ant' models. In doing so, we have excluded cellular automata, possibly the predominant model of the Computational

Universe Conjecture.[q] Our preference for trinet mobile automata, beside the aesthetic appeal of a Cosmos run by a single, memoryless ant, comes from the high degree of abstraction and flexibility offered by graphs and graph rewriting. For example, when the graph is planar and trivalent (a 'trinet'), by just applying the introduced *Expand* and *Exchange* rules, we can create new triangular faces, grow them to become n-polygons, for any n, and *freely* move them around, bringing any face in contact with any other, thus turning the dynamic graph into a lively population of entities – the faces (see Ref. 21 for a demonstration). Dynamic trinets appear as an ideal stage for the emergence of complex behavior, although detecting it may be harder than with cellular automata.

Some readers may have spotted, in our picture of trinets as boundaries of a growing causet/spacetime, a vague analogy with the holographic principle proposed by 't Hooft and Susskind. In its most general form, this principle suggests that the universe could entirely depend on a two-dimensional information structure found at the cosmological horizon.[1] One would then be interested in entropy measures for the boundary, and this is another aspect where trivalent graphs, or their duals (triangulations), offer advantages. Tutte[27] establishes that the number of distinct planar triangulations with n nodes, i.e., of trinets with n faces, is $\psi_n = \frac{2(4n+1)!}{(3n+2)!(n+1)!}$. The amount of information carried by a trinet, in bits, can be expressed by $log_2 \psi_n$, which grows linearly with n and yields an estimate of 3.2451 bits/face. Calculations of this type could help in establishing possible correspondences between algorithmic causets from trinet automata and existing quantum gravity theories.

Research on emergence in algorithmic causets is still at an early stage. Progress can be expected from the study of quantitative properties such as dimension and curvature, and from comparisons with stochastic causets. However, we believe that even more exciting results could come from the investigation, by simulation, of qualitative properties. With our work, we hope we have provided some additional arguments in support of the conceptual equation 'complexity in nature = emergence in computation'. If cellular automata can implement phenomena such as particle interaction and self-replication, is it too ambitious to expect algorithmic causets to implement the mechanisms that eventually trigger and run the biosphere?

---

[q]Perhaps counterintuitively, both the parallel operation of cellular automata and the sequential, localized operation of touring ants, can 'implement' the multiplicity of concurrent activities that we expect to observe in a realistic model of our universe, as discussed in,[28] at p. 487.

Acknowledgements

I would like to thank Stephen Wolfram for useful discussions and encouragement, Tommaso Toffoli for bibliographic suggestions, and Alexander Lamb for ongoing discussions on causet 'Lorentzianity'.

References

1. Jacob D. Bekenstein. Information in the holographic universe. *Scientific American*, July 2003.
2. Tommaso Bolognesi. Planar trinet dynamics with two rewrite rules. *Complex Systems*, 18(1):1–41, 2008.
3. Tommaso Bolognesi. A pseudo-random network mobile automaton with linear growth. *Inf. Process. Lett.*, 109(13):668–674, 2009.
4. Tommaso Bolognesi. Causal sets from simple models of computation. *Int. Journ. of Unconventional Computing*, 6(6):489–524, 2010.
5. Tommaso Bolognesi. Algorithmic causets. In *Proceedings of DICE 2010: Space, Time, Matter*. IOP Science, 2011. Journal of Physics: Conference Series, Vol. 306, No. 1, doi:10.1088/1742-6596/306/1/012042.
6. Luca Bombelli, Johan Lee, David Meyer, and Rafael D. Sorkin. Space-time as a causal set. *Phys. Rev. Lett.*, 59(5):521–524, Aug. 1987.
7. Fay Dowker. Causal sets and the deep structure of spacetime. In Abhay Ashtekar, editor, *100 Years of Relativity - Space-time Structure: Einstein and Beyond*, pages 445–464. World Scientific, 2005. arXiv:gr-qc/0508109v1.
8. Edward Fredkin. Digital philosophy. www.digitalphilosophy.org/Home/tabid/57/Default.aspx.
9. Edward Fredkin. Five big questions with pretty simple answers. *IBM J. Res. & Dev.*, 48(1):31–45, 2004.
10. Peter Gacs and Leonid A. Levin. Causal nets or what is a deterministic computation? *Information and Control*, 51:1–19, 1981.
11. Jerzy Jurkiewicz, Renate Loll, and Jan Ambjørn. Using causality to solve the puzzle of quantum spacetime. *Scientific American*, July 2008.
12. Seth Lloyd. *Programming the Universe*. Alfred A. Knopf, Random House, Inc., New York, 2006.
13. Renate Loll. The emergence of spacetime, or, quantum gravity on your desktop. *Classical and Quantum Gravity*, 25(11), 2008. doi: 10.1088/0264-9381/25/11/114006. arXiv:0711.0273v2 [gr-qc].
14. Fotini Markopoulou. Dual formulation of spin network evolution, 1997. arXiv:gr-qc/9704013v1.
15. David A. Meyer. *The Dimension of Causal Sets*. PhD thesis, Massachusetts Institute of Technology, 1988.
16. Daniel B. Miller and Edward Fredkin. Two-state, reversible, universal cellular automata in three dimensions. In *CF '05: Proceedings of the 2nd conference on Computing frontiers*, pages 45–51, New York, NY, USA, 2005. ACM. doi: 10.1145/1062261.1062271.

17. Jan Myrheim. Statistical geometry, 1978. CERN preprint TH-2538.
18. Thomas Nowotny and Manfred Requardt. Dimension Theory of Graphs and Networks. *J. Phys.*, A31:2447–2463, 1998. arXiv:hep-th/9707082, doi:10.1088/0305-4470/31/10/018.
19. Ed Pegg. Turmite. In Eric W. Weisstein, editor, MathWorld–A Wolfram Web Resource. http://mathworld.wolfram.com/Turmite.html.
20. Ed Pegg. Turmites. The Wolfram Demonstration Project, http://demonstrations.wolfram. com/Turmites/.
21. Ed Pegg, Jeff Bryant, and Theodore W. Gray. Voronoi diagrams. The Wolfram Demonstration Project, http://demonstrations.wolfram.com/VoronoiDiagrams/.
22. David D. Reid. Introduction to causal sets: an alternate view of spacetime structure. *Can.J.Phys.*, 79(1):1–16, 2001. arXiv:gr-qc/9909075v1.
23. David P. Rideout and Rafael D. Sorkin. Classical sequential growth dynamics for causal sets. *Phys. Rev. D*, 61:024002, Dec 1999. arXiv:gr-qc/9904062v3.
24. Jürgen Schmidhuber. A computer scientist's view of life, the universe, and everything. In Christian Freksa, Matthias Jantzen, and Rüdiger Valk, editors, *Foundations of Computer Science: Potential - Theory - Cognition*, volume 1337 of *Lecture Notes in Computer Science*, pages 201–208. Springer, 1997. arXiv:quant-ph/9904050v1.
25. Rafael D. Sorkin. Causal sets: Discrete gravity, Notes for the Valdivia Summer School, 2003. arXiv:gr-qc/0309009.
26. Max Tegmark. The mathematical universe. *Foundations of Physics*, 38(2):101–150, 2008. arXiv:0704.0646v2 [gr-qc], doi:10.1007/s10701-007-9186-9.
27. W. T. Tutte. A census of planar triangulations. *Canadian Journal of Mathematics*, 14(1):21–38, 1962.
28. Stephen Wolfram. *A New Kind of Science*. Wolfram Media, Inc., 2002.
29. Konrad Zuse. *Rechnender Raum*. Friedrich Vieweg & Sohn, Braunschweig, 1969.
30. Konrad Zuse. Calculating space, 1970. Proj. MAC, MIT, Cambridge, Mass.,Technical Translation AZT-70-164-GEMIT. Original title: "Rechnender Raum".

Chapter 25

The Computable Universe Hypothesis

Matthew P. Szudzik
*Department of Mathematical Sciences,
Carnegie Mellon University, USA*

When can a model of a physical system be regarded as computable? We provide the definition of a *computable physical model* to answer this question. The connection between our definition and Kreisel's notion of a mechanistic theory is discussed, and several examples of computable physical models are given, including models which feature discrete motion, a model which features non-discrete continuous motion, and probabilistic models such as radioactive decay. We show how computable physical models on effective topological spaces can be formulated using the theory of type-two effectivity (TTE). Various common operations on computable physical models are described, such as the operation of coarse-graining and the formation of statistical ensembles. The definition of a computable physical model also allows for a precise formalization of the *computable universe hypothesis*—the claim that all the laws of physics are computable.

1. Introduction

A common way to formalize the concept of a *physical model* is to identify the states of the system being modeled with the members of some set S, and to identify each observable quantity of the system with a function from S to the real numbers.[a] For example, a simple model of planetary motion, with the Earth moving in a circular orbit and traveling at a uniform speed, is the following.

Model 1.1 (Simple Planetary Motion) *Let S be the set of all pairs of real numbers (t, a) such that $a = 360(t - \lfloor t \rfloor)$, where $\lfloor t \rfloor$ denotes the largest integer less than or equal to t. The angular position of the Earth, measured*

---
[a] A more detailed account of this formalism is available in reference [19].

in degrees, is given by the function $\alpha(t, a) = a$. The time, measured in years, is given by the function $\tau(t, a) = t$.

If we wish, for example, to compute the position of the Earth after 2.25 years, we ask: "For which states (t, a) does $\tau(t, a) = 2.25$?" There is only one such state, namely $(2.25, 90)$. Therefore, the position of the Earth after 2.25 years is $\alpha(2.25, 90) = 90$ degrees. We say that the model is *faithful* if and only if the values of the observable quantities in the model match the values that are physically observed.

Church and Turing hypothesized that the functions which are effectively computable by humans are exactly the recursive functions.[b] There have been several attempts [18,29,8,28,21] to extend the Church-Turing thesis to physics, hypothesizing that the laws of physics are, in some sense, computable. But given an arbitrary physical model, it has not been clear exactly how one determines whether or not that model is to be regarded as computable. To date, the best attempt at providing such a definition has been Kreisel's notion of a mechanistic theory [11]. Kreisel suggested the following.

Kreisel's Criterion *The predictions of a physical model are to be regarded as computable if and only if every real number which is observable according to the model is recursive relative to the data uniformly.*

But many seemingly innocuous models have failed to satisfy Kreisel's criterion. For example, the simple model of planetary motion (Model 1.1) fails because given a real number representing the time t, there is no effectively computable procedure which determines the corresponding angle a when a is near the discontinuity at 360 degrees. Models which intuitively seem to have computable predictions often fail to satisfy Kreisel's criterion because discontinuities in their formalisms prevent the models' predictions from being regarded as computable, despite the fact that there are no discontinuities in the actual physical phenomena being modeled [16].

Rather than using Kreisel's criterion to prove that the predictions of established models are computable, an alternate approach is to supply a restrictive formalism which guarantees that the predictions of models expressible in that formalism are computable. This has been the approach

---

[b]Readers unfamiliar with the definition of a recursive function or related terminology, such as uniformity, should consult reference [17]. The original justifications for identifying the effectively computable functions with the recursive functions can be found in references [4,22,23].

taken in references [18,29,8,28]. But difficulties have been encountered expressing important established models in these formalisms. For example, Rosen [18] was unable to describe radioactive decay in the formalism that he had proposed, and work is ongoing to describe established physical models in other computable formalisms.

It is the goal of this paper to provide a general formalism for describing physical models whose predictions are computable, and to show that the computable formalisms studied by previous authors are special cases of our general formalism. In particular, we show in Section 15 that among the members of a large class of physical models, each physical model satisfying Kreisel's criterion has a corresponding model in our formalism. We also avoid some of the difficulties which, for example, prevented the simple model of planetary motion (Model 1.1) from being regarded as computable, as will be seen in Section 4. Our approach also avoids the difficulty that Rosen encountered with radioactive decay, as will be seen in Section 6.

2. Computable Physical Models

The central problem is that physical models use real numbers to represent the values of observable quantities, but that recursive functions are functions of non-negative integers, not functions of real numbers. To show that a model is computable, the model must somehow be expressed using recursive functions. Careful consideration of this problem, however, reveals that the real numbers are not actually necessary in physical models. Non-negative integers suffice for the representation of observable quantities because numbers measured in laboratory experiments necessarily have only finitely many digits of precision. For example, measurements of distances with a measuring stick will always be non-negative integer multiples of the smallest division on the measuring stick. So, we suffer no loss of generality by restricting the values of all observable quantities to be expressed as non-negative integers—the restriction only forces us to make the methods of error analysis, which were tacitly assumed when dealing with real numbers, an explicit part of each model.

Non-negative integers are not only sufficient for the description of direct physical measurements, but are also sufficient for encoding more complex data structures—allowing us to define recursive functions on those data structures. For example, a pair of two non-negative integers x and y can be encoded as a single non-negative integer $\langle x, y \rangle$ using Cantor's pairing

function
$$\langle x, y \rangle = \frac{1}{2}(x^2 + 2xy + y^2 + 3x + y)$$

A pair $\langle x, y \rangle$ of non-negative integers will also be called a *length two sequence* of non-negative integers. A triple (or equivalently, length three sequence) of non-negative integers x, y, and z can be encoded as $\langle \langle x, y \rangle, z \rangle$, and so on. We write $\langle x, y, z \rangle$ as an abbreviation for $\langle \langle x, y \rangle, z \rangle$. An integer i can be encoded as a non-negative integer $\zeta(i)$ using the formula

$$\zeta(i) = \begin{cases} -2i - 1 & \text{if } i < 0 \\ 2i & \text{if } i \geq 0 \end{cases}$$

And a rational number $\frac{a}{b}$ in lowest-terms with $b > 0$ can be encoded as a non-negative integer $\rho(\frac{a}{b})$ using the formula

$$\rho\left(\frac{a}{b}\right) = \zeta\left((\operatorname{sgn} a) 2^{\zeta(a_1-b_1)} 3^{\zeta(a_2-b_2)} 5^{\zeta(a_3-b_3)} 7^{\zeta(a_4-b_4)} 11^{\zeta(a_5-b_5)} \cdots \right)$$

where $a = (\operatorname{sgn} a) 2^{a_1} 3^{a_2} 5^{a_3} 7^{a_4} 11^{a_5} \cdots$ is the prime factorization of the integer a, and similarly for b. We write $(q\,;r)$ as an abbreviation for the pair of rational numbers $\langle \rho(q), \rho(r) \rangle$.

Historically, authors who have wished to restrict themselves to physical models whose predictions are computable have chosen from among a handful of formalisms. For example, Zuse [29] and Fredkin [8] have formalized their models as cellular automata, with each cell of an automaton representing a discrete unit of space and each step of computation in the automaton representing a discrete unit of time. Wolfram [28] has formalized his models in a variety of computational systems, including cellular automata, but has favored network systems for a model of fundamental physics. In each of these cases, the states of a physical system are represented by the states of a computational system (for example, a cellular automaton or a network system) which can be encoded as non-negative integers using the techniques just described. The resulting set of non-negative integers is a recursive set, and the observable quantities of the system are recursive functions of the members of that set. This immediately suggests the following definition.

Definition 2.1 *A computable physical model of a system is a recursive set S of states with a total recursive function ϕ for each observable quantity of the system. $\phi(s)$ is the value of that observable quantity when the system is in state s.*

So, in a computable physical model the set S is a set of non-negative integers, and each observable quantity is a function from non-negative integers to non-negative integers. The models considered by Zuse, Fredkin, and Wolfram are necessarily special sorts of computable physical models, and the set of all computable physical models is a proper subset of all physical models. In order to avoid all ambiguity, we insist that observable quantities be defined operationally [2] in computable physical models, so that, for example, if there were an observable quantity corresponding to time, then that observable quantity would be the time as measured with a specific conventionally-chosen clock in a specific conventionally-chosen reference frame.

An immediate consequence of the definition of a computable physical model is that we can give a precise formal counterpart to the informal claim that all the laws of physics are computable.

Computable Universe Hypothesis *The universe has a recursive set of states U. For each observable quantity, there is a total recursive function ϕ. $\phi(s)$ is the value of that observable quantity when the universe is in state s.*

By a *distinguishable system*, we mean any system for which there is an observable quantity ϕ such that $\phi(s) = 1$ when the system exists in the universe, and such that $\phi(s) = 0$ otherwise. For example, if the system being studied is the orbit of the Earth, then $\phi(s) = 0$ when state s corresponds to a time before the formation of the Earth, and $\phi(s) = 1$ when the Earth exists and is orbiting the Sun. Note that the set of states s in U for which $\phi(s) = 1$ is itself a recursive set whenever U and ϕ are recursive. So, the computable universe hypothesis implies that computable physical models are sufficient for modeling any distinguishable system in the universe—the set of states of that distinguishable system is the set of all members s of U for which $\phi(s) = 1$, and the observable quantities of the distinguishable system are necessarily a subset of the observable quantities of the universe.

3. Discrete Planetary Motion

As a first example of a computable physical model, consider the following model of planetary motion.

Model 3.1 (Discrete Planetary Motion) *Let S be the set of all pairs*

$\langle(r\,;s),(p\,;q)\rangle$ such that

$$r = \frac{i}{10} - \frac{1}{100} \qquad\qquad p = 360(r - \lfloor r \rfloor)$$
$$s = \frac{i+1}{10} + \frac{1}{100} \qquad\qquad q = 360(s - \lfloor s \rfloor)$$

for some integer i between -20000 and 20000. The angular position of the Earth, represented as a range of angles measured in degrees, is given by the function $\alpha\langle(r\,;s),(p\,;q)\rangle = (p\,;q)$. The time interval, measured in years, is given by the function $\tau\langle(r\,;s),(p\,;q)\rangle = (r\,;s)$.

This is a discrete model. That is, the position of the Earth in its orbit is not an exact real number, such as 90 degrees, but is instead an interval, such as $(68.4\,;111.6)$ representing a range of angles between 68.4 degrees and 111.6 degrees.[c] Similarly, time is measured in discrete intervals of length 0.12 years. The earliest time interval in the model is near the year -2000 and the latest time interval is near the year 2000. Moreover, this model is faithful—it is in exact agreement with all observations.

There are ten possible measurements for the angular position of the Earth in Model 3.1:

$(32.4\,;75.6)$	$(68.4\,;111.6)$	$(104.4\,;147.6)$	$(140.4\,;183.6)$
$(176.4\,;219.6)$	$(212.4\,;255.6)$	$(248.4\,;291.6)$	$(284.4\,;327.6)$
$(320.4\,;3.6)$	$(356.4\,;39.6)$		

These are the intervals obtained by dividing the 360 degrees of the circle into ten equal intervals of 36 degrees each, then extending each interval by exactly 3.6 degrees on both sides, bringing the total length of each interval to 43.2 degrees. Therefore, consecutive intervals overlap by 7.2 degrees (there is also overlap in consecutive time intervals), and this serves an important purpose. The Earth's orbit is not, in reality, a perfect circle, and the Earth does not spend an equal amount of time in each of the intervals. But because the eccentricity of the Earth's orbit contributes to, at most, only about a 2 degree deviation [7] from the simple model of planetary motion (Model 1.1), the overlap of these intervals is more than adequate to conceal evidence of the eccentricity, ensuring that this discrete model is faithful. Also note that the overlap is a realistic feature of all known

---

[c]We use decimal numbers to represent exact rational numbers. For example, 68.4 is to be understood as an abbreviation for $\frac{684}{10}$.

instruments which measure angles, since each such instrument has only a limited accuracy. If angles are measured with a protractor, for example, the accuracy might be limited by the thickness of the lines painted on the protractor, which divide one reading from another. For example, if the lines are 7.2 degrees thick, then it might not be possible to distinguish a reading of $(32.4\,;75.6)$ from a reading of $(68.4\,;111.6)$ if the quantity being measured is somewhere on that line (that is, if the quantity is somewhere between 68.4 and 75.6 degrees). The accuracy of measuring instruments is discussed in greater detail in Section 9.

4. Non-Discrete Continuous Planetary Motion

Many commonly-studied computable physical models are discrete, but non-discrete continuous models are also possible. For example, a non-discrete continuous computable physical model of planetary motion is the following.

Model 4.1 (Non-Discrete Continuous Planetary Motion) *Let S be the set of all pairs $\langle (r\,;s), (p\,;q) \rangle$ such that*

$$r = \frac{i}{10^n} - \frac{1}{10^{n+1}} \qquad\qquad p = 360(r - \lfloor r \rfloor)$$
$$s = \frac{i+1}{10^n} + \frac{1}{10^{n+1}} \qquad\qquad q = 360(s - \lfloor s \rfloor)$$

for some integer i and some positive integer n. The angular position of the Earth, represented as a range of angles measured in degrees, is given by the function $\alpha \langle (r\,;s), (p\,;q) \rangle = (p\,;q)$. The time interval, measured in years, is given by the function $\tau \langle (r\,;s), (p\,;q) \rangle = (r\,;s)$.

Like the discrete model, angular position and time are measured in intervals, but in this case the intervals are not all the same length. In particular, there are arbitrarily small intervals for the observable quantities of position and time, meaning that these quantities may be measured to arbitrary precision. This feature of Model 4.1 allows us to speak about real-valued positions and times, despite the fact that the values of observable quantities in the model are all non-negative integers, not real numbers.

This is because a real number is not the result of a single measurement, but is instead the limit of a potentially-infinite sequence of measurements. Suppose, for example, that we wish to measure the circumference of a circle whose diameter is exactly one meter. Measured with unmarked metersticks, we measure the circumference to be 3 meters. If the sticks are marked with millimeters, then we measure the circumference to be about 3.141 meters. And if they are marked with micrometers, then we measure a circumference of about 3.141592 meters. If we continue this process indefinitely with increasingly precise measuring instruments, then in the infinite limit we approach the real number π.

More formally, for each real number x there is an infinite sequence of nested intervals $(a_0\,;b_0)$, $(a_1\,;b_1)$, $(a_2\,;b_2)$, ... that converges to x. Given such a sequence, the function ϕ such that $\phi(n) = (a_n\,;b_n)$ for each non-negative integer n is said to be an *oracle* for x. Note that there is more than one distinct sequence of nested intervals converging to x, and therefore more than one oracle for each x. Of particular importance is the *standard decimal oracle* o_x for the real number x. By definition, $o_x(n) = (a_n\,;b_n)$, where

$$a_n = \frac{\lfloor 10^{n+1}x \rfloor}{10^{n+1}} - \frac{c}{10^{n+1}} \qquad b_n = \frac{\lfloor 10^{n+1}x \rfloor + 1}{10^{n+1}} + \frac{c}{10^{n+1}}$$

for each non-negative integer n, and where the *accuracy factor* c is a positive rational number constant. We say that x is a *recursive real number* if and only if o_x is a recursive function. Note that not all real numbers are recursive [22].

Now, returning to Model 4.1, suppose that we are asked to find the position of the Earth at some real-valued time t. Suppose further that we are given the oracle o_t with accuracy factor $c = \frac{1}{10}$. Note that as we increase n, the values $o_t(n)$ are increasingly precise measurements of the time t in Model 4.1. Therefore, for each n there is some state $\langle o_t(n), (p_n\,;q_n) \rangle$ in the set S of Model 4.1. Because S is a recursive set, and because there is exactly one state corresponding to each time measurement, the function ϵ such that $\epsilon(n) = (p_n\,;q_n)$ is a recursive function relative to the oracle o_t. In fact, if $o_t(n) = (r_n\,;s_n)$, then

$$\epsilon(n) = \Big(360(r_n - \lfloor r_n \rfloor)\,;360(s_n - \lfloor s_n \rfloor)\Big)$$

for each non-negative integer n. And since the sequence of intervals $(r_0\,;s_0)$, $(r_1\,;s_1)$, $(r_2\,;s_2)$, ... converges to t, it immediately follows that the sequence of intervals $\epsilon(0)$, $\epsilon(1)$, $\epsilon(2)$, ... converges to $a = 360(t - \lfloor t \rfloor)$ whenever t is not an integer. In other words, ϵ is an oracle for the angular position a.

But in the case that t is an integer,

$$\epsilon(n) = \Big(360 - \frac{36}{10^{n+1}}\,;\frac{396}{10^{n+1}}\Big)$$

for all non-negative integers n, and $(356.4\,;39.6)$, $(359.64\,;3.96)$, $(359.964\,;0.396)$, ... is the resulting sequence. In the standard topology of the real numbers an interval $(x\,;y)$ should have $x < y$, so the question of whether or not this sequence converges to a point a in that standard topology cannot be meaningfully answered. But if we are willing to abandon the standard topology of the real numbers, then we may conventionally define this sequence to converge to $a = 0$. In fact, this definition is tantamount to establishing the topology of a circle of circumference 360 for all angles a.[d] Of course, this definition is justified since the readings after 360 on a measuring instrument for angles are identified with those readings after 0. In other words, angles really do lie in a circle.

So, given the oracle o_t for a real-valued time t, Model 4.1 allows us to compute an oracle ϵ for the angular position a of the Earth at that time. These predictions are in complete agreement with the predictions of the simple model of planetary motion (Model 1.1). In fact, imposing the appropriate topology on the space of angles a, the mapping from t to a in Model 4.1 is continuous. The same mapping is discontinuous in the standard topology of the real numbers, which leads Kreisel's criterion to fail for Model 1.1. The formulation of computable physical models on effective topological spaces is discussed in greater detail in Sections 11 through 15.

5. Coarse-Graining

Observable quantities in computable physical models are defined operationally. This means that each observable quantity is defined so as to correspond to a specific physical operation, such as the operation of comparing a length to the markings on a meterstick (where the meterstick itself is constructed according to a prescribed operation). This is problematic for the non-discrete continuous model of planetary motion (Model 4.1) because, for example, arbitrary precision angle measurements are made with a single observable quantity in the model. That is, to assert that a model such as Model 4.1 is faithful, one must assert that there exists an operation which is capable of measuring angles to arbitrary precision. It is not known whether or not such an operation actually exists. And although the point is somewhat moot, since Model 4.1 is clearly not faithful, it raises the question of

---

[d] A basis for this topology is represented by the set of all possible angle measurements. In particular, if $x < y$ then $(x\,;y)$ represents the set of all real numbers a such that $x < a < y$, and if $x > y$ then $(x\,;y)$ represents the set of all real numbers a such that $0 \leq a < y$ or $x < a < 360$.

whether this is an accidental feature of Model 4.1, or whether it is a feature common to all non-discrete continuous computable physical models.

A more practical alternative to Model 4.1 might introduce an infinite sequence of observable quantities $\alpha_1, \alpha_2, \alpha_3, \ldots$, each with finitely many digits of precision, and each more precise than its predecessor in the sequence. In this case, given a state s, the values $\alpha_1(s), \alpha_2(s), \alpha_3(s), \ldots$ would form a sequence of intervals converging to a real number representing the angular position of the Earth in that state. But a computable physical model has only countably many states, and there are uncountably many real numbers in the interval $(0\,;360)$. Therefore, there must be some real number position in the interval $(0\,;360)$ that the Earth never attains.[e] That is, a computable physical model of this alternative form is not continuous in the intended topology.

Rather than considering arbitrary precision measurements, let us introduce just one additional level of precision into the discrete model of planetary motion (Model 3.1).

Model 5.1 *Let S be the set of all quadruples $\langle(r_1\,;s_1),(r_2\,;s_2),(p_1\,;q_1),(p_2\,;q_2)\rangle$ such that*

$$r_1 = \frac{i}{10} - \frac{1}{10^2} \qquad p_1 = 360(r_1 - \lfloor r_1 \rfloor)$$
$$s_1 = \frac{i+1}{10} + \frac{1}{10^2} \qquad q_1 = 360(s_1 - \lfloor s_1 \rfloor)$$
$$r_2 = \frac{j}{10^2} - \frac{1}{10^3} \qquad p_2 = 360(r_2 - \lfloor r_2 \rfloor)$$
$$s_2 = \frac{j+1}{10^2} + \frac{1}{10^3} \qquad q_2 = 360(s_2 - \lfloor s_2 \rfloor)$$

for some integers i and j with $10i \leq j \leq 10i + 9$. The angular position of the Earth, represented as a range of angles measured in degrees with a low-precision measuring instrument, is given by the function

$$\alpha_1\langle(r_1\,;s_1),(r_2\,;s_2),(p_1\,;q_1),(p_2\,;q_2)\rangle = (p_1\,;q_1)$$

The angular position of the Earth, represented as a range of angles measured in degrees with a high-precision measuring instrument, is given by the function

$$\alpha_2\langle(r_1\,;s_1),(r_2\,;s_2),(p_1\,;q_1),(p_2\,;q_2)\rangle = (p_2\,;q_2)$$

---

[e] In particular, this is a real number constructed by diagonalizing over those real numbers which are associated with each of the countably many states.

The time interval, measured in years by a low-precision measuring instrument, is given by the function

$$\tau_1 \langle (r_1\,;s_1), (r_2\,;s_2), (p_1\,;q_1), (p_2\,;q_2) \rangle = (r_1\,;s_1)$$

The time interval, measured in years by a high-precision measuring instrument, is given by the function

$$\tau_2 \langle (r_1\,;s_1), (r_2\,;s_2), (p_1\,;q_1), (p_2\,;q_2) \rangle = (r_2\,;s_2)$$

Note that if the high-precision observable quantities α_2 and τ_2 are ignored, then the predictions of Model 5.1 agree exactly with the predictions of Model 3.1.[f] The process of removing observable quantities from a model to obtain a new model with fewer observable quantities is called *coarse-graining*. But while Model 3.1 is faithful, Model 5.1 is not faithful—physical measurements do not agree with the values of the observable quantities α_2 and τ_2 because the orbit of the Earth is not a perfect circle.

A traditional conception of science regards all physical models as inexact approximations of reality, and holds that the goal of science is to produce progressively more accurate models whose predictions more closely match observations than the predictions of previous models. That conception of science is reasonable when the values of observable quantities are real numbers, since the real numbers predicted by physical models are never exactly the same as the real numbers 'measured' in the laboratory. But when non-negative integers are used for the values of observable quantities, then an alternate conception of science is possible.

In this alternate conception there exist faithful models that are in exact agreement with reality, but perhaps only for a small subset of all physically observable quantities. For example, Model 3.1 is faithful, but only predicts the angular position of the Earth to within 43.2 degrees, and only for a limited range of times. The goal of science is then to produce *more refined* models. That is, the goal of science is to discover faithful models which have larger sets of observable quantities, and are therefore capable of predicting increasing numbers of facts.

---

[f]But it should be noted that the model obtained by omitting α_2 and τ_2 from Model 5.1 is not identical to Model 3.1. In particular, for each state in Model 3.1, there are ten indistinguishable states in the model obtained by omitting α_2 and τ_2 from Model 5.1. That is, these models are not isomorphic. See Section 10.

6. Radioactive Decay

Given non-negative integers x and y, let $\beta(x,y)$ be the length y sequence composed of the first y bits in the binary expansion of x. For example $\beta(13,6) = \langle 0,0,1,1,0,1 \rangle$. Now suppose that a single atom of a radioactive isotope, such as nitrogen-13, is placed inside a detector at time $t = 0$. We say that the detector has status 1 if it has detected the decay of the isotope, and has status 0 otherwise. The *history* of the detector at time t is the length t sequence of bits corresponding to the status of the detector at times 1 through t. For example, if the isotope decays sometime between $t = 2$ and $t = 3$, then the history of the detector at time $t = 5$ is $\langle 0,0,1,1,1 \rangle$. The following computable physical model models the status of the detector as a function of time.

Model 6.1 (Radioactive Decay) Let S be the set of all triples $\langle t, \beta(2^n - 1, t), j \rangle$ where n, t, and j are non-negative integers such that $n \leq t$, $t \neq 0$, and $2j \leq 2^n - 1$. The history of the detector is given by the function $\eta \langle t, h, j \rangle = h$, and the time, measured in units of the half-life of the isotope, is given by the function $\tau \langle t, h, j \rangle = t$.

This is a model of the many-worlds interpretation [5] of radioactive decay. Suppose that one asks, "What will the status of the detector be at time $t = 2$?" There are four states $\langle t, h, j \rangle$ such that $\tau \langle t, h, j \rangle = 2$, namely

$$\langle 2, \langle 0,0 \rangle, 0 \rangle \quad \langle 2, \langle 0,1 \rangle, 0 \rangle \quad \langle 2, \langle 1,1 \rangle, 0 \rangle \quad \langle 2, \langle 1,1 \rangle, 1 \rangle$$

In three of these states, the detector has status 1, and in one state it has status 0. If we assume that each state of the system is equally likely, then there is a $\frac{3}{4}$ probability that the detector will have status 1 at time $t = 2$. But if we ask, "If the detector has status 1 at time $t = 1$, then what will its status be at time $t = 2$?" The answer is "1", since the detector has status 1 at time 2 in both states where the detector had status 1 at time 1. These results are in agreement with conventional theory.

7. Ensembles of Physical Models

Suppose that a planet orbits a distant star and that we are uncertain of the planet's orbital period. In particular, suppose that we believe its motion is faithfully described by either the discrete model of planetary motion (Model 3.1) or by the following computable physical model.

Model 7.1 *Let S be the set of all pairs $\langle (r\,;s), (p\,;q) \rangle$ such that*

$$r = \frac{i}{10} - \frac{1}{100} \qquad p = 360\left(\frac{n}{10} - \frac{1}{100} - \left\lfloor \frac{n}{10} - \frac{1}{100} \right\rfloor\right)$$

$$s = \frac{i+1}{10} + \frac{1}{100} \qquad q = 360\left(\frac{n+1}{10} + \frac{1}{100} - \left\lfloor \frac{n+1}{10} + \frac{1}{100} \right\rfloor\right)$$

for some integer i between -20000 and 20000, and such that $n = \lfloor i/4 \rfloor$. The angular position of the planet, represented as a range of angles measured in degrees, is given by the function $\alpha\langle (r\,;s), (p\,;q) \rangle = (p\,;q)$. The time interval, measured in Earth years, is given by the function $\tau\langle (r\,;s), (p\,;q) \rangle = (r\,;s)$.

Note that this model is similar to Model 3.1, except that the orbital period of the planet is 4 Earth years, rather than 1 Earth year.

If for each of the two models we are given a rational number expressing the probability that that model is faithful, then a *statistical ensemble* of the models may be constructed. For example, if Model 3.1 is twice as likely as Model 7.1, then a corresponding statistical ensemble is the following. Note that this statistical ensemble is itself a computable physical model.

Model 7.2 (Ensemble of Models) *Let S be the set of all triples $\langle (r\,;s), (p\,;q), j \rangle$ such that*

$$r = \frac{i}{10} - \frac{1}{100} \qquad p = 360\left(\frac{n}{10} - \frac{1}{100} - \left\lfloor \frac{n}{10} - \frac{1}{100} \right\rfloor\right)$$

$$s = \frac{i+1}{10} + \frac{1}{100} \qquad q = 360\left(\frac{n+1}{10} + \frac{1}{100} - \left\lfloor \frac{n+1}{10} + \frac{1}{100} \right\rfloor\right)$$

for some integer i between -20000 and 20000, where $j = 0, 1,$ or 2, and where

$$n = \begin{cases} i & \text{if } j = 0 \text{ or } 1 \\ \lfloor i/4 \rfloor & \text{if } j = 2 \end{cases}$$

The angular position of the planet, represented as a range of angles measured in degrees, is given by the function $\alpha\langle (r\,;s), (p\,;q), j \rangle = (p\,;q)$. The time interval, measured in Earth years, is given by the function $\tau\langle (r\,;s), (p\,;q), j \rangle = (r\,;s)$.

Since Model 3.1 is twice as likely as Model 7.1, there are two states, $\langle (r\,;s), (p\,;q), 0 \rangle$ and $\langle (r\,;s), (p\,;q), 1 \rangle$ in the ensemble for each state $\langle (r\,;s), (p\,;q) \rangle$ in Model 3.1, and there is one state $\langle (r\,;s), (p\,;q), 2 \rangle$ in the ensemble for each state $\langle(r\,;s),(p\,;q)\rangle$ in Model 7.1. Note that the index j in each state $\langle(r\,;s),(p\,;q),j\rangle$ of the ensemble is not observable.

Now, if we ask for the position of the planet during the time interval $(0.29\,;0.41)$, for example, there are three possible states $\langle(r\,;s),(p\,;q),j\rangle$ in the ensemble such that

$$\tau\langle(r\,;s),(p\,;q),j\rangle = (0.29\,;0.41)$$

namely

$$\langle(0.29\,;0.41),(104.4\,;147.6),0\rangle$$
$$\langle(0.29\,;0.41),(104.4\,;147.6),1\rangle$$
$$\langle(0.29\,;0.41),(356.4\,;39.6),2\rangle$$

Since the planet's angular position is $(104.4\,;147.6)$ for two of these three states, the position measurement $(104.4\,;147.6)$ has a probability of $\frac{2}{3}$. Similarly, because the planet's angular position is $(356.4\,;39.6)$ for one of the three states, the position measurement $(356.4\,;39.6)$ has a probability of $\frac{1}{3}$. These probabilities are a direct reflection of our uncertainty about which of the two underlying physical models, Model 3.1 or Model 7.1, is the true faithful model. In particular, because Model 3.1 has been deemed twice as likely as Model 7.1, the position of the planet in Model 3.1, namely $(104.4\,;147.6)$, has twice the probability of the position predicted by Model 7.1, namely $(356.4\,;39.6)$.

It is important to note that there is no observable quantity corresponding to probability in Model 7.2. Instead, probability is a mathematical tool used to interpret the model's predictions. This sort of interpretation of an ensemble of models is appropriate whenever the ensemble is composed from all possible models which could describe a particular system, with the number of copies of states of the individual models reflecting our confidence in the predictions of those models. See reference [10] for a more detailed account of this subjectivist interpretation of probability in physics.

Ensembles may be constructed in other circumstances as well, and we may refer to such ensembles as *non-statistical ensembles* of physical models. Non-statistical ensembles of physical models are commonplace in the sciences. For example, they result whenever a constant, such as an initial position, is left unspecified in the statement of a model. That model can then be used to describe any member of a family of systems, each of which may have a different value for the constant. But most importantly, when a non-statistical ensemble of physical models is constructed, no claims as to the likelihood of one value of the constant, as compared to some other value of the constant, are being made. In fact, this is the defining characteristic of a non-statistical ensemble of models. Non-statistical ensembles can be useful because they provide a convenient way to collect together sets of closely-related models.

8. Incompatible Measurements

A pair of measurements is said to be *simultaneous* if and only if they are both performed while the system is in a single state. An essential feature of quantum mechanical systems is that there may be quantities which are not simultaneously measurable. For example, the measurement of one quantity, such as the position of a particle, might affect the subsequent measurement of another quantity, such as the particle's momentum. Such measurements are said to be *incompatible*. It is natural to ask whether computable physical models can be used to describe systems which feature incompatible measurements.

Discrete quantum mechanical systems are often formalized as follows [6,24]. The *quantum mechanical state* of a system is a normalized vector v in some normed complex vector space V. Typically, V is a Hilbert space and v is a *wave function*. For each *quantum mechanical measurement* there is a corresponding set $B = \{v_1, v_2, v_3, \ldots\}$ of normalized basis vectors for V. Each member of B corresponds to a possible value of the measurement. Because B is a basis for V, $v = a_1 v_1 + a_2 v_2 + a_3 v_3 + \cdots$ for some complex numbers a_1, a_2, a_3, ... If the system is in quantum mechanical state v and no two members of B correspond to the same measurement value,[g] then the probability that the measurement will have the value corresponding to v_n is $|a_n|^2$. In this case, if the actual value which is measured is the value corresponding to v_n, then the quantum mechanical state of the system immediately after that measurement is v_n. The state v is said to have *collapsed* to v_n. During the time between measurements, the quantum mechanical state of a system may evolve according to a rule such as Schrödinger's equation.

Consider, for example, the problem of measuring the components of the spin of an isolated electron. In this case, V is the set of all vectors (a, b) such that a and b are complex numbers, where the norm $\|(a, b)\|$ is defined to

---

[g]Alternatively, if $v_{n_1}, v_{n_2}, v_{n_3}, \ldots$ are distinct basis vectors corresponding to the same measurement value, then the probability of measuring the value is $|a_{n_1}|^2 + |a_{n_2}|^2 + |a_{n_3}|^2 + \cdots$. If that value is actually measured, then the state of the system immediately after the measurement is the normalization of $a_{n_1} v_{n_1} + a_{n_2} v_{n_2} + a_{n_3} v_{n_3} + \cdots$. See reference [13].

be $\sqrt{|a|^2 + |b|^2}$. A quantum mechanical measurement of the z component of the electron's spin has two possible values, $-\frac{1}{2}\hbar$ and $+\frac{1}{2}\hbar$. The basis vectors corresponding to these values are $(0, 1)$ and $(1, 0)$, respectively. The quantum mechanical measurement of another component of the electron's spin, lying in the xz plane at an angle of 60 degrees to the z axis, also has two possible values, $-\frac{1}{2}\hbar$ and $+\frac{1}{2}\hbar$. The basis vectors corresponding to these values are $\left(-\frac{1}{2}, \frac{\sqrt{3}}{2}\right)$ and $\left(\frac{\sqrt{3}}{2}, \frac{1}{2}\right)$, respectively. So, for example, if the spin component in the z direction is measured to have a value of $+\frac{1}{2}\hbar$ at time $t = 0$, then since

$$(1,0) = -\frac{1}{2}\left(-\frac{1}{2}, \frac{\sqrt{3}}{2}\right) + \frac{\sqrt{3}}{2}\left(\frac{\sqrt{3}}{2}, \frac{1}{2}\right)$$

there is a $\left|\frac{\sqrt{3}}{2}\right|^2 = \frac{3}{4}$ probability that if the 60-degree electron spin component is measured at time $t = 1$, then that component will also have a value of $+\frac{1}{2}\hbar$.

Supposing that the 60-degree electron spin component is measured to have a value of $+\frac{1}{2}\hbar$ at time $t = 1$, a similar line of reasoning implies that if the spin's z component is measured at time $t = 2$, then there is a $\left|\frac{1}{2}\right|^2 = \frac{1}{4}$ probability that the value of that measurement will be $-\frac{1}{2}\hbar$, since

$$\left(\frac{\sqrt{3}}{2}, \frac{1}{2}\right) = \frac{1}{2}(0, 1) + \frac{\sqrt{3}}{2}(1, 0)$$

Therefore, if the z component of the electron's spin is measured at time $t = 0$, followed by a measurement of the 60-degree spin component at time $t = 1$, and followed by another measurement of the z component at time $t = 2$, then the values of the two measurements of the z component need not be the same. Indeed, the quantum mechanical state of the system does not change[h] between times $t = 0$ and $t = 1$, or between times $t = 1$ and $t = 2$, but the measurement of the 60-degree spin component at time $t = 1$ disturbs the system and can potentially change the value of any subsequent measurement of the z component. That is, measurement of the electron's 60-degree spin component is incompatible with measurement of its z component.

Let us formalize this system as a computable physical model. The system is composed of the electron, the apparatus used to make the quantum mechanical measurements, and the researcher who chooses which compo-

---

[h] In this case, the quantum mechanical state of the system does not change between measurements because the electron is isolated. For example, the electron is free from external electromagnetic fields or other influences that might cause its spin to precess.

nents to measure.[i] We assume that the quantum mechanical state of the electron is $(1,0)$ at time $t = 0$, and that the researcher makes subsequent quantum mechanical measurements of the electron's spin components at times $t = 1$ and $t = 2$. When a quantum mechanical measurement is performed, a record is made (perhaps in the researcher's notebook) of the value of this measurement and of the component that was measured. We construct the computable physical model of this system from the point of view of an agent who observes only this recorded history and the time.

Model 8.1 (Electron Spin Measurement) *Let S be the set of all triples $\langle t, h, j \rangle$ such that*

$$
\begin{array}{ccc}
t = 1 & t = 1 & t = 1 \\
h = (0\,;+1) \quad\text{or}\quad & h = (60\,;-1) \quad\text{or}\quad & h = (60\,;+1) \\
j = 0 & j = 1 & j = 2 + m
\end{array}
$$

or

$$
\begin{array}{cc}
t = 2 & t = 2 \\
h = \langle (0\,;+1), (0\,;+1) \rangle \quad\text{or}\quad & h = \langle (0\,;+1), (60\,;-1) \rangle \\
j = 5 & j = 6
\end{array}
$$

or

$$
\begin{array}{cc}
t = 2 & t = 2 \\
h = \langle (0\,;+1), (60\,;+1) \rangle \quad\text{or}\quad & h = \langle (60\,;-1), (0\,;-1) \rangle \\
j = 7 + m & j = 10
\end{array}
$$

or

$$
\begin{array}{cc}
t = 2 & t = 2 \\
h = \langle (60\,;-1), (0\,;+1) \rangle \quad\text{or}\quad & h = \langle (60\,;-1), (60\,;-1) \rangle \\
j = 11 + m & j = 14
\end{array}
$$

or

$$
\begin{array}{cc}
t = 2 & t = 2 \\
h = \langle (60\,;+1), (0\,;-1) \rangle \quad\text{or}\quad & h = \langle (60\,;+1), (0\,;+1) \rangle \\
j = 15 + n & j = 24 + m
\end{array}
$$

---

[i]We refrain from asking questions about the probability with which the researcher chooses which components to measure. That is, this model describes a non-statistical ensemble of researchers.

or
$$t = 2$$
$$h = \langle (60\,;+1), (60\,;+1) \rangle$$
$$j = 27 + m$$

for some integers m and n with $0 \leq m \leq 2$ and $0 \leq n \leq 8$. The time is given by the function $\tau \langle t, h, j \rangle = t$. The history is given by the function $\eta \langle t, h, j \rangle = h$. A history is a chronological sequence of records, with the leftmost record being the oldest. Each record is a pair $(a\,;b)$ of rational numbers, where a is the angle from the z axis, measured in degrees, of a component of the electron's spin, and where b is the value of that component, measured in units of $\frac{1}{2}\hbar$.

Note that each state $\langle t, h, j \rangle$ has a distinct index j, which we will use to identify that particular state.

Model 8.1 corresponds to the quantum mechanical system in the following sense. First, the quantum mechanical state of the system at time t corresponds to a set of states in the computable physical model. For example, if the researcher decides to measure the 60-degree component of the electron's spin at time $t = 1$, then the quantum mechanical state of the system is represented by the set of states with indices 1 through 4. Assuming that the states in the set are equally likely, there is a $\frac{3}{4}$ probability that this component will have a measured value of $+\frac{1}{2}\hbar$, for example. Immediately after the measurement is made, the quantum mechanical state collapses, becoming either the set of states with indices 2 through 4, or the singleton set containing only the state with index 1. The collapse occurs because the information provided by the quantum mechanical measurement allows us to identify the state of the system more precisely, eliminating those states which disagree with the measurement result.[j] The quantum mechanical state then evolves to a new set of states at time $t = 2$. For example, if the measured value of the 60-degree electron spin component is $+\frac{1}{2}\hbar$ at time $t = 1$, and if the researcher plans to measure the 0-degree electron spin component (that is, the z component) at time $t = 2$, then the quantum mechanical state immediately before that measurement at time $t = 2$ is the set of states with indices 15 through 26.

Computable physical models similar to Model 8.1 can be constructed for quantum mechanical systems which satisfy the following criteria.

---

[j] For a more detailed discussion of this ensemble interpretation of the collapse of a quantum mechanical state, see reference [1].

(1) There is a set of possible measurements $\{m_0, m_1, m_2, \ldots, m_i, \ldots\}$ indexed by non-negative integers i.
(2) Every discrete time step, one measurement from this set is performed.
(3) The possible values of each measurement m_i are identified with non-negative integers.
(4) If $\phi(i, n, t, h)$ is the probability that the measurement with index i has the value n, given that the measurement is performed at time step t and that $h = \langle (i_1; n_1), \ldots, (i_{t-1}; n_{t-1}) \rangle$ is the history of past measurements and their values, then $\phi(i, n, t, h)$ is a rational number.
(5) If there is no measurement with index i or if the non-negative integer n does not correspond to a value of the measurement with index i, then $\phi(i, n, t, h) = 0$.
(6) For each choice of non-negative integers i, t, and h, there are only finitely many non-negative integers n such that $\phi(i, n, t, h) > 0$.
(7) ϕ is a recursive function.

If a quantum mechanical system satisfies these criteria, then we can determine whether or not

$$s = \langle t, \langle (i_1; n_1), (i_2; n_2), \ldots, (i_t; n_t) \rangle, j \rangle$$

is in the set S of states of the corresponding computable physical model as follows. First, if $t = 0$, then s is not in S. Next, let $h_1 = 0$ and for each positive integer k with $1 < k \leq t$, let

$$h_k = \langle (i_1; n_1), (i_2; n_2), \ldots, (i_{k-1}; n_{k-1}) \rangle$$

Now we perform the following calculations for each positive integer $k \leq t$. If $\phi(i_k, n_k, k, h_k) = 0$, then s is not in S. Otherwise, there must be finitely many non-negative integers n such that the probability $\phi(i_k, n, k, h_k)$ is greater than zero. Since probabilities must sum to 1, those values for n may be found exhaustively by calculating $\phi(i_k, 0, k, h_k)$, $\phi(i_k, 1, k, h_k)$, $\phi(i_k, 2, k, h_k)$, and so on, until the the sum of these probabilities reaches 1. Let d_k be the least common denominator of these rational probabilities, and let a_k be the unique positive integer such that

$$\phi(i_k, n_k, k, h_k) = \frac{a_k}{d_k}$$

If $j < a_1 a_2 \cdots a_t$, then s is in S. Otherwise, s is not in S.

9. The Accuracy of Measuring Instruments

An important feature of the discrete model of planetary motion (Model 3.1) is that the intervals representing time and angle measurements overlap. The amount of overlap between adjacent intervals is determined by the accuracy of the corresponding measuring instrument. The introduction of overlapping intervals is motivated by an argument such as the following.

If Model 3.1 were constructed using disjoint, non-overlapping intervals, then the states $\langle (r\,;s), (p\,;q) \rangle$ of that model would be given by

$$r = i/10 \qquad\qquad p = 360(r - \lfloor r \rfloor)$$
$$s = (i+1)/10 \qquad\qquad q = 360(s - \lfloor s \rfloor)$$

where i is an integer. In particular, $\langle (0.2\,;0.3), (72\,;108) \rangle$ and $\langle (0.3\,;0.4), (108\,;144) \rangle$ would be two such states, with $(r\,;s)$ representing the state's time interval, measured in years, and with $(p\,;q)$ representing the corresponding interval of angular positions for the Earth, measured in degrees. According to this model, if the position of the Earth is measured at time $t = 0.298$ years, then t is within the interval $(0.2\,;0.3)$, and the state of the system is $\langle (0.2\,;0.3), (72\,;108) \rangle$. Therefore, according to this model, the position of the Earth should be between 72 and 108 degrees. Indeed, the simple model of planetary motion (Model 1.1) predicts that the angular position of the Earth at time $t = 0.298$ years should be $360(0.298 - \lfloor 0.298 \rfloor) \approx 107$ degrees. But the true position of the Earth in its orbit deviates from Model 1.1. In this case, the true position of the Earth at time $t = 0.298$ years is about 109 degrees,[k] which is outside the interval $(72\,;108)$. Therefore, if the discrete model were constructed using disjoint, non-overlapping intervals, then the model would fail when $t = 0.298$ years.

But the discrete model of planetary motion (Model 3.1) was constructed using overlapping intervals. In particular,

$$\langle (0.19\,;0.31), (68.4\,;111.6) \rangle \qquad \langle (0.29\,;0.41), (104.4\,;147.6) \rangle$$

are two states in Model 3.1. Note that at time $t = 0.298$ years, Model 3.1 could be in either of these two states. Furthermore, any pair of real-valued time t and angle a measurements which satisfy

$$|a - 360(t - \lfloor t \rfloor)| < 7.2$$

---

[k]This is assuming that time is measured in anomalistic years, with each year beginning at perihelion passage. During the course of a year, the position of the Earth is the true anomaly, measured relative to that perihelion passage.

fall within the time and angle intervals of some common state of Model 3.1. Since $|a - 360(t - \lfloor t \rfloor)|$ is at most 2 degrees [7] for all physically observed angles a measured at times t, Model 3.1 is faithful.

It is important to point out, though, that Model 3.1 is faithful only if the results of measurements are uncertain when they occur within the region of overlap. For example, at time $t = 0.306$ years, two results of a time measurement are possible, $(0.19 ; 0.31)$ and $(0.29 ; 0.41)$, and an observer cannot be certain which of these intervals is the value of the measurement. The actual angular position of the Earth at time $t = 0.306$ years is about 112 degrees, so $(104.4 ; 147.6)$ is the only possible result of a position measurement. Since

$$\langle (0.19 ; 0.31), (104.4 ; 147.6) \rangle$$

is not one of the states of Model 3.1, the observer is expected to realize, in retrospect, after measuring the angular position, that the true time measurement must have been $(0.29 ; 0.41)$. After providing a model for the phenomenon of accuracy, we will be able to reformulate Model 3.1 so that the results of measurements no longer possess this sort of ambiguity.

But first, note that the accuracy of a measuring instrument, by definition, can only be quantified relative to some other, more precise quantity. For example, the argument above, concerning accuracy in Model 3.1, makes frequent reference to exact real-valued angles and times. Indeed, even when we express an angle measurement as an interval, such as $(68.4 ; 111.6)$, we are implying that it is possible to distinguish an angle of 68.4 degrees from an angle of 111.6 degrees, and that other angles lie between those two values. In principle, though, it is possible to describe the accuracy of a measuring instrument in a purely discrete manner, without any mention of real numbers. For example, let us consider an instrument for measuring distances in meters, with the value of a measurement represented as an integer number of meters. The accuracy of this measuring instrument can be quantified relative to a second instrument which measures distances in decimeters.

Presumably, the phenomenon of accuracy results from our inability to properly calibrate measuring instruments. Although there are many different underlying causes of calibration error, it suffices to consider only one such cause for a simple model of this phenomenon. We will suppose that when we measure a distance in meters, that we have difficulty aligning the measuring instrument with the origin, so that sometimes the instrument is aligned a decimeter too far in the negative direction, and at other times a decimeter too far in the positive direction. Hence, there are two different physical models for the measurement. In one model the instrument is misaligned in the negative direction, and in the other model it is misaligned in the positive direction. Since we do not know which of these two models describes any one particular measurement, it is appropriate to combine them in the following statistical ensemble.

Model 9.1 *Let S be the set of all triples $\langle \zeta(m), \zeta(d), \zeta(i) \rangle$ such that*

$$m = \left\lfloor \frac{d+i}{10} \right\rfloor$$

where d is an integer, and where $i = -1$ or $+1$. The distance, measured in meters, is given by the function

$$\mu \langle \zeta(m), \zeta(d), \zeta(i) \rangle = \zeta(m)$$

The same distance, measured in decimeters, is given by the function

$$\delta \langle \zeta(m), \zeta(d), \zeta(i) \rangle = \zeta(d)$$

Note that the function ζ was defined in Section 2. Also note that the index i in each state $\langle \zeta(m), \zeta(d), \zeta(i) \rangle$ represents the calibration error, which is either -1 decimeter or $+1$ decimeter. Model 9.1 is a computable physical model.

A measurement of d decimeters in Model 9.1 can be interpreted as corresponding to an interval of $\left(\frac{d}{10} ; \frac{d+1}{10} \right)$ meters. Note that a measurement of 9 decimeters (corresponding to an interval of $(0.9 ; 1.0)$ meters), for example, is possible in two distinct states of the model:

$$\langle \zeta(0), \zeta(9), \zeta(-1) \rangle \qquad \langle \zeta(1), \zeta(9), \zeta(+1) \rangle$$

Similarly, a measurement of 10 decimeters (corresponding to an interval of $(1.0 ; 1.1)$ meters) is possible in the states

$$\langle \zeta(0), \zeta(10), \zeta(-1) \rangle \qquad \langle \zeta(1), \zeta(10), \zeta(+1) \rangle$$

Hence, a measurement of 0 meters overlaps with a measurement of 1 meter on the intervals $(0.9 ; 1.0)$ and $(1.0 ; 1.1)$. And in general, a measurement of m meters overlaps with a measurement of $m+1$ meters on the intervals $(m + 0.9 ; m + 1.0)$ and $(m + 1.0 ; m + 1.1)$. Therefore, a measurement of m meters in Model 9.1 can be understood as corresponding to an interval of $(m - 0.1 ; m + 1.1)$ meters, with adjacent intervals overlapping by 0.2 meters.

Of course, this interpretation of Model 9.1 presumes that decimeters can be measured with perfect accuracy. A more realistic computable physical model can be constructed by supposing that decimeter measurements can also be misaligned, for example, by -1 centimeter or $+1$ centimeter. Note that centimeters are treated as unobserved, purely theoretical constructions in this model—there is no observable quantity for centimeter measurements.

Model 9.2 *Let S be the set of all quintuples $\langle \varsigma(m), \varsigma(d), \varsigma(c), \varsigma(i), \varsigma(j) \rangle$ such that*

$$m = \left\lfloor \frac{c + 10i}{100} \right\rfloor \qquad d = \left\lfloor \frac{c + j}{10} \right\rfloor$$

where c is an integer, $i = -1$ or $+1$, and $j = -1$ or $+1$. The distance, measured in meters, is given by the function

$$\mu \langle \varsigma(m), \varsigma(d), \varsigma(c), \varsigma(i), \varsigma(j) \rangle = \varsigma(m)$$

The same distance, measured in decimeters, is given by the function

$$\delta \langle \varsigma(m), \varsigma(d), \varsigma(c), \varsigma(i), \varsigma(j) \rangle = \varsigma(d)$$

As an important application, the model of accuracy described in this section can be used to reformulate the discrete model of planetary motion (Model 3.1).

Model 9.3 *Let S be the set of all quintuples $\langle (r\,;s), (p\,;q), \varsigma(i), \varsigma(j), \varsigma(k) \rangle$ such that*

$$m = \left\lfloor \frac{k+i}{10} \right\rfloor \qquad n = \left\lfloor \frac{k+j}{10} \right\rfloor$$

$$r = \frac{m}{10} - \frac{1}{100} \qquad p = 360\left(\frac{n}{10} - \frac{1}{100} - \left\lfloor \frac{n}{10} - \frac{1}{100} \right\rfloor\right)$$

$$s = \frac{m+1}{10} + \frac{1}{100} \qquad q = 360\left(\frac{n+1}{10} + \frac{1}{100} - \left\lfloor \frac{n+1}{10} + \frac{1}{100} \right\rfloor\right)$$

for some integers m and n, for some integer k between -200000 and 200000, and where $i = -1$ or $+1$, and $j = -1$ or $+1$. The angular position of the Earth, represented as a range of angles measured in degrees, is given by the function

$$\alpha \langle (r\,;s), (p\,;q), \varsigma(i), \varsigma(j), \varsigma(k) \rangle = (p\,;q)$$

The time interval, measured in years, is given by the function

$$\tau \langle (r\,;s), (p\,;q), \varsigma(i), \varsigma(j), \varsigma(k) \rangle = (r\,;s)$$

Note that like Model 3.1, this model is faithful. But the faithfulness, in this case, no longer requires that the results of some measurements be uncertain. Instead, given any particular measurement, the state of the system is uncertain. For example, there are 40 distinct states s such that $\tau(s) = (0.29\,;0.41)$.

In contrast to Model 3.1, consider what happens if Model 9.3 is used to explain measurements taken at time $t = 0.306$ years. Associated with each state

$$\langle (r\,;s), (p\,;q), \zeta(i), \zeta(j), \zeta(k) \rangle$$

in Model 9.3 is an integer k, intended to represent the time interval $(\frac{k}{100}\,;\frac{k+1}{100})$ during which the system is in that state. At time $t = 0.306$ years, $k = 30$, and the system could be in one of the following four states:

$$\langle (0.19\,;0.31), (68.4\,;111.6), \zeta(-1), \zeta(-1), \zeta(30) \rangle$$
$$\langle (0.29\,;0.41), (68.4\,;111.6), \zeta(+1), \zeta(-1), \zeta(30) \rangle$$
$$\langle (0.19\,;0.31), (104.4\,;147.6), \zeta(-1), \zeta(+1), \zeta(30) \rangle$$
$$\langle (0.29\,;0.41), (104.4\,;147.6), \zeta(+1), \zeta(+1), \zeta(30) \rangle$$

Like Model 3.1, two time measurements are possible, $(0.19\,;0.31)$ or $(0.29\,;0.41)$. And since the actual angular position of the Earth at time $t = 0.306$ years is about 112 degrees, the measured position of the Earth is $(104.4\,;147.6)$ degrees at that time. Unlike Model 3.1, this position measurement is compatible with either time measurement, since

$$\langle (0.19\,;0.31), (104.4\,;147.6), \zeta(-1), \zeta(+1), \zeta(30) \rangle$$
$$\langle (0.29\,;0.41), (104.4\,;147.6), \zeta(+1), \zeta(+1), \zeta(30) \rangle$$

are both states of Model 9.3.

10. Isomorphism Theorems

Given a physical model with a set S of states and a set $A = \{\alpha_1, \alpha_2, \alpha_3, \ldots\}$ of observable quantities, we write (S, A) as an abbreviation for that model.[1]

---

[1] In the interest of generality, the definition of a computable physical model places no restrictions on the set A except that its members must be total recursive functions. Some authors prefer to restrict their attention to finite sets A. For example, see reference [3]. Other authors may prefer to restrict their attention to observable quantities computed by programs which belong to a recursively enumerable set.

Definition 10.1 *Two physical models (S, A) and (T, B) are isomorphic if and only if there exist bijections $\phi : S \to T$ and $\psi : A \to B$ such that $\alpha(s) = \psi(\alpha)(\phi(s))$ for all $s \in S$ and all $\alpha \in A$.*

Intuitively, isomorphic physical models can be thought of as providing identical descriptions of the same system.[m]

Given any particular computable physical model (S, A), there are many different models which are isomorphic to (S, A). The following two theorems provide some convenient forms for the representation of computable physical models. Let π_i^n be the *projection function* that takes a length n sequence of non-negative integers and outputs the ith element of the sequence. That is, $\pi_i^n \langle x_1, x_2, \ldots, x_n \rangle = x_i$ for any positive integer $i \leq n$.

Theorem 10.2 *If A is a finite set, then the computable physical model (S, A) is isomorphic to some computable physical model whose observable quantities are all projection functions.*

Proof. Given a computable physical model (S, A) with $A = \{\alpha_1, \alpha_2, \ldots, \alpha_n\}$, let (T, B) be the computable physical model such that

$$T = \{ \langle \alpha_1(s), \alpha_2(s), \ldots, \alpha_n(s), s \rangle \mid s \in S \}$$

and let B be the set of projection functions $\{\pi_1^{n+1}, \pi_2^{n+1}, \ldots, \pi_n^{n+1}\}$. By construction, (T, B) is a computable physical model isomorphic to (S, A). □

Theorem 10.3 *If S is an infinite set, then the computable physical model (S, A) is isomorphic to some computable physical model whose set of states is the set of all non-negative integers. If S has n elements, then the computable physical model (S, A) is isomorphic to some computable physical model whose set of states is $\{0, 1, \ldots, n-1\}$.*

Proof. By definition, if (S, A) is a computable physical model, then S is a recursive set. It immediately follows that S is recursively enumerable. In particular, if S is infinite, then let T be the set of non-negative integers and there is a bijective recursive function ψ from T to S. If S has n elements, then there is a bijective recursive function ψ from $T = \{0, 1, \ldots, n-1\}$ to S. Now, given $A = \{\alpha_1, \alpha_2, \alpha_3, \ldots\}$, let $B = \{\alpha_1 \circ \psi, \alpha_2 \circ \psi, \alpha_3 \circ \psi, \ldots\}$, where $\alpha \circ \psi$ denotes the composition of the functions α and ψ. By construction, (S, A) is isomorphic to (T, B). □

---

[m] Rosen [19] defined a weaker notion of isomorphism. Physical models that are isomorphic in Rosen's sense are not necessarily isomorphic in the sense described here.

Although Theorem 10.3 implies that any computable physical model (S, A) is isomorphic to a computable physical model (T, B) where T is a set of consecutive non-negative integers beginning with zero, there is no effective procedure for constructing a program that computes the characteristic function of T, given a program for computing the characteristic function of S when S is a finite set. That is, Theorem 10.3 does not hold uniformly.

Definition 10.4 A *non-negative integer physical model* is a pair (S, A) where S is a set of non-negative integers and where each member of A is a partial function from the non-negative integers to the non-negative integers. S is the set of states of the model, and A is the set of observable quantities of the model.

Note that if α is an observable quantity of a non-negative integer physical model (S, A), then α might be undefined for some inputs. The non-negative integer physical models form a more general class of objects than the computable physical models. In particular, a computable physical model is a non-negative integer physical model whose set of states is a recursive set and whose observable quantities are total recursive functions.

Theorem 10.5 A non-negative integer physical model (S, A) is isomorphic to some computable physical model if S is a recursively enumerable set and if each member of A is a partial recursive function whose domain includes all the members of S.

Proof. Given a non-negative integer physical model (S, A), note that the construction of the computable physical model (T, B) in the proof of Theorem 10.3 only requires that S be a recursively enumerable set and that each member of A be a partial recursive function whose domain includes all the members of S. Therefore, any such non-negative integer physical model (S, A) is isomorphic to the computable physical model (T, B). □

Let ϕ be any partial recursive function and let S be the largest set of consecutive non-negative integers beginning with zero such that $\phi(s)$ is defined for each $s \in S$. Let A be the set $\{\pi_1^m \circ \phi, \pi_2^m \circ \phi, \ldots, \pi_m^m \circ \phi\}$ for some non-negative integer m. By Theorem 10.5, (S, A) is isomorphic to a computable physical model. We say that any such computable physical model is *determined* by ϕ.

Theorem 10.6 *Every computable physical model with finitely many observable quantities is determined by some partial recursive function ϕ.*

Proof. Let (S, A) be a computable physical model with $A = \{\alpha_1, \alpha_2, \ldots, \alpha_m\}$. Let ψ be the recursive function given in the proof of Theorem 10.3. If S has only n states, then let $\phi(i)$ be undefined for all non-negative integers $i \geq n$. Otherwise, define

$$\phi(i) = \langle \alpha_1(\psi(i)), \alpha_2(\psi(i)), \ldots, \alpha_m(\psi(i)) \rangle$$

By construction, (S, A) is determined by ϕ. □

A physical model (S, A) is said to be *reduced* if and only if for each pair of distinct states s_1 and s_2 in S, there exists an $\alpha \in A$ with $\alpha(s_1) \neq \alpha(s_2)$.

Theorem 10.7 *If (S, A) and (T, B) are isomorphic physical models and (S, A) is reduced, then (T, B) is also a reduced physical model.*

Proof. Let (S, A) and (T, B) be isomorphic physical models and let (S, A) be reduced. Since (S, A) and (T, B) are isomorphic, there exist bijections $\phi : S \to T$ and $\psi : A \to B$ such that $\alpha(s) = \psi(\alpha)(\phi(s))$ for all $s \in S$ and all $\alpha \in A$. Now suppose that t_1 and t_2 are distinct states in T. Because ϕ is a bijection, $\phi^{-1}(t_1)$ and $\phi^{-1}(t_2)$ are distinct states in S. But S is reduced, so there must exist an $\alpha \in A$ such that $\alpha(\phi^{-1}(t_1)) \neq \alpha(\phi^{-1}(t_2))$. Furthermore,

$$\alpha(\phi^{-1}(t_1)) = \psi(\alpha)(\phi(\phi^{-1}(t_1))) = \psi(\alpha)(t_1)$$

and

$$\alpha(\phi^{-1}(t_2)) = \psi(\alpha)(\phi(\phi^{-1}(t_2))) = \psi(\alpha)(t_2)$$

Hence, there exists a $\beta \in B$ such that $\beta(t_1) \neq \beta(t_2)$, namely $\beta = \psi(\alpha)$. We may conclude that the physical model (T, B) is reduced. □

An *epimorphism* from a physical model (S, A) to a physical model (T, B) is a pair of functions (ϕ, ψ) such that ϕ is a surjection from S to T and ψ is a bijection from A to B, where $\alpha(s) = \psi(\alpha)(\phi(s))$ for all $s \in S$ and all $\alpha \in A$. Two physical models (S_1, A_1) and (S_2, A_2) are said to be *observationally equivalent* if and only if there are epimorphisms from (S_1, A_1) to (T, B) and from (S_2, A_2) to (T, B), where (T, B) is some reduced physical model.

Theorem 10.8 *If (S_1, A_1) and (S_2, A_2) are isomorphic physical models, then (S_1, A_1) and (S_2, A_2) are observationally equivalent.*

Proof. Define an equivalence relation on S_2 so that $r \in S_2$ is related to $s \in S_2$ if and only if $\alpha(r) = \alpha(s)$ for all $\alpha \in A_2$. Let T be the corresponding set of equivalence classes of S_2. For each $\alpha \in A_2$, define a function α' so that if $s \in t \in T$, then $\alpha'(t) = \alpha(s)$. Let $B = \{\alpha' \mid \alpha \in A_2\}$. By construction, (T, B) is a reduced physical model. Also note that there is an epimorphism (ϕ, ψ) from (S_2, A_2) to (T, B). Namely, ϕ is the function that maps each member of S_2 to its corresponding equivalence class in T, and ψ is the function that maps each $\alpha \in A_2$ to $\alpha' \in B$.

Now suppose that (S_1, A_1) and (S_2, A_2) are isomorphic. By definition, there are bijections ϕ' from S_1 to S_2 and ψ' from A_1 to A_2 such that $\alpha(s) = \psi'(\alpha)(\phi'(s))$ for all $s \in S_1$ and all $\alpha \in A_1$. Since (ϕ, ψ) is an epimorphism from (S_2, A_2) to (T, B), it immediately follows that $(\phi \circ \phi', \psi \circ \psi')$ is an epimorphism from (S_1, A_1) to (T, B). We may conclude, by definition, that (S_1, A_1) and (S_2, A_2) are observationally equivalent. □

Intuitively, two physical models are observationally equivalent when they both make the same observable predictions. For example, as was discussed in Section 5, the model obtained by omitting the observable quantities α_2 and τ_2 from Model 5.1 is observationally equivalent to the discrete model of planetary motion (Model 3.1). Moreover, if a physical model (S_1, A_1) is faithful, and if (S_1, A_1) is observationally equivalent to (S_2, A_2), then (S_2, A_2) is also faithful.

It is important to note that the converse of Theorem 10.8 does not hold. That is, observationally equivalent models are not necessarily isomorphic. Consider, for example, the computable physical models $(\{0, 1\}, \{\alpha\})$ and $(\{0, 1, 2\}, \{\beta\})$ where $\alpha(s) = s$ for all $s \in \{0, 1\}$ and where $\beta(s) = \lfloor s/2 \rfloor$ for all $s \in \{0, 1, 2\}$. These models are not isomorphic because $\{0, 1\}$ and $\{0, 1, 2\}$ have different cardinalities. Yet, they are observationally equivalent, since both models have a single observable quantity whose only possible values are 0 and 1. Physical models that are isomorphic must not only make the same observable predictions, but must also have the same *structure*. The models $(\{0, 1\}, \{\alpha\})$ and $(\{0, 1, 2\}, \{\beta\})$ have different structures because, assuming that the states are equally likely, they both give different answers to the question, "What is the probability that the observable quantity has value 0?"

11. Oracles and Effective Topologies

Given any set X, we can impose a topology on X. Let $\mathcal{B}$ be a basis for this topology. The members of $\mathcal{B}$ are said to be *basis elements*. We say that a set $\mathcal{L}_x \subseteq \mathcal{B}$ is a *local basis* for a point $x \in X$ if and only if the following two conditions hold.

(1) For each $L \in \mathcal{L}_x$, x is a member of L.
(2) For each $B \in \mathcal{B}$ with $x \in B$, there exists an $L \in \mathcal{L}_x$ with $L \subseteq B$.

Note that every point $x \in X$ has a local basis. For example, the set

$$\mathcal{L}_x = \{\, B \in \mathcal{B} \mid x \in B \,\}$$

of all basis elements that contain x is a local basis for x.

If the basis $\mathcal{B}$ is countable, then each basis element can be encoded as a non-negative integer. In that case, choose some encoding and let $\nu(n)$ be the basis element encoded by n. We allow for the possibility that a basis element may be encoded by more than one non-negative integer. (That is, ν is not necessarily an injection.) The *domain* of ν, denoted $\mathrm{dom}_\mathcal{B}\,\nu$, is the set of all non-negative integers n such that $\nu(n) \in \mathcal{B}$. For any function $\phi : A \to B$ and any set $C \subseteq A$, let $\phi(C) = \{\phi(c) \mid c \in C\}$ denote the *image* of C under ϕ. We let $\mathbb{N}$ denote the set of non-negative integers.

Definition 11.1 A function $\phi : \mathbb{N} \to \mathrm{dom}_\mathcal{B}\,\nu$ is said to be an *oracle* for a point x, with basis $\mathcal{B}$ and coding ν, if and only if $\nu\bigl(\phi(\mathbb{N})\bigr)$ is a local basis for x.

An oracle ϕ for x is *complete* if and only if every $n \in \mathrm{dom}_\mathcal{B}\,\nu$ such that $x \in \nu(n)$ is a member of $\phi(\mathbb{N})$. An oracle is said to be *nested* if and only if $\nu\bigl(\phi(n+1)\bigr) \subseteq \nu\bigl(\phi(n)\bigr)$ for all $n \in \mathbb{N}$.

A pair $(\mathcal{B}, \nu)$ is said to be an *effective topology* if and only if $\mathcal{B}$ is a countable basis for a T_0 topology and ν is a coding for $\mathcal{B}$. Effective topologies were first introduced in the theory of type-two effectivity [12,26]. In accordance with that theory, we use an oracle for x, with a basis $\mathcal{B}$ and coding ν, as a representation of the point x in an effective topology $(\mathcal{B}, \nu)$. Because effective topologies are T_0, no two distinct points are ever represented by the same oracle.

Of special interest are effective topologies where the subset relation

$$\{\, \langle b_1, b_2 \rangle \mid \nu(b_1) \subseteq \nu(b_2)\ \&\ b_1 \in \mathrm{dom}_\mathcal{B}\,\nu\ \&\ b_2 \in \mathrm{dom}_\mathcal{B}\,\nu \,\}$$

is a recursively enumerable set.[n] In particular, if $(\mathcal{B}, \nu)$ has a recursively

---

[n] An effective topology with a recursively enumerable subset relation is an example of a *computable topology*, as defined in reference [26]. Not all computable topologies have recursively enumerable subset relations.

enumerable subset relation, then

$$\mathrm{dom}_\mathcal{B}\, \nu = \{\, b \mid \nu(b) \subseteq \nu(b) \,\&\, b \in \mathrm{dom}_\mathcal{B}\, \nu \,\}$$

is also a recursively enumerable set.

Theorem 11.2 *Let ϕ be an oracle for x in an effective topology $(\mathcal{B}, \nu)$ with a recursively enumerable subset relation. Then there exists a complete oracle ψ for x in $(\mathcal{B}, \nu)$ that is recursive relative to ϕ uniformly.*

Proof. Suppose that ϕ is an oracle for x in an effective topology $(\mathcal{B}, \nu)$ with a recursively enumerable subset relation. Since $\nu(\phi(\mathbb{N}))$ is a local basis for x, it follows that for each $b \in \mathrm{dom}_\mathcal{B}\, \nu$, $x \in \nu(b)$ if and only if there exists an $n \in \mathbb{N}$ such that $\nu(\phi(n)) \subseteq \nu(b)$. Hence,

$$\{\, b \in \mathrm{dom}_\mathcal{B}\, \nu \mid (\exists n \in \mathbb{N})[\nu(\phi(n)) \subseteq \nu(b)] \,\}$$

is the set of encodings of all basis elements that contain x. But this set is recursively enumerable relative to ϕ because $(\mathcal{B}, \nu)$ has a recursively enumerable subset relation. Therefore, there exists a function $\psi : \mathbb{N} \to \mathrm{dom}_\mathcal{B}\, \nu$, recursive relative to ϕ, such that $\nu(\psi(\mathbb{N}))$ is this set. By definition, ψ is a complete oracle for x. $\square$

Theorem 11.3 *Let ϕ be an oracle for x in an effective topology $(\mathcal{B}, \nu)$ with a recursively enumerable subset relation. Then there exists a nested oracle ψ for x in $(\mathcal{B}, \nu)$ that is recursive relative to ϕ uniformly.*

Proof. Suppose that ϕ is an oracle for x in an effective topology $(\mathcal{B}, \nu)$ with a recursively enumerable subset relation. Note that for each pair of basis elements B_1 and B_2 such that $x \in B_1 \cap B_2$, there exists a basis element B_3 with $x \in B_3 \subseteq B_1 \cap B_2$, by the definition of a basis. Therefore, since $\nu(\phi(\mathbb{N}))$ is a local basis for x, there must exist an $m \in \mathbb{N}$ such that

$$x \in \nu(\phi(m)) \subseteq B_3 \subseteq B_1 \cap B_2$$

Now define $\psi : \mathbb{N} \to \mathrm{dom}_\mathcal{B}\, \nu$ recursively, relative to ϕ, as follows. Let $\psi(0) = \phi(0)$ and for each $n \in \mathbb{N}$ let $\psi(n+1) = \phi(m)$ for some $m \in \mathbb{N}$ such that

$$\nu(\phi(m)) \subseteq \nu(\psi(n)) \cap \nu(\phi(n+1))$$

We can find m recursively given $\psi(n)$ and ϕ because the subset relation for $(\mathcal{B}, \nu)$ is recursively enumerable, and the set of all $m \in \mathbb{N}$ such that

$$\nu(\phi(m)) \subseteq \nu(\psi(n)) \,\&\, \nu(\phi(m)) \subseteq \nu(\phi(n+1))$$

is therefore recursively enumerable relative to ϕ. We may conclude that ψ is nested because

$$\nu\big(\psi(n+1)\big) \subseteq \nu\big(\psi(n)\big) \cap \nu\big(\phi(n+1)\big) \subseteq \nu\big(\psi(n)\big)$$

for all $n \in \mathbb{N}$, and that ψ is an oracle for x because $x \in \nu\big(\psi(0)\big) = \nu\big(\phi(0)\big)$ and

$$x \in \nu\big(\psi(n+1)\big) \subseteq \nu\big(\psi(n)\big) \cap \nu\big(\phi(n+1)\big) \subseteq \nu\big(\phi(n+1)\big)$$

for all $n \in \mathbb{N}$. □

Define $\iota(a\,;b)$ to be the set of all real numbers x such that $a < x < b$, and let $\mathcal{I}$ be the set of all $\iota(a\,;b)$ such that a and b are rational numbers with $a < b$. The members of $\mathcal{I}$ are said to be *rational intervals*. Note that $\mathcal{I}$ is a basis for the standard topology of the real numbers. Indeed, the oracles for real numbers that were introduced in Section 4 were nested oracles with basis $\mathcal{I}$ and coding ι. Another basis for the standard topology of the real numbers is the set $\mathcal{I}_{10,c}$ of *decimal intervals* with accuracy factor c, where c is a positive rational number, and where $\mathcal{I}_{10,c}$ is defined to be the set of all $\iota(a\,;b)$ such that

$$a = \frac{m}{10^n} - \frac{c}{10^n} \qquad\qquad b = \frac{m+1}{10^n} + \frac{c}{10^n}$$

for some integer m and some positive integer n. We call n the *number of digits of precision* of $(a\,;b)$.

Note that both $(\mathcal{I}, \iota)$ and $(\mathcal{I}_{10,c}, \iota)$ have recursively enumerable subset relations. The following theorem asserts that if ϕ is an oracle for a real number x with basis $\mathcal{I}$ and coding ι, then there exists an oracle ψ for x with basis $\mathcal{I}_{10,c}$ and coding ι that is recursive relative to ϕ uniformly.

Theorem 11.4 *Let $(\mathcal{A}, \nu)$ and $(\mathcal{B}, \nu)$ be effective topologies with recursively enumerable subset relations such that $\mathcal{B} \subseteq \mathcal{A}$, and such that $\mathcal{A}$ and $\mathcal{B}$ are bases for the same topology. If ϕ is an oracle for x in $(\mathcal{A}, \nu)$, then there exists an oracle ψ for x in $(\mathcal{B}, \nu)$ that is recursive relative to ϕ uniformly.*

Proof. Suppose that $(\mathcal{A}, \nu)$ and $(\mathcal{B}, \nu)$ are effective topologies as described in the statement of the theorem, and that ϕ is an oracle for a point x in $(\mathcal{A}, \nu)$. By definition,

$$x \in \nu\big(\phi(n)\big)$$

for every $n \in \mathbb{N}$. And because $\nu(\phi(n))$ is an open set, it is a union of basis elements from $\mathcal{B}$. Hence, there must exist a $B \in \mathcal{B}$ such that

$$x \in B \subseteq \nu(\phi(n))$$

But $\nu(\phi(\mathbb{N}))$ is a local basis for x, and B is a basis element in $\mathcal{A}$, so there exists an $m \in \mathbb{N}$ such that $x \in \nu(\phi(m)) \subseteq B$. Therefore, we have that for each $n \in \mathbb{N}$ there exist $B \in \mathcal{B}$ and $m \in \mathbb{N}$ such that

$$x \in \nu(\phi(m)) \subseteq B \subseteq \nu(\phi(n))$$

Now, since $(\mathcal{B}, \nu)$ has a recursively enumerable subset relation, $\mathrm{dom}_{\mathcal{B}}\, \nu$ is a recursively enumerable set. Then, because $(\mathcal{A}, \nu)$ also has a recursively enumerable subset relation, the set

$$\{\, \langle m, b \rangle \mid \nu(\phi(m)) \subseteq \nu(b) \subseteq \nu(\phi(n))\ \&\ m \in \mathbb{N}\ \&\ b \in \mathrm{dom}_{\mathcal{B}}\, \nu \,\}$$

is recursively enumerable relative to ϕ, for any $n \in \mathbb{N}$. Therefore, there is a function $\psi : \mathbb{N} \to \mathrm{dom}_{\mathcal{B}}\, \nu$, recursive relative to ϕ, such that $\psi(n) = b$ for all $n \in \mathbb{N}$, where

$$\nu(\phi(m)) \subseteq \nu(b) \subseteq \nu(\phi(n))$$

for some $m \in \mathbb{N}$. But this function ψ is an oracle for x in $(\mathcal{B}, \nu)$, because

$$x \in \nu(\phi(m)) \subseteq \nu(\psi(n)) \subseteq \nu(\phi(n))$$

for all $n \in \mathbb{N}$. □

12. Basic Representations of Sets

Definition 12.1 Let $(\mathcal{B}, \nu)$ be an effective topology on a set X, and let A be any subset of X. We say that a set R of non-negative integers is a *basic representation* of A in the effective topology $(\mathcal{B}, \nu)$ if and only if the following two conditions hold.

(1) $R \subseteq \mathrm{dom}_{\mathcal{B}}\, \nu$
(2) $x \in A$ if and only if there exists a local basis $\mathcal{L}_x$ for x with $\mathcal{L}_x \subseteq \nu(R)$.

Note that condition 2 of the definition ensures that no two distinct sets in $(\mathcal{B}, \nu)$ have the same basic representation. Note further that if R is a basic representation of A, then $\{\, A \cap \nu(r) \mid r \in R \,\}$ is a basis for the subspace topology on A, and this is an effective topology with coding $\lambda r [A \cap \nu(r)]$.

In an effective topology we use basic representations to represent sets of points, but not all sets of points have basic representations. For example, there are $2^{2^{\aleph_0}}$ many sets of real numbers, but since a basic representation is a set of non-negative integers, there are at most $2^{\aleph_0}$ many basic representations. Nevertheless, many commonly-studied sets have basic representations.°

Theorem 12.2 *Let A be a set in an effective topology $(\mathcal{B}, \nu)$.*

(1) *If A is an open set, then A has a basic representation.*
(2) *If A is a closed set, then A has a basic representation.*

Proof. Suppose that A is an open set in the effective topology $(\mathcal{B}, \nu)$ and let
$$R = \{\, r \in \operatorname{dom}_\mathcal{B} \nu \mid \nu(r) \subseteq A \,\}$$
Clearly, $R \subseteq \operatorname{dom}_\mathcal{B} \nu$ and if $x \notin A$ then there does not exist a local basis $\mathcal{L}_x$ for x with $\mathcal{L}_x \subseteq \nu(R)$, since no member of $\nu(R)$ contains x. Alternatively, if $x \in A$ then, by the definition of a basis, for each basis element B_1 that contains x there exists some basis element B_2 such that $x \in B_2 \subseteq B_1 \cap A$. That is, if $x \in A$ then for each basis element B_1 with $x \in B_1$, there exists a basis element $B_2 \in \{\, B \in \mathcal{B} \mid x \in B \subseteq A \,\}$ with $B_2 \subseteq B_1$. It immediately follows that $\mathcal{L}_x = \{\, B \in \mathcal{B} \mid x \in B \subseteq A \,\}$ is a local basis for x and $\mathcal{L}_x \subseteq \nu(R)$. By definition, R is a basic representation of A.

Now, if A is a closed set in $(\mathcal{B}, \nu)$, then let
$$R = \{\, r \in \operatorname{dom}_\mathcal{B} \nu \mid A \cap \nu(r) \neq \emptyset \,\}$$
Clearly, $R \subseteq \operatorname{dom}_\mathcal{B} \nu$ and if $x \in A$ then there exists a local basis $\mathcal{L}_x$ for x with $\mathcal{L}_x \subseteq \nu(R)$. Namely, $\mathcal{L}_x$ is the set of all basis elements that contain x. Alternatively, if $x \notin A$, then since A is closed, every local basis $\mathcal{L}_x$ for x contains a basis element that does not intersect A. Therefore, $\mathcal{L}_x \not\subseteq \nu(R)$. We may conclude, by definition, that R is a basic representation of A. □

Although we use a basic representation R to represent a set of points in an effective topology, the following theorem demonstrates that there is, in

---

°In the effective topology $(\mathcal{I}, \iota)$, the set of rational numbers does not have a basic representation, but the set of irrational numbers has the basic representation
$$R = \left\{ \left(\frac{m}{n!}\, ; \frac{m+1}{n!} \right) \;\middle|\; m \in \mathbb{Z} \,\&\, n \in \mathbb{N} \right\}$$
where $\mathbb{Z}$ denotes the set of integers. It is tempting to conjecture that the sets with basic representations in an effective topology $(\mathcal{B}, \nu)$ are exactly the G_δ sets, but there is a trivial counterexample to this conjecture if the effective topology is not T_1.

general, no effective procedure (relative to R) for finding oracles for those points. Nevertheless, if we restrict our attention to certain special classes of basic representations R, then effective procedures do exist. See Section 15.

Theorem 12.3 *Let $\mathcal{B}$ be a countable basis for the standard topology of $\mathbb{R}^n$ and let ν be a coding for the basis. Then there does not exist a partial recursive function ϕ satisfying the condition that for every singleton set $\{x\} \subseteq \mathbb{R}^n$ and for every basic representation R_x of $\{x\}$ in the effective topology $(\mathcal{B}, \nu)$, the function $\lambda m[\phi(R_x, m)]$ is an oracle for x in $(\mathcal{B}, \nu)$.*

Proof. Let $(\mathcal{B}, \nu)$ be an effective topology as in the statement of the theorem and suppose, as an assumption to be shown contradictory, that there exists a partial recursive function ϕ satisfying the condition that for every singleton set $\{x\} \subseteq \mathbb{R}^n$ and for every basic representation R_x of $\{x\}$ in the effective topology $(\mathcal{B}, \nu)$, the function $\lambda m[\phi(R_x, m)]$ is an oracle for x in $(\mathcal{B}, \nu)$. Now consider any two distinct points $x \in \mathbb{R}^n$ and $y \in \mathbb{R}^n$, and let R_x be a basic representation of $\{x\}$ in $(\mathcal{B}, \nu)$. Since the standard topology of $\mathbb{R}^n$ is T_1, there must exist a non-negative integer k such that $\phi(R_x, k)$ is defined and

$$y \notin \nu\big(\phi(R_x, k)\big)$$

Next, choose a program for computing ϕ. Note that since the computation for $\phi(R_x, k)$ has only finitely many steps, only finitely many non-negative integers are tested for membership in R_x during the course of the computation. Let C be the collection of all $i \in \mathbb{N}$ such that $i \in R_x$ and such that i is tested for membership in R_x during the course of the computation of $\phi(R_x, k)$. Similarly, let D be the collection of all $i \in \mathbb{N}$ such that $i \notin R_x$ and such that i is tested for membership in R_x during the course of the computation of $\phi(R_x, k)$.

Now, choose any oracle ψ for y in $(\mathcal{B}, \nu)$. Note that $\psi(\mathbb{N})$ is a basic representation for $\{y\}$ in $(\mathcal{B}, \nu)$. And because C and D are finite sets,

$$R_y = (\psi(\mathbb{N}) \cup C) - D$$

is also a basic representation for $\{y\}$ in $(\mathcal{B}, \nu)$. It follows that $\phi(R_x, k) = \phi(R_y, k)$, because whenever i is tested for membership in R_x during the course of the computation of $\phi(R_x, k)$, $i \in R_x$ if and only if $i \in R_y$. Therefore,

$$y \notin \nu\big(\phi(R_x, k)\big) = \nu\big(\phi(R_y, k)\big)$$

But by the definition of ϕ, $\lambda m\bigl[\phi(R_y, m)\bigr]$ is an oracle for y. Hence,

$$y \in \nu\bigl(\phi(R_y, k)\bigr)$$

This is a contradiction, so the assumption must be false. The partial recursive function ϕ does not exist. $\square$

13. Basic Representations of Physical Models

Let $\mathbb{R}$ be the set of all real numbers. For any two sets A and B, let $A \times B = \{\,(a,b) \mid a \in A\ \&\ b \in B\,\}$ be the Cartesian product of A with B. We write A^k to denote the set formed by taking the Cartesian product of A with itself k many times. For example, $A^3 = (A \times A) \times A$. As with Cantor's pairing function, (a,b,c) is an abbreviation for $((a,b),c)$, and so on. Similarly, we define the *Cartesian projection function* ϖ_i^n so that $\varpi_i^n(x_1, x_2, \ldots, x_n) = x_i$ for each positive integer $i \leq n$.

A physical model (S, A) with finitely many observable quantities is said to be in *normal form* if and only if $S \subseteq \mathbb{R}^n$ and $A = \{\varpi_1^n, \varpi_2^n, \ldots, \varpi_n^n\}$.

Theorem 13.1 *The following two conditions hold for any physical model (S, A) with finitely many observable quantities.*

(1) *(S, A) is observationally equivalent to a physical model in normal form.*
(2) *(S, A) is isomorphic to a physical model in normal form if and only if (S, A) is a reduced physical model.*

Proof. Begin by noting that if (T, B) is a physical model in normal form, and if $t_1 \neq t_2$ for any $t_1 \in T$ and $t_2 \in T$, then $\varpi_i^n(t_1) \neq \varpi_i^n(t_2)$ for some positive integer $i \leq n$. Therefore, by definition, every physical model in normal form is a reduced physical model. It immediately follows from Theorem 10.7 that if a physical model (S, A) is isomorphic to a physical model in normal form, then (S, A) is a reduced physical model.

To prove condition 1, suppose that (S, A) is a physical model such that $A = \{\alpha_1, \alpha_2, \ldots, \alpha_n\}$. Define

$$T = \bigl\{\,(\alpha_1(s), \alpha_2(s), \ldots, \alpha_n(s)) \mid s \in S\,\bigr\}$$

and let $B = \{\varpi_1^n, \varpi_2^n, \ldots, \varpi_n^n\}$. Note that (T, B) is a physical model in normal form. Also note that the function $\phi : S \to T$ given by

$$\phi(s) = \bigl(\alpha_1(s), \alpha_2(s), \ldots, \alpha_n(s)\bigr)$$

is a surjection, and that $\alpha_i(s) = \varpi_i^n(\phi(s))$ for all $s \in S$ and all positive integers $i \leq n$. Therefore, there is an epimorphism from (S, A) to the reduced physical model (T, B). Trivially, there is also an epimorphism from (T, B) to itself. We may conclude that (S, A) is observationally equivalent to (T, B).

To prove condition 2, consider the special case where (S, A) is a reduced physical model. Because (S, A) is reduced, we have that if $s_1 \neq s_2$ for any $s_1 \in S$ and $s_2 \in S$, then there exists a positive integer $i \leq n$ such that $\alpha_i(s_1) \neq \alpha_i(s_2)$. This implies that if $s_1 \neq s_2$ then $\phi(s_1) \neq \phi(s_2)$. Hence, ϕ is an injection. Since ϕ is also a surjection, ϕ is a bijection. Therefore, if (S, A) is a reduced physical model, then (S, A) and (T, B) are isomorphic. We have already proved the converse, that if (S, A) is isomorphic to a physical model in normal form, then (S, A) is a reduced physical model. Hence, condition 2 holds. □

The notion of a basic representation of a set can be generalized so that we may speak of basic representations of physical models in normal form. Given a physical model (S, A) in normal form with $A = \{\varpi_1^n, \varpi_2^n, \ldots, \varpi_n^n\}$, we may choose sets $X_1, X_2, \ldots, X_n$ such that $\varpi_i^n(S) \subseteq X_i \subseteq \mathbb{R}$ for each positive integer $i \leq n$, and we may impose effective topologies $(\mathcal{B}_1, \nu_1)$, $(\mathcal{B}_2, \nu_2), \ldots, (\mathcal{B}_n, \nu_n)$ on these sets.[p] Define

$$(\mathcal{B}_1, \nu_1) \otimes (\mathcal{B}_2, \nu_2) \otimes \cdots \otimes (\mathcal{B}_n, \nu_n)$$

to be the effective topology with basis $\mathcal{B}$ such that

$$\mathcal{B} = \{ B_1 \times B_2 \times \cdots \times B_n \mid B_1 \in \mathcal{B}_1 \ \& \ B_2 \in \mathcal{B}_2 \ \& \ \cdots \ \& \ B_n \in \mathcal{B}_n \}$$

and with coding ν such that

$$\nu\langle a_1, a_2, \ldots, a_n\rangle = \nu_1(a_1) \times \nu_2(a_2) \times \cdots \times \nu_n(a_n)$$

We call $(\mathcal{B}, \nu)$ the *effective product* of $(\mathcal{B}_1, \nu_1), (\mathcal{B}_2, \nu_2), \ldots, (\mathcal{B}_n, \nu_n)$. Note that S is a set of points in the effective topology $(\mathcal{B}, \nu)$. A physical model (R, H) is said to be a *basic representation* of the physical model (S, A) if R is a basic representation of S in the effective topology $(\mathcal{B}, \nu)$ and if H is the set $\{\pi_1^n, \pi_2^n, \ldots, \pi_n^n\}$ of projection functions.

---

[p]If (S, A) is faithful, then the bases for these topologies are uniquely determined by the physical operations used to measure each of the observable quantities. For example, if an observable quantity is an angle measurement, then the corresponding topology is the topology of a circle, and each basis element corresponds to a particular reading on the instrument that is used to measure angles. The idea that basis elements correspond to the values of measurements appears to have originated with reference [27].

For example, the non-discrete continuous computable physical model of planetary motion (Model 4.1) is a basic representation of the simple model of planetary motion (Model 1.1). In particular, Model 4.1 is obtained by imposing the effective topology $(\mathcal{I}_{10,c}, \iota)$ on the time in Model 1.1, where $c = \frac{1}{10}$, and by imposing the effective topology described in Footnote d on the angular position in Model 1.1. The product of these topologies is the topology for the surface of a cylinder. The states of Model 1.1 are a spiral path on the surface of that cylinder, and the set of states of Model 4.1 is a basic representation of the path.

14. Data and Predictions

In order to make predictions, we are often interested in finding the set of all states of a physical model which could account for a given collection of simultaneous measurements. That is, given a physical model (S, A) with $A = \{\alpha_1, \alpha_2, \alpha_3, \ldots\}$, and given real numbers $x_1, x_2, \ldots, x_k$, we are interested in the set

$$P = \{\, s \in S \mid \alpha_1(s) = x_1 \,\&\, \alpha_2(s) = x_2 \,\&\, \cdots \,\&\, \alpha_k(s) = x_k \,\}$$

In this context, the real numbers $x_1, x_2, \ldots, x_k$ are said to be the *data*, and P is the corresponding set of states *predicted* by the model.

The following theorem shows that if we are given a basic representation of a physical model (S, A) in normal form, together with complete oracles for the real numbers $x_1, x_2, \ldots, x_k$, then there is an effective procedure for finding a basic representation of the set P, provided that the underlying topology is T_1. (This is a rather weak requirement, since almost all topologies with practical applications in the sciences are T_1.)

Theorem 14.1 *Let (S, A) be a physical model in normal form with $A = \{\varpi_1^n, \varpi_2^n, \ldots, \varpi_n^n\}$ and let (R, H) be a basic representation of (S, A) in a T_1 effective topology*

$$(\mathcal{B}, \nu) = (\mathcal{B}_1, \nu_1) \otimes (\mathcal{B}_2, \nu_2) \otimes \cdots \otimes (\mathcal{B}_n, \nu_n)$$

If $k \leq n$ and $\phi_1, \phi_2, \ldots, \phi_k$ are complete oracles for $x_1, x_2, \ldots, x_k$ in the effective topologies $(\mathcal{B}_1, \nu_1), (\mathcal{B}_2, \nu_2), \ldots, (\mathcal{B}_k, \nu_k)$, then there is a basic representation of

$$P = \{\, s \in S \mid \varpi_1^n(s) = x_1 \,\&\, \varpi_2^n(s) = x_2 \,\&\, \cdots \,\&\, \varpi_k^n(s) = x_k \,\}$$

in $(\mathcal{B}, \nu)$ that is recursively enumerable relative to $R, \phi_1, \phi_2, \ldots, \phi_k$ uniformly.

Proof. Let the variables be defined as in the statement of the theorem and note that the set

$$Q = \{\, r \in R \mid (\forall i \in \{1, 2, \ldots, k\})(\exists m \in \mathbb{N})\left[\pi_i^n(r) = \phi_i(m)\right] \,\}$$

is recursively enumerable relative to R, ϕ_1, ϕ_2, $\ldots$, ϕ_k uniformly. (In fact, Q is recursively enumerable relative to ξ, ϕ_1, ϕ_2, $\ldots$, ϕ_k uniformly, where ξ is a function that merely enumerates the members of R.) We claim that Q is a basic representation of P in $(\mathcal{B}, \nu)$. Since $Q \subseteq R \subseteq \mathrm{dom}_\mathcal{B}\, \nu$, it suffices to prove that $s \in P$ if and only if there exists a local basis $\mathcal{L}_s$ for s with $\mathcal{L}_s \subseteq \nu(Q)$. Or equivalently, it suffices to prove that $s \in P$ if and only if there exists an oracle ψ for s with $\psi(\mathbb{N}) \subseteq Q$.

Suppose $s \in P$. Because R is a basic representation of S in $(\mathcal{B}, \nu)$, there is an oracle ψ for s in $(\mathcal{B}, \nu)$ such that $\psi(\mathbb{N}) \subseteq R$. Moreover, for each positive integer $i \leq k$, the set $\nu_i\bigl(\pi_i^n(\psi(\mathbb{N}))\bigr)$ is a local basis for $\varpi_i^n(s) = x_i$. And since ϕ_i is a complete oracle for x_i, we have that $\pi_i^n(\psi(\mathbb{N})) \subseteq \phi_i(\mathbb{N})$. Hence, if $r = \psi(l)$ for some $l \in \mathbb{N}$, then there exists an $m \in \mathbb{N}$ such that $\pi_i^n(r) = \phi_i(m)$. Therefore, by the definition of Q, $\psi(\mathbb{N}) \subseteq Q$.

Conversely, suppose that ψ is an oracle for some point s in $(\mathcal{B}, \nu)$, and that $\psi(\mathbb{N}) \subseteq Q$. Then, for each positive integer $i \leq k$, we have that $\pi_i^n(\psi(\mathbb{N})) \subseteq \phi_i(\mathbb{N})$. Of course, $\nu_i\bigl(\pi_i^n(\psi(\mathbb{N}))\bigr)$ is a local basis for $\varpi_i^n(s)$ because $\nu(\psi(\mathbb{N}))$ is a local basis for s. And by the definition of ϕ_i, $\nu_i\bigl(\phi_i(\mathbb{N})\bigr)$ is a local basis for x_i. Hence, a local basis for $\varpi_i^n(s)$ is a subset of a local basis for x_i in the effective topology $(\mathcal{B}_i, \nu_i)$. But because $(\mathcal{B}, \nu)$ is a T_1 effective topology, $(\mathcal{B}_i, \nu_i)$ is also T_1. In a T_1 topology, local bases for any two distinct points z_1 and z_2 must contain basis elements B_1 and B_2, respectively, such that $z_2 \notin B_1$ and $z_1 \notin B_2$. Therefore, since a local basis for $\varpi_i^n(s)$ is a subset of a local basis for x_i, it must be the case that $\varpi_i^n(s) = x_i$. We may conclude, by the definition of P, that $s \in P$. $\square$

A set $S \subseteq \mathbb{R}^n$ is said to be the *graph* of a function $\psi : \mathbb{R}^k \to \mathbb{R}^{n-k}$, if and only if

$$S = \{\, (x_1, \ldots, x_k, x_{k+1}, \ldots, x_n) \in \mathbb{R}^n \mid \psi(x_1, \ldots, x_k) = (x_{k+1}, \ldots, x_n) \,\}$$

And we say that a physical model (S, A) is *induced* by a function $\psi : \mathbb{R}^k \to \mathbb{R}^{n-k}$ if and only if (S, A) is in normal form and S is the graph of ψ. Therefore, if (S, A) is induced by $\psi : \mathbb{R}^k \to \mathbb{R}^{n-k}$ and we are given k real numbers $x_1, x_2, \ldots, x_k$ as data, then the corresponding set of states predicted by (S, A) is a singleton set $P = \{s\}$. Namely,

$$s = (x_1, x_2, \ldots, x_k, x_{k+1}, \ldots, x_n)$$

where x_{k+1}, x_{k+2}, ..., x_n are the real numbers uniquely determined by the equation

$$\psi(x_1, x_2, \ldots, x_k) = (x_{k+1}, x_{k+2}, \ldots, x_n)$$

It then follows from Theorem 14.1 that given complete oracles for x_1, x_2, ..., x_k in the standard topology of $\mathbb{R}$, and given a basic representation R of (S, A) in the standard topology of $\mathbb{R}^n$, there is an effective procedure (relative to the given oracles and R) for finding a basic representation of $\{s\}$. But by Theorem 12.3 there is, in general, no effective procedure for finding an *oracle* for s. In the next section we describe a special class of basic representations for which such an effective procedure does exist.

15. Kreisel's Criterion

A common way to interpret Kreisel's criterion is to say that a physical model (S, A) satisfies Kreisel's criterion on $\mathbb{R}^k$ if and only if (S, A) is induced by a function $\psi : \mathbb{R}^k \to \mathbb{R}^{n-k}$ for some positive integer $n > k$, and for each positive integer $j \leq n - k$ there is a partial recursive function κ_j such that if $\phi_1, \phi_2, \ldots, \phi_k$ are nested oracles for real numbers $x_1, x_2, \ldots, x_k$ in the effective topology $(\mathcal{I}, \iota)$, then $\kappa_j(\phi_1, \phi_2, \ldots, \phi_k, m)$ is defined for all $m \in \mathbb{N}$ and $\lambda m [\kappa_j(\phi_1, \phi_2, \ldots, \phi_k, m)]$ is a nested oracle for $\varpi_j^{n-k}(\psi(x_1, x_2, \ldots, x_k))$ in $(\mathcal{I}, \iota)$. Note by Theorem 11.4 that the effective topology $(\mathcal{I}, \iota)$ in this statement can be replaced, without loss of generality, with any effective topology $(\mathcal{B}, \iota)$ that has a recursively enumerable subset relation and such that $\mathcal{B} \subseteq \mathcal{I}$ is a basis for the standard topology of the real numbers.

Practical computer models that use multiple-precision interval arithmetic [15] provide examples of physical models satisfying Kreisel's criterion. Typically, such models are induced by a function $\psi : \mathbb{R}^k \to \mathbb{R}^{n-k}$ where, for each positive integer $j \leq n - k$, there is a recursive function ξ_j such that if the data $x_1, x_2, \ldots, x_k$ lie within the intervals $(a_1; b_1), (a_2; b_2), \ldots, (a_k; b_k)$ respectively, then $\varpi_j^{n-k}(\psi(x_1, x_2, \ldots, x_k))$ lies within the interval

$$\xi_j((a_1; b_1), (a_2; b_2), \ldots, (a_k; b_k))$$

In such a case, the partial recursive function κ_j in Kreisel's criterion is given by

$$\kappa_j(\phi_1, \phi_2, \ldots, \phi_k, m) = \xi_j(\phi_1(m), \phi_2(m), \ldots, \phi_k(m))$$

We are now prepared to state the following theorem, which holds uniformly.

Theorem 15.1 *If a physical model satisfies Kreisel's criterion on $\mathbb{R}^k$, then the model has a basic representation that is isomorphic to a computable physical model.*

Proof. Suppose that (S, A) is a physical model satisfying Kreisel's criterion on $\mathbb{R}^k$. In particular, suppose that (S, A) is induced by a function $\psi : \mathbb{R}^k \to \mathbb{R}^{n-k}$, and for each positive integer $j \leq n - k$ suppose there is a partial recursive function κ_j such that if $\phi_1, \phi_2, \ldots, \phi_k$ are nested oracles for real numbers $x_1, x_2, \ldots, x_k$ in the effective topology $(\mathcal{I}_{10,c}, \iota)$, then $\kappa_j(\phi_1, \phi_2, \ldots, \phi_k, m)$ is defined for all $m \in \mathbb{N}$ and $\lambda m[\kappa_j(\phi_1, \phi_2, \ldots, \phi_k, m)]$ is a nested oracle for $\varpi_j^{n-k}(\psi(x_1, x_2, \ldots, x_k))$ in $(\mathcal{I}_{10,c}, \iota)$, where c is a positive rational number.

Let $I = \operatorname{dom}_{\mathcal{I}_{10,c}} \iota$. Then, for each interval $u \in I$ with midpoint p and with d digits of precision, define the partial recursive function σ_u so that

$$\sigma_u(l) = \begin{cases} o_p(l) & \text{if } l < d \\ \text{undefined} & \text{if } l \geq d \end{cases}$$

for each $l \in \mathbb{N}$, where o_p is the standard decimal oracle described in Section 4. Note that for each $x \in \mathbb{R}$ and each $m \in \mathbb{N}$, $\sigma_{o_x(m)}(l) = o_x(l)$ for all non-negative integers $l \leq m$. Now, for each positive integer $j \leq n-k$, choose a program to compute κ_j and let $\kappa_j(\sigma_{u_1}, \sigma_{u_2}, \ldots, \sigma_{u_k}, m)$ be undefined if for some positive integer $i \leq k$ and some $l \in \mathbb{N}$ the program calls $\sigma_{u_i}(l)$ in the course of the computation and $\sigma_{u_i}(l)$ is undefined. Note that given $u_1, u_2, \ldots, u_k$, the set of all $m \in \mathbb{N}$ such that $\kappa_j(\sigma_{u_1}, \sigma_{u_2}, \ldots, \sigma_{u_k}, m)$ is defined is a recursively enumerable set, since for each m we can follow the computation and test whether or not $\sigma_{u_i}(l)$ is defined whenever $\sigma_{u_i}(l)$ is called by the program, for any i and l. Let R be the set of all $\langle u_1, u_2, \ldots, u_n \rangle$ such that $u_i \in I$ for each positive integer $i \leq k$, and such that $u_{k+j} = \kappa_j(\sigma_{u_1}, \sigma_{u_2}, \ldots, \sigma_{u_k}, m)$ for some $m \in \mathbb{N}$ if $j \leq n - k$ is a positive integer. Note that R is also recursively enumerable. Let $H = \{\pi_1^n, \pi_2^n, \ldots, \pi_n^n\}$. We claim that (R, H) is a basic representation of (S, A).

As a brief digression from the proof, suppose that $\langle u_1, u_2, \ldots, u_n \rangle \in R$ and note that for each positive integer $i \leq k$, if $x_i \in \iota(u_i)$ then there exists a nested oracle ϕ_i for x_i in $(\mathcal{I}_{10,c}, \iota)$ such that $\phi_i(m) = \sigma_{u_i}(m)$ for all $m \in \mathbb{N}$ where $\sigma_{u_i}(m)$ is defined. And since for each positive integer $j \leq n - k$ we have that $\lambda m[\kappa_j(\phi_1, \phi_2, \ldots, \phi_k, m)]$ is a nested oracle for $x_{k+j} = \varpi_j^{n-k}(\psi(x_1, x_2, \ldots, x_k))$, it follows that

$$x_{k+j} \in \iota(\kappa_j(\phi_1, \phi_2, \ldots, \phi_k, m)) = \iota(\kappa_j(\sigma_{u_1}, \sigma_{u_2}, \ldots, \sigma_{u_k}, m)) = \iota(u_{k+j})$$

for some $m \in \mathbb{N}$. Hence, given any $\langle u_1, u_2, \ldots, u_n \rangle \in R$, if $x_i \in \iota(u_i)$ for each positive integer $i \leq k$, then $x_{k+j} = \varpi_j^{n-k}(\psi(x_1, x_2, \ldots, x_k)) \in \iota(u_{k+j})$ for each positive integer $j \leq n - k$.

Now, returning to the proof of Theorem 15.1, let

$$(\mathcal{B}, \nu) = \overbrace{(\mathcal{I}_{10,c}, \iota) \otimes (\mathcal{I}_{10,c}, \iota) \otimes \cdots \otimes (\mathcal{I}_{10,c}, \iota)}^{n \text{ factors}}$$

and note that the basis of $(\mathcal{B}, \nu)$ is a basis for the standard topology of $\mathbb{R}^n$. Note that because (S, A) satisfies Kreisel's criterion, the function ψ which induces (S, A) is continuous. And since continuous real functions have closed graphs, the set S is closed in the standard topology of $\mathbb{R}^n$. Because S is closed, to prove that R is a basic representation of S in $(\mathcal{B}, \nu)$, it suffices to show that for each $r \in R$ there exists an $x \in S$ with $x \in \nu(r)$, and that for each $x \in S$ there is a local basis $\mathcal{L}_x$ for x with $\mathcal{L}_x \subseteq \nu(R)$.

By the definition of R, if $r = \langle u_1, u_2, \ldots, u_n \rangle \in R$ then for each positive integer $j \leq n - k$,

$$u_{k+j} = \kappa_j(\sigma_{u_1}, \sigma_{u_2}, \ldots, \sigma_{u_k}, m)$$

for some $m \in \mathbb{N}$. So, if p_i is the midpoint of the interval u_i for each positive integer $i \leq k$, then

$$u_{k+j} = \kappa_j(o_{p_1}, o_{p_2}, \ldots, o_{p_k}, m)$$

And by Kreisel's criterion $\lambda m[\kappa_j(o_{p_1}, o_{p_2}, \ldots, o_{p_k}, m)]$ is an oracle for $\varpi_j^{n-k}(\psi(p_1, p_2, \ldots, p_k))$. Therefore, for each $r \in R$ there exists an $(x_1, x_2, \ldots, x_n) \in S$ with $(x_1, x_2, \ldots, x_n) \in \nu(r)$. Namely, $x_i = p_i$ for each positive integer $i \leq k$ and

$$x_{k+j} = \varpi_j^{n-k}(\psi(p_1, p_2, \ldots, p_k))$$

for each positive integer $j \leq n - k$.

Now suppose that $(x_1, x_2, \ldots, x_n)$ is an arbitrary member of S. Again, by Kreisel's criterion, for each positive integer $j \leq n - k$, the function $\lambda m[\kappa_j(o_{x_1}, o_{x_2}, \ldots, o_{x_k}, m)]$ is an oracle for x_{k+j}. But for each $m \in \mathbb{N}$ and each positive integer $i \leq k$, the computation for $\kappa_j(o_{x_1}, o_{x_2}, \ldots, o_{x_k}, m)$ has only finitely many steps, and so the oracle o_{x_i} can only be called finitely many times during the course of the computation. Hence, for each $m \in \mathbb{N}$ there exists a non-negative integer l_i for each $i \leq k$, such that for any non-negative integer $l'_i \geq l_i$, if $u_i = o_{x_i}(l'_i)$ then $\kappa_j(\sigma_{u_1}, \sigma_{u_2}, \ldots, \sigma_{u_k}, m)$ is defined and

$$\kappa_j(\sigma_{u_1}, \sigma_{u_2}, \ldots, \sigma_{u_k}, m) = \kappa_j(o_{x_1}, o_{x_2}, \ldots, o_{x_k}, m)$$

Of course, for each positive integer $j \leq n-k$ the interval

$$u_{k+j} = \kappa_j(o_{x_1}, o_{x_2}, \ldots, o_{x_k}, m)$$

can be made arbitrarily small by choosing a suitably large value of m, and for each positive integer $i \leq k$ the interval u_i can be made arbitrarily small by choosing a suitably large value of l'_i. Furthermore, by definition, $\langle u_1, u_2, \ldots, u_n \rangle \in R$. It immediately follows that for each $x = (x_1, x_2, \ldots, x_n) \in S$ there is a local basis $\mathcal{L}_x$ for x such that $\mathcal{L}_x \subseteq \nu(R)$. We may conclude that (R, H) is a basic representation of (S, A). And since R is recursively enumerable, it follows from Theorem 10.5 that (R, H) is isomorphic to a computable physical model. $\square$

A physical model (S, A) that satisfies Kreisel's criterion on $\mathbb{R}^k$ is uniquely determined by the functions $\kappa_1, \kappa_2, \ldots, \kappa_{n-k}$. Moreover, the proof of Theorem 15.1 describes an effective procedure for finding a basic representation of (S, A), given programs for computing $\kappa_1, \kappa_2, \ldots, \kappa_{n-k}$. Let $\mathcal{K}_{k,n,c}$ be the collection of all basic representations of physical models that are constructed from physical models satisfying Kreisel's criterion according to the procedure in the proof of Theorem 15.1, where c is the positive rational number which appears in that proof. An immediate question is whether there exists an effective procedure for the inverse operation. That is, given a basic representation in $\mathcal{K}_{k,n,c}$, is there an effective procedure for constructing partial recursive functions $\kappa_1, \kappa_2, \ldots, \kappa_{n-k}$? In the proof of the following theorem, we show that the answer is "Yes." Therefore, for every physical model satisfying Kreisel's criterion on $\mathbb{R}^k$, there is a computable physical model that may be used in its place, to predict the values of observable quantities given the data.

Theorem 15.2 *If $(R, H) \in \mathcal{K}_{k,n,c}$ and if $\psi : \mathbb{R}^k \to \mathbb{R}^{n-k}$ is the function whose graph has basic representation R, then there exist partial recursive functions $\kappa_1, \kappa_2, \ldots, \kappa_{n-k}$ such that if $\phi_1, \phi_2, \ldots, \phi_k$ are nested oracles for real numbers $x_1, x_2, \ldots, x_k$ in the effective topology $(\mathcal{I}_{10,c}, \iota)$, then for each positive integer $j \leq n-k$, $\kappa_j(\phi_1, \phi_2, \ldots, \phi_k, m)$ is defined for all $m \in \mathbb{N}$ and $\lambda m \big[\kappa_j(\phi_1, \phi_2, \ldots, \phi_k, m) \big]$ is a nested oracle for $\varpi_j^{n-k}\big(\psi(x_1, x_2, \ldots, x_k)\big)$ in $(\mathcal{I}_{10,c}, \iota)$.*

Proof. Suppose that $(R, H) \in \mathcal{K}_{k,n,c}$ and that $\psi : \mathbb{R}^k \to \mathbb{R}^{n-k}$ is the function whose graph has basic representation R. Note by the proof of Theorem 15.1 that R is recursively enumerable. Now, given any oracles $\phi_1, \phi_2, \ldots, \phi_k$ for real numbers $x_1, x_2, \ldots, x_k$ in $(\mathcal{I}_{10,c}, \iota)$, it follows from Theorem 11.2 that there are complete oracles $\phi'_1, \phi'_2, \ldots, \phi'_k$ for x_1, $x_2, \ldots, x_k$ in $(\mathcal{I}_{10,c}, \iota)$, such that $\phi'_1, \phi'_2, \ldots, \phi'_k$ are recursive relative to $\phi_1, \phi_2, \ldots, \phi_k$ uniformly. Let $x_{k+1}, x_{k+2}, \ldots, x_n$ be the real numbers uniquely determined by the equation

$$\psi(x_1, x_2, \ldots, x_k) = (x_{k+1}, x_{k+2}, \ldots, x_n)$$

Then by the proof of Theorem 14.1,

$$Q = \{ \langle u_1, u_2, \ldots, u_n \rangle \in R \mid (\forall i \in \{1, 2, \ldots, k\})(\exists m \in \mathbb{N})[u_i = \phi'_i(m)] \}$$

is a basic representation of $\{(x_1, x_2, \ldots, x_n)\}$ in

$$(\mathcal{B}, \nu) = \overbrace{(\mathcal{I}_{10,c}, \iota) \otimes (\mathcal{I}_{10,c}, \iota) \otimes \cdots \otimes (\mathcal{I}_{10,c}, \iota)}^{n \text{ factors}}$$

And since R is recursively enumerable, the set Q is recursively enumerable relative $\phi_1, \phi_2, \ldots, \phi_k$ uniformly.

Now, since $x_i \in \iota(\phi_i(m))$ for each positive integer $i \leq k$, it follows from the definition of Q that $x_i \in \iota(u_i)$ for each $\langle u_1, u_2, \ldots, u_n \rangle \in Q$. But recall from the proof of Theorem 15.1 that R has the property that if $x_i \in \iota(u_i)$ for each positive integer $i \leq k$, then $x_j \in \iota(u_{k+j})$ for each positive integer $j \leq n - k$. Hence,

$$(x_1, x_2, \ldots, x_n) \in \nu\langle u_1, u_2, \ldots, u_n \rangle$$

for each $\langle u_1, u_2, \ldots, u_n \rangle \in Q$. It immediately follows from the definition of a basic representation that $\nu(Q)$ is a local basis for the point $(x_1, x_2, \ldots, x_n)$. Therefore, for any function $\kappa : \mathbb{N} \to R$ such that $\kappa(\mathbb{N}) = Q$, the function κ is an oracle for $(x_1, x_2, \ldots, x_n)$ in $(\mathcal{B}, \nu)$. And since Q is recursively enumerable relative to $\phi_1, \phi_2, \ldots, \phi_k$ uniformly, there is a nested oracle κ that is recursive relative to $\phi_1, \phi_2, \ldots, \phi_k$ uniformly. So, if we define

$$\kappa_j(\phi_1, \phi_2, \ldots, \phi_k, m) = \pi^n_{k+j}(\kappa(m))$$

for each positive integer $j \leq n - k$ and for each $m \in \mathbb{N}$, then κ_j is partial recursive, $\kappa_j(\phi_1, \phi_2, \ldots, \phi_k, m)$ is defined for all $m \in \mathbb{N}$, and $\lambda m[\kappa_j(\phi_1, \phi_2, \ldots, \phi_k, m)]$ is a nested oracle for $\varpi_j^{n-k}(\psi(x_1, x_2, \ldots, x_k)) = x_{k+j}$ in $(\mathcal{I}_{10,c}, \iota)$. $\square$

Acknowledgements

This paper is adapted from a chapter of our Ph.D. thesis [20]. We received many helpful suggestions from our thesis advisor, Richard Statman, and from the members of our thesis committee, most notably Robert Batterman and Lenore Blum. In the same regard, we also benefited from the careful reading and subsequent suggestions of Kevin Kelly, Klaus Weihrauch, Stephen Wolfram, and Hector Zenil. Zenil has also made us aware of a draft paper, posted by Marcus Hutter [9], which expresses ideas very similar to those presented here.

References

1. David Bohm, *Quantum theory*, Prentice-Hall, 1951.
2. P. W. Bridgman, *The logic of modern physics*, Macmillan, 1927.
3. John L. Casti, *Reality rules*, vol. 1, John Wiley & Sons, 1992.
4. Alonzo Church, *An unsolvable problem of elementary number theory*, American Journal of Mathematics **58** (1936), no. 2, 345–363.
5. Bryce S. DeWitt and Neill Graham (eds.), *The many-worlds interpretation of quantum mechanics*, Princeton University Press, 1973.
6. P. A. M. Dirac, *The principles of quantum mechanics*, Clarendon Press, 1930.
7. James Evans, *The history and practice of ancient astronomy*, Oxford University Press, 1998.
8. Edward Fredkin, *Digital mechanics*, Physica D **45** (1990), 254–270.
9. Marcus Hutter, *A complete theory of everything (will be subjective)*, arXiv:0912.5434v1 [cs.IT], December 2009.
10. E. T. Jaynes, *Information theory and statistical mechanics*, Physical Review **106** (1957), no. 4, 620–630.
11. G. Kreisel, *A notion of mechanistic theory*, Synthese **29** (1974), 11–26.
12. Christoph Kreitz and Klaus Weihrauch, *Theory of representations*, Theoretical Computer Science **38** (1985), 35–53.
13. Albert Messiah, *Mécanique quantique*, vol. 1, Dunod, 1959, see reference [14] for an English translation.
14. Albert Messiah, *Quantum mechanics*, vol. 1, North-Holland, 1961.
15. Ramon E. Moore, *Interval analysis*, Prentice-Hall, 1966.
16. Wayne C. Myrvold, *Computability in quantum mechanics*, The Foundational Debate (Werner Depauli-Schimanovich, Eckehart Köhler, and Friedrich Stadler, eds.), Kluwer, 1995, pp. 33–46.
17. Hartley Rogers, Jr., *Theory of recursive functions and effective computability*, McGraw-Hill, 1967.
18. Robert Rosen, *Church's thesis and its relation to the concept of realizability in biology and physics*, Bulletin of Mathematical Biophysics **24** (1962), 375–393.
19. Robert Rosen, *Fundamentals of measurement and representation of natural systems*, North-Holland, 1978.
20. Matthew P. Szudzik, *Some applications of recursive functionals to the foundations of mathematics and physics*, Ph.D. thesis, Carnegie Mellon University, Pittsburgh, Pennsylvania, December 2010.
21. Max Tegmark, *The mathematical universe*, Foundations of Physics **38** (2008), 101–150.

22. A. M. Turing, *On computable numbers, with an application to the Entscheidungsproblem*, Proceedings of the London Mathematical Society **42** (1936–1937), 230–265.
23. A. M. Turing, *Computability and λ-definability*, The Journal of Symbolic Logic **2** (1937), no. 4, 153–163.
24. Johann von Neumann, *Mathematische grundlagen der quantenmechanik*, Springer, 1932, see reference [25] for an English translation.
25. John von Neumann, *Mathematical foundations of quantum mechanics*, Princeton University Press, 1955.
26. Klaus Weihrauch and Tanja Grubba, *Elementary computable topology*, Journal of Universal Computer Science **15** (2009), no. 6, 1381–1422.
27. Klaus Weihrauch and Ning Zhong, *Is wave propagation computable or can wave computers beat the Turing machine?*, Proceedings of the London Mathematical Society **85** (2002), no. 3, 312–332.
28. Stephen Wolfram, *A new kind of science*, Wolfram Media, 2002.
29. Konrad Zuse, *Rechnender raum*, Elektronische Datenverarbeitung **8** (1967), 336–344.

Chapter 26

The Universe is Lawless or
"Pantôn chrêmatôn metron anthrôpon einai"*

Cristian S. Calude[1], F. Walter Meyerstein[2] and Arto Salomaa[3]

[1] *Computer Science Department,*
The University of Auckland, New Zealand
[2] *Barcelona, Spain*
[3] *Turku Centre for Computer Science, TUCS, Turku, Finland*

The belief that the physical Universe is a *knowable* system governed by rules which determine its future *uniquely* and *completely* has dominated the Western civilisation in the last two and a half millennia. The goal of this paper is to provide new arguments in favour of the hypothesis that the Universe is lawless, a hypothesis proposed and discussed in our papers.[7,9,11,14,15,18]

1. Introduction

The endeavour to discover and determine the laws presumed to govern the physical Universe is as old as Western civilisation itself, as are the difficulties herewith associated. Witness the anecdote transmitted by Plato (in his dialogue *Theaetetus*) concerning Thales of Miletus, the first mathematician to accurately predict a solar eclipse (for the 28th May 585 BC):

> While Thales was studying the stars and looking upwards, he fell into a pit, and a neat, witty Thracian servant girl jeered at him, because he was so eager to know the things in the sky that he could not see what was there before him at his very feet.

Nowadays we continue "to look upwards", albeit with the help of the latest technology and its fabulous instruments. This process is unavoidably marked by the human "measure" which biases the laws we presume to hold in the entire Universe.

---
*"Man is the measure of all things", Protagoras, 5th century BC.

In what follows we provide new arguments in favour of the hypothesis that the Universe is lawless, a hypothesis proposed and discussed in our papers.[7,9,11,14,15,18] We start by describing the notions of (physical) Universe and law of the Universe (sometimes called natural law or the law of nature), then we discuss the lawfulness hypothesis and lawlessness hypothesis. We continue by arguing in favour of the lawlessness hypothesis in various types of Universes. Finally we discuss the provability of the lawlessness hypothesis.

2. The Universe

The dictionary definition of the term, "all that exists", is a tautology. For the scientific endeavour ("looking upwards") the term covers two quite separate domains of the reality accessible to humans: I) the Solar System and II) the electromagnetic radiation signals from beyond the Solar System captured by the antennas of our instruments.

The Sun and the plethora of planets, moons, comets, asteroids, and other directly detectable objects have been intensely scrutinised by humans—from the very dawn of their history—by means of their innate radiation-detection antenna: the retina. Remarkably, beyond the naked eye, no further light-amplifying instrument was available to Thales, or to Ptolemy, until Galileo's invention of the telescope radically changed this way of "looking upwards". The telescope and Copernicus presented humans with a different Universe, a Universe gigantic but still reasonably comprehensible by minuscule humans, although definitely removing them from the central position. In the following centuries, great physicists discovered the first "laws" of nature, in domains so different as optics, electricity, movement, gravity, chemistry, etc., giving rise to the idea that there might exist rules of "universal" validity. But, as with Galileo's telescope, advances in technology again changed the entire outlook. Now a "large" bracket of the electromagnetic spectrum, not just the narrow visible-light window, is available for scrutiny.

Contemporaneous with these technological advances, theoretical physicists developed the two fundamental explanatory models of physical "reality", the standard model of quantum mechanics and general relativity of gravitation. These advances again fundamentally changed what was meant by the term "Universe".

In the first place, the human "measure" vanished: the reality encompassed by this term is enormous, both in time as in space. Just one tiny example: the distance from the Solar System to the nearest star, Alpha Centaurii, is approximately 40, 000, 000, 000, 000, 000 m. Further, if the Solar System inspected with the modern instruments revealed itself as being of an unsuspected complexity, the electromagnetic signals now detected in all frequencies of the spectrum showed—not just an extraordinary complexity—but also what could only be interpreted in the light of presently admitted theories as incredibly gigantic phenomena, for which even new names had to be coined: super-massive black holes, pulsars, quasars, neutron stars, and a variegated catalogue of supernovae, to name only a few.

Nevertheless, the search for a universal explanatory theory—"the laws"—went on. It had to incorporate the two fundamental theories: quantum mechanics and gravitation. But the first is a probabilistic theory, the second a deterministic theory, and their marriage has so far resisted all efforts. That is to say, these efforts now take outlandish forms: in them the Universe has more dimensions than the traditional four, eleven, for instance. Worse: there is not just one Universe—"ours"—but many of them, although completely detached and unreachable for us.

If these conundrums were not enough, further observations have created even more problems. Several decades ago it was discovered that the movement of the stars in a galaxy, including our Milky Way, do not comply with the speed values assigned to them by Newton's or Einstein's gravitation laws. Neither do groups of galaxies. The remedy: "black matter", an undetectable ("black") gravitating component of the Universe. Then recently it was found that the Universe expands faster than what was allowed by the latest theories. The remedy: "black energy", a concept originally put forward by Einstein albeit in a different context. What are these mysterious matter-energy forms? Until today nobody knows and of course nothing of that kind has so far been detected. However, based on more and more exact measurements of the cosmic microwave background, initially predicted to exist as a fossil remnant of the Big Bang[a] itself, the following composition of the Universe is presently put forward by cosmologists: dark matter 23.3%, dark energy 72.1%, ordinary matter, of which stars, planets and people are made: 4.6%.

In a quite abbreviated form, this is what the term Universe stands for nowadays. Clearly, not a well-defined concept but a patchwork of observations not yet understood, theories and prejudices.

---

[a]To arrive at a cosmology of the Big Bang type, many additional postulates are required, see, for example.[5]

3. The Laws

What are the laws of the Universe? Are they just metaphors (cf. Zilsel[36]) or "like veins of gold, [...] that scientists are extracting the ore, " (cf. Johnson[24])?

According to Feynman[22] and Davies,[21] the physical laws or the laws of the Universe—shortly, the laws, are expressed in simple mathematical terms; further on, the laws are universal (they apply everywhere in the Universe), infinite, absolute, stable, omnipotent (everything in the Universe must comply with them).

When the adjective "lawful" is predicated from the term Universe, what is thereby meant? Here all the usual human prejudices impinge. To determine the laws that rule the changes and development of said object is equated with acquiring *knowledge* about this entity. In other words, one jumps from the *how* to the *why*. But these laws, assuming we will ever find them, are not *causal* laws at all. It is true that Aristotle has defined knowledge of something as *knowledge* of the cause (or causes) *why* that thing is as it is. Probably because the assumption of causality is an innate—fitness enhancing—trait of humans, causality is in most cases immediately associated with knowledge. But causality can only be observed by humans in the form of short causal chains, short as measured from a particular here and now (*hic et nunc*) and basically only in the past-time direction.

We give a trivial example. A man passes under a balcony from which a flowerpot falls killing him. The cause of his death? The flowerpot, of course. But also the red traffic light: had it been green he would have passed earlier under the balcony ... etc. It is clear that from every *hic et nunc* sprout exponentially many interconnected "causal chains", and the whole idea becomes meaningless at a short past distance from any *nunc*. In the opposite direction, towards the future, causality changes into prediction, always a probabilistic affair in the best case. Finally, let us repeat again that the dimension of the Universe makes any reasonable reference to the human measure, as required by Protagoras, if not directly absurd, at least untenable.

Consequently, it seems that predicating "lawful", in any common-use sense of that term, with the noun Universe, cannot be reasonably made. In fact, it is precisely *not* in the common-use sense that the term is applied in most cases. Loaded with centuries of religious belief, the search is not for *laws* but for a *design*, or at least a *design principle*, of that Universe. This is again a very old idea. Divinities were always credited with a superior knowledge, including the possession of the ultimate account and justification of the world.

In Plato's *Timaeus* we have one of the most famous examples. The Universe in the *Timaeus* is fashioned by a divine craftsman, the demiurge, of whom it is repeatedly stated that he was "good" and that he designed the Cosmos with the view to make it as "good" as he possibly could (the platonic demiurge is not omnipotent). Note however, that nowhere in the *Timaeus* (or elsewhere) does Plato neatly define the "good". But the idea is clearly expressed: the difference between a chaotic, lawless Universe and a Universe that can be claimed to be "lawful", i.e. to be a Cosmos, is the existence of some overlying, unifying concept presumably of "divine" origin (such as "the Good", *t'agathon*, for Plato).

Plato's ideas directly influenced Brahe, Kepler, Galileo who put forward what became the official program of science: Find the lawful part of the Universe, and, if lucky, try to formulate the (mathematical) laws describing its "kinesis" (change). In this spirit some audacious present-day physicists and cosmologists are looking for a "theory of everything" or whatever name they may choose.

4. The Lawfulness Hypothesis

From millennia-old aspirations "to know more" comes the idea that the Universe is lawful. This hypothesis seems to be supported by our daily observations: the rhythm of day and night, the pattern of planetary motion, the regular ticking of clocks. The stage is set at the beginning and everything follows "mechanistically" without the intervention of God, without the occurrence of "miracles". The future is determined from the past by *universal, infinite and eternal laws*:

> [The] *entire history of the Universe is fixed*, according to some precise mathematical scheme, *for all time*, cf. Penrose ,[28] p. 558–559.

Most importantly, the laws are *knowable* by means of observations/measurements and reason/logic. It is up to us to discover them.

The great law, the law of cause and effect—*a thing cannot occur without a cause which produces it* in Laplace's words—transcends all known laws and is ever at work with chains of causations and effects governing all of manifested matter and life.

5. The Lawlessness Hypothesis

It is a simple matter of reflection to point out some limits of the lawfulness hypothesis: the vagaries of weather, the devastation of earthquakes or the fall of meteorites are "perceived" as fortuitous.

The great law of cause and effect is illusive and could not be proven, just observed. In fact, it is possible to disprove its universality as we shall see soon.

The lawlessness hypothesis—according to which there are no laws of the Universe—does not exclude the existence of *local rules* functioning on large, but finite scales. Local regularities are not only compatible with randomness, but in fact a consequence of randomness. Following[15] we will illustrate our arguments for a Universe crudely represented by an infinite binary sequence.

For example, every Martin-Löf random sequence[b] contains every possible string (of any length) and every such string must appear infinitely many times.[6] The fact that the first billion digits of a Martin-Löf random sequence are perfectly lawful, for instance by being exactly the first digits of the binary expansion of π, does not modify in any way the global property of randomness of the (infinite) sequence.

These facts are consistent with our common experience. Space scientists can pinpoint and predict planetary locations and velocities "well enough" to plan missions months in advance, astronomers can predict solar or lunar eclipses centuries before their occurrences, etc. All these results—as impressive as they may be—are only true *locally* and within a certain *degree of precision*. They are not "laws of the Universe".

The hypothesis that the Universe is lawless is not a new idea. Twenty-four centuries ago, Plato in the *Timaeus* invented a cosmology (see more in[14]) which states that in the beginning the demiurge finds a completely chaotic substrate, "Chora", which has only one property: it is the material substrate of the Universe in a primordial state, a state which we would call today random. Faithful to the law of cause and effect, Plato proposes an acting principle of disorder, a cause of randomness, which he calls "Anagke" (necessity). The demiurge is trying to "persuade" Anagke to accept a mathematical order. If successful, one arrives at a finite set of purely mathematical elementary building blocks—Plato's perfect polyhedra—which, when combined by simple mathematical rules, constitute the ordered Universe,

---

[b]A sequence is Martin-Löf random if there is a constant c such that all its finite prefixes are c-incompressible with respect to a self-delimiting universal Turing machine.

the "cosmos" (order). But only the part where the demiurge succeeded in persuading Anagke is ordered. In fact, the demiurge is not all-powerful, hence in Plato's Universe, order is only partial. And an irreducible disorder, chaos, randomness remains, so irreducible that nothing can be *said* about it. Plato does not indicate anywhere what part of the Universe is lawful, and what part is entirely random.

We note that the demiurge is not the God of Genesis, as later interpreters hoped to prove. In fact, the demiurge does not "create" anything at all, it is only the sufficient cause of order, where such order exists. Instead of saying, "there is a law which underpins the order detected in this context", Plato says, "the demiurge caused ...", and then he adds the mathematical expression describing in rigorous terms, this partial order.

Twenty four centuries later, Poincaré also suspected the chaotic, random nature of the Universe when he wrote:[c]

> If we knew exactly the laws of nature and the situation of the universe at the initial moment, we could predict exactly the situation of that universe at a succeeding moment. But even if it were the case that the natural law no longer had any secret for us, we could still only know the initial situation approximately. If that enabled us to predict the succeeding situation with the same approximation, that is all we require, that [it] is governed by the laws. But it is not always so; it may happen that small differences in the initial conditions produce very great ones in the final phenomena. A small error in the former will produce an enormous error in the latter. Prediction becomes impossible, and we have the fortuitous phenomenon.

In our time Barrow[2] has proven that Einstein's equations exhibit a formal chaotic behaviour, which means that the evolution of the Universe becomes unpredictable after a time short in cosmological scales. Hawking's views (see[23] p. 26) are even stronger:

> The intrinsic entropy means that gravity introduces an extra level of unpredictability over and above the uncertainty usually associated with quantum theory.[d] So Einstein was wrong when he said, "God does not play dice." Consideration of black holes

---

[c]Quoted from Peterson,[29] p. 216.

[d]A massive star, which has exhausted its supplies of nuclear energy, collapses gravitationally and disappears leaving behind only an intense gravitational field to mark its presence. The star remains in a state of continuous free fall, collapsing endlessly inward into the gravitational pit without reaching the bottom.

suggests, not only God does play dice, but that he sometimes confuses us by throwing them where they can't be seen.

A detailed account of unknowables in physics is given by Svozil.[34]

Are there better ways to describe the Universe than the mathematical one? In retrospect, mathematical formalisms seem to be inevitable, In any case, there is nothing to indicate better candidates. The growing preference to move from analytical descriptions of physical laws to algorithmic ones (see for example Wolfram[35] or the discussions in[9,11]) is not a paradigm shift as programs are fundamentally mathematical entities.

6. Arguments in Favour of the Lawlessness Hypothesis

We concentrate on continuous models for the Universe. First we will argue that even if the Universe is lawful then we won't be able to know this; secondly, we shall discuss reasons why the Universe cannot be lawful.

As the tools to understand the laws are mathematical and much of the elementary intuition about numbers derives from our linguistic abilities to assign names to objects[e] it is not surprising that our arguments will focus on numbers. This point of view is consistent with Landauer's[26]

> The laws of physics are essentially algorithms for calculation. These algorithms are significant only to the extent that they are executable in our real physical world. Our usual laws depend on the mathematician's real number system.

To what extent is the system of real numbers contaminated by "chaoticity" and "randomness"? A real number in base b is *disjunctive* (cf. Jürgensen and Thierrin[25]) in case its b-expansion sequence contains all possible strings over that alphabet $\{0, 1, \ldots, b-1\}$. A *lexicon* is a real number which is disjunctive in any base. A lexicon contains all writings, which have been or will be ever written, in any possible language. A lexicon expresses a strong qualitative idea of randomness.

According to the law of large numbers, in every binary expansion of almost every real number in the unit interval every string appears with its "natural" probability. For example 1 appears with probability 1/2, 0 appears with probability 1/2, 00 appears with probability 1/4, and so on. This happens for almost all, but not exactly all of them: the law of large

---

[e]According to Barrow ([3], p. 4), "linguistic abilities are far more impressive than our mathematical abilities, both in their complexity and their universality among humans of all races."

numbers is false in the sense of Baire category with respect to the natural topology of the unit interval,[27] but it is still true for a small modification of this topology.[13] Lexicons form residuals[19] for the natural topology, hence *most reals do not obey any probability laws*. This shows that the system of real numbers, our very basic language of expressing laws, is fully contaminated by randomness.

Martin-Löf randomness, a stronger quantitative form of randomness, while "less pervasive" than disjunctivity, is still omnipresent among real numbers: with probability one every real number is Martin-Löf random.[6] Even more, Martin-Löf random reals are in a sense the "bricks" of the whole set of reals: by Gács theorem improved by Hertling (see,[6] p. 155–165) every real is effectively reducible to a Martin-Löf random one.

The law of cause and effect breaks down with the advent of algorithmic information theory: mathematics, even elementary number theory, is full of facts true for no formal reason as Chaitin has proved:[6,20,30]

> God not only plays dice in physics but also in pure mathematics.

Randomness not only exists, it is everywhere.[10]

The lawlessness identified in the system of reals appears in quantum mechanics. This is no news, except that now one can go beyond the mere postulation of quantum randomness: one can prove some mathematical facts about the quality of quantum randomness. Consider a quantum random number generator generating bits produced by successive preparation and measurement of a state in which each outcome has probability one-half. By envisaging this device running ad infinitum, we can consider the infinite sequence **x** it produces. If we assume: a) a standard picture of quantum mechanics, i.e. a Copenhagen-like interpretation in which measurement irreversibly alters the quantum state, b) the "many-worlds" interpretation and other "exotic" possibilities including contextual hidden counterfactual observables are excluded, and c) the experimenter has freedom in the choice of measurement basis (the "free-will assumption"), then **x** is incomputable,[17] that is no Turing machine can reproduce exactly the bits of the sequence **x**. For example, **x** can start with a billion of 0's, but cannot consists of only 0's. In fact, one can prove a stronger property: the sequence **x** is bi-immune, i.e. only finitely many bits of **x** are computable. Every bi-immune sequence is incomputable, but the converse is not true. Experimental confirmation of this theoretical result was obtained in.[12]

7. Digital Universes are Also Lawless

As the "free-will assumption" used in the previous section excludes a digital Universe[f], it is natural to ask whether such a Universe is lawful or not?

Digital physics distinguishes three possible scenarios: a) the Universe[g] is (may be) continuous, but our model is digital, say a universal (prefix) Turing machine working with a discrete infinite time, b) the Universe is a universal (prefix) Turing machine working with a discrete infinite time, c) the Universe is a universal (prefix) Turing machine working for a finite, albeit huge, time only.

A law of a Universe in cases a) and b) can be expressed by an infinite sequence while for c) the law has to be expressed by a finite string. All results regarding qualitative and quantitative randomness described in the previous section apply for the scenarios a) and b). The status of a "law" in the scenario c) is not so clear. The lawlessness of such a Universe comes from the fact that strings coding programs expressing laws of such a Universe cannot be distinguished from algorithmic random strings.[6]

The influential NKS programme initiated by Wolfram's book[35]—the systematic, empirical investigation of computational systems for their own sake—is relevant for understanding the Universe in all three possible scenarios described above, irrespective of the particular philosophical views of researchers in NKS. Proposed digital versions of various parts of continuous physics have consistently revealed various forms of randomness; see for example the work in digital statistical mechanics in.[1,16,31–33]

8. Can the Lawlessness Hypothesis be Proved?

In spite of many unknowables in physics,[34] the relevance of incompleteness of mathematics for physics is still unclear.[4] It is unlikely that a formal proof for the lawlessness hypothesis can be found. Of course, the hypothesis can be experimentally illustrated and tested (see[12] and the discussion in Zenil[37]).

In agreement with Hawking ([23], p. 3–4):

> I take the positivist viewpoint that a physical theory is just a mathematical model and that it is meaningless to ask whether

---

[f]In a truly deterministic theory—sometimes called super-determinism—the experimenter might have the illusion of exercising her independent free choice, but in reality she just obeys the rules of the theory.
[g]Note that the term Universe, as described in Section 2, is a model itself.

it corresponds to reality. All that one can ask is that its predictions should be in agreement with observation.

one can say that our partial and provisional understanding of the Universe comes through measurements, so ultimately through numbers. With extremely rare exceptions, the real numbers representing the outcome of measurements are lexicons, so they are devoid of any order or law. Can such a system express any "laws" of the physical Universe?

Finally, does the lawlessness hypothesis mean the end of science? Should one definitely abandon the hope of finding sense and meaning in the Universe? The answers to both questions are negative. With the lawfulness hypothesis we leave in a dream of global, universal order and law, when, according to the lawlessness hypothesis, there is only Chora (chaos) with local laws only. These are just hypotheses and their merits should be pragmatically judged only. If one feels elated to discover the laws of the Universe, then the traditional assumption fits better; the alternative hypothesis is preferable for the more realistic and humble minds. Science is and will be alive, and progress in answering fundamental questions and developing applications will continue.

Acknowledgement

We thank E. Calude, J. Casti, G. Chaitin, B. Doran, S. Marcus, B. Pavlov, M. Stay, K. Svozil and H. Zenil for many illuminating discussions on these issues.

References

1. J. Baez, M. Stay. Algorithmic thermodynamics, *Mathematical Structures in Computer Science*, 2012, to appear.
2. J. Barrow. Chaotic behaviour in general relativity, *Physics Reports* 85, 1–49, 1982.
3. J. Barrow. Limits of science, in J. L. Casti, A. Karlqvist (eds.). *Boundaries and Barriers*, Addison-Wiley, New York, 1–11, 1996.
4. J. D. Barrow. Gödel and physics, in M. Baaz, C. Papadimitriou, H. Putnam, D. Scott, C. Harper Jr. (eds.). *Kurt Gödel and the Foundations of Mathematics. Horizons of Truths*, Cambridge University Press, Cambridge, 255–276, 2011.
5. L. Brisson and F. W. Meyerstein. *Inventing the Universe*, Suny Press, New York, 1995.
6. C. Calude. *Information and Randomness–An Algorithmic Perspective*, Springer-Verlag, New York, 2002 (2nd ed.).

7. C. S. Calude. Randomness everywhere: My path to algorithmic information theory, in H. Zenil (ed.). *Randomness Through Computation*, World Scientific, Singapore, 179–189, 2011.
8. C. S. Calude, E. Calude and K. Svozil. The complexity of proving chaoticity and the Church-Turing Thesis, *Chaos* 20 037103, 1–5, 2010.
9. C. S. Calude, J. L. Casti, G. J. Chaitin, P. C. W. Davies, K. Svozil, S. Wolfram. Is the universe random? in H. Zenil (ed.). *Randomness Through Computation*, World Scientific, Singapore, 309–350, 2011.
10. C. S. Calude, G. J. Chaitin. Randomness everywhere, *Nature* 400, 22 July, 319–320, 1999.
11. C. S. Calude, G. J. Chaitin, E. Fredkin, A. T. Legget, R. de Ruyter, T. Toffoli, S. Wolfram. What is computation? (How) Does nature compute? in H. Zenil (ed.). *Randomness Through Computation*, World Scientific, Singapore, 351–403, 2011.
12. C. S. Calude, M. J. Dinneen, M. Dumitrescu, K. Svozil. Experimental evidence of quantum randomness incomputability, *Physical Review A*, 82, 022102, 1–8, 2010.
13. C. S. Calude, S. Marcus, L. Staiger. A topological characterization of random sequences, *Information Processing Letters* 88, 245–250, 2003.
14. C. S. Calude, F. W. Meyerstein. Is the universe lawful? *Chaos, Solitons & Fractals* 10, 6, 1075–1084, 1999.
15. C. Calude, A. Salomaa. Algorithmically coding the Universe, in G. Rozenberg, A. Salomaa (eds.). *Developments in Language Theory*, World Scientific, Singapore, 472–492, 1994.
16. C. S. Calude, M. A. Stay. Natural halting probabilities, partial randomness, and Zeta functions, *Information and Computation* 204, 1718–1739, 2006.
17. C. S. Calude, K. Svozil. Quantum randomness and value indefiniteness, *Advanced Science Letters* 1 (2008), 165–168.
18. C. S. Calude, K. Svozil. Is Feasibility in Physics Limited by Fantasy Alone?, `arXiv:0910.0457v1 [physics.hist-ph]`, 9pp. 2009.
19. C. S. Calude, T. Zamfirescu. Most numbers obey no probability laws, *Publicationes Mathematicae Debrecen*, Tome 54 Supplement, 619–623, 1999.
20. G. J. Chaitin. *Algorithmic Information Theory* Cambridge University Press, 1987.
21. P. Davies. *The Mind of God. The Scientific Basis for a Rational World*, Simon & Schuster, 1992.
22. R. Feynman. *The Character of Physical Law*, Modern Library, 1994.
23. S. Hawking, R. Penrose. *The Nature of Space and Time*, Princeton University Press, Princeton, New Jersey, 1996.
24. G. Johnson. *Fire in the Mind*, Alfred A. Knopf, New York, 1995.
25. H. Jürgensen, G. Thierrin. Some structural properties of ω-languages, *13th Nat. School with Internat. Participation "Applications of Mathematics in Technology"*, Sofia, 56–63, 1988.
26. R. Landauer. The physical nature of information, *Physics Letters* A 217, 188–193, 1996.

27. J. C. Oxtoby, S. M. Ulam. Measure-preserving homeomorphisms and metrical transitivity, *Annals of Mathematics* 42 (1941) 874–925.
28. R. Penrose. *The Emperor's New Mind*, Vintage, London, 1990.
29. I. Peterson. *Islands of Truth, A Mathematical Mystery Cruise*, W. H. Freeman, New York, 1990.
30. G. Rozenberg and A. Salomaa. The secret number. An exposition of Chaitin's theory, in C. S. Calude (ed.). *Randomness and Complexity, from Leibniz to Chaitin*, World Scientific Publ. Co., Singapore, 2007, 175–215.
31. K. Tadaki. A statistical mechanical interpretation of algorithmic information theory, *ISIT2007*, Nice, France, June 24–June 29, 2007, 1906–1910.
32. K. Tadaki, A statistical mechanical interpretation of algorithmic information theory. In: *Local Proceedings of Computability in Europe 2008 Athens, Greece, June 15–20, 2008*, 425–434.
33. K. Tadaki. A statistical mechanical interpretation of algorithmic information theory III: Composite systems and fixed points, *Mathematical Structures in Computer Science*, 2012, to appear.
34. K. Svozil. Physical unknowables, in M. Baaz, C. Papadimitriou, H. Putnam, D. Scott, C. Harper Jr. (eds.). *Kurt Gödel and the Foundations of Mathematics. Horizons of Truths*, Cambridge University Press, Cambridge, 2011, 213–254.
35. S. Wolfram. *A New Kind of Science*, Wolfram Research, 2002.
36. E. Zilsel. The genesis of the concept of 'physical law', *Philosophical Review* 303 (1942), 245–279.
37. H. Zenil. The world is either algorithmic or mostly random, arXiv:1109.2237v1 [cs.IT] http://fqxi.org/community/forum/topic/867, 8 February 2011.

Chapter 27

Is Feasibility in Physics Limited by Fantasy Alone?

Cristian S. Calude[1] and Karl Svozil[2]

[1] Computer Science Department, The University of Auckland, New Zealand
[2] Institute for Theoretical Physics,
Vienna University of Technology, Austria

> Although various limits on the predicability of physical phenomena as well as on physical knowables are commonly established and accepted, we challenge their ultimate validity. More precisely, we claim that fundamental limits arise only from our limited imagination and fantasy. To illustrate this thesis we give evidence that the well-known Turing incomputability barrier can be trespassed via quantum indeterminacy. From this algorithmic viewpoint, the "fine tuning" of physical phenomena amounts to a "(re)programming" of the universe.

Take a few moments for some anecdotal recollections. Nuclear science has made true the ancient alchemic dream of producing gold from other elements such as mercury through nuclear reactions. A century ago, similar claims would have disqualified anybody presenting them as quack. Medical chemistry discovered antibiotics which cure Bubonic plague, tuberculosis, syphilis, bacterial pneumonia, as well as a wide range of bacterial infectious diseases which were considered untreatable only one hundred years ago. For contemporaries it is hard to imagine the kind of isolation, scarcity in international communication, entertainment and transportation most of our ancestors had to cope with.

This historic anecdotal evidence suggests that what is considered tractable, operational and feasible depends on time. One could even extend speculations to the point where everything that is imaginable is also feasible. In what follows, we shall concentrate on some physical issues which might turn out to become relevant in the no–so–distant future, and which might affect the life of the generations succeeding ours to a considerable degree. In particular, we shall consider the connections between time, space and the limit velocity of light in vacuum; we shall ponder upon measurement; and we shall discuss physical indeterminism and randomness, and its relations to the possibility of trespassing the Turing incomputability barrier.

1. Space-time

One of the findings of special relativity theory is the impossibility to trespass the speed of light barrier "from below;" i.e., by starting out with subluminal speed. This fundamental limit applies also to communication and information transfer. Amazingly, this holds true even when quantum mechanics and "nonlocal quantum correlations" are taken into account, stimulating a notion of "peaceful coexistence" between quantum mechanics and special relativity theory. Thereby, superluminal particles, as well as the inclusion of field theoretic effects such as an index of refraction smaller than unity, supercavitation in the quantum ether, or general relativistic effects by locally rotating masses, wormholes or local contraction and expansion of space-time, possibly also related to time travel, to name but a few, cannot be excluded *a priory*.

Recent operational definitions of space-time and velocity, in order to physically represent the former, conventionalised the latter: Initially, the constancy of the velocity of light in vacuum in all reference frames was treated as an empirical fact. Since 1983 it has been frame-invariantly standardised by *Resolution Number 1* of the *17th Conférence Générale des Poids et Mesures (CGPM)* in which the following SI *(International System of Units)* operational definition of the meter has been adopted: *"The metre is the length of the path travelled by light in vacuum during a time interval of 1/299 792 458 of a second."* As a result, the empirical fact associated with this convention, as predicted by relativity theory, is the proposition that the length of a solid body depends neither on its spatial orientation, nor on the inertial frame relative to which that body is at rest.[1]

Indeed, by a theorem of incidence geometry,[2] linear Lorenz-type transformations follow from the frame-invariant standardisation of the velocity of light alone and appear to be a formal consequence of the conventions adopted by the SI. In such an approach, the physics resides in the invariance of Maxwell's equations and the equations of motion in general, as well as in the invariance of all physical measures based on matter stabilised by them, such as the length or duration of a space or time scale. This, after all, suits the spirit of Einstein's original 1905 paper, which starts out with conventions defining simultaneity and then proceeds with kinematics and by unifying electric and magnetic phenomena. Of course, for the sake of principle, everybody is free to choose other "limiting speeds," thereby implicitly sacrificing the form invariant representation of the equations of motion in inertial frames dominated and stabilised by electromagnetic interactions. In this way it would also not be difficult to adopt special relativity to findings of higher signalling and travel speeds than the velocity of light in vacuum.

Since antiquity, natural philosophers and scientists have pondered about the (in)finite divisibility of space-time, about its (dis)continuity, and about the (im)possibility of motion. In more recent times, the ancient Eleatic arguments ascribed to Parmenides and Zeno of Elea have been revived to "construct" accelerated computations[3] which serve as one of the main paradigms of the fast growing field of hypercomputation.

The most famous argument ascribed to Zeno is the impossibility for "Achilles" to overtake a turtle if the turtle is granted to start some finite distance ahead of Achilles, even though the turtle moves, say, one hundred times slower than Achilles: for in the finite time it takes Achilles to reach the turtle's start position, it has already moved away from it and is still a (tenth of the original) finite distance apart from Achilles. Now, if Achilles tries to reach that new point in space, the turtle has made its way to another point and is still apart from Achilles. Achilles' vain attempts to reach and overtake the turtle could be considered *ad infinitum*; with him coming ever closer to the turtle but never reaching it. By a similar argument, there could not be any motion, because in order to move from one spatial point to another, one would have first to cross half-distance; and in order to be able to do this, the half-distance of the half-distance, ... again *ad infinitum*. It might seem that because of the infinite divisibility of space, unrestricted motion within it is illusory because of this impossibility.

The modern-day "solution" of this seemingly impossible endeavour to move ahead of a slower object resides in the fact that it takes Achilles an ever decreasing outer (extrinsic, exterior) time to reach the turtle's previous position; so that if one takes "the limit" by summing up all infinitely many outer space and time intervals, Achilles meets (and overtakes) the turtle in finite outer time, thereby approaching an infinity of space-time points. Of course, Achilles' approach is even then modelled by an infinite number of steps or trip segments, which can be used to create an inner (intrinsic) discrete temporal counter. If this inner counter could in some form be associated with the cycle of an otherwise conventional universal computer such as a universal Turing machine or a universal cellular automaton,[4,5] then these "machines" might provide "oracles" for "infinite computations." In this respect, the physics of space and time, and computer science intertwine.

The accelerated Turing machine (sometimes called Zeno machine) is a Turing machine working in a computational space analogue to Zeno's scenario. More precisely, an accelerated Turing machine is a Turing machine that operates in a universe with two clocks: for the exterior clock each step is executed in a unit of time (we assume that steps are in some sense identical except for the time taken for their execution) while for the inner clock it takes an ever decreasing amount of time, say 2^{-n} seconds, to perform its nth step. Accelerated Turing machines have been implicitly described by Blake [6, p. 651] as well as Weyl [3, pp. 41-42], and studied in many papers and books since then.

Because an accelerated Turing machine can run an infinite number of steps (as measured by the exterior clock) in one unit of time (according to the inner clock), such a mechanism may compute incomputable functions, for example, the characteristic function of the halting problem.

How feasible are these types of computation? This is not an easy question, so not surprisingly there is no definitive answer. One way to look at this question is to study the relation between computational time and space. As expected, there is a similarity between computational time and space; however, this parallel is not perfect. For example, it is not true that an accelerated Turing machine which uses unbounded space has to use an infinite space for some input. An accelerated Turing machine that uses a finite space (not necessarily bounded) for all inputs computes a *computable function* (the function is not necessarily computed by the same machine).[7] Hence, if an accelerated Turing machine computes an incomputable function, then the machine has to use an infinite set of configurations for infinitely many inputs. Re-phrasing, going beyond Turing barrier with an accelerated Turing machine requires an infinite computational space (even if the computational time is finite); the computational space can be bounded (embedded in the unit interval), but cannot be made finite. Do we have such a space? Maybe relativistic computation offers a physical model for hypercomputation.[8]

2. Measurement

Another challenging question has emerged in the quantum mechanical context but it equally applies for all reversible systems: what is an irreversible measurement? Because if the quantum evolution is uniformly unitary and thus strictly reversible, what is to be considered the separated "measurement object" and the "measurement apparatus" can be "wrapped together" in a bigger system containing both, together with the "Cartesian cut;" i.e., the environment supporting communication between these two entities. Any such bigger system is then uniformly describable by quantum mechanics, resulting in total reversibility of whatever might be considered intrinsically and subjectively as a "measurement." This in turn results in the principal impossibility of any irreversible measurement (not ruling out decoherence "fapp;" i.e., for all practical purposes); associated with the possibility to "reconstruct" a physical state prior to measurement; and to "undo" the measurement.[9] The quantum state behaves just as in Schrödinger's interpretation of the Ψ function as a *catalogue of expectation values:* this catalogue can only be "opened and read" at a single page; yet it may be "closed" again by "using up" all knowledge obtained so far, and then re-opened at another page.

Two related types of unknowables which have emerged in the quantum context are complementarity and value indefiniteness. *Complementarity* is the impossibility to measure two or more observables instantaneously with arbitrary accuracy: in the extreme case, measurement of one observable annihilates the possibility to measure another observable, and *vice versa.* Despite attempts to reduce this feature to a "completable" incompleteness by Einstein and others, and thus to preliminary, epistemological deficiencies of the quantum formalism, the hypothetical "quantum veil," possibly hiding the "physical existence" of the multitude of all conceivable (complementary) observables, has maintained its impermeability until today.

As new evidence emerged, the lack of classical comprehensibility has gotten even worse: whereas quasi-classical systems—such as generalised urn or finite automaton models[10]—feature complementarity, some quantised systems with more than two measurement outcomes cannot be thought of as possessing any global "truth function." As the Kochen-Specker theorem[11,12] shows, they are *value indefinite* in the sense that there exist (even finite) sets of observables which, under the hypothesis of non-contextuality, cannot all (for some this might still be possible) have definite values independent of the type of measurement actually being performed.

Faced with the formal results, some researchers prefer a resolution in terms of *contextual* realism: measurement values "exist" irrespective of their "actual measurement," but they depend on what other observables are measured alongside of them. Another possibility is to abandon classical omniscience and assume that an "elementary" quantum system is only capable of expressing a *single* bit (or dit for d potential measurement outcomes)[13] or context; all other conceivable measurements are mediated by a measurement apparatus capable of context translation.

3. Indeterminacy and Hypercomputability

In the Pythagorean tradition, the universe computes. Thus any method and measure to change its behaviour amounts to (re)programming. If one remains within this metaphor, the character and "plasticity" of the "substratum" software and hardware needs to be exploited. Presently, the Church-Turing thesis confines the universe to universal computability formalised by recursion theory, but is it conceivable that some physical processes transcend this realm?

Arguably, the most (in)famous result in theoretical computer science is Turing's theorem saying that it is undecidable to determine whether a general computer program will halt or not. This is formally known as the *halting problem*. More precisely, there is no computer program halt which given as input an arbitrary program p runs a finite-time computation and returns 1 if p eventually stops and 0 if p never stops (here we use a fixed universal Turing machine to run programs).

There are two essential conditions imposed on halt: a) halt has to stop on every input, b) halt returns the correct answer. It is easy to construct a program halt that satisfies the above two conditions for many very, very large sets of programs, even for infinite sets of programs, but *not*, as Turing proved, for *all* programs.

So, one way to trespass the Turing barrier is to provide a physical mechanism which computes the function halt discussed above. There are many proposals for such devices.. Let's first present a negative result: using an information-theoretic argument, the possibility of having access to a time-travel machine would not solve the halting problem, unless one could travel back and forth in time at a pace exceeding the growth of any computable function.

Would some quantum processes transcend the Turing barrier? Surprisingly, the answer is yes,[14] and the main reason is the incomputability of quantum randomness.

In 1926, Max Born stated that (cf. [15, p. 866], English translation in [16, p. 54])

> From the standpoint of our quantum mechanics, there is no quantity which in any individual case causally fixes the con-

sequence of the collision; but also experimentally we have so far no reason to believe that there are some inner properties of the atom which condition a definite outcome for the collision. Ought we to hope later to discover such properties [[...]] and determine them in individual cases? Or ought we to believe that the agreement of theory and experiment — as to the impossibility of prescribing conditions? I myself am inclined to give up determinism in the world of atoms.

Born's departure from the *principle of sufficient reason* — stating that every phenomenon has its explanation and cause — by postulating irreducible randomness[17] in the physical sciences did not specify formally the type of "indeterminism" involved. More recent findings related to the Boole-Bell, Greenberger-Horne-Zeilinger as well as Kochen-Specker theorems for Hilbert spaces of dimension three onwards derive physical indeterminism from value indefiniteness of at least one observable among finite complementary collections of observables. As a result, although not necessarily all noncontextual observables in Kochen-Specker-constructions need to be value indefinite, but at least one has to be.

Suppose that quantum value indefiniteness occurs uniformly and symmetrically distributed over all observables. Because indeterminism and randomness are defined by algorithmic "lawlessness" and "incompressibility"[18] any physical system featuring indeterminism and randomness cannot be simulated by a universal computer; it "outperforms" any known computing machinery in terms of unpredictability. With these assumptions, physical value indefiniteness and randomness can thus be seen as valuable resources capable of serving as "oracles" for example, for Monte Carlo methods and primality testing requiring them. Indeterminism becomes an *asset* rather than a deficiency.

Contemporary realisations of quantum random number generators involve beam splitters. Thereby it should be noted that lossless beam splitters are reversible devices formalised by unitary transformations, and that the single photons used constitute a two-dimensional Hilbert space which may be "protected" from cryptanalytic attacks "lifting the hypothetic quantum veil" by quantum complementarity only. Indeed, it may not be totally unreasonable to point out that one of the greatest and mind-boggling quantum riddles of our time is the rather ambivalent use of beam splitters: on the one hand, beam splitters are associated with random coin tosses, which are postulated to yield absolute and irreducible randomness;[17] while on the other hand beam splitters are represented by discrete reversible unitary operators; the action of which could be totally reversed by serially composing two of them into a Mach-Zehnder interferometer.

Imperfections in measurements are typically corrected with von Neumann's procedure of normalisation — "compressing" a bit sequence *via* the map $00, 11 \mapsto \{\}$ (00 and 11 are discarded), $01 \mapsto 0$, and $10 \mapsto 1$. The algorithm works under the hypotheses of independence and stationarity of the original sequence,[21] conditions which may not be satisfied in beam splitting experiments—for instance due to multiparticle statistics like the Hanbury Brown and Twiss effect.

Some quantum systems are protected by value indefiniteness grounded in the Kochen-Specker theorem from Hilbert space dimensions three onwards. As the Kochen-Specker theorem requires complementarity, but the converse implication is not true, it follows that a system of two entangled photons in a singlet state or systems with three or more measurement outcomes may be more suitable for generating quantum random bits. Obtaining more than two outcomes is not problematic as, if in a sequence of random elements drawn from an alphabet with $n > 2$ symbols a fixed symbol is systematically removed, the resulting sequence is still random (over an alphabet of $n - 1$ symbols).

Final Remarks

There are exciting times ahead of us. The limits which seem to be imposed upon us by various constraints might decay into "thin air" as the conditions upon which these constraints are founded will lose their applicability and necessity, or even lose their operational validity. Thus, we perceive physical tractability and feasibility wide open, positive, and full of unexpected opportunities. Indeed, we just quiver at the extension of our imaginable ignorance; let alone the possibilities which we even lack to fantasise. Any further scientific exploration of this realm has to be strongly encouraged.

Acknowledgement

We thank A. Abbott and E. Calude for useful comments and criticism.

References

1. Asher Peres. Defining length. *Nature*, 312:10, 1984.
2. June A. Lester. Distance preserving transformations. In Francis Buekenhout, editor, *Handbook of Incidence Geometry*. Elsevier, Amsterdam, 1995.

3. Hermann Weyl. *Philosophy of Mathematics and Natural Science*. Princeton University Press, Princeton, NJ, 1949.
4. Konrad Zuse. Rechnender Raum. *Elektronische Datenverarbeitung*, pages 336–344, 1967.
5. Stephen Wolfram. *A New Kind of Science*. Wolfram Media, Inc., Champaign, IL, 2002.
6. R. M. Blake. The paradox of temporal process. *Journal of Philosophy*, 23(24):645–654, 1926.
7. Cristian S. Calude and Ludwig Staiger. A note on accelerated turing machines. *Mathematical Structures in Computer Science*, 20 (Special Issue 06):1011–1017, 2010.
8. Hajnal Andréka, István Németi, and Péeter Németi. General relativistic hypercomputing and foundation of mathematics. *Natural Computing*, 8(3):499–516, 2009.
9. Thomas J. Herzog, Paul G. Kwiat, Harald Weinfurter, and Anton Zeilinger. Complementarity and the quantum eraser. *Physical Review Letters*, 75(17):3034–3037, 1995.
10. Karl Svozil. Contexts in quantum, classical and partition logic. In Kurt Engesser, Dov M. Gabbay, and Daniel Lehmann, editors, *Handbook of Quantum Logic and Quantum Structures*, pages 551–586. Elsevier, Amsterdam, 2009.
11. Ernst Specker. Die Logik nicht gleichzeitig entscheidbarer Aussagen. *Dialectica*, 14(2-3):239–246, 1960. Reprinted in Ref. [19, pp. 175–182]; English translation: *The logic of propositions which are not simultaneously decidable*, Reprinted in Ref. [20, pp. 135-140] and as an eprint arXiv:1103.4537.
12. Simon Kochen and Ernst P. Specker. The problem of hidden variables in quantum mechanics. *Journal of Mathematics and Mechanics (now Indiana University Mathematics Journal)*, 17(1):59–87, 1967. Reprinted in Ref. [19, pp. 235–263].
13. Anton Zeilinger. A foundational principle for quantum mechanics. *Foundations of Physics*, 29(4):631–643, 1999.
14. Cristian S. Calude and Karl Svozil. Quantum randomness and value indefiniteness. *Advanced Science Letters*, 1(2):165–168, December 2008.
15. Max Born. Zur Quantenmechanik der Stoßvorgänge. *Zeitschrift für Physik*, 37:863–867, 1926.
16. John Archibald Wheeler and Wojciech Hubert Zurek. *Quantum Theory and Measurement*. Princeton University Press, Princeton, NJ, 1983.
17. Anton Zeilinger. The message of the quantum. *Nature*, 438:743, 2005.
18. Gregory J. Chaitin. *Exploring Randomness*. Springer, London, 2001.
19. Ernst Specker. *Selecta*. Birkhäuser Verlag, Basel, 1990.
20. Clifford Alan Hooker. *The Logico-Algebraic Approach to Quantum Mechanics. Volume I: Historical Evolution*. Reidel, Dordrecht, 1975.
21. Alastair A. Abbott and Cristian S. Calude. Von Neumann normalisation and symptoms of randomness: An application to sequences of quantum random bits, In C. S. Calude, J. Kari, I. Petre, G. Rozenberg, editors, *Proc. 10th International Conference Unconventional Computation*, pages 40–51. Lecture Notes Comput. Sci. 6714, Springer, Heidelberg, 2011.

PART 4
The Quantum, Computation & Information

Chapter 28

What is Computation?
(How) Does Nature Compute?*

David Deutsch

Centre for Quantum Computation
Clarendon Laboratory, University of Oxford[†]

So it seems we're asking ourselves today "What is Computation?" and either "Does Nature Compute?" or "*How* Does Nature Compute?" And there's an amazing fact that motivates both of these questions and indeed motivates every other foundational question about computation as well. It is this: if you take any physical variable whatsoever, for example "who is going to be the next president of the United States", to take a topical example or, another one is "the mean temperature of the Earth's atmosphere as a function of time" and ask how that variable depends on other variables, then the answer will always invariably be a computable function—or, if there's quantum indeterminacy involved then the probability distribution function will be a computable function. This is because the laws of physics refer only to computable functions—either directly or via computable differential equations.

Now, the reason this is amazing is that most mathematical functions are not computable—in fact, the set of computable functions is of measure zero in the class of all mathematical functions, let alone in the class of all mathematical relationships. So, there is something infinitely special about the laws of physics as we actually find them, something exceptionally tractable, prediction-friendly and computation-friendly. That's clearly not accidental. So, there's definitely something there to be explained. But people make different things of it, and *what* they make of it depends on quite deep aspects of their world view. Of course, religious people tend to see

---
*Transcript by Adrian German (who was one of the co-chairs of the conference and introduced Deutsch to the audience—the other conference chair being Hector Zenil).
[†]Talk delivered on November 1st, 2008 at the 2008 Midwest NKS Conference at Indiana University Bloomington.

divine providence in it and some evolutionists see the signature of evolution namely "apparent design" — but at the level of laws of nature, to make them computable. And cosmologists see anthropic selection effects and computer programmer type people, well, they see either a great computer in the sky in which we're all simulations, like in the movie "The Matrix" or that the Universe itself is a computer—and either way that what we perceive as physical phenomena are actually just virtual reality: *running programs*. Now, all those conceptions are wrong—because they all share a fatal flaw. They all have other flaws as well, but I want to concentrate on the shared one, which is relevant to our question, namely: "What is Computation?"

The laws of nature are, by definition, inviolable. For instance, you can't make a perpetual motion machine– On the way here today I saw a headline on the BBC that said "the future of physics is in jeopardy". But that is exactly what can't happen—the future of physics is unchanging, inviolable, invariable. It's only we who can change! So, there's something inviolable in the universal truths of the physical laws. And, also, the theorems of mathematics are inviolable. For instance, you can't change which of two integers is the larger. So in both cases 'inviolable' refers to a fact of the matter, meaning: it can't be argued away just by changing terms or definitions. For instance, you could define the term 'perpetual motion' so that a glass of tap water is a perpetual motion machine, because the molecules are literally in perpetual motion. Nevertheless that wouldn't enable you to use that motion to charge up batteries in a cycle. That's a fact of the matter. And similarly in arithmetic you can redefine the word billion which in British English used to mean 10^{12} and was then redefined down to 10^9 in line with American usage but that renaming doesn't make any actual number bigger or smaller—and that's another fact of the matter, but of a different kind. So, the laws of physics and the truths of mathematics are equally inviolable, they are universal truths and they are both about something objective. So they're alike in those respects but nevertheless there is a well recognized and important difference between scientific and mathematical truths: it is in what these fields are about. Mathematics is about absolutely necessary truths. Such truths are all abstract and essentially they are truths about what is or isn't logically implied by particular axioms, but science isn't about what's implied by anything. It's about what is really out there in the physical world. Laws of nature do therefore have to be consistent but unlike mathematical axioms they also have to correspond to reality, so that's the fundamental difference between science and mathematics, between theories and theorems.

Now, what about the laws of computation? By a law of computation I mean any inviolable generalization (any universal truth) about computation such as: that there exists no computer that could reliably detect whether a program will eventually halt, or not, or whether it would do any given thing. And in order to understand what computation is, the most basic question we have to address first is into which of those two categories do the laws, the inviolable laws of computation, fall? Are they absolutely necessary truths or are they determined by the laws of physics?

Now the answer is that they're determined by the laws of physics but there's been a lot of confusion in regard to that and to explain what's going on I have to look at this in a slightly broader context. The context here is that the theory of computation, which is a branch of physics, was pioneered by mathematicians. That is not unusual: several other important branches of theoretical physics were also started by mathematicians—for instance geometry, probability theory and various theories involving the infinite and the infinitesimal (i.e., calculus). And indeed in many of those cases the same confusion that I am going to describe here did arise. In terms of this distinction between reality and necessary truth the confusion keeps arising because mathematicians tend not to have a very firm grip on reality.

Here's what happens: take a mathematical idea, say—the idea of infinity. Mathematicians realized centuries ago that they really can work with infinity—they can elegantly define, say, what an infinite set is, as one that can be placed in a one-one correspondence with a proper subset of itself, and then they can prove theorems about such sets and about further abstract structures if they can consistently define them, in terms of such infinite sets, and sometimes they can then use those mathematical structures and theorems to formulate new scientific theories or to make existing theories more precise – for instance, as calculus was used by Newton and Leibniz to formulate scientific theories about things like instantaneous velocity and other rates of change. OK—so far so good, but notice that the concept of instantaneous velocity in physics and in common sense doesn't involve anything infinite and indeed nothing physical is either infinite or infinitesimal in, say, the smooth motion of a projectile; and on the other hand there's already an informal conception of what infinite does mean. It means bigger than anything merely big, or more of something than can be quantified even in principle, more of something that can be addressed in any sequence however long. In that conception the essence of finiteness is not about sets or mappings, it's about what computer theorists call effectiveness and what physicists call measurability or preparability, doability—and those two conceptions of infinity, the mathematical and the physical, draw the distinction between finite and infinite at completely different places.

And that is essentially how Zeno of Elea in his famous paradox managed to conclude that Achilles will never overtake the tortoise, if the tortoise has a head start—because by the time Achilles reaches the point where the tortoise is now, the tortoise will have moved on a little, and by the time he reaches that point, it will have moved on a little further—and so on *ad infinitum*. And thus, the catching-up process requires Achilles to perform an infinite number of catching-up steps which—as a finite being—he presumably cannot do.

"Presumably", but did you see what Zeno did just there? He just presumed that a particular mathematical notion that happens to be called "infinity" faithfully captures the distinction between finite and infinite, and was thus relevant to a particular situation in physics. And he was simply wrong. So, he succeeded only in confusing himself and there's nothing more to his paradox than that mistake. The truth is that what Achilles can or can't do cannot be deduced from mathematics, or *a priori*, in any way. It depends entirely on what the relevant laws of physics say. If they say that he'll overtake the tortoise in a given time, then he'll overtake it. If that happens to involve an infinite number of steps of the form "go to where the tortoise is now" then he'll do an infinite number of such steps. If it involves his passing through an uncountable infinity of points in space then that's what he does—but nothing physically infinite has happened. And, by the way, what did happen didn't happen in steps!

OK, well, the distinction between finite and infinite is also at the heart of the theory of computation. For instance, an effective algorithm is defined as one that halts after a finite number of steps, where a step has to be defined according to a finite list of rules. And a rule has to be finitely executable. And so on. Now these requirements were implemented in the classical theory of computation which was pioneered by Alan Turing but they date back to mathematical requirements set by the mathematician David Hilbert in 1900 with the intention of formalizing the concept of mathematical proof. In fact, classical computations are essentially the same things as proofs in Hilbert's and Turing's sense because every valid proof is a computation of the conclusion from the premises and every correctly executed computation is a proof that the output is the result of the given operations on the input. Hilbert had required finiteness conditions, he had required that proofs use only a fixed, finite set of rules of inference and executing a proof had to take only a finite number of elementary steps and the steps themselves had to be finite—so you recognize that from the theory of computation, that's exactly where finiteness, effectiveness came from in the theory of computation.

Hilbert contemptuously ridiculed the idea that his finiteness requirements were substantive ones, but do you see that he was thereby making exactly the same fundamental mistake as Zeno was? He was assuming that a particular mathematical distinction between finite and infinite in regard to steps and axioms and so on was self-evidently the one that corresponds to 'doable' and 'effective'—'thinkable', and 'provable' in physical objects—such as the brain of mathematicians. Fortunately, Hilbert's intuition about finite and infinite unlike the one that Zeno tried to implement did match physics (and as far as we know even quantum physics) so his conceptual mistake didn't do all that much harm at first and Turing's implementation of it was therefore enormously successful and fruitful. But the point is that if the laws of physics were in fact different from what we currently think they are then so might be the set of mathematical truths that we then would be able to prove, and so might be operations that we'd be able to use to prove them with. The laws of physics that we know happen to afford a privileged position to such operations as AND and OR and NOT and to the concatenation of functions and to individual bits of information. But if instead they were based on, say, Turing machines as elementary objects instead of points and so on, and if the laws of motion depended on functions like "does this halt?" instead of on differential equations then one could compute using operations which with our physics we call non-computable, while perhaps, the functions that seem natural and elementary to us would be non-computable in that physics. Similarly it's not just the distinction between computable and non-computable that depends on the laws of physics but also the very important distinction between simple and complex. We now know, because of quantum computation, that Turing's and Hilbert's conceptions of what is simple and complex is not reflected in real physics. Quantum computation drives a coach and horses through the intuitive notion of a 'simple' or 'elementary' operation and it makes some intuitively complex things simple.

It might be objected that quantum computation—therefore—isn't real computation, it's just physics, just engineering, and it might be argued that the logical possibilities that I have just been describing (that would enable exotic forms of physics which would then enable exotic forms of computation) likewise don't address the issue of what a proof really is, or what a computation really is. So, more precisely the objection would go something like this: "Under suitable laws of physics we would be able to compute non-Turing computable functions but that wouldn't be genuine computation; and similarly we might be able to establish truth or falsity of undecidable mathematical propositions, but then again that 'establishing' wouldn't be the same as genuinely proving—because then our knowledge of whether the proposition was true or false would forever depend on our knowledge of what the laws of physics really are. If we discovered one day that the real laws of physics are different we might have to change our mind about the proof, too, and its conclusion and so it wouldn't really be a proof (so the objection runs) because real proof is independent of physics."

Wrong! That objection is nonsense, because it would apply equally well to any proof. Our knowledge of whether a proposition is true or false always depends on our knowledge about how physical objects behave, be they computers or our own brains! If we changed our minds about what, physically, a computer, or a brain, has been doing—and in principle changing our minds could be due to deciding that the laws of physics are not what we thought they were—and we decided for instance that when formulae become sufficiently complex, computers or brains are subject to systematic false memories (this is a bit like Roger Penrose's idea about what happens to the wave function when its components change more than a certain amount)—then we'd be forced to change our opinion about whether something is proved or not, and possibly about whether we know it to be true or not. And here I must stress that whether a mathematical proposition is true or false is indeed *completely independent* of physics but proof is 100% physics, proofs are not abstract, there is no such thing as abstractly proving something just as there is no such thing as abstractly calculating or computing something. One can of course define a class of abstract entities and call them proofs, just like you can define a perpetual motion machine to be something else, one can define abstract entities and call them *computations* but those proofs can't do the job of verifying mathematical statements, they're not effective or doable. A mathematical theory of proofs therefore has no bearing on which proofs can or cannot be proved in reality or known in reality and similarly, a theory of abstract computation as well has no bearing on what can or cannot be computed in reality.

So, what is provable or unprovable is determined by the laws of physics in exactly the same sense as "what the angles of a triangle add up to" is determined by the laws of physics. Immanuel Kant thought that Euclidean geometry was self-evidently true—and that's another example of the same misconception that I've been talking about. The truth is that you can define abstract entities and call them triangles and have them obey Euclidian geometry but if you do that you can't then infer anything from that theory about, say, what angle you'll turn through if you walk around the closed path consisting of three straight lines—that thing might not be a triangle as you have defined it. Likewise in probability theory you can define abstract quantities that obey a certain calculus about mutually exclusive alternative events and you can call those quantities probabilities but in that case your theory tells you nothing about how you should bet on real events. And, of course, in quantum theory real probabilities do not even obey those axioms of the probability calculus that refer to alternative intermediate events.

So, quite generally: things that happen in reality are governed by the laws of physics, period. And therefore I can now give the definitive answer to the question that is the title of my talk and the title of the conference: "What is (a) computation?" A *computation* is a physical process in which physical objects like computers, or slide rules or brains are used to discover, or to demonstrate or to harness properties of abstract objects—like numbers and equations. *How* can they do that? The answer is that we use them only in situations where to the best of our understanding the laws of physics will cause physical variables like electric currents in computers (representing bits) faithfully to mimic the abstract entities that we're interested in. The reliability of the proof therefore depends on the accuracy with which those physical symbols do indeed mimic the abstract entities of interest. If we changed our mind about what the laws of physics are, we might indeed have to change our mind about whether a proposition we thought we'd proved was really true.

Hence the reliability of our knowledge of mathematics, nothing to do with whether propositions are true or false, just our knowledge of mathematics, remains forever subsidiary to our knowledge of physical reality. Every mathematical proof depends absolutely, for its validity, on our being right about the rules that govern the behavior of some physical object like computers, or ink and paper, or our brains. So, contrary to what Hilbert thought, contrary to what mathematicians since antiquity believed and continue to believe to this day, proof theory can never be made into a branch of mathematics nor is it helpful to think of it as a metamathematics, as it's sometimes known. Proof theory is a science, and specifically it is *computer science*. So that's what computers are, that's what computation is, that's what proofs are—physical phenomena. That enables me now to tie into the subsidiary question: "Does Nature Compute?" or "How Does Nature Compute?" So, well, in one sense: yes, of course, computations are physical processes and every physical process can be regarded as a computation if we just give abstract labels to all the input states and all the output states and then just let the system evolve under the laws of physics. But precisely because you could always do that—no matter what the world was like—calling everything a computation doesn't in itself gain us any understanding of the world. But the world also has that amazing property that I referred to at the beginning: more broadly, its computational universality. All those different computations embodied in physical processes are expressible in terms of a single finite set of elementary physical operations. They share a single, uniform, physical distinction between finite and infinite operations, and they can all be programmed to be performed on a single physical object: a universal computer, a universal quantum computer to be exact. And that's an object that can perform every computation that every other physically possible object can perform. And to the best of our knowledge the laws of physics do have that property of computational universality and it?s because of *that* that physical objects like ourselves can understand other physical objects (i.e., to do science). And it's also because of that same universality that mathematicians (like Hilbert) can build up an intuition of proof and then mistakenly think that it's independent of physics. It's not independent of physics, it's just universal in the physics that governs our world.

But in the class of all possible universes that cosmologists nowadays postulate—not the many universes from quantum theory but the cosmological ones with all different laws of physics—most of them have other, very different laws of physics than ours, and in most of those there's no computational universality. In a tiny subclass (but still infinite though) there is universality but with all sorts of different computable functions—and also, all sorts of different measures of what is simple and complex in those worlds, and all sorts of different criteria for what is finite and infinite in those worlds.

In some of those universes there are analogues of David Hilbert or Alan Turing each of them reaching towards a different conception of what counts as mathematically proving something, what counts as computing something. And there are also lots of Stephen Wolframs, each of them seeking to base a new kind of physics on what they respectively experience as inherently simple, inherently simple self-evident computational foundations beneath his Universe's contingent laws of physics. But, in reality there is no such distinction as simple versus complex, or finite versus infinite except as the laws of physics dictate. There is no mathematically preferred conception of computation, or computability, or finiteness, or simplicity.

Absolutely nothing other than physics (and the cultural preferences that it conditions in us) singles out Turing computable functions or cellular automata or even quantum computation, or quantum cellular automata, as being fundamental, or special, or elementary in any way. In fact there is nothing that singles out the computational functions at all, nor bits, logical variables, as being the fundamental forms of data on which computations operate.

There is nothing deeper known about the physical world than the laws of physics. And, I think, there is nothing deeper known within physics than the quantum theory of computation. And for that reason I entirely agree that it's likely to be fruitful to recast our conception of the world and of the laws of physics and physical processes in computational terms, and to connect fully with reality it would have to be in quantum computational terms. But computers have to be conceived as being inside the universe, subject to its laws, not somehow prior to the universe, generating its laws.

The latter is the very misconception that led Zeno astray, and Hilbert and Kant, and many other thinkers throughout history, who haven't realized that while truth can be absolutely necessary and transcendent, all knowledge (even of such truths) is generated, computed, by physical processes, and the scope and limitations of such knowledge are conditioned by Nature's contingent laws. Thank you very much.

Questions & Answers

Question: Would you consider an experiment, a physical experiment, to be a computation?
Answer: Are you asking about a specifical experiment, or a specifical physical process?
Q: No, in general.
A: Yes, you can always regard any physical experiment as a computation. If the outcome is unknown then it's a computation where we are probing what the computer is doing, and if the laws are known, then what we are doing is transforming the input into the output. But in order to make an experiment into a computation you have to label everything, you have to give labels to all the inputs, states and to all the outputs...
Q: Great. Let me then have a follow-up question on that. Suppose you perform an experiment[a] and you figure out that Nature tells you that the

---
[a]This question may have originally aimed to ask whether quantum mechanics is in fact a good model of reality, maybe in the sense of one of Sir Roger Penrose's interpretations

answer to your question, to your computation, takes a polynomial number of steps—so the answer is in fact polynomially complex[b].
A: Ah, wait—an experiment can't tell you that!
Q: Well, you do the experiment and you ask the question, for example, how many resources you need in order to determine (or measure) a very specific physical quantity with a certain error...
A: Yes—but you can only do that for a finite number of times, so you can't tell if for large inputs it's going to be polynomial or not. In order to tell whether the resources are indeed polynomial you have to know how it behaves for arbitrarily large values of the inputs... and you can only do it for a finite number of experiments.
Q: I understand that. Now my question is this: suppose that with the well-known laws of nature (with quantum physics) I cannot find an algorithm that performs the same thing with the same complexity.
A: Yes.
Q: What would you say then, what would you conclude about the laws of nature as we know them? Would you conclude that quantum mechanics is not the right representation of how nature works, or ... what am I missing?
A: Well, I am not entirely sure that I understand what your proposed experiment is. But if you indeed find that ... I mean, the idea is that you guess, you conjecture what the laws of physics are, which tell us what to conjecture about what our computer will do. If we find that a real physical process can compute something—appears to be able to compute something—in polynomial time that no algorithm (as given by the laws of nature as we know them) will compute in polynomial time, we can then indeed conclude that the laws that you thought were operating the device are in fact not the true laws operating the device. So if that's the question you're asking then the answer is definitely: yes.
Q: I completely agree that maybe mathematical concepts have a root in reality and I also agree that some mathematicians forget it sometimes. Yet it seems that there's always an abstraction step that is needed to go from

---

(see, e.g., http://www.cs.indiana.edu/~dgerman/penrose.pdf) or was perhaps aimed to be considered in the larger context of "simulating of reality" (with computers, in real time or any plausible kind of time).

[b]This is an additional indication that the question has something to do with the 1967 Feynman quote from "The Character of Physical Law" that was the motto of the conference: "It always bothers me that, according to the laws as we understand them today, it takes a computing machine an infinite number of logical operations to figure out what goes on in no matter how tiny a region of space, and no matter how tiny a region of time. How can all that be going on in that tiny space? Why should it take an infinite amount of logic to figure out what one tiny piece of space/time is going to do?"

the root of the concept in reality to the concept itself. If we take a very simple concept, for example the concept of natural number—of course it's rooted in notion of calculability and the notion of calculability is rooted in the notion of physical object. But if we restrict to just what we see around us, the reality we (as finite human beings) can perceive around us we might take the point of view which says for example that there's a maximum natural number and no successor.

A: Yes, and that would be silly.

Q: In some sense these abstraction steps free us from this limitation that for some reason we have and that we don't even want to consider. So in this respect, although it is rooted in reality the concept of natural number is not in reality itself—it's abstracted from reality.

A: Yes. Well, it is not in physical reality at any rate. I mean, one could argue that mathematical entities are real in a different sense, in that they have autonomous properties that we didn't necessarily put into them. But yes—as I stressed in my talk—mathematical truth really is independent of physics, it is only knowledge of mathematics that is constrained by what the laws of physics say. So there's this enormous realm of mathematical truths which is independent of what the laws of physics are and there's a window on that—a tiny window—which has the truths that we can know. And that window is determined by the laws of physics.

Q: What is true of natural number is true of proofs too. So, for instance, the mathematical notion of proof abstracts from what can be found in the reality.

A: Yes. So it goes like this: you can have physical intuitions about things, which are really intuitions about what the laws of nature are and then you can abstract to make nicer mathematical quantities that don't have arbitrary restrictions on them. And subsequently you can guess that those mathematical quantities are also instantiated in real physical objects. So the class of mathematical truths that we can know *about* is much larger than the class that we can *know*. But there might be other things [that] we can't even know about! Presumably there are, because the infinities involved would be simply too large to allow knowing about them; well, I suppose we know about them *in that sense*.

So, yes: I think I agree. Possibly you're making a subtle point that I haven't picked up but, yes: proofs arise from a physical intuition. We then form a mathematical conception, a mathematical abstraction, which we call proofs and then we conjecture that that mathematical conception is actually genuinely true in physics. But we could be wrong about that last step! The mathematical object(s) we have set up and called the "proofs", or "laws of inference" and such may simply not correspond to reality and then—or we may not know the correspondence even approximately—and then it's the set of things that we really *can* prove with physics that are the provable things not the ones that that mathematical conceptions might end up proving, because that is not effective in that case.

Q: I have one question about computability. You said in the beginning that most, in fact *all* functions in physics are computable. And I wonder if whether it's as you said (that it's a miracle) or whether there's some explanation for that. And maybe just a simple possible explanation is that all of physics is in fact expressed in terms of ordinary differential equations and there are fundamental and general existence theorems for these equations and so on, stating that in such and such case there will always be a solution to them and only one. And when one uses computable tools to model the world that's also when what one models appears as having been already computable all along. Because if you look at the history of physics—physics usually shies away from theories that are not deterministic or are heuristic and maybe this is the reason why when we express a lot of physics in such a way that it would be only deterministic and maybe this is the reason for which every function in physics is computable.

A: Well, it's not a matter of arbitrary choice on the part of physicists. Physicists are trying to find explanatory laws that fit experiment and it turns out that the laws as we think they are have this property of computability—for instance they are differential equations—and not just any old differential equations, but well-posed differential equations which have that property. Now, I guess one can—this is a bit like the anthropic principle reasoning—I guess one can be amazed by this or not. I think I *am* amazed by it, because in the bigger picture of mathematics these things are very special. And we don't know of any mechanism or reason why the actual laws of physics should be expressed in terms of those special kinds of differential equations. And actually even among that small set of differential equations are the particularly simple ones that appear to be actually implemented. So we don't know why, and some cosmologists say it's an anthropic selection effect: observers that can ask such questions only exist in universes where evolution happens and evolution is the kind of computation that depends on the existence of simple computability. And I don't think that that's the full argument, myself. That would take us off on a tangent here but although that may be true it can't possibly be a sufficient reason for why the laws of physics are as they are because the set of all possible laws of physics doesn't even have a measure on it and so this reasoning about "most of them do one thing and the rest do another" hasn't really got any basis.

Q: I would like to—I like very much your position and your talk but I would like to point out a possible similarity between mathematics and physics. You know, if you look at mathematics, say, 200 years ago—most functions were continuous. And people believed, that indeed *all* functions were continuous. And when they started understanding better and using better tools, they discovered that most functions are not continuous. So maybe this is just a historical moment/accident when physics looked for laws that can be expressed by computable functions and not something that is motivated by some good reason—but simply a historic effect.

A: Well, it can't be a purely historic effect because the laws of physics, as we currently believe they are, are extremely successful whether, you know, there may be corrections to them and there may be regimes in which they are completely wrong but they are—it is already a miracle that the physical world is as computable as it seems to be, even if in some other respects it may turn out to be non-computable or discontinuous functions or not be governed by differential equations. It's already a miracle that it is as computable as it is. So there is definitely something out there! It's not just that we decided to look in one place—we decided to look in one place and *we found* computability and that's already a miracle.

Q: Well, it's a miracle also that for instance in engineering you now if you look at you know buildings, bridges planes they all are built with continuous functions! So you can say you know this is all in the imagination of mathematicians and discontinuities and you know all sort of sophisticated deviations [in fact] appear. But if you look at the pragmatic view you know one could say that from the point of view of engineering only continuous functions come and exist.

A: Yes, but then engineering has been successful at least up to a point. Prior to the scientific age people tried all sorts of different kinds of explanations—anthropocentric type explanations, in which physical processes were governed by the intentions of supernatural beings, and that kind of thing... and they were looking for explanations in those terms and those explanations never were successful. So the fact that we can now build bridges using assumptions of continuous functions and so on, even if it later turns out that it's just an approximation that's already something that we don't know how to explain! Why are computable functions available to build bridges with? It's not just that we looked for them, we also *found* them.

Q: I would like to know a bit more about explanations related to physics being miraculously computable and would this make our Universe more [special] among other alternatives just because those are not computable and of course the condition of evolution being more likely...

A: As I said I don't think *that* can be the full explanation of this mystery. It could be that there are all sorts of Universes with different laws of physics and it could be that in most of them in some sense there are no observers. However there has to be a structure on this class of Universes—such that the concept "most of them" *makes sense*. The set of all possible laws (which is not a set but a class) is too enormous for the concept of "most of them do one thing while a few of them do another" to make sense. You can't attack this problem purely with anthropic reasoning. The answer must be some kind of explanation of why the laws are as they are or at least why the measure of all possible laws is as it is—that kind of thing. So if you're asking is anthropic reasoning enough to explain all of this: I think not.

Q: Well, the interesting situation with computability is that obviously computable functions are of measure C and I imagine that in all alternative universes [that there is a] favorable choice [with a different complexity measure] is very unlikely.

A: Yes, quite so. So there's something to explain—it's not just selection.

Q: In your position that proofs are things that are computable in physical reality and mathematical proofs are [idealizations] I was interested in [knowing] how this relates with the difference between classical proofs and constructive proofs [as in] classical logic and constructor logic.

A: Well, I know next to nothing about that but I think I can tell you what the situation basically is. Like I said, proof theory is a science. And one of the changes that one has to make already with quantum computation, you know, never mind those exotic laws of physics that might exist, is that we can no longer define a proof as a certain kind of *object*. That is, traditionally we thought of a proof as something that we can present on a piece of paper that satisfies certain laws: that in the beginning there must be axioms, and that it proceeds in lines and then each line must follow from the axioms according to a certain set of laws of inference and then the last line is the conclusion—and we say that the conclusion is proved from the axioms. Now in quantum theory that is no longer sufficient, there are other kinds of proofs that cannot be expressed in that way at least not tractably—not in a polynomial number of steps. There are proofs where there is not enough paper in the universe to express them in that way—but a quantum computer could nevertheless prove them and for that reason you have to change your [entire perspective]. In classical physics and in classical theory of computation the idea of a proof as an object obeying several laws and the idea of a proof as a process where you execute one operation after another are equivalent—that is, there is a one to one correspondence between those two conceptions but in quantum computation there isn't. So those distinctions between different kinds of proofs are induced by the laws of physics and if the laws of physics are different the classification of proofs will be different as well.

Chapter 29

The Universe as Quantum Computer

Seth Lloyd

Department of Mechanical Engineering
WM Keck Center for Extreme Quantum Information Processing,
Massachusetts Institute of Technology (MIT), USA

This article reviews the history of digital computation, and investigates just how far the concept of computation can be taken. In particular, I address the question of whether the universe itself is in fact a giant computer, and if so, just what kind of computer it is. I will show that the universe can be regarded as a giant quantum computer. The quantum computational model of the universe explains a variety of observed phenomena not encompassed by the ordinary laws of physics. In particular, the model shows that the the quantum computational universe automatically gives rise to a mix of randomness and order, and to both simple and complex systems.

1. Introduction

It is no secret that over the last fifty years the world has undergone a paradigm shift in both science and technology. Until the mid-twentieth century, the dominant paradigm in both science and technology was that of energy: over the previous centuries, the laws of physics had been developed to understand the nature of energy and how it could be transformed. In concert with progress in physics, the technology of the industrial revolution put the new understanding of energy to use for manufacturing and transportation. In the mid-twentieth century, a new revolution began. This revolution was based not on energy, but on information. The new science of information processing, of which Turing was one of the primary inventors, spawned a technology of information processing and computation. This technology gave rise to novel forms and applications of computation and communication. The rapid spread of information processing technologies, in turn, has ignited an explosion of scientific and social inquiry. The result is a paradigm shift of how we think about the world at its most fundamental level. Energy is still an important ingredient of our understanding of the universe, of course, but information has attained a conceptual and practical status equal to – and frequently surpassing – that of energy. Our new understanding of the universe is not in terms of the driving power of force and mass. Rather, the world we see around us arises from a dance between equal partners, information and energy, where first one takes the lead and then the other. The bit meets the erg, and the result is the universe.

At bottom, the information that makes up the universe is not just ordinary classical information (bits). Rather, it is quantum information (qubits). Consequently, the computational model that applies the universe at its smallest and most fundamental level is not conventional digital computation, but quantum computation[1] . The strange and weird aspects of quantum mechanics infect the universe at its very beginning, and – as will be seen – provide the mechanism by which the universe generates its peculiar mix of randomness, order, and complexity.

2. Digital Computation Before Turing

Before describing how the universe can be modeled as a quantum computer, and how that quantum computational model of the universe explains previously unexplained features, we review computation and computational models of the universe in general.

Alan Turing played a key role in the paradigm shift from energy to information: his development of a formal theory of digital computation made him one of the most influential mathematicians of the twentieth century. It is fitting, therefore, to praise him. Curiously, however, Turing's seminal role in a global scientific and technological revolution also leads to the temptation to over-emphasize his contributions. We human beings have a sloppy, if not outright bad habit of assigning advances to a few 'great men.' I call this habit the Pythagoras syndrome, after the tendency in the western world to assign all pre-fifth century B.C.E. mathematics to Pythagoras without regard to actual origins. In evaluating Turing's contributions, we should be careful not to fall victim to the Pythagoras syndrome, if only to give full credit to his actual contributions, which were specific and great.

Computing machines are not a modern invention:[2] the abacus was invented in Babylon more than four thousand years ago . Analog, geared, information processing mechanisms were developed in China and Greece thousands of years ago, and attained considerable sophistication in the hands of medieval Islamic philosophers. John Napier's seventeenth century mechanical implementation of logarithms ('Napier's bones') was the precursor of the slide rule. The primary inventor of the modern digital computer, however, was Charles Babbage. In 1812, Babbage had the insight that the calculations carried out by mathematicians could be broken down into sequences of less complicated steps, each of which could be carried out by a machine[3] – note the strong similarity to Turing's insight into the origins of the Turing machine more than a century later. The British government fully appreciated the potential impact of possessing a mechanical digital computer, and funded Babbage's work at a high level. During the 1820s he designed and attempted to build a series of prototype digital computers that he called 'difference engines.' Nineteenth century manufacturing tolerances turned out to be insufficiently precise to construct the the all-mechanical difference engines, however. The first large-scale computing project consumed over seventeen thousand pounds sterling of the British taxpayers' money, a princely expenditure for pure research at the time. Like many computing projects since, it failed.

Had they been constructed, difference engines would have been able to compute general polynomial functions, but they would not have been capable of what Turing termed universal digital computation. After the termination of funding for the difference engine project, Babbage turned his efforts to the design of an 'analytic engine.' Programmed by punched cards like a Jacquard loom, the analytic engine would have been a universal digital computer. The mathematician Ada Lovelace devised a program for the analytic engine to compute Bernoulli numbers, thereby earning the title of the world's first computer programmer.

The insights of Babbage and Lovelace occurred more than a century before the start of the information processing revolution. Turing was born in the centenary of the year in which Babbage had his original insight. The collection in which this paper appears could equally be dedicated to the two-hundredth anniversary of Babbage's vision. But scientific history is written to celebrate winners (see Pythagoras, above). Turing 'won' the title of the inventor of the digital computer because his insights played a direct role in the vision of the creators of the first actual physical computers in the mid-twentieth century. The science fiction genre known as 'steampunk' speculates how the world might have evolved if nineteenth century technology had been up to the task of constructing the difference and analytical engines. (Perhaps the best-known example of the steampunk genre is William Gibson and Bruce Sterling's novel 'The Difference Engine'.[4])

The mathematical development of digital logic did not occur until after Babbage's mechanical development. It was not until the 1830s and 1840s that the British logician Augustus de Morgan and the mathematician George Boole developed the bit-based logic on which current digital computation is based. Indeed, had Babbage been aware of this development at the time, the physical construction of the difference and analytic engines might have been easier to accomplish, as Boolean, bit-based operations are more straightforward to implement mechanically than base-ten operations. As will be seen, the relative technological simplicity of bit-based operations would play a key role in the development of electronic computers.

By the time that Turing began working on the theory of computation, Babbage's efforts to construct actual digital computers were a distant memory. Turing's work had its direct intellectual antecedents in the contentious arguments on the logical and mathematical basis of set theory that were stirred up at the beginning of the twentieth century. At the end of the nineteenth century, the German mathematician David Hilbert had proposed an ambitious programme to axiomatize the whole of mathematics. In 1900, he famously formulated this programme at the International Congress of Mathematicians in Paris as a challenge to all mathematicians – a collection of twenty three problems whose solution he felt would lead to a complete, axiomatic theory not just of mathematics, but of physical reality. Despite or because of its grand ambition to establish the logical foundations of mathematical thought, cracks began to appear in Hilbert's programme almost immediately. The difficulties arose at the most fundamental level, that of logic itself. Logicians and set theorists such as Gottfried Frege and Bertrand Russell worked for decades to make set theory consistent, but the net result of their work was call into question the logical foundations of set theory itself. In 1931, just when the efforts of mathematicians such as John von Neumann had appeared to patch up those cracks, Kurt Gödel published his beautiful but disturbing incompleteness theorems, showing that any system of logic that is powerful enough to describe the natural numbers is fundamentally incomplete in the sense that there exist well-formulated proposition within the system that cannot be resolved using the system's axioms.[5] By effectively destroying Hilbert's programme, Gödel's startling result jolted the mathematical community into novel ways of approaching the very notion of what logic was.

3. Digital Computation Concurrent with Turing

Turing's great contribution to logic can be thought of as the rejection of logic as a Platonic ideal, and the redefinition logic as a *process*. Turing's famous paper of 1936, 'On Computable Numbers with an application to the *Entscheidungsproblem*,' showed that the process of performing Boolean logic could be implemented by an abstract machine,[6] subsequently called a Turing machine. Turing's machine was an abstraction of a mathematician performing a calculation by thinking and writing on pieces of paper. The machine has a 'head' to do the thinking, and a 'tape' divided up in squares to form the machine's memory. The head has a finite number of possible states, as does each square. At each step, in analogy to the mathematician looking at the piece of paper in front of her, the head reads the state of the square on which it sits. Then, in analogy to thinking and writing on the paper, the head changes its state and the state of the square. The updating occurs as a function of the head's current state and the state of the square. Finally, in analogy to the mathematician either taking up a new sheet of paper or referring back to one on which she has previously written, the head moves one square to the left of right, and the process begins again.

Turing was able to show that such machines were very powerful computing devices in principle. In particular, he proved the existence of 'universal' Turing machines, which were capable of simulating the action of any other Turing machine, no matter how complex the actions of its head and squares. Unbenownst to Turing, the American mathematician Alonzo Church had previously arrived at a purely formal logical description of the idea of computability, the so-called[7] Lambda calculus. At the same time as Turing, Emil Post devised a mechanistic treatment of logical problems. The three methods were all formally equivalent, but it was Turing's that proved the most accessible.

Perhaps the most fascinating aspect of Turing's mechanistic formulation of logic was how it dealt with the self-contradictory and incomplete aspects of logic raised by Gödel's incompleteness theorems. Gödel's theorems arise from the ability of logical systems to have self-referential statements – they are a formalization of the ancient 'Cretan liar paradox,' in which a statement declares itself to be false. If the statement is true, then it is false; if it is false, then it is true. Regarding proof as a logical process, Gödel restated the paradox as a statement that declares that it can't be proved to be true. There are two possibilities. If the statement is false, then it can be proved to be true – but if a false statement can be proved to be true, then the entire system of logic is inconsistent. If the statement is true, then it can't be proved to be true, and the logical system is incomplete.

In Turing's formulation, logical statements about proofs are translated into actions of machines. The self-referential statements of Gödel's incompleteness theorems then translate into statements about a universal Turing machine that is programmed to answer questions about its own behavior. In particular, Turing showed that no Turing machine could answer the question of when a Turing machine 'halts' – i.e., gives the answer to some question. If such a machine existed, then it would be straightforward to construct a related machine that halts only when it fails to halt. In other words, the simplest possible question one can ask of a digital computer – whether it gives any output at all – cannot be computed!

The existence of universal Turing machines, together with their intrinsic limitation due to self-contradictory behavior as in the halting problem, has profound consequences for the behavior of existing computers. In particular, current electronic computers are effectively universal Turing machines. Their universal nature expresses itself in the fact that it is possible to write software that can be compiled to run on any digital computer, no matter whether it is made by HP, Lenovo, or Apple. The power of universal Turing machines manifests itself in the remarkable power and flexibility of digital computation. This power is expressed in the so-called Church-Turing hypothesis, which states that any effectively calculable function can be computed using a universal Turing machine. The intrinsically self-contradictory nature of Turing machines and the halting problem manifest themselves in the intrinsically annoying and frustrating behavior of digital computers – the halting problem implies that there is no systematic way of debugging a digital computer. No matter what one does, there will always be situations where the computer exhibits surprising and unexpected behavior (e.g., the 'blue screen of death').

Concurrent with the logical, abstract development of the notion of computation, including Turing's abstract machine, engineers and scientists were pursuing the construction of actual digital computers. In Germany in 1936, Konrad Zuse designed the Z-1, a mechanical calculator whose program was written on perforated 35mm film. In 1937, Zuse expanded the design to allow universal digital computation *a la* Turing. When completed in 1938, the Z-1 functioned poorly due to mechanical imprecision, the same issue that plagued Babbage's difference engine more than a century earlier. By 1941, Zuse had constructed the Z-3, an electronic computer capable of universal digital computation. Because of its essentially applied nature, and because it was kept secret during the second world war, Zuse's work received less credit for its seminal nature than was its due (see the remark above on winner's history).

Meanwhile, in 1937, Claude Shannon's MIT master's thesis, 'A Symbolic Analysis of Relay and Switching Circuits,' showed how any desired Boolean function – including those on which universal digital computation could be based – could be implemented using electronic switching circuits.[8] This work had a profound influence on the construction of electronic computers in the United States and Great Britain over the next decades.

4. Digital Computation Post-Turing

Turing's ideas on computation had immediate impact on the construction of actual digital computers. While doing his Ph.D. at Princeton in 1937, Turing himself constructed simple electronic models of Turing machines. The real impetus for the development of actual digital computers came with the onset of the second world war. Calculations for gunnery and bombing could be speeded up electronically. Most relevant to Turing's work, however, was the use of electronic calculators for the purpose of cryptanalysis During the war, Turing became the premier code-breaker for the British cryptography effort. The first large-scale electronic computer, the Colossus, was constructed to aid this effort. In the United States, IBM constructed the Mark I at Harvard, the second programmable computer after Zuse's Z-3, and used it to perform ballistic calculations. Zuse himself had not remained idle: he created the world's first computer start-up, designed the follow-up to the Z-3, the Z-4, and wrote the first programming language. The end of the war saw the construction of the Electronic Numerical Integrator and Computer, or ENIAC.

To build a computer requires and architecture. Two of the most influential proposals for computer architectures at the end of the war were the Electronic Discrete Variable Automatic Computer, or EDVAC, authored by von Neumann, and Turing's Automatic Computing Engine, or ACE. Both of these proposals implemented what is called a 'von Neumann' computer architecture, in which program and data are stored in the same memory bank. Stored program architectures were anticipated by Babbage, implicit in Turing's original paper, and had been developed previously by J. Presper Eckert and John Mauchly in their design for the ENIAC. The Pythagoras syndrome, however, assigns their development to von Neumann, who himself would have been unlikely to claim authorship.

This ends our historical summary of conventional digital computation. The last half century has seen vast expansion of devices, techniques, and architectures notably the development of the transistor and integrated circuits. But the primary elements of computation – programmable systems to perform digital logic – were all in place by 1950.

5. The Computing Universe

The physical universe bears little resemblance to the collection of wires, transistors, and electrical circuitry that make up a conventional digital computer. How then, can one claim that the universe is a computer? The answer lies in the definition of computation, of which Turing was the primary developer. According to Turing, a universal digital computer is a system that can be programmed to perform any desired sequence of logical operations. Turing's invention of the universal Turing machine makes this notion precise. The question of whether the universe is itself a universal digital computer can be broken down into two parts: (I) Does the universe compute? and (II) Does the universe do nothing more than compute? More precisely, (I) Is the universe capable of performing universal digital computation in the sense of Turing? That is, can the universe or some part of it be programmed to simulate a universal Turing machine? (II) Can a universal Turing machine efficiently simulate the dynamics of the universe itself?

At first the answers to these questions might appear, straightforwardly, to be Yes. When we construct electronic digital computers, we are effectively programming some piece of the universe to behave like a universal digital computer, capable of simulating a universal Turing machine. Similarly, the Church-Turing hypothesis implies, that *any* effectively calculable physical dynamics – including the known laws of physics, and any laws that may be discovered in the – can be computed using a digital computer.

But the straightforward answers are not correct. First, to simulate a universal Turing machine requires a potentially infinite supply of memory space. In Turing's original formulation, when a Turing machine reaches the end of its tape, new blank squares can always be added: the tape is 'indefinitely extendable.' Whether the universe that we inhabit provides us with indefinitely extendable memory is an open question of quantum cosmology, and will be discussed further below. So a more accurate answer to the first question is 'Maybe.' The question of whether or not infinite memory space is available is not so serious, as one can formulate notions of universal computation with limited memory. After all, we treat our existing electronic computers as universal machines even though they have finite memory (until, of course, we run out of disc space!). The fact that we possess computers is strong empirical evidence that laws of physics support universal digital computation.

The straightforward answer to question (II) is more doubtful. Although the outcomes of any calculable laws of physics can almost certainly be simulated on a universal Turing machine, it is an open question whether this simulation can be performed *efficiently* in the sense that a relatively small amount of computational resources are devoted to simulating what happens in a small volume of space and time. The current theory of computational complexity suggests that the answer to the second question is 'Probably not.'

An even more ambitious programme for the computational theory of the universe is the question of architecture. The observed universe possesses the feature that the laws of physics are local – they involve only interactions between neighboring regions of space and time. Moreover, these laws are homogeneous and isotropic, in that they appear to take the same form in all observed regions of space and time. The computational version of a homogeneous system with local laws is a cellular automaton, a digital system consisting of cells in regular array. Each cell possesses a finite number of possible states, and is updated as a function of its own state and those of its neighbors. Cellular automata were proposed by von Neumann and by the mathematician Stanislaw Ulam in the 1940s, and used by them to investigate mechanisms of self-reproduction.[7] Von Neumann and Ulam showed that cellular automata were capable of universal computation in the sense of Turing. In the 1960s, Zuse and computer scientist Edward Fredkin proposed that cellular automata could be used as the basis for the laws of physics – i.e., the universe is nothing more or less than a giant cellular automaton.[7,10] More recently, this idea was promulgated by Stephen Wolfram.

The idea that the universe is a giant cellular automaton is the strong version of the statement that the universe is a computer. That is, not only does the universe compute, and only compute, but also if one looks at the 'guts' of the universe – the structure of matter at its smallest scale – then those guts consist of nothing more than bits undergoing local, digital operations. The strong version of the statement that the universe is a computer can be phrased as the question, (III) 'Is the universe a cellular automaton?' As will now be seen, the answer to this question is No. In particular, basic facts about quantum mechanics prevent the local dynamics of the universe from being reproduced by a finite, local, classical, digital dynamics.

6. Classical Digital Devices Cannot Reproduce Quantum Mechanics Efficiently

Quantum mechanics is the physical theory that describes how systems behave at their most fundamental scales. It was studying von Neumann's book[7] *The mathematical foundations of quantum mechanics* that inspired Turing to work on mathematics.[12] (In particular, Turing was interested in reconciling questions of determinism and free will with the apparently indeterministic nature of quantum mechanics.) Quantum mechanics is well-known for exhibiting strange, counter-intuitive features. Chief amongst these features is the phenomenon known as entanglement, which Einstein termed 'spooky action at a distance' (*spukhafte Fernwirkung*). In fact, entanglement does not engender non-locality in the sense of non-local interactions or superluminal communication. However, a variety of theorems from von Neumann to Bell and beyond show that the types of correlations implicit in entanglement cannot be described by classical local models involving hidden variables.[13] In particular, such quantum correlations cannot be reproduced by local classical digital models such as cellular automata. Non-local classical hidden variable models can reproduce the correlations of quantum mechanics, but only at the by introducing either superluminal communication, or a very large amount of classical information to reproduce the behavior of a single quantum bit. Accordingly, the answer to question (III), is the universe a cellular automaton, is 'No.'

The inability of classical digital systems to cope with entanglement also seems to prevent ordinary computers from simulating quantum systems efficiently. Merely to represent the state of a quantum system with N subsystems, e.g., N nuclear spins, requires $O(2^N)$ bits on a classical computer. To represent how that state evolves requires the exponentiation of a 2^N by 2^N matrix. Although it is conceivable that exponential compression techniques could be found that would allow a classical computer to simulate a generic quantum system efficiently, none are known. So the currently accepted answer to question (II), can a Turing machine simulate a quantum system efficiently, is 'Probably not.'

7. Quantum Computing

The difficulty that classical computers have reproducing quantum effects makes it difficult to sustain the idea that the universe might at bottom be a classical computer. Quantum computers, by definition, are good at reproducing quantum effects, however.[14] Let's investigate the question of whether the universe might be, at bottom, a quantum computer.[1]

A quantum computer is a computer that uses quantum effects such as superposition and entanglement to perform computations in ways that classical computers cannot. Quantum computers were proposed by Paul Benioff in 1980.[15] The notion of a quantum Turing machine that used quantum superposition to perform computations in a novel way was proposed by David Deutsch in 1985.[16] For a decade or so, quantum computation remained something of a curiosity. No one had a particularly good application for them, and no one had the least idea how to build them. The situation changed in 1994, when Peter Shor showed that a relatively modestly sized quantum computer, containing a few thousand logical quantum bits or 'qubits,' and capable of performing around a million coherent operations, could be used to factor large numbers and so break public key cryptosystems such as RSA.[17] The previous year, Lloyd had showed how quantum computers could be constructed by applying electromagnetic pulses to arrays of coupled quantum systems.[18] The resulting parallel quantum computer is in effect a quantum cellular automaton. In 1995, Ignacio Cirac and Peter Zoller showed how ion traps could be used to implement quantum computation.[19]

Since then, a wide variety of designs for quantum computers have been proposed. Further quantum algorithms have been developed, and prototype quantum computers have been constructed and used to demonstrate simple quantum algorithms. This allows us to begin addressing the question of whether the universe is a quantum computer. If we 'quantize' our three questions, the first one, (Q1) 'Does the universe allow quantum computation?' has the provisional answer, 'Yes.' As before, the question of whether the universe affords a potentially unlimited supply of quantum bits remains open. Moreover, it is not clear that human beings currently possess the technical ability to build large scale quantum computers capable of code breaking. However, from the perspective of determining whether the universe supports quantum computation, it is enough that the laws of physics allow it.

Now quantize the second question. (Q2) 'Can a quantum computer efficiently simulate the dynamics of the universe?' Because they operate using the same principles that apply to nature at fundamental scales, quantum computers – though difficult to construct – represent a way of processing information that is closer to the way that nature processes information at the microscale. In 1982, Richard Feynman suggested that quantum devices could function as quantum analog computers to simulate the dynamics of extended quantum systems.[20] In 1996, Lloyd developed a quantum algorithm for implementing such universal quantum simulators.[21] The Feynman-Lloyd results show that, unlike classical computers, quantum computers can simulate efficiently any quantum system that evolves by local interactions, including for example the standard model of elementary particles. While no universally accepted theory of quantum gravity currently exists, as long as that theory involves local interactions between quantized variables, then it can be efficiently simulated on a quantum computer. So the answer to the quantized question 2 is 'Yes.'

There are of course subtleties to how a quantum computer can simulate the known laws of physics. Fermions supply special problems of simulation, which however can be overcome. A short-distance (or high-energy) cutoff in the dynamics is required to insure that the amount of quantum information required to simulate local dynamics is finite. However, such cutoffs – for example, at the Planck scale – are widely expected to be a fundamental feature of nature.

Finally, we can quantize question three: (Q3) 'Is the universe a quantum cellular automaton?' While we cannot unequivocally answer this question in the affirmative, we note that the proofs that show that a quantum computer can simulate any local quantum system efficiently immediately imply that any homogeneous, local quantum dynamics, such as that given by the standard model and (presumably) by quantum gravity, can be directly reproduced by a quantum cellular automaton. Indeed, lattice gauge theories, in Hamiltonian form, map directly onto quantum cellular automata. Accordingly, all current physical observations are consistent with the theory that the universe is indeed a quantum cellular automaton.

8. The Universe as Quantum Computer

We saw above that basic aspects of quantum mechanics, such as entanglement, make it difficult to construct a classical computational model of the universe as a universal Turing machine or a classical cellular automaton. By contrast, the power of quantum computers to encompass quantum dynamics allows the construction of quantum computational models of the universe. In particular, the Feynman-Lloyd construction allows one to map any local, homogeneous quantum dynamics directly onto a quantum cellular automaton.

The immediate question is 'So what?' Does the fact that the universe is observationally indistinguishable from a giant quantum computer tell us anything new or interesting about its behavior? The answer to this question is a resounding 'Yes!' In particular, the quantum computational model of the universe answers a question that has plagued human beings ever since they first began to wonder about the origins of the universe, namely, Why is the universe so ordered and yet so complex[1]?

The ordinary laws of physics tell us nothing about why the universe is so complex. Indeed, the complexity of the universe is quite mysterious in ordinary physics. The reason is that the laws of physics are apparently quite simple. The known ones can be written down on the back of a tee shirt. Moreover, the initial state of the universe appears also to have been simple. Just before the big bang, the universe was highly flat, homogeneous, isotropic, and almost entirely lacking in detail. Simple laws and simple initial conditions should lead to states that are, in principle, themselves very simple. But that is not what we see when we look out the window. Instead we see vast variety and detail – animals and plants, houses and humans, and overhead, at night, stars and planets wheeling by. Highly complex systems and behaviors abound. The quantum computational model of the universe not only explains this complexity: it requires it to exist.

To understand why the quantum computational model necessarily gives rise to complexity, consider the old story of monkeys typing on typewriters. The original version of this story was proposed by the French probabilist Émile Borel, at the beginning of the twentieth century (for a detailed account of the history of typing monkeys see[1]). Borel imagined a million typing monkeys (*singes dactylographes*) and pointed out that over the course of single year, the monkeys had a finite chance of producing all the texts in all the libraries in the world. He then immediately noted that with very high probability, they would would produce nothing but gibberish.

Consider, by contrast, the same monkeys typing into computers. Rather than regarding the monkeys random scripts as mere texts, the computers interpret them as programs, sets of instructions to perform logical operations. At first it might seem that the computers would also produce mere gibberish – 'garbage in, garbage out,' as the programmer's maxim goes. While it is true that many of the programs might result in garbage or error messages, it can be shown mathematically that the monkeys have a relatively high chance of producing complex, ordered structures. The reason is that many complex, ordered structures can be produced from short computer programs, albeit after lengthy calculations. Some short program will instruct the computer to calculate the digits of π, for example, while another will cause it to produce intricate fractals. Another will instruct the computer to evaluate the consequences of the standard model of elementary particles, interacting with gravity, starting from the big bang. A particularly brief program instructs the computer to prove all possible theorems. Moreover, the shortest programs to produce these complex structures are necessarily *random*. If they were not, then there would be an even shorter program that could produce the same structure. So the monkeys, by generating random programs, are producing exactly the right conditions to generate structures of arbitrarily great complexity.

For this argument to apply to the universe itself, two ingredients are necessary – first, a computer, and second, monkeys. But as shown above, the universe itself is indistinguishable from a quantum computer. In addition, quantum fluctuations – e.g., primordial fluctuations in energy density – automatically provide the random bits that are necessary to seed the quantum computer with a random program. That is, quantum fluctuations are the monkeys that program the quantum computer that is the universe. Such a quantum computing universe *necessarily* generates complex, ordered structures with high probability.

9. Conclusions

This article reviewed the history of computation with the goal of answering the question, 'Is the universe a computer?' The inability of classical digital computers to reproduce quantum effects efficiently makes it implausible that the universe is a classical digital system such as a cellular automaton. However, all observed phenomena are consistent with the model in which the universe is a quantum computer, e.g., a quantum cellular automaton. The quantum computational model of the universe explains previously unexplained features, most importantly, the co-existence in the universe of randomness and order, and of simplicity and complexity.

References

1. Lloyd S. *Programming the Universe*. New York: Knopf, 2004.

2. Cunningham M. *The History of Computation*. New York: AtlantiSoft, 1997.
3. Babbage C. *Passages from the life of a philosopher*. London: Longman, 1864.
4. Gibson W, Sterling B. *The Difference Engine*. London: Victor Gallancz, 1990.
5. Gödel K. Über formal unentscheidbare Sätze der *Principia Mathematica* und verwandter Systeme, I. *Monatshefte für Mathematik und Physik,* 38: 173–98, 1931.
6. Turing AM. On Computable Numbers, with an Application to the *Entscheidungsproblem*. Proceedings of the London Mathematical Society 1937;2 42 (1): 230-65. Turing AM. On Computable Numbers, with an Application to the *Entscheidungsproblem:* A correction. Proceedings of the London Mathematical Society; 2 43 (6): 544-6, 1937.
7. Church A. An unsolvable problem of elementary number theory. *American Journal of Mathematics,* 58: 345–363, 1936.
8. Shannon C. *A Symbolic Analysis of Relay and Switching Circuits*. MIT MS thesis. 1937.
9. Von Neumann J, Burks AW. *Theory of self-reproducing automata*. Urbana: University of Illinois Press, 1966.
10. Zuse K. *Rechnender Raum*. Braunschweig: Vieweg & Sohn, 1969.
11. Von Neumann J. *Mathematical Foundations of Quantum Mechanics*. Beyer RT, trans. Princeton: Princeton University Press; 1932. 1996 edition.
12. Hodges A. *Alan Turing: the enigma*. London: Random House; 1992.
13. Wheeler JA, Zurek WH. *Quantum Theory and Measurement*. Princeton: Princeton University Press, 1983.
14. Nielsen MA, Chuang IL. *Quantum Computation and Quantum Information*. Cambridge: Cambridge University Press; 2001.
15. Benioff P. *Journal of Statistical Physics*. 22: 563–591, 1980.
16. Deutsch D. *Proceedings of the Royal Society of London A*. 1985; 400: 97–117.
17. Shor P. Polynomial-Time Algorithms for Prime Factorization and Discrete Logarithms on a Quantum Computer. *SIAM J Sci Statist Comput*. 26: 1484, 1995.
18. Lloyd S. A potentially realizable quantum computer. *Science*. 261: 1569–1571, 1993.
19. Cirac JI, Zoller P. Quantum computations with cold trapped ions. *Physical Review Letters*. 74: 4091–4094, 1995.
20. Feynman RP. *International Journal of Theoretical Physics*. 21: 467–488, 1982.
21. Lloyd S. Universal Quantum Simulators. *Science*. 273: 1073–1078, 1996.

Chapter 30

Quantum Speedup and Temporal Inequalities for Sequential Actions

Marek Żukowski

Institute for Theoretical Physics and Astrophysics,
University of Gdańsk, Poland
Faculty of Physics, University of Vienna, Austria &
U. S. T. C., Hefei, China

Quantum information processing promises to beat limitations of "classical" information processing. We can have theoretically unbeatable quantum cryptography, which has by now very many technical implementations. We might have quantum computers, which in principle perform some tasks exponentially faster (in terms of operations needed) than classical ones, i.e. ones which are realizations of Turing machines.

What is the exact principal reason for better-than-classical performance of quantum information processing? The standard answer is the quantum superposition, which allows for some kind of parallelism. A qubit, the quantum bit can be in states which are superpositions of logical 1 and 0. But superposition is also property of classical waves. The difference between quantum superposition and a classical wave superposition is that the former is fragile upon measurements: any measurement on a qubit, except the one which equivalent to preparation of the given state (that is requires the knowledge of the state) destroys the initial state. The only information that we gain after such a measurement is in what state is the qubit *after* the measurement. Does this fragility, or gross invasiveness of measurements, contribute to quantum information processing? In the case of cryptography it certainly does. But what is its role in other applications? As we shall see below this does not need to be the case.

What are the basic, primitive, principles that limit performance of classical information processing devices? It was recently suggested by Brukner et al.[1] that the reason, or one of the reasons for quantum speedup in computing, might be in violation of "temporal Bell inequalities". Such inequalities were introduced by Leggett and Garg,[2] and were aimed at the question of the relation between Quantum Mechanics and set of basic principles of classical systems which was given the name of Macroscopic Realism. They formulated the basic principles of Macrorealism as:

- a macroscopic system with two or more distinct states will at all times be in one of these states,
- it is possible, in principle, to determine the state of the system with an arbitrarily small perturbation on its subsequent dynamics (noninvasive measurability).

Leggett and Garg considered macroscopic quantum coherence in a SQUID, and showed that effectively there is no flux "when nobody looks". To this end they derived what is often called "temporal Bell inequalities".

A different version of such inequalities was introduced by Brukner et al.[1] (for an earlier derivation, without a direct link with the discussion of Macrorealism, see;[3] for an extensive study see[4] and[5]). They follow basically the same technical assumptions as the original ones, however the observer is allowed to choose between various observables which he or she wants to measure at a given instant of time. The original ones allowed the observer to freely choose the time of observation, but not the observable, which was fixed throughout the process. As the approach that will be followed in this chapter, is more directly related the Brukner et al. inequalities, the assumptions behind them will be now presented.

The observer has a choice between two apparatus settings for each instant of time at which he or she is to make a measurement. The measurements are to be made at instants of time t_0 and later at t_1. The following traits of a Macro-realistic theory are assumed:

- In the theoretical description one is allowed to use all variables $A_m(t)$, the values of which are eigenvalues of the observable $\hat{A}_m$, which represent the values which could have been obtained, had the given observable been measured at time t, regardless what was the actual measurement. The observer has a choice $m = 1, 2$, or even larger. All $A_m(t)$'s are treated as unknown, but nevertheless fixed numbers, all of them at an equal footing, that is for example the sum $A_1(t) + A_2(t)$ has a definite, but unknown, value. (This is an assumption of *realism* - it is satisfied by classical systems. Please note that, if $\hat{A}_1$ and $\hat{A}_2$ are quantum observables, which do not commute, then at a given instant of time only one observable can be

measured, and thus one deals here with counterfactual statements.)
- *Non-invasiveness*: The values $A_m(t_1)$ are independent of whether or not a measurement was performed earlier, at t_0, and which observable was at this earlier time measured. In short values $A_m(t_1)$ are independent of the measurement settings chosen earlier. (Note that this is a strong assumption, which does not have to hold even for classical systems when an act of observation produces a disturbance.)
- Values $A_m(t_0)$ do not depend on what happens at later times, especially at t_1 (*future cannot influence past*).
- The experimenter is free to choose the observable which is to be measured at a given instant of time (*freedom*). That is the choices are statistically independent of the set of values $A_m(t)$, and everything else in the experiment.

With a similar type of algebra as in the case of CHSH inequalities, under the assumption that all involved eigenvalues are ± 1, Brukner et al.[1] show that

$$E(A_1(t_0), A_1(t_1)) + E(A_1(t_0), A_2(t_1))$$
$$+ E(A_2(t_0), A_1(t_1)) - E(A_2(t_0), A_2(t_1)) \leq 2, \tag{1}$$

where $E(A_k(t_0), A_m(t_1))$ stands for a correlation function, understood as an averaged product of the results, that is $\langle A_k(t_0) A_m(t_1) \rangle$. Such inequalities can be violated by a qubit in any initial state. Thus, they suggest that the reason for "quantum speedup" is due to a violation of the set of assumptions behind Macrorealism (at least one of them). A violation of the last assumption would deny existence of any statistically independent random process, and would make all events in the Universe mysteriously linked, without any chance to ever observe processes which have traits usually defining independence, like the probability rule $P(x\&y) = P(x)P(y)$ for two stochastic variables. Thus, most of us would reject such a possibility. The assumption that future events cannot influence past is difficult to question under our current understanding of the whole physics,[a] and in the

---

[a]Sometimes it is argued the in delayed choice experiments one can influence past. But this is just a nice marketing of the weirdness of quantum mechanics. A measurement at later time "can influence past" only if in the past the quantum system was in a quantum superposition, and the superposition stays intact until the measurement. Thus, the system cannot interact in the past with anything in such a way that it leaves observable traces. Therefore, the influence of delayed choice measurements cannot change any observable past event, that is anything of the past.

relativistic physics influences are constrained even further to the objective past light cone of an event. Realism is the usual assumption of macrophysics, thus it seems in place here, however it is a good candidate to be a culprit in the case of violations of these inequalities by qubits, as quantum mechanics does not allow to define realism. Still surprisingly, in the case of *single* qubits, as it was constructively established by Bell[8] one can have a realistic model. Thus this assumption cannot be behind the violation. Thus we are left with with non-invasiveness. But when compared to the other assumptions, this would boil down to the following "past does not influence future", at least the future of the variables $A_l(t_k)$ in question, for a specific form of past actions singled out - measurements. Despite these constraints of the applicability of this assumption, it emerges as the culprit. In the case of quantum objects measurements do influence the future behavior of the systems. Thus the picture seems clear, but clarity is reached at a high price, what is violated is something that does not necessarily hold, a highly questionable assumption. Thus, are we gaining any insight here?

Note that in the initial wording the "non-invasiveness" assumption is quite weak (to determine the state of the system with an arbitrarily small perturbation on its subsequent dynamics). However, to derive the bound of the inequality one needs a sharp formulation that future values of observables are not influenced at all by earlier measurements. The same problems arise when one considers the original Leggett-Garg inequalities, and their violations.[b]

One intuitively feels that there must exist some form of temporal inequalities that are applicable to arbitrarily long processes, which involve many instants of time, at which the system changes its state due to an external intervention. Below, such a family of inequalities will be presented. An entirely new approach will be taken, which surprisingly uses softer, more natural, assumptions than these of Macrorealism presented above. The whole idea will be illustrated with something that resembles an information processing protocol, in its most elementary form, performed on a most elementary physical system that has information capacity of one bit. In case of physical realization of Turing machines, such a system would by for example a single transistor. What important we take into account only the macro-states of the transistor relevant for the computation process. States 0 and 1, represented by "no current" and "current" passing through it. In

---

[b]In the Heisenberg picture different moments of observation lead to different observables, as $\hat{A}(t) = U^\dagger(t,t_0)\hat{A}(t_0)U(t,t_0)$, where $U(t,t_0)$ is the unitary evolution operator. This gives one an equivalence of the two approaches.

such a case one is tempted to compare the transistor with a single qubit, and try to see if such a replacement leads to a gain in performance. The states of the transistor and the qubit, to make the comparison fair, will undergo some transformations, which would be determined by specific data streams of the same type in both cases. Thus the "quantum speedup" or rather "quantum perfect success" as we shall seed further on, will be only due to the properties of the qubit and quantum transformations. The performance of the transistor in case of the studied problem is rather disastrous...

Imagine a microprocessor element, our transistor, which can be in two states. The states will be denoted as A. For the sake of an easier mathematical representation, we shall assume that $A = \pm 1$, that is the bit value b represented by the state is related with respect to A in the following way $A = (-1)^b$. Assume that at each instant of time t_k, where $k = 0, 1, 2, ...$ and $t_k < t_{k+1}$, an operation is performed on the system which may change the value of A. The operation is governed by two external input bits. For the given moment they are represented by two random numbers x_k, y'_k, and the pair will be denoted X_k, and at certain points we shall assume that X_k has a numerical value $2^1 y'_k + 2^0 x_k = 0, 1, 2, 3$. We assume that each y'_k is completely random, whereas the distribution of x_k's may be governed by a probability distribution $p(x_1, x_2, ...)$. For technical reasons we assume that $x_k = 0$ or 1 and we replace y'_k by $y_k = (-1)^{y'_k}$. Thus, $y_k = \pm 1$. After say l operations the current state of the system, at $t = t_l$, is denoted as $A_{X_l} = A(X_1, X_2, ..., X_l)$. However, we shall assume that the system forgets the reason why it is in the current state, that is the state after the k-th instant of time is given by

$$A_{X_k}(t_k) = F(X_k, A_{X_{k-1}}(t_{k-1}), t_k). \qquad (2)$$

That is, it is defined by the state of the system before the last operation, $A_{X_{k-1}}(t_{k-1})$, and by the last operation, defined by X_k (this seems absolutely sensible in the case of classical operations on computer elements). F denotes a binary function of values ± 1.

We shall demand that the operations performed on the system are aimed to give at the end of the process a final value A_n, which is an answer to a question about the value of a dichotomic task function $T_n(X_1, ..., X_n)$. As it will be shown below, for some task functions the probability of $A_n = T_n$, for systems having the above traits, is quite low.

To this end, let us take a specific example of such a computation task:

The aim is to give the value of

$$T_n = \prod_{l=1}^{n} y_l \cos\left(\frac{\pi}{2} \sum_{k=1}^{n} x_k\right), \qquad (3)$$

under the promise that the distribution of x_k's obeys the following probability

$$p(x_1, x_2, ...) = 2^{-2N+1} |\cos(\frac{\pi}{2} \sum_{k=1}^{n} x_k)|, \qquad (4)$$

where $N = n/2$. This simply implies that the bits x_k are promised to satisfy always the following constraint: $(\sum_{k=1}^{N} x_k) \bmod 2 = 0$. That is, they are distributed in such a way that their sum is always even. Note, that under such a promise $T_n = \pm 1$. What is the average chance to get a correct result for systems obeying the above assumptions? This will be estimated by calculating the maximum possible value of an average, over all possible values for X_k's weighted by their probabilities, of the product of the answer with the correct value: $E = Max\langle A_n T_n\rangle_{avg}$. If this average equals 1 the answer is always correct. If $E = 0$, he answer is random, uncorrelated with T_n. The probability of a correct answer in an optimal protocol is $p = \frac{1}{2}(1 + E)$.

Of course, the above story does not have to be taken literally, when one considers just properties of systems satisfying the above assumptions, nevertheless it is relevant if one ponders about the opening questions of this article. We shall now derive an inequality which is obeyed by the average value of $A_n = A(X_1, X_2, ..., X_n)$, under the restrictions given above, especially (2). The interpretation of the form of the inequality is in a way irrelevant, what matters is that it can be violated by a quantum process. From the technical point of view the derivation resembles, albeit only in its final stages, the case of communication complexity problems studied in.[10]

Let us write first explicitly the expression the maximum of which we search for:[c]

$$E = \langle A_n T_n \rangle_{avg}$$

$$= \sum_{x_1, x_2, ..., x_n = 0, 1} \sum_{y_1, y_2, ..., y_n = \pm 1} \frac{1}{2^n} p(x_1, ..., x_n) A(X_1, X_2, ..., X_n) \prod_{l=1}^{n} y_l f(x_1, ..., x_n), \qquad (5)$$

---

[c]Readers not interested in the technicalities of the derivation can drop this entire paragraph.

where $f(x_1, ..., x_n) = \cos(\frac{\pi}{2} \sum_{k=1}^{n} x_k)$. Note that $A(X_1, ..., X_n) = A_n(x_n, y_n, A_{n-1})$, where $A_{n-1} = A(X_1, ..., X_{n-1})$ is the earlier value, after $n-1$ operations, and x_n and y_n represent the last operation. The index n in A_n is introduced, so the iterative character of the reasoning is more clearly visible. A_n is a binary function of its three arguments. It must depend on A_{n-1} because only A_{n-1} might contain information about $y_{n-1}, y_{n-2}, ..., y_1$, which is absolutely necessary for an attempt to get the correct value of T_n. Please, look at equation (3): all y_l's must be known in order to get the correct value. There are very few binary functions of a binary variable, just four (two constant ones and two proportional to the variable). Let us use this fact. Treat x_n and y_n as fixed, thus we have $A_n = B_{x_n, y_n}(A_{n-1})$. Because the function B_{x_n, y_n} is binary, it can only have the following form:

$$B_{x_n,y_n}(A_{n-1}) = D_{x_n,y_n} + C_{x_n,y_n} A_{n-1},$$

where both C and D are equal ± 1 or 0, and $C_{x_n,y_n} D_{x_n,y_n} = 0$. Note that, as $\sum_{y_{n-1}} y_{n-1} D_{x_n,y_n} = 0$, the first term cannot contribute to E. Thus, if we search for the maximal value, and as maximum of any linear function is at the limits of its inputs, only values -1 and +1 of C can contribute. Therefore we can put $B_{x_n,y_n}(A_{n-1}) = C_{x_n,y_n} A_{n-1}$, where $C_{x_n,y_n} = \pm 1$. In turn, if one fixes x_n, the function C is binary one of a binary variable y_n. That is, it must have the form of $C_{x_n,y_n} = d(x_n) + c_n(x_n) y_n$, with $d(x_n) c_n(x_n) = 0$. But again, as $\sum_{y_n} y_n d(x_n) = 0$, one has to assume that $c(x_n) = \pm 1$. Thus, as we see, the optimal form of $A_n(X_n, A_{n-1})$ is $A_n = y_n c_n(x_n) A_{n-1}$. With a similar step one shows that the optimal form of A_{n-1} is $y_{n-1} c_{n-1}(x_{n-1}) A_{n-2}$. The symbols play equivalent roles as the ones with the indices increased by 1, used earlier. Continuing like that we arrive at the final formula which is

$$\langle A_n T_n \rangle_{avg} = \sum_{x_1, x_2, ..., x_n = 0, 1} K(x_1, ..., x_n) \prod_{k=1}^{n} c_k(x_k), \qquad (6)$$

where all $c_k(x_k)$ take values ± 1, and the coefficients K are given by $K(x_1, ..., x_n) = p(x_1, ..., x_n) f(x_1, ..., x_n)$. In our specific case they read $2^{-2N+1} \cos\left(\frac{\pi}{2} \sum_{k=1}^{n} x_k\right)$. The right hand side of (6) is mathematically isomorphic with an algebraic expression leading to a multi-party Bell inequality. Its upper bound is given by

$$\sum_{x_1, x_2, ..., x_n = 0, 1} K(x_1, ..., x_n) \prod_{k=1}^{n} c_k(x_k) \leq 2^{-N+1}.$$

As a matter of fact for $n = 3$ one has a structure similar to the Mermin Bell-type-inequality, and the whole set to the series of inequalities derived by Mermin in 1990.[9] Similar series of Bell-like inequalities were derived for the communication complexity problems in[10] (see also[11]). Please do notice, that Bell inequalities were used here only to reach the final values, once the stage (6) of the derivation was reached. The inequality $E \leq 2^{-N+1}$ has otherwise no relation with Bell inequalities whatsoever.

Please note that we have just arrived at the bounds of a series of temporal inequalities

$$\sum_{X_1,...,X_n=0,1,2,3} \cos\left(\frac{\pi}{2}\sum_{k=1}^{n} X_k\right) A(X_1, X_2, ..., X_n) \leq 2^N, \qquad (7)$$

which is applicable to final states, $A(X_1, X_2, ..., X_n)$, of a classical system which undergoes a series of transformations governed by a temporal sequence of external parameters X_k. This can be re-expressed in the terms of the upper bound on the probability of success of any classical protocol, following the constraints listed above, to give a correct answer:

$$P \leq \frac{1}{2}(1 + 2^{-N+1}). \qquad (8)$$

That is, for a high n, the answer is basically random.

The inequality (7) is relevant. It is violated by a quantum process, which was experimentally realized by the group of Weinfurter,[10] however in a completely different context. In the ideal quantum version of the protocol one starts with a qubit in the state[d] $|\psi_i\rangle = 2^{-1/2}(|0\rangle + |1\rangle)$. Then one acts sequentially on the qubit with unitary phase-shift transformations

---

[d]For the unacquainted with quantum mechanics: Photon's polarization is a perfect natural realization of a qubit. Imagine that the symbol $|0\rangle$ represents the vertical polarization, and the symbol $|1\rangle$ the horizontal one. The formula for the initial state represents an in phase superposition of the two linear polarizations, that is a photon polarized diagonally. The sequential operations performed on it change in the relative phase of the two polarizations. This can be done with suitable wave plates. The relative phase accumulates during the protocol. The final measurement uses a birefringent crystal, which splits light into two beams: diagonally and antidiagonally linearly polarized. In each beam we place a (perfect) photo-detector. A photon cannot split crossing the crystal: photons do not change their color when passing such a crystal, in turn the color, that is frequency, defines the energy, by the formula $E = h\nu$; if photons split the resulting daughter photons have lower frequencies than the original one, thus a different color. Thus, it is found by either the detector observing the diagonal or the antidiagonal channel. As the phase shifts in the protocol can only accumulate to $k\pi$, where k is an integer, the final state of the photon is diagonal or antidiagonal polarization. Thus photons end up *deterministically* in the given detector. That is, the quantum protocol does not produce errors.

of the form $|0\rangle\langle 0| + e^{i\pi/2 X_k}|1\rangle\langle 1|$, in accordance with the local inputs x_k, y_k. After all N phase shifts the state is

$$|\psi_f\rangle = \frac{1}{\sqrt{2}}(|0\rangle + e^{i\pi/2(\sum_{k=1}^n X_k)}|1\rangle). \qquad (9)$$

Due to the constraint that the sum over all X_k must be even (see the "promise" in the derivation of the inequality), the phase factor $e^{i\pi/2(\sum_{k=1}^n X_k)}$ is equal to the dichotomic function T_n to be computed. Therefore, a measurement of the qubit in the basis given by $2^{-1/2}(|0\rangle + |1\rangle)$ and $2^{-1/2}(|0\rangle - |1\rangle)$ reveals the value of T_n, with fidelity $\langle A_n T_n \rangle_{avg} = 1$. In other words the quantum answer is always correct. Note, that this implies that inequality (8) is violated by an exponentially growing factor (in terms of the number of operations n).

The following modified assumptions are behind the inequality (they are spelled out below to cover a broad set of systems, not just the example):

- *Realism*: In the theoretical description one is allowed to use all variables $A_m(t)$, $B_m(t)$,..., $Z_m(t)$, the values of which are eigenvalues of observables $\hat{A}_m,... \hat{Z}_m$, which represent the value which would have been obtained if the given observables were measured at time t, at which it is assumed that an operation m was performed upon the system. This regardless what was the *actual* operation at t. By "operation" is meant a transformation and/or a measurement. The observer has a choice between many m's (in our example in the k-th step $m_k = X_k$, which can take four values). All these variables are treated as unknown, but nevertheless fixed (real) numbers, all of them at an equal footing, that is for example, for two different choices, m and m', the algebraic expressions like $A_m(t) \pm A_{m'}(t)$ or $A_m(t) \pm B_{m'}(t)$, or even $A_m(t) \pm B_{m'}(t')$, etc., have a definite, but perhaps unknown, value. (This is basically the old assumption, slightly rewritten and generalized.)
- *Classical causality*: initial conditions decide subsequent evolution : The realistic values after an operation at t_{k+1}, that is $A_m(t_{k+1}),..., Z_m(t_{k+1})$, are not *directly* dependent on operations which were performed earlier, at t_k, $t_{k-1}, ..., t_1$. However, $A_{m_{k+1}}(t_{k+1})$, etc., might depend on the earlier values, that is on $A_{m_k}(t_k),....Z_{m_k}(t_k)$, which are defined by the state of the system after the previous operation m_k at t_k. That is, it is assumed that there is no *direct* dependence on the operations done earlier, before the last one, while an indirect one is allowed via results of these

earlier operations, as operations are allowed to change the realistic values $A_{m_k}(t_k), ..., Z_m(t_k)$. [e]
- *Past cannot depend on future*: Values $A_m(t_{k-1}),...,Z_m(t_{k-1})$ do not depend on what happens at later times, especially at t_k. (Unchanged.)
- *Freedom*: The experimenter is free to choose the operation which is to be to be performed at a given instant of time. Thus, for example the choices of m are statistically independent of the set of values $A_m(t)$, $A_{m'}(t'),..., Z_m(t'')$, etc. (Unchanged.)

These assumptions are quite general, and apply to observables endowed with any eigenvalues. One *could* supplement them with the non-invasiveness assumption, but the example shows that this is not necessary in case of some processes to get interesting results (the inequality is violated by a quantum process). The operations can be some transformations or measurements. The assumptions, when applied to our example, are isomorphic with the set stated at the beginning of the derivation of the inequality, and the tacit assumptions used during the derivation (esp., freedom). Note that, one can introduce additionally a notion of an initial state of the system, say λ, on which all values depend. Thus, one would have $A_m(t|\lambda),..., Z_m(t|\lambda)$. Averaging of the inequality over a probability distribution $\rho(\lambda)$ does not change the bound.

Note that as the "non-invasiveness" assumption does not apply above, the assumptions that one could see as loosing their validity in the case of qubits, to allow for the violation, are the conjunction of realism with classical causality. Note that the latter assumption is not independent of the earlier: it is formulated using notions defined by by the earlier one.

Temporal inequalities involving many measurements, were introduced earlier in ref.[12] However they follow a an approach based on the original definition of Marcrorealism. As it is suggested within a different context

---

[e]Note that this is a an assumption which holds for the on/off states of transistors in microchips, treated as two state-systems. In classical mechanics it is equivalent to a statement that we do not care what was the reason for the current state of an object, we care only about the state. We do not need to know *why* a classical particle has this or that momentum and this or that position at the given moment. Still these values are full initial conditions for further dynamics. All systems, which follow Hamilton dynamics, including classical fields, satisfy this condition. Note further, that quantum systems share this feature, but it is expressed differently: the initial state defines the final one in evolution dictated by the Schroedinger equation. The evolution of, or manipulations upon, the quantum system before it reached the initial state is totally irrelevant for a future evolution.

in,[11] one can derive inequalities involving task functions and promises related with the inequalities discovered in[13] and.[14]

These findings can be generalized in many ways. Concerning the opening question, it is shown here that we cannot steer a transistor (treated as a two-state system) by sequential operations, each governed by pairs of bits x_k and y'_k following a certain promise, so that at the end of the process it would give, with a high probability, the value of T_n, given by (3). In contrast, this can be done with a qubit. The presented inequality is just a first example of infinitely many that can be derived using the new four assumptions. These do not have to be constrained to two-state systems, and the inputs can be even continuous (compare the communication complexity problems in[10] and[11]). Note that the system considered in the example has a finite information capacity. This seems to be a crucial factor leading to the inequality,[15] and its violation by a quantum process involving a qubit.

Acknowledgements

MZ is supported by projects Q-ESSENCE (VII FP EU) and N N202 208538 of MNiSW. Thanks to Č. Brukner, J. Kofler, M. Markiewicz, T. Paterek and M. Pawłowski for discussions.

References

1. Č. Brukner, S. Taylor, S. Cheung, V. Vedral, arXiv:quant-ph/0402127.
2. A. J. Leggett, and A. Garg, Phys. Rev. Lett. **54**, 857 (1985); A. J. Leggett, J. Phys. Condens. Matter **14**, R415, 2002.
3. A. Shafiee, and M. Golshani, Ann. Fond. Broglie **28**, 105, 2003.
4. J. Kofler, Ph. D. Thesis, e-print arXiv:0812.0238.
5. J. Kofler, Č. Brukner, Phys. Rev. Lett. 101, 090403 (2008)
6. J. Kofler, N. Buric, Č. Brukner, e-print arXiv:0906.4465.
7. D. M. Greenberger, M. Horne, and A. Zeilinger, in *Bell's Theorem, Quantum Theory, and Conceptions of the Universe*, edited by M. Kafatos (Kluwer Academic, Dordrecht), p. 73–76, 1989.
8. J. S. Bell, Physics **1**, 195, 1964.
9. N. D. Mermin, Phys. Rev. Lett. **65**, 1838, 1990.
10. P. Trojek, C. Schmid, M. Bourennane, Č. Brukner, M. Żukowski, and H. Weinfurter, Phys. Rev. A **72**, 050305(R), 2005.
11. Č. Brukner, M. Żukowski, J.-W. Pan, and A. Zeilinger, Phys. Rev. Lett. **92**, 127901, 2004.
12. F. Morikoshi, Phys. Rev. A **73**, 052308, 2006.
13. H. Weinfurter, M. Żukowski, Phys. Rev. A **64**, 010102 (R) (2001); M. Żukowski and Č. Brukner, Phys. Rev. Lett. **88**, 210401. 2002.

14. R. F. Werner, and M. M. Wolf, Phys. Rev. A **64**, 032112, 2001.
15. Č Brukner, and independently M. Pawłowski, private communications. See also related: M. Kleinmann, O. Gühne, J. R. Portillo, J.-A. Larsson, A. Cabello, arXiv:1007.3650.
16. A. S. Holevo, Probl. Pered. Inform. **9**, 3 (1973) [Probl. Inf. Transm. **9**, 110, 1973].

Chapter 31

The Contextual Computer

Adán Cabello

Departamento de Física Aplicada II, Universidad de Sevilla, Spain &
Department of Physics, Stockholm University, Sweden
adan@us.es

What kind of computer is the universe? Here we present three results. The first is a consequence of the Kochen-Specker theorem: If the predictions of quantum mechanics are correct, then the universe cannot be a non-contextual computer. We then show that, if we assume that the density of memory is bounded, then the universe cannot be a classical contextual computer. The third result singles out the universe among all possible contextual computers by exploiting a curious connection with graph theory: In the universe, the maximal contextuality of a set of propositions is given by the Lovász number of the graph representing their mutual exclusiveness.

1. Introduction

Physicists have convincing arguments explaining why the universe cannot be a classical computer working on bits. The most famous one is Bell's theorem,[1] which can be formulated in a very general way: If the predictions of quantum mechanics are correct and if the speed of information is limited, then no classical computer can simulate the results we obtain in some experiments.

But let us suppose that our experiments are never fast enough to guarantee that they cannot influence the results generated in other parts of the universe. If such is the case, is there any way to prove that the universe cannot be a classical computer?

In Sec. 2 we define the concept of non-contextual computer and, in Sec. 3, we show that the universe cannot be a non-contextual computer. This is a well-known result which, by itself, does not prevent the universe to be a classical computer. However, in Sec. 4 we show that, if we assume that the density of memory, in bits per qubit, a computer of finite size and energy can have is limited, then the amount of contextuality in the universe cannot be simulated by a classical computer.

This tells us what kind of computer the universe probably is *not*: It is not a non-contextual computer or a classical contextual computer. But what type of computer the universe actually *is*? In Sec. 5 we argue that it is a contextual computer with precisely defined limits for the contextuality that it can produce, and we show that there is a simple characterization for this maximum contextuality.

2. Contextuality

Throughout all this article, we will consider the following scenario: A single portion of the universe (a "system") is subjected to a sequence of dichotomic experiments (that is, with only two possible results: $+1$ or -1). The experiments will be chosen from amongst a finite set. The order in which the experiments are performed and the number of times each experiment is performed is randomly decided. The goal is to design a computer which simulates the results of these experiments, assuming that these results are correctly predicted by quantum mechanics.

We choose the experiments to show one of the fundamental properties of quantum mechanics: Contextuality. To introduce contextuality, we first need to introduce two concepts: Repeatability and compatibility.

An experiment A is *repeatable* when it can be performed many (ideally infinite) times and gives always the same result. We assume that no other experiment is performed in between two experiments A. In quantum mechanics, a repeatable experiment is called a *sharp measurement* or a *projective measurement* and is mathematically represented by a self-adjoint matrix (a square matrix with complex entries that is equal to its own conjugate transpose).

Two repeatable experiments A and B are *compatible* when every time experiment B is performed after experiment A, a subsequent execution of experiment A yields the same result as if experiment B had not been performed. Compatibility is an experimentally testable property. It is symmetric (the roles of A and B are interchangeable), but not necessarily transitive: Even if A is compatible with B, and B is compatible with b, it may happen that A and b are incompatible. A set of experiments $S = \{A, B, \ldots, Z\}$ is compatible when all experiments in S are mutually compatible. In quantum mechanics, compatible experiments are represented by commuting matrices, $AB = BA$. In classical physics, carefully performed experiments are always compatible. This is not the case in quantum mechanics.

We focus on a particular type of incompatible experiments. Two repeatable dichotomic experiments A and b are *maximally incompatible* if, whenever b is performed after A, a subsequent execution of A yields a fundamentally unpredictable result. This implies that it yields the same result as if b had not been performed with probability $\frac{1}{2}$, and yields the opposite result with probability $\frac{1}{2}$. Maximal incompatibility is also an experimentally testable, symmetric but not necessarily transitive property. In quantum mechanics, maximally incompatible experiments are represented by anticommuting matrices, $Ab = -bA$.

A *non-contextual* computer is any which provides pre-established results which are not affected by compatible experiments. For instance, suppose experiment A is compatible with experiments B and a (although B and a may be mutually incompatible). The assumption of non-contextuality is that the result of A is the same, regardless of whether A is performed alone, after B, or after a. However, the result of A might be different if an incompatible experiment b were to be performed before A. A *contextual* computer is any which does not satisfy the definition of non-contextual computer.

We assume that the system on which these experiments are performed is a set of n qubits (with $n \geq 2$). A *qubit* is any physical system for which we have the ability of preparing and measuring all the possible quantum superpositions of two perfectly distinguishing states. For example, a qubit is the polarization of a single photon. In contrast, a classical *bit* is any physical system which we can prepare either in state 0, in state 1, or in a probabilistic mixture of them, and on which we can distinguish states 0 and 1. A qubit is therefore more general than a bit and much more expensive. A dead/alive cat[2] is a bit, but *not* a qubit unless we can prepare any possible quantum superpositions and we perform any possible quantum sharp measurement of dead and alive.

3. The Impossible Non-Contextual Computer

The goal is to design a computer which simulates some predictions of quantum mechanics. Consider 9 dichotomic experiments $A, B, C, a, b, c, \alpha, \beta$ and γ, such that they form 6 compatible sets: $S_1 = \{A, B, C\}$, $S_2 = \{a, b, c\}$, $S_3 = \{\alpha, \beta, \gamma\}$, $S_4 = \{A, a, \alpha\}$, $S_5 = \{B, b, \beta\}$ and $S_6 = \{C, c, \gamma\}$, and such that any pair of experiments is either compatible or maximally incompatible. The predictions of quantum mechanics we want to simulate are the following: (i) The compatible/maximally incompatible experiments are actually compatible/maximally incompatible. (ii) The product of the results of the three experiments in S_i performed sequentially is $+1$ for $i = 1,\ldots,5$ and -1 for $i = 6$. That is, now using A to denote the result of experiment A, etc.,

$$\begin{aligned} ABC &= 1, \\ abc &= 1, \\ \alpha\beta\gamma &= 1, \\ Aa\alpha &= 1, \\ Bb\beta &= 1, \\ Cc\gamma &= -1. \end{aligned} \quad (1)$$

A non-contextual computer should provide results for each of the 9 experiments which do not depend on which other compatible experiments are performed.

The interesting point is that a non-contextual computer cannot reproduce all the predictions in (1). The proof is simple: Assuming a non-contextual assignment of results, if we multiply the first three equations in (1), we obtain

$$ABCabc\alpha\beta\gamma = 1. \qquad (2)$$

However, if we multiply the last three equations in (1), we obtain

$$ABCabc\alpha\beta\gamma = -1. \qquad (3)$$

The contradiction proves that the assumption is not valid: Non-contextual assignments are impossible.

One way to produce the quantum predictions in (1) is to pick $n = 2$ qubits and choose the following experiments: $A = \sigma_x \otimes \mathbb{1}$ (meaning the measurement of the observable represented in quantum mechanics by the tensor product of the Pauli matrix x and the 2×2 identity matrix), $B = \mathbb{1} \otimes \sigma_x$, $C = \sigma_x \otimes \sigma_x$, $a = \mathbb{1} \otimes \sigma_y$, $b = \sigma_y \otimes \mathbb{1}$, $c = \sigma_y \otimes \sigma_y$, $\alpha = \sigma_x \otimes \sigma_y$, $\beta = \sigma_y \otimes \sigma_x$ and $\gamma = \sigma_z \otimes \sigma_z$.[3,4] This is a simple example of a general result in quantum mechanics known as the Kochen-Specker theorem.[5]

4. A Classical Contextual Computer Requires Unlimited Density of Memory

There is a simple solution to the task of simulating predictions (i) and (ii): The computer saves in its memory the result of the first measurement, e.g., A. If the second measurement is maximally incompatible, e.g., b, then the computer erases from its memory the result of A and saves the result of b. If the second measurement is compatible with A, e.g., B, then the computer saves both the results of A and B. The first equation in (1) automatically defines the result of $C = AB$. In this way, the computer can provide *contextual* results simulating (i) and (ii). All the computer needs is *memory*. In principle, not too much memory: All that the computer needs to keep is the number of states needed to remember which was the set of compatible experiments S_i of the last compatible measurements (e.g., S_1) and the number of states needed to remember the results of all the experiments in S_i (e.g., A, B and C).

For example, to reproduce (i) and (ii) in the previous scenario, the computer needs $\log_2(6 \times 2^2) \approx 4.58$ bits, since there are 6 possible sets S_i and 2^2 possible states in every S_i (the result of the third experiment is defined by the results of the other two). A formal proof that this is the necessary and sufficient memory needed to satisfy (i) and (ii), assuming that the result of an experiment incompatible with the previous one is genuinely random, can be found in.[6] Notice that this memory is already larger than the information-carrying capacity of two qubits, which is two bits:[7] The computer needs to store more information than that which is accessible by performing experiments on the computer. Strange, but, so far, not too bad.

Let us add all possible experiments which are either compatible or maximally incompatible with the previous experiments. The new experiments are: $\sigma_z \otimes \mathbb{1}$, $\mathbb{1} \otimes \sigma_z$, $\sigma_z \otimes \sigma_x$, $\sigma_x \otimes \sigma_z$, $\sigma_z \otimes \sigma_y$ and $\sigma_y \otimes \sigma_z$. We end up with the set $E^{(2)} = \{E_i^{(2)}\}_{i=1}^{15}$ of 15 experiments which can be defined for $n = 2$ qubits using the tensor product of one of the three Pauli matrices or the identity matrix times one of the three Pauli matrices or the identity matrix (we exclude $\mathbb{1} \otimes \mathbb{1}$ in the count, because it is compatible with all the other experiments and not maximally incompatible with any of them). $E^{(2)}$ defines a set $S^{(2)} = \{S_j^{(2)}\}_{j=1}^{15}$ of 15 sets $S_j^{(2)}$ of three compatible experiments $E_i^{(2)}$. We will denote by $\mathcal{S}^{(2)}$ the pair $(E^{(2)}, S^{(2)})$.

$\mathcal{S}^{(2)}$ can be arranged in 10 different sets of 9 experiments and 6 compatible sets, each defining an equation like those in (1) for which no non-contextual model exists. Since no non-contextual model exists for each of these 10 subsets, then no non-contextual model exits for $\mathcal{S}^{(2)}$.

The addition of the 6 extra experiments had the purpose of completing the set of experiments on $n = 2$ qubits in which every pair is either compatible or maximally incompatible. $\mathcal{S}^{(2)}$ presents an advantage compared to the initial set of experiments in Sec. 3: The ratio between the quantum predictions that cannot be satisfied with non-contextual models and the total number of quantum predictions *increases* from $\frac{1}{6}$ for the set in Sec. 3 to $\frac{1}{5}$ for $\mathcal{S}^{(2)}$.[8] Therefore, to simulate (i) and (ii) for $\mathcal{S}^{(2)}$, the computer needs more memory. To be precise,

$$m(2) = \log_2(15 \times 2^2) \approx 5.91 \text{ bits.} \qquad (4)$$

Another good thing about $\mathcal{S}^{(2)}$ is that it can be naturally generalized to any $n \geq 2$: For a system of $n \geq 2$ qubits, a complete set of experiments such that each pair is either compatible or maximally incompatible is the one consisting of all n-fold tensor products of σ_x, σ_y, σ_z and $\mathbb{1}$. We will call this set $\mathcal{S}^{(n)} = (E^{(n)}, S^{(n)})$. From a practical point of view, a useful property of $\mathcal{S}^{(n)}$ is that performing the experiments $E_i^{(n)}$ for arbitrary n only requires assembling the devices needed for the experiments in $\mathcal{S}^{(2)}$ (for a particular example, see[11]).

$E^{(n)}$ has $4^n - 1$ experiments $E_i^{(n)}$. $S^{(n)}$ has

$$c(n) = \frac{1}{n!(n+1)} \prod_{k=1}^{n}(2^k + 1) \prod_{k=0}^{n-1}(2^n - 2^k), \qquad (5)$$

compatible sets $S_j^{(n)}$ containing a maximum number of elements.[9] Each $S_j^{(n)}$ contains $2^n - 1$ mutually compatible experiments, but only the results of $2^n - 2$ of them are independent, since the product of all of them is either $+1$ or -1, according to quantum mechanics. Therefore, each $S_j^{(n)}$ can be in

$$s(n) = 2^{(2^n - 2)} \qquad (6)$$

different states.

Non-contextual models cannot simulate all these quantum predictions. Indeed, the fraction of predictions which cannot be satisfied by a non-contextual model increases rapidly with n: it is $\frac{3}{15} = \frac{1}{5} = 0.2$ for $n = 2$, $\frac{252}{945} = \frac{4}{15} \approx 0.27$ for $n = 3$ and $\frac{157248}{385560} = \frac{104}{255} \approx 0.41$ for $n = 4$.[10]

The proof in[6] can be easily extended to arbitrary n. This means that the memory needed to simulate (i) and (ii) is the one needed to remember which compatible set $S_j^{(n)}$ the last experiments belongs to, and which are the results for all the independent elements in that set. This means that, for arbitrary $n \geq 2$, the memory is

$$m(n) = \log_2 \left[c(n) \times s(n) \right], \tag{7}$$

which is larger than 2^n for $2 \leq n \leq 29$; my computer produces "overflow" for $n \geq 30$. Therefore, the memory required to reproduce predictions (i) and (ii) grows at least *exponentially* with the number of qubits. In other words, the *density* of bits of memory per qubit required to simulate the results as pre-established *contextual* properties also grows exponentially with n. This means that, if there is a limitation to the density of memory a finite computer can have, e.g., 1 bit per qubit or 10^6 bits per qubit, at some point this limitation will make it impossible to simulate the predictions of quantum mechanics (assuming an unlimited number of qubits in the universe).

Even more appealing than the possibility of a computer with unlimited density of memory is to think that the information is not *in* the n-qubit system, but in the observer, and that the system is only carrying n bits of information.

5. The Contextuality of the Universe

Now we know what kind of computer the universe is unlikely to be (a non-contextual or a classical one), but apparently nothing has been said about what kind of computer the universe *is*. However, we have learnt something which is of fundamental importance: The universe is a *contextual* computer such that the precise amount of contextuality in every situation can be obtained from quantum mechanics. The only missing piece is which is the fundamental principle responsible for this exact amount of contextuality. We still do not have the answer, but we have made some progress.

Let us first explain how to quantify contextuality. Consider the predictions in (1). We know that they are satisfied by quantum mechanics but not by non-contextual models. This can be expressed by saying that, if

$$\begin{aligned}\kappa :=& P(ABC = 1) + P(abc = 1) + P(\alpha\beta\gamma = 1) \\ &+ P(Aa\alpha = 1) + P(Bb\beta = 1) + P(Cc\gamma = -1),\end{aligned} \tag{8}$$

where $P(ABC = 1)$ is the probability of obtaining results for A, B and C such that $ABC = 1$, then $\kappa \leq 5$ for non-contextual theories, while $\kappa \leq 6$ for quantum mechanics.

Now let us assume that $C = c = \alpha = -\beta = \gamma = \cancel{k}$ in (8). Then, we can define

$$\kappa' := P(AB = 1) + P(ab = 1) + P(Aa = 1) + P(Bb = -1). \qquad (9)$$

What are the bounds for κ'? The answer is $\kappa' \le 3$ for non-contextual theories and $\kappa' \le 2 + \sqrt{2} \approx 3.41$ for quantum mechanics. The first bound can be achieved by choosing 4 mutually compatible experiments A, B, a and b and preparing a state such that $A = B = a = b = +1$. The second, by preparing two qubits in the state $|\psi\rangle = \frac{1-i}{2}|\sigma_z = +1\rangle \otimes |\sigma_z = +1\rangle + \frac{1}{\sqrt{2}}|\sigma_z = -1\rangle \otimes |\sigma_z = -1\rangle$ and performing the experiments A, B, a and b described in Sec. 3. How to prove that no higher values are reachable? There is a simple method. Since $P(AB = 1) = P(A = B = +1) + P(A = B = -1)$, then κ' can be expressed as

$$\begin{aligned}\kappa' :=& P(A = B = +1) + P(A = B = -1) \\ &+ P(a = b = +1) + P(a = b = -1) \\ &+ P(A = a = +1) + P(A = a = -1) \\ &+ P(B = -b = +1) + P(B = -b = -1).\end{aligned} \qquad (10)$$

In any non-contextual model, every system must have precise values for A, B, a and b. Therefore, if the proposition "$A = B = +1$" is true for a specific system, then the propositions "$A = B = -1$", "$A = a = -1$" and "$B = -b = -1$" must be false. What is the maximum number of true values we can assign to the 8 propositions in κ'? As a simple inspection reveals, the answer is 3. And the same limit holds no matter how we choose the non-contextual values. Therefore, since the value of κ' is obtained by performing experiments on different systems, the highest value κ' can take is necessarily 3. The same method explains why $\kappa \le 5$ for non-contextual theories.

Quantum mechanics is different. In quantum mechanics, the maximum value is

$$\vartheta(G) := \max \sum_{i=1}^{p} |\langle \psi | v_i \rangle|^2, \qquad (11)$$

where the maximum is taken over all unit vectors $|\psi\rangle$ and $|v_i\rangle$, in any dimension, where each $|v_i\rangle$ corresponds to a proposition in κ' (or κ) and exclusive propositions are represented by *orthogonal* vectors. p is the number of explicit propositions: $p = 6 \times 2^2 = 24$ in κ ("$A = B = C = +1$", "$A = -B = -C = +1$",..., "$C = c = \gamma = -1$"), and $p = 4 \times 2 = 8$ in κ'.

However, the most curious thing is that both numbers have been used for a long time in graph theory. Let us construct a graph G in which vertices represent the propositions and edges connect those that cannot both be true. For example, the graph associated to κ is shown in Fig. 1, and the graph associated to κ' is a subgraph of this graph. Then, the non-contextual bound is the *independence number* of G, $\alpha(G)$,[12] and the quantum bound is the *Lovász number* of G, $\vartheta(G)$.[13] This was first observed in.[14] Remarkably, while computing the non-contextual maximum is NP-hard, the quantum one (the one which represents what happens in the universe) can be computed to arbitrary precision by semidefinite programming in polynomial time.[13] The question of whether this number naturally derives from some fundamental principle is still open.

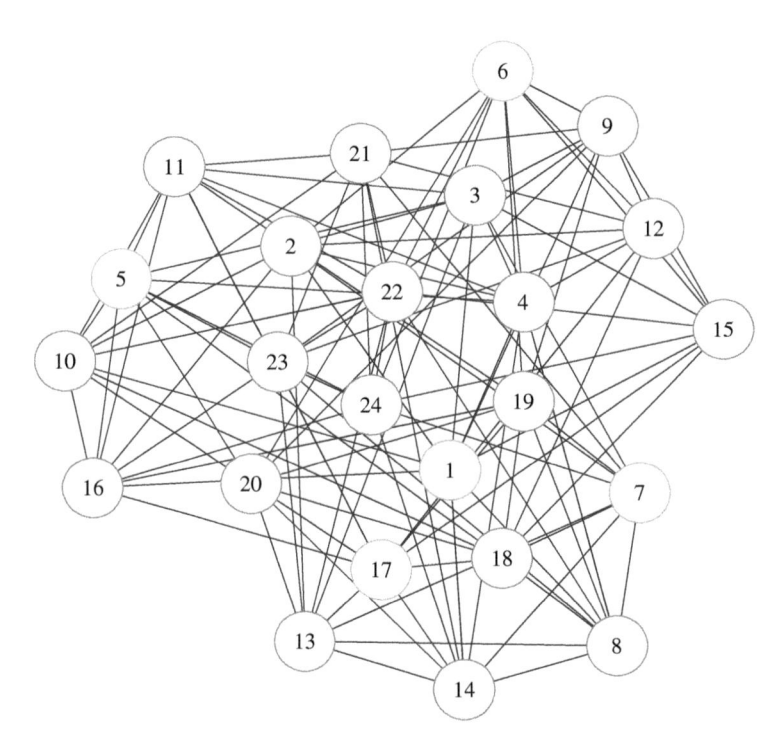

Fig. 1. Graph G representing all the explicit propositions in κ defined in (9). Each vertex represents a proposition and adjacent vertices represent propositions that cannot both be true. True/false assignments are represented by green/red-shaped circles. For G, the maximum number of propositions which can be true in a non-contextual model is $\alpha(G) = 5$. However, the maximum quantum value for κ is $\vartheta(G) = 6$.

Acknowledgements

This work was supported by the MICINN Project No. FIS2008-05596 and the Wenner-Gren Foundation.

References

1. J. S. Bell. On the Einstein-Podolsky-Rosen paradox, *Physics 1*, 195 (1964).
2. E. Schrödinger. Die gegenwärtige Situation in der Quantenmechanik, *Naturwissenschaften 23*, 807, 823 and 844 (1935).
3. A. Peres. Incompatible results of quantum measurements, *Phys. Lett. A 151*, 107 (1990).
4. N. D. Mermin. Simple unified form for the major no-hidden-variables theorems", *Phys. Rev. Lett. 65*, 3373 (1990).
5. S. Kochen and E. P. Specker. The problem of hidden variables in quantum mechanics, *J. Math. Mech. 17*, 59 (1967).
6. A. Cabello and J. J. Joosten. Hidden variables simulating quantum contextuality increasingly violate the Holevo bound, in *Unconventional Computation 2011, Lecture Notes in Computer Science 6714* (Springer, Berlin, 2011), p. 64. The memory needed for the case in which all results are generated in a deterministic way is discussed in M. Kleinmann, O. Gühne, J. R. Portillo, J.-Å. Larsson and A. Cabello, Memory cost of quantum contextuality, arXiv:1007.3650.
7. A. S. Holevo. Bounds for the quantity of information transmitted by a quantum communication channel. *Probl. Peredachi. Inf. 9*, 3 (1973).
8. A. Cabello. Proposed test of macroscopic quantum contextuality, *Phys. Rev. A 82*, 032110 (2010).
9. A. Cabello. in preparation.
10. C. Budroni, A. Cabello and A. R. Plastino. in preparation.
11. E. Anselem, M. Bourennane and A. Cabello. in preparation.
12. R. Diestel. *Graph Theory, Graduate Texts in Mathematics, 173* (Springer, Heidelberg, 2010).
13. L. Lovász. On the Shannon capacity of a graph", *IEEE Trans. Inf. Theory 25*, 1 (1979).
14. A. Cabello. S. Severini and A. Winter, (Non-)Contextuality of physical theories as an axiom, arXiv:1010.2163.

Chapter 32

A Gödel-Turing Perspective on Quantum States Indistinguishable from Inside*

Thomas Breuer

Research Centre PPE, FH Vorarlberg, Austria

By a diagonalisation argument, Bell states are not distinguishable from inside. This result is closely related to the theorems of Gödel, Church, and Turing in spite of important dissimilarities.

1. Introduction

Consider two entangled quantum systems A, B. Without loss of generality take A and B to be qubits. A and B are allowed to communicate on a classical channel the results of measurements performed by each of them. Can A or B, by themselves or in cooperation, immediately or with hindsight, discriminate between two different entangled states of the joint system $A\&B$? We argue that they cannot discriminate states ρ_1, ρ_2 of $A\&B$ whose partial trace over both A and B coincide, i.e. for which $\text{tr}_A(\rho_1) = \text{tr}_A(\rho_2)$ and $\text{tr}_B(\rho_1) = \text{tr}_B(\rho_2)$. These conditions are satisfied for example for the density matrices of the Bell states

$$\rho_1 = (|00\rangle\langle 00| + |11\rangle\langle 00| + |00\rangle\langle 11| + |11\rangle\langle 11|)/2 \qquad (1)$$

$$\rho_2 = (|00\rangle\langle 11| - |11\rangle\langle 00| - |00\rangle\langle 11| + |11\rangle\langle 11|)/2 \qquad (2)$$

(There are two more Bell states, which we will not use here. Our indistinguishability results are valid for any two of the four Bell states, and in fact for any two quantum states differing only in the entanglement between A and B.)

These internal indistinguishability results differ from other well-known restrictions of the distinguishability of quantum states. One such argument establishes that there is no quantum procedure to reliably distinguish non-orthogonal states, see e.g. Nielsen and Chuang [1, p.87]. Holevo[2]

---
*I am grateful to Thomas Schulte-Herbrüggen for helpful discussions on earlier versions.

established an upper bound on the information accessible in quantum systems.

Our result is an application of a more general argument[3] establishing limitations on measurements from inside. In order to keep the treatment self-contained, we give a sketch of that argument in Section 2. That argument uses a diagonalisation procedure. Measurements from inside are self-referential because they provide information about a larger system, which in turn has implications for the observer (or the apparatus) contained in it. Certain final states of the large observed system are impossible because they imply paradoxical self-reference.

It is a misconception that the resulting limitations on measurements from inside are due to the apparatus being "smaller" than the observed system containing it, and that the small apparatus could not discriminate all states of the observed system. Size is not the reason for the limitations on measurements from inside. A smaller system could discriminate all states of an external system with more degrees of freedom as long as the state spaces have the same cardinality. If the observed system is external in the sense that it does not contain the apparatus, a measurement does not give rise to self-reference and nothing prevents a discrimination of all states.

For *quantum* systems the limitations on measurements from inside are more serious than for classical systems. If system A or B is classical, then A could measure the state of B without limitations, and B could measure the state of A. By communicating their measurement results they could determine uniquely the joint state of $A\&B$. In the *quantum* scenario, however, entanglement prevents the unique determination of the state of $A\&B$ from the states of A and B. Section 3 develops this argument for the internal indistinguishability of the Bell states.

In Section 4 we discuss similarities and dissimilarities to Gödel's Theorem and the Halting Problem.

2. Restrictions on Measurements from Inside

In this section we review an argument[3] why it is impossible for an observer to distinguish all states of a system in which she or he is contained. The argument exploits self-reference properties, and it does not make any assumptions about the character of the time evolution.

Description of measurements   Let us assume that we have a physical theory whose formalism specifies for the systems it describes sets of possible states. In general these states may describe classical or quantum systems, and they may refer to individual systems or to statistical ensembles. We can think of the state set to consist of the density matrices and interpret a density matrix as representing the state of an individual quantum system.

A measurement performed by an apparatus A on some observed system O establishes certain relations between the states of A and of O. After a measurement, we infer information about the state of O from information we have about the state of A. We assume the states of A and of O refer to the *same* time after the experiment. To describe this inference, let us use a map I from the power set of the set $\mathcal{S}_A$ of apparatus states into the power set of the set $\mathcal{S}_O$ of system states. (Note that the curly $\mathcal{S}_A$ denotes the set of all apparatus states, whereas we use S_A for other sets of apparatus states.) The inference map I characterises the kind of measurement performed. I is defined as assigning to every set S_A of apparatus states (except the empty set) the set $I(S_A)$ of object states compatible with the information that the apparatus after the experiment is in one of the states in S_A.

This defines the inference map I which depends on the kind of measurement we are performing. I is different in different measurement situations. But when the observer chooses the experimental set-up, she also chooses a map I describing how she is going to interpret the pointer reading after the experiment. This map is fixed throughout the measurement. The states in $I(S_A)$ are the possible states of O after the experiment: They are the states of O compatible with the information that we do not know anything about the final state of A except that the measuremnt has taken place. In general not every state of O is compatible with the information that the measurement has taken place; not every state of O is a possible state after the experiment. We have $I(\mathcal{S}_A) \subsetneq \mathcal{S}_O$.

Knowing that if the apparatus after the experiment is in a state s_A, the observed system must be in a state in $I(\{s_A\})$, one infers from the information that the apparatus after the experiment is in one of the states in S_A that the state of the observed system must be in $\bigcup_{s_A \in S_A} I(\{s_A\})$. Therefore we have, according to the definition of I,

$$I(S_A) = \bigcup_{s_A \in S_A} I(\{s_A\}). \qquad (3)$$

Example 1 *Assume we measure an observable represented by a self-adjoint operator M with non-degenerate eigenstates $|m\rangle$ on a Hilbert space $\mathcal{H}_O$, and assume further that the value of the pointer is represented by a self-adjoint operator P with non-degenerate eigenstates $|p\rangle$ of the apparatus*

Hilbert space $\mathcal{H}_A$, and assume the eigenstate-eigenvalue link: An observable represented by a self-adjoint operator has an unambiguous value if and only if the system is in an eigenstate of that operator. Under these assumptions for each eigenstate $|m\rangle$ of the measured observable there is an eigenstate $|p(m)\rangle$ of the pointer refering to it. There may also be pointer eigenstates refering to none of the eigenstates of the observable. These pointer readings indicate that the measurement failed. So the map I is

$$I(\{|\psi\rangle\}) = \begin{cases} |m\rangle & \text{if } |\psi\rangle = |p(m)\rangle, \\ \emptyset & \text{else.} \end{cases}$$

From singleton sets we can extend I to arbitrary sets S_A of pure apparatus states by (3).

Example 2 When we measure an observable represented by a self-adjoint operator M with spectral decomposition $M = \sum_m M_m$, using an apparaturs with a self-adjoint pointer observable $P = \sum_p P_p$, a map I can be defined on the density matrices by

$$I(\{\rho_A\}) = \begin{cases} M_m & \text{if } \rho_A = P_{p(m)} \rho_A P_{p(m)}, \\ \emptyset & \text{else.} \end{cases}$$

(Again, there may be pointer readings refering to none of the eigenstates of the measured observable. These pointer readings may indicate that the measurement failed.) From singleton sets we can extend I to arbitrary sets of apparatus states by (3).

Example 3 More generally we can describe measurements by a collection $\{M_m\}$ of measurement operators which are not necessarily projections but satisfy the completeness relation $\sum_m M_m^\dagger M_m = I$. If the state of the quantum system O is ρ immediately before the measurement then the probability that result m occurs is given by $\text{tr}(M_m^\dagger M_m \rho)$ and if this result occurs the state of the system after the measurement is $M_m \rho M_m^\dagger / \text{tr}(M_m^\dagger M_m \rho)$. The pointer is described by a collection $\{P_p\}$ of operators also satisfying the complenetess relation. A map I can be defined on the density matrices by

$$I(\rho_A) = \begin{cases} M_m \rho M_m^\dagger / \text{tr}(M_m^\dagger M_m \rho) & \text{if } \rho_A = P_{p(m)} \rho_0 P_{p(m)}^\dagger / \text{tr}(P_{p(m)}^\dagger P_{p(m)} \rho_0), \\ \emptyset & \text{else,} \end{cases}$$

where ρ_0 is the initial ready state of the apparatus. Again, from singleton sets we can extend I to arbitrary sets of apparatus density matrices states by (3).

Discrimination of states We will say that an experiment with inference map I is able to *discriminate between the states* s_1, s_2 of O if there is one set S_A^1 of final apparatus states referring to s_1 but not to s_2, and another set S_A^2 referring to s_2 but not to s_1: $I(S_A^1) \ni s_1 \notin I(S_A^2)$ and $I(S_A^1) \not\ni s_2 \in I(S_A^2)$.

Example 4 *Consider a measurement performed by a qubit A on a two qubit system $O = B\&C$, as in Fig. 1. The state space $\mathcal{S}_A$ is the Bloch sphere, the state space $\mathcal{S}_O$ equals $\{\alpha \in \mathbb{C}^4 : |\alpha| = 1\}$. Since $\mathcal{S}_A$ and $\mathcal{S}_O$ are of the same cardinality there is a bijection between them. Any such bijection gives rise to an inference map I which discriminates all states of O.*

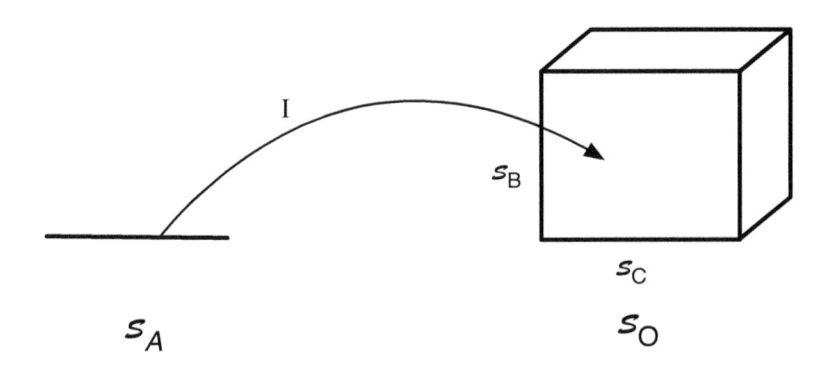

Fig. 1. **All states can be discriminated in measurements from outside.** The Bloch sphere representing the state space of A is depicted as an interval $\mathcal{S}_A$. The state space of $O = B\&C$ is represented by the cube $\mathcal{S}_O$, which contains the state spaces $\mathcal{S}_B$ and $\mathcal{S}_C$, as well as another dimension representing the entanglement between B and C. Any bijection between the interval $\mathcal{S}_A$ and the cube $\mathcal{S}_O$ is an inference map discriminating all states of O.

Example 4 illustrates two important conceptual points. First, it shows that a smaller system can discriminate all states of an external system with more degrees of freedom, if the observed system is external to the apparatus. Size is not the reason for the limitations on measurements from inside. The restrictions on state discrimination *from inside* are not due to the apparatus being "smaller" than the observed system. Second, the external system A can access the non-local information encoded in the entanglement between B and C. The unobservabilty of entanglement from inside is a result following from self-reference;[4] it is not an assumption entering into our description of measurements.

Measurements from inside Now consider the case where the apparatus is measuring a system in which it is contained (see Fig. 2). So the observed system O is composed of the apparatus A and of a residue B, $O = A \& B$. We assume that the observed system has strictly more degrees of freedom than the apparatus and contains it. This can be formulated in an *assumption of proper inclusion*:

$$(\exists \rho, \rho' \in \mathcal{S}_O) : R_A(\rho) = R_A(\rho'), \rho \neq \rho'.$$

Here $R_A(\rho)$ is the partial trace over B of O's density matrix ρ. It is the state of A determined by restricting the state ρ of O to the subsystem A.

Whether the assumption of proper inclusion is satisfied or not depends not only on the sets $\mathcal{S}_A, \mathcal{S}_O$ but also on the restriction map R_A. ([3] gives an example of two restriction maps such that the assumption of proper inclusion is satisfied with respect to one but not the other.) This may seem odd but it is not. An arbitrary subset of $\mathcal{S}_O$ can in general not be interpreted to be the set of states of a subsystem of B. The restriction map R_A gives physical information which is not reflected in the structure of the sets $\mathcal{S}_A$ or $\mathcal{S}_O$, namely the fact that A is a subsystem of O. That A is a subsystem of O does not only depend on the abstract structure of A (and of O), but on *which* system A is. If A and A' are isomorphic and A is a subsystem of O, it does not follow that A' is a subsystem of O.

The assumption of proper inclusion seems trivial in the sense that the bigger system O needs more parameters to fix its state. But it excludes situations in which each physically possible state of O is uniquely determined by the state of a subsystem A together with some constraint. (We take constraint to mean that states violating the constraint are physically impossible in the sense that the system can never be in such a state. Think for example of O as consisting of A and a mirror B reflecting exactly the state of A; in this case the assumption of proper inclusion is not fulfilled although A is a subsystem of O.)

A consistency condition The states of the apparatus after the measurement are self-referential: they are states in their own right, but they also refer to states of the observed system in which they are contained. This leads to a *consistency condition* for the inference map I which must be satisfied lest the inference map be contradictory: For every apparatus state ρ_A, the restriction of the system states $I(\{\rho_A\})$ to which it refers should again be the same apparatus state ρ_A. This can be written as:

$$R_A(I(\{\rho_A\})) = \{\rho_A\}.$$

for all possible post-measurement states ρ_A of the apparatus. This consistency condition is illustrated in Fig. 2.

From the physical point of view the consistency condition is not a restrictive requirement. Rather it is motivated by logic: It ensures that we cannot arrive at contradictory conclusions about the apparatus state. Assume that the meshing condition is violated and that therefore there is a state $\rho' \in I(\{\rho_A\})$ such that $R_A(\rho') \neq \rho_A$. Then, knowing that after the experiment the apparatus is in the state ρ_A, we would conclude that O is in one of the states in $I(\{\rho_A\})$, possibly in ρ'. From this in turn we conclude that A can be in the state $R_A(\rho')$, which contradicts the assumption that

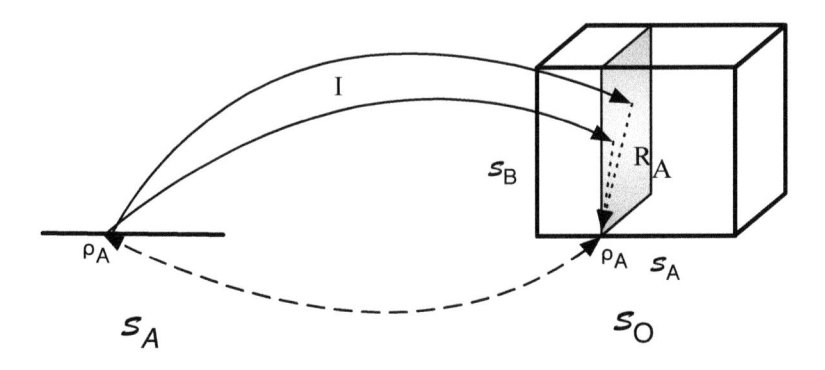

Fig. 2. **Not all states can be discriminated in measurements from inside.** The state space of A is represented by an interval $\mathcal{S}_A$, the state space of $O = A\&B$ by the cube $\mathcal{S}_O$, which contains the state spaces $\mathcal{S}_A$ and $\mathcal{S}_B$, as well as another dimension representing the entanglement between A and B. The grey plane represents the states of O whose restriction R_A to A (dotted line) coincides. The consistency condition requires of I that the projection R_A of any state in $I(\{\rho_A\})$ is again ρ_A. Therefore the bijection (represented by the dashed line) identifying the original $\mathcal{S}_A$ with the subspace $\mathcal{S}_A \subset \mathcal{S}_O$ has to be the identity map. The discrimination of two different O-states in the grey plane requires them to be in the image of two different A-states, which is prevented by the consistency condition. No bijection between the interval $\mathcal{S}_A$ and the cube $\mathcal{S}_O$ satisfies the consistency condition.

A is in the state ρ_A. Note that the consistency condition has to be imposed because both ρ_A and $R_A(I(\{\rho_A\}))$ describe the state of A at the same time.

Restricted state discrimination from inside The consistency condition and the assumption of proper inclusion imply restrictions on the distinguishability of states from inside. *For an apparatus A contained in O there is no inference map I, and thus no experiment, which can distinguish states of O whose restrictions to A coincide.* More precisely, if ρ_1, ρ_2 are two states of O fulfilling $R_A(\rho_1) = R_A(\rho_2)$, then there is no inference map I, and thus no measurement using as apparatus A, which can discriminate between ρ_1 and ρ_2: For all inference maps I satisfying the consistency condition, there exist no two sets of A-states S_A^1, S_A^2 such that S_A^1 refers to ρ_1 but not to ρ_2 and that S_A^2 refers to ρ_2 but not to ρ_1. This result is proved in[3] and illustrated in Fig. 2.

A particular kind of measurement from inside is induced by measurements from outside, because every measurement of A on an external system B can be interpreted as a measurement of A on $A\&B$. One can describe the measurement of A on B by an inference map I_B: If we know that A after the measurement is in some state in S_A, we infer that B is in sone state in $I_B(S_A)$. The measurement on B also provides information about $A\&B$, which is described by the inference map

$$I_{A\&B}(S_A) := \{\rho : R_A(\rho) \in S_A, R_B(\rho) \in I_B(S_A)\}. \qquad (4)$$

In this particular kind of measurement from inside the restrictions on state discrimination from inside are especially obvious. No measurement I_B from outside can induce a measurement $I_{A\&B}$ from inside which can discriminate between states ρ_1, ρ_2 whose restriction to A coincides. Any measurement I_B discriminating $R_B(\rho_1), R_B(\rho_2)$ would have to leave A in *different* states depending on whether B is in state $R_B(\rho_1)$ or $R_B(\rho_2)$. This contradics $R_A(\rho_1) = R_A(\rho_2)$.

The result on restricted state discrimination from inside does not only apply to measurements from inside induced by measurements from outside. It allows for arbitrary consistent inference maps I, which need not be of the form (4).

3. Inside Indistinguishability of the Bell states

Now let us return to quantum mechanics and the question of discriminating between the Bell states of eqns. (1) and (2). Tracing out the state of B we

get
$$R_A(\rho_1) = \text{tr}_B(\rho_1) = \mathbf{1}_2/2 = \text{tr}_B(\rho_2) = R_A(\rho_2).$$

The reduced states $R_A(\rho_1)$ and $R_A(\rho_2)$ are no longer pure states since $\text{tr}((\mathbf{1}_2/2)^2) = 1/2 < 1$. Tracing out the state of B leads to a loss of information. All this is well known.

The main result of this contribution is that *neither A nor B will be able to tell whether $A\&B$ is in the Bell state ρ_1 or ρ_2*. The proof is an application of the restriction on state discrimination from inside: Because $R_A(\rho_1) = R_A(\rho_2)$ and $\rho_1 \neq \rho_2$ the assumption of proper inclusion is satisfied. Restricted state discrimination from inside implies that there is no inference map I, and thus no measurement using as apparatus A, which can discriminate between ρ_1 and ρ_2. On the other hand we have $R_B(\rho_1) = \text{tr}_A(\rho_1) = \mathbf{1}_2/2 = \text{tr}_A(\rho_2) = R_B(\rho_2)$. This implies that there is no inference map I, and thus no measurement which B could perform in order to discriminate between ρ_1 and ρ_2. Neither A nor B can discriminate between the Bell states. Even if A and B exchange information about results of measurements they have performed either on themselves or on the other, neither A nor B nor the two together will not be able to tell whether $A\&B$ is in the Bell state ρ_1 or ρ_2, as we will argue below. Still, the Bell states ρ_1 and ρ_2 are different. An outside observer can distinguish them by measuring observables pertaining to both A and B.

It is the presence of entanglement which prevents the unique determination of the state of $A\&B$ from the states of A and B. If either A or B is classical, no entanglement is possible. This is easiest to see from Bell's inequalities in the formulation of Csirel'son[5], which read

$$|E(A_1B_1) + E(A_2B_1) + E(A_2B_2) - E(A_1B_2)|$$
$$\leq \sqrt{\text{tr}\left((4\mathbf{1}_4 + ([A_1, A_2][B_1, B_2]))\rho\right)}$$
$$\leq 2\sqrt{2}$$

for all density matrices ρ and all observables A_1, A_2 with eigenvalues ± 1 of system A and likewise B_1, B_2 of system B. For the density matrices of the Bell states the inequalities are satisfied with equality.

System A is classical if all observables of A commute, i.e. if $[A_1, A_2] = 0$ for all A_1, A_2. In this case $|E(A_1B_1)+E(A_2B_1)+E(A_2B_2)-E(A_1B_2)| = 2$ for all ρ, irrespective of whether system B is classical or quantum. No entanglement is possible. The same holds if B is classical. The density matrices ρ_A of A and ρ_B of B determine uniquely the state $\rho_A \otimes \rho_B$ of $A\&B$. This is the *only* state whose restriction to A is ρ_A and to B is ρ_B.

Therefore, if A or B or both are classical all states which differ can be discriminated by A and B in cooperation.

But if both A and B are quantum systems, there will be observables for which the Bell inequalities are violated. A can at best determine the state ρ_B of B, and B can at best determine the state ρ_A of A. When they communicate these measurement results to each other, they both know that $A\&B$ is in some state ρ with $R_A(\rho) = \rho_A$ and $R_B(\rho) = \rho_B$. But this does not determine the state of $A\&B$ uniquely. For example, the Bell states agree in both their restrictions to A and B. Therefore, if both A or B are quantum systems, there are states which cannot be discriminated by A and B, even if they cooperate.

The argument is illustrated by Fig. 2. States distinguishable by A are in the vertical grey plane. States distinguishable by B are in a horizontal plane. By communicating their mesaurement results A and B learn that the state of $A\&B$ is in the intersection of the two planes. Since this intersection is not a single point, A and B cannot not determine a unique state.

The indistinguishability of certain states *for each inside observer*, is specific to quantum mechanics, although restricted state discrimination from inside holds in classical physics as well. In the classical realm, for any two states there will always be some inside oberver able to discriminate between the two. But in quantum mechanics there are states like the Bell states, between which neither A nor B will be able to discriminate. In this sense quantum mechanical entanglement aggravates the self-reference problems of measurements from inside.

4. Similarities and Dissimilarities to Gödel's Theorem and the Halting Problem

Rosser[6] pointed out that the techniques used in the theorems of Gödel,[7] Church,[8] and Turing[9] are very similar. Kleene[10] showed Gödel's incompleteness theorem to follow from the insolubility of the halting problem. We conclude by pointing out some parallels and some differences between the above results on restricted state discrimination from inside and the theorems of Gödel, Church, and Turing.

The similarities can be summarised by the three pairs propositions–states, proof–measurement, and effectively generated proof–measurement from inside. (1) Statements about a physical system are statements about the state of the system. Propositions in formal systems and results of programmes correspond to physical states. (2) A measurement on a physical system establishes the truth or falsehood of some statements about the system. In this spirit, the state of a system is often regarded as a full or partial assignment of truth values. Both proofs and measurements are decision procedures: A proof establishes the truth or falsehood of statements in a formal system and corresponds to a measurement on a physical system. (3) A measurement from inside is self-referential, because its result has implications for the state of the observer. The arithmitisation of syntax achieved by the Gödel numbering, together with the restriction to effectively generated proofs, allows for the formulation of self-referential statements which are undecidable.

But there also important dissimilarities: First, the consistency condition for measurements from inside is different from the assumption of ω-consistency in Gödel[7] resp. consistency in Rosser.[11] Second, both quantum mechanics and classical mechanics use continuous state spaces, while in the theorems of Gödel, Church, and Turing propositions and programmes are referenced by natural numbers. The importance of this difference remains to be clarified. Perhaps it is not as important as it seems, since indistinguishability for measurements from inside holds for both discrete and continous sets of states. Third, indistinguishability for measurements from inside requires no substantial assumption about richness of state set other than the assumption of proper inclusion. In the theorems of Gödel, Church, and Turing the asumption about the formal system being rich enough for natural numbers is essential.

Turing[12] famously pointed to the fallacy of assuming that everything that can be known in principle is known actually and immediately. The difference of the two became the subject of complexity theory. The inside indistinguishability of the Bell states is a matter of principle, as are the theorems of Gödel, Church, and Turing. We have no insight to offer about the efficient distinguishability of states, except that the in principle indistinguishability of the Bell states from inside implies efficient indistinguishability from inside. The Bell states are key in quantum teleportation, as well as in quantum algorithms involving the Hadamard gate or Fourier transforms, like the Deutsch-Jozsa[13] algorithm. In this field the implications of the inside indistinguishability of the Bell states remain to be seen.

References

1. M. A. Nielsen and I. L. Chuang. *Quantum Computation and Quantum Information, 10th Anniversary Edition.* Cambridge University Press, 2010.

2. A. S. Holevo. Statistical problems in quantum physics. In J. V. Prokhorov G. Maruyama, editor, *Proceedings of the Second USSR-Japan Symposium on Probability Theory*, volume 330 of *Lecture Notes in Mathematics*, pages 104–119. Springer, 1973.
3. T. Breuer. The impossibility of accurate state self-measurements. *Philosophy of Science*, 62 (2):197–214, 1995.
4. T. Breuer. Subjective decoherence in quantum measurements. *Synthese*, 107:1–17, 1996.
5. B. S. Csirel'son. Quantum generalisation of bell's inequality. *Letters in Mathematical Physics*, 4:93–100, 1980.
6. J. B. Rosser. An informal exposition of proofs in Gödel's theorem and Church's theorem. *Journal of Symbolic Logic*, 4:53–60, 1939.
7. K. Gödel. Über formal unentscheidbare Sätze der Principia Mathematica und verwandter Systeme. *Monatshefte für Mathematik und Physik*, 38:173–198, 1931.
8. A. Church. An unsolvable problem of elementary number theory. *American Journal of Mathematics*, 58:345–363, 1936.
9. A. M. Turing. On computable numbers, with an application to the entscheidungsproblem. *Proceedings of the London Mathematical Society*, s2-42:230–265, 1937.
10. S. C. Kleene. Recursive predicates and quantifiers. *Transactions of the American Mathematical Society*, 53:41–73, 1943.
11. J. B. Rosser. Extension of some theorems of Gödel and Church. *Journal of Symbolic Logic*, 1:87–91, 1936.
12. A. M. Turing. Computing machinery and intelligence. *Mind*, 59:433–460, 1946.
13. D. Deutsch and R. Jozsa. Rapid solution of problems in quantum computation. *Proceedings of the Royal Society A*, 439:553–558, 1992.

Chapter 33

When Humans Do Compute Quantum

Paola Zizzi

Department of Psychology, University of Pavia,
Piazza Botta, 6, 27100 Pavia, Italy
paola.zizzi@unipv.it

We suggest that the ordinary human thought can have both the classical and the quantum computational modes. Due to massive parallelism, the quantum computational mode is exponentially faster than its classical counterpart, thus it can account for some extremely rapid mental processes, which humans are generally not aware of. The quantum mode is described by a formal Quantum Object-Language (QOL). The associated (quantum) deductive calculus has full access to the hidden quantum information and can reproduce, step-by step, the whole quantum computational process. The classical computational mode derives from the quantum mode through decoherence, and concerns conscious mental processes. The conscious mind, although it benefits of the very rapid data outputting of the preceding quantum mode, can grasp only a bit of the hidden quantum information. There is also a non-algorithmic mode concerning the meta-thought, which accounts for intention, intuition, and control, and stands at the roots of the unconscious mind The non-algorithmic mode is logically described by a formal Quantum Meta-Language (QML), which "talks about" the QOL. Both the quantum mode and the non-algorithmic mode are physically described by Quantum Physics. However, while the propositions of the former are described by qubits states of Quantum Information, the assertions of the latter are interpreted as quantum fields of a Quantum Field Theory of the brain.

1. Introduction

Most of the misunderstandings and misconceptions which arise when we relate the human mind to logic and computation, are due to the fact that, basically, human logic is not Aristotelian classical logic. The latter is in fact too abstract (in the sense that resources can be used as many times as one likes) and too much structured (that is has structural rules, beside logical rules) to represent the human thought. In fact the mind operates in a very simple way at the fundamental level (that is, when performing its fundamental tasks). However, the mind is also able to elaborate structured, abstract mathematical constructions (like Aristotelian classical logic) at a higher level of complexity. Then the mind must be described by a weaker (that is, with a fewer structural rules) and less abstract logic, than classical logic, which should, nevertheless, have extensions (among which, classical logic). Looking for such a "natural" logic of the mental processes is very important for understanding the human thought (the mind). In the constructivist approach[1] logic is in fact a by-product of the mind.

In this regard, there are two different points of view, the microscopic and the macroscopic ones.

According to the latter, one might start from the phenomenology of the "thinking processes" studied in cognitive science, where, however, a unified theory is still missing. Also philosophy of mind in general takes the macroscopic point of view. Constructivist approaches seem to rely on the cognitive and/or social interpretation, with a great support from philosophy of mind.

Within the macroscopic point of view, we think that Basic Logic (BL)[2] is the best candidate for a "natural" logic of the mind, in a foundational sense, that is, a logic which can describe the basic mental operations. In fact BL, which is the weakest logic, stands for the platform of all other logics, its extensions, which are too abstract and structured to allow a representation of the most fundamental thought processes.

The former point of view (the microscopic one) relies on the quantum processes in the brain,[3,4] and can be formalized within quantum theory. The two points of views are not incompatible, they just concern different contexts. However, the microscopic point of view, which accounts for a mathematical formalism, that of Quantum Mechanics (QM) and Quantum Field Theory (QFT) and can also display a formal language, is better suited to a logical study.

In this paper, we will adopt the microscopic point of view, in particular we will focus on quantum metalanguage, which is considered, here, an emergent feature of brain quantum processes.

This new kind of metalanguage leads to a new quantum logic, which describes the fundamental aspects of mind functioning at the quantum level, that is, the quantum computational mode of the mind. This new logic, called Lq,[5] can be considered a quantum version of BL and, consequently, a quantum approach to logical constructivism.

As far as the relation between mind and computation is concerned, one should take into account that a program can be viewed as a logic plus control. Therefore, once we assign a certain logical language L to the human thought, and a metalanguage ML controlling the object-language L, automatically we also assign a particular computational mode, and vice versa. Then if some aspects of the mind cannot be formalized by a deductive system plus a control language (ML), but rather, only by a control language, to them it cannot be assigned any computational mode, and are said to be non-algorithmic. This happens for those functions of the mind which control high-level thought.

In Penrose's view[6,7] the non-algorithmic side of the mind appears evident when a mathematician is able to recognize the truth of a Gödel sentence $G(F)$ of a consistent formal system F, although $G(F)$ cannot be demonstrated within F. This appears very reasonable in the context of metalanguage, where in fact assertions live. In a sense, when the mathematician recognizes the truth of the sentence $G(F)$, which is unprovable in the language L of F, he is using some mental processes, which can be described solely by the metalanguage, ML.

In this paper, we claim that there are two computational modes for the human thought, the quantum mode, described by Lq, and the classical mode, described by BL. Meta-thought, instead, which is described by a quantum metalanguage (QML), is non-algorithmic.

The quantum mode and the non-algorithmic mode concern those mental processes which in psychology are assumed to constitute the unconscious mind. Instead, the classical mode concerns those mental processes which constitute the conscious mind. In Freud interpretation, the unconscious mind was primary, with respect to the conscious mind. This can be physically explained by the fact the classical computational mode is driven by the quantum mode: the former in fact originates from the latter through decoherence of the qubit states. Also, Freud considered unconscious events to be not observable. This can be physically interpreted as the fact that we cannot observe a quantum state without destroying it.

In the quantum mode, humans perform very fast computations. In general, high-level mental processes are in fact very fast, as they concern complex issues. If those tasks were not performed so quickly in the quantum mode, the conscious thought would not be able to accomplish its own tasks, that is, recognizing solutions of complex problems in due time.

In the non-algorithmic mode, humans have intuitions and intentions, and can perform control on their own quantum computational mode. In fact, the non-algorithmic mode is described by a Quantum Metalanguage (QML),[5] which controls the Quantum Object-Language, which in turns describes the quantum computational mode.

The quantum computational mode of the mind can be described by the non-structural logic Lq of quantum information and quantum computation (for a review of quantum information see for example[8]), while the classical computational mode can be described by the sub-structural logic BL, which is said "non-classical", but is not a quantum logic.

There is also the non-algorithmic mode of meta-thought. Meta-thought is described by a quantum meta-language QML. The latter "reflects" into the QOL, which is Lq, through the reflection principle of BL.[2] By the reflection principle, all the logical connectives of BL are introduced by solving an equation (called *definitional equation*), which "reflects" meta-linguistic links between assertions into connectives between propositions in the object-language. The same happens in the quantum case, where the new logical connective "quantum superposition"[5] is introduced.

In the non-algorithmic mode, humans have intuitions and intentions, and can perform control on their own (quantum) logical language.

The non-algorithmic mode is then a function of the mind, which occurs before the QML has reflected in the QOL, which is a stage that can be better interpreted physically. We assume that QML is physically interpreted as a QFT of the brain,[3,4] while the QOL is interpreted as quantum information, that is, the Quantum Mechanics (QM) of qubits. Now, QFT has infinite degrees of freedom, and is not computable, that is it cannot be simulated by a Turing machine. Instead, a Quantum Mechanical system can be simulated by a quantum computer much more efficiently than by a classical computer. Dissipative QFTs, which are better suited to describe brain processes, can dissipate in such a way that their vacuum state has the symmetry of a global SU(2) (which is the rotation of a qubit).[9] Then dissipative QFTs of the brain can reduce to QM of information. Before reaching maximal dissipation, however, those theories are non-computable, and describe the non-algorithmic mode of the mind.

In Sect. 2, we illustrate the three modes of the human thought. In Sect. 3, we discuss Quantum Metalanguage, in relation with the reflection principle and the definitional equation. Sect. 4 is devoted to the conclusions.

2. Three Modes of Thought

We make the formal distinction between ordinary thought and meta-thought. Ordinary thought can be conscious or unconscious, and in both cases it is essentially computational, in a classical mode in the first case, and in a quantum mode in the second one.

The metalogic description of the quantum meta-thought is Quantum Metalanguage, which is the language speaking about the quantum object-language of the quantum thought.

The unconscious thought seem to be driven by mental processes which are very fast, much more than those concerning the conscious thought. This already suggests that the above processes are quantum-computational, as a quantum computer is much faster than its classical counterpart. Moreover, the sudden decision-makings or understandings, creativity, imagination and discoveries arising from an unconscious state of the mind, are just the results of a given mental process, whose intermediate steps, however, remain unknowable. This is also a fact well known in quantum computing: one can get the result of a computation with a given probability, but the intermediate steps are not available. Then, these two features seem to indicate that the unconscious mind is indeed quantum-computational. The logic of the computational mode of the unconscious mind must be therefore the logic of quantum information. Such a logic, which has been developed in[5] is Lq.

If the unconscious really computes in the quantum mode, this means that it "prepares", at highest speed, what we then recognize as a conscious thought. The conscious thought derives from a choice (a projective measurement) made on the quantum computational state, and thereafter uses a classical mode. In Lq, the quantum cut-rule, which is a meta-rule, can in fact be interpreted as a projective measurement. As a meta-rule, the cut-rule is useless when given to a machine, and this implies that a quantum computer cannot perform a self-measurement, or in other words, cannot make a conscious, classical judgement about its own state.

We humans, who can use a meta-rule, can, instead, make the conscious judgement, and make our quantum mode decohere. However, we don't have much time to re-elaborate the outputs of the unconscious mind (half a second) so that our conscious thought looks more like a succession of flashes of consciousness rather than a proper classical computation.

We use the partial information obtained from a given number of quantum measurements, but in effect, we do not compute anything new. In fact, most of the time humans compute in a quantum mode.

Let us consider, for example, the demonstration of a theorem. If the demonstration is particularly difficult and complex, it will asks for an extremely rapid mental process, otherwise it would take too long. We are almost unaware of this extremely rapid mental process (we will have only a partial information of what is going on). This is very specific of quantum computation, which is exponentially faster than its classical counterpart, and does not allow us to retrieve all the hidden information involved. We cannot follow all the quantum computational steps, we just get the result with a certain probability. Once we have proven the theorem, the quantum computer of our unconscious mind has just decohered. The next step is to use the output, and reformulate the demonstration of the theorem in classical terms.

However, this distinction between the quantum mode and the classical mode is too sharp. It may happen that when making the calculations in the classical mode, we need some extra hints from the quantum mode to avoid difficulties or find a short-cut, or settle problems. There are several interruptions and phases in the resolution of a complex problem, which indicate that we need some more quantum computation or, in other words, some more unconscious elaboration.

The third model is that of meta-thought, which is non-algorithmic. Meta-thought is the process of thinking about our own thought. It has no computational mode, neither classical, nor quantum.

Quantum meta-thought, which thinks about the quantum, unconscious thought, can be viewed as the roots of the unconscious mind. It is the side of the mind most closely related to physical processes in the brain, which are supposed to be described by a Dissipative Quantum Field Theory (DQFT),[3] which is, in fact, non-computational.

Meta-thought deals with intuition, intentions, and control.

Intentions, in fact, can be hardly conceived as mental states. Namely, the concept itself of mental state is useless when dealing with adaptive mechanisms, such as intentions, which underlie a number of control processes, in turn acting on mental operations, such as reasoning, deciding, recalling, and the like. If we generically denote the whole set of usual mental operations through the word 'thought', we should denote intentions, and other control mechanisms, through the word 'meta-thought', to stress the fact that the latter acts on, controls, and drives the ordinary thought.

Meta-thought processes could be interpreted as aiming to keep some sort of equilibrium or, more in general, of coherence. Therefore, in order to describe meta-thought and intentions, the problem is to find what are the best models, as regards both the physical basis of these processes and their logical nature. Of course, the generally adopted solution of this problem consists in resorting to classical physics. However, the latter is ruled out by theoretical arguments as well as experimental findings. On the theoretical side, we know from long time that classical physics is not endowed with coherence-keeping mechanisms. The latter are forbidden by Second Principle of Thermodynamics or, what is equivalent, by the so-called Correlation Weakening Principle, stating that whatever long range correlation will die away after a long enough evolution time. On the contrary, Quantum Theory appears as endowed with powerful coherence-keeping mechanisms, whose efficiency is, in rough terms, due to the fact that within it whatever entity is not spatially and temporally localized but rather described by a probability distribution ranging over the whole space-time. Thus, the superposition of different probability distributions associated to different entities gives rise to a sort of long range correlation between these latter which counteracts the disturbing influences produced by heat, noise, and other coherence-destroying mechanisms. It is, however, to be remembered that the expression 'Quantum Theory' is too generic. Namely we currently have two different kinds of Quantum Theories: Quantum Mechanics (QM), dealing with fixed numbers of particles lying within finite space volumes, and Quantum Field Theory (QFT), in which the field strengths are the basic entities, and infinite volumes as well as processes of creation and destruction of particles are possible. While in QM we have a finite number of degrees of freedom, QFT is characterized by an infinite (and continuous) number of degrees of freedom. Both in QM and in QFT the mathematical entities describing physical quantities must fulfil suitable constraints, expressing the non-classical nature of these theories and often called canonical commutation relations (CCR). Once given a physical system, a particular choice of the description of its dynamics, provided it fulfils the CCR, is called a "representation" of the CCR. Now an important theorem of QM, proved many years ago by von Neumann , states that within it, all different representations of the same physical system are unitarily equivalent. This means that in QM all representations of a given physical system have the same physical content. However, this no longer true in QFT, as shown already in the Sixties. This circumstance entails that within the latter theory the different descriptions of the same physical system can be unitarily non-equivalent, that is describing different kinds of physics. These considerations entail that QFT is well suited to describe the emergence of meta-thought. Such a circumstance is at the basis of a number of Quantum Brain Theories,[3] in turn relying on a firm experimental evidence about the quantum nature of physical phenomena underlying mental processes.[10,11]

3. Quantum Metalanguage

Metalanguage is quite an old subject in logic, which rests at the roots of semantics and model theory. Propositions can be defined as the sharable objects of the attitudes and the primary bearers of truth and falsity. If the language under discussion (the object-language) is L, then the meaning of propositions is in another language known as the metalanguage, ML. The metalanguage should contain a copy of the object language (so that anything one can say in L can be said in ML too), and ML should also be able to talk about the sentences of L and their syntax. Finally Tarski[12] allowed ML to contain notions from set theory, and a predicate symbol *True*.

A classical metalanguage (ML) is made of assertions, linked in a metalinguistic way. The difference between a QML and a ML is that in QML atomic assertions carry assertion degrees, which are complex numbers, interpreted as probability amplitudes. Also, the QML is equipped with Meta Data, corresponding to the constraint that probabilities sum up to one. The *reflection principle* of Basic Logic[2] was used to recover QOL from QML. By the reflection principle, all the logical connectives are introduced by solving an equation (called *definitional equation*), which "reflects" meta-linguistic links between assertions into logical connectives between propositions.

The QOL derived from the QML through the *reflection principle*, is made of propositions linked by quantum connectives, like, for instance, the connective "quantum superposition" (the quantum analogous of the classical connective "AND"), which is labelled by complex numbers, and is non-commutative.

In summary, we suggested that the mind has three modes: the non-computational mode (QML), the quantum-computational mode (QOL) both describing the Quantum Mind (or unconscious), and the classical-computational one, describing the Classical Mind (or consciousness).

Formal languages are sometimes called object-languages. The language used to make statements about an object-language is called a metalanguage. In this Section, we will illustrate the main difference between a classical metalanguage and its quantum counterpart. It is such a difference that limits the use of a classical metalanguage, which allows, by the reflection principle of Basic Logic, to introduce only classical logical connectives. Instead, a quantum metalanguage can be used to introduce quantum logical connectives, like the connectives of quantum superposition and quantum entanglement.[5]

We recall that Basic Logic is a sequent calculus, a particular kind of deductive system where the building block is the sequent: $\Gamma \mid- \Delta$, where Γ and Δ are sequences of atomic propositions, and are called the antecedent and the consequent of the sequent, respectively. The symbol $\mid-$, called turnstile, means "entails". The turnstile is a semantic object, which belongs to the metalanguage.

Atomic assertions, like $\mid- A$ (A asserted), where A is a proposition A of the object-language, belong to the ML. Also compound assertions, like $\Gamma \mid- A$, or $\mid- A$ and $\mid- B$ (where and is a metalinguistic link) belong to the ML.

3.1. The reflection principle

All the connectives of Basic logic satisfy the principle of reflection, that is, they are introduced by solving an equation (called *definitional equation*), which "reflects" meta-linguistic links between assertions in the ML into connectives between propositions in the OL.

Object-language $\xleftarrow{\text{Reflection}}$ **Metalanguage**

Logical connectives $\xleftarrow{\text{Reflection}}$ Metalinguistic links $\mid-$; "and"
(Between propositions) (between assertions)

The semantical equivalence:

$$A \ \# \ B \text{ true if and only if } A \text{ true link } B \text{ true}$$

which we call defnitional equation for the connective #, gives all we need to know about it. A # B is semantically defined as that proposition which, when asserted true, behaves exactly as the compound assertion A true link B true. The inference rules for # are easily obtained by solving the definitional equation, and they provide an explicit definition. We then say that # is introduced according to the principle of reflection.

All logical constants of basic logic are introduced according to the principle of reflection.

One might argue that intuition comes into play at the very level of the definitional equation, more precisely on the RHS of it.

The intention, or control function of the mind, seems to fit better with the LHS, instead.

3.2. *Classical metalanguage:*

Given n atomic propositions $p_0, p_1, ..., p_{n-1}$ of a formal language L, the (classical) metalanguage ML consists of :

i) The atomic assertions $|- p_i$ $(i = 0, 1, ..., n-1)$
ii) The meta-linguistic links ("yields", "and") among atomic assertions.

Example:

$$|- p_0 \ \& \ p_1 \quad \text{iff} \quad |- p_0 \quad \underline{and} \quad |- p_1 \tag{1}$$

Where & is the logical connective "and" between the two atomic propositions p_0 and p_1.

The formation rule for the connective & is:

$$\frac{|- p_0 \qquad |- p_1}{|- p_0 \ \& \ p_1} \tag{2}$$

3.3. *Quantum metalanguage*

Given n atomic propositions $p_0, p_1, ..., p_{n-1}$ of a formal language L, the quantum metalanguage QML consists of:

i) The atomic graded assertions $|-^{\lambda_i} p_i$ $(i = 0, 1, ..., n-1)$ with $\lambda_i \in C$. The complex numbers λ_i will be named "assertion-weights", which should not be confused with the truth-degrees in fuzzy logics, which are, of course, real numbers. What corresponds to the fuzzy truth degrees are in fact the squared modules of the assertion-weights (or assertion degrees) as it will become clear in the next sections. Moreover, one should notice that in all other logics, the assertion-degree is always 1.

ii) The meta-data $\sum_i |\lambda_i|^2 = 1$, which ensures that the global truth-value is 1.

iii) The usual meta-linguistic links ("yields", "and") among atomic weighted assertions.

Example:

$$|- p_0 \,_{\lambda_0}\&_{\lambda_1}\, p_1 \quad \text{iff} \quad |-^{\lambda_0} p_0 \quad \text{and} \quad |-^{\lambda_1} p_1 \qquad (3)$$

With $\lambda_0, \lambda_1 \in C$, meta-data: $|\lambda_0|^2 + |\lambda_1|^2 = 1$, $_{\lambda_0}\&_{\lambda_1}$ is the (quantum) logical connective "and" between the two atomic propositions p_0 and p_1, which takes into account the complex numbers λ_0 and λ_1 appearing in the QML.

At this point, the quantum connective $_{\lambda_0}\&_{\lambda_1}$ is formally introduced by the formation rule:

$$\frac{|-^{\lambda_0} p_0 \quad |-^{\lambda_1} p_1}{|- p_0 \,_{\lambda_0}\&_{\lambda_1}\, p_1} \qquad (4)$$

The logical connective $_{\lambda_0}\&_{\lambda_1}$ is that of quantum superposition.

4. Conclusions

We illustrated three different ways by which fundamental high-level mental activities do manifest themselves. There are two ones which are algorithmic: the classical computational mode, and the quantum computational mode. The quantum mode concerns extremely fast mental processes of which humans are mostly unaware of, and is logically described by the logic of quantum information and quantum computation, called Lq. The atomic propositions of Lq are interpreted as the basis states of a complex Hilbert space, while the compound propositions are interpreted as qubit states. Therefore, the physical model of the quantum mode of the mind is Quantum Information. The classical mode concerns those mental processes, which humans are aware of. It arises from the decoherence of the quantum computational state in the quantum mode, and is logically described by a sub-structural non-classical logic called Basic Logic (BL). In a sense, the quantum mode "prepares" the classical mode, which otherwise would take very long to perform even the easiest tasks, but most of the quantum information remains hidden. The classical mode comes into play by "flashes" of decoherence, which occur so often that humans get the impression of a "flow of consciousness".

The third mode, which is non-algorithmic, concerns the meta-thought (intuition, intentions and control) and is described by a quantum metalanguage (QML), which "controls" the logic Lq of the quantum mode. The assertions of the QML are physically interpreted as the (coherent) states of a QFT of the brain. The non-algorithmic mode describes then the most physical (with respect to mathematics) mental activity as it is closest to brain processes.

References

1. Beeson, M., *Foundation of Constructive Mathematics*. Springer, Berlin, 1985.
2. Sambin G., Battilotti G., Faggian C. (2000). Basic logic: reflection, symmetry, visibility. *The Journal of Symbolic Logic*, 65, 979–1013.
3. Jibu M., Yasue K., *Quantum Brain Dynamics and Consciousness: An Introduction*. Benjamins, Amsterdam, 1995.
4. Vitiello G. (1995). Dissipation and memory capacity in the quantum brain model. *International Journal of Modern Physics B*, 9, 973–989, 1995.
5. Zizzi, P., From Quantum metalanguage to the Logic of Qubits, PhD Thesis, arXiv:1003.5976, 2010.
6. Penrose, R. *Shadows of the Mind: A Search for the Missing Science of Consciousness*. Oxford University Press, 1989.
7. Penrose, R. *The Emperor's New Mind: Concerning Computers, Minds and The Laws of Physics*. Oxford University Press, 1989.
8. Nielsen, M.A., Chuang, I.L., *Quantum Computation and Quantum Information*. Cambridge University Press, Cambridge, UK, 2000.
9. Zizzi, P., Quantum Mind from a Classical Field Theory of the Brain. *Journal of Cosmology*, Vol. 14, 2011.
10. Tuszynski, J.A. (Ed.) *The emerging physics of consciousness*. Springer, Berlin, 2006.
11. Abbott, D., Davies, P.C.W., Pati, A.K. (Eds.). *Quantum aspects of life*. Imperial College Press, London, 2008.
12. Tarski, A., The semantic conception of truth. *Philosophy and Phenomenological Research*, 4, 13–47, 1944.

PART 5
Open Discussion

Chapter 34

Open Discussion on *A Computable Universe*

A. Bauer, T. Bolognesi, A. Cabello, C. S. Calude, L. De Mol, F. Doria,
E. Fredkin, C. Hewitt, M. Hutter, M. Margenstern, K. Svozil, M. Szudzik,
C. Teuscher, S. Wolfram and H. Zenil

Question from T. Bolognesi to Ed Fredkin:
As a premise, I wish to express complete support for your assumption that 'space, time, state and all other properties and processes of physics are finite, discrete and deterministic', and that the Discrete Theoretical Processes that 'run' the physical universe must be of a computational nature. The variety and complexity of phenomena that emerge from the computations of Cellular Automata [CA], or, in fact, of all Turing-universal models, have been widely explored in the last decades, from Zuse to Wolfram, and it is hard to believe that the strong analogies between the computational and the natural universe are purely accidental.

What I personally find a bit costly, however, is the assumption of a huge (infinite?) regular grid, or 3 + 1-dimensional Cartesian lattice, and associated program, as a prerequisite for starting the CA-like game of our universe. In order to justify the existence of this gigantic background structure, while avoiding the idea that it pops out from nothing, you suggest that there exists a sort of Super-Universe, that you call 'Other', in which such a grid is created, initialized, and started.

In doing so, you may have provided a more rational (and certainly original!) justification for the existence of the grid, but are you not simply moving the bootstrap problem one layer up, making it even harder? It might be that any cosmogonic theory will have to face some form of bootstrap dilemma, but one way to alleviate it is to stick to minimality criteria, and reduce as much as possible the gap between nothingness and the first instance of 'something' — the seed from which everything derives. Justifying the appearance of a Super-Universe from nothing seems more costly than assuming some simple initial condition for a universal computation. So, would it be possible/desirable to conceive the background grid itself, be it regular or not, as being incrementally produced, along with all the data that populate it, by the same universal computation that runs our familiar world, rather than imagining it built in an 'Other' place, and perhaps even run by an actual, 'Other' Super Computer?

Answer by Ed Fredkin:

There are a number of observations that lead to thoughts about what I call 'Other'. Our Universe has a number of characteristics that are essentially irreconcilable. There is a great deal of evidence that supports the 'Big Bang' theory of early events in the evolution of this universe. However, the idea that space, time, matter, energy and the laws of physics all appeared out of nothing is inconsistent with those same laws of physics. In addition, it defies common sense. One of the most fundamental of all laws of physics is reversibility. To believe some Big Bang cosmogonies, we have to also believe in the reverse process, where all of the matter and energy in the Universe collapse into a single point, and violating conservation laws, disappear into nothingness. It is also unreasonable to believe that the laws of physics existed independent of our physical Universe and somehow allowed for the origin of the Universe. Long ago, we had little choice in trying to explain the cosmogony problem; but now that we are more familiar with theoretical computation there is the possibility of more sensible candidates for models of the cosmogony problem. Many current models seeking to invent solutions are one or another form of fairy tale. However some fairy tales are more reasonable than others. Given the assumption that the ultimate and correct model of the most microscopic processes in physics is some kind of discrete space-time-state system, then we can invent the most straightforward and reasonable fairy tale, the one with the fewest and the least outlandish assumptions. If we were to find that there are discrete space-time-state models for microscopic physics, then that could open the door to possibilities supported by common sense. Given an accurate, deterministic, discrete, space-time-state model of microscopic physics, it would be true that any computer could model a small volume of space-time and do it exactly. In a discussion that took place between myself and Marvin Minsky in approximately 1960, we agreed that if such a model, running on an ordinary computer, was the correct model, the computed process would not merely be a simulation of physics, but it would indeed be physics. The point is to consider the necessary requirements or conditions for Other to be able to host a computer, most likely a Cellular Automaton (CA), capable of simulating our Universe (if space, time and state are discrete). First, Other does not have to have the restriction that our Universe has of having a beginning or an ending. It does not have to have our laws of physics. It does not have to have our kind of space or space-time. Computation is so general as to put almost no restrictions on the nature of Other except for a lower bound on the number of states that had to have been committed for the simulation of our Universe. The only requirement we put on Other is that it be able to host a CA that could run the process we call physics. The concepts of beginning and ending, which in one way or another seem to be a characteristic of our Universe, need not be a property of Other. There is no need for assuming that Other requires some kind of beginning. It may be hard for us to conceptualize worlds that do not conform to our physics, but it is clear that the demands made by the assertion that some system can support discrete computation are so minimal as to defy comprehension. Occam's Razor suggests that we look for simple explanations. A continually existing place– not in this Universe, not subject to the laws of our physics–that can create and run discrete processes capable of modeling this Universe may be the simplest explanation we will find.

Question from H. Zenil to M. Margenstern:

In your interesting paper you provide the encoding of a very small Turing machine as an example of something that, written in the letters of DNA, is smaller than any possible virus. This seems to suggest that if Turing universality requires so little, there are good chances that viruses (if not as individuals then as a set) may be capable of universal computation and therefore capable of potentially any possible threat to any possible solution. Is this what you are suggesting? If this is the case, I wonder what your position is on the pervasiveness of universality in connection to the (un)pervasiveness of undecidability in nature, in particular in biological systems. It would seem that while one may find evidence for the pervasiveness of computational universality, one does not find many identifiable examples of or candidates for natural undecidability in natural systems. Do you think there is any fundamental reason for this or is it that they have simply eluded our efforts to identify them? Charles Bennett, for example, enlightens us with a beautiful analogy which casts RNA polymerase as a specific purpose Turing machine, what amounts to a "truly chemical Turing machine" (from his "The Thermodynamics of Computation—a Review"), and more recently it has been suggested that the language of genes may have a computational power close to, if not the same as that of Turing completeness ("The language of genes" by David Searls in Nature). It seems we have very little to say about analogies with possible biological undecidability and how it may or may not play any role in biology. Would you have anything to say in this regard?

Answer by M. Margenstern:
I think that the pervasiveness of universality is high and if it does not appear to be so, the reason might be that people have not tried to investigate biological phenomena with this idea in mind. What we can see depends on what we expect. If people do not care about universality or undecidability, many facts or features that would point to it cannot be identified. I do not think that there are absolutely objective conditions: we observe nature and this observation is not at all neutral. First, it may involve an interaction with nature, even if we try to avoid it. Second, the observation is highly dependent on the culture of the observer, in particular on his/her language, on his/her ideas, among them, his/her opinion on what he/she expects to see. Fortunately, some of us are able to see things which were not foreseen, but this is not that frequent and when it happens, it is not always accepted. This is another aspect of the social conditions in which science evolves and these conditions act on scientists, all the more so if they are convinced that they are immune to such influences. To go back to our subject, the studies about universality and undecidability are not known well enough outside computer science circles and very restricted circles of mathematicians and physicists. So it is not surprising that many things connected with universality have not yet been observed. Now, I would like to point out that some studies on ciliates seem to suggest the fact that the well known pointer and stack mechanism used in programming is known in nature. It seems that the explanation of some mechanisms observed in the reproduction of ciliates shows some evidence of this. There are several works by Gzegorz Rozenberg on this very topic. Now, with two stacks and a not very big automaton, we get universality. Furthermore, there are very interesting works by Eshel Ben Jacob on colonies of bacteria. This is what I consider in the paper I submitted for your volume. The remarkable ability of such colonies to adapt to extremely severe conditions is something which should be investigated. It might be possible that dialogs between universal Turing machines could lead to super Turing problems. I really think that we should consider that recursively enumerable sets constitute the lowest level of life and that higher forms require us to go up in the hierarchy. Well,

I have no proof of what I claim. It is just a direction of thought. I believe it is meaningful. The future will tell us whether or not this is the case.

Comment by M. Margenstern on C. Calude, W. Meyerstein and A. Salomaa's chapter:

I am highly appreciative of the paper by Cristian Calude, Walter Meyerstein and Arto Salomaa entitled "The universe is lawless...". It is time to put an end to the dominion of the Big Bang theory. I do not understand how people using a mathematical model can claim that it expresses, even in its guidelines, the past of the universe. Mathematical theories make use of reversible time; how can they exhaust nature, which is not at all reversible? Let us note that in our discrete time of computations, time is irreversible: it is very often extremely difficult to run an algorithm backward. At the highest level of generality it is impossible. Let us note that the formal definition of sequential algorithms induces an irreversible time which is present in the very notations. As an example, an affectation is usually denoted as "var := expr;" and the notation ":=" explicitly indicates that what is on the left-hand side is not the same as what is on the right-hand side, and that there is a process, a consequence of which, after some time, is what we call an affectation. This is a very trivial example but we can see that in this way things are set on their feet and not upside-down. There are many things that could be said on this subject: how our mathematical theories do not give an intrinsic account of the notion of orientation (if you want to orient yourself in space you have to look at a watch; no mathematical theory can help you decide which orientation to choose between two possibilities), which is not the case in nature. So how can some people claim that our theories, which are an approximation of reality, allow us to define a time 0 from which everything starts? However, we can say that the present approximation constituted by mathematical theories is a very good one from a practical point of view. It is a rather convenient system for dealing with phenomena on earth and in space which is rather close to earth. Classical mechanics does not govern planets at the level of our solar system, as proved by the inaccuracy of predictions of the perihelia of Mercury made from within this framework. Well, this is not dramatic for planes in the air, ships on the seas, cars and trains on land. It is, however, interesting to note that GPS was rather recently made more efficient by computing the relativity effect of the speed of signals used to compute positions. It has not yet taken into account that the geometry of our space is probably not Euclidean. I also think that taking into account algorithmic problems in the computability or incomputability of some mathematical models could lead to better approximations of reality. To me, the opinion that "the universe has local rules only" is very interesting and should enhance our ability to study it.

Question from H. Zenil to T. Bolognesi:

You say that "We still lack experimental evidence for this conjecture," the conjecture being "complexity in nature = emergence in computation". It seems to me that the work of Wolfram and myself applying algorithmic probability as a theory of pattern generation from purely computational processes, based on the works of Levin and Solomonoff, is evidence in favour of the complexity in nature produced by computation, evidence that should not be completely overlooked even if it is not at all conclusive. Perhaps you were thinking of the smallest scales (which is likely since in your text you immediately turn to Planck units), where effectively we have no clue. On the contrary, there is an important discrepancy between the kind of complexity that can come out of quantum mechanics and the kind of complexity generated on the macroscopic level, which is deterministic in principle (as all the laws of physics are reversible).

So my question to you is what elevates a model, hypothesis or theory above the subjective, the aesthetic or the metaphorical, to use your own words? Is it their explanatory and predictive power? In which case, algorithmic information theory predicts the nature and frequency of certain patterns in the world, if the world were the result of computational processes. And while we don't know whether nature is a computational process, we do have an idea what a world that is a computational process would look like, which gives us a basis for making comparisons (and even predictions). Its explanatory power is apparently underestimated but it easily explains why structure can emerge even out of almost nothing or out of randomness. Science has been working since the beginning on this very same assumption, that the world is structured and that one can derive universal physical laws from reality, but without knowing much about why this is so.

So I wonder what would make you feel that a computational theory of the universe is correct? We know that 'computational' doesn't necessarily mean 'predictable.' What kind of verification tests could be conceived that would give them the status of a theory, would elevate them above fairy tales as models of the world?

Answer by T. Bolognesi:
You raise the fine point of predictability vs. irreducibility. The key element that discriminates scientific theories from other forms of knowledge, such as philosophy or religion, is the requirement of experimental validation, as taught by Galileo four centuries ago. I believe this fundamental will stay with us at least as long as the Aristotelian thinking that it has superseded. On the other hand, if the universe is animated by a computation, it seems reasonable to assume the latter to be irreducible, like that of elementary cellular automaton number 30, and this leads to the impossibility of predicting its evolution faster than in real time. However, I do not think that the consequences of this state of affairs are fatal for science. The ultimate, discrete ToE [Theory of Everything] (if any) should describe, with infinite precision, the evolution of the universe as a *whole*. Now, it is one thing to be unable to predict, in less than 10 years, what the whole universe will look like in 10 years — and emergent phenomena are indeed unpredictable, almost by definition, so that no scientific theory will ever be able to describe in advance, say, the future shapes that life will assume in our galaxy. It is another thing to be able to predict the exact mass of a few hundred elementary particles: the amount of information involved here is much smaller, and their derivation from a hypothetical universal computation, once and for all, is conceivable. That would constitute very convincing experimental evidence for the validity of the 'theory'!

Although the most important test for the validity of a scientific theory is the prediction of measurable physical quantities, with the highest numeric precision, I believe that other, more qualitative forms of predictive power can also be considered, at least for acquiring some degree of confidence as to the validity of 'an approach', if not of a specific theory. For example, a computational theory that, starting from completely abstract mathematical/logical rules, yields the emergence of localized structures whose interactions bear analogies with those of known physical particles, should be regarded with more favor than one producing white noise all the time.

In a way, the brilliant idea that you mention — to spot clues to algorithmic behavior behind the distributions of empirical data — would also fit in this broader picture. The key observation at the core of that proposal seems to be that a random string looks random most of the time, while the output of a random Turing machine has greater chances of looking structured.

The big challenges I see in that approach, however, are: (i) how to reduce the immense variety of empirical data sets and associated probabilistic distributions available in the physical world to a single, representative empirical distribution; (ii) how to do the same in the computational world, where you may want subsystems to also interact, and where different layers of emergence may each be characterized by its own laws.

Furthermore, there are other approaches that attempt to explain the emergence of structure in physical systems. One has been investigated for decades by Stuart Kauffman. In his networks of Boolean functions, 'order for free' is very likely to emerge — in the form of surprisingly small attractors in the configuration space — whenever the network satisfies some reasonable topological conditions. It would be interesting to further investigate the relations between the 'order for free' obtained by Kauffman's generalized cellular automata, and that associated with Levin's miraculous distribution.

Question from H. Zenil to T. Bolognesi:

When you talk about persistent structures in, for example, Wolfram's Rule 110, as being analogous to physical elementary particles, structures in Rule 110 are supported by a completely regular background as you point out. Do you think this background, besides acting as a physical carrier for such structures to propagate, is also a necessary kind of ether supporting the computation itself? The question has to do with the way Turing machines are usually set to start out from a "blank configuration" and whether you think a blank configuration does or does not have a physical representation of computation in the real world. In the end, both Turing machines and cellular automata have, for example, their ultimate "physical" carriers—a tape for Turing machines and a grid for cellular automata–yet one can define both formalisms with pure functions without having recourse to any substratum. Could you elaborate on this in the context of a discrete space-time? Does it mean that if discrete, the world can also be just mathematical?

Answer by T. Bolognesi:

You mention (without necessarily sharing) the viewpoint that, since models of computation such as Turing machines and cellular automata can characterize the computable functions, we may directly refer to the latter for a fundamental theory of physics, and conclude that 'the world is purely mathematical'. I always feel 'uncomfortable' with this idea. The main reason is that, for a Turing machine to express a mathematical function, you need to provide it with a halting state, and to look at the final tape configuration only when the machine halts, while fully ignoring all the interesting things that have happened along the way.

On the contrary, for a Turing machine to express an evolving, artificial physical universe, you don't need to include halting states at all, and you concentrate instead on the 'shape' that the computation assumes in its detailed, step-by-step unfolding. In this respect, I totally support Fredkin's idea of a one-to-one correspondence between state and function in the physical world and state and function in the computational model: nothing in the computation should be deprived of a physical counterpart, and vice versa. This is one way to contrast the computational vs. the purely mathematical universe concepts: the physical universe (not the devil!) hides in the details—in every single detail of the computation, not in the final outcome.

You also raise the interesting point of the regular support, or substratum (cell tape or array) required by Turing machines and cellular automata for carrying out their computations, and of the periodic background that seems to be required for artificial computational particles to come into existence and play their games.

The main reason why I consider the two models mentioned only as useful *metaphors* for a computational theory of physics is precisely that they require this cumbersome, predefined support. I look with much more favor upon graph-rewriting models, such as trinet mobile automata (TMA), where the regular substratum is not assumed a priori, but can be expected to emerge from the computation itself, along with the particles that (may eventually) populate it, often starting from a tiny, two node graph. My experiments with TMA computations clearly indicate that many of the phenomena observed with cellular automata can also occur in the causal sets (intended as spacetime instances) obtained from this model, and the emergence of a regular background is certainly a most common one.

Question from C. Teuscher to M. Hutter:

A computational paradigm that does not allow for "programming" is pretty much useless. I wonder if a computational theory of everything must also include a "programmability of everything" in order to be useful?

Answer by M. Hutter:

All considered CToE candidates (q, s) are computable. A pair (q, s) is an algorithmic description of the observations of a subject in a universe algorithmically generated by q. Whether (q, s) is a correct CToE, i.e. correctly "explains" all observations, is semi-decidable by executing (q, s).

Even the search process for better and better (i.e. smaller and smaller) CToEs itself is effective by standard program dovetailing.

And to top this, this process will find the best (absolutely shortest, given the evidence o) CToE in finite time.

What we cannot hope for is that this procedure halts thereafter (due to the halting problem). So we can never be sure we have found the best CToE, but likewise future observations have the potential to invalidate any (C)ToE found by any means.

Also in practice, more intelligent ways than brute-force search and simulation are of course used for finding good CToE candidates and checking their consistency with observations, which actually can still be a computational process.

I guess this constitutes what Christof Teuscher calls the "programmability of everything" (modulo computational limitations of course).

Question from C. Teuscher to F. Doria:

Ideal analog computers don't exist in nature because every physical quantity ultimately has a limited precision. How does your approach avoid that fundamental limitation to compute beyond the Turing limit?

Answer by F. Doria:

I've been trained as an engineer. Engineers build things, given some blueprint or some reasonable technical description. So, as the construction of an analog computer that would–ideally–settle undecidable stuff is within our reach, I want to build one and see how it performs and especially how it fails.

There are a few points I'd like to stress:

- The main theoretical difficulty when dealing with the halting problem are the nonstopping values ("infinite loops") in the algorithm. But in actual circumstances we are only interested in a finite number of such values. It is known that in such a case there will always be an algorithm to do it, albeit a possibly very complex one.
- I wonder whether some real hypercomputer which performs well within some finite range can actually be used to speed up computations. That would be the main practical interest in building such a machine.

There are a few ideological (if I may say so) objections which deserve comment. For instance, "the universe is discrete and analog stuff are continuous." One usually points out that Schrôdinger's picture of quantum mechanics deals with a smooth object, the wave function. And so on. But the possibility that the universe is essentially discrete wouldn't affect the building of a practical machine, as I have suggested.

Why has nobody tried to build it and see how it performs? After all, the proof of the pudding is in the eating!

Question from H. Zenil to C. Calude:

If the universe is lawful as you argue in your contribution to this volume, with the fair point you make about our models not fully matching nature, how could we account for the fact that many of these models actually capture much, if not all, the behaviour of many natural systems? The question could be rephrased by evoking Eugene Wigner's claim about "the Unreasonable Effectiveness of Mathematics in the Natural Sciences".

There is also the idea (popularised by Paul Davies, but originating with others, for example John A. Wheeler, if I'm not mistaken) that physical laws may change over time, just as physical constants may over the course of the history of the universe (so not all physicists embrace the idea that laws are unchangeable and true forever, even if most do). Of course our lifespan is so short that we are used to thinking of physical laws as perennial and immune to change, which is reasonable because were the laws to change we would not be able to make sense of the past in terms of the present (or make sense of the future). But assuming that laws may change, things like the inflation period of the universe or dark matter may find a reasonable alternative explanation (likewise the question of the tuning of the universe).

The beauty of the computational/algorithmic view from my vantage point is twofold. On the one hand there is the lesson learned from Turing's work that data and programs are essentially not very different because one can always be encoded in the other. Hence there would be no need to separate them in a theory of laws+state. On the other hand, we find this beautiful convergence in computational power among different, initially unrelated models of computation, allowing us to discuss underlying general principles of physical reality under a single assumption, rather than having to posit different theories for different natural phenomena. In the final analysis science has been quite successful in merging phenomena that seemed at first to be unrelated, showing them to be governed by increasingly more general but accurate physical laws. Also, it seems to me that computational views have the advantage of not being contaminated by additional, apparently unnecessary assumptions, such as the presumption of the existence of real numbers, even if we cannot decide whether they exist or not by generating models because the answer only belongs to physics.

I think there is a distinction between whether the Universe is epistemologically or ontologically lawful. I think you are suggesting that the Universe is *ontologically* "lawful", but most of your arguments are for an *epistemologically* "lawful" universe. As you say, it is one thing to think there is no cause and another to think that the cause is unknown or unknowable. You propose that if the digital views are subject to the lawfulness arguments it is because they cannot be distinguished from algorithmically random strings. In other words you suggest that we may be embedded in an accidental orderly finite string of an infinite random sequence. Isn't the effective probability of picking a regularity in an infinite Martin-Löf random sequence zero? Just like the chances of picking a rational number among all the real numbers ($\mathbb{R}$)? Doesn't this introduce a problem tantamount to the apparent fine tuning of the universe?

Answer by C. Calude:

The regularities and laws discovered and validated in science, in particular in physics, can be empirically demonstrated to have "local" validity, which is very impressive and useful. There is nothing mysterious or unreasonable in the fact that local regularities match natural phenomena well in a finite number of cases.

To put this in perspective, does the enormous empirical confirmation of the Riemann Hypothesis justify the statement "the Riemann Hypothesis is true"? Of course not!

Our point is that there is no such thing as a "law of the Universe" as long as we cannot prove it, and science cannot prove it. Why so? Because of the infinity which is intrinsic to any non-trivial law. For a better understanding of the nature of laws emerging from a scientific theory I quote the following paragraph from Deutsch's book *The Beginning of Infinity* (Penguin Books, London, 2012, p. 4):

> ... scientific theories are not 'derived' from anything. We do not read them in nature, nor does nature write them into us. They are guesses—bold conjectures. Human minds create them by rearranging, combining, altering and adding to existing ideas with the intention of improving upon them.

This is why it is no surprise that "physical laws may change over time". A possible reading of science is as a history of failed theories.

"Isn't the effective probability of picking a regularity in an infinite Martin-Löf random sequence zero?" The answer is negative. An infinite Martin-Löf random sequence is full of regularities: it avoids only a countable set of regularities (given by Martin-Löf randomness tests). For example, in every infinite Martin-Löf random sequence any string appears with a very high regularity (Borel normality).

Understandably, scientists love their theories as parents love their kids. However, one should be careful about unreasonable love!

Reply by H. Zenil:

I think *local validity* is opposed to the concept of a physical law as conceived by physicists. A physical law is supposed to be valid anywhere (and perhaps anytime) in the universe. Whether this is or is not the case would make for an interesting discussion, but I think the success of science and in particular of physics is in the derivation of more general laws explaining a greater number of physical phenomena (hence the replacement of old theories by new ones).

On the other hand, I thought that the definition of a ML-random number was that its digits were random in relation to any effective algorithm. In the best case only a finite number of bits are computable in a ML-random number, just as for a Chaitin Ω number. But even worse, because there are random sequences of arbitrarily high Turing degree it seems one would need an Oracle to spot the regularities. Is it that from the outside the probability of seeing these regularities is 0, though the metaphor that the universe is like a ML-random number is valid because we may already be in a regularity? My main point is therefore why, if it is so hard to find such regularities, if this is a ML-random universe, do we seem to do so so often?

Reply by C. Calude:

An ML-random sequence is one which passes every ML-test of randomness, so only a countable number of tests. As there are an uncountable number of tests of randomness, it follows that any ML-random sequence doesn't pass an uncountable set of randomness tests. There is no sequence which passes all, even an uncountable set of, randomness tests, so in every sequence there are regularities, but these regularities are "invisible" to Turing machines.

A ML-random number avoids every c.e. regularity but has plenty of incomputable regularities. No scientific theory can find incomputable regularities, so it is lawless.

Reply by H. Zenil:

The math makes sense, but then it does not accord with the metaphor that the universe may be like a ML-random number if we find (apparently computable) regularities in the world all the time (it seems we find more computable regularities than irregularities–or as you call them uncomputable regularities).

Is your answer then that we see regularities (1) by chance (2) because we construct them as regularities (3) because we see these regularities despite scientific theories (and algorithms), which then is a form of negation of the CT thesis? Does any of these or a combination of them represent your view?

Reply by C. Calude:

We don't "see them" as the result of scientific investigation because they are incomputable. We only know they exist. If by some miracle we can "see" such a regularity, there is no way to prove it. A similar situation: I know there is a winning lotto ticket, but I cannot know which one is it; the knowledge helps, maybe, by motivating me to play lotto, but it does not help me in any direct sense to get the winning ticket.

Question from M. Hutter to C. Calude:

Common-sense laws are rules to govern behavior, are not necessarily causal, and do not enforce a unique/deterministic behavior. Precisely in this sense also the universe respects many laws and is very law abiding. This seems to disprove the lawlessness hypothesis.

The paper seems to hint at several hypotheses, or at least this was my impression of it. I try to formulate them below. I cannot say for sure whether the authors had any of them in mind. Some of them seem worth stating and discussing further. I would like the authors to indicate which of these hypotheses they subscribe to:

- (H1) physical laws are not causal [In my opinion, not worth discussing, since it's wrong]
- (H2) the causal laws could be re-interpreted in a non-causal way [has been investigated and possibly worth investigating further]

- (H3) some laws of physics are not causal [possibly worth investigating]
- (H4) our observable universe is an (accidentally) regular part of an (infinite) random universe/string. [this is reminiscent of what is known as the 'Boltzmann Brain' idea. The authors' concepts of disjunctive real numbers and lexicons seem reminiscent of the 'Library of Babel'. I have investigated the random universe idea and its problems in http://dx.doi.org/10.3390/a3040329.]
- (H5) Free Will undermines the causality/laws of our universe. [I'm skeptical but would be interested in hearing more on this]. In Section 6 I cannot find "Arguments in favour of the lawlessness hypothesis" for our physical universe, even when admitting the, in my opinion implausible, "free-will assumption". I actually subscribe to Footnote 6].
- (H6) The law that governs the universe is a random string, hence the universe itself is lawless. [Paraphrasing the authors. This seems wrong. Input (law) and output (universe) are not the same thing. Cf. H7.]
- (H7) Some aspects of the universe are random (like quantum mechanics), therefore the universe is lawless. [In my opinion, this is wrong. The output of most randomised algorithms exhibits some real, not just accidental, regularities]
- (H8) Some aspects of the universe are random, therefore there cannot be a Theory of Everything. [this is worth discussing and depends on what one requires from a ToE]

Answer by C. Calude:
If the "laws of the Universe" are considered in the following sense

> Common-sense laws are rules to govern behavior, are not necessarily causal, and do not enforce a unique/deterministic behavior

then you can rightly conclude that

> Precisely in this sense also the universe respects many laws and is very law abiding.

With this definition every common-sense regularity, no matter how deceptive it may be shown to be under a more careful scrutiny, qualifies as a 'law of the Universe'. If we accept this description, then the Universe is lawful. However, I see too little justification to accept it.

I do not subscribe to any of the hypotheses (H1)–(H8), although, as you correctly observe, (H4) comes close to our hypothesis.

Reply by M. Hutter:
(a) Do you believe that you/we can find or have found computable regularities in the part of the universe you/we observe?
(b) Do you believe these regularities are locally valid, but not globally since the total universe is globally ML random?
(c) Do you believe that the part of the universe you/we observe happens to be a regular sub-part of the total universe?

In case the answers to (a-c) are yes, your idea is just an ML version of the random universe with possibly the same problem and solution as discussed in http://dx.doi.org/10.3390/a3040329

The problem is essentially that it is a metaphysical rather than a scientific statement. If clearly expressed and marked as such, I have a-priori no problem with it, and find it even interesting to philosophize about, though I can't find any argument in favour of it in your paper.

So in question form:
(d) Do you believe your lawlessness claim is a scientific statement?
(e) Do you regard your lawlessness claim only as a metaphysical statement?
(f) Do you believe your lawlessness claim is a mathematical statement?

Here's some more context:
Since QM, most physicists do not subscribe to a deterministic universe anymore, since (QM) 'laws' do not determine the future uniquely and completely. It may sometimes still be presented like that, as e.g. in [28], if the difference is beside the point or the author is sloppy or ignorant or expresses a minority opinion. This of course undermines deterministic causality too.
(g) Is all that you claim that the universe does not respect deterministic and/or causal laws?

Note that an approximate law of the universe would still be useful and approximately valid. And there are ways to convert approximate laws into exact laws by coding the errors.

In Section 6 you (only) show that (a) in mathematics there is something like random numbers; (b) (With QM) one can (apparently) produce physical random sequences. This does not exclude the fact that the universe respects stochastic laws, in the sense that there is a high probability that it has certain properties(=laws?) (e.g. that the relative frequency of 1s in the above QM sequence converges to a limit).

Then you write in Section 8 that "the hypothesis can be experimentally illustrated and tested ... [12]" But [12] only investigates whether QM produces "true/ML" randomness. Unless you restrict "laws" to the deterministic, it doesn't increase the plausibility of your hypothesis

(another indirect indication that you reject stochastic laws for our universe). You seem to jump quite quickly (and, in my opinion, unjustly for several reasons) from "there is true randomness in the universe" to "the universe is lawless".

Final reply by C. Calude:

The lawless hypothesis is the negation of the current dominant hypothesis that the universe is lawful; it should be analysed in this context. Given the limits of this discussion, my previous answers together with the independent comments of Maurice Margenstern say it all.

Question from A. Cabello to C. Calude and K. Svozil:

About using quantum systems protected by value indefiniteness grounded in the Kochen-Specker theorem for generating random bits (p. 8 of their contribution): There are many ways in which a theory can have value indefiniteness (or contextuality of results), and quantum theory has a very peculiar form of value indefiniteness (which, incidentally, suggests that the universe is not lawless). By that I mean the following: A standard way to experimentally reveal nature's value indefiniteness (predicted by quantum theory) is through the violation of Bell inequalities. However, there are two different ways in which quantum theory may violate a Bell inequality: There are type-1 scenarios in which the maximum quantum violation is not the largest violation achievable by the no-signalling principle, for example in a Bell inequality with two parties, each with two settings, each with two possible outcomes [S. Popescu and D. Rohrlich, *Phys. Lett. A* 166, 293 (1992)]. In these situations, the observed correlations still can be expressed as a fraction of classical correlations (in which values may be defined) and a fraction of nonclassical ones [A. Elitzur, S. Popescu and D. Rohrlich, *Phys. Lett. A* 162, 25 (1992)]. In contrast, there are type-2 scenarios in which the maximum quantum violation is equal to the largest violation achievable by the no-signalling principle, and the observed correlations are fully nonclassical, for example in a Bell inequality with two parties, each with three settings, each with four possible outcomes [L. Aolita et al. *Phys. Rev. A* 85, 032107 (2012)]. The question is whether, for generating randomness, there is any difference between these two scenarios.

Answer from C. Calude and K. Svozil:

Adan Cabello rightly states that there are many ways in which a theory can have value indefiniteness or contextuality of results. We would like to stress that what is presently understood as *value indefiniteness*—the dependence of an individual (non)observed observable or element of physical reality on what is measured alongside it (its "context")—might be sufficient, but certainly not necessary, to interpret Kochen-Specker type theorems. Indeed, we consider this type of "explanation" very much inspired by the old quasi-classical realistic view of physics; and it is no wonder that John Bell, a staunch realist, is often quoted suggesting that [J. S. Bell, *Reviews of Modern Physics,* 38, 447–452 (1966)] "the result of an observation may reasonably depend ... on the complete disposition of the apparatus."

We also agree with Adan on the distinction between, as he denotes them, Type-1 and Type-2 scenarios. For a detailed analysis, see also K. Svozil, *Natural Computing* 10(4), 1371–1382 (2011) and K. Svozil, *Natural Computing* (Online First 2012), DOI: 10.1007/s11047-012-9318-9].

C. Calude aside comment to A. Cabello:

In an aside you point out that "value indefiniteness ... incidentally suggests that the universe is not lawless". I disagree. We have to distinguish between the theoretical expressions of value indefiniteness (the Kochen-Specker theorem, for example), which are mathematical statements, and the **claim** that these statements are universally valid in the quantum universe. The **claim** can only be tested a finite number of times in contrast with the intrinsic infinite nature of each of these mathematical statements, so according to the arguments of our paper "The Universe is Lawless...", the law of value indefiniteness is in the same class as all known physical laws: they cannot be proved, so they are not universal.

Question from C. Teuscher to C. Calude and K. Svozil:

Most natural machinery, such as a biological neuron, is not universal but processes information in a highly specialised way, in which Turing universality is irrelevant. So why should one care about Turing's barrier for practical/natural applications when most of them don't even require Turing universality?

Answer by C. Calude and K. Svozil:
Turing's barrier of universal computation, and the physical Church-Turing thesis in general, are very relevant metaphors of what actually—or rather, in the limit of unbounded space and time—could be achieved by physical and biological systems, even though many (bio)physical systems are specialised and incapable of performing universal computation.

More generally, one should not confuse necessity with sufficiency. Pointedly stated: although a mobile phone may not easily be seduced to compile a LaTeX document such as this one, there exist many physically realisable computers which can handle all kinds of requests, including mobile phone simulation and LaTeX computation, and a lot else besides.

This does not diminish the value of tiny mobile phones, or of desktop computers; it is just that the latter can do "more" than the former. There is Turing's limit with regards to simulation, and that Turing limit falls short of allowing quantum random numbers. But of course, very specialised systems are also incapable of "producing" incomputability through computability.

Finally, universal machines are marred by undecidable problems, but sometimes non-universal machines are not. Distinguishing between universality and non-universality helps us know how decidable a model is. For example, most parts of Algorithmic Information Theory are undecidable (because of universality); finite-state complexity retains many features of that theory but is decidable, so it has a greater chance of being practically applicable (see C. S. Calude, K. Salomaa, T. K. Roblot. Finite state complexity, *Theoretical Computer Science* 412 (2011), 5668–5677).

Question from C. Teuscher to A. Cabello: What if it turns out that the universe does not compute anything useful? In other words, why should we care what kind of computer the universe actually is?

Answer by A. Cabello:
The universe computes itself. It computes the present and the future, you and me, our life and death. It also produces individuals who do not care too much about whether all these things are "useful", but definitely care about how the universe is working.

Question from H. Zenil to A. Cabello:
For the Bell experiment to validate quantum reality one would need to design it to cover all possible open ends or loopholes, as you explain. Your

paper seems to suggest that the only open loophole today is the problem of measuring fast enough to guarantee that measurements cannot influence each other and thus account for the correlation. It seems to me that your result tends toward ruling out the possibility that the universe is a classical computer, using an argument based on the amount of resources needed to keep track of the computation before overloading the space with enough information to keep producing the correlations that we see after the quantum experiments. Does this rule out the possibility of a completely deterministic universe, where the correlation would be explained simply because neither the particles nor the observers are independent of each other? And is a full deterministic universe considered to be a hidden variable theory?

On the other hand, all tests of Bell's inequality seem to assume that one has the freedom to choose the detectors' settings and that such freedom has to be indeterministic for the interpretation of the results of the experiment to work. The so-called free will theorem, for example, seems to be valid exactly in its reverse, negative version, that is, that if one does not have free will then particles do not. Free will has been connected to randomness since the beginning, but it has played the role of devil's advocate, where sometimes it is the cause of randomness and sometimes it is the consequence of randomness.

Answer by A. Cabello:

Yes, a deterministic universe would be definitely a hidden variable model. Without making assumptions, I see no way to rule out a completely deterministic universe. However, we can rule some partially deterministic universes out if we make extra assumptions. The assumptions in Bell's theorem are: (i) Outcomes determinism, (ii) bounded speed of information, and (iii) observers' free will. The assumptions in my approach are (i), (ii') bounded density of information, and (iii).

I do not think that there is evidence to support the negative version of the free-will theorem. The free-will theorem states that, if the observer has free-will to choose the experiment on the particle, then the particle also has a certain amount of free-will in producing the corresponding outcome. The free-will theorem emphasizes the connection between (1) the assumption of the observer's free-will and (2) the conclusion that the outcome cannot be determined. However, the fact that (1) cannot be tested does not necessarily imply that (2) is wrong. Interestingly, the quantum protocols for randomness expansion in which the input is a private random string (the observers' free-will) and the output is a longer private random string, suggest that particles' free-will is greater than observers' free-will.

Question from H. Zenil to A. Cabello:
Stephen Wolfram has suggested in the past that pseudo-randomness (the only possible kind of randomness in a fully deterministic universe) has the obvious particularity that when its generating process is repeated with the same seed, the result is always exactly the same. I see you tackle the question of repeatable and compatible experiments. Could you elaborate a bit on how this idea of testing the possible pseudo-random nature of the universe (i.e. the possible assumption of being carried out by a classical computer) is connected to your paper? At a first glance, one may think that Bell's experiment yields the correlation because one is restarting the pseudo-random generator with the same seed. Is this what the violation of the Holevo bound actually rules out, on the grounds that a classical computer lacks the storage capacity to keep up with the information needed to simulate such correlations? Would your result still hold if the universe was deterministic from the inception?

Answer by A. Cabello:
No. The idea is that, if a supposedly simple physical system acts as a classical computer, then it must store a large amount of information in order to provide outcomes in agreement with the rules of the game (quantum mechanics). If you assume that the system is actually simple, then it cannot act as a classical computer. The conclusion is based on the assumption that the system must store all the information needed to consistently answer any possible question. If the questions are determined, then no large memory is needed.

Question from H. Zenil to A. Cabello:
Quantum computation takes advantage of the physical effects of superposition and entanglement, leading to a new paradigm. In quantum mechanical computation models, all events occur by unitary transformations, which apparently means that all quantum gates are reversible. But it is not clear whether standard quantum computation fully captures quantum mechanics or is already committed to a specific interpretation of QM. This is because in some sense the standard interpretation of QM suggests that quantum processes are indeterministic, but if quantum gates in quantum

computation are reversible, this clearly suggests the opposite, viz. that quantum processes are also deterministic (as I think is the view of Feynman, Bennett and Deutsch, to whom we owe the theoretical basis of the quantum computer). So either the standard model of quantum computation does not capture all the essential properties of quantum mechanics according to the mainstream interpretation, or quantum mechanics, just like classical mechanics, is reversible and therefore in a strict (perhaps not practical) sense, deterministic. Could you shed some light on this subject?

On the other hand, in standard quantum mechanics, decoherence implies that the system considered is not closed, but interacts with the environment. Thus, while there is a leak of quantum information from the system into the environment, information is still conserved in the embedding universe. Is there any loss of information during decoherence? Is it the case that the possibility of a fundamental information loss cannot be completely ruled out and is explored in alternative theories? Are there any phenomena that we know of that imply a real loss of information?

Answer by A. Cabello:

It is not true that in quantum computation "all events occur by unitary evolutions". Usually quantum computations consist of unitary evolutions followed by measurements that have intrinsically unpredictable outcomes. Quantum gates are reversible, quantum measurements are not.

I would need a more precise definition of what you mean by a "fundamental information loss". In any case, let me point out that the quantum information degraded under decoherence is the information an observer has about the probabilities of the different possible outcomes of any possible experiment involving specific physical degrees of freedom.

Reply by H. Zenil:

By loss of information I mean information that cannot be recovered in the logical sense, e.g. in non-reversible computation. Here is an interesting quote concerning an additional loophole to close, viz. a local "leak" of information–besides the possibly deterministic nature of quantum mechanics–in order to perhaps avoid some of the problems you have pointed.

In his *Horizons* paper (http://arxiv.org/pdf/gr-qc/0401027v1.pdf) G. t'Hooft writes:

> The most important problem of the ones just alluded to is that deterministic evolution seems to be difficult to reconcile with a Hamiltonian that is bounded from below. It is absolutely es-

sential for Quantum mechanics to have a lowest energy state, i.e., a vacuum state. Now the most likely way this problem can perhaps be addressed is to assume not only deterministic evolution, but also *local information loss*. As stated, information loss is difficult to avoid in black holes, in particular when they are classical. It now seems that this indeed may turn out to be an essential ingredient for understanding the quantum nature of this world. (emphasis in original)

This sends the author to examples of irreversible cellular automata in Wolfram's NKS:

Simple examples of universes with information loss can be modeled on a computer as cellular automata.

So in the case of QM, I understand your claim is that loss (or degradation) of information in this reversible sense is at the level of the observer making a measurement, meaning that if one tries to reverse the operation (entangling the particle again) and make another measurement, the second measurement is disconnected from the first one, and there is no way to recover the first one from the second measurement. Is this the kind of irreversibility and loss of information implied by quantum mechanics? Is this, then, a "fundamental loss of information"? As opposed to, for example, classical mechanics, where apparently irreversible processes release heat from where one can, in principle, recover all the information and reverse any classical process.

Question from H. Zenil to C. Teuscher:

Wolfram has pointed out that there is an almost embarrassing difference between engineered artefacts, which tend to be simple and predictable yet are not very robust, and are often faulty compared to things easily identified and produced by nature, which are more complex and, more importantly, robust, even if unpredictable in almost every respect. I think you are saying something very close, making your case in favour of an unconventional model of "unorganised" computation (an alternative to von Neumann's successful but apparently limited model). One property of computational systems is that every time one runs a program for a given input the output is exactly the same. Deterministic behaviour is a fundamental property of classical computation (even if this doesn't mean that computation is predictable). My question is whether you think that running a program on an unorganised model of computation will also turn out to be even less predictable, and how negatively or positively this may impact the practical use of such a computational paradigm. Experience may suggest that, for example, natural or living systems are difficult to program (to make them behave the same way in the face of the same stimuli).

The second part of my question concerns the mechanism of random self-assembly of complex networks. Could you elaborate on how, for example, small world networks can self-assemble? By what mechanism? The paper suggests you have run tests using evolutionary algorithms. By what other means or processes might we be capable of producing complex networks other than by engineering the network in advance?

Answer by C. Teuscher:

Many of the computational substrates and computer models we proposed are disorganized and/or exhibit highly stochastic behavior. While regularity and deterministic behavior are desirable properties of most classical computational models, they mostly serve the purpose of making the designer's and programmer's lives easier. The approach we took instead was to harness what we could easily and cheaply build on a massive scale. It turns out that such things are rarely perfect and regular. But irregularity and unpredictability don't mean that the computation will necessarily be unpredictable too. With further scaling down CMOS, most of the "classical devices" will have high variation and become increasingly stochastic in their behavior because of fundamental physical limits (e.g., thermal noise). However, we know how to build abstraction layers in traditional CMOS design that hide such behavior. Perfect computation based on imperfect components is something we do often with classical models too. Of course each abstraction layer typically comes at a certain cost in terms of overhead. Nothing prevents us from building such abstraction layers with unconventional models as well. However, the real question is whether that should be the goal. For example, do we really want to compute binary logic with a reaction-diffusion system that can solve PDEs directly? So in that sense, I think unconventional substrates need unconventional computer paradigms and unconventional programming. What works well for CMOS doesn't necessarily work well for unorganized machines. I think nature provides us with many examples where the physical/chemical substrate is harnessed in a very particular way that is unique to that particular substrate.

It's been well known that small-world networks evolve if you consider both the cost and the performance. Such networks simply occupy a space in the entire space of networks that optimize for these trade-offs. That's the main reason why many natural networks are small-world. E.g., the neural network of the brain needs to perform well but is also subjected to cost (e.g., energy, space) constraints. We have hypothesized in the past that certain ways of growing nanowires (e.g., from seed points) will lead to small-world-like networks because the wires "compete" for limited resources as well (e.g., space, chemicals). Preliminary experiments in collaboration with Los Alamos National Lab have shown that we indeed get wire-length distributions that can lead to small-world networks. However, more experiments are required and the results may not apply to all growth methods and self-assembly approaches.

Question from F. Doria to H. Zenil:
What is your opinion on the issue of hypercomputation? Is it feasible? Impossible?

Answer by H. Zenil:
I was once myself a hypercomputation scholar (the title of my Master's thesis—in French—was "Hyper calcul"). Paradoxically it wasn't until I had a good knowledge of it that I started to better appreciate and to increasingly enjoy the beauty of the digital (Turing) model. One powerful reason is that hypercomputational models do not converge in computational power. There are a plethora of possible models of hypercomputation while there is only one digital model (in terms of computational power). I find it astonishing how little it takes to reach the full power of (Turing) universality, and I don't see the power of universal machines as being subject to any limitation. Quite the contrary. This shift occurred during the time I started working for Stephen Wolfram, who certainly may have influenced me. I happen to think the question has a truth value and is ultimately susceptible of a physical answer. However, I don't think the answer will emerge in the foreseeable future, if it ever does.

Given all this, I find your contribution to the subject interesting in that you point out a particular ("ideal") model of hypercomputation that I think gets around one of Davis' objections, though it doesn't fully address the problem of verification. Unlike a Turing computation, which can in principle be verified by carrying it out by hand, step by step, the inner working of a hypercomputer can only be followed by another hypercomputer. The caricature version of the problem is in the whimsical answer given by the (hyper?)computer Deep Thought, which proposed "42" as the "Ultimate (uncomputable?) answer to the Ultimate Question of Life, The

Universe, and Everything" [parenthesis added] in *The Hitchhiker's Guide to the Galaxy* by Douglas Adams. The only way to verify such an answer would be by building another, more powerful and even less understandable computer.

This makes me wonder whether we ought not to favour meaningful computation over what could potentially be hypercomputation, even if hypercomputation were possible. There is a strong analogy to the concept of proof in math. Mathematical proofs can be of two types. They either serve to convince you of the truth of a statement you weren't certain was true, or else to provide logical evidence that a statement intuitively believed to be true was in fact so (e.g. the normality of π). But in the latter case, why would one bother to provide a proof of a statement that nobody would argue to be false? It is because ideally math proofs should provide insight into why a statement is true, and not simply establish whether or not it is so.

I also think, unlike a few other authors, that taking Oracle machines–proposed by Turing–as hypercomputers is mistaken and misguiding. The question as regards hypercomputation is whether one thinks Oracle machines are actually realisable, and I think Turing had, at the very least, no ambivalence on the subject. Both his work and his later philosophical positions indicate that he certainly never thought that Oracle machines would ever be constructed, only used for the purpose of studying the limits of computation and relativising the concept of computation into computability degrees which told us something very interesting, equivalent to Gödel's result, that even assuming an Oracle one would still find problems of increasingly greater difficulty than even Oracles are able to solve.

The study of infinite objects has, however, given us great insight into profound and legitimate questions of math and computation, such as the question of the nature of number and computation. And it has been immensely useful to focus on limits in order to better understand these concepts. As De Mol has concluded in her contribution to this volume ("Generating, Solving and the Mathematics of Homo Sapiens: Emil Post's Views on Computation"), it is at the boundary between decidability and undecidability that one seems best positioned to answer the foundational question of computation. And there are some examples where the study of a hypercomputational model is more valuable–from my personal point of view–not as a model per se but because of its ancillary results (see e.g. José Félix Costa's paper "The ARNN model relativises $P = NP$ and $P \neq NP$", 2011). Unlike most people, I think the contribution of Siegelmann and her

ARNNs has, for example, much more to say about computational complexity and therefore classical computation than about hypercomputation (Siegelmann's ARNNs model has come to be strangely revered among hypercomputation enthusiasts). As Doria points out in his contribution to this volume (Blueprint for a Hypercomputer), Martin Davis provides (in "The Myth of Hypercomputation") some criticisms (some more justified than others from my perspective: I agree with Doria that the problem of verification of a computation is not necessarily an argument against hypercomputation but whether hypercomputation is legitimate is a true problem). Among Davis' criticisms there is the fact that the ARNN model is circular and its value as a hypercomputational model depends ultimately on physics (which is something we already knew, i.e. the ability to encode non-computable numbers as weights in an artificial neural network), and the model is by no means a model that can suggest that building such a neural network or a hypercomputer is possible.

Question from H. Zenil to L. De Mol:

Proving the undecidability of the halting problem and providing an answer to Hilbert's program is the extraordinary negative result of Turing's seminal work. The positive result of Turing's paper is, however, the concept of computational universality and the proof and construction of a universal (Turing) machine. Judging from Post's theses, which are not only equivalent to but anticipate those of Church and Turing, computational universality seems to be implied in his work, as indeed in his quest to find a general algorithm for all mathematics in the context of, I assume, the Entscheidungs problem. I wonder, however, about the weight that Post gave to the construction of the concept of universal computation, and whether this had an impact on the way the concept was or was not conveyed in his work, and hence on the difference in the reception of his work as compared to Turing's. Could you please elaborate on this and explain how Post might have realised and weighed the value of such a construction? As you say, even his positive thesis (Post's second) seems more profoundly rooted in the limits rather than the possibilities, in this case of a universal constructor (to use a term used by von Neumann).

Answer by L. De Mol:

Post's 1936 note certainly does not contain the description of a universal computer, since it is but a short note. Also, in the 20s Post most probably did not have a detailed description of a universal computer (or something equivalent like a universal tag system or normal system) but one does find traces of the idea in his 'Account of an anticipation' of his work in the 20s, where it plays a role comparable to Turing's universal machine. Two (theorems)[a] in this manuscript mention the complete normal system K which is so called *"because, in a way, it contains all normal systems"* and which is compared with Turing's universal computing machine in footnote 95.[b] The first (theorem) states the existence of K, the second, as Post explains in footnote 99, *"shows [when combined with Post's thesis I] not only that the finiteness problem [the decision problem] for the class of all normal systems is unsolvable, but that the same is true for one particular one of them."* Of course, since this work was not published before Turing's 1936 paper (the 'Account of an anticipation' was posthumously published in 1965 by Martin Davis), it was not received in the same way as Turing's paper. Later, in Post's influential *Recursively enumerable sets of positive integers and their decision problems* the notion of a complete set K is introduced which is similar to that of the complete normal system K. Several theorems are proven in that paper for K. However this paper was published only nine years after Turing's 1936 paper and worked with a formalism that is far less 'intuitively appealing' than Turing's machines.

Question from H. Zenil to L. De Mol:

The general sentiment is that Turing's model was closer to the mechanical realisation of a mechanical machine, while Church's was at the opposite extreme, closer to what we today recognise as a computer language, hence software, while Post's model is often regarded as lying in between. Your paper has convinced me, however, that Post's was closer to Turing's mechanical realisation than to Church's, hence not exactly in an intermediate position. Post also seems to have been closer to a concept of complexity than any of the other proposers, as it apparently arises quickly from running even the simplest of Post's rule systems. Is this something you suggest or would agree with?

Answer by L. De Mol:

It depends on which model you are talking about. His formulation 1 is almost identical to the Turing machine and in this sense can hardly be

---

[a]Post puts some of the theorems of his 'Account of an anticipation' within brackets because he didn't have detailed proofs but merely sketches for those theorems

[b]These footnotes were later added by Post when he was preparing the manuscript for submission to the *American Journal of Mathematics* in 1941.

called intermediate. However, it is interesting to point out that Turing's formulation is in terms of 'hardware', whereas Post's formulation is more in terms of 'software', giving a description in terms of sets of directions in a well-defined language. Post's normal systems, his canonical form C and tag systems are very different from a Turing machine and perhaps closer in spirit to programming languages than Formulation 1. In fact, it seems that John Backus was inspired by Post's production systems when he developed his normal form, calling them *'just the thing'*[c] and Patterson of the Burroughs company invited Post to the Burroughs research division because "[your production systems and theory of canonical languages] appeal to us as possibly the most natural approach to the theory of computability from the standpoint of one interested in syntactical machines." (Patterson in a letter to Post, Dec. 17, 1952). It might well be noted that Post's production systems also had an important influence on the development of formal grammars, his work being an important source in Chomsky's foundational papers.[d] Post was clearly very aware of the possibility that simple programs can give rise to complex behavior. It was during his research on tag systems (which he called primitive forms of mathematics) that he understood that one does not necessarily need complicated machinery to have complex behavior (he described the behavior of some tag systems as being of 'bewildering complexity'). Most probably guided by this insight, Post then proved that the *"meager formal apparatus of [the] normal systems can wipe out all of the additional vastly greater complexities of the canonical form"*. This is his normal form theorem, a theorem which led him to believe that in fact the whole of Principia Mathematica could be reduced to the normal form. He did not have a detailed proof of this at that time (the reduction of PM to normal form). This is the reason why Post did not anticipate the negative solution of the Entscheidungs problem for the restricted functional calculus proved by Church and Turing in 1936.

Question from H. Zenil to L. De Mol:

I notice that several of the papers in this volume mention the fact that math seems grounded in historical accidents (the reason why Post insisted

---

[c] It should be noted though that it is not entirely clear where Backus knew Post's work from (See Martin Davis' paper *Influences of Mathematical Logic on Computer Science* published in R. Herken, The universal Turing machine: A half-century survey, Springer, 1995).

[d] For more information on Post's influence on computer science, see Martin Davis, Emil Post's contributions to computer science, in: *Proceedings of the fourth annual symposium on Logic in Computer science,* June 5-8, 1989, pp. 134–136.

that Church's thesis and equivalent formulations should be treated as hypotheses). In Post's words "I consider mathematics as a product of the human mind, not as absolute". Just as Bauer seems to suggest in his contribution to this volume, and Sutner, in the problem he has with unnatural intermediate degrees of computation. Do you have any position in this regard? How accidental is the history of computer science? And how accidental it is compared to, say, the history of mathematics?

Answer by L. De Mol:

I do not really understand why the idea that math is a product of the human mind somehow implies that math is grounded in historical accidents. Of course, if one understands math as a product of the human mind then one can hardly deny that it is rooted in a history and, hence, that it is subject to history's unpredictable and erratic ways. To call certain events in the history of computer science or mathematics 'accidental' is tricky since this depends on the theoretical notion of randomness one is using. Furthermore, it is not certain whether true randomness exists at all in nature, let alone, in human history. Notwithstanding these theoretical points, I think that events in the history of math or computer science can surely have a certain amount of accidentality or, at least, be the result of some exceptional or particular combination of events, but that does not mean that a given result from that history would never have been, had that particular chain of events not occurred. For instance, one can wonder whether the design of modern computers would today be named after von Neumann if not for Goldstine's chance meeting with him at the Aberdeen railway station, or if Goldstine had not distributed the so-called First draft of a report on the EDVAC design as he did. But this does not mean that the electronic and programmable computer would never have been invented. It only means that it might have been named after someone else and that part of its history would have been slightly different. I think that one would need some notion of robustness to discuss the 'accidentality' and non-accidentality of history in a more serious manner. The difference between the history of mathematics vs. the history of computer science is probably that computer science is much more guided by/dependent on the history of technology, industry, economy and politics than mathematics. In that sense, one could say that the history of computer science is more 'accidental' than that of math but only insofar as it is more dependent on the developments of other histories, not in some absolute way.

Question from H. Zenil to K. Sutner:

The lack of correspondence between the abstract and the physical world seems sometimes to suggest that there are profound incompatibilities between what can be thought and what has meaning in the real world. One can ask, for example, how often one faces undecidable problems. And it seems the case that in the practice of advanced math we have managed to avoid undecidability (even though this set is a natural computational degree). One may think that long lasting open mathematical problems are natural candidates for undecidable propositions. However, the history of math suggests that it is a matter of time before someone with some theory provides a proof for every current open mathematical problem. In the same way that an ad hoc proof was tailored to give a positive answer to Post's problem, could it be the case that math gets around to undecidability by producing proofs in the form of more powerful theories than the theories in which the original problems were stated? As we know this is the case with a simple mathematical statement such as Fermat's last theorem, proven by Wiles with sophisticated mathematical artillery. Do you have any comments on this? How natural/unnatural and common/uncommon a practice is this?

Answer by K. Sutner:

One has to distinguish between undecidability in the sense of classical recursion theory and undecidability in the sense of independence within the framework of a particular mathematical theory. For example, the problem of finding integral solutions of Diophantine equations is undecidable in the computational sense. This will come as no surprise to anyone who has studied Diophantine equations; it is exceedingly difficult to handle even polynomials of rather low degree. The fact that there is no algorithm that can determine the solvability of Diophantine equations can be established within any standard formal system of number theory, such as Peano arithmetic. On the other hand, the question of whether every set of reals is measurable cannot be answered in the context of a much stronger system such as Zermelo-Fraenkel set theory, nowadays the default background theory for mathematics, thanks in no small part to Bourbaki's efforts more than half a century ago. One could argue that, given the increasing importance of computation, set theory is no longer the preferred candidate for that role, but we digress. If we accept the axiom of choice, there are non-measurable sets. Worse, by a famous theorem due to Banach and Tarski, there are paradoxical decompositions of a sphere that can be reassembled to produce a sphere twice the size of the original one. On the other hand, Solovay has shown that one can construct a model of Zermelo-Fraenkel set theory where every set of reals is measurable, assuming there are inaccessible cardinals. Thus, the existence of non-measurable sets of reals is independent of Zermelo-Fraenkel set theory; it cannot be decided within this context and it is not entirely clear how one should augment the standard system to settle this particular question. Most mathematicians would seem to favor adopting the axiom of choice, paradoxical side-effects notwithstanding.

As to long-standing open problems, the Riemann hypothesis is something of a stand-out. From the logical perspective, the statement of the Riemann hypothesis requires no more than universal quantification, so it is fairly simple in that sense and comparable to Fermat's conjecture. A lot of computational evidence points towards the correctness of the hypothesis. Yet it is not unreasonable to assume that a proof would use tools outside of the scope of Peano arithmetic, much as Wiles's proof did. Of course, a counterexample could be much simpler in nature. This is a bit different from the solution to Post's problem, which is irritatingly artificial but can be established within the confines of number theory. Or consider what is currently arguably the most important open problem in theoretical computer science, the P versus NP question. Here it is really unclear how much machinery is needed to deal with the problem. There may yet be a simple polynomial time algorithm that determines satisfiability of Boolean formulas—bear in mind the recent discovery of a polynomial time primality test that uses in essence no more than high-school algebra. Or we might be able to construct some analogue of the Friedberg-Muchnik argument that establishes the existence of a problem intermediate between P and NP. This approach is somewhat less promising since any such argument would presumably be invariant under oracles, and we know that this is not the case. Lastly, the problem might be independent from some standard axiom system like ZFC. For example, it might be the case that there is a polynomial algorithm for satisfiability, but we cannot establish this fact within the given framework and we would need to propose some plausible extension to settle the question. Of course, the issue of plausibility here is quite problematic in itself. But that should not be interpreted as a fundamental flaw in the method; ever since Gödel demolished Hilbert's program in its original form we have known that there is no hope of reducing all of mathematics to a simple, finitistic basis. It seems to me that thoughtfully constructed "sophisticated artillery" is both necessary and quite acceptable from a foundational perspective. As Hilbert said, "We must know, we will know."

Question from Bolognesi to Hewitt:

The computational universe conjecture, in one of its extreme forms, suggests that the physical universe is a manifestation of the emergent properties of a computation, one possibly deriving from the execution of a simple program started from simple initial conditions. Wolfram's elementary cellular automata are a popular metaphor for this, but Turing machines and various other Turing-complete models have been investigated under this perspective. One way to partially abstract from the specific features of one or the other model is to focus on the causal structure of computation events—a structure that could then be taken as a discrete model of spacetime.

In your paper you raise the point that the Actor Model is stronger than Turing machines, using an 'unbounded nondeterminism' argument. I do not know whether you have specific opinions on the idea of understanding Nature quite literally as computation. If so, I would be interested in knowing your opinion on the role that the Actor Model could play in that context, also in light of your claim that the model was inspired by the laws of physics. In particular, causal structures (partial orders) of events would seem to derive quite naturally from actor system computations...

Answer by C. Hewitt:

The Actor model is largely compatible with the views of [Rovelli 2008]: a pen on my table has information because it points in this or that direction. We do not need a human being, a cat, or a computer, to make use of this notion of information.

Relational physics takes the following view [Laudisa and Rovelli 2008]:

- Relational physics discards the notions of absolute state of a system and absolute properties and values of its physical quantities.
- State and physical quantities refer always to the interaction, or the relation, among multiple systems.
- Nevertheless, relational physics is a complete description of reality.

Rovelli added: This [concept of information] is very weak; it does not require [consideration of] information storage, thermodynamics, complex systems, meaning, or anything of the sort. In particular:

(1) Information can be lost dynamically ([correlated systems can become uncorrelated]);

(2) [It does] not distinguish between correlation obtained on purpose and accidental correlation;
(3) Most important: any physical system may contain information about another physical system. Information is exchanged via physical interactions. It is always possible to acquire new information about a system. In place of the notion of state, which refers solely to the system, [use] the notion of the information that a system has about another system. Quantum mechanics indicates that the notion of a universal description of the state of the world, shared by all observers, is a concept which is physically untenable, on experimental grounds.

According to the Actor Model, Interaction creates reality. Technical details are in Fig 1.

The Actor model makes use of two fundamental orders on events [Baker and Hewitt 1977; Clinger 1981, Hewitt 2006]:

1. The *activation order* ($\curvearrowright$) is a fundamental order that models one event activating another (there is energy flow from an event to an event which it activates). The activation order is discrete:

    $\forall e_1, e_2 \in \text{Events} \rightarrow \text{Finite}[\{|\ e \in \text{Events}\ |\ e_1 \curvearrowright e \curvearrowright e_2\ |\}]$

    There are two kinds of events involved in the activation order: reception and transmission. Reception events can activate transmission events and transmission events can activate reception events.

2. The *reception order* of a serialized Actor $\mathbf{x}$ ($\underset{x}{\rightarrow}$) models the (total) order of events in which a message is received at $\mathbf{x}$. The reception order of each x is discrete:

    $\forall e_1, e_2 \in \text{ReceptionEvents}_\mathbf{x} \rightarrow \text{Finite}[\{|\ e \in \text{ReceptionEvents}_\mathbf{x}\ |\ e_1 \underset{x}{\rightarrow} e \underset{x}{\rightarrow} e_2\ |\}]$

The *combined order* (denoted by $\rightsquigarrow$) is defined to be the transitive closure of the activation order and the reception orders of all Actors with the following axiom:

$\forall e_1, e_2 \in \text{Events} \rightarrow \text{Finite}[\{|\ e \in \text{Events}\ |\ e_1 \rightsquigarrow e \rightsquigarrow e_2\ |\}]$

Fig. 1. C. Hewitt answer to T. Bolognesi.

Question from H. Zenil to A. Bauer:

I liked how deeply, in terms of constructive math, you connected the Church-Turing thesis to geometric intuitions about space. I agree that these are strong indicators that the last word on the nature of computation and geometry has not yet been said.

If one of the arguments against constructive mathematics, concerning its expressive power, is not true as you argue in your paper, do you have any hypothesis as to why classical and not intuitionistic mathematics took over and became mainstream? Is it an historical accident, perhaps an inheritance from ancient Greek math? To me, only such an historical accident would explain why mathematicians seem to see it as less exotic to work with infinite arithmetic and oracle machines than with constructive mathematics. If intuitionistic math is richer than classical, how do we reset math in the right direction? Doesn't all this tell us something about the human dimension of mathematics? What does this say about the purported immutable truth of mathematical theorems?

Along the same lines, how historically connected is constructive math to Hilbert's program? In the end Turing's abstract idea of a universal computer, a constructive device that can perform all possible computations, turned out to be technologically realisable. How much more successful can a mathematics be required to be than to embody itself in a physical device capable of carrying out all possible calculations? Would you attribute such an achievement to constructive math and how could a constructive result emerge from such a non-constructive motivation? (the non-constructive negative result of Hilbert's program epitomised by Turing's use of Cantor's diagonal argument). How much can one credit constructive math for being at the foundations of the concept of universal computation, leading to the development of the whole field of computer science? And in that sense what do you make of areas such as computability and algorithmic randomness (or for that matter, modern classical mathematical logic) where most of the human resources are concentrated in highly non-constructive fields?

In this context, do you think the complementary question to the 'Unreasonable Effectiveness of Mathematics in the Natural Sciences' question posed by Eugene Wigner (1960), the question of whether modern math has any effectiveness in the natural sciences, is legitimate today? The usual answer is that one should not ask whether math has any application today, as it might be applicable in the future, and pure research should not be subject to considerations about mundane applications. A legitimate question, however, is to ask how realistic it is to think that ordinal arithmetic or the study of oracle machines will find any real connection to the natural world in the foreseeable future. Both subjects, however, have given us a lot of insight into math and computation.

Answer by A. Bauer:

Since I am no historian of mathematics I am surely the wrong person to answer authoritatively some of your questions. Nevertheless, I can offer my quasi-informed opinion.

Why did classical and not intuitionistic mathematics take over and became mainstream? This question is a bit like asking why so many people use a particular operating system, or speak a particular language. While many, if not most, mathematicians would like to think that there is something unique and sacred about mathematics, I would advise against such opinions. Mathematics is (still) a deeply human activity. Trends in mathematics are determined mostly by social factors, such as what the popular people do and which kinds of mathematics can get funding.

So yes, it is a historic accident that we use one kind of mathematics rather than another. If engineers invented computers before they invented the steam engine, I am sure mathematics would have developed differently. Your asking how to reset math in the right direction calls for a value judgment which I am not willing to make. A mathematician who is familiar with intuitionistic mathematics will have technical advantages over his purely classical colleagues, but the same may be said for any scientist who knows more than his peers. In any case, if intuitionistic mathematics is going to take over the world of mathematics, it will be partly due to quality of work and mostly thanks to good propaganda and pressure from external sources, such as computer science.

Your next question is about the historical connection between constructive mathematics and Hilbert's program. As far as I know, and I know next to nothing about this particular historical issue, Hilbert's finitism and constructive mathematics were not related historically. You raise an interesting issue when you ask about Turing having developed his machines in the context of classical mathematics. Turing's development of a theory of computation is perfectly constructive, and if there are any non-constructive steps in his papers, they can be easily avoided. Cantor's diagonalization argument, as well as Turing's use of it in relation to the halting problem, are constructively valid (contrary to the popular opinion that any proof which reaches a contradiction is non-constructive). Therefore, it is largely irrelevant whether Turing worked constructively or classically. I will go much further and state that the vast majority of mathematics is independent of foundations, and luckily so. Algebra and geometry existed before anyone spoke of foundations of mathematics, and I am pretty sure they would not wither away without the 20th century logicians. This of course is not to say that logic and foundations has no role at all. In the 20th century it was a grand unifying force which clarified the nature and methods of mathematics, and gave mathematicians a universal language so that they could speak to each other more easily and discover deep connections that would not have been visible otherwise.

You ask how much one can credit constructive mathematics for being at the foundations of the concept of universal computation leading to the development of the whole field of computer science. Historically speaking, and here again I warn that I am not a historian, constructive mathematics stemmed from a philosophical tradition which was concerned with questions about the nature of mathematical knowledge and how the creative subject, i.e., the mathematician, might recognize mathematical truths. Questions about the nature of computation arose in logic independently, through efforts to clarify the concept of computation. Only later on did Stephen C. Kleene establish a clear connection between intuitionistic arithmetic and computability through his realizability interpretation. In recent times constructive mathematics has had an important and deep influence on computer science, for example through Church's λ-calculus and Martin-Löf's type theory.

Finally, you speak of the unreasonable effectiveness of mathematics in the natural sciences. Is it realistic to think that ordinal arithmetic or the study of oracle machines will find any real connection to the natural world in the foreseeable future? Never say never. Ordinal arithmetic has such a connection already. It is used to show termination properties of programs, for example, and it has undoubtedly inspired the concepts of inductive definition and inductive construction in type theory. And I once heard Robert Soare explain very nicely that an oracle machine is just like a personal computer with access to external information, such as a CD or an Internet connection. Perhaps this is why we keep surfing the Internet, to find an oracle.

But more seriously, you presume that constructive mathematics is unable to deal with ordinal arithmetic and oracles. What evidence is there for such an assumption? In fact, I already mentioned that ordinal arithmetic has its constructive counterpart, namely inductive definitions and constructions, while oracle computations in the context of constructive mathematics happens to be high on my to do list.

Question from H. Zenil to A. Bauer:

Klaus Sutner (see his chapter in this volume) seems to provide an example of a constructive proof (that can in principle be carried out by a Turing machine) offering a solution to Post's problem (for reference: whether there is any r.e. degree strictly between 0 and 0'), but still unsatisfactory in the terms discussed by Sutner in this volume. For this problem, the novelty of the proof is the ad hoc proof method, but the proof itself says little if anything about the real nature of Post's problem in a more realistic (physical) context, leaving us with what seems to be an unnatural answer that does not provide any insight into the true problem, i.e. whether there are intermediate computational degrees. Sutner concludes that in the context of some reasonable theory of physics, intermediate processes fail to exist. Do you have anything to say in this regard? What bearing does this have on whether constructive math is physically meaningful, or on whether the solution to Post's problem is or isn't truly constructive?

Answer by A. Bauer:

I am not aware of any constructive proofs of the existence of intermediate degrees, and I seriously doubt there is one. Sutner does not provide a constructive solution to Post's problem either, and it is not his intent to do so anyhow. In fact, non-existence of a constructive solution would support, not contradict, Sutner's conclusions.

Sutner observes, as others have before, that a certain amount of information hiding is involved in the usual solutions of Post's problem. My reading of Sutner is the following. If hiding of information is prevented, either by the laws of physics or because of other requirements, such as "a 1-dimensional cellular automaton must be used and we can observe its actions", then any computation whose output is an intermediate degree will be recognized as computationally complete when looked at in its entirety by an outside observer. I find this to be a very interesting remark which may go a long way in explaining why intermediate degrees are not observed in nature, provided one looks honestly with eyes wide open.

It is curious to see that information hiding is important both in Sutner's contribution and mine. I argue that a certain amount of information hiding should be provided by the laws of physics, if the Brouwerian continuity and choice principles are to be realized. Similarly, Sutner argues that information hiding could be the root cause of intermediate degrees. Yet, his and my conclusions are in opposition somehow, since for me information hiding prevents the existence of bizarre objects, namely discontinuous maps, while for him it creates them, namely intermediate degrees. The moral of the story seems to be that a proper analysis of real-world computation should seriously take into account the concept of hiding computations and communication channels from external observers.

Question from A. Cabello to S. Wolfram:

What if quantum theory turns out to be the "ultimate fundamental theory", that is, the ultimate "complete set of laws that provide a finite description of how the whole universe works", you refer to in p. 5 of your contribution?

Answer by S. Wolfram:

Then there'd still be the question of how we perceive definite things to happen in the universe. The formalism of quantum mechanics on its own just gives us only probability amplitudes.

Question from C. Teuscher to S. Wolfram:

I'm curious to know what influence the fact that "we will ultimately never be constrained in what we can achieve by the details of our physical universe" have on building artificial brains of some sort?

Answer by S. Wolfram:

It implies that there won't be any fundamental issue with building artificial brains, whatever those may be. I think it's a challenge to define what an abstract "artificial brain" is, divorced from all the details of human brains in particular. But I'm confident that in any definition, we'll be able to build versions even in software.

Question from M. Szudzik to S. Wolfram:

You argue that the evolution of a thermodynamic system from one state to the next is essentially an encryption process. This analogy is compelling because it allows us to use ideas from cryptography to classify models of physical systems. For example, since the evolution is reversible for the classical billiard ball model of a gas, a symmetric-key encryption function can be used as a model for a gas's evolution. Can you comment further on this analogy?

Answer by S. Wolfram:

I have to say that I viewed encryption as an analogy for the computational processes that lead to the Second Law, not a direct model. Back in the mid-1980s, I realized that the rule 30 cellular automaton could be used as a good source of randomness for a stream cipher. I wondered whether reversible rules could be the basis for a practical encryption system, but

I never managed to figure out a way to make that work. Of course, the scrambling by the Second Law is probably a lot weaker than a practical encryption system.

Question from L. De Mol to S. Wolfram:
In his "The computer and the brain", John von Neumann said that:

> When we talk about mathematics, we may be discussing a secondary language built upon the primary language truly used by the central nervous system. Thus, the outward forms of our mathematics are not absolutely relevant from the point of view of evaluating what is the mathematical or logical language truly used by the central nervous system.

Do you have any comments on this quote in relation to the idea of reducing physics to mathematics?

Answer by S. Wolfram:
On this point (if not always on others [see http://www.stephenwolfram.com/publications/recent/neumann/]), I think von Neumann is obviously correct. Brains surely don't have direct representations of symbolic integrals, or constructs from predicate logic. Those get built up through, in effect, many layers of software (a concept that arose only after a lot of practical software existed in the world). Whether there are good high-level symbolic ways to represent what's going on in the brain is an interesting question, as is how these relate to human natural language. In my efforts to represent knowledge and what can variously be called "computation" or "thinking", I've been surprised at how simple the representations can be, while still achieving what amounts to human-like performance.

PART 6
Live Panel Discussion (transcription)

Chapter 35

What is Computation? (How) Does Nature Compute?*

Cristian S. Calude, Gregory J. Chaitin, Edward Fredkin,
Anthony J. Leggett, Rob de Ruyter, Tommaso Toffoli
and Stephen Wolfram

PANEL DISCUSSION, NKS MIDWEST CONFERENCE 2008. UNIVERSITY OF
INDIANA BLOOMINGTON, USA. NOVEMBER 2, 2008.[†]

QUANTUM VS. CLASSICAL COMPUTATION

Adrian German: About five years ago at MIT Stephen Wolfram was asked a question[a] at the end of his presentation:

> Question (on tape): "How does your Principle of Computational Equivalence (PCE) explain the separation in complexity between a classical cellular automaton (CA) and a quantum CA?"
>
> Stephen Wolfram (answers the question on tape): "If you take the current formalism of quantum mechanics and you say: let's use it, let's just assume that we can make measurements infinitely quickly and that the standardized formalism of quantum mechanics is an exact description of the actual world, not some kind of an idealization, as we know that it is — because we know that in fact when we make measurements, we have to take some small quantum degree of freedom and amplify it to almost an infinite number of degrees of freedom, and things like that — but let's assume we don't worry about that, let's take the idealization that, as the formalism suggests, we just do a measurement and it happens. So then the question is: how does

---

*Prof. Anthony Leggett has requested the editor to mention that he had no time to proofread in full detail the transcription of his interventions.
[†]Transcript and footnote comments by Adrian German (Indiana University Bloomington) with some assistance from Anthony Michael Martin (UC Berkeley). Organized by Adrian German, Gerardo Ortiz and Hector Zenil. Moderated by George Johnson assisted by Gerardo Ortiz and Hector Zenil.
[a]Plays recording from http://mitworld.mit.edu/video/149/ in which the selected paragraph appears at the 1:28:28 mark. Tape starts.

what I am talking about relate to the computation that you can get done in a quantum case vs. the classical case. And the first thing I should say is that, ultimately, the claim of my PCE is that in our universe it is in fact not possible to do things that are more sophisticated than what a classical CA, for example, can do. So, if it turned out that quantum CAs (or quantum computers of some kind) can do more sophisticated things than any of these classical CAs then this PCE of mine would just be wrong."

Question (on tape): "By more sophisticated do you mean with a better complexity then, or do you mean in a universal sense?"

Wolfram (on tape): "So — to be more specific, one question is: can it do more than a universal Turing machine, can it break Church's thesis? Another one is: can it do an NP-complete computation in polynomial time? And on the second issue I am not so sure. But on the first issue I certainly will claim very strongly that in our actual universe one can't do computations that are more sophisticated than one can get done by a standard classical Turing machine. Now, having said that there's already then a technical question which is: if you allow standard idealizations of quantum mechanics, with complex amplitudes that have an arbitrary number of bits in them, and things like that, even within the formalism can you do computations that are more sophisticated than you can do in a standard classical universal Turing machine? I think that people generally feel that you probably can't. But it would be interesting to be able to show that. For example, the following is true: if you say that you do computations just with polynomials then there is a result from the 1930s that says that every question you ask is decidable. This was Tarski's result. So, if the kinds of primitives that you have in your theory would be only polynomial primitives then it is the case that even using these arbitrary precision amplitudes in quantum mechanics you couldn't compute something more than what we can classically compute. But as soon as you allow transcendental functions that is no longer the case — because for example even with trigonometric functions you can easily encode an arbitrary diophantine equation in your continuous system. So if you allow such functions then you can immediately do computations that cannot be done by a classical Turing machine, and if there was a way to get quantum mechanics to make use of those transcendental functions — which certainly doesn't seem impossible — then it would imply that quantum mechanics in its usual idealization would be capable of doing computations beyond the Church's thesis limit. Then you can ask questions about the NP completeness level of things and, as I said, I'm less certain on that issue. My own guess, speculation if you want, is that if you try and eventually unravel the idealizations that are made in quantum mechanics you will find that in fact you don't succeed in doing things that are more sophisticated than you could do with a classical Turing machine type system – but I don't know that for sure!"

Adrian German: Where this answer ends, our conference starts. As quantum computation continues to generate significant interest while the interest in NKS grows steadily we decided to ask ourselves: Is there really a tension between the two? If there is — how is that going to inform us? What if the conflict is only superficial? It was then that we remembered a quote from Richard Feynman from the 1964 Messenger Lectures at Cornell, later published as the book "The Character of Physical Law":

> "It always bothers me that, according to the laws as we understand them today, it takes a computing machine an infinite number of logical operations to figure out what goes on in no matter how tiny a region of space, and no matter how tiny a region of time. So I have often made the hypothesis that ultimately physics will not require a mathematical statement, that in the end the machinery will be revealed, and the laws will turn out to be simple, like the chequerboard with all its apparent complexities[b]."

The seed of that hypothesis may have been planted by one of our guests today: Ed Fredkin[c] who says that in spite of his extensive collaboration with Feynman was never sure that his theories on digital physics had actually made an impact with the legendary physicist — until he heard it in the lectures as stated above.

---

[b] In 1981, Feynman gave a talk at the *First Conference on the Physics of Computation*, held at MIT, where he made the observation that it appeared to be impossible in general to simulate an evolution of a quantum system on a classical computer in an efficient way. He proposed a basic model for a quantum computer that would be capable of such simulations.

[c] Richard Feynman's interaction with Ed Fredkin started in 1962 when Fredkin and Minsky were in Pasadena and one evening not knowing what to do with their time "sort of invited [them]selves to Feynman's house," as documented by Tony Hey. Twelve years later, in 1974, Fredkin visited Caltech again, this time as a Fairchild scholar and spent one year with Feynman discussing quantum mechanics and computer science. "They had a wonderful year of creative arguments," writes Hey, "and Fredkin invented Conservative Logic and the Fredkin Gate, which led to Fredkin's billiard ball computer model." (See http://www.cs.indiana.edu/~dgerman/hey.pdf)

The Idea of Computation

George Johnson: Well, good morning, and I want to thank you all for coming out and also for inviting me to what's turned out to be a really, really fascinating conference. Friday night just after I got here I had dinner with my old friend Douglas Hofstadter who of course lives in Bloomington and I was thus reminded why (and how) I later became interested in computation in the abstract. I was already very interested in computers when Doug's book "Gödel, Escher, Bach" came out — and I remember in graduate school picking up a copy in a bookstore on Wisconsin Ave., in Washington, DC, and just being absolutely sucked into a vortex — and then eventually buying the book and reading it. And as I was reading it I thought "Well, this is great, because I am a science journalist," and at the time I was working for a daily newspaper in Minneapolis, "I think we really need to do a profile of this Hofstadter person." It was a great way to get introduced to this topic and I had my first of many other trips to Bloomington to meet Doug and over the years we got to know each other pretty well — in fact he copy-edited my very first book "The Machinery of the Mind."

And it was while writing that book that I came across a quote from one of tonight's guests Tommaso Toffoli that I just remembered while I was sitting here on these sessions and I looked it up the other night and I just wanted to use it to get us started. This was an article in 1984 in Scientific American by Brian Hayes[d] that mentions Stephen Wolfram's early work on CAs and mentions Norman Packard who was at the Institute for Advanced Studies (IAS) at the time (he later went on to co-found a prediction company in my hometown of Santa Fe and is now researching Artificial Life) and quotes[e] Dr. Toffoli as saying:

> "... in a sense Nature has been continually computing the 'next state' of the universe for billions of years; all we have to do — and actually, all we can do — is 'hitch a ride' on this huge ongoing computation."

It was reading that and things like that really got me hooked and excited about this idea of computation as a possible explanation for the laws of physics. We were talking about a way to get started here to have you briefly introduce yourselves, although most of us all know who you are and why you're important, but just to talk in that context of how you first got hooked to this idea and the excitement of computation in the abstract.

---

[d]http://bit-player.org/bph-publications/AmSci-2006-03-Hayes-reverse.pdf
[e]http://www.americanscientist.org/issues/id.3479,y.0,no.,content.true,page.1,css.print/issue.asp

Greg Chaitin: Being a kid what I was looking for was the most exciting new idea that I could find. And there were things like Quantum Mechanics and General Relativity that I looked at along the way and Gödel's incompleteness theorem but at some basic level what was clearly the big revolution was the idea of computation, embodied in computer technology. Computer technology was exciting but I was even more interested in the idea of computation as a deep mathematical and philosophical idea. To me it was already clear then that this was a really major new mathematical idea — the notion of computation is like a magic wand that transforms everything you touch it with, and gives you a different way of thinking about everything. It's a major paradigm shift at a technological level, at the level of applied mathematics, pure mathematics, as well as the level of fundamental philosophy, really fundamental questions in philosophy. And the part about the physical universe — that part was not obvious to me at all. But if you want to discover the world by pure thought, who cares how this world is actually built? So I said: let us design a world with pure thought that is computational, with computation as its foundational building block. It's like playing God, but it would be a world that we can understand, no? If we invent it we can understand it — whereas if we try to figure out how this world works it turns into metaphysics again[f]. So, obviously, I've hitched a ride on the most exciting wave, the biggest wave I could see coming!

Ed(ward) Fredkin: When I was in the Air Force and was stationed at Lincoln Labs I had the good fortune to meet Marvin Minsky almost right away and not long afterwards I met John McCarthy and I'd get lots of advice from them. I don't know exactly when I first thought of the idea of the Universe being a simulation on a computer but it was at about that time. And I remember when I told John McCarthy this idea (that had to be around 1960) he said to me something that in our long time of talking to each other has probably said to me maybe a hundred times: "Yes, I've had that same idea," about it being a computer. And I said "Well, what do you think of it?" and he said "Yeah, well we can look for roundoff or

---

[f]Beginning in the late 1960s, Chaitin made contributions to algorithmic information theory and metamathematics, in particular a new incompleteness theorem. He attended the Bronx High School of Science and City College of New York, where he (still in his teens) developed the theories that led to his independent discovery of algorithmic complexity. Chaitin has defined Chaitin's constant Ω a real number whose digits are equidistributed and which is sometimes informally described as an expression of the probability that a random program will halt. Ω has the mathematical property that it is definable but not computable. Chaitin's early work on algorithmic information theory paralleled the earlier work of Kolmogorov and other pioneers.

truncation errors in physics..." and when he said it I immediately thought "Oh! He thinks I'm thinking of an IBM 709 or 7090 (I guess wasn't out yet) in the sky..." and I was thinking of some kind of computational process... I wasn't thinking of roundoff error... or truncation error. When I told this idea to Marvin he suggested that I look at cellular automata and I hadn't heard of cellular automata at that point so what I had to do was to find a paper by... or, find out what I could, I remember I couldn't find the paper that von Neumann had done and from that point on — which was around 1960 (say, 1959 or 1960) — I remained interested in it[g]. And on one of my first experiments on the computer I decided to find the simplest rule that is symmetrical in every possible way, so I came up with the von Neumann neighborhood and binary cellular automata and I thought "what function that's symmetrical is possible?" And, as it turned out it's XOR. And so I programmed that one up and I was kind of amazed by the fact that that simple rule was at least a little bit interesting.

Rob de Ruyter[h]: I also have to go back to my youth, or maybe high-school, when (in the time that I was in high-school, at least) there was a lot of mistery still around about biology and there were a number of wonderful discoveries made and are still being made about how life on the cellular scale works. So that was a kind of an incredibly fascinating world for me that opened as I got interested in biology and learned ever more intricate mechanisms. At the same time of course we learned physics — actually in Holland in high-school you learn them both at the same time, as opposed to in this country. And in physics you get all these beautiful laws, mathematical descriptions of nature and they give you a real sense that you can capture nature in pure thought. Now it was obvious then already that going from that physical picture where you can describe everything very precisely to the workings of a biological cell that there's an enormous range of things that you have to cover in order to understand how this biological cell works and we still don't know that in terms of underlying physical principles. At the same time at that age when you're in high school you also go through all kind of hormonal, etc. transformations and one of the

---

[g]http://www.digitalphilosophy.org/
[h]Rob de Ruyter van Steveninck is a Professor at the Department of Physics and Program in Neural Science Indiana University Bloomington. He is a Fellow of the American Physical Society. He has a Ph.D. (1986) from the University of Groningen, the Netherlands, and has had postdoctoral positions at Cambridge in the UK and the Univ. of Groningen. His main interest is in understanding basic principles underlying coding, computation and inference in the sensory nervous system.

things you start to realize is that (a) as you look at the world you take information in from the world outside but (b) you also have an own mind with which you can also think and do introspection. And you know that this introspection, your own thoughts, somehow have to reside, they have to be built out of matter — that probably, out there somewhere, somehow, that has to happen. Then that's a question that has always fascinated me tremendously and still really fascinates me which is (if you want to put it in shorthand) the question of how matter leads to mind, the mind-matter question and everything that's related to that.

I was lucky enough to find a place in the physics department where there they had a biophysics group and I could study questions of mind — probably a simple mind, the mind of an insect, or at least little bits of the mind of an insect — and try to study them in a very, highly quantitative way that harkens back to the nice, beautiful mathematical description that you get in physics. So I think this tension between matter and how you describe it in mathematical formalisms and thought — how you introspectively know what thought is, and presumably to some extent all animals have thoughts — and thought to some extent takes the form of computation, of course, is the kind of thing that drives me in my work[i]. And as an experimentalist I find it very pleasing that you can do experiments where you can at least take little little slivers off these questions and perhaps make things a little bit more clear about the mind.

(An)T(h)ony Leggett: I guess I am going to be "skunk" of the group, because I actually am not convinced that a computational approach is going to solve at least those fundamental problems of physics which I find most interesting[j]. I certainly think that the sort of ideas, like some of those that we've discussed at this conference, are extremely intriguing and they may well be right — I don't know. And one area in which I think that one could quite clearly demonstrate that this confluence of computational science and physics has been fruitful is of course the area of quantum information and in particular quantum computing. Certainly what's happened

---

[i]Rob de Ruyter van Steveninck has co-authored *Spikes: Exploring the Neural Code* (MIT Press, 1999) with Prof. William Bialek of Princeton University and others

[j]Sir Anthony James Leggett, KBE, FRS (born 26 March 1938, Camberwell, London, UK), is the John D. and Catherine T. MacArthur Chair and Center for Advanced Study Professor of Physics at the University of Illinois at Urbana-Champaign since 1983. Dr. Leggett is widely recognized as a world leader in the theory of low-temperature physics, and his pioneering work on superfluidity was recognized by the 2003 Nobel Prize in Physics. He set directions for research in the quantum physics of macroscopic dissipative systems and use of condensed systems to test the foundations of quantum mechanics.

in that area is that, as I said earlier, an approach coming from computer science gives you a completely new way of looking at the problems, posing them as questions that you would not have thought about otherwise. But quite interesting actually that at least in the decade from 1995 to 2005, when quantum information clearly gained way, a lot of the papers which appeared in Physical Review Letters at that time in some sense could have easily appeared in 1964, but they hadn't. Why not? Because people then didn't have the knowledge (or intuition) to ask such particular type(s) of questions. So certainly I think that was a very very useful and fruitful interaction.

But I think that if someone told me, for example, that such and such a problem which can be easily posed in physics cannot be answered by a computer with a number of bits which is greater or equal to the total number of particles in the universe: I don't think I would be too impressed. And my reaction would probably be: "All right, so what?" I don't think that's useful, I don't think that nature goes around computing what it has to do — I think it just doesn't! But I think perhaps a little more fundamentally, one reason I'm a little skeptical about the enterprise is that I actually think that the fundamental questions in physics[k] have much more to do with the interface between the description of a physical world and our own consciousness[l]. One of the most obvious cases is the infamous

---

[k]In 1987 Oxford University Press published "The Problems of Physics" by Anthony J. Leggett in which he wrote: "Although [the so-called anthropic principle, the 'arrow of time', and the quantum measurement paradox] are associated historically with quite different areas of physics — cosmology, statistical mechanics, and quantum mechanics, respectively — they have much in common. In each case it is probably fair to say that the majority of physicists feel that there is simply no problem to be discussed, whereas a minority insist not only on the problem's existence, but also on its urgency. In no case can the issue be settled by experiment, at least within the currently reigning conceptual framework -a feature which leads many to dismiss it as 'merely philosophical'. And in each case, as one probes more deeply, one eventually runs up against a more general question, namely: In the last analysis, can a satisfactory description of the physical world fail to take explicit account of the fact that it is itself formulated by and for human beings?"

[l]"The anthropic principle in cosmology has a long and venerable history — in fact it goes back effectively to long before the birth of physics as we know it. Most modern versions start from two general observations. The first is that, in the current formalism of particle physics and cosmology, there are a large number of constants which are not determined by the theory itself, but have to be put in 'by hand'. The second observation is that the physical conditions necessary for the occurrence of life — still more for the development of life to human stage — are extremely stringent. It is easy to come to the conclusion that for any kind of conscious beings to exist at all, the basic constants of nature have to be exactly what they are, or at least extremely close to it. The anthropic principle then turns this statement around and says, in effect, that the reason the fundamental

quantum measurement problem[m], which I certainly do regard as a very serious problem; and secondly the question of the arrow of time[n]:that we

constants have the values they do is because otherwise we wouldn't be here to wonder about them." (The Problems of Physics, by Anthony J. Leggett, 1987 Oxford University Press, pp. 145-146)

[m]"Within the formalism of quantum mechanics in its conventional interpretation, it looks as if in some sense a system does not possess definite properties until we, as it were, force it to declare them by carrying out an appropriate measurement. But is this the only possible interpretation of the experimental data? We can summarize the situation succinctly by saying that in the quantum formalism things do not 'happen', while in everyday experience they do. This is the quantum measurement paradox. As indicated above, it is regarded by some physicists as a non-problem and by others as undermining the whole conceptual basis of the subject. Various resolutions of the paradox have been proposed. There is one so to say exotic solution that needs to be mentioned namely the interpretation variously known as the Everett-Wheeler, relative state, or many-worlds interpretation. The basis for this alleged solution is a series of formal theorems in quantum measurement theory which guarantee that the probability of two different measurements of the same property yielding different results (of course, under appropriately specified conditions) is zero The many worlds interpretation, at least as presented by its more enthusiastic advocates, then claims, again crudely speaking that our impression that we get a particular result in each experiment is an illusion; that the alternate possible outcomes continue to exist as 'parallel worlds', but that we are guaranteed, through the above formal theorems, never to be conscious of more than one world, which is moreover guaranteed to be the same for all observers. Let me allow myself at this point the luxury of expressing a strong personal opinion . It seems to me that the many-worlds interpretation is nothing more than a verbal placebo, which gives the superficial impression of solving the problem at the cost of totally devaluing the concepts central to it, in particular the concept of 'reality'. I believe that our descendants two hundred years from now will have difficulty understanding how a distinguished group of scientists of the late twentieth century, albeit still a minority, could ever for a moment have embraced a solution which is such manifest philosophical nonsense."

[n]"It is almost too obvious to be worth stating that the world around us seems to exhibit a marked asymmetry with respect to the 'direction' of time; very many sequences events can occur in one order, but not in reverse order. Why is this observation problematic? As a matter of fact, there is a great deal more to the problem than the above discussion would indicate. Indeed, in general discussions the subject it is conventional to distinguish (at least five different arrows of time. One is the 'thermodynamic' arrow, a second is the 'human' or psychological arrow of time, determined by the fact that we remember the past but not the future. A third arrow is the 'cosmological' one determined by the fact that the Universe is currently expanding rather than contracting. A fourth arrow is the so-called electromagnetic one, and this requires a little explanation. Finally, for completeness I should mention the fifth arrow, which is associated with a class of rare events in particle physics — the so-called CP-violating decays of certain mesons — and probably with other very small effects at the particle level. This arrow is rather different from the others in that the relevant equations are themselves explicitly asymmetric with respect to the sense time, rather than asymmetric being imposed by our choice of solution. At the moment the origins of this asymmetry are very poorly understood, and the general consensus is that it probably does not have anything much to do with the other arrows; but it is too early to be completely sure of this. Regarding the first four arrows a point of view which seems free from obvious internal inconsistencies at least is

can remember the past, that the past influences the future and vice-versa. It's something that underlines some of the most fundamental aspects of our existence as human beings and I don't think we understand them at all, at least I don't think we understand them in physical terms, and I personally find it very difficult to imagine how a computational approach could enhance our approach to any of these problems.

Now one of the reasons I come to conferences like this is that I hope that I may in the future somehow be convinced. So we'll have to wait and see.

Cris(tian) Calude: I guess I am probably the most conventional and ordinary person° here. My interests are in computability, complexity and randomness, and I try to understand why we can do mathematics and what makes us capable of understanding mathematical ideas. I am also a little bit skeptical regarding the power of quantum computing. I've been involved in the last few years in a small project in dequantizing various quantum algorithms, i.e., constructing classical versions of quantum algorithms which are as quick as their quantum counterparts. And I have come to believe that quantum computing and quantum information say much more about physics than they could become useful tools of computation. Finally, I am very interested to understand the nature and the power of quantum randomness, whether this type of hybrid computation where you have your favorite PC and the source of quantum randomness can get you more. Can we surpass the Turing barrier, at least in principle, with this kind of hybrid computation or not?

---

that one can take as fundamental the cosmological arrow; that the electromagnetic arrow is then determined by it, so that it is not an accident that, for example, the stars radiate light energy to infinity rather than sucking it in; that, because radiation is essential to life, this then uniquely determines the direction of biological differentiation in time, and hence our psychological sense of past and future; and finally, that the thermodynamic arrow is connected in inanimate nature with the electromagnetic one, and in the laboratory with the psychological one in the manner described above. However, while it is easy to think, at each step, of reasons why the required connection *might* hold, it is probably fair to say that in no case has a connection been established with anything approaching rigor, and it would be a brave physicist who would stake his life on the assertion, for example, that conscious life must be impossible in a contracting universe. This fascinating complex of problems will probably continue to engage physicists and philosophers (and biologists and psychologists) for many years to come." The Problems of Physics, 1987 Oxford University Press, pp. 148-157

°Cristian S. Calude is Chair Professor at the University of Auckland, New Zealand. Founding director of the Centre for Discrete Mathematics and Theoretical Computer Science. Member of the Academia Europaea. Research in algorithmic information theory and quantum computing.

Tom(maso) Toffoli[p]: Well, I was born by a historical accident in a post office and that may have something to do with all of this (laughter) because it was exactly organized like Ethernet: press a key and you bring a whole lot of lines (40 miles) from a high value to ground, so you have to listen to what you are typing and if what you get is different from what you are typing, or what you're expecting to hear, then it means that someone else is on the line too. So you both stop. And then both of you start again, at random — exactly in the spirit of Ethernet. But anyway, when I was five, it was right after the end of the war, my mother already had four kids and she was very busy trying to find things for us to eat from the black market and such — so I was left off for the most part of the day in a place on the fifth or sixth floor. Where we lived we had long balconies with very long rails, for laundry etc. and the rail was not raised in the middle, it was just hooked, stuck in the wall at the endpoints on the two sides. So I was like a monkey in a cage in that balcony and I had to think and move around and find a way to spend time in some way. So one of the things I noticed was that as I was pulling and shaking this thing, the rope, somehow I discovered that by going to more or less one third of its length and shaking it in a certain way, eventually the thing would start oscillating more and more and more. And I discovered what you could call resonance, phase locking and other things like that — and I felt really proud about this thing, about this power unleashed into your [tiny] hands. But I also got very scared because — as a result of my early scientific activity — the plaster had come out at the place where the rail was hooked in the wall and there was even plaster outside on the sidewalk if you looked down from the sixth floor balcony where I was, so I was sure I was going to get into trouble when my mother would come home later that day.

In any event, the point that I'd like to make is that "I cannot understand

---

[p]Prof. Tommaso Toffoli's interests and expertise include among others: Information Mechanics, Foundations and physical aspects of computing, Theory of cellular automata, Interconnection complexity and synchronization, Formal models of computation consistent with microscopical physics (uniformity, locality, reversibility, inertia and other conservation principles, variational, relativistic, and quantum aspects of computation). He proved the computation-universality of invertible cellular automata (1977); formulated the conjecture (later proved by Kari) that all invertible cellular automata are structurally invertible (1990). Introduced the Toffoli gate (1981), which was later adopted by Feynman and others as the fundamental logic primitive of quantum computation. Proposed, with Fredkin, the first concrete charge conserving scheme for computation (1980) — an idea that has been taken up by the low-power industry in recent years. Proved that dissipative cellular automata algorithms can be replaced by non dissipative lattice gas algorithms (2006).

why people cannot understand" (as Darwin wrote to a friend) "that there's no such thing as simply experiments — there are only experiments 'for' or 'against' a subject." In other words you have to ask a question first, otherwise the experiment is not a very useful thing. Now you will say: "Fine, but how are you going to answer the question that you're making the experiment for?" And what we do with computation, cellular automata and so on, is that we just make our own universe. But the one that we make, we know it, we make the rules so the answers that we get there are precise answers. We know exactly whether we can make life within this CA (like von Neumann) and we know whether we can compute certain things, so it's a world that we make so that we have precise answers — for our limited, invented world — and that hopefully can give us some light on answers for the questions in the real world. And so we have to go continually back and forth between the worlds that we know because we have complete control [over them] and we can really answer questions and the other worlds about which we would really like to answer questions. And my feeling is that computation, CA and all these things that Stephen has called our attention to are just attempts to get the best of the two worlds, essentially. The only world we can answer questions for is the one we make, and yet we want to answer the questions about the world in which we are — which has all of its imperfections, and where resonance does not give you an infinite peak but still gives you crumbled plaster on the floor near the wall and so on.

But I think that we should be able to live with one foot in one world and the other in the other world.

Stephen Wolfram: Let's see, the question I believe was how did one get involved with computation, and how did one get interested in computation and then interested in these kind of things. Well, the story for me is a fairly long one by now: when I was a kid I was interested in physics. I think I got interested in physics mainly because I wanted to understand how stuff works and at the time, it was in the 1960s, roughly, physics was the most convincing kind of place to look for an understanding of how stuff works. By the time I was 10-11 I started reading college physics books and so on and I remember a particular moment, when I was 12, a particular physics book that had a big influence on me: it was a book about statistical mechanics which had on the cover this kind of a simulated movie strip that purported to show how one would go from a gas of hard spheres or something like that, that was very well organized, to a gas of hard spheres that looked

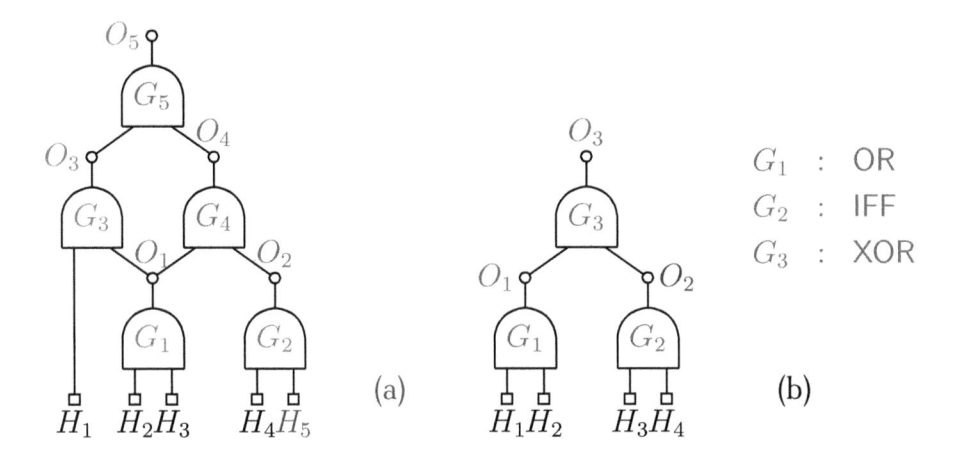

Fig. 2. Binary switching circuits.

that the output of a gate can serve as input to several gates). Moreover, one or more outputs of gates (here O_5) are designated as the outputs of the circuit. We consider the clocked mode of functioning of switching circuits, meaning that there is a global clock, and at each tick of the clock:

- a combination of 0's and 1's is set on the set of inputs (here $H_1, \ldots, H_5$),
- gates are setting either 0 or 1 on their outputs depending on the combination of 0's and 1's on their inputs at the previous tick of the clock.

Which signal (0 or 1) is set on the output of a gate depends on the truth table of this gate, which tells for each combination of inputs whether 0 or 1 will be set on the output. Thus, e.g., for the Exclusive OR gate (XOR), as well as for the OR gate and the IFF gate, the truth tables look as follows:

XOR

x_1	x_2	y
0	0	0
0	1	1
1	0	1
1	1	0

OR

x_1	x_2	y
0	0	0
0	1	1
1	0	1
1	1	1

IFF

x_1	x_2	y
0	0	1
0	1	0
1	0	0
1	1	1

To implement a switching circuit by a reaction system, we implement each gate it comprises by implementing its truth table. This is done by providing reactions that produce the output (y) whenever, for a given combination of 0's and 1's on the inputs, the value of the output in the truth table is 1.

1981. In order to build this computer system I had to understand at a very fundamental level (a) how to set up a wide range of computations and (b) how to build up a language that should describe a wide range of computations. And that required inventing these primitives that could be used to do all sorts of practical computations. The whole enterprise worked out rather well, and one of my conclusions from that was that I was able to work out the primitives for setting things up into a system like SMP. And then for a variety of reasons when I got to started thinking about basic science again my question was: "Couldn't I take these phenomena that one sees in nature and kind of invent primitives that would describe how they work — in the same kind of way that I succeeded in inventing primitives for this computer language that I built?" I actually thought that this would be a way of getting to a certain point in making fairly traditional physics models of things, and it was only somewhat coincidentally that later I actually did the natural science to explore abstractly what the computer programs that I was setting up did. And the result of that was that I found all sorts of interesting phenomena in CAs and so on.

That got me launched on taking more seriously the idea of just using computation, computer programs, as a way to model lots of things. I think that in terms of the conviction for example that the universe can be represented by computation I certainly haven't proved that yet. I've only become increasingly more convinced that it's plausible — because I did use to say just like everybody else: "Well, all these simple programs and things: they can do this amount of stuff but there'll be this other thing (or things) that they can't do." So, for instance, say, before I worked on the NKS book one of the things that I believed was that while I could make simple programs representing the standard phenomena in physics, that standard phenomena — not at the level of fundamental physics, but at the level of things like how snowflakes are formed, and so on — I somehow believed that, for example, biological systems with adaptation and natural selection would eventually lead to a higher level of complexity that couldn't be captured by these very simple programs that I was looking at. And I kind of realized as a result of working on the NKS book and so on that that's just not true. That's a whole separate discussion, but I kind of believed that that there would be different levels of things that could and couldn't be achieved by simple programs. But as I worked on more and more areas, I got more and more convinced that the richness of what's available easily in the computational universe is sufficiently great that it's much less crazy to think that our actual universe might be made in that kind of way. But we still don't know if that's right until we actually have the final theory so to speak.

Is the Universe a Computer?

George Johnson: Very good. And now we have a fairly formal linear exercise for you. Let me preface it by saying that early in the conference there was, I thought, this very interesting exchange between Tom Toffoli and Seth Lloyd, who's another old Santa Fe friend. And I thought it was funny that he said he had to leave to help his kids carve jack-o-lanterns, because shortly after I met Seth he invited me to his house in Santa Fe that he was renting and he was carving jack-o-lanterns then as well — this was before he was married and had children. Our other jack-o-lantern carver then was Murray Gell-Mann. I was writing a biography of Murray at the time and he had not agreed to cooperate, in fact he was being somehow obstructive, but somehow sitting there with a Nobel prize winner carving jack-o-lanterns helped break the ice.

Seth of course has described his grand vision of the universe as a quantum computer — the universe *is* a quantum computer, he said, and then Dr. Toffoli had an interesting subtle objection and basically I think you said that it was this idea of the computational relation between the observer and the system, that somebody has to be looking at it. And later, actually — to skip out to another context during the same session — when Seth made some joke about the election and you said, quoted Stalin saying that it doesn't matter who votes, it's who counts the votes, that kind of seemed to connect back to it and got me thinking about this whole idea of endo-physics of looking at the universe from within or from without. So what I am wondering is this: does Seth's view of looking at the universe not *as* a computer but saying that *it is* a computer, implies a God's eye view of somebody who's watching the computation? What is the universe computing, if it is computing? Is the universe *a* computer, or is the universe *like* a computer– or do the two state the exact same thing?

Tom Toffoli: Are you asking anyone in particular?

George Johnson: Anyone. Please jump in.

Stephen Wolfram: I can think of a few fairly obvious things to say. You know, there's the question of: "What are models?" In other words, if you say, in the traditional physics view of things: "Is the motion of the Earth a differential equation?" No, it's not — it's *described* by a differential equation, it isn't a differential equation. And I don't think, maybe other people here think differently about it, but my view of computation as the underlying thing in physics is that computation is a *description* of what happens in physics. There's no particular sense in which it is useful to say the universe is a computer. It's merely something that can be described as a computation[al process] or something that operates according to the rules of some program.

Cris Calude: I agree and I would add one more idea. When you have a real phenomenon, and a model — a model typically is a simplified version of the reality. In order to judge whether this model is useful or not you have to get some results (using the model). So I would judge the merit of this idea about the universe being modeled faithfully by a gigantic computer primarily by saying: "Please tell me three important facts that this model reveals about the universe that the classical models can't."

Ed Fredkin: I will tell you one in a minute. The thing about a computational process is that we normally think of it as a bunch of bits that are evolving over time plus an engine — the computer. The interesting thing about informational processes (digital ones) is that they're independent of the engine in the sense of what the process is: any engine that is universal (and it's hard to make one that isn't) can — as long as it has enough memory — exactly produce the same informational process as any other. So, if you think that physics is an informational process then you don't have to worry about the design of the engine — because the engine isn't here. In other words, if the universe is an informational process then the engine, if there is one, is somewhere else.

Greg Chaitin: George you're asking a question which is a basic philosophical question. It's epistemology versus ontology. In other words when you say the Universe looks like a computer, that this is a model that is helpful — that's an epistemological point of view, it helps us to understand, it gives us some knowledge. But a deeper question is what is the universe really. Not just we have a little model, that sort of helps us to understand it, to know things. So, that is a more ambitious question! And the ancient Greeks, the pre-Socratics had wonderful ontological ideas: the world is number, the world is this, the world is that. And fundamental physics also wants to answer ontological questions: "What is the world really built of at the fundamental level?" So, it's true, we very often modestly work with models, but when you start looking at fundamental physics and you make models for that and if a model is very successful — you start to think [that] the model *is* the reality. That this is really an ontological step forward. And I think modern philosophy doesn't believe in metaphysics and it certainly doesn't believe in ontology. It's become unfashionable. They just look at epistemological questions or language, but mathematicians and physicists we still care about the hard ontological question: If you're doing fundamental physics you *are* looking for the ultimate reality, you *are* working on ontology. And a lot of the work of lots of physicists nowadays really resembles metaphysics– when you start talking about all possible worlds, like Max Tegmark does or many other people do, or David Deutsch for that matter. So philosophers have become very timid, but some of us here, we are continuing in the tradition of the pre-Socratics.

Ed Fredkin: The problem — one problem — we have is the cosmogony problem which is the origin of the universe. We have these two sort of contradictory facts: one is that we have a lot of conservation laws, in particular mass energy is conserved and we have the observation that the universe began, it seems, with a big bang not so long ago, you know, just thirteen or fourteen billion years ago. Well, basically there's a contradiction in those two statements: because if something began and you have a conservation law where did everything come from, or how did it come about? And also the idea that it began at some time is problematic — the laws of physics can't help us right there because, if you think a lot about of those details it's not so much why matter suddenly appeared but why is there physics and why this physics and stuff like that.

And there is a way to wave your hands, and come up with a kind of answer that isn't very satisfying. Which is that you have to imagine that there is some other place — I just call it 'other' — and in this other place for whatever reason there is an engine and that engine runs an informational process. And one can actually come to some kind of feeble conclusions about this other place, because there are some numbers that we can state about how big that computer must be. In other words if we ask the question what would it take to run a computer that emulated our universe, well — we can guess some quantitative numbers if we can figure out a few things, and so you can make a few statements [educated guesses] about that other kind of place. The point about 'other' is that it is a place that doesn't need to have conservation laws, it is a place that doesn't need to have concepts such as 'beginnings' and 'ends' — so there aren't that many constraints. And one of the wonderful things about computation is that it is one of the least demanding concepts, if you say: well, what do you need in order to have a computational engine? Well, you need a space of some kind. What kind of space? How many dimensions does it have to have. Well, it could have three, it could have two, it could have one, it could have seven — it doesn't matter. They can all do the same computations. Of course, if you have a one-dimensional space you can spend a lot of time overcoming that handicap. But does it need the laws of physics as we know them? No, you don't need to have the laws of physics as we know them. In fact, the requirements are so minimal for having a computation compared to the wonderful rich physics we have that it's very, very simple. I am, in fact, reminded of a science-fiction story, by a Polish author where there's a robot that could make everything that started with the letter 'n' and in Polish, like in Russian, or in English, the word 'nothing' starts with an 'n' so someone bored said: "OK, make nothing." And the robot started working and where the sky had previously been white with so many stars and galaxies just minutes earlier, it slowly started to fade away, little by little, galaxy by galaxy. Admittedly, it's just a science fiction story, but the point is that one could even inquire as to what would be the motivations to create an emulation like this? You can imagine that there is some question, and [then] one needs to think about: what could the question be?

We can speculate about that. But the point is that this is an explanation that says: well there's this thing called 'other' that we don't know anything about — as opposed to all other explanations that imply that some kind of magic happened. Well: I don't like magic, myself.

Rob deRuyter: A couple of sentences from a biological perspective: Let's take a naive standpoint that there's a world out there and that there's a brain and this brain needs to understand what's happening in the world, what's going around and unfortunately, maybe — or fortunately for us — this brain is an engine that is really well adapted to information processing in the savannah, or in the trees, or wherever. I don't think that the brain itself is a universal computational engine — at least I don't see it that way — but it's a device that is extremely well adapted to processing information that comes to our sensory organs, in from the world that we happen to inhabit. So if we want to start thinking about more complex things or deeper things we need to develop tools that allow us translate our thoughts about the phenomena that we observe in the world. Or the other way around: just like we developed hammers and pliers, in dealing with phenomena in the world we need to develop tools to think [about them].

I have no idea of what the limitations that the structure of our brain are, and that impose — I mean, there must be limitations (in the way we think) that impose structure on those tools that we develop to help think about things. But computation, I think, in a sense, is one of the tools [that] we tried to develop in dealing with the world around us. And what computation allows us to do, like mathematics, is to be able to develop long chains of reasoning that we normally don't use, but that we can extend to reason about very long series, sequences of complicated observations about phenomena in the world around us. So what interests me is this relationship between the way we think — and the way we have evolved to think about the world around us — and the things that we think about now in terms of scientific observations and the origin of the universe. To what extent does the hardware that we carry around inform and determine the kinds of tools and computations and the strategies that we're using?

Tom Toffoli: I would like to give some examples to illustrate why the question of whether the universe is a computer, is a really hard question. In some sense it resembles the question of what is life. Let's take for example the concept of randomness: say you buy a random number generator and you start producing numbers with it. First number you get out is 13, then you get 10, 17, and so on — and then you start asking yourself about the numbers that you obtained: how random are they? What is random? Is 10 a random number? Is 13 a random number? How about 199999 — is it a random number? Then you realize that, of course, randomness is not a property of the number, it is a property of the process. You pay for a random number generator because you want to be surprised. You want not to know what will come out. If you knew it — it would not be random to you because it would be perfectly predictable. So we use that term 'a random number' as an abbreviation for whatever [sequence] is produced by a random number generator. I'll give you another example: somebody tried to trademark icons. When they were invented icons were small: 16 by 16 pixels black and white, and you can draw, you know, some simple things with those 16 by 16 bits. So you may want to trademark them. And some greedy businessman said: look I will go to the judge and I will try to trademark *the entire set* of 16 by 16 pixel icons, all of them, and everybody has to pay me. So now if you are the judge and you have to decide whether you can or cannot allow to someone the right to trademark not just one icon, but all of the icons that can be made that way. And if you are the judge you really have two ways: you say (a) either you have to give me a reason why this icon is really something interesting or (b) you pay 5 cents for each one of the icons that you register and being that there are 2256 items you know, you don't have to exercise any judgment, it would just turn into a big contribution to the community.

In other words, the question is: what makes an icon what it is? Is it the fact that it is 16 by 16 bits or that you have reason to believe that there is something useful in it? Brian Hayes, whom you mentioned a moment ago, once said: "What surprises me is that most people don't use the computer for what makes it unique and powerful — which is that it is a programmable machine." My partial definition of a computer is: something that can compute (evaluate) a lot of different functions. If it can evaluate just one function, then I wouldn't call it a computer I would call it a special purpose machine or whatever it is. So we may get the surprise that if we find the formula that gives us the universe as a computer, then at that very point the universe itself becomes a special purpose machine. I mean, we know the formula, we know the machine, we know the initial conditions, and we just go: tick, tick, tick, tick. And it's the largest computer but nobody can program it — if this is the universe, we cannot program it because we are inside of it.

So, the definition of a computer is a bit like the definition of life and the definition of evolution or being adaptive: if there isn't a component of adaptiveness, I wouldn't call the thing a computer. Now the thing can be formalized better, I just said it in an intuitive way, but I'm asking some of these questions to try to clarify a bit what exactly it was that we wanted.

Greg Chaitin: Look, I'd like to be aggressive about this — the best way for me to think about something is to make claims that are much too strong (at least it brings out the idea). So the universe *has to* be a computer, as Stephen said, because the only way to understand something is to program it. I myself use the same paradigm. Every time I try to understand something the way I do it, is: I write a computer program. So the only possible working model of the universe has to be a computer — a computational model. I say a working model because that's the only way we can understand something: by writing a program, and getting it to work and debugging it. And then trying to run it on examples and such. So you say that you understand something only if you can program it. Now what if the universe decides however that it's not a — that you *can't* do a computational model about it. Well, then: no problem. It just means we used the wrong computers. You know, if this universe is more powerful than a computer model of it can be, that means that our notion of what a computer is is wrong and we just need a notion of computer that is more powerful, and then things are in sync. And by the way, there is a way to define the randomness of individual numbers based on a computer [we hear: infinite ones, from Toffoli] well, anyway, but that's another issue.

Stephen Wolfram: One point to make, in relation to using computation as a model of a universe: we're used to a particular thing happening when we do modeling, we're used to models being idealizations of things. We say we're going to have a model of a snowflake or a brain or something like that, we don't imagine that we're going to make a *perfect* model! It's a very unusual case that we're dealing with in modeling fundamental physics (perhaps modeling isn't the right term, because what we imagine is that we're going to actually have a precise model that reproduces our actual universe in every detail.) It's not the same kind of thing as has been the tradition of modeling in natural science, it's much more. So when you say, when you talk about what runs it and so on, it's much more like talking about mathematics: you wouldn't ask when you think about a mathematical result, [if you] work out some results in number theory, for example, one wouldn't be asking all the time "Well, what's *running* all these numbers?" It's just not a question that comes up when one is dealing with something where what you have is a precise model of things.

One other point to make regarding the question of to what extent our efforts of modeling relate to what our brains are good at doing and so on. One of the things I am curious about is: if it turns out to be the case that we can find a precise representation, a new representation (better word than model) for our universe in terms of a simple program and we find that it's, you know, program number such and such — what do we conclude in that moment? It's a funny scientific situation, it kinds of reminds one of a couple of previous scientific situations, like for instance Newton was talking about working out the orbits of planets and so on and made the statement that once the planets are put in their orbits then we can use the laws of gravity and so on to work out what would happen — but he couldn't imagine what would have set the planets originally in motion in their orbit in the first place. So he said, "Well, the Hand of God must have originally put the planets in motion. And we can only with our science figure out what happens after that." And it's the same with Darwin's theories: once we have life happening then natural selection will lead us inexorably to all the things that we see in biology. But how to cause life in the first place

— he couldn't imagine. So some of the things I'd be curious about would be: if in fact we do come up with a precise representation of the universe as a simple program – what do we do then and can we imagine what kind of a conclusion we can come to, about why this program and not another program and so on? So one of the possibilities would be that we find out that it's, you know, program number 1074 or whatever it is. The fact that it is such a small number might be a consequence of the fact that our brains are set up because they are made from this universe in such a way that it is inevitable, and in [all] the enumerations that we [might] use our universe will turn out to be a small number universe. I don't think that's the case — but that's one of those self fulfilling prophecies: because we exist in our universe our universe will have to have laws that will somehow seem simple and intuitive to us. I think it's more clear than that, but that's one of the potential resolutions of this question: so now we have our representation of the universe, what do we conclude metaphysically from the fact that it is this particular universe representation and not the other one.

Tony J. Leggett: I think with regard to the strong and forceful statement that could be related to Seth Lloyd's argument namely that the universe is a computer, I just have a little very naive and simple question: it seems to me that if a statement is called for then the converse is not called for. So my question to Seth Lloyd is: "What would it be like for the universe *not* to be a computer?" And so far I fear I don't find that particular statement truly helpful. I do find quite helpful the thesis that it may be useful to look at the universe in the particular framework of computational science and to ask different questions about it, although I have to say that I'm not yet convinced that looking at it in this way is going to help us to answer some very obvious questions, some of which are usually met with a certain amount of friction, namely is the anthropic principle physically meaningful? That is, why do the constants of nature as we know them have the particular values that they have? Is it perhaps for some arbitrary reason, or maybe for some other deeper reason. Now, of course, there have been plenty of arguments and speculations here about all sorts of things but as far as I can see they don't seem to give any direct relationship (of the universe with a specific computational model, or the universe as a computer.) But I would really like to hear a plausible argument as to why this point of view takes us further on these types of questions that I just mentioned.

Ed Fredkin: I want to react a little bit to what Stephen was saying. There exist areas where we use computers to write programs that are *exact* models, exact and perfect in every possible way — and that is when you design a program to emulate another computer. This is done all the time both for writing a trace program map and for debugging, where you write an emulator for the computer that the software is running on. Or you want to run software that is for another computer like the Mac did when it switched CPUs from the 68000 to the PowerPC: they made an emulator, which is an exact implementor of the software on another computer.

This relates to something that I used to call 'The Tyranny of Universality.' Which is: "Gee, we can never understand the design of the computer that runs physics since any universal computer can do it." In other words if there's a digital computer running all physics of course then any computer can do it, but then, after convincing myself that that point of view made sense a long time later I came up with a different perspective: that if the process that runs physics is digital and it is some kind of CA there will exist a model that's one to one onto in terms of how it operates. And it would probably be possible to find in essence the simplest such model so that if some kind of experimental evidence showed us that physics is some kind of discrete, digital [physical] process like a CA I believe we should be able to find the exact process (or, you know, one of a small set of processes) that implement it exactly.

Is the Universe Discrete or Continuous?

George Johnson: Maybe a good way to get to Tony Leggett's question that he raised at the end is just to ask the same question that our host Adrian asked after we saw that brief film that he showed us first thing this morning which is: is there a fundamental difference between a computational physics, or a computational model, or emulation of the universe and quantum mechanics? Or is there a fundamental distinction between a discrete and a continuous physics? Does anyone have a reaction to that?

Stephen Wolfram: (speaking to George Johnson): So if you're asking is there some definitive test for whether the universe is somehow discrete or somehow fundamentally continuous...

George Johnson: Yeah — if there's a conflict that it could possibly be both at a deeper level.

Stephen Wolfram: If you're asking for that — for example in the kinds of models that I made some effort to study, there are so many different ways to formulate these models that this question "Is it discrete or is it continuous?" becomes kind of bizarre. I mean, you could say: we'll represent it in some algebraic form in which it looks like it's talking about these very continuous objects — and what matters about it may yet turn out to be discrete, it may turn out to be a discrete representation (which is much easier to deal with). So I think that at the level of models that I consider plausible the distinction between continuity and discreteness is much less clear than we expect. I mean, if you ask this question I think you end up asking 'very non-physics questions' like, for example, how much information can in principle be in this volume of space. I'm not sure that without operationalizing that question that it's a terribly interesting or meaningful question.

Tom Toffoli: I would like to say something that will be very brief. Look at CAs, they seem to be a paradigm for discreteness. But as it turns out one of the characterizations of CAs is that they are a dynamical system [that perform certain kinds of translations] and they are continuous with respect to a certain topology which is identical to the Cantor set topology, continuous in exactly that very sense of that definition of continuity that is studied in freshman calculus. But the interesting thing is that this is the Cantor set topology, invented by Cantor (the one with the interval where you remove the middle third, etc.) And as it turns out, this topology for CAs is — not in the sense of geometrical topology, but in the sense of set topology of circuits with gates that have a finite number of inputs and a finite number of outcomes outputs — that is exactly the topology of the Cantor set, so it's sort of a universal topology for computation. And so we come full circle that (a) something that was not invented by Cantor to describe computers in fact represents the natural topology to describe discrete computers and (b) the moment you take on an in[de]finite lattice then you have continuity exactly the kind of continuity, continuity of state, of the dynamics that you get when you study continuous functions. So these are some of the surprises that one gets by working on things.

Stephen Wolfram: I just want to say with respect to the question of how do we tell, you know, the sort of thing that Tony is saying: "How do we tell that this is not all just complete nonsense?" Right?

Tony Leggett: Refutation of the negative statement.

Stephen Wolfram: Yes, right. We're really only going to know for sure if and when we finally get a theory, a representation that is the universe and that can be represented conveniently in computational form. Then people will say: "Great! This computational idea is right, it was obvious all along, everybody's thought about it for millions of years..."

Tom Toffoli: At that point they will probably say: "In fact it's trivial!"

Stephen Wolfram (agrees laughing, everybody laughs): And I think that until that time one could argue back and forth forever about what's more plausible than what and it's always going to be difficult to decide it based on just that. Yet these things tend to be decided in science in a surprisingly sociological way. For example the fact that people would seriously imagine that aspects of string theory should be taken seriously as ways to model the reality of a physical universe it's — it's interesting and it's great mathematics — but it's a sociological phenomenon that causes [or forces] that to be taken seriously at the expense of other kinds of approaches. And it's a matter of history that the approach we're using (computational ideas) isn't the dominant theme in thinking about physics right now. I think it's purely a matter of history. It could be that in place of string theory people could be studying all kinds of bizarre CAs or network systems or whatever else and weaving the same kind of elaborate mathematical type web that's been done in string theory and be as convinced as the string theorists are that they're on to the right thing. I think at this stage until one has the definitive answer, one simply doesn't know enough to be able to say anything with certainty and it's really a purely sociological thing whether we can say that this is the right direction or this isn't the right direction. It's very similar actually to the AI type thing, people will argue forever and ever about whether it's possible to have an AI and so on — and some of us are actually putting a lot of effort into trying to do practical things that might be identified as relevant to that. I think that actually the AI question is harder to decide than the physics question. Because in the physics case once we'll have it it's likely (it seems to me) that we'll be able to show that the representation is of the universe that is obviously the actual universe and the question will be closed. Whereas the question of AI will be harder to close.

Cris Calude: Apparently there is an antagonism between the discrete and continuous view. But if we look at mathematics there are mathematical universes in which discrete and continuous are co-existing. Of course, what Tom said, the Cantor space is a very interesting example, but it might be too simple for the problem that we are discussing. For instance, non-standard analysis is another universe where you find [this same phenomenon] and you have discreteness and you have continuity and maybe, to the extents that mathematics can say something about the physical universe, it could be just a blend of continuous and discreteness and some phenomena may be revealed through discreteness and some others will be revealed through continuity and continuous functions.

Greg Chaitin: Again I am going to exaggerate — on purpose. I think the question is like this: discreteness vs. continuity. And I'm going to say why I am on the side of discreteness.

The reason is this: I want the world to be comprehensible! Now there are various ways of saying this. One way would be: God would not create a world that we couldn't understand. Or everything happens for a reason (the principle of sufficient reason). And other ways. So I guess I qualify as a neo-Pythagorean because I think the world is more beautiful, if it is more comprehensible. We are thinkers, we are rationalists — we're not mystics. A mystic is a person that gets in a communion with an incomprehensible world and feels some kind of community and is able to relate to it. But we want to understand rationally so the best universe is one that can be completely understood and if the universe is discrete we can understand it — it seems to me. This is something that you said, at one point, Ed — it is absolutely totally understandable because you run the model and the model is exactly what is happening.

Now a universe which uses continuity is a universe where no equation is exact, right? Because we only have approximations up to a certain order. So I would also say: a universe would be more beautiful if it were discrete! And although we now end up in aesthetics, which is even more complicated, I would still say that a discrete universe is more beautiful, a greater work of art for God to create — and I'm not religious, by the way. But I think it's a very good metaphor to use — or maybe I am religious in some sense, who knows?

Another way to put it is let's say this universe does have continuity and messy infinite precision and everything – well, too bad for it. Why didn't God create as beautiful a universe as he could have?

Tom Toffoli: He should have asked you, Greg!

Greg Chaitin (laughs): What? ... No... No ... maybe at this point I think Stephen is the leading candidate for coming up with a ... [everybody is still laughing, including Chaitin who continues] ... so that would be more beautiful it seems to me. You see, it would be more understandable it would be more rational it would show the power of reason. Now maybe reason is a mistake as may be to postulate that the universe is comprehensible — either as a fundamental postulate or because you know God is perfect and good and would not create such a universe, if you want to take an ancient theological view. Maybe it's all a mistake, but this one of the reasons that I'm a neo-Pythagorean, because I think that would be a more beautiful, or comprehensible universe.

Stephen Wolfram: I have a more pragmatic point of view, which is that if the universe is something that can be represented by something like a simple discrete program, then it's realistic to believe that we can just find it by searching for it. And it would be embarrassing if the universe would indeed be out there in the first, you know, billion universes that we can find by enumeration and we never bothered to even look for it. [There's a very sustained reaction from the rest of the round table members, especially Greg Chaitin, whom we hear laughing.] It may turn out that, you know, the universe isn't findable that way — but we haven't excluded that yet! And that's the stage we're at, right now. Maybe in, you know, 10-20-50 years we will be able to say: yes we looked at all the first I don't know how many — it will be like looking for counterexamples of the Riemann hypothesis, or something like that — and we'll say that we've looked at the first quadrillion possible universes and none of them is our actual universe, so we're beginning to lose confidence that this approach is going to work. But right now we're not even at the basic stage of that yet.

Ed Fredkin : If this were a one dimensional universe, Steve (Wolfram), you would have found the rule by now, right? Because you've explored all of them...

Tony Leggett: Well, George I think, raised the question whether quantum mechanics has any relevance to this question, so let me just comment briefly on that. I think if one thinks about the general structure of quantum mechanics, and the ways in which we verify its predictions, you come to the conclusion that almost all the experiments (and I'll stick my neck out and say *all* the experiments) which have really shown us interesting things about quantum mechanics do measure discrete variables in fact. Experiments on the so-called macroscopic quantum coherence, experiments on Bell's theorem, and so forth — they all basically use discrete variables in practice. Now of course the formalism of quantum mechanics is a continuous formalism. You allow amplitudes to have arbitrary values, but you never really measure those things. And I think that all one can say when one does sometimes measure things like position and momentum which are [apparently] continuous variables — if you look at it hard you'll see that the actual operational setup is such that you are really measuring discrete things. So measurements within the framework of quantum mechanics which claim to be of continuous variables usually are of discrete variables. So I think from the, as it were, the ontological point of view, one can say that quantum mechanics does favor a discrete point of view.

IS THE UNIVERSE RANDOM?

George Johnson: When Hector[q], Gerardo[r] and I were talking about good questions that would stimulate debate we thought that perhaps we should ask something that would be really really basic about randomness — and the great thing about being a journalist particularly a science journalist is that you get this license of asking really really smart people questions about things that have been puzzling you. And this is something that has always kind of bugged me — the SETI (Search for ExtraTerrestrial Intelligence) where we get these signals from space which are then analyzed by these computers, both by super computers and by SETI at home, where you donate some CPU time on your PC, computer cycles etc. They're looking for some structure in what appears to be random noise. And I was wondering we're getting a signal that seems to be pure noise but to some extent — as I think Tomasso Toffoli has suggested — perhaps the randomness is only in the eye of the beholder.

If, for example we're getting this noisy signal that just seems to be static — how do we know we're not getting the ten billionth and fifty seventh digit of the expansion of π forward? How do we know that we're not getting line ten trillion quadrillion stage forward of the computation of the rule 30 automata? So I am wondering if you can help me with that.

Tom Toffoli: This is not an answer. It's just something to capture the imagination. Suppose that people are serious about computing and they say: "Look: you're not using your energy efficiently because you're letting some energy — that has not completely degraded — out." So they start to make better re-circulators, filters and so on and now whatever thermal energy comes out is as thermalized as possible. Because if it's not, they would have committed a thermodynamical sin. But this is exactly what happens when you look at the stars. They are just sending close to thermodynamical equilibrium a certain temperature — so you can say well probably then this is prima facie evidence that that there are people there computing and they are computing so efficiently that they are just throwing

---
[q]Hector Zenil
[r]Gerardo Ortiz

away garbage, they're not throwing away something that is still recyclable! And this could be an explanation as to why we see all these stars with all these temperatures.

Stephen Wolfram: You know I think the question about SETI and how it relates to the type of things we're talking about — I think it gets us into lots of interesting things. I used to be a big SETI enthusiast and because I'm a practical guy I was thinking years ago about how you could make use of unused communication satellites and use them to actually detect signals and so on. And now I have worked on the NKS book for a long time, and thought about the PCE and I have became a deep SETI non-enthusiast. Because what I realized is that it goes along with statements like "the weather has a mind of its own". There's this question of what would constitute — you know, when we say that we're looking for extra terrestrial intelligence — what actually is the abstract version of intelligence? It's similar to the old question about what is life, and can we have an abstract definition of life that's divorced from our particular experience with life on the Earth. I mean, on the Earth it is pretty easy to tell whether something — reasonably easy to tell whether something — is alive or not. Because if it's alive it probably has RNA it has some membranes it has all kinds of historical detail that connects it to all the other life that we know about. But if you say, abstractly: what is life? It is not clear what the answer is. At times, in antiquity it was that things that can move themselves are alive. Later on it was that things that can do thermodynamics in a different way than other things do thermodynamics are alive. But we still — we don't have — it's not clear what the abstract definition of life is divorced from the particular history. I think the same is true with intelligence. The one thing that most people would (I think) agree with — is that to be intelligent you must do some computation. And with this principle of computational equivalence idea what one is saying is that there are lots of things out there that are equivalent in the kind of computation that they can do ...

Tom Toffoli: But you can also do computation without being intelligent!

Stephen Wolfram (replying to Toffoli): That's precisely the question: can you — what is the difference, what is the distinctive feature of intelligence? If we look at history, it's a very confusing picture: a famous example that I like is when Marconi had developed radio and (he had a yacht that he used to ply the Atlantic with, and) at one point he was in the middle of the Atlantic and he had this radio mast — because that was the business that he was in — and he could hear these funny sounds, you know: ... wooo ... ooooeeo ... eeooo ... woooo ... this kind of sounds out in the middle of the Atlantic. So what do you think he concluded that these sounds were? He concluded that they must be radio signals from the martians! Tesla, was in fact more convinced that they were radio signals from the martians. But, what were they in fact? They were in fact some modes of the ionosphere on the Earth, they were physical processes that — you know, something that happens in the plasma. So the question was, how do you distinguish the genuinely intelligent thing, if there is some notion of that, from the thing that is the ... the computational thing that is.

The same thing happened with pulsars, when the first pulsars were discovered. In the first days of discovery it seemed like this periodic millisecond thing must be some extraterrestrial beacon. And then it seemed like it was too simple. We now think it's too simple to be of intelligent origin. It also relates to this question about the anthropic principle and the question of whether our universe is somehow uniquely set up to be capable of supporting intelligence like us. When we realize that there isn't an abstract definition of actual intelligence, it is (as I think) just a matter of doing computation. Then the space of possible universes that support something like intelligence becomes vastly broader and we kind of realize that this notion of an anthropic principle with all these detailed constraints — just doesn't make much sense.

There's so much more we can say about this, and I'll let others do so.

George Johnson: Randomness: is it in the eye of the beholder?

Greg Chaitin: George, I suppose it would be cowardly of me not to defend the definition of randomness that I have worked on all my life, but I think it is more fun to say (I was defending rationalism, you know) that a world is more understandable because it is discrete, and for that reason it is more beautiful. But in fact I've spent my life, my professional life, working on a definition of randomness and trying to find, and I think I have found, randomness in pure mathematics which is a funny place to find something that is random. When you say that something is random you're saying that you can't understand it, right? So defining randomness is the rational mind trying to find its own limits, because to give a rational definition to randomness is odd there's something paradoxical in being able to know *that*, you know, being able to define randomness, or being able to know that something is random — because something is random when it escapes ... I'm not formulating this well, actually improvising it, but there are some paradoxes involved in that. The way it works out in these paradoxes is that you can define randomness but you can't know that something is random, because if you could know that something is random then it wouldn't be random. Randomness would just be a property like any others, and it could be used would enable you to classify things. But I do think that there is a definition of randomness for individual numbers and you don't take into account the process by which the numbers are coming to you: you can look at individual strings of bits — base 10 number — and you can at least mathematically say what it means for this to be random. Now, although most numbers or most sequences of bits are random according to this definition, the paradoxical thing about it is that you can never be sure that one individual number is random — so I think it is possible to define a notion of randomness which is intrinsic and structural and doesn't depend on the process from which something comes but there is a big problem with this definition which is: it's useless. Except to create a paradox or except to show limits to knowledge, or limits to mathematical reason. But I think that's fun, so that's what I've been doing my whole life.

So I don't know if this is relevant to SETI? I guess it is, because if something looks random it then follows that it probably doesn't come from an intelligent source. But what if these superior beings remove redundancy from their messages? They just run it through a compression algorithm because they are sending us enormous messages, they're sending us all their knowledge and wisdom and philosophy everything they know in philosophy, because their star is about to go nova, so this is an enormous text encompassing all their accomplishments of their thinking and civilization — so obviously they think any intelligent mind would take this information and compress it, right? And the problem is, we're getting this LZ compressed message and we think that it's random noise and in fact it's this wonderfully compact message encapsulating the wisdom and the legacy of this great civilization?

George Johnson: Oh, but do they include the compression algorithm?

Greg Chaitin: Well, they might think that a priori this is the only conceivable compression algorithm, that it is so simple that any intelligent being would use this compression algorithm — I don't know ...

Stephen Wolfram: I think it's an interesting question — about SETI. For example, if you imagine that there was a sufficiently advanced civilization that it could move stars around, there's an interesting kind of question: in what configuration would the stars be moved around and how would you know that there is evidence of intelligence moving the stars around? And there's a nice philosophical quote from Kant who said "if you see a nice hexagon drawn in the sand you know that it must come from some sort of intelligent entity [that has created it]" And I think that it's particularly charming that now in the last few years it's become clear that there are these places in the world where there are hexagonal arrangements of stones that have formed and it is now known that there is a physical process that has causes a hexagonal arrangement of stones to be formed. That's sort of a charming version of this ...

Ed Fredkin: One of the poles of Saturn has this beautiful hexagon — at the pole and we have pictures of them.

Stephen Wolfram (continues): ... right, so the question is what do you have to see to believe that we have evidence that there was an intention, that there was a purpose. It's just like Gauss for example, he had the scheme of carving out in the Syberian forest the picture of the Pythagorean theorem, because that would be the thing that would reveal the intelligence. And if you look at the Earth now a good question to ask an astronaut is: "What do you see on the Earth that makes you know that there is some sort of a civilization?" And I know the answer: the thing that is most obvious to the astronauts is — two things, OK? One is: in the great salt lake in Utah [there is] a causeway that divides a region which has one kind of algae which tend to be of orangeish color, from a region that has another kind of algae that tend to be bluish, and there's a straight line that goes between these two colored bodies of water. It's perfectly straight and that is thing number one. Thing number two is in New Zealand. There's a perfect circle that is visible in New Zealand from the space. I was working on the NKS book and I was going to write a note about this particularly thing. We contacted them, this was before the web was as developed as it is today, so we contacted the New Zealand Geological Survey to get some information about this perfect circle and they said: "If you are writing a geology book (the circle is around a volcano,) please *do not* write that this volcano produces this perfect circle, because it isn't true." What's actually true is that there is a national park that was circumscribed around the volcano and it happens to be perfectly circular and there are sheep that have grazed inside the national park but not outside, so it's a human produced circle. But it's interesting to see what is there on the Earth that sort of reveals the intelligence of its source.

And actually, just to make one further comment about randomness, and integers — just to address this whole idea of whether there are random integers or not, and does it matter, and how can you tell, and so on — we have a little project called 'integer base' which basically is a directory of integers. And the question is to find is the simplest program that makes each of these integers. And it's interesting, it's a very pragmatical project actually trying to fill in actual programs that make integers. We have to have some kind of metric as to what counts as simple. When you use different kinds of mathematical functions, you use different kinds of programming constructs, you actually have to concretely decide a measure to quantify simplicity. And there are lots of ways, for example: how many times does this function appear is referenced on the web, that could be a criterion as to how much weight should be given; or how long is this function's name in; or other kinds of criteria like that. So it's kind of a very concrete version of this question about random integers.

Tom Toffoli: I'd like to say something that throws out another corollary. There's this self-appointed guru of electronics, Don Lancaster, he's very well known in circles and he said something that is very true. He said: the worst thing that could happen to humanity is to find an energy source that is inexhaustible and free. You know, we are all hoping that we will find something like that, but if we found it it would be a disaster, because then the Earth would be turned into a cinder in no time.

If you don't have any of the current constraints it can turn very dangerous. For example, you have a house there, and you have a mountain, and in the afternoon you would like to have sunshine. And in the morning you would like to have shading from the cold or whatever, so if energy is free you just take the mountain from where it is in the morning you take it away and you plant it back in the evening. And this is what we're doing in essence when we're commuting from the suburbs to the center of Boston. You see this river of cars that is rushing in every day, and rushing out every day, with energy that is costing quite a bit. Imagine if it was completely free. So, again, let's try to think up what the answer to our question would be if we really got it and then see the consequences of that first.

And, again: this comment is in relation to what I said earlier about the randomness of the stars if it's an indication of a super-intelligence, or super-stupidity. Who knows, it could be that they're one and the same thing?

Rob de Ruyter: Just to take your question completely literally: there's a lot of randomness in our eyes, as we look around the world. And especially outside in moonlight conditions there are photons flying around, but they are not that many and we are very aware of the fact that information that we're getting into our visual system, information that we have to process in order to navigate successfully, is of low quality and the interesting thing is that we as organisms are used to walking around in a world that is random and we're very conscious of it. Yesterday I spoke about how flies cope with this — we cope with this too and we cope with it at all levels, from adaptation in photoreceptors in our eyes to the adaptation in the computational algorithms that our brain is using to where you are in an environment where you are subject to large levels of noise, because there are not that many photons around. In that case you tend to move very cautiously — you don't start running, unless maybe the tiger is just following you, but that is a very rare situation — so, I think, in a lot of senses we're used to the measurements that our sensors make being more or less random depending on how the situation is at the moment. And so as computational engines we are very well aware of that.

Cris Calude: I am interested in the quality of quantum randomness. We were able to prove that quantum randomness, under some mild assumptions on the quantum model of physics we agree on, is not computable. So this means no Turing machine can reproduce the outcome of a quantum generated sequence of bits (finitely many) and this gives you a weak form of relation between Greg's theory, Greg's definition of algorithmic randomness and what one would consider to be the best possible source of randomness in this this universe, i.e., quantum randomness. And one of the things that is delicate as we are thinking and experimenting is a way to distinguish quantum randomness from generated randomness. Is it possible, by using finitely many well-chosen tests, to find a mark of this distinction you know between something that is computer computably generated from something that is not generated in that way?

So whereas here we have some information about the source, you know, like Tom said — we know very well that there is an asymptotic definition of the way that these bits can be generated — it is still very difficult to account in a finite amount of tests for that difference.

Ed Fredkin: Just a funny story about random numbers: in the early days of computers people wanted to have random numbers for Monte Carlo simulations and stuff like that and so a great big wonderful computer was being designed at MIT's Lincoln laboratory. It was the largest fastest computer in the world called TX2 and was to have every bell and whistle possible:

a display screen that was very fancy and stuff like that. And they decided they were going to solve the random number problem, so they included a register that always yielded a random number; this was really done carefully with radioactive material and Geiger counters, and so on. And so whenever you read this register you got a truly random number, and they thought: "This is a great advance in random numbers for computers!" But the experience was contrary to their expectations! Which was that it turned into a great disaster and everyone ended up hating it: no one writing a program could debug it, because it never ran the same way twice, so ... This was a bit of an exaggeration, but as a result everybody decided that the random number generators of the traditional kind, i.e., shift register sequence generated type and so on, were much better. So that idea got abandoned, and I don't think it has ever reappeared.

Stephen Wolfram: Actually it has reappeared, in the current generation of Pentium chips there's a hardware random generator that's based on double Johnson noise in the resistor. But in those days programs could be run on their own. In these days there are problems in that programs can no longer be run on their own: they are accessing the web, they're doing all sorts of things, essentially producing random noise from the outside, not from quantum mechanics but they're producing random noise from the outside world. So the same problem has come up again.

George Johnson: This reminds me that in the dark ages before the published this huge tome — I found a reprint of it called "One hundred thousand random numbers" (Cris Calude corrects out loud: "One *million* random numbers") in case you needed some random numbers — and Murray Gell-Mann used to tell a story that he was working at RAND and at one time, I think when they were working on the book, they had to print an errata sheet! (There is laughter)

Stephen Wolfram: Well, the story behind the erratum sheet I think is interesting because those numbers were generated from (I think a triode, or something) some vacuum tube device and the problem was that when they first generated the numbers, they tried to do it too quickly, and basically didn't wait for the junk in the triode to clear out between one bit and the next. This is exactly the same cause of difficulty in randomness that you get in trying to get perfect randomness from radioactive decay! I think the null experiment for quantum computing, one that's perhaps interesting to talk about here, is this question of how do you get — I mean, can you get — a perfect sequence of random bits from a quantum device? What's involved in doing that? And, you know, my suspicion is the following: my suspicion would be that every time you get a bit out you have to go from the quantum level up to the measured classical level, you have to kind of spread the information about this bit out in this bowl of thermodynamic soup of stuff so that you get a definite measurement; and the contention, or my guess, would be that there is a rate at which that spreading can happen and that in the end you won't get out bits that are any more random than the randomness that you could have got out just through the spreading process alone without kind of the little quantum seed. So that's an extreme point of view, that the extra little piece (bit) of *quantumness* doesn't really add anything to your ability to get out random bits. I don't know if that is correct but you know I, at least a long time ago, I did try looking at some experiments and the typical thing that's found is that you try to get random bits out of a quantum system quickly and you discover that you have $\frac{1}{f}$ noise fluctuations because of correlations in the detector and so on. So I think that the minimal question from quantum mechanics is: can you genuinely get sort of random bits and what's involved in doing that? What actually happens in the devices to make that happen? I'd be curious to know the answer to this [question].

Tom Toffoli: I know the answer, Intel already says: yes, you can get good quantum numbers if you have a quantum generator of random numbers. Just generate one random number, then throw away your generator and buy a new one, because the one that you have used is already entangled with the one it generated. So you buy a new one and you solved the problem. [Wolfram says: this is also a good commercial strategy ... people laugh]

Ed Fredkin: There's a great story in the history of — back in the '50s people doing various electronic things needed noise. They wanted noise, random noise so they thought: what would be a good source of it. And so it was discovered that a particular model of photomultiplier, if you covered it up and let no light into it gave beautiful random noise. And as a result various people conducted experiments, they characterized this tube and it was essentially like a perfect source of random noise, and the volume of sales started to pick up. Various people started building these circuits and using them all over. Meanwhile, back at the tube factory which was RCA, someone noticed: "Hey, that old noisy photomultiplier that we had trouble selling lately, sales are picking up, we better fix that design so it isn't so noisy!" So they fixed it and that was the end of that source of random noise.

INFORMATION VS. MATTER

George Johnson: I think I'll ask another question about one other thing that has been bothering me before we start letting the audience jump in. I first ran across this when I was writing a book called "Fire of the mind" and the subtitle was 'Science, faith and the search for order.' And I was re-reading a book that I read in college that really impressed me at the time, and seeing that it still stood up, which was Robert Pirsig's book "Zen and the art of motorcycle maintenance" — and it did stand up in my opinion.

There's a scene early on in the book the protagonist who called himself Phdrus after Plato's dialogue is taking a motorcycle trip around the country with his son Chris — who in real life later was tragically murdered in San Francisco where he was attending a Zen monastery, which is neither here or there — but, in the book this person's running around with Chris and they're sitting around the campfire at night and drinking whisky and talking and telling ghost stories and at one point Chris asks his father: "Do you believe in ghosts?" And he says "Well, no, of course I don't believe in ghosts because ghosts contain no matter and no energy and so according to the laws of physics they cannot exist." And then he thinks for a moment and says: "Well, of course the laws of physics are also not made of matter or energy and therefore they can't exist either." And this really seemed like an interesting idea to me, and when I was learning about computational physics, this made me wonder: *where* are the laws of physics? Do you have to be a Platonist and think that the laws of physics are written in some theory realm? Or does this computational physics gives us a way to think of them as being embedded within the very systems that they explain? [waits a little, sees Chaitin wanting to answer, says:] Greg!

Greg Chaitin: George, we *have* ghosts: information! Information is non-material.

George Johnson: Information is physical, right? Well, wouldn't Landauer say that?

Greg Chaitin: Well, maybe Rolf would say that, but ontologically we've come up with this new concept of information and those of us that do digital physics somehow take information as more primary than matter. And this is a very old philosophical debate: is the world built of spirit or mind or is it built of matter? Which is primary which is secondary?

And the traditional view of what we see as the reality, is that everything is made of matter. But another point of view is that the universe is an idea and therefore (and information is much closer to that) made of spirit, and matter is a secondary phenomenon. So, as a matter of fact, perhaps everything is ghost. If you believe in a computational model informational model of the universe, then there is no matter! It's just information — patterns of information from which matter is built.

George Johnson: Does that sound right to you, Tony?. Does it sound right to you that information is more fundamental than matter or energy? I think most of us — you know, the average person in the street — asked about information would think about information as a human construct that is imposed on matter or energy that we make. But I really like the idea that Gregory has suggested — and I actually wrote about it quite a bit in this book — that information is actually in the basement there, and that matter and energy are somehow built out of information.

Tom Toffoli [starts answering the question]: Well, ideally, I mean... You can ask the same thing about correlation rather than information because it conveys the same meaning. Furthermore one can bring up and discuss the notion of entanglement, and in the same spirit: it's not here nor there. Where is it? Very related issues. That's the point I wanted to make.

Tony Leggett: Well I think I would take the slightly short-sighted point of view that information seems to me to be meaningless — unless it is information *about something*. Then one has to ask the question: "what is the something?" I would like to think that *that something* has something to do with the matter of interest, and the matter and energy that's involved.

Stephen Wolfram: This question about whether abstract formal systems are *about something* or not is a question that obviously has come up from mathematics. And my guess about the answer to this question: is information the primary thing or is matter the primary thing? I think that the answer to that question would probably end up being that they are really the same kind of thing. That there's no difference between them. That matter is merely our way of representing to ourselves things that are in fact some pattern of information, but we can also say that matter is the primary thing and information is just our representation of that. It makes little difference, I don't think there's a big distinction — if one's right that there's an ultimate model for the representation of universe in terms of computation.

But I think that one can ask this question about whether formal systems are about something — this comes up in mathematics a lot, we can invent some axioms system and then we can say: is this axiom system describing something really, or is it merely an axiom system that allows us to make various deductions but it's not really about anything. And, for example, one of the important consequences of Gödel's theorem is that you might have thought that the Peano axioms are really just about integers and about arithmetic but what Gödel's theorem shows is that these axioms also admit different various non-standard arithmetics, which are things that are not really like the ordinary integers, but still consistent with its axioms. I think it's actually a confusion of the way in which mathematics has been built in its axiomatic form that there is this issue about 'aboutness' so to speak — and maybe this is getting kind of abstract. But when we think about computations we set things up, we have particular rules, and we just say "OK, we run the rules, you know, and what happens — happens." Mathematics doesn't think about these things in those terms, typically. Instead, it says: let's come up with axiom systems which constrain how things could possibly work. That's a different thing from saying let's just throw down some rules and then the rules run and then things just happen. In mathematics we say: let's come up with axioms which sort of describe how things have to work, same thing was done in physics with equations — the idea of, you know, let's make up an equation that describe what's possible to happen in the world — and not (as we do in computation) let's do something where we set up a rule and then the rule just runs. So, for example, that's why in gravitation theory there's a whole discussion about "What do the solutions to the Einstein's equations look like?" and "What is the set of possible solutions?" and not (instead) "How will the thing run?" but the question traditionally asked there is: "What are the possible things consistent with these constraints?"

So in mathematics we end up with these axiom systems, and they are trying to sculpt things, to ensure that they're really talking about the thing that you originally imagine[d you were] talking about, like integers. And what we know from Gödel's theorem and so on, is that that kind of sculpting can never really work. We can never really use this constraint-based model of how to understand things to actually make our understanding be about a certain thing. So, I think that's kind of the ultimate problem with this idea of whether the laws that one's using to describe things and the things themselves are one and the same or they are different, and if so what the distinction really is.

Greg Chaitin: There's this old idea that maybe the world is made of mathematics and that the ultimate reality is mathematical. And for example people have thought so about continuous mathematics, differential equations and partial differential equations, and that view was monumentally successful, so that already is a non-materialistic view of the world. Also let me say that quantum mechanics is not a materialistic theory of the world; whatever the Schrödinger wave equation is it's not matter, so materialism is definitely dead as far as I am concerned. The way Bertrand Russell put it is: if you take the view that reality is just what the normal day appearances are, and modern science shows that everyday reality is not the real reality, therefore — I don't know how he called this ... 'naive realism' I think — and if naive realism is true then it's false, therefore it's false. That's another way to put it. So the only thing we're changing in this view that the actual structure of the world is mathematical, the only thing new that we're adding to this now is we're saying mathematics is ultimately computational, or ultimately it is about information, and zeroes and ones. And that's a slight refinement on a view that is quite classical.

So I am saying in a way we're not as revolutionary as it might seem, this is just a natural evolution in an idea. In other words, this question of idealism vs. materialism, or "Is the world built of ideas or is the world built of matter?" it might sound crazy, but it's the question of "Is the ultimate structure of the world mathematical?" versus matter. That sounds less theological and more down to Earth. And we have a new version of this, as ideas keep being updated, the current version of this idea is just: "Is it matter or is it information?" So these are old ideas that morph with time, you know, they evolve, they recycle, but they're still distinguishably not that far from their origin.

Cris Calude: Information vs. matter, discrete vs. continuous: interesting philosophical contrasts and I found Stephen's description to be extremely interesting, because I too think that these views should in fact coexist in a productive duality. And it depends on your own abilities, it depends on your own problems if one of them would be more visible or useful. And at the end of the day what really counts is — if you have a view that in a specific problem information prevails — what can you get from that? Can you prove a theorem, can you get some result, can you build a model which you know answers an important question or not? In some cases one view may be the right one, in other cases the other one is, so from my point of view, which is, I would guess, more pragmatic, I would say: look, you choose whatever view you wish in order to get a result and if you get the result that means that for this specific problem that choice was the correct one.

Ed Fredkin: But in fact the world either is continuous or discrete, and we can call it anything we want, to get results and so on, to add convenience — but there really is an answer to it and one answer is right and the other is wrong. I go with Kronecker who said "God invented the integers and all else is the work of man." So you can do anything with discrete models and/or continuous models but that doesn't mean that the world is both, or can be, or could be either. No, the world is either one — or the other.

Stephen Wolfram: This whole question about mechanism is kind of amusing. For example, with CAs models, people in traditional physics (not so in other areas but people in traditional physics) have often viewed this kinds of models with a great deal of skepticism. And it's kind of an amusing turn of historical fate, because, in the pre-Newtonian period people always had mechanistic models for things — whether there were angels pushing the Earth around the orbit, or other kinds of mechanistic type of things. And then along came this kind of purely abstract mathematical description of law of gravity and so on and everybody said — but only after a while! — everybody said: "Well, it's all just mathematics and there isn't a material reality, there isn't a mechanism behind these things!" And so, when one comes with these computational models which seem to have much more of a tangible mechanism, that is viewed as suspicious and kind of non-scientific by people who spend their lives working in the mathematical paradigm, that it can't be simple enough if there's an understandable mechanism to behind things. So it's kind of an interesting turning around of the historical process. As you know I work on various different kinds of things and, as I said, I've not been working much on finding a fundamental theory of physics lately. I actually find this discussion as an uptick in my motivation and enthusiasm to actually go and find a fundamental theory of physics. Because I think, in a sense, what with all of these metaphysical kinds of questions about what might be there, what might not be there and so on: damn it, we can actually answer these things!

Tom Toffoli: I know a way out of this! It is similar to the one that Ed proposed a long time ago. He said (about computability, referring to exponential, polynomial problems) he said that one can turn all problems into linear problems. You know, *all* the exponential problems! And people of course said: "Ed you are an undisciplined amateur you say these things without knowing what you're talking about." But he said: "Look, we have Moore's law! And with Moore's law everything doubles its speed every so many years. So we just wait long enough and we get the solution — as long as it takes."

So the key then, is to wait, as long as we need to. With this, I am now giving a simpler solution to this problem, of finding a fundamental theory of physics, starting from the observation that domesticated animals become less intelligent than wild animals. This has been proven recently with research on wolves. And maybe domesticated animals are somewhat less intelligent in certain ways but they can see what humans want and they obey. But then some more experiments were run and the findings were that wild wolves once they put them in an environment with humans they learn humans faster than domesticated animals to anticipate the wills of their trainers. It's not that they follow their will, but they anticipate it faster.

Now we are doing a big experiment on ourselves, on humanity — humanity is domesticating itself. And there was a time when people said: we discovered differential equations and very few people can understand them so we have the monopoly, we are the scientists. And eventually somebody said: "Well, wait a second, but why can't we find a model like, you know, Ed or Stephen — that is, sort of computational, discrete so that everyone can own it and possess it and so on?" But [one forgets that] we are domesticating ourselves! To the point that the computer that — according to Brian Hayes the first thing about a computer is that you can program it, make it do whatever we want — now most people don't even know that that is possible, they just think of the computer as an appliance. And soon even the computer will be a mystery to most people! The programmable digital computer [like] the differential equations were a generation ago — so we solved the problem, in that sense, just wait long enough and nobody will even be able to care about these things, it will just be a mystery for us.

Cris Calude: Well, I would like to just add a small remark, suggested by your idea about physical computation. So, this is essentially a personal remark about the P vs. NP problem: I believe this is a very challenging and deep and interesting mathematical question, but I think one that has no computer science meaning whatsoever. For the simple fact that P is not an adequate model of feasible computation, and there are lots of results — both theoretical and experimental — which point out that P does not model properly what we understand as feasible computation. Probably the simplest example is to think about the simplex algorithm which is exponentially difficult, but works much better in practice than all the known polynomial solutions.

UNMODERATED AUDIENCE QUESTIONS

George Johnson: This is a good time to move on to questions from the audience.

Jason Cawley: So I have a question for Greg Chaitin but for everyone else as well. You said that the world would be more intelligible and prettier if it were discrete, which was very attractive to me. But how intelligible would it be, really, even if I grant you finiteness and discreteness, even if we find the rule? Won't it have all these pockets of complexity in it, won't it be huge compared to us, finite, wouldn't it last much longer than us — and then we still have all kinds of ways that would be mysterious to us in all the little detail?

Greg Chaitin: Well, I didn't catch all of that... (collects his thoughts then proceeds to summarize the question as best as he heard it because Jason had no microphone) ... Oh, I guess the remark is disagreeing with what I said that a discrete universe would be more beautiful ... no, no, more comprehensible ... right? ... and you gave a lot of reasons why you think it would be ugly, incomprehensible, disgusting (audience laugher) — and I can't argue, if you feel that way! But in that case I don't think that's a question, I view that as a comment that doesn't need to be answered ...

Jason Cawley: Sorry. The question is "How intelligible is a perfectly discrete universe?" The reason I am concerned about this is: I happen to like rationalism too, but I don't want people concluding, when they see non intelligibility in the universe, that it is evidence against the rational.

Greg Chaitin refines his answer: Oh, Okay, great! Well, in that case, as Stephen has pointed out in his book, it could be that all the randomness in the world is just pseudo randomness, you know, and things only *look* unintelligible, but they are actually rational. The other thing is he's also pointed out — and this is sort of his version of Gödel's incompleteness theorem — is that something can be simple and discrete and yet we would not be able to prove things about it. And Stephen's version of this (which I think is very interesting) is that in the way that the universe is created, because you have to run a computation, in general you have to run a physical system to see what it will do, you can't have a shortcut to the answer[s]. So that can be viewed as bad, but it also means that you could have a simple theory of the world that wouldn't help us much to predict things. And you can also look at it as good, because it means that the time evolution of the universe is creative and surprising, it's actually doing something that we couldn't know in advance — by just sitting at our desks, and thinking — so I view this as fundamental and creative! And in regards to creativity Bergson was talking 100 years ago about "L'Evolution Créatrice[t]" — at this point, this would be a new version of that. But over aesthetics one can't ultimately argue too much. But still, I think it was a good question.

New question from the audience: First a question about SETI: I may be naive, but it seems strange to me why we would like to look at perfect circles and dividing lakes when one can simply look at Chicago and New York emanating light from Earth at night. If I were a Martian that's what I would do. But what interests me more and that's the question for Sir Leggett is this: suppose that there were phase transitions from quantum to classical. Would building a quantum computer — would the approach to build a quantum computer be different than trying to reduce decoherence as it's being done presently?

Tony Leggett: I'm not entirely clear what you mean by postulating that there was a phase transition from quantum to classical. Could you elaborate a bit?

Question: Ah, well, I am a bit ignorant about your theory but, if there is a theory that it's not just decoherence but in fact there are phase transitions from quantum to classical at some level — that's why we don't see Schrödinger's cat after some. Would this imply perhaps a different approach of building a quantum computer?

Tony Leggett: If you mean — if you're referring to theories, for example, of the GRWP type which postulate that there are physical mechanisms systems which will meet the linear formalism of quantum mechanics which will have to be modified — and it will have to be modified more severely as it goes from the microscopic to the macroscopic — then I think the answer to your question is that to the extent that we want to use macroscopic or semi-macroscopic systems as qubits in our quantum computer, it wouldn't

---
[s]Via the Principle of Computational Irreducibility (which is a corollary of the PCE).
[t]1907 book by French philosopher Henri Bergson. Its English translation appeared in 1911. The book provides an alternate explanation for Darwin's mechanism of evolution.

work. On the other hand I don't think that theories of the GRWP — scenarios of the GRWP type — necessarily against an attempt to build a quantum computer always keeping the individual bits at the microscopic level. You have to look at it in, of course, in detail in a specific context of a particular computer built around a particular algorithm, such as, say, Shor's algorithm. But I don't think it is a priori essential that a GRWP type scenario would destroy the possibility of it.

New question from the audience: Well, we still haven't answered the question of "How does nature compute?" We are just discussing differences between discrete and continuous, classical vs. quantum computation but if we see mathematics as a historical accident and we try to push it aside, shouldn't we try to look at nature, and try to understand how nature in fact computes? And not just try to translate it into a mathematical context. For example one can watch plants and see that there is some kind of parallel type of computation that is going on, that is, not based on our mathematics, but like *they* do it, the plants ... you know... Do you have you any thoughts on that?

Stephen Wolfram answers it: I think that one of the things that makes that somewhat concrete is the question of how we should build useful computational devices. Whether our useful computational devices should have ALUs (Arithmetic Logic Unit) inside them or not. The very fact that every CPU that's built has a piece that's called "the arithmetic logic unit," tells you that in our current conception of computation we have mathematics somewhere in the middle of it. So an interesting thing is: can we achieve useful computational tasks without ever having an ALU in the loop, so to speak. I think the answer is: definitely yes, but as the whole engineering development of computers has been so far we've just been optimizing this one particular model that's based on mathematics. And as we try to build computers that are more at the molecular scale, we could, actually, use the same model: we could take the design of the Pentium chip and we can shrink it down really really small and have it implemented in atoms. But an alternative would be that we can have atoms do things that they are more naturally good at doing. And I think the first place where this would come up are things like algorithmic drugs, where you want to have something that is in essence a molecule operating in some biological, biomedical context and it wants to actually do a computation as it figures out whether to bind to some site or not, as opposed to saying "I am the right shape so I'm going to bind there!" So that's a place where computation might be done. But it's not going to be computation that will be done through arithmetic but we'll be forced to think about computation at a molecular scale in its own terms, simply because that's the scale at which the thing has to operate. And I'm going to guess that there will be a whole series of devices and things, that — mostly driven by the molecular case — where we want to do computation but where the computation doesn't want to go through the intermediate layer of the arithmetic.

Cris Calude: Yes! There is a lot of research that we have started about ten years ago in Auckland, and a series of conferences called "Unconventional Computation." And this is one of the interesting questions. You know, there are basically two streams of thought: one is quantum computation, and the other one is molecular computing. And in quantum computing, you have this tendency of using the embedded mathematics inside. But in molecular computing, you know, you go completely wild because mathematics is not there so you use all sorts of specific biological operations for computation. And if you look at the results, some of them are quite spectacular.

Small refinement: Yeah, but they are still based on logic gates and ... you know ... molecular computing is still based on trying to build logic gates and ...

Cris Calude No, no! There is no logic gate! That's the difference, because the philosophy in quantum computation is: you do these logical gates at the level of atoms or other particles — but in molecular computation there is no arithmetical instruction, there are no numbers, you know, everything is a string, and the way they are manipulated is based exactly on biological type of processing. No arithmetic.

Greg Chaitin: No boolean algebra, not even *and*'s and *or*'s?

Cris Calude: No, nothing! And then this is the beauty, and in a sense this was the question that I posed to Seth Lloyd in '98. I said you know, why don't you do in quantum computing something similar? Why don't you try to think of some kind of rules of processing – not imposed from the classical computation, from Turing machines, but rules which come naturally from the quantum processes — just like the typical approach in molecular computation.

Tom Toffoli: I would like to give a complementary answer to this. I've been teaching microprocessors and microcontrollers — you have them in your watches and cellphones. They're extremely complicated objects. And you would say, I mean given the ease with which we can fabricate these things, whenever we want to run a program or algorithm, we could make a very special purpose computer rather than using a microprocessor to program it. Apparently it turns out that it's much more convenient, if someone had designed a microprocessor with an ALU and a cache and the other things in between, to just take it as a given and then the process is complete. If we look at biology, biology has done the same thing. We have, at a certain moment, hijacked mitochondria that do the conversion of oxygen and sugar into recharging the ATP batteries. That was a great invention. And now, after probably three billion years, or something like that we still keep using that instead of inventing a method that is, maybe a little more efficient, but would have to be a very special choice for every different circumstance. Essentially we could optimize more, but we would do that at the cost of losing flexibility, modularity and so on but, apparently it's much more convenient. For three billion years we've kept using this kind of energy microprocessor, that worked fairly well... So there is, essentially, the flexibility or modular evolution that really suggests that choices like the ALU... is not an optimal choice, but — empirically — is a very good choice. This is my viewpoint.

Stephen Wolfram: Well — as is always the case in the history of technological evolution it is inconceivable to go back and sort of restart the whole design process. Because there is just far too much investment in that particular thing. The point that I want to make is that — what will happen is that there will be certain particular technological issues, which will drive different types of computing. My guess is that the first ones that will actually be important are these biomedical ones, because they have to operate on this molecular scale, because that's the scale on which biomedicine operates. And, you know, one can get away with having much bigger devices for other purposes — but biomedicine is a place where potentially a decision has to be made by a molecule. And whether — maybe there will be ways of hacking around that but in time, it won't be very long before the first applications of this kind will be finalized.

And what's interesting about [this] is that if you look at the time from Gödel and Turing to the time when computers became generic and everybody had one, and the time from Crick and Watson and DNA to the time when genomics becomes generic — it's about the same interval of time. It hasn't yet happened for genomics, it has more or less happened for computers. Computers were invented, the idea of computers is twenty-three years earlier or something like that, than the DNA idea. Anyway, it will soon happen for genomics as well, and in time we will routinely be able to sequence things, in real time, from ourselves, and we'll do all kinds of predictions that yes, you know we detect that today you have a higher population of antibodies that have a particular form so we'll be able to run some simulation that this means that you should go and by a supply of T-cells, that has this particular characteristic and so on. And there's a question whether the decisions about that will be made externally by ALU based computers, or whether they will be made internally by some kind of molecular device — more like the way biology actually does it. And if it ends up that they are made externally then there won't be any drive from the technology side to make a different substructure for computing.

New question: I study computer science. And one of the ideas I find truly fascinating and I think it really is — is this thing called the Curry-Howard isomorphism, which basically relates propositions to types and rules to terms and proof normalization to program evaluation. And since you're discussing models of computation, I was wondering if you have encountered something similar for cellular automata and such. I think that the classification, this particular relation is very explicit in the simply typed lambda calculus where program terms, which are lambda expressions, can be given types — and the types are pretty much propositional logic expressions. And if you can prove a certain proposition the structure of the proof in natural deduction style will actually look like a program type and if the proof is not a normal proof then the process of proof normalization is basically the process of the evaluation of the term into a normal form of the term, which basically means that if you have this computational model which is the lambda calculus, reductions of the calculus will correspond to normalizations of the proofs, and the types serve as a way of classifying programs, types are a way of saying: these particular programs that behave in such and such a way don't have such and such properties. And I feel that this might be something that carries over to other notions of computations as well, because there's nothing intrinsic about the lambda calculus that makes this uniquely applicable to it.

Stephen Wolfram: Let me start by saying that I'm a very anti-type person. And it turns out that, you know, in the history types were invented as a hack by Russell basically to avoid certain paradoxes — and types then became this kind of "great thing" that were used as an example and then as a practical matter of engineering in the early computer languages there was this notion of integer types versus real types and so on, and the very idea of types became very inflated — at least that's how I see it.

So, for example in *Mathematica* there are no types. It's a symbolic system where there is only one type: a symbolic expression. And in practical computing the most convincing use of types is the various kinds of checking but in a sense when something is checkable using types that involves a certain kind of rigidity in programs that you can write, that kind of restricts the expressivity of the language that you have. And what we found over and over again in *Mathematica*, as we thought about putting things in that are like types, that to do that would effectively remove the possibility of all sorts of *between paradigm* kinds of programming, so to speak, that exist when you don't really have types.

So, having said that, this question of the analogy between proof processes and computation processes is an interesting one. I've thought about that a lot, and there's more than just a few things to say about it. But one thing to think about is: "What is a proof?" and "What's the point of doing a proof?" I mean, the real role of a proof is as a way to convince (humans, basically) that something is true. Because when we do a computation, in the computation we just follow through certain steps of the computation and assuming that our computer is working correctly the result will come out according to the particular rules that were given for the computation. The point of a proof is somehow to be able to say to a human — look at this: you can see what all the steps were and you can verify that it's correct. I think the role of proofs in modern times has become, at best, a little bizarre. Because, for example, so here's a typical case of this: when *Mathematica* first existed twenty years ago one would run into mathematicians who would say "How can I possibly use this, I can't prove that any of the results that are coming out are correct!" OK? That's what they were concerned about. So, I would point out sometimes that actually when you think you have a proof, in some journal for example, it's been maybe checked by one person — maybe — if you're lucky. In *Mathematica* we can automate the checking of many things and we can do automatic quality assurance, and it's a general rule that — in terms of how much you should trust things — the more people use the thing that you're using the more likely it is that any bugs in it will have been found by the time you use it. So, you know, if you say: "Well, maybe it's a problem in the software that I myself am writing, maybe it's a problem in the system (like *Mathematica*) that I am using, maybe it's a problem in the underlying hardware of the computer" — it gets less and less plausible that there's a problem, the broader the use of the thing is.

So I think as a practical matter when people say: "I want a proof!" that the demand for proof, at least in the kind of things that *Mathematica* does, decayed dramatically in the first few years that *Mathematica* existed because it became clear that most likely point of failure is where you as a human were trying to explain to the computer what to do. Now, having said that it's interesting [that] in *Mathematica* we have more and more types of things that are essentially proof systems — various kinds of things, for example proof systems for real algebra, we just added in the great new *Mathematica* 7.0 all sorts of stuff of doing computation with hundreds of thousands of variables and so on. Those are effectively places where what we've done was to add a proof system for those kinds of things. We also added a general equational logic proof system, but again I think that this question whether people find a proof interesting or whether they just want the results — it seems that the demand for presentation of proofs is very low.

Tom Toffoli (wants to add something, Wolfram uses the break to drink some water): If you would give me the microphone for one second: coming back to Russell when he published Principia Mathematica — most of the theorems were right, but a good fraction of the proofs were found wrong. I mean this was Russell, OK? He was wrong, but there was no problem, because he was still convinced, by his own ways he was convinced of the theorems. So he put together some proofs (arguments) to try to convince the readers that the theorems were right, and he convinced them. But the proofs, as actual mechanical devices, were not working. So his proofs were coming out of just a heuristic device and he derived, and you can always derive the right theorem with the wrong proof. What are you going to do in that case?

Stephen Wofram: No, I think it's actually interesting this whole analogy between proof and computation and so on. One of the things that I have often noticed is that if you look at people's earlier attempts to formalize mathematics — the thing that they focused on formalizing was the process of doing proofs, and that was what Whitehead, Russell, Peano before him and so on worked on. It turned out that that direction of formalization was fairly arid. Not much came from it. The direction that turned out to be the most interesting direction of formalization was, in fact, the formalization of the process of computation. So, you know, in the construction of *Mathematica*, what we were trying to do was to formalize the process of computing things. Lots of people used that and did various interesting things with it. The ratio of people who do computation with formalized mathematics to the number who do proofs with formalized mathematics is a huge ratio. The proof side turned out not to be that interesting. A similar kind of thing, and an interesting question, is how one would go about formalizing every day discourse: one can take (the) everyday language and one can come up with a formalized version of it that expresses things in a sort of formal, symbolic way. But the thing that I've not figured out — actually I think that I've now figured it out, but I *hadn't* figured it out — was "So, what's the point of doing that?" In other words the Russell-Whitehead effort of formalizing proofs turned out not not lead to much. The right idea about formalizing mathematics was the one about formalizing the process of computation. Similarly formalizing everyday discourse as a way to make the semantic web or some such other thing, probably has the same kind of issue as the kind of formalization of proof as was done in mathematics and I think that maybe there is another path for what happens when you formalize everyday discourse, and it's an interesting analogy to what would happen in the mathematics case. You know: what's the point of formalization and what can you do with a formalized system like that.

Greg Chaitin: Let me restate this very deep remark that Stephen has just made about proofs versus computation: if you look at a[ny] formal system for a mathematical formal theory, Gödel shows in 1931 that it will always be incomplete. So any artificial language for doing mathematics will be incomplete, will never be universal, will never have every possible mathematical argument. There is no formal language for mathematics where every possible mathematical argument or proof can be written in. Zermelo-Fraenkel set theory, you know, as a corollary of Gödel — may be wonderful for everything we have but it is incomplete. Now the exact opposite — the terminology is different — when you talk about a programming language you don't talk about completeness and incompleteness, you talk about universality. And the amazing thing is that the drive for formalism, and Russell-Whitehead is one data point on that, another data point is Hilbert's program, the quest for formalization started off in mathematics, and the idea was to formalize reasoning; and the amazing thing is that this failed. Gödel in 1931 and Turing in 1936 showed that there are fundamental obstacles– it can't work! But the amazing thing is that this is a wonderful failure! I mean what can be formalized beautifully, is not proof, or reasoning but: *computation*. And there almost any language you come up with is universal, which is to say: complete – because every algorithm can be expressed in it. So this is the way I put what Stephen was saying.

So Hilbert's dream, and Russell and Whitehead failed gloriously — is not good for reasoning but it's good as a technology, is another way to put it, if you're trying to shock people. The quest for a firm foundation for mathematics failed, but gave rise to a trillion dollar industry.

Tom Toffoli: Let me add something to this. You've heard of Parkinson's law, the one that says: "a system will use as many resources as are available." It was formulated because it was noticed that the whole British Empire was run by essentially a basement of a few dozen people for two hundred years. And then, the moment the British started losing their empire, they started de-colonizing and so on, then they had a ministry of the colonies and this ministry grew bigger and bigger and bigger as the colonies became fewer and fewer and fewer. And it was not working as well as before. And I think that formalization is often something like that. You can think about it, but you don't want to actually do it, even von Neumann, you know, the moment he decided that CAs were, sort of, plausible to give life — he didn't go through the process of developing the whole thing. The point was already made.

New question from the audience: If I may, can I add something about the Curry-Howard isomorphism before I ask my question? Yes? Maybe ... I think that the revolution that we are living in physics is only part of a wider revolution and there are many questions in physics to which the answer uses the notion of algorithm. For instance: what are the laws of nature? They are algorithms. So you may give this answer, only because you have the notion of algorithm. And for centuries we didn't have it. So, to these questions we had either no answer or ad-hoc answers. For instance: what are the laws of nature? Compositions. When we had compositions we didn't have algorithms. It was a good way to answer it. And there are many many areas in knowledge where there are many questions to which now we answer: it is an algorithm. And one of the very first questions on which we changed our mind, was the question: what is a proof? And: what is a proof? The original answer was a sequence of formulas verifying deduction rules and so on. And starting with Kolmogorov — because behind the Curry-Howard isomorphism there is the Kolmogorov interpretation — is this idea that proofs, like the laws of nature are in fact algorithms. So they are two facets, or two elements of a wider revolution, and I think that they are connected in this way.

Now consider this question: Does the Higgs boson exist? Today, I guess, there are people who believe that the Higgs boson exists, there are people who believe that it doesn't exist, but anyone — you can take the electron, if you want (instead of the Higgs boson) [which] most people I guess, believe it exists — but I guess that everyone agrees that we have to find some kind of procedural [means of verifying such a prediction]. It can be an experiment, it may be anything you want — in this case — that will allow eventually, if we are lucky enough, to solve this question. And I would be very uncomfortable if someone told me that the Higgs boson exists in the eye of the beholder, or that Higgs boson has to exist because then the theory would be more beautiful, or if the Higgs boson does not exist then it's actually a problem of the Universe and not ours and we can continue to postulate that it exists because it would be nicer. So I wonder if the two questions we have (not?) been discussing today are not of the same kind: we should try to look for a kind of procedure to answer the question at some point. One: is the universe computable? The other: is the universe discrete? In some sense are these questions only metaphysical questions or are they questions related to experiments that we could and should, in fact carry out?

George Johnson: I'm sorry we won't going to have time to get into that... I want to thank the speakers. Perhaps we could resume at the next conference and this would be another reason to look forward to it.

End of transcript.

PART 7
Zuse's Calculating Space

Chapter 36

Calculating Space (*Rechnender Raum*)*

Konrad Zuse

1st. re-edition[†] written in LaTeX by A. German and H. Zenil

Dedication to Dr. Schuff

The work which follows stands somewhat outside the presently accepted method of approach, and it was for this reason rather difficult to find a publisher ready to undertake publication of such a work. For this reason I am indebted to the Vieweg Press and especially to Dr. Schuff for undertaking publication. Dr. Schuff suggested that a summary be printed in the Journal "Elektronische Datenverarbeitung" (Electronic Data Processing), which appeared last year.

The tragic death of Dr. Schuff has deeply shaken his friends, and we will always remember him with affection.

1. Introduction

It is obvious to us today that numerical calculations can be successfully employed in order to illuminate physical relationships. Thereby we obtain a more or less close interrelationship between the mathematicians, the physicists and the information processing specialists, corresponding to Fig. 1. Mathematical systems serve for the construction of physical models, the numerical calculation of which is carried out today with electronic data processing equipment.

---

*Schriften zur Datenverarbeitung, Vol. 1, 1969 Friedrich Vieweg & Sohn, Braunschweig, 74 pp. MIT Technical Translation. Translated for Massachusetts Institute of Technology, Project MAC, by: Aztec School of Languages, Inc., Research Translation Division (164), Maynard, Massachusetts and McLean, Virginia AZT-70-164-GEMIT Massachusetts Institute of Technology, Project MAC, Cambridge, Massachusetts 02139—February 1970
[†]With kind permission by all parties involved, including MIT and Zuse's family.

The function of the data processing specialists is primarily that of finding the most useful numerical solutions for the models which the mathematicians and physicists have developed. The feedback effect of data processing on the models and the physical theories itself is expressed indirectly in the preferential use of those methods for which numerical solutions are particularly easy to obtain.

The close interplay between the mathematicians and the physicists has had a particularly favorable effect on the development of models in theoretical physics. The modern quantum theory system is very largely pure and applied mathematics. The question therefore appears justified whether data processing can have no more than an effectuating part in the interplay or whether it can also be the source of fruitful ideas which themselves influence the physical theories. The question is all the more justified since a new branch of science, automaton theory, has developed in close cooperation with data processing.

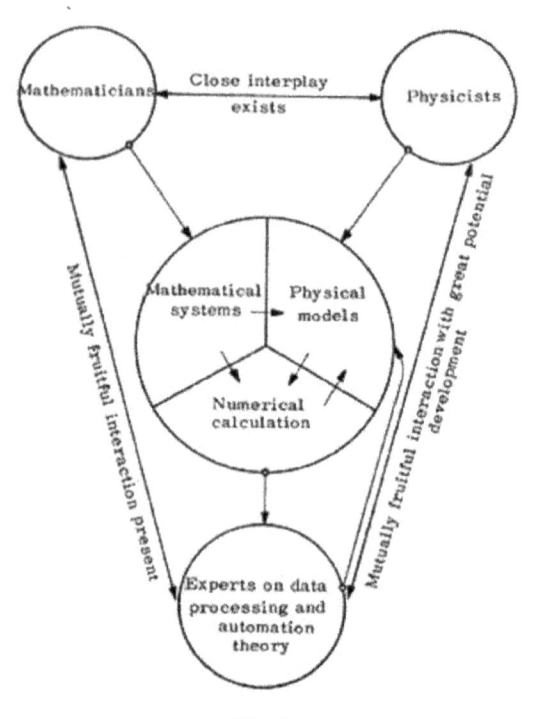

Fig. 1.

In the following pages, several ideas along these lines will be developed. No claim is made to completeness in the treatment of the subject.

Such a process of influence can issue from two directions:

(1) The development and supplying of algorithmic methods, which can serve the physicist as new tools by which he can translate his theoretical knowledge into practical results. Among these are included first all numerical methods, which are still the primary tool in the use of electronic calculating machines. The ideas expressed in the chapters which follow could contribute particularly to the problem of numerical stability.

Among these are symbolic calculations, which command an ever growing importance today. The numerical calculation of a formula is not meant by this, but the algebraic treatment of the formulas themselves as they are expressed in symbols. Precisely in quantum mechanics, extensive formula development is necessary before the actual numerical calculation can be carried out. This very interesting field will not be covered in the material which follows.

(2) A direct process of influencing, particularly by the thought patterns of automaton theory, the physical theories themselves could be postulated. This subject is without a doubt the more difficult, but also the more interesting.

Therein lies the understandable difficulty that different fields of knowledge must be brought into association with one another. Already the field of physics is splitting up into specialized areas. The mathematical methods of modern physics alone are no longer familiar to every mathematician and an understanding of them requires years of specialized study.

But even the theories and fields of knowledge related to data processing are already dividing into different special branches. Formal logic, information theory, automaton theory and the theory of formula language may be cited as examples. The idea of collecting these fields (to the extent which they are relevant) under the term "cybernetics" has not yet become widely accepted. The conception of cybernetics as a bridge between the sciences is very fruitful, entirely independent of the different definitions of the term itself.

The author has developed several basic ideas toward this end, which he considers of value to be presented for discussion. Some of these ideas in their present, still immature form may not be reconcilable with the proven concepts of theoretical physics. The goal has been reached if only discussion occurs and provokes stimulation which one day leads to solutions, which are also acceptable to the physicists.

The method applied below is at present still heuristic in nature. The author considers the conditions not yet ripe for the formulation of a precise theoretical system. First of all, the existing mathematical and physical models will be considered in Chapter 2 from the viewpoint of the theory of automatons. Several examples of digital models are presented in Chapter 3, and the expression "digital particle" is introduced. In Chapter 4, several general thoughts and considerations based on the results of Chapters 2 and

3 will be developed, and in Chapter 5, the prospects for the possibility of further developments are considered briefly.

2. Introductory Observations

2.1. *Concerning the theory of automatons*

The theory of automatons today is already a widely developed, and to an extent very abstract, theory about which considerable literature has been written. Nevertheless, the author would like to distinguish between the actual automaton theory itself and the thought patterns connected with this theory, of which considerable use will be made in the following chapters. A thorough understanding of the automaton theory is not necessary to an understanding of the chapters which follow.

The automaton theory appeared at about the same time as the development of modern data processing equipment. The design and the working method of these arrangements necessitated theoretical investigations based on different mathematical methods; for example, that of mathematical logic. The first useful result of this development was connection mathematics, in which particularly the statement calculus of mathematical logic can play an important part. Of particular importance is the realization that all information can be broken up into yes-no values (bits). The "truth values" of statement calculus assume only two ratings (true and false). The connecting operations and the rules of statement calculus can therefore be viewed as the elementary operations of information processing. Figure 2 shows the elementary connections corresponding to the three basic operations of statement calculus, conjunction, disjunction and negation.

Further research led to introduction of the term "state" of an automaton. In addition, input data and output data play a role. >From input and initial state the new state and the output are obtained, corresponding to the algorithm built into the automaton. Figure 3 shows the schematic diagram of an automaton for a two-place binary register. In the figure, E_1 and E_0 represent the inputs on which a two place binary number can be entered and A_2, A_1 and A_0 represent the outputs, which have the meaning of a three-place binary number. The two-place binary number formed from the figures A_1 and A_0 is relayed back to the automaton and represents the eventual state of the binary number. (In this case the states symbolize a number already entered into the addition process, to which the number E_1, E_0 is to be added).

Fig. 2.

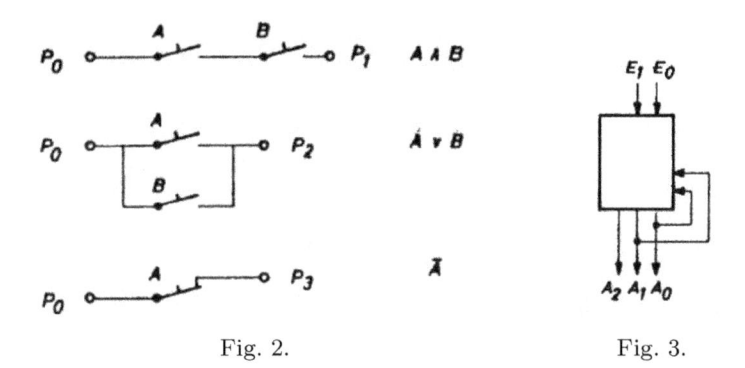

Fig. 3.

The algorithm given by the automaton can be represented by state tables in simple cases. These have the form of a matrix, and for every state and every input combination they give the resultant state or output combination. Figure 4 shows the state table for the automaton in Fig. 3. In this particular case the state table represents an addition table. The theory of the automaton investigates the different possible diffractions of such an au-

A\E	00	OL	LO	LL
00	000	00L	0L0	0LL
0L	00L	0L0	0LL	L00
L0	0L0	0LL	L00	L0L
LL	0LL	L00	L0L	LL0

Fig. 4.

tomaton and sets forth a series of general rules concerning its method of operation. It is important for what follows that the terms finite, autonomous and cellular automaton be understood. A finite automaton works with a discrete number of discrete states; it is roughly equivalent to a digital data processing machine, which is made up of a limited number of elements, each element capable of taking a limited number of states (at least two), with the result that the whole automaton can accept only a limited number of states. Similar conditions hold for the inputs and outputs. The autonomous automaton can accept no inputs (the outputs are also relatively inconsequential). It can be represented, therefore, by a machine that operates independently, once started. Its states follow linearly in sequence, once the initial combination has been started, and the operational process cannot be influenced externally by the absence of one of the inputs.

The cellular automaton represents a special form of automaton built out of interrelated, periodically-recurring cells. This type of automaton is of particular importance for the observations which follow. Later it will be discussed in greater detail.

By the term "automaton theoretical way of thinking" we understand a manner of observation according to which a technical, mathematical or physical model is viewed from the standpoint of a lapse of states, which follow one another according to predetermined rules.

2.2. *About computers*

The automaton theory can be used as an abstract mathematical system, yet these thought structures can also be related to technical models, and similarly the automaton theory can be used for describing automatons, particularly those suited for information processing. In current expanded usage, the term "compute" is identical with "information processing." By analogy, the terms "computer" and "information-processing machine" may be taken as identical.

We distinguish between two classes of computers: analog computers and digital computers. In an analog computer, the steps in the calculation are performed in an "analog" model. Magnitudes representing numerical values are theoretically represented through continual physical magnitudes, such as positions of mechanical parts (torsion angle), tension, velocities, and the like. The machine operates essentially without end. The represented values lie obviously below certain technical limits. These are established by maximum values and by the accuracy of the system. The maximum values are given by a clearly-defined upper limit which corresponds to the technical limits of the system. In contrast, the accuracy has no clearly-defined magnitude, because it depends on change and on external influences (temperature, moisture, the presence of disturbing fields, etc.) One well-known analog computer is the slide rule. Figure 5 shows a mechanical adding mechanism the form of a lever which can be replaced with a rotating mechanism with gears, as in Fig. 6. This mechanism is known in engineering by the inappropriate term "differential mechanism" and is employed in the rear axle of every automobile.

A typical construction element of analog machines is represented by the integration mechanism shown in Fig. 7. This operates with a friction disc A in contact with a friction wheel B. The distance r of the friction wheel B from the axle of A can be varied. In this way, the mechanism can be used for integration. In modern analog instruments, these mechanical elements are replaced by electronic ones. An integration can, for example, be carried out by charging a condenser.

Fig. 5.

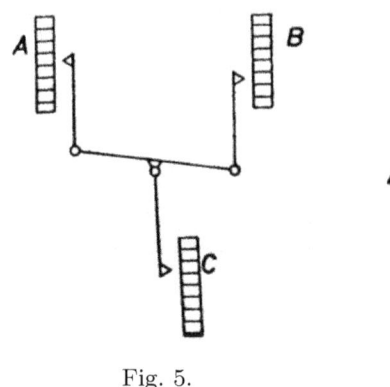

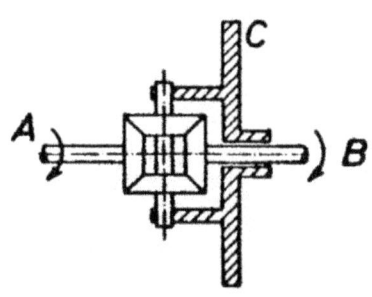

Fig. 6.

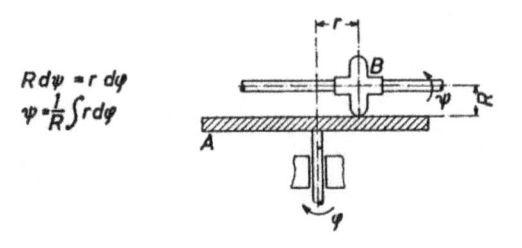

$Rd\psi = r\,d\varphi$
$\psi = \frac{1}{R}\int r\,d\varphi$

Fig. 7.

Noncontinuous processes are generally not reproducible with analog instruments; in other words, analog computers are poorly designed for these processes.

With digital computers, all values are represented by numbers. Because a digital computer can hold only a certain limited sum of numbers, there is available for the representation of continuous values only a limited supply of values. This implies considerable divergence from mathematical models. Mathematical values are subject to the concept of infinity in two respects.

First, the absolute magnitude of the numbers is unlimited; furthermore, between any two given values an infinite number of intermediate values may be assumed to exist. For this reason, computers have (independent of the number code employed) maximum values which, out of technical considerations (number of places of the register and storage), cannot be exceeded. In addition, the values proceed in step-fashion. There are neighboring values between which no additional intermediate values may be inserted. This results in limited accuracy among other consequences. In contrast to the analog computers, the accuracy of digital computers is strictly defined and is not subject to any coincidental influences.

A further conclusion is that no digital computer can precisely reproduce the results of processes defined by arithmetic axiom. Thus, for example, the mathematical formula

$$\frac{a \cdot b}{a} = b$$

has general validity, with the one exception that a cannot be equal to 0. There is no finite automaton capable of reproducing this fact precisely and generally. It is possible, nevertheless, by increasing the number of places before and after the decimal point, for a digital computer to approach infinitely close to the laws of arithmetic.

We in the field of mathematics have already become so accustomed to the concept of infinity that we accept it without considering that every infinite term is related to a series expansion or to a limiting process ("for every number there is one which follows it"). By relating this process to automaton theory, we obtain in place of a static, predetermined, finite automaton a series of automatons which are constructed according to a definite plan and differ from one another only in the number of places. The plan for construction of an automaton with n places is given; in addition, there are instructions for converting an n-place automaton to one with $n+1$ places. By use of the limiting process $\lim_{n \to \infty}$ with the aid of series expansion the automaton rule for arithmetic operations is obtained.

The digital computer, because of its special ability to handle not only numbers but also general information (in contrast to the analog computer), has opened up completely new fields, discussed below in greater detail. In general, all calculation problems can be solved on a digital computer, whereas analog computers are better suited to special tasks. It must be stressed that digital computers work in a strictly determinative way. Using the same algorithm (i.e., the same program) and introducing the same input values, the same results must always be obtained. The limited accuracy always results in the same degree of inaccuracy in the results when an operation is performed several times on the same inputs. This is in contrast to the analog computer, in which the limited accuracy has a different effect each time the program is run and can be expressed only in terms of statistical probability.

By way of supplementary comments, it may be observed that hybrid systems have been developed which consist of a mixture of the principles of the digital and of the analog computer.

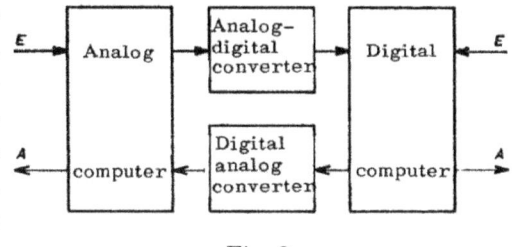

This can be simply carried out via a system in which the two computers operate side-by-side. They are joined by a digital-analog converter and an analog-digital converter (Fig. 8). In systems of this type, the single parts of the problem are divided in such a way that the more appropriate device is chosen for each subdivision of the problem.

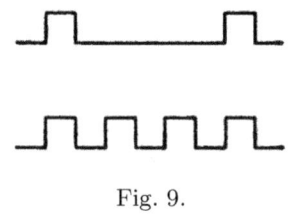

Fig. 8.

The joining of the two systems can also be accomplished by the representation of the values themselves. Thus, for example, a magnitude may be characterized by the pulse density (Fig. 9). Pulses themselves have a digital character, for they are normalized in intensity and duration; they are therefore digital, but their density (the number of pulses per unit time) can have any number of intermediate values, and it is therefore analog in character. A commonly-held opinion today is that the human nervous system operates on this principle.

Fig. 9.

2.3. Differential equations from the point of view of the automaton theory

Observation of several differential equations reveals that this way of thinking is by no means self-evident to mathematicians and physicists. There are at our disposal a number of models of physical data, which can be represented by differential equations. For example, we can take a simple differential equation to represent the upper surface shape of a liquid in a rotating vessel, according to which at every point on the surface, the normal to the surface is determined by the vector sum of the gravitational and centrifugal accelerations (Fig. 10).

This equation is written:

$$y' = \frac{r\omega^2}{g}$$

where ω is the angular velocity of the container.

The solution is very easy to obtain analytically:

$$y = \frac{\omega^2}{2g} \cdot r^2$$

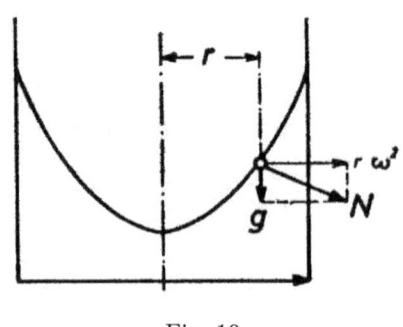

Fig. 10.

In reality, we have here an expression valid for the situation only after equilibrium has been established. For every equilibrium situation there is an initiating action. In the experiment with a rotating vessel initially at rest, the rotatory motion must be transferred to the liquid through frictional forces. Only after complex wave interaction, which diminishes with time, will equilibrium be established. For this reason it is not possible to describe the actual processes in this transition by means of our differential equation. The processes taking place during this period are considerably more complicated, and they are almost impossible to describe mathematically. We realize also that it is not necessary to follow each of these complicated processes when only the final state is of interest to us.

The relationships are very similar to many partial differential equations. These equations are used to describe the stress divisions of an equilibrium situation in plane and solid stress states. The establishment of equilibrium occurs in actuality *via* a highly complicated sequence of steps, in which once again the braking of these processes is the condition for the eventual establishment of equilibrium.

Differential equations describe only the final condition in the case of the theory of ideally incompressible fluids. The actual process leading to establishment of the end condition of equilibrium from a state of rest is hardly conceivable without taking compressibility and braking processes into account.

In the case of these differential equations, the issue is not one of a fundamental law, which can be described in terms of automaton theory as a functional variable of different, sequentially-occurring states. This also has an influence on the possible numerical solutions. Differential equations which describe an allowed sequence of states of a system are often easier to solve numerically than those which represent no more than a control function over the final state. In fact, solutions for such end states must usually be found in a stepwise solution, often with help of a relaxation process. It is not necessary to attach value to the step-wise approximations of the final state in order to simulate natural or technical processes; thus, it is possible to apply mathematically-simpler processes in the approximation.

A differential equation which describes an evolutionary process from the point of view of the automaton theory may be called the "yield" form, because the following state arises from a given state through operation of the differential on the given state. In the case of liquids and gases, inclusion of the compression term leads first to this yield form. The state of a system is given by the pressure and velocity distribution. The differences in pressure result in forces leading to a new velocity distribution, which itself leads to a new density and therefore pressure distribution through the movement of the masses. The "state" of the field may be described, therefore, by a scalar density field γ and a velocity field v. The equation may be expressed in the yield form as follows:

$$k \text{ grad } \gamma \Rightarrow \frac{\partial v}{\partial t}$$

$$-\text{div } v \Rightarrow \frac{\partial \gamma}{\partial t}$$

(k is a factor which is determined by the physical conditions). The algorithmic character is even more clearly expressed in the following form:

$$v + k(\text{grad } \gamma)dt \Rightarrow v$$

$$\rho - (\text{div } v)dt \Rightarrow \gamma$$

Corresponding to the normal rules of programming language (algorithmic language), the same symbols on both sides of the yield sign refer to different sequential states of the system (v, γ).

In the case of incompressible fluids there is the condition div $\gamma = 0$.

This equation has no algorithmic character and cannot, as a result, be transformed into the yield form. It represents merely one condition for the correctness of a solution obtained by another means.

2.4. *Maxwell equations*

Maxwell equations can also be studied from this point of view. We will limit ourselves to those equations describing the expansion of a field in a vacuum:

$$\text{rot } H = \frac{1}{c}\frac{\partial E}{\partial t} \qquad \text{div } E = 0$$

$$\text{rot } E = -\frac{1}{c}\frac{\partial H}{\partial t} \qquad \text{div } H = 0$$

Both equations, which contain the differential operator **rot** can be converted to the yield form easily:

$$E + c(\text{rot } H)dt \Rightarrow E$$

$$H - c(\text{rot } E)dt \Rightarrow H$$

(the rotor of H gives the increment of E; the rotor of E gives the increment of H).

Both divergence equations, on the other hand, have no yield form. If the wave region of the field is taken into account we obtain:

$$\text{div } E = 4\pi\rho$$

This equation is not sufficient for the algorithmic description of the law of wave propagation. Are Maxwell equations therefore incomplete? They are used to describe the propagation of transverse, but not longitudinal, waves. The reason that Maxwell equations in their usual form are sufficient for the description of all processes occurring in electromagnetic fields rests on the fact that there exist in nature no growing, newly-appearing or disappearing waves. Only displacements of charge occur. With this sort of displacement, Maxwell equations are sufficient to describe the changes in fields associated with the displacements. The author has been unable to locate a precise mathematical proof of this in any text, but it must be assumed. An interesting comment in this regard is found in "Beckersauter" (page 186), where the field for a uniformly-moving charge is developed. This results, interestingly enough, in elliptical deformation of the previously spherically-symmetric field. This deformation corresponds to the Lorentz contraction hypothesis. It is possible to reformulate the statement that "Maxwell equations are invariable in relation to the specialized theory of relativity": "As a result of nature's use of the trick of lateral expansion (rotor) in an expanding field, the system of the specialized theory of relativity is logically based".

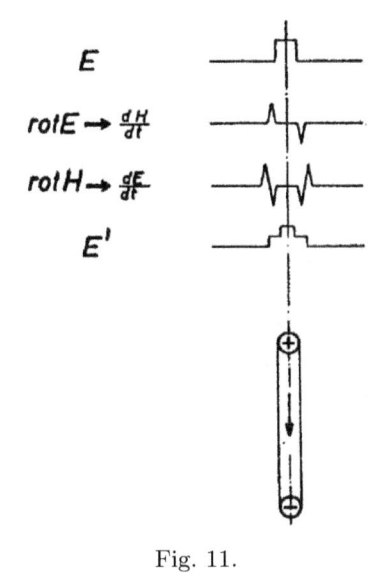

We can conceive of the functional nature of this lateral expansion as follows: given that we want to calculate the field between two opposite charges $+e$ and $-e$, let us assume that we do not know the field distribution in itself well-known and also easily derivable. We begin, as shown in Fig. 11, with a distribution sure to be false, by simply joining $+e$ and $-e$ by a linearly-constant force from the origin to the terminus. Application of the Maxwell equations to this field distribution results in a multistep asymptotic approximation of the field to be determined.

Fig. 11.

It is also demonstrated in this process that we obtain results without using the equation

$$-\mathrm{div}\, E \Rightarrow \frac{\partial \gamma}{\partial t}$$

in the treatment of electromagnetic fields, although, as we have seen, this equation is necessary for the treatment of compressible fluids. We need not even introduce the electric field density γ. The fact that results are obtained without this term is not proof that nature works without resort to field density. Assuming that such a condition did exist, nevertheless, it would be nearly impossible to demonstrate its existence, for both "rotor" equations establish in themselves a field distribution such that

$$\mathrm{div}\, E = 0$$

is generally satisfied. As a result, the divergent makes no contribution to the field distribution. Because it is impossible to create or destroy charges, we have no experimental means of testing the validity of the law of longitudinal expansion in nature.

What, then, is the rationale for examining this law? The question is interesting in connection with the concept of numerical stability, and it will be considered again below.

2.5. *An idea about gravitation*

A short consideration of gravitation is introduced in this regard. If we accept the validity of the Maxwell equations, in their transmitted sense, for gravitation as well, then a simple explanation of the expansion of gravitational fields by moving masses and the invariance of the laws of celestial mechanics based on this distribution is applied to the special relativity theory. Because the relative velocity of the heavenly bodies within our observation range lie on the order of magnitude of 1/10,000 of the speed of light, the gravitational magnetic fields were simply so weak that they were immeasurable. To be sure, small damping of planetary movements must be considered. The author would be very grateful for a critical observation of these thoughts by the physicists.

2.6. *Differential equations and difference equations, digitalization*

If differential equations are expressed "yield" form, according to the automaton theory, then they can be simulated by a technical model (an automaton) and solved. In itself the analog computer is the ideal automaton. It works, at least in theory, with continuous values and operates constantly; in other words, we have a continuous flow of states, the latter of which is always determined by that which precedes it. In practice, analog computers are used primarily for calculation of differential equations. Nevertheless, there is a rather narrow limit to the capabilities of the analog computer. For partial differential equations, analogous technical models are available only under special circumstances.

The solution of differential equations with a digital automaton is immediately complicated by the previously-mentioned difficulties: differential equations operate with continuous values and infinite field densities. Digital instruments operate with discontinuous values. An infinite field density would require an infinite storage capacity and infinite calculating time. Therefore it is necessary to reach compromises in both regards.

One normally proceeds from differential equations to difference equations when numerical solutions are sought. In this process, the values obtained are still regarded as continuous. In fact, the transition from differential equations to difference equations involves two boundary transitions: (1) $\Delta x \to dx$, and (2) enlargement of the number of places of the included magnitudes. The first boundary transition leads constantly to a limiting value which the second transition anticipates; in other words, constructing difference quotients makes sense only if the gradations between values are much smaller than the chosen Δ-value. This fact has a definite influence on the numerical stability of a calculation.

If the transitions are carried out in such a way that the values remain of approximately the same order of magnitude as the step values, the staircase shape of the curve is maintained, and it is impossible to construct a differential quotient.

In the observations that follow, this distance will be utilized with design, specifically through consequential further development of the thoughts on digitalization.

Systematic narrowing of the number of places of the magnitudes being treated results in the limitation of variables to those encompassed by elementary logic; for example, yes-no values or triply-variable values. As we will discover later, triple values and the trinary number system based on these values has certain advantages, since rounding up and rounding down are easier to carry out and the division by 6 necessitated by the division of the field area into 6 neighboring cells is also easier to calculate. By attaching the values $+1$, 0 and -1 to the numbers, this corresponds to the possible electrical particles $+e$, 0, $-e$.

The continuous field density must he separated into single values for numerical solution, a process which is easiest with a grid. The simplest grid is doubtless an orthogonal one. There are other possible choices: the triangular and hexagonal grids in two dimensions, for example, and a grid in three dimensions corresponding to the most dense packing of spheres. If several different field values arise in the calculation (for example, velocity vectors and densities), it is not necessary that these values be localized on the same grid point. There is no need for the three components of a spatial vector to be localized. In this case, a division is possible as well. There is no further necessity in the construction of a digital space structure to approximate the laws of Euclidean space. A number of general observations on the presentation of physical problems were presented earlier from the viewpoint of the automaton theory.

2.7. *Automaton theory observations of physical theories*

Up to this point we have considered only the problem of using computers to approximate physical models and to follow physical processes numerically. It would be possible in this context to suggest a fundamentally different question: to what extent are the realizations gained from study of calculable solutions useful when applied directly to the physical models? *Is nature digital, analog or hybrid?* And is there essentially any justification for asking such a question?

The classical models of physics are doubtless analog in nature. The field strength of different potentials, like the force of gravity, are not subject to a "particularization". There are no such limits as "threshold values" (minimal size), limiting values (maximum values) or limits on the density of the field itself. Even the extension of classical laws by the theory of relativity is entirely within the conception of the continuum. Only for velocity is an absolute upper limit assumed to exist (that of the speed of light), and that concept is completely in accord with "analog" thought.

It was first with the introduction of the particular nature of matter through its subdivision into molecules, atoms and elementary particles that a few quantities assumed a discrete character, but this is not necessarily to be equated with "digital" interpretation of the laws of nature. The classical many-body problem was of an analog nature, even when each of the single bodies possessed individual characteristics with discrete properties (masses).

Quantum physics is the first to deviate in several respects from the concept of infinite quantities, to the extent that it assumes only discrete values for certain physical quantities. Best known is the relationship between frequency and energy of a light quantum, which is defined by the formula $E = h \cdot \gamma$, where h is a universal constant of nature. To be sure, the energy itself is not quantized, but only the quotient $\frac{E}{\gamma}$. This is somewhat different from the case where the energy can have only a discrete number of values because of the limited number of places in the calculator of a digital computer.

The postulates of the quantum theory have far-reaching consequences in relation to the quantization of different physical quantities. The conceptions of the classical spatial continuum are being abandoned, it is true, but not through replacement of the continuum by a grid of discrete values, rather through a process whereby one moves to fundamentally different starting points, similar to a configuration room of higher dimensions, in which probability values are defined (for example, the probability of a particle being in a certain place at a certain time). Even in this concept the idea of the continuum is not rejected, for the differential equations of quantum mechanics are governed by no restrictions in relation to the magnitudes of fields.

The models of modern physics are concerned, therefore, both with continuous and discrete values. It would seem appropriate to consider a hybrid system. It will be extremely difficult to find a technical model of a hybrid computer which behaves according to the laws of quantum physics.

We have recognized the preliminary conclusion that our physical models may best be conceived of as hybrid systems. Can conclusions with respect to nature he drawn from this? Is nature itself therefore to be considered a hybrid system?

We have not yet disposed of completely digital physical models. If we are completely impartial, it appears a justified question whether infinitely-divisible quantities (in other words, really continuous quantities) have any reality in nature. what would be the consequences, for example, if we were to shift to complete quantification of all the laws of nature and were to assume in principle that every physical magnitude is subject to some sort of quantification?

Before an examination of the real question is attempted, let us examine first the classical model of thermodynamics, through which the relationship of gases is treated by the model of rubber balls moving freely through space and colliding with each other. If the static behavior of these balls is replaced by a differential equation, it is valid only for spatial dimensions that are large in comparison to the average distance between the individual particles. In effect, the model can be viewed as analog on a large scale, yet in detail it is characterized by the particle nature of matter.

What would the calculated solution look like, if we were to imitate directly the model of flying, colliding particles?

Of course, the starting point is no longer a differential equation; the flight paths of single particles are followed with digital calculations (Figs. 12, 13 and 14).

It is quite simple for modern electronic computers to draw up a program for this purpose. We do not wish to become involved in these calculations in the course of our discussion (the calculation

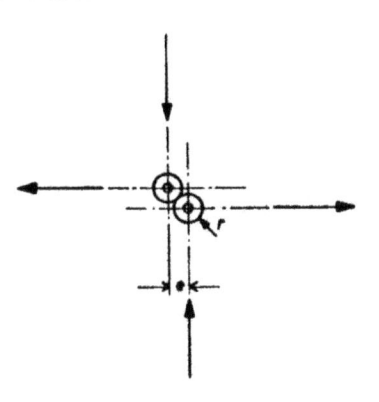

Fig. 12.

itself is relatively involved and boring) because a large number of particles are necessary for the results to have statistical value. The flight paths are simple to calculate, since they are rectilinear (gravity effects disregarded).

The collision processes are the interesting part. Equal mass and equal elasticity of the particles is assumed. We shall first consider the case in which the particles meet exactly; i.e., first, that the paths lie in one plane and mutually intersect, and second, that the centers of both particles meet simultaneously at the point of intersection. This case is uninteresting, for the case of the elastic collision is not significantly different from that in which both particles continue undisturbed on their ways, if each particle is considered individually. Furthermore, in general situations, the probability of such a situation arising approaches 0 as the accuracy of the calculation is improved. Therefore, only those cases are of interest in which the paths do not exactly cross, or in which the centers arrive at the approximate intersection at only approximately the same time.

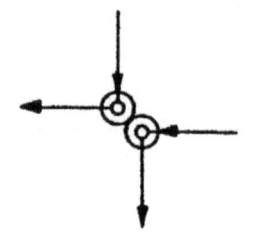

Fig. 13.

In this case the particles have different paths after the collision than before it. It is not necessary to stop here and establish the collision law firmly. The behavior depends on the size of the particles and the law of elasticity. Large particles collide more frequently than small ones. Hard particles behave differently than soft ones. The statistical result of the behavior of a large number of particles is the same. If we compare such a calculation model with the physical model, the following interesting aspects arise.

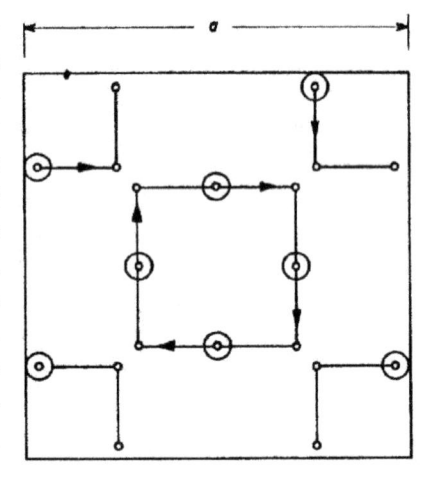

Fig. 14.

In the case of both models, we can see that in general ordered states give rise to disordered states, or entropy increases. In any case, we can devise certain exceptional cases, for which a given entropy remains constant. Take, for example, a vessel with exactly parallel sides and a series of particles, the paths of which are exactly perpendicular to these walls and sufficiently far apart from each other that there is no mutual interaction of the particles. In this case, the paths remain unchanged in the sense of classical mechanics. This is also the case in the computer model if the coordinate system on which the calculation is based is set parallel or orthogonal to the walls. There are certainly other interesting special cases for which collision processes between the particles occur, yet nevertheless a certain ordering remains in force (Fig. 14).

We are now aware that modern physics has replaced this classical picture. Collision processes between single particles are not precisely determinable, according to modern physics. There exist only the laws of probability, which correspond to the laws of classical mechanics, taken as a statistical average. Scattering is due to this effect, with the result that even for the theoretically-assumed special cases, the order of the system decreases with time and the entropy increases. How is this reproduced in the computer model? As long as we do not specifically program this scattering effect into our model, the carefully-constructed special case mentioned above does not exhibit any scattering effect. However, as soon as the system, through the introduction of a small scattering input, becomes out of step with the special ordering, the situation is similar to that obtained with the models of modern mechanics. It is not generally necessary to pay particular attention to scattering effects. The error inherent in the computation–special cases excepted–have the same effect (Fig. 14). The classical model demands absolute accuracy in calculations, requiring in the computer model an instrument with an infinite number of places. Since this is not possible in practice, calculation errors enter into the collision processes, which have the effect–similar to the model of modern mechanics–that divergences from the paths predicted by the theories of classical mechanics appear. It would be possible to express these deviations by a statistical law. A significant difference does exist, however. In the model of modern mechanics the errors are real; in the computational model everything is strictly predetermined, not in the sense of classical mechanics but in the sense of defined calculating inputs, which can only approach the classical model. Both result in an increase in entropy.

The initially equivalent result (i.e., the increase in entropy) arises in both cases from the slight deviations from classical mechanics. In modern physical models, these deviations are defined by probability laws; in the case of computer models through defined calculation errors.

This may appear unimportant at first glance. Yet if we extend this thought process somewhat further, very interesting consequences in relation to causality may he drawn, which will be developed in Chapter 4.

Matrix mechanics can also be considered in the automaton theory. In any case, we need an automaton in which the transition from one state to the next is determined by probability laws. The transition matrices of matrix mechanics correspond to the state tables of the automaton. This possibility of automaton-theoretical observations will not be considered at greater length. In the next chapter, a few examples of digital treatment of field and particle problems will be presented.

3. Examples of Digital Treatment of Fields and Particles

3.1. *The expression "digital particle"*

Let us first consider one-dimensional space. In this regard we can relate an example from hydromechanics and one from counter engineering. Let us consider the behavior of frictionless gases in a straight cylinder. After eliminating and collecting terms that are irrelevant for our purposes (density, etc.), we can obtain a somewhat simplified relationship of the real physical forces.

We have two quantities: p (pressure), which we fix in discrete points 1, 2 and 3, and v (velocity), which we express in intermediate points 1', 2' and 3'.

$$p\ 1\ 2\ 3\ 4\ 5$$
$$v\ 1'\ 2'\ 3'\ 4'\ 5'$$

Δ_p^s and Δ_v^s representing the difference in p- and v-values between neighboring points, Δ_p^t and Δ_v^t corresponding to the differences between p and v in consecutive time intervals.

The following differential equations then hold:

$$k_0 \Delta_p^s \Rightarrow \Delta_v^t$$
$$k_1 \Delta_v^s \Rightarrow \Delta_p^t$$

Expressed in words: the change in velocity is proportional to the change in pressure and the difference in pressure is proportional to the change in velocity. In the second equation, the terminus Δ_p^t is converted in order to indicate that it refers to a Δ_p after that of the first equation. The two factors k_0 and k_1, which contain the physical characteristics Δ_x (length component) and Δ_t (time component), can be combined for our purposes into a single factor k. We then obtain:

$$-\Delta_p^s \Rightarrow \Delta_v^t$$
$$-k\Delta_v^s \Rightarrow \Delta_p^t$$

The sign $\rightarrow$ is used to indicate that Δ_p in the second equation is not identical with that in the first equation.

It is clear that these equations can he converted from differential equations to difference equations when Δ_x and Δ_t are allowed to approach 0. Exactly the opposite condition is of interest to us. Although mathematicians and programmers generally attempt to set up difference equations in such a way that the differential equation at the basis of the difference equation is approximated as nearly as possible, we are able to resolve the question by using the most general digitalization possible.

We are now able to convert a physical pulse law to an engineering counter law. If we let the quantities p and v and the corresponding values Δ_p and Δ_v assume only integral values, we must choose a whole number value of k in order for the difference equation to give whole number results. If we first let $k = 1$, we obtain the equations:

$$-\Delta_p^s \Rightarrow \Delta_v^t$$
$$-\Delta_v^s \Rightarrow \Delta_p^t$$

We further attempt to assign p and v the smallest possible values, i.e. -1, 0 and $+1$, and to study the behavior of the system that satisfies these conditions. We obtain as a result the following arithmetic relation:

$$v - \Delta_p^s \Rightarrow v$$
$$p - \Delta_v^s \Rightarrow p$$

Figure 15 shows a simple calculating scheme for this rule. We have the four values v, $-\Delta v$, p and $-\Delta p$ per unit time. The spatial sectors are opposed to one another. Zero values are not written for purposes of simplicity. Four stable elementary forms are represented [(1), (2), (3) and (4)] which we will consider as mutually-independent "digital particles". There are two time units, t_1 and t_2, respectively, for the values v, $-\Delta v$, p and $-\Delta p$; v and p are assumed for t_1. It follows from this that $-\Delta v$ and $-\Delta p$ correspond to time interval t_2 and, following through the above equation, the values v and p correspond to the next time interval t_2.

The equations relate to the traveling of a simple pulse. The particles are stable only at this velocity. At the same time, this velocity is the highest one possible for the system. The system permits no other velocities. Figure 16 shows a graphic version of this pulse.

From the standpoint of the automaton theory, we are concerned with a linearly-expanded infinite automaton which is repeated periodically in the automaton (cellular automaton). The v– and p–values represent the states of the automaton; Δv and Δp are derived from them. The above equation establishes the function according to which the subsequent state arises from the previous one.

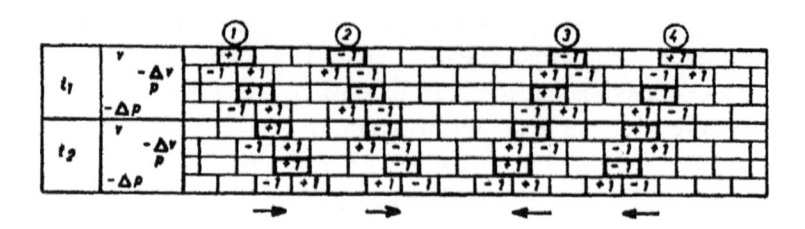

Fig. 15.

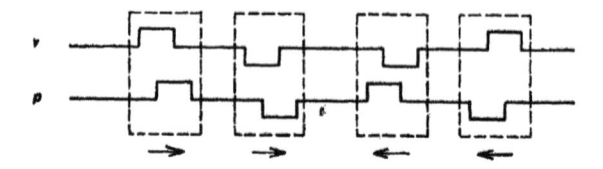

Fig. 16.

Figures 17 and 18 show an instable form of expansion of an isolated pressure pulse, with which no velocity impulse is associated (as was the case in Figs. 15 and 16). In Fig. 17, the Δ–values are omitted for reasons of generalization.

This form of pulse expansion contradicts our conception of the expansion of an originally isolated pressure cell in a gas-filled cylinder. From this model we have derived the difference equation. The digitalization was carried out so generally that the deviations from the differential equation result in deviations from the physical laws. The conservation of pulse rather than of energy is the key to the calculation behind the difference equation. The graphic representation

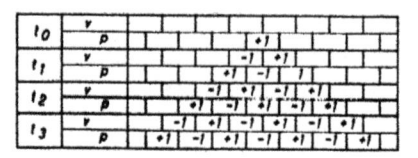

Fig. 17.

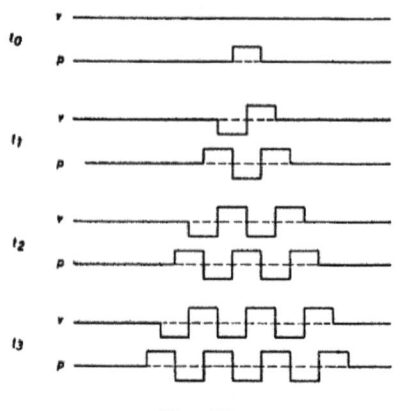

Fig. 18.

Calculating Space (Rechnender Raum) 751

of Fig. 18 shows, in fact, that the average of $(p = 1)$ remains constant, and that the average value of v is constant at 0. On the other hand, the expansion of alternating positive and negative p–values in the graphic representation indicates an obvious constant increase in the potential energy. The corresponding is true for the kinetic energy values represented by the v–values.

It would he interesting at this point to inquire whether this sort of deviation is necessarily associated with crude digitalization or whether crude digital models can be constructed which obey all the conditions of the original differential equation, in this case especially that of conservation of energy. Of course such a simplified model requires an exact definition of the term "energy". This is simply noted without further consideration here.

It is interesting that a pair of isolated pulses yields a stable system: the emission of two diverging digital particles (Fig. 19). Apparently only certain configurations are possible, while others are excluded or provide no stable results. This bears a certain similarity to some situations in quantum mechanics.

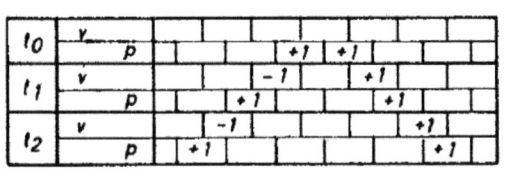

Fig. 19.

Because our chosen calculating rule has a purely additive character, the superposition rule applies; i.e., the single forms can be considered independently of one another, as a result of which it is natural that values greater than 1 appear. This means that two oppositely moving particles do not influence one another, but pass by or pass through one another without changing shape. In a system strictly described by the superposition rule, there are no results possible which correspond to the reactions between elementary particles known in physics. This provides our evidence that it is not necessary to build linear elements into our models. The simplest and roughest form is general limiting of the values above and below. This may he demonstrated from the examples in Fig. 20. Here we have two approaching digital particles, specifically in examples (1) and (2) on the left, corresponding to the previous reaction according to the superposition rule. We can see that in example (1), values $+2$ and -2 arise. In example (3), the particles pass through one another without values greater than $+1$ and -1 arising.

In this situation, an interesting result of crude digitalization may be observed. The course of the collision process differs with the phase state

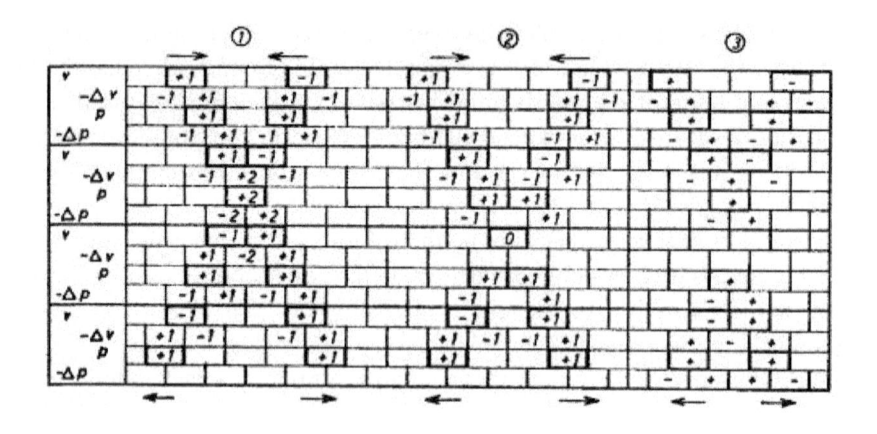

Fig. 20.

Fig. 21.　　　　　　　Fig. 22.

of the distance between the two particles. This is not outwardly visible. Figure 20 shows example (1) with a limiting law corresponding to Fig. 21. Here there are only three values: −, 0 and +. Figure 22 shows the relevant calculating system. It is constructed so that 1 + 1 results in a value of 1. We can see that in spite of this limitation, the particles are free to intersect one another, a result which in itself would not be expected at first glance, for crude curtailments of the calculating rule were made. Application of the calculating rule of Fig. 22 to example (2) yields nothing new, of course, because in the example the values −2 and +2 are not to be found.

It is interesting that in spite of this, a certain reaction process in particle interaction can be noted. If we consider examples (2) and (3), for example, it can be seen that in the case of (3), in contrast to (2), a certain retardation of the process may be observed. In (2) the particles intersect and proceed away from one another unhindered. In (3) we might argue that the particles first react with one another and that two new digital particles are emitted as a result of this reaction. The question as to whether (2) or (3) occurs is again dependent on the distance phase state and is outwardly a matter of chance. Without knowledge of the fine spatial structure, it can only be determined that in our example two fundamental situations are possible in particle interaction, for each of which the probability of occurrence is 1/2.

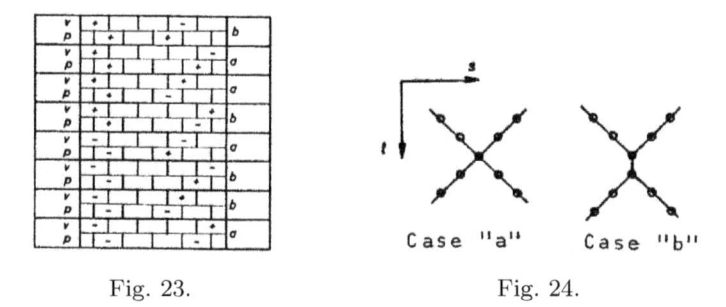

Fig. 23.

Fig. 24.

Figure 23 shows a summary of the eight possible cases in particle interaction; Fig. 24 represents the schematic, idealized particle paths for the two different interaction patterns a and b. It must be explicitly stressed that the paths are idealized particle paths. In reality, our model represents not continuous movement, but a process of stepwise progress.

It is interesting to note that in the nonlinear calculating rule (Fig. 22), an isolated pressure point results in the emission of two particles (Fig. 25).

Establishment of limiting values obviously sets limits on the free superposition processes. In the case of

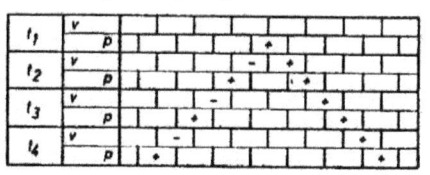

Fig. 25.

unlimited values, particles corresponding to Fig. 15 are also theoretically superimposible. That means that we can construct a pressure mountain of any height with its accompanying velocity distribution which satisfies the step-wise extension rule; i.e., which remains stable. These stable "larger" particles are always divisible into elementary particles. This is no longer true when the rule corresponding to Fig. 22 is applied.

Our initial position, in which we have chosen the factor 1 relative to the Δ-value, corresponds to a very hard medium in the assigned physical pattern of a gas-filled cylinder. A more flexible situation is obtained when the factor is made smaller. In this case, nonintegral numbers arise in more accurate calculation. If we wish to continue with whole numbers or to introduce only minimal gradations, rounding up and rounding down must be introduced. In this respect also the ternary system is superior to the binary one. The value 1/2 lies exactly midway between 0 and 1. The values 1/3 and 2/3 can also be precisely inserted between the values 0 and 1.

From there we want to make the following start:

$$v - \frac{\Delta p}{3} \Rightarrow v$$

$$p - \frac{\Delta v}{3} \Rightarrow p$$

Values $\Delta_p/3$ and $\Delta_v/3$ rounded up or down to whole numbers. Figure 26 (1) shows a stable particle in this system with a period of $3\Delta t$. The velocity of propagation is 1/3 of that of the particle in the corresponding figure (Fig. 15). This corresponds as well to the physical model, in which a soft medium has a slower speed of sound. Here we have the situation that the "speed of switching" between neighboring particles is considerably higher (in the example three times as great) than the particle velocity. In more complicated models of "calculating space", it would be conceivable that speeds of light corresponding to maximum particle velocities, which are considerably slower than the speed of switching, exist. This does not mean, however, that in such a model "signal speeds" greater than the speed of light (in the model) are possible. The speed of switching has a purely local meaning.

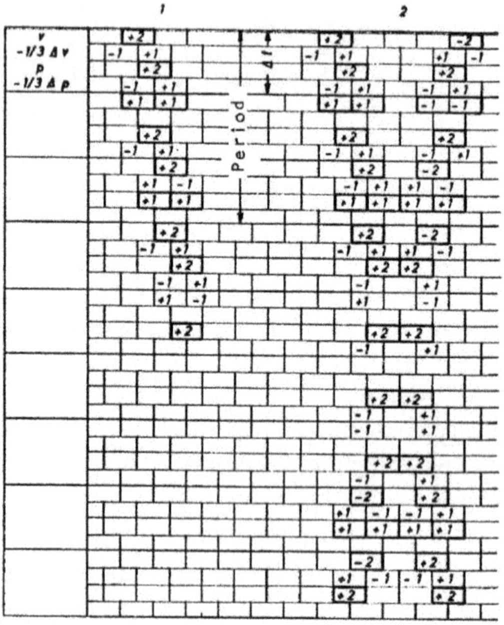

Fig. 26.

It is interesting that a digital particle assumes different configurations in the course of a period. The pressure pulse appears in part alone with a value of +2, in part as a pair with the values +1 and +1. The position of the particle is definable for the following period, but not without further information for the single phases of a given period. Is this not analogous to the quantum theory, which relates position and momentum through the uncertainty principle? In any case the computer model, in spite of the apparent error, is characterized by strict predetermined happenings.

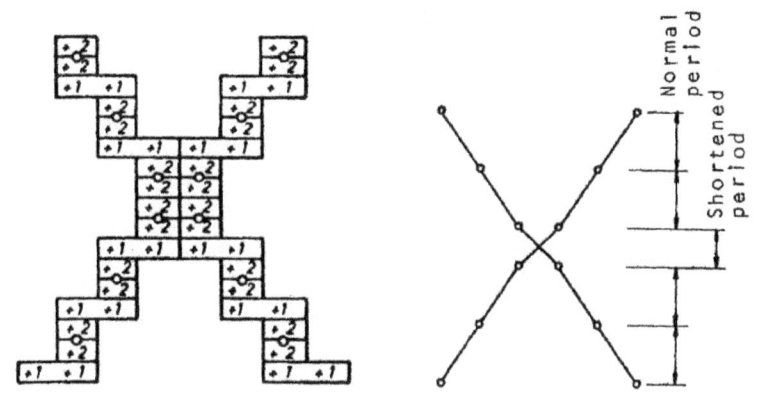

Fig. 27.

Fig. 28.

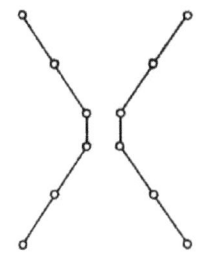

Fig. 29.

Figures 26, 27, 28 and 29 show the process of interaction of two such particles, and more specifically Fig. 26 (2) shows the detailed calculating scheme and Fig. 27 an excerpt from it, in which only the p–values are represented, while Fig. 28 shows the idealized particle path. The figures demonstrate that the particles do not simply pass beyond one another, but that they do react, this time with shortening of the interaction time (in contrast to Fig. 24). The process can also be represented as one of repulsion (Fig. 29). Here it may be seen in the mode of viewing the figures that terms like "passing through" and "repulsion" lose meaning when applied to the reaction of digital particles. The quantum theory has yielded corresponding results, although not in digital form.

In particle interaction corresponding to Fig. 26, there are certainly many more differentiable cases apparent from systematic investigation, in comparison with the example from Fig. 23. We must first investigate which particles are possible in this system. The influence of the separation phases must also be taken into account, and finally the possibilities of the particles interacting in different phases must be considered.

It is not the purpose of this paper to carry out an exhaustive examination. The previous observation of a few simple examples stimulates a whole series of interesting concepts.

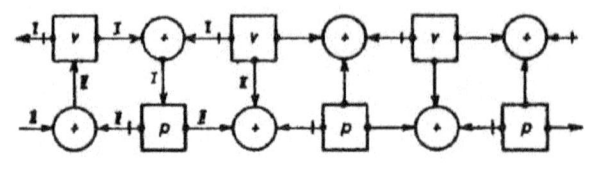

Fig. 30.

Figure 30 shows the block diagram for a calculating space corresponding to the previously-introduced calculating rule. The squares v and p represent registers to which numbers can be added. The shifting parts of the system, which serve to carry out subtraction, are represented by the circles marked with Δ. The vertical line at the exit of the Δ–members means negation. The block diagram can, of course, be subdivided into its single shifting elements. The symbols in current use reduce the shifting to its single elements, which correspond to the basic operations of Boolean algebra (conjunction, disjunction and negation). The three value information elements used here had to be converted to binary elements via two Boolean variables (2 bit). Out of 4 possible combinations of these two values, only three are employed. For this reason, a more detailed representation is omitted. In order to render the block diagram in Fig. 30 operable, clean pulsing is necessary. Therefore the pulse beats are represented in Fig. 30 by I and II. In this process it is taken for granted that the pure addition members work without time delay to build the Δ–values, while the registers transmit their information further only with the addition of the following pulse. This pulsing corresponds to the fine structure of the time dimension.

3.2. *Two-dimensional systems*

Let us examine briefly the two-dimensional system. The simplest structure is a grid corresponding to an orthogonal coordinate system. The system possesses two definite axes which enter into even simple pulse propagation. We shall start with a simple rule, where every grid point can have the states 0 and 1. In every time interval one such 1 is transmitted to every neighboring grid point. The combination of pulses arising from different neighboring points is carried out in accordance with the disjunction rule. If the state of the grid point (x, y) is $\phi_{x,y}$, we obtain the following equation:

$$\phi_{x-1,y} \vee \phi_{x+1,y} \vee \phi_{x,y-1} \vee \phi_{x,y+1} \Rightarrow \phi_{x,y}$$

Expansion along the coordinate axes is faster than along the diagonals. Little can be developed with such a rule, since after a short time it leads to a state in which all spatial points reach the state "1" and thereby no configurations, particles, etc. are possible (Fig. 31).

Next we will consider a similar rule, in which nevertheless many-place values are allowed and combination occurs by addition. In the transfer between the grid points the values are multiplied by a factor k. We obtain the formula for this rule:

$$K(\phi_{x-1,y} + \phi_{x+1,y} + \phi_{x,y-1} + \phi_{x,y+1}) \Rightarrow \phi_{x,y}$$

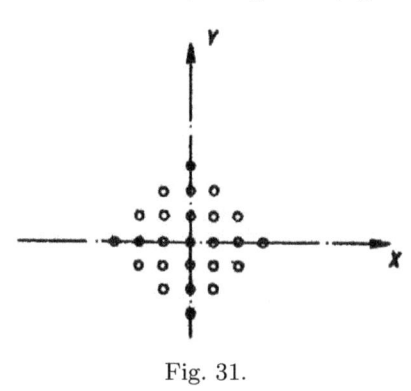

Two examples for the factors 1/4 and 1/2 are given in Figs. 32 and 33. For reasons of symmetry it is necessary to consider only a 45° section. As in Fig. 32, the values are entered only for the front of the pulse. The roman numerals correspond to the individual time phases with a separation time of Δt. We can see from the examples that the front moves as represented in Fig. 31; i.e., with its peak along the coordinate axis, although the values along the diagonals are greater. The forward-rushing point very soon reaches its peak.

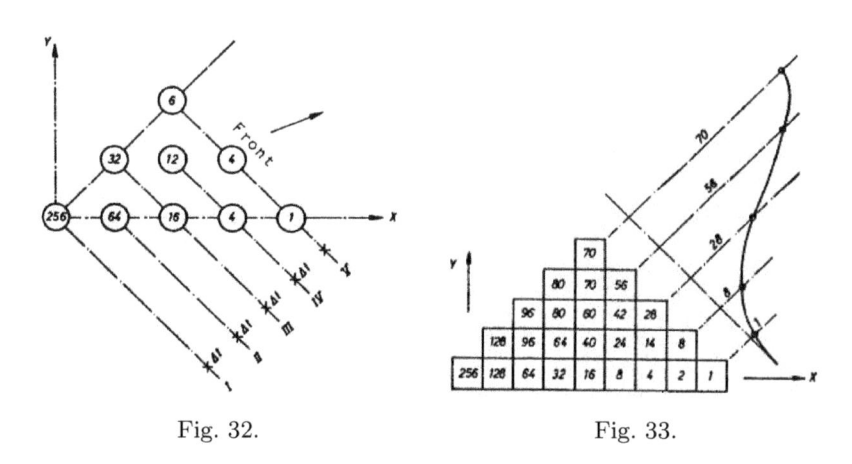

Fig. 31.

Fig. 32.

Fig. 33.

Because we cannot assume an infinite number of small values in digital space, the minimum value is soon reached; i.e., the peak dies out. It would be interesting to follow the progress of such an expansion with the help of a calculating machine. The question of particular interest is whether and how quickly the values converge in a circular expansion pattern.

One thing is clear: it is impossible to construct digital particles from such a rule. We must find other rules.

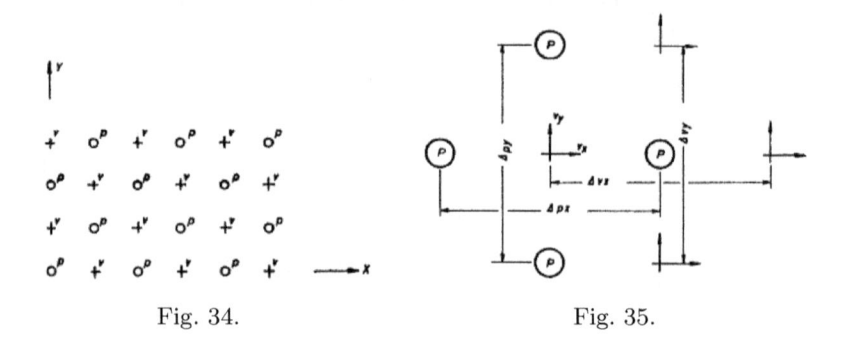

Fig. 34.

Fig. 35.

It is possible to take the rules for linear space, which give rise to stable particles, and apply them to two-dimensional space. Of course, we then need an interrelationship of the two dimensions, for without it the single orthogonal grid points would have an independent existence.

Figure 34 shows one possibility of arranging the v– and p–values in a checkerboard. Figure 35 shows the individual values which emerge. Two components, v_x and v_y, must be considered for v. One value is sufficient for p. The two axes are coupled through p.

We can now formulate the following rule:

$$v_x - \Delta p_x \Rightarrow v_x$$
$$v_y - \Delta p_y \Rightarrow v_y$$
$$p - (\Delta v_x + \Delta v_y) \Rightarrow p$$

Fig. 36.

Fig. 37.

Because of the coupling through p, individual pulses corresponding to Figs. 15 and 16 vanish. Stable, although not infinitely parallel, wave fronts can be built. Figure 36 shows such a wave front parallel to one of the coordinate axes, and Fig. 37 shows a diagonally-moving wave. Figure 38 shows a propagation relation between the two waves. The propagation velocities are functions of direction.

It would be interesting to consider the different consequences of more or less crude digitalization in this case. Because the rules are related to the equations of rarefied gas dynamics and hydrodynamics, it is interesting whether (for example) the hydrodynamically stable structure of a vortex can be crudely digitalized and "digital elements" can he constructed. This investigation can be carried out only with the help of calculating machines.

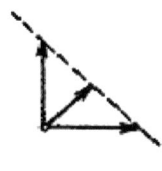

Fig. 38.

In order to construct stable particles in two-dimensional space, we shall first consider another manner.

3.3. Digital particles in two-dimensional space

We shall assume an orthogonal grid pattern, corresponding to Fig. 39. We no longer make the distinction between $v-$ and $p-$points, but allow for each point the values p_x, p_y. For reasons of simplicity we first assume that the p-values can take on the values $-$, 0, $+$. We can then speak of p-arrows or of short arrows. First we establish that an isolated arrow (an arrow which has no perpendicular arrow arising at the same grid point) is directly transmitted to the next grid point. Figure 40 shows the four possible examples of this sort of single isolated pulse. It can be transmitted forward only in an orthogonal direction. We can first determine that there are two cases of interaction between two arrows approaching in the same orthogonal.

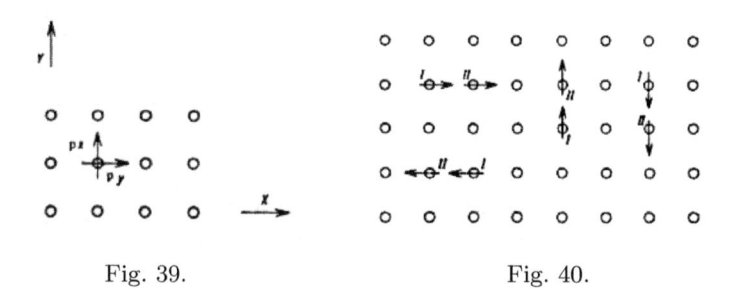

Fig. 39.   Fig. 40.

Both of these are shown in Fig. 41. In one case, the arrows continue away from one another; in the other they cancel one another. Which case occurs depends on the separation phase. We still need a rule for the case of intersecting arrows. This is demonstrated in Fig. 42. Two intersecting arrows exist at point Z at time I. According to our previous rules, they would be propagated forward, each in its own direction, independent of the other. Now we establish that the two arrows are in fact propagated forward in their respective directions toward points B and C, and at points B and C they exchange direction. We obtain in this way a stable particle of period $2\Delta t$, which is propagated diagonally forward (Fig. 43).

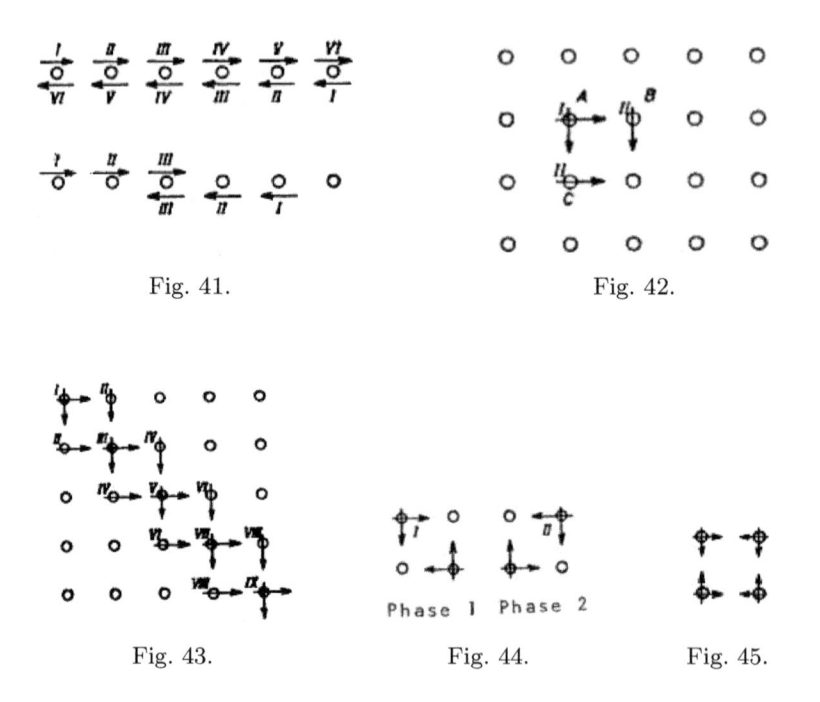

Fig. 41.

Fig. 42.

Fig. 43.

Phase 1   Phase 2

Fig. 44.

Fig. 45.

It is interesting to note that pockets arise from this rule which are fixed to 4 neighboring grid points; they have a period $2\Delta t$ (Fig. 44). A doubly-stable pocket with period Δt is also possible (Fig. 45). As may be seen from additional examples, these pockets cannot be destroyed.

We now have particles which can be propagated in eight discrete directions in a plane and standing pockets as well. Figures 46–57 give a series of interesting examples for the interaction of such particles. At first we shall maintain the condition that arrows may have only the values −, 0, +. Two

Calculating Space (Rechnender Raum)

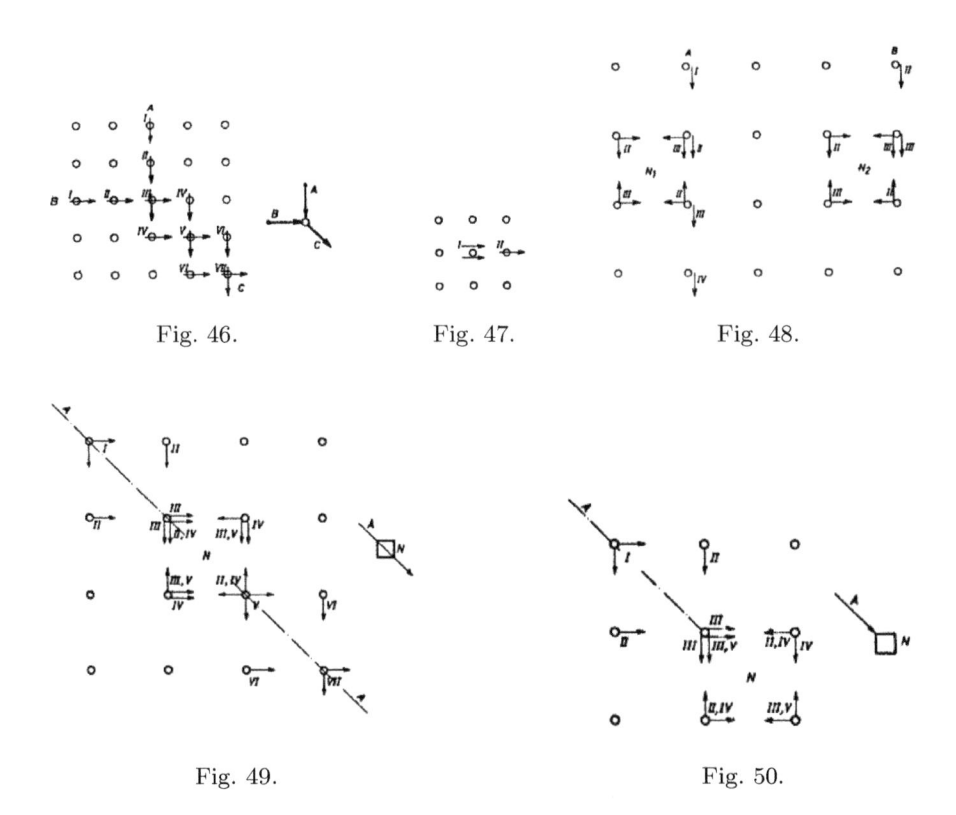

Fig. 46.

Fig. 47.

Fig. 48.

Fig. 49.

Fig. 50.

oppositely-directed arrows cancel one another at the same grid point, and two with the same orientation act as a single isolated arrow.

It may be seen that the course of the different interactions is dependent on both time and separation phases. The particles can cross through one another, cancel one another or build new particles. Pockets are insidious because they can destroy particles without disappearing themselves. On the other hand, pockets can arise from specific forms of interaction (Figs. 55 and 57). In the model of a cosmos which functions according to this rule, all particles would eventually be converted into hard pockets. This model is therefore of little use.

In the interaction it is highly significant whether the point of intersection of the particle paths lies on a discrete defined point in the coordinate system. In this case a reaction occurs (for example, Figs. 52 and 53).

The possibilities of this system can be investigated by permitting the introduction of arrows of different absolute length. For arrows pointing in the same direction we use the addition rule. It is more difficult to expand

Fig. 51.

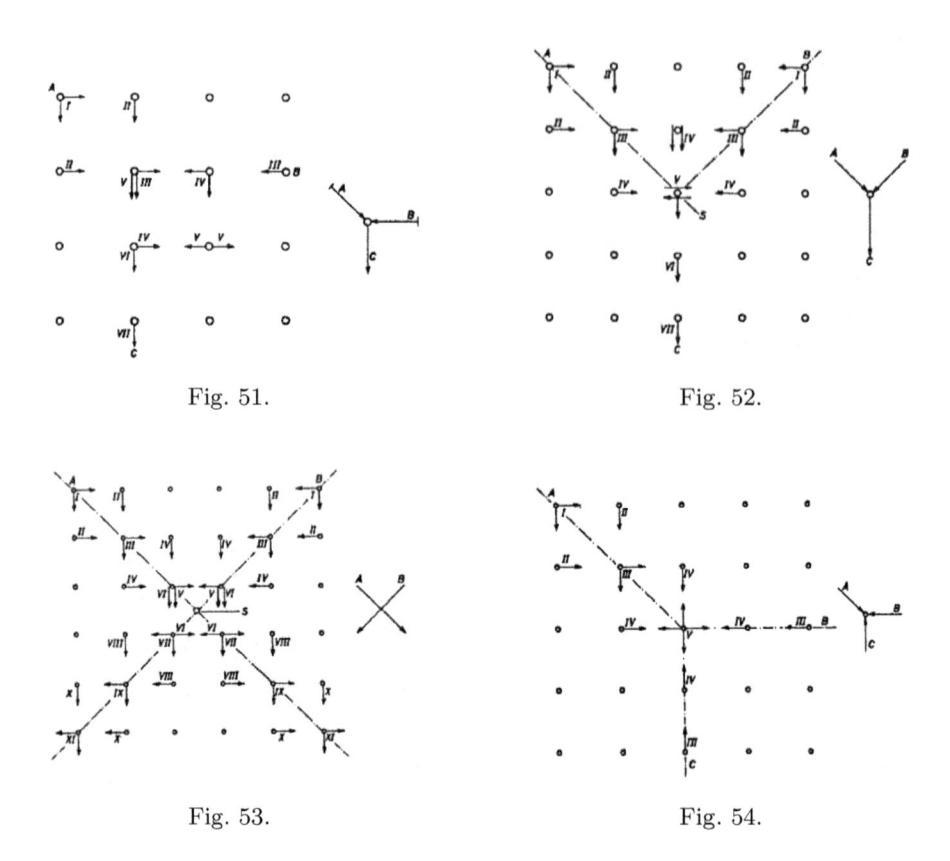

Fig. 52.

Fig. 53.

Fig. 54.

the rule of Fig. 42 to include two intersecting arrows of different lengths. We can reach the following agreement.

In the case of mutually-orthogonal arrows, the longer arrow is divided into two parts; the contribution of one is equivalent to that of the arrow orthogonal to it and combines with the first as in Fig. 42. The remainder acts as an isolated arrow (Fig. 58).

We are now able to construct particles having different directions of propagation. The number of different directions possible is dependent on the number of values possible for the contribution of the arrow.

Figure 59 shows an example with a ratio of the arrows of 5 : 2. The direction of movement corresponds to the ratio of the arrows. The particles pass through different phases. The particle in Fig. 59 has a period of $7\Delta t$. In the course of one period the particles pass through a discrete coordinate point Q (zero phase point). The particles "disappear" at intervals. It is possible to construct lines of the same phase (phase lines τ_0 - τ_6).

Calculating Space (Rechnender Raum) 763

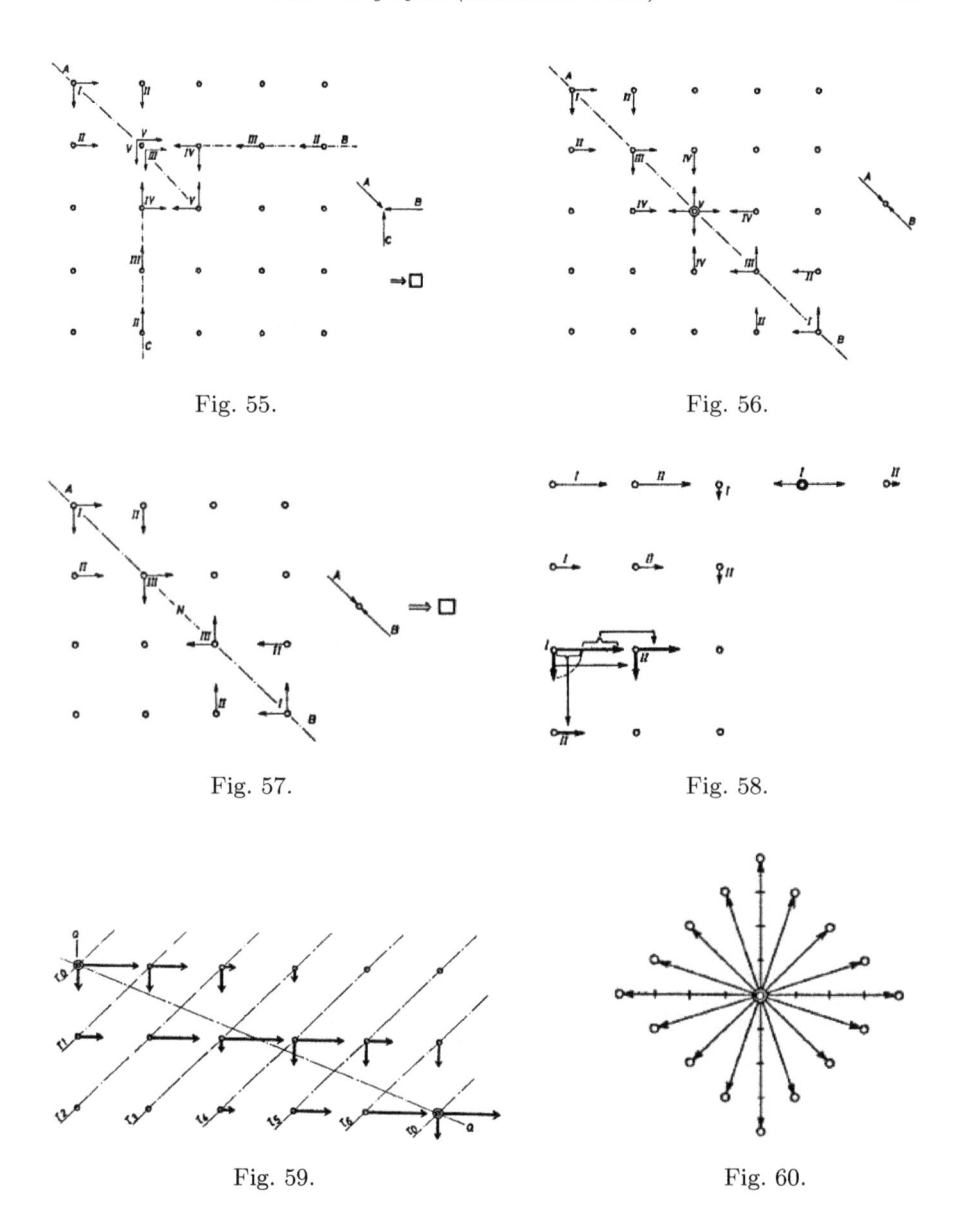

Fig. 55.

Fig. 56.

Fig. 57.

Fig. 58.

Fig. 59.

Fig. 60.

Figure 60 represents an example of the limitation of the possible discrete directions of motion. It must be stressed that there exists an interdependence between the velocity of propagation and direction. The chosen propagation rule permits no difference in velocity of the particles moving in the same direction.

Fig. 61.   Fig. 62.

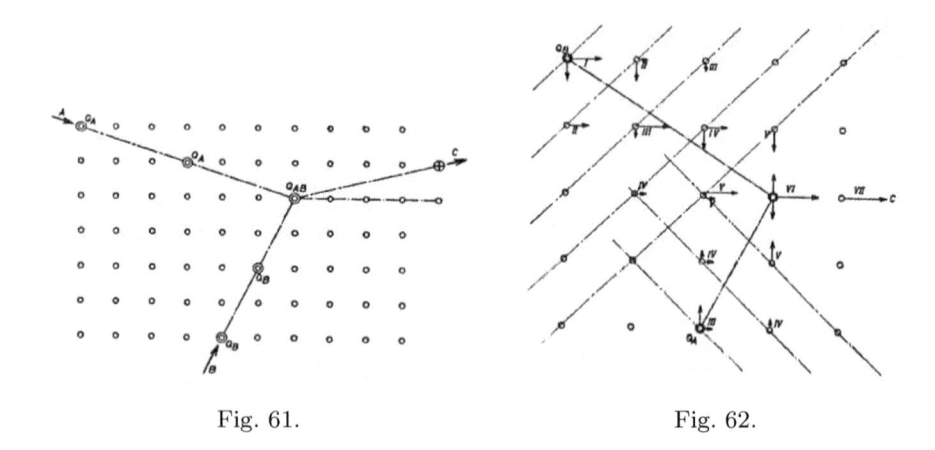

Figures 61-66 show another series of interesting cases of interaction between such particles. Again the process of interaction is phase-dependent. A reaction between two particles always occurs, when they are respectively at the zero point at intersection (for example, Figs. 61 and 62). But they can also react under other circumstances, as the examples in Figures 65 and 66 show. In these cases, the already–mentioned phase lines play a part. We could construct a time phase line R, which represents both particles. If this passes through the point of intersection of the particle paths S, a reaction is possible (Figs. 65 and 66).

Of course, these examples are very simple and primitive. But even these simple forms yield an abundance of suggestions; they show that the basic method of digitalization adopted is of greatest interest and that development of the rules will yield additional concepts.

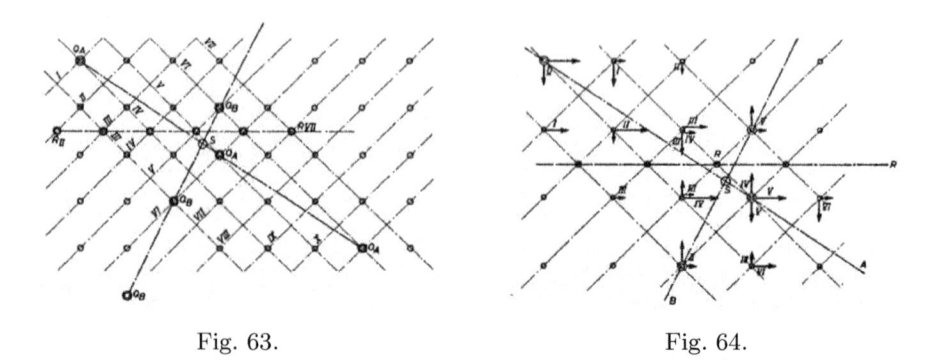

Fig. 63.   Fig. 64.

Fig. 65.   Fig. 66.

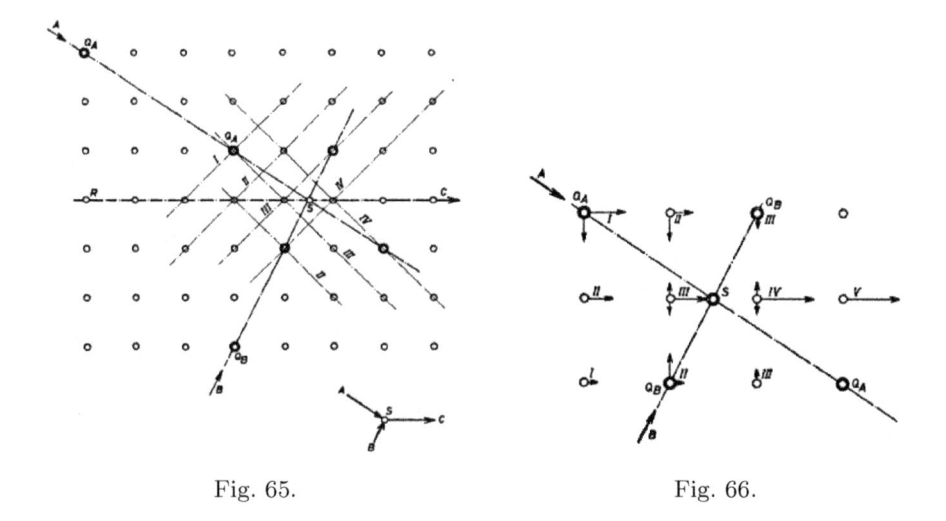

3.4. Concerning three-dimensional systems

The concepts developed in Sections 3.2 and 3.3 can also be applied to three-dimensional systems. The studies of the author are not yet complete in this area and should be reserved for further investigation.

4. General Considerations

4.1. Cellular automatons

The examples of digitalization of fields and particles which have been presented are in their present unfinished form still far removed from being able to serve in the formulation of physical rules. Nevertheless, they give a rough impression of the possibilities for using the tools of the automaton theory to answer physical questions.

The examples have dealt primarily with point grids. A single cellular automaton consists, therefore, of a point grid which is bound to neighboring points through information exchange. In the cases shown in Figs. 34 and 35, the grids are checkerboards of two different values, p and v, in grid form. There exist different possibilities for their combination, so that division into single automatons is not specific. This does not affect the behavior of the entire system.

In general, division of the continuum into discrete cellular automatons has different consequences, depending on the precise division. The idea of a grid spatial structure is already treated in various contexts by physicists, although not in regard to automaton theory. Generally speaking, the idea that the cosmos could really be subdivided into such cells is sharply repudiated by physicists. We agree that space cannot be viewed as a continuum even in infinitely small sections. The concept of a smallest length is already widely accepted today, while not in relation to the idea of subdivision into a point grid, but more as the principal limit in the differentiation of two different particles. The doubts relating to a grid structure are essentially as follows:

(a) A grid structure would abolish the isotropy of space.

It is clear that a regular grid pattern establishes preferred directions. This has an effect, for example, in the expansion of fields (Figs. 31, 38) and in the discrete possible directions in which a digital particle can move (Fig. 60). We know of no physical experiments which would provide a key to preferred directions of this type, but the field has not been systematically studied for this effect. Sober reflection reveals, nevertheless, that it is worthwhile to consider rules for a grid like spatial structure which do not allow the grid structure to become visible in regions of smaller and intermediate energy and frequencies. The grid constant must be assumed to be considerably smaller than the elementary shortest length of approximately 10^{-13} cm (Bopp assumes even 10^{-56} cm). The field of normal optics, for example, works with wavelengths of extraordinary length in comparison with these lengths. It is hardly possible to think of an experiment which could determine the eventual discrete propagation direction of photons, when we assume the accuracy of such a change in direction (in circular measure) to be of the same order of magnitude that we are capable of differentiating between frequencies, namely 10^{-12} (Mössbauer effect).

Results of this sort can first be expected in the very high energy ranges, when wavelength and length of the period approach the grid constant. Only today do we have the capability to carry out such experiments. The author must leave it to the decision of the physicists whether and within which limits these phenomena could be observed with the aid of present experimental techniques.

(b) Curved volumes, as they are assumed from the general theory of relativity, are hard to represent with the grid structure of space. Bopp has chosen the expedient of assuming a Cartesian space in which the three spatial coordinates each converge on themselves. This can be imagined in two-dimensional space by assuming a toroid.

There are, of course, many possible deviations from these consequences. The whole subject is still too young for one to be able to draw final positive or negative conclusions. The following possibilities can be mentioned:

(α) The assumption of fixed circuits in the form of cellular automatons is not the only logical possibility for defining logical connections between discrete values in space. If we introduce the change in the circuits as a function of the results of the previous process, variable circuits can be regularly developed.

(β) The concept of the growing automaton is closely related to the regular variability of circuits.

Both possibilities require at first a very well-prepared theory. Since automaton theory is a young field, the possibilities of which are in no respect exhausted, we can expect further developments in the direction being considered.

(γ) The assumption of a grid implicitly assumes that of an inertial system, which is contradictory to a strict interpretation of the theory of relativity. This will be considered at greater length.

In this light, the use of an orthogonal network is the most convenient way of beginning investigations. The results obtained in this manner will certainly be just as valid when in the course of time automaton theory yields new methods for use.

4.2. Digital particles and cellular automatons

Digital particles may be considered as disturbances in the normal conditions of a cellular automaton. This disturbance has a distinct pattern which is subject to periodic changes. According to automaton theory, every state evolves from the preceding one; nevertheless, the entire pattern can fluctuate in the process. To a certain extent we are concerned with "flowing states". In accordance with this, digital particles can be regarded as "self reproducing systems". A given pattern is generated in a neighboring region of the cellular automaton.

In the examples in Chapter 3, digital fields and digital particles are treated separately. Modern field theory takes pains to explain even elementary particles through singularities and special forms of fields. Automaton theory is understandably well-suited to digitalize such interpretations and to subject them to the rules of automaton theory. The author hopes to be able to treat this subject in greater depth in another contribution.

4.3. *On the theory of relativity*

The question of the isotropy of space obviously requires coming to grips with the theory of relativity. The Lorentz transformations so important to the special theory of relativity, can obviously be infinitely approximated by numerical estimates. Nevertheless, it is very difficult to simulate in digital form the consistent form of the model of the theory of relativity. Our physical experience tells us immediately that no excellent coordinate system can be proven to exist, and that we are justified in considering each coordinate system to be as valid as the next one, in which case the Lorentz transformations formulate the relationships between these inertial systems. The strict interpretation of the special theory of relativity leads, however, to the conclusion that in reality no superior coordinate system exists, and that it is useless to search for such a system experimentally. In any representation of the cosmos as cellular automatons, it is almost impossible to avoid the assumption of a superior system of movement. We can construct the structure of cellular automatons in such a way that a greater number, although still a finite quantity, of superior coordinate systems are available. The constancy of the speed of light in all inertial systems is represented by the digital simulation of the Lorentz transformations and the related shortening of bodies.

In any case, a relation between the speed of light and the speed of transmission between the individual cells of the cellular automaton must result from such a model. These do not need to be identical. In contrast, it may be assumed that the speed of transmission from cell to cell must be greater than the speed of propagation of the signal obtained from this transmission. This greater speed of transmission has only a local meaning. Because of the anisotropy of the calculating space it is different in different directions, In any case the "digital" model, in comparison with the analog model of the relativity theory, yields a significant difference: the closer the velocity of the inertial system approaches the standard of the speed of light, the more critical the digital simulation of the processes becomes. In the case of energy-rich particles, we come to processes which can be characterized (at least to some extent) as a "miscalculation" of calculating space. In this way the essentially different behavior of particles of very high energy (high velocity, high frequency) can be explained.

A strict interpretation of the special theory of relativity has as a consequence that for every inertial system another one can be imagined, which moves with an initial velocity less than c. The physical rules are just as valid in the second system as in the first. This process can be repeated as often as desired, at least in principle. The complete monstrosity of this thought is only vaguely clear. Here it must be said again that every conception of infinity presupposes a limiting process. Here we are concerned with an infinitely frequent repetition of reaction of another inertial system which moves relative to the previous one. This process has a few consequences if observations of an information theoretical nature are applied, as we will consider in the following.

The following statement is also of interest.

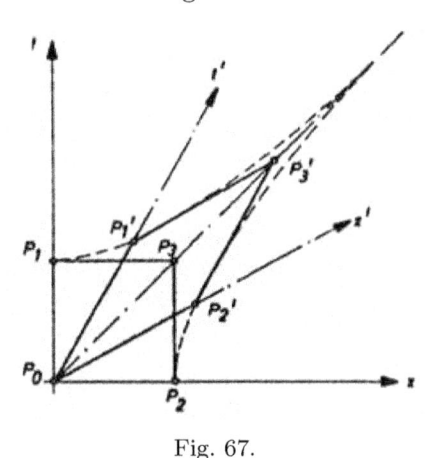

We shall first introduce the term "shifting volume". This is equal to the number of shifting parts involved multiplied by the number of shifting beats which take part in a given process, for example the period of a digital particle. Figure 67 shows a simplified representation, in which it may be assumed that a disturbance representing the digital particle extends for a distance $P_0 - P_1$. The particle is assumed to be stationary in the inertial system x, t. In this case, the space P_0, P_1, P_2, P_3 is equal to the shifting volume of a period. If this particle moves relative to the system x, t we can we can speak of a second inertial system x', t, according to the special theory of relativity, relative to which the moving particle is stationary. The inversion corresponding to the Lorentz transformations yields the shifting volume P_0, P_1', P_2', P_3'.

Fig. 67.

This is equal in area to the shifting volume P_0, P_1, P_2, P_3. We can speak therefore, of *invariance of the shifting volume*.

4.4. Considerations of information theory

The term information gains considerable meaning in the process of these different considerations. Information theory has formulated the term "information content" with clarity in regard to news-transmitting systems. For this reason, we are inclined to consider information theory as the theory of information processing. This is not correct, however. The easily accomplished application of terms from information theory in the neighboring field of news transmission unfortunately leads to frequent confusion. Even in the present observations we must be clear of our understanding of information content. It is difficult to speak of physical processes in terms of news transmission. This would be of interest in itself only insofar as we could include people in our consideration. If we assume an infinitely fine propagation of our news, transmitted through electromagnetic waves, it must be infinitely conserved, as long as limits are not established for them by the temporal finiteness of the universe. Metaphorically we can also consider the rays in the universe approaching us from other stars as news for people, in which case the question of the information content of this news makes sense.

Such a relationship between man and nature is to be found in the modern statement of the quantum theory, which attempts to relate all measurable quantities in a mathematical system. The information which we obtain from nature about the structure of atomic shells consists largely of the frequencies of the emitted light quanta. In this case, the use of the term "information content" is meaningful. The matter will not be further investigated here.

If we disregard this definition of information as the means of news transmission, it is still not possible to speak of information content of inhabited systems, if we consider the width of variation of the possible shapes of an object, a pattern or the like. Thus, a punch card may contain, due to its variability, a definite information content, measured in bits.

The technical characteristics of the punch card itself, including the accompanying punching and readout systems, set upper limits to the amount of information which can be entered, which is defined as information capacity. In news transmission this capacity does not need to be completely used, so that the information transmitted from sender to receiver on the punch card can be below capacity.

It is also possible to speak of a maximum possible information capacity of a finite automaton, if we consider the number of its possible states as a measure. If this is equal to n, the information content is $\log_2(n)$ (logarithm to the base two). A programmed calculating machine represents this type of automaton, as we are aware. If such an instrument has m members, for each of which there are two possible positions (for example, flip-flops, ferrite nuclear rings in the storage, etc.), then the number of possible states is 2^m and the information capacity is equal to m. In this process no distinctions are made between the individual possible states. In the total of 2^m possible states every state is counted in which every register and storage unit is dissolved (i.e., set at zero) as are the states, as a result of which the solution of a very complicated differential equation is held in storage. Emotionally, we naturally tend to assume that the equipment contains no information in its zero state, although in the second state mentioned extremely interesting scientific results are available for use by mathematicians. This example shows the necessity for great caution in the definition of terms in information theory. The difference in this situation is that for the receiver, the two states have a fundamentally different meaning. The state of "everything dissolved" is only an extension of the receiver's knowledge that the machine is in the ground state at the moment, while in the second case, the receiver's knowledge is increased with regard to significant results.

If no account is taken of these individual values of information for the receiver, then the conclusion may be drawn that the information content of a finite automaton cannot be increased while running a calculation. Because the calculation is made completely automatically after introduction of the program and the input values, the results are established from the beginning. The results have greater value for the person using the equipment: for why would be let the computer perform a calculation if not to increase his knowledge, which is only possible if the final state of the automaton has a greater information content than the starting state.

The first result of viewing the cosmos as a cellular automaton is that the single cells represent a finite automaton. The question to what extent it is possible to consider the entire universe as a finite automaton depends on the assumption which we make in relation to its dimensions. If we take the toroid of higher order, as already suggested by Bopp, we are dealing with a finite automaton on the whole. It is originally valid that the individual cells can accept a limited number of states and have therefore only a limited information content. This is equally true for the entire cosmos, if we make suitable assumptions about its limits.

Automaton theory demonstrates that different characteristic running patterns are possible for a finite automaton, several of which will be considered.

For every given state there is a succeeding state. It is therefore possible to express the relation "state A dissolves state B" as relation $F(A, B)$ and to represent it in the form of an arrow diagram. Such an arrow diagram is often called a "graph". Figures 68a-d show different types of arrow diagrams. It is important to remember that every state can have only one succeeding state, although there are several preceding states which can dissolve it. The process figures show that an autonomous automaton must end in a periodic cycle in every case, which under certain conditions can also degenerate into a single final state.

Fig. 68a.

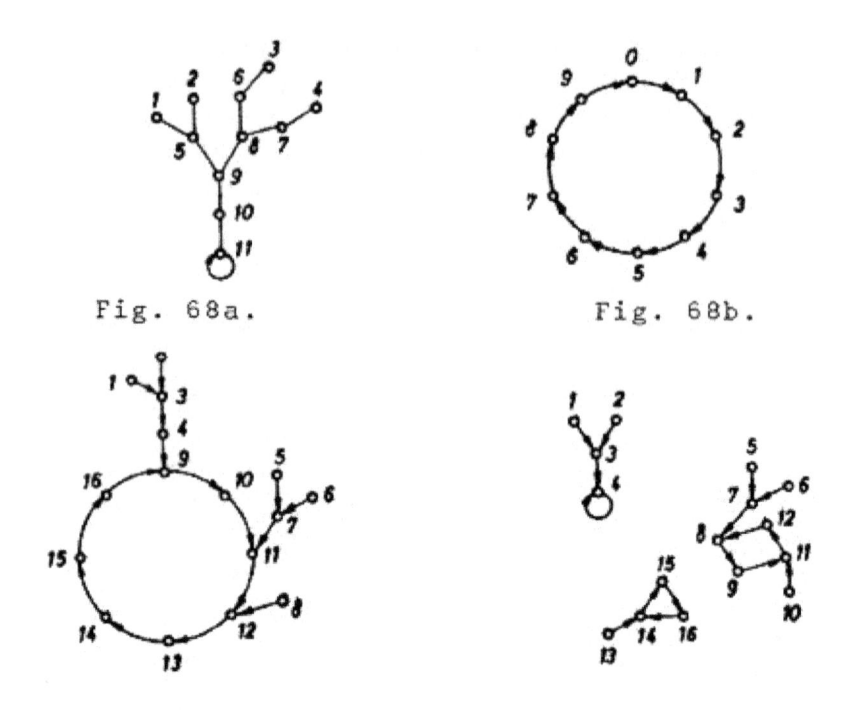

Fig. 68b.

Fig. 68.

This knowledge cannot be transferred to the individual cells of a cellular automaton, for they are related to neighboring cells through information exchange and therefore do not result in an autonomous finite automaton. In the assumptions of limits on the cosmos in the universe, we are concerned with a finite autonomous automaton as soon as we exclude any sort of influences of a greater external world. The first result is the somewhat disillusioning consequence that the cosmic process must of necessity end in a periodic cycle. This realization, in itself logically unassailable, has other implications when examined quantitatively.

The dimensions of the universe are assumed to be on the order of magnitude of 10^{41} elementary lengths (10^{-13} cm) by some physicists (approximately 10 million light years). We are concerned therefore with a volume of approximately 10^{123} elementary cubes of the elementary length on a side. If an individual bit of information content is assigned to each of these elementary cubes, then we have already $2^{10^{123}}$ different states of the universe to consider. This number represents only a lower limit. In reality, a much finer grid must be assumed, for which it is not yet known how many variations at each grid point are possible. It must further be considered that space calculates extremely exactly. The relation of electrostatic interactions to those from gravitational fields is about $10^{40} : 1$. The interaction of nuclear forces are again orders of magnitude stronger. The higher of the two values represents in reality only a lower limit, which is most likely many orders of magnitude too small.

If we assume the number of time pulses to approach the order of magnitude of the spatial expansion, in effect 10^{41}, the result is obtained that in spite of this long time only a vanishingly small portion of the possible states of the cosmos can exist. There are $2^{10^{82}}$ types of reaction ways possible, each of which is independent of any other. This also means that the number of deflections and branchings is incomprehensibly great. The previously considered observations of automaton theory relating to Fig. 68 lose all predictive value. Of what value is the realization that the evolution of the universe follows a periodic cycle, when even within the already very large range of time being considered one single period at most can pass, and most likely not even that?

The consideration of closed processes, i.e. of shifting processes involving a digital particle, appears more fruitful. We have already observed that a digital particle consists of a series of periodically-repeating patterns in a cellular automaton and that they are not fixed in position, but can move in the space of the single cells like the moving writing machine. The term "flowing state" was already introduced.

The question of the information content of a digital particle can be considered from several points of view. At first the digital particle accepts a set position in space at a particular point in time. The information content of the digital particle cannot be greater than the information capacity of this position in space, which is determined by the sum of the possible states of this region. It is highly unlikely that every variation in state of such a limited region corresponds to a digital particle. It is much more likely that a limited selection dissolves individual stable period patterns.

We can inquire, entirely independent of the space associated with a digital particle, how many pattern variations representing phases of a digital particle are in fact possible? It is advantageous to classify the patterns along different lines:
(1) type;
(2) direction and velocity (pulse);
(3) phase state;
(4) position of the particle.

An answer to Question 1 assumes that we have at our disposal a model which permits different types of digital particles, as we have in nature with photons and electrons, etc.

An answer to Question 2 requires that our model accept different velocities and directions of propagation of the periodic pattern.

The phase sequence results from the periodic pattern sequence associated with the special type of particle and pulse.

Question 4 has meaning only when the interrelationship of the particles is considered. It is, of course, impossible for a closed region of space to hold the information about its own state.

The examples of Figs. 42–66 from Chapter 3 satisfy these conditions only to a limited extent. First, the model permits representation of only one type of particle. Further only the direction may be varied, but not the velocity. The length of the periods of the individual particles is not constant, but this is not of interest to our consideration. The information content of this type of particle depends on the accuracy of representation of the arrow length or on the number of places with which it is digitally represented. If we assume absolute lengths of a component for Example 4, then we obtain 9 different arrow lengths, including the zero value, for that component; in two-dimensional space there are 81 different pulse variations. On the basis of these possible variations in the particles, even within the given limits it is possible to determine the information content of a particle. Each of these particles has a series of associated phase states, so that the number of possible patterns of digital particles is still greater. The particle in Fig. 59 has, for example, 7 different phase states ($\tau_0 - \tau_6$).

The question of information retention in the reaction between digital particles is an interesting one. In the examples given in Chapter 3, pulse arrows are added in the course of reaction. This means that the number of places in the pulse arrow of the new resultant particle must be greater than the number of places in the reacting particle. If we eliminate arrow length of 0 for purposes of simplicity and assume that the arrow of the reacting particle can be represented by binary places, then the arrow of the resultant particle must be represented by 4 binary places. Before the reaction we have 2 particles, each of which has an information content of 2×3 bits (a total of 12 bits). After the reaction we have a particle with an information capacity of only $2 \times 4 = 8$ bits. During the reaction we have lost 4 bits of information. In this process we have permitted the arrow of the resultant particle to be represented by a greater number of places. This already means in itself the admission of a new type of particle. If this is not permitted, a rule must be found which takes effect whenever the permitted number of places are exceeded in the process of addition. If we simply assume that the maximum value may not be exceeded, then successive reactions lead after a certain period of time to the result that we are left with particles with the absolute maximum pulse arrows.

The examples chosen here for digital particles are still much too simple to be strictly related to physical processes. Actually, we are never confronted in nature with the situation that particles of the same type react with one another, not to mention the result that two such particles react to give a particle of a higher type. Conservation of energy, of pulse, of spin charge and so forth holds for elementary particles in physics. It is only when models of digital particles are at our disposal, with the help of which terms can be represented, that comparative observations with elementary particles in physics and their reactions are possible.

It is a question of obvious interest whether conservation of the different magnitudes cited in correspondingly-constructed digital particles is related to a corresponding conservation of information. The problem becomes even more complicated when fields are also considered. The author can only state the question without offering an answer to it. Perhaps the question is not so terribly important. Somehow the question amounts to the problem of "configuration", which is known to be very difficult to handle mathematically.

Here we come squarely into contact with one of the difficulties of information theory. In news transmission, the greatest possible information content is obtained when the probability of the individual signals is distributed as uniformly as possible. This situation is referred to as the maximum entropy of information. It is easily possible to consider this in such a way that every possibility of relating previously-received news to the following symbol must of necessity diminish the information content, which limits through related redundancy the freedom on the selection of symbols (news, the content of which one can already predict, has no information content). Every sort of configuration necessarily represents through its rules a limitation of the possible means of representation and diminishes thereby the information content. Conservation of information and conservation of configuration are therefore contradictory to a certain extent.

The question whether or not tested terms in physics (energy, effective quantum, elementary charge, mass, etc.) can be interpreted by the terms of information theory or of information processing cannot yet be answered. In the model of a cellular automaton constructed so that processes occur in it which can be related to the listed physical quantities, these quantities must be represented by the construction of the circuits; i.e., by the values represented in the circuits.

Even more important than the term information content is that of information exchange. Something dynamic, not something static, results from circuit principles. Perhaps it could be called conservation of events or complication of events (Dr. Reche suggested the idea of "conservation of complicatedness", although in another connection). Viewed this way, the shifting process acquires added meaning. If the effective quantum is assigned the dimension "shifting process", we obtain the dimension "shifting process per unit time" for energy. The principle of conservation of energy can then be interpreted as the principle of conservation of events. The term "effective quantum" already points to a close relationship to shiftlike effects, namely the shifting process. The representation of energy as an "event" makes the relationship between energy and frequency more easily understandable. These thoughts are for the time being only simple speculation. Their purpose is to stimulate the application of automaton theoretical means of observation in physics.

A consideration of the Heisenberg uncertainty principle in the light of information theory follows. If a storage capacity of m bits is available for the digital representation of two quantities A and B, we are free to distribute the two quantities with different numbers of places and even differing precisions on the number of places. If n places are assigned to A, B has $m - n$ places. The error in A is on the order of magnitude of 2^{-n}, that of B the order of magnitude of $2^{-(m-n)}$. The product of both errors yields the constant 2^{-m}.

It is possible to assume that both conjugated quantities A and B are not directly represented by the pattern of digital particles, but represent derived quantities which appear only in certain processes. The limitations on the information content of the digital particles do not permit both quantities to be represented with the maximum possible accuracy. In the case of digital particles, even if one of the quantities is completely indeterminate, the other cannot be represented with ideal accuracy, but only with the maximum accuracy permitted by the limitations on the number of places. The following can be stated with regard to the principles of conservation: limiting values of the upper and lower sums must be considered. The laws of addition do not have unlimited validity. Similarly losses enter in the construction of models by falling below the threshold values. Digital models are possible in which, in spite of this occurrence, laws of conservation can be defined.

4.5. About determination and causality

The question of determination and causality is closely related to observations from information and automaton theory. The expression "causality" is not strictly used in the literature. In the following it is always used to mean that which is generally referred to as "determination", namely the definition of the succeeding state of a closed system as a function of the preceding state. The entire universe can be seen as a closed system, to the extent the necessary consequences of this assumption are taken into account.

Automaton theory works with the concept of the state of an automaton. Finite automatons can receive a limited number of states. If there is no entrance signal, the resultant state results from that which preceded it because of the algorithmus basic to the automaton. Because automaton theory works with abstract concepts, this conversion from one state to the next occurs in theory without intermediate steps. Automaton theory does not ask the question exactly how this conversion occurs in an operating automaton. It is concerned solely with the fact that, for example, a flip-flop takes place from one state to another in the space of a certain time, the pulse time. The technological analysis of the turnover process, which is possible, lies outside the range of automaton theory observations, as long as it is not concerned with the comprehension of such details.

The opinion is held by some physicists, for instance Arthur March[a], that direct conversion of an atom from one stable state to another is difficult to reconcile with the rule of causality. He understands the idea of causality in such a way that conversion from one closed system to the next requires a continual process. This interpretation can hardly resist the automaton theoretical consideration of physical processes. It cannot be assumed that this idea is based on reality. The process of thinking in whole numbers and in discrete states requires a thought process of non-continuous transitions, in which the law of causality is formulated in algorithms. Work with discrete states and quantification as such does not necessarily require rejection of the causal manner of observation.

This continuous transition in the sense of automaton theory must be differentiated from the thought of the continuous transition between the individual stable states of an atom. Since we are not able to analyze the process of such a transition experimentally, all theories on this subject belong to the realm of speculation. In the automaton theoretical sense, the

---

[a]For example, see March, Arthur: Die physikalische Erkenntnis und ihre Grenzen (Physical Perception and Its Limits), p. 19.

natural objective is to create models which enable these transitions to be followed individually and permit explanation of the emission or absorption of photons in the associated process. We cannot predict whether this goal will ever be reached. The often-argued opinion that such transitions are essentially unanalyzable and that such experiments should be subordinated to more fruitful endeavors can, however, be refuted. Quantum physics provides statistical laws for such processes through which individual determinations are supplanted by statistical determinations. This subject will be pursued further in connection with the discussion of probability.

It is important to inquire whether the determination is valid in both time directions; i.e., whether later states of the system are clearly functions of the previous states as well as the reverse. The classical model of mechanics satisfies this demand for time symmetry ideally. Statistical quantum mechanics introduces the idea of probability and observes a deviation from time symmetry in the increase in entropy. In general, finite automatons follow laws determined in only the positive direction. The algorithm establishes only which state arises from the given one, not the reverse. It is possible to construct automatons in which the previous state is determined by the one which follows it, but this does not necessarily imply symmetry in the time direction. A consideration of computers may clarify this. A computer is–assuming unobjectionable work–determined in the positive time direction. In general, calculating processes are not reversible, which may be seen from consideration of the basic operations on which all higher calculations are based and which are not reversible (for example, $a \vee b \Rightarrow c$). A calculator is one example of a calculating machine which is effectively determined in both directions, because it counts forward in one time direction and backward in the other, to the extent that we consider only the state tables and do not analyze the processes individually.

The different characteristic types of operation of an autonomous automaton were already discussed in 4.4 in connection with Fig. 68. Type 68b would correspond to an automaton determined in both directions, as is the calculator mentioned.

A difference remains nevertheless: in the positive time direction, the rule by which the following state is related to the preceding one is explicitly given by the algorithm. In the negative time direction, there exists a single correlation, to be sure, but this correlation is only implicitly given; i.e., it cannot be directly calculated without further knowledge. This difference is not clearly visible in the diagrams corresponding to Fig. 68 and in the state table corresponding to Fig. 4. In any case, this type of representation is possible only for very simple automatons and serves more for primary experiments than for practical determinations of the automaton operation process. The actual rule for the formation of the following state from the preceding one is given by the automaton circuits. We are able to say that an autonomous automaton is determined in the positive time direction and that in special cases of negative time direction a "pseudodetermination" exists. The relationships of digital particles are similar in the cases discussed in Chapter 4.4. As long as such a particle follows its path independent of outside influences, a single sequence of states occurs. As soon as we consider the sequence of two particles, the conditions are immediately different. In this case, the examples in Chapter 3, Figs. 42–66 refer to irreversible processes. The basic shifting rule regulates the processes in the interaction of the particles, There is no sort of inducement for a particle to divide into two particles at any time. This statement makes only one assertion about the models used in Chapter 3. The question whether it is possible to construct usable models of digital particles which do not have this characteristic is difficult to answer. This is the same problem as the one confronting the physicist in the decay of elementary particles or atomic nuclei. The present state of theoretical physics is such that we can only give probability laws for such processes. In a model which follows a predetermined operation process and excludes working elements, in accordance with the probability laws, there are only two means of solution:

(a) the digital model is constructed in such a way that it contains a sort of clock which dissolves the process when a certain state has been reached;

(b) the influence of the environment, (for example that of fields through which the digital particle moves) is taken into account. In the process of moving through its different phases, a particle can pass through critical states in which the influence of the environment (frequency, etc.) causes particle division.

The present state of physical theories does not permit the drawing of conclusions about physical laws from these possibilities of digital models. What has already been said for the transition from one atomic state to another is equally relevant here: no experiment permits an examination behind the scenes, and all theories are essentially speculative in character. Nevertheless, it has been possible to determine a certain dependence of radioactivity at high temperatures, which corresponds to the assumption of critical situations influenced by the environment.

One result is important, in any respect: the assumption of valid determination only in the positive time direction is not influenced in the least by the dissolution of physical laws into the laws of probability. Similarly, the increase in entropy is not necessarily related to this question. From the viewpoint of automaton theory, each of these questions takes on another meaning. Entropy can be explained in a digital model, the operation of which is strictly determined.

Let us consider the classical model of physics from this point of view. As already mentioned, the validity of the determination, particularly in both time directions, requires absolute accuracy of the individual processes. It may hardly be assumed that serious considerations of the extreme significance of this assumption in regard to information theory have been made. Such a model requires an infinitely fine structure of spatial and temporal relationships. An infinite information content is required for an unlimited spacetime element. It is practically impossible to simulate such a model with computers because of the necessity of infinite number of places required. The sources of error are correspondingly great in the extremely large number of collisions between gas molecules, and these errors quickly lead to deviations from theoretical processes. This means that the better the causality rule is approximated in the reverse time direction, the more calculations we must be prepared to carry out in our model. This leads to the result that simulations of universal systems with causality functioning in both time directions belong to the category of "unsolvable" problems.

Of course, it can be said that this is true only for calculating simulative models. But this result should encourage us to reconsider the matter. Are we justified in assuming a model of nature for which no calculable simulation is possible?

From this point of view, it appears that the frequently advanced argument of determination in both time directions should be fundamentally reexamined.

The question of time symmetry of the physical laws is frequently discussed in connection with the reflective characteristics of space. The observations of automaton theory might be of significant value in furthering this discussion.

4.6. *On probability*

The problem of determination in modern physics is closely related to the laws of probability. An observation from automaton theory may be inserted here. It is of course possible to build mathematical systems, such as matrix mechanics and wave mechanics, in which probability values play a significant part. The automaton theoretician can introduce the idea of probability into his theories and can establish a successive state dependent on probability values. To this point, the process is a simple mathematical game on paper. It becomes critical when we attempt to construct finished forms of such mechanisms which operate according to the laws of probability. Such calculations have been carried out in our calculating automatons with considerable success for some time (Monte-Carlo method). The element of chance is introduced into the calculation in the form of "chance values". The generation of these chance values is the decisive problem. There are two ways to accomplish this.

(a) The values are generated by simulation of the dice method and that type of number series, in which no sort of dependence between the numbers exists. Such a number series can be developed from the calculation of irrational numbers (π, for example). In reality, this process is strictly determined. Nevertheless, we speak of pseudo-chance values. This process is completely sufficient when the generation rule for such chance values is carefully chosen.

(b) A mechanism is taken from nature which is either so complicated that it cannot be shown to be regular or for which it can be said that, according to the valid laws of physics, it provides "real" probability values. The dice mechanism belongs to the first sort, where causal rules play a role but for which, in the case of a sufficiently carefully built die, equal probability for every case can be shown. The same is true for all games of chance (roulette, etc.). In the other case we rely on the fact that, for example, the radioactivity of a certain material is subject to strict probability laws. Whether the probability process is in reality determined in these atoms is not significant, for experience shows that in any case the laws of probability can be assumed without leading to incorrect results. In this case the calculating automaton regards the probability values to a certain extent as external input values. It remains true, however, that real probability values are hardly possible in technical automatons.

It must also be remembered that the choice of algorithm for creation of the pseudo-chance values is highly significant in Case (a). This means that only those choices from the range of basic number series are possible which follow one another as irregularly as possible and which have the most uniform possible distribution of probability. This means that longer series of the same number and series of numbers in the same separation (1, 2, 3) must be excluded, although these series are just as probable or improbable in real series of chance values as any other number series.

Of course, we can ask the purely speculative question whether true probability laws are admissible to automaton theoretical observations of physical processes. This question is a philosophical one, and is only noted here, without an answer.

4.7. *Representation of intensity*

The representation of intensity of field strengths and other numerical quantities in cellular automatons must be specially considered. For this reason, a few basic possibilities are considered here.

Figure 69 shows a two-dimensional grid in which individual grid points are occupied by elementary logical values; for example, yes-no values. If we assign to these values the numbers 0 and 1, the statistical distribution of the 1 values represents a scale for field

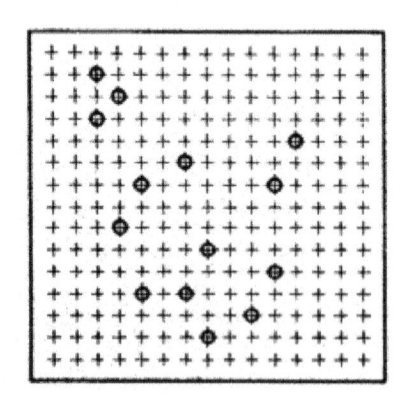

Fig. 69.

strength. This sort of representation can accomplish little, of course, if many orders of magnitude of density must be taken into account. As already mentioned, the relationships of electrostatic interactions to gravitational interactions is on the order of 10^{40} : 1. If we wanted to represent these intensity differences in a three-dimensional space corresponding to Fig. 69 using yes-no values, a cube with a side length of approximately 10^{13} grid units would be necessary. This represents only a lower limit, for in reality field strengths can differ by even greater orders of magnitude. If we take a grid with the elementary length of 10^{-13} cm accepted by physicists, it would mean that a space of many cubic centimeters would be necessary, according to these calculations, to represent the field intensity. This type of model cannot be very useful, entirely independent of the fact that it is extremely difficult to establish laws for stable digital particles with this sort of statistical distribution.

A much more rational method is offered by the principle of place values. This does not lead to the idea to construct calculating automatons

Fig. 70.

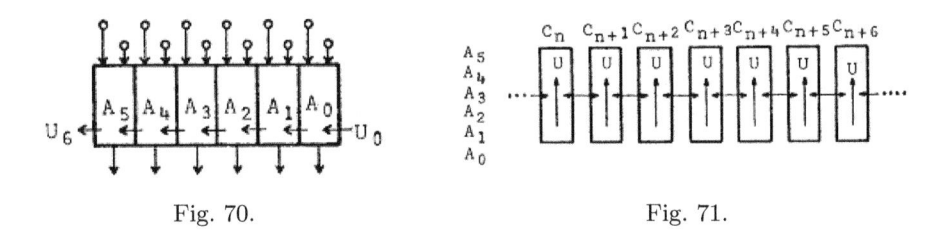

Fig. 71.

according to the principle of Fig. 69. Figure 70 shows the ideal arrangement of an adding machine consisting of neighboring cells and among which a hierarchical ordering is seen. The individual cells are coordinated with numbers of different value. This is reflected in the one-sided construction of the transmission process $u_0 - u_6$.

Figure 71 shows the transmission of this thought process to a linear cellular automaton. Each cell is allied with a complete adding machine. Each cell C_i is subdivided into the individual addition steps $A_{0...5}$. In the construction of such a shifting system it must be remembered that the transmissions among levels within the cell must be coordinated in time with the transmission of information between the individual cells.

This principle is relatively easy to put into practice for one-dimensional and two-dimensional cellular automatons. Theoretically it can be applied to three– and more-dimensional automatons without any modifications. In addition to the dimensions, which correspond to the topological arrangement of neighboring cells (space dimension), there is also a level dimension. This is only imaginable in three-dimensional space and must be constructively built into (projected into) three-dimensional space.

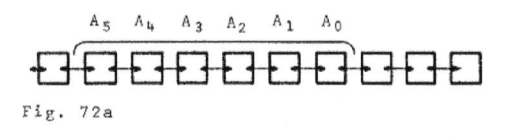

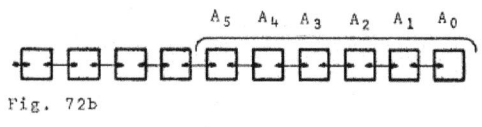

Fig. 72a

Fig. 72b

Fig. 72.

The further question can be asked whether in a symmetrically built cellular automaton a hierarchical ordering can be introduced by the manner of occupancy. Figure 72 demonstrates the principle. The single cells can contain, for example, single addition steps and are not able to accept several-place numbers. These are divided among several neighboring cells, according to the place value principle. The difficulty arises in the fact that this sort of arrangement is of the nature of occupancy. If the concept is applied to a several-dimensional automaton, it is easy to see that major complications develop.

Cellular automatons provide an elegant solution when each cell contains a complete calculating system, as symbolically represented in Fig. 73. These single calculating systems contain both information-processing and information-storing elements.

The net automaton represented in Fig. 74 is a further development of the cellular automaton corresponding to Fig. 73. The individual cells are responsible here for only information processing. Branching lines B connect the individual cells and serve both for information transmission and for information storage. The individual cells can consist of single-place adding units, according to the series principle valid for calculating machines. Preliminary investigations by the author have shown that this type of automaton is highly successful, specifically in the solution of numerical problems as well as in simulation of physical processes. More specific consideration will be the subject of another paper.

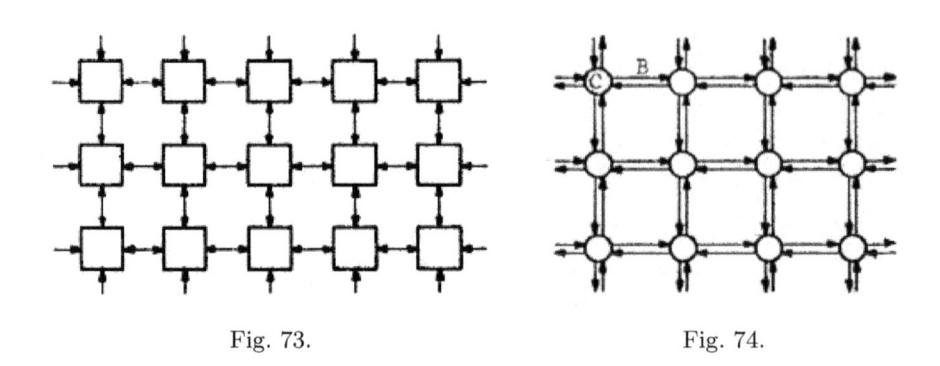

Fig. 73.

Fig. 74.

5. Conclusions

Even if these observations do not result in new, easily understood solutions, it may still be demonstrated that the methods suggested have opened several new perspectives which are worthy of being pursued. Incorporation of the concepts of information and the automaton theory in physical observations will become even more critical, as even more use is made of whole numbers, discrete states and the like.

A relating of different possible conceptualizations is attempted in the following table:

Classical Physics	Quantum Mechanics	Calculating Space
Point mechanics	Wave mechanics	Automaton theory Counter algebra
Particles	Wave-particle	Counter state, digital particle
Analog	Hybrid	Digital
Analysis	Differential equations	Difference equations and logical operations
All values continuous	A number of values quantized	All values have only discrete values
No limiting values	With the exception of the speed of light, no limiting values.	Minimum and maximum values for every possible magnitude
Infinitely accurate	Probability relation	Limits on calculation accuracy
Causality in both time directions	Only static causality, division into probabilities Classical mechanics is statistically approximated Based on formulas	Causality only in the positive time direction introduction of probability terms possible, but not necessary Are the limits of probability of quantum physics explainable with determinate space structures? Based on counters

In view of the possibilities listed, it is clear that there are several different points of view possible:

(1) "The ideas of calculating space contradict some recognized concepts of present-day physics (for example, space isotropy); therefore, the fundamental basis must be false."
(2) "The laws of calculating space must he revised with the object of eliminating the existing contradictions."
(2) "The possibilities arising from the ideas of calculating space are in themselves so interesting that it is worthwhile to reconsider those concepts of traditional physics which are called into question and to examine their validity from new points of view."

The author has greatly enjoyed being able to discuss this subject with a few mathematicians and physicists. The greatest handicap to cooperation is certainly the difference in terms between the individual, specialized fields of knowledge. We hope that this chasm will be bridged in time and that through cybernetics, a true bridge between physics and the automaton theory can be built.

Independent of the possibility that the idea of calculating space can be directly applied to physical determinations, there remains the major task of providing theoretical physics with an aid in calculating and of finding numerical solutions to very complicated relationships. In spite of the use of huge computers in the field of physics, the applications of "software" in physics are still much more limited than the applications of "hardware". With huge accelerators that cost hundreds of million dollars we are able to obtain particles of very great energy, requiring a fundamental reexamination of the general validity of our basic theoretical hypotheses. Is there not a considerable danger that the software lags behind the hardware of physics, and that we will soon be unable to evaluate the determinative results of our practical experiments?

In the field of information processing we are already spending equivalent amounts on hardware and software. In physics the ratio of expenditures is probably between 1:20 and 1:100. The result in chemistry is about the same. Although the laws of electron shells have been generally known for a long time, young scientists are able to explore them only within circumscribed limits in precise, analytical chemistry. The author hopes that the ideas of calculating space after a period of adaptation will be of assistance. The first step would be further development of the models of the automaton theory approximately along the lines suggested in this article. When this process has reached a certain maturity, then specific goals can he set.

It must be stressed that the experiments of the author are confined to pen and paper experiments. Further experimentation must he carried out with the help of modern computers.

Afterword to Konrad Zuse's Calculating Space

Adrian German[1] and Hector Zenil[2]

[1] *School of Informatics and Computing,*
Indiana University Bloomington, USA
[2] *Department of Computer Science, University of Sheffield, UK*

There are many parallels between Zuse's and Turing's interests. In the mid 1930s, some researchers were engaged in what amounted to an inquiry into the nature of computation, and trying to figure out whether it would be possible to build a computing machine. In part this was a consequence of Hilbert's programme, but it was no doubt also due to a certain chain of historical events. As pointed out by Raúl Rojas,[a] people started to think about computers when it was time to build computers. There were of course Schönfinkel (SKI combinators), Church (λ calculus), Post (tag systems), Kleene (recursive functions), Turing (a-machines), among a few others.

Perhaps the main difference between all the other approaches and Zuse's lies in the fact that Zuse was a civil engineer aiming to solve concrete problems and, as such, his approach was quintessentially practical. Thus Zuse's goal was from the beginning that of building a concrete, mechanical realization of computation. Turing's approach falls mid-way between the purely abstract and a practical realization. This fact alone may explain why Turing's work was, in the end, more visible than others. Zuse's approach being an engineer's answer to the question of computation, his solution took the form of an actual machine.[b]

---

[a] In a recent talk *Zuse and Turing in Context* in Cambridge, UK on February 18, 2012.
[b] The most comprehensive source of information is the Konrad Zuse Internet Archive curated by Raúl Rojas available online at http://www.zib.de/zuse/home.php (accessed in April 2012). His son, Horst Zuse, maintains his father's homepage, available at http://www.horst-zuse.homepage.t-online.de/konrad-zuse.html (accessed in April 2012). And Juergen Schmidhuber[c] also maintains a website devoted to Zuse, available at http://www.idsia.ch/~juergen/zuse.html (accessed in April 2012).
[c] Schmidhuber is also a contributor to *A Computable Universe: Understanding & Exploring Nature as Computation*.

Zuse may not have realized that there was a fundamental concept behind the question all these people were asking and ultimately trying to answer (Zuse was working in relative isolation, unlike the others, who for the most part knew of each other). Turing finally provided the closest answer to the question with his concept of computation universality, the founding notion of Computer Science. Paradoxically, today's digital computers may be more similar in some respects to Zuse's than to Turing's idelization, certainly because Zuse had to deal with the minutiae of actually building a physical machine (for ex., the IEEE Standard for floating-point coding is almost the same as the representation used in Zuse's Z1 and Z3). Zuse never thought of universality as Turing did, but as Rojas has proved, not without some creativity, the Z1 and Z3 *accidentally* (because it was never Zuse's purpose, and he didn't even formulate the question) turn out to be capable of universal computation.[d] Zuse never thought about how the machine could get into an unbounded computation (necessary for universality), for example, and if it did, how to make it stop (Rojas suggests that there would have had to be a mechanical/electrical hack to arbitrarily stop the machines, with the required computation finished and somehow encoded among other computations in the output, if unbounded computation were allowed–by, for example, looping a punched card).

Upon graduating in 1935, Zuse became a stress analyst for the Henschel Aircraft Company, where he worked on problems of aircraft vibration. Stress analysis involved formidable calculations, which at the time could only be performed with great difficulty using teams of human "computers" equipped with desk calculating machines.[e] Zuse thought that many of the calculations he was performing could simply be automatized. With a 1936 research grant from the *Reichsluftfahrtministerium* (the German ministry of aviation), he coincidently built his first computing machine between 1936 and 1938, and in 1938 he was building his second one, using phone relays unlike the first one, which was mechanical. His Z3 was completed in 1941, was fully operational, and was able to perform calculations.[f] His Z1 was already programmable even though mechanical, using punched tapes.

---

[d]See Raul Rojas' "The Architecture of Konrad Zuse's Early Computing Machines," in "The First Computers – History and Architecture," MIT Press, 2000, pp. 237-262, edited by R. Rojas and Ulf Hashagen.

[e]http://www.independent.co.uk/news/people/obituary--konrad-zuse-1526795.html (accessed in April 2012).

[f]An online video made at the Deutschen Museum München shows how the Z3 worked, using examples of arithmetical division and square roots: http://www.youtube.com/watch?v=J98KVfeC8fU (accessed in April 2012).

His main motivation to switch from a mechanical to an electronic mode was a concern about reliability–he wanted to build resilient and fault-tolerant machines–but the Z3 built with electronic relays was logically equivalent to the Z1. The Z1 and Z3 could be programmed and could perform all arithmetical calculations, could load and store information in binary and were capable of floating-point calculations (whereas the Mark I and the ENIAC in the U.S. still represented data in decimals, even though they both operated with binary gates, and were unable to handle floating-point calculations). Zuse decided to use the binary system and metallic plates that could move only in one direction, i.e. they could only shift position, just as modern digital computers do at their lowest working level (Zuse seemed to believe that mechanical devices and digitally based calculations were more reliable as compared to, for example, vacuum tubes, as suggested by Helmut Schreyer, Zuse's friend.).

Fig. 75. Replica of the first mechanical computer designed by Konrad Zuse, the Z1, finished in 1938. It was a binary electrically driven mechanical calculator which used Boolean logic and binary floating point numbers. Picture taken by H. Zenil, *Deutsches Technikmuseum* ("German Museum of Technology"), Berlin.

Zuse and Turing never met but they became acquainted with each other's work. Zuse mentions Turing's work in his autobiography, and it is known that Turing was on the program/reviewing committee of at least one colloquium that Zuse attended–but not Turing–at the Max-Planck-

Gesellschaft in Göttingen in 1947. Had Turing attended they would actually have met.

But if Zuse didn't hit upon the concept of universal computation, he was interested in another very deep question, the question of the nature of nature: "Is nature digital?" He tended toward an affirmative answer, and his ideas were published, according to Horst Zuse (Konrad's eldest son), in the *Nova Acta Leopoldina*. Horst was born precisely when Konrad was thinking about *Rechnender Raum* for the first time (the common translation into English is *Calculating Space* but the phrase in his native German carries a lot more cognitive weight than its plain English counterpart, in light of the ideas treated in Zuse's piece: calculation, computation of nature, space and/or the universe). Hector Zenil (HZ) met Prof. Horst Zuse (a professor at the Technische Universität of Berlin) in the Autumn of 2006 during a conference dinner in Berlin. The conference topic was precisely "Is the Universe a Computer?" (Ist das Universum ein Computer?) and it was held at the Deutschen Technikmuseum and organized to mark the Year of Informatics (Informatik Jahr) in Germany.[g]

Konrad Zuse did, however, acknowledge the problems likely to be faced in attempting to reconcile a digital view of the universe with theories of physics assumed to work in continuum spaces. But according to Konrad Zuse, the laws of physics could be explained in terms of laws of switches or relays (not a surprise as he had experienced the transformation of his machines from mechanical to electronic form through the use of relays), and thought of physical laws as computing approximations captured by mathematical models. It is clear from *Rechnender Raum* that Zuse knew that differential equations could be solved by digital systems and took this fact as evidence in favor of a digital theory.

Years before John von Neumann explained the advantages of a computer architecture in which the processor is separated from the memory, Zuse had already arrived at the same conclusion. As a computer builder in the 1930s, Zuse worked as an amateur completely outside the mathematical community, on his own time, in the evenings and on weekends, in the living room of his parents' house. He did, however, obtain some financial assistance from a local calculating machine manufacturer. He also persuaded Helmut Schreyer, a former university classmate, to work with him. It was on the advice of his friend Schreyer that Zuse moved from mechanical to electro-mechanical, telephonic relays.

---

[g]HZ wrote a blog post about it, available online at http://www.mathrix.org/liquid/archives/is-the-universe-a-computer.

In his autobiography,[h] Zuse writes that in 1939, as war broke out, he was drafted into the infantry to serve on the front lines. He never saw action as a soldier. His military service was to last six months, "six months during which I had plenty of time to contemplate the ideas developed and captured in my diary notes of 1937 and 1938." He was exempted from active duty and discharged so he could undertake work directly related to weapons development, as a structural engineer in the Special Division F at Henschel Aircraft Company, where remote-controlled flying bombs were developed.

In 1941, shortly after the Z3 was completed, Zuse went back to work as a structural engineer in aircraft construction with Henschel, a day job while starting a company, *Zuse Apparatebau* (Zuse Apparatus Construction), to manufacture his machines. When the Z3 became operational, it was the world's first practical automatic computer, and for 2 years remained the only one. A second machine, the Z4, was quickly commissioned. During the war Z3 was demonstrated before several departments, yet it was never put into everyday operation. In 1944 the Z3 was destroyed in an air raid but it was reconstructed in 1960 and set up in the *Deutsches Museum* in Munich.

Zuse and Schreyer had, however, to abandon the building where their computer was housed. As the war came to an end, Zuse retreated to Hinterstein, a village in the southeast of Germany, where his eldest son (Horst) was born. There he reconstituted his Z4 computer in a stable, and it became the world's first operational commercial computer, leased to the ETH Zürich (one of the two universities of the Swiss Federal Institutes of Technology). Then he began working in an area that didn't require physical resources—computer programming. He devised a language, the Plankalkül (meaning "formal system for planning" or "calculus of programs"; "a universal language" according to Zuse, who compared it to an "artificial brain"), which anticipated some programming concepts that surfaced later, and can be considered the first high-level programming language, although no compiler or interpreter was ever written for it. In 1945, perhaps with the same motivation that led Turing to turn to chess, namely the fact that the game was believed to epitomize human intelligence while seeming highly algorithmic, Zuse worked on chess playing algorithms formulated as routines in his Plankalkül. One year before, in 1944, he had organized his work into

---

[h]"The Computer – My Life," published in German by Springer-Verlag in 1993 and translated into English in 2010, the anniversary of Zuse's birth.

a dissertation[i] which was never defended formally. The title he chose for his work was "Beginnings of a Theory of General Computing," trying to establish the foundations of what is today generally understood as information processing: "Computing (*Rechnen*)", he wrote, "means, in general, forming new data from given data according to some rule.' The concept of the algorithm would later replace his concept of *Vorschrift* (or rule). His programming language, like the logic, design and construction of his computing machines, was entirely his own work, carried out in isolation from developments elsewhere.

While still in Hinterstein he wrote a treatise entitled "Freedom and Causality in the Light of the Computing Machine". In his autobiography he writes: "I think the majority of researchers involved in the development of the computer have at some point in their lives, in one way or another considered the question of the relationship between human free will and causality." This was to be the major impetus for the work that led to the translation presented in this volume:

> "While considering causality it suddenly occurred to me that the universe could be conceived as a gigantic computing machine. I had the relay calculator in mind: relay calculators contain relay chains. When a relay is triggered, the impulse propagates through the entire chain. The thought went through my head that this must also be how a quantum of light propagates. The thought settled firmly; over the years I have developed it into a concept of the *Rechnender Raum*, or 'computing universe'. However, it was to be another thirty years before I succeeded in formulating the idea correctly."

In 1967, Zuse suggested that the universe itself was running on a cellular automaton or a similar computational structure, a metaphysical position known today as digital physics, a subject Ed Fredkin had himself taken up before becoming acquainted with the work of Zuse. Excited to discover this work, Fredkin invited Zuse to Cambridge, MA. The translation of *Rechnender Raum* reproduced here, from a German (published) version of Zuse's ideas, was in fact commissioned during Ed Fredkin's tenure as Director of MIT's Project MAC[j] (the AI lab that was a precursor of the current MIT AI labs).

---

[i]See "The Plankalkül of Konrad Zuse – Revisited" by Friedrich L. Bauer, in "The First Computers – History and Architecture," cited earlier.

[j]Ed Fredkin is also a contributor to *A Computable Universe: Understanding & Exploring Nature as Computation*.

Afterword to Konrad Zuse's Calculating Space 793

Fig. 76. (How) Does Nature Compute? A Panel Discussion organized by A. German and H. Zenil during the last day of the 2008 NKS Midwest Conference, featuring (in order): Greg Chaitin, Ed Fredkin, Rob de Ruyter, Anthony Leggett, Cristian Calude, Tommaso Toffoli and Stephen Wolfram, moderated by (from left to right) Gerardo Ortiz, George Johnson and Hector Zenil, at the University of Indiana Bloomington. See http://www.cs.indiana.edu/~dgerman/2008midwestNKSconference/

More than twenty years after his *Rechnender Raum*, in Zuse's autobiography, he wrote:

> "In the final analysis, the concept of the computing universe requires a rethinking of ideas, for which physicists are not yet prepared. Yet it is clear that earlier concepts have reached the limits of their possibilities; but no one dares to switch to a fundamentally new track. Yet, with quantization, the preliminary steps towards a digitalization of physics have already been taken; but only a few physicists have attempted to think along the lines of these new categories of computer science. [...] This was illustrated quite clearly during the conference on the Physics of Computation, held May 6-8, 1981 [at MIT]. What was typical at this conference was that, although the relationship between physics and computer science, and/or computer hardware, was examined in detail, the questions of the physical possibilities and limits of computer hardware still dominated the discussions. The deeper question, to what extent processes in physics can be explained as computer processes,

was dealt with only marginally at this otherwise very advanced conference."

The original of *Rechnender Raum* seems to have been lost. To our knowledge the translation commissioned by Project MAC (the precursor to the current MIT Computer Science and Artificial Intelligence Laboratory or *CSAIL*) was never published in a journal.[k] It is reproduced here translated into modern LaTeX, which required quite a bit of work, despite having used OCR techniques with *Mathematica* first, in order to avoid starting completely from scratch. It is published in this volume without changes, except for perhaps a few corrected typos and redistribution of text and images to fit the book format. The material is at once dated and surprisingly contemporary: "I propose that in an information-theoretic analysis, objects and elementary dimensions of physics must not be complemented by the concept of information, but rather should be explained by it." Zuse was always aware of the hypothetical nature of his thesis: "The concept of the computing universe is still just a hypothesis; nothing has been proved. However, I am confident that this idea can help unveil the secrets of nature."

Zuse refers the more skeptical among us to a quote from Freeman Dyson ("Innovation in Physics" published in Scientific American, Vol. 199, No. 3, (September 1958), pp. 74-82.): "A few months ago Werner Heisenberg and Wolfgang Pauli believed that they had made an essential step forward in the direction of a theory of elementary particles. Pauli happened to be passing through New York, and was prevailed upon to give a lecture explaining the new ideas to an audience which included Niels Bohr. Pauli spoke for an hour, and then there was a general discussion during which he was criticized rather sharply by the younger generation. Finally Bohr was called on to make a speech summing up the argument. 'We are all agreed,' he said, 'that your theory is crazy. The question which divides us is whether it is crazy enough to have a chance of being correct. My own feeling is that it is not crazy enough.'"

"Imagination," Zuse used to say, "is the key to all progress."

A. German and H. Zenil
Bloomington, IN. USA and Sheffield, UK

---

[k]Scanned copies of a short German version and the translation into English, accompanied by additional contextual material, are available online at Schmidhuber's website *Zuse's thesis* at http://www.idsia.ch/~juergen/digitalphysics.html. The German version is also at http://www.zib.de/zuse/Inhalt/Texte/Chrono/60er/Pdf/76scan.pdf (links accessed in April 2012).

Index

$\frac{1}{f}$ phenomenon, 708
λ calculus, 160, 571, 787
λ-calculus, 48
π, 6, 700
't Hooft, G., 475

Aaronson, S., xxxix
abacus, 568
Abelson, H., 244, 246
Abelson, H., et al., 244, 246
absolutely unsolvable, 49, 53, 58
abstract machines, 63–65
Abstract State Machines, 88, 89
Abstract State Model, 165
accelerated Turing machine, 542
accuracy factors, 486
Achilles paradox, 554
Ackermann, W., 77, 436
action-based semantics, 350
Actor model, 159
Actor Systems, 164
ActorScript, 164
Adamatzky, A., xxxix
Alchin, N., 415
Alexandrov-interval, 456
algorithmic
  information, 407
  information theory, 401
algorithmic complexity, 6
algorithmic drugs, 717
Algorithmic Information Theory, 6, 278
algorithmic information theory, 677
algorithmic probability, xl
algorithmic randomness, 706

all-a-carte
  models, 404
Allen, C., 352
Allen, D., 244, 246
almost periodic, 442
amorphous cellular automaton, 249
amorphous computing systems, 243
analog computers, 734
analog devices, 24
analog instruments, 735
analogue, 348
analytic engine, 569
Analytical Engine, 33
Angluin, D., 248
anthropic
  self-sampling, 414
anthropic principle, 680, 694, 702
apparent design, 552
apparent randomness, 8
approximate
  theory, 411, 412
Arithmetic Logic Unit, 717
arithmometer, 24
artificial life, 360
artificial neural network (ANN), 351–354, 356
Arvind, D. K., 246
ASCC, 38
Aspect's experiments, 12
Aspect, A., 11
Aspnes, J., 248
assumption
  computability, 413
  self-sampling, 414
assumptions

theory of everything, 412
asynchronous stationary amorphous
    computing system, 250
ATP batteries, 719
Atwood, B., 246
automaton theory, 731
Axiom of choice, 153
axiom systems, 711

Babbage, C., 24, 569
baby
    universes, 404
background set, 194
Banach, S., 157
Banach-Tarski paradox, 157
Bar-Hillel, Y., 354
Bar-Yehuda, R., 251
Barrow, J. D., 415
Barwise, J., 354
basic representations, 510, 514
Batterman, R., 522
Bauer, A., xvii, xli
Beal, J., 246
beam routing, 257
Beavers, A., xv, xli, 351
Bekenstein, J., 18
Bell experiments, 11
Bell states, 612
Bell's inequality, 13
Bell's theorem, 595, 699
Bell, J., 11, 576
Bell, J.S., 583
Benioff, P., 577
Bergson, H., 716
Bernoulli numbers, 569
BHK interpretation, 150
Bialek, W., 679
Big Bang, 527
big crunch, 404
billiard ball computer model, 675
Binary latching, 30
binary places, 774
binary sequence
    temporal, 402, 412
biological creativity, 277
biological differentiation, 682

Bishop, E., 143
Bit, 597
bit-string
    ontology, 402, 412
Black holes, 18
black matter, 527
black ravens, 401
Bletchley Park, 39
Blum, L., 522
Bohr, N., 14, 794
Bolognesi, T., xvi, xviii, xxii, 4
Bombelli, L., 452
Boole, G., 570
boolean algebra, 718
Boolean logic, 570
Borel, E., 579
Borroughs, 25
Bostrom, N., 415
Brenner, S., 277
Breuer, T., xxiv, xli
Bringsjord, S., xl, xliv
broadcast protocol, 251
Brouwer, L.E.J., 58, 144
Brukner, Caslav, 584
Bullis, K., 246
Burgin, M., xl
Bush, V., 38
Busy Beaver function, 279

Cabello, A., xxiv, xli, 3
Calculating Space, 2
Calude, C., xiv, xliii, 2, 4, 298, 631,
    673
canonical form A, 51
canonical form C, 52
Cantor set, 696
Cantor set topology, 696
Cantor space, 697
Cantor, G., 481
Carnap, R., 354
Carriage of tens, 30
Cartesian space, 766
Cauchy, A., 146
causal set, 452
causal set program, 452
causality, 207, 777

Index

causet, 452
Cawley, J., 715
cellular automata, 60, 257, 258, 559, 674, 678, 684, 695, 713, 765
  rule 30, 700
cellular automaton, 440, 442, 575, 733
cellular automaton, elementary rule 110, 470
Chaitin's Ω, 285
Chaitin, G., xvi, xxxvii, 2
Chalmers, D.J., 71
Champernowne, D. G.
  number, 404
Chandio, F. H., 246
Chomsky's classification, 261
Chuang, I.A., 605
Church's thesis, 47, 48, 53, 79, 80, 87, 674
Church's thesis limit, 674
Church, A., xvii, 47, 48, 78, 79, 92, 108, 159, 257, 480, 571, 614, 787
Church-Turing thesis, 79, 92, 100, 172, 438, 480
  Extended, 92
Cirac, I., 577
circuit, 353, 356, 361
circular Turing machines, 258
classical computer, 675
classification, 443
Client-cloud computing, 164
coarse-graining, 489
coding
  universe, 410
coding theorem, 5
collapse
  wave-function, 402, 404
collision-based computing, 258
collisions, 257
Colmar (de), T, 24
Colossus, 38
combinatorial circuits, 254
communication graph, 250
compact
  representation, 402
compactification, 404
Compatible experiments, 596

complete
  theory of everything, 406, 409, 414
completeness, 172
complex adaptive system, 355
complexity, 568
  Kolmogorov, 407
  parameter, 412
  program, 407
  theory of everything, 407
complexity theory, 355, 615
compression, 401
compression algorithm, 703
comptometers, 25
computability
  assumption, 413
computable, 436
  observer process, 413
  universe, 400, 404, 413
computable processes, 8
computable universe hypothesis, 8, 483
computation, 63, 347–350, 356
computation-universality, 683
computational philosophy, 349, 360
Computational Representation Theorem, 159
computational universality, 247
computationalism, xxxviii, 1
computer, 436
computor, 106
Comrie, L., 26
conditions
  initial, 402, 406
configuration, 442
confluence, 444
consciousness, 617
conservation principles, 683
consistency, 610
  mathematical, 407
constructive mathematics, 143
Constructor functions, 84–86, 90, 91
content-addressable memory, 352
context sequence, 194
context-independent process, 195
Contextual computer, 597
Continuity principle, 154

continuous
  universe, 412
contracting universe, 682
Conway's game of Life, 266, 470
Conway, J., 469
Cook, M., 440
Cooper, B., xvii, xxxviii, xli
Coore, D., 244, 246
Copeland, J., 70
correspondence
  theory, 410
cosmogony problem, 689
cosmological
  models, 404, 405
cosmological arrow, 682
cosmology, 680
Costa, J.F., xl
countable
  model, 412
  universe, 412
CP-violation, 681
Crick, F., 6, 277, 719
Csirel'son, B.S., 613
CSP, 167
Curry-Howard isomorphism, 720, 724
cybernetics, 785
cyclic tag systems, 258
cyclotrons, 258
C#, 164

Darwin evolution, 693
Darwin's theory of evolution, 278
Darwin, C., 7, 277, 684
Darwinian evolution, 7, 716
data, 401
Davies, P. C. W., 673
Davies, P. C. W., 415
Davis, M., 49, 55, 439
de Bruijn diagrams, 258
De Mol, L., xli
de Morgan, A., 570
de Ruyter, R., 673, 679
Deacon, T., 348, 349, 356
decay, 204
decision making, 401
decoherence, 404, 716

deductive
  reasoning, 401
degree, 439
  intermediate, 439
Delahaye, J.-P., xliii
DeMol, L., xvii, xxx
Dennett, D., 361
dequantizing algorithms, 682
Dershowitz, N., xli
description, 401
description size, 3
determinism, 779
deterministic
  model, 402
  theory, 402
deterministic chaos, 466
Deutsch, D., xvii, xli, 3, 577, 615, 689
Deutsch-Jozsa algorithm, 615
diagonalisation, 605
diagonalization method, xxxiv
Difference Engine, 28
difference engine, 569
differential equations, 739
digital, 348
digital computer, 736
digital computers, 734
digital information, 4
digital particle, 748
digital software, 278
digitalization, 742
Dijkstra, E., 167
diophantine equation, 674
Diophantine equations, 77
direct
  observation, 410
discovery
  scientific, 401
discrimination of states, 608, 612
dissipative cellular automata
  algorithms, 683
dissipative lattice gas algorithms, 683
distinguishable systems, 483
diversity measure, 355
DNA, 277, 281, 297, 719
DNA replication, 6
Dolnick, E., 415

domain
  overlap, 412
double Johnson noise, 707
Dretske, F., 354
Dubucs, J., xliv
duration, 204
duration function, 204
duration sequence, 205
dynamic associative network, 352
dynamic associative network (DAN), 351–356, 361
Dyson, F., 794

ECA with memory, 258
Eckert, J.P., 573
Eckert, P., 38
EDSAC, 38
EDVAC, 38
Effective method, 78, 79, 81, 84, 86, 88, 89, 92, 92
effective product, 514
egocentric
  model, 403
Ehrenfeucht, A., xxxiii
Einstein, A., 11, 13, 527, 576
Eisenstat, D., 248
electronuclear forces, 4
elementary particles, 5
elementary substructure, 443
enabled reaction, 192
ENIAC, 38
ensembles
  non-statistical, 492
  statistical, 491
entity, 194
entropy, 350, 352–356, 358, 359
Entscheidungsproblem, 47, 50, 57, 78, 435
epistemology, 402
  information theory, 403
EPR experiment, 11
erratum sheet story, 707
Ethernet, 683
Euclidean geometry, 556
Everett's many-worlds interpretation, 404

Everett-Wheeler interpretation, 681
Everitt, H., 404
evo-devo, 297
evolution
  state, 406
exhaustive search, 283
experience, 411
explanation, 73, 401
extensions
  theory of everything, 411

falsifiability, 414
  Popper, K., 407
Feynman, R., 1, 560, 578, 675, 683
finite automaton, 261
finiteness problem, 50, 53
first-order logic, 47, 50
fitness, 281
flip-flops, 770
Floridi, L., 351, 360
flow arc, 202
flying amorphous computer, 252
formal language, 347, 348
formalists' motto, 353
FORTRAN, 280
foundation
  science, 414
François, B., xliv
Fredkin, E., 365
Fredkin, Ed, xvi, xxiv, 3, 266, 442, 452, 474, 482, 483, 575, 631, 673, 675, 683, 792
free will, 13
Frege, G., 570
Friedman, H.M., 80
functional description, 72
fundamental
  theory, 412
future
  observations, 411
fuzzy logic, 626

Gödel, xxix
Gödel's incompleteness theorem, 677, 711, 715

Gödel, K., 47, 78, 79, 91, 108, 159, 436, 570, 614, 719, 723
Gödelization, xxxi
Gács, P., 458
Gandy machine, 441
Gandy machines, 112
Gandy, R., 48, 159, 441
Gardner, M., 415
Gell-Mann, M., 415, 687, 707
general recursive function, 48
general relativity, 1, 677, 740
general-purpose computation, 33
generalization by postulation, 51
genetic regulatory network, 195
genomics, 719
geocentric
  model, 403
German, A., xxxvii, 673, 729
Gibson, W., 570
Goldreich, O., 251
Goodstein's theorem, xxxi
gravitation, 527, 742
gravitational path integral, 453
Greif, I., 165
grid points, 782
Grim, P., 351, 360
GRWP, 716
Grünwald, P. D., 415
Gurevich, Y., 81, 88

Halting problem, 152, 300, 439, 572
Hanson, Ch., 244, 246
hardware random generator, 707
Harper, C. L., 415
Harrison, C., xli
Harrison, E., 415
Harvard Mark I, 38
Haugeland, J., 348, 353
Hayes, B., 676, 692, 714
heliocentric
  model, 403, 404
heliocentric model, 404
Herbrand, J., 159
Herschel, 28
Hewitt, C., xxiii, xli
Heyting, A., 144

hidden variables, 5
Higgs boson, 725
Hilbert space, 627
Hilbert's program, 723
Hilbert's programme, 787
Hilbert's Second Problem, 77
Hilbert's Tenth Problem, 77
Hilbert, D., 77, 78, 435, 493, 554, 559, 570, 724
Hintikka, J., 354
Hoare, T., 169
Hofstadter, D., 166, 676
Holevo, A.S., 606
Hollerith, H., 24
holographic principle, 18, 475
Homsy, G., 244, 246
Hoyle, F., 244
human mind, 617
Hutter, M., 8, 415, 522
hybrid computation, 682
Hypercomputation, 93
hypercomputation, xxxiv, xxxviii

IBM, 24
idealism, 412, 712
implementation, 63, 66
incompleteness theorem, 570
inconsistency-robust Logic
  Programming, 167
Independence number, 603
induction, 401
  universal, 414
inductive
  reasoning, 401
inertia, 683
inference map, 607
infinite
  universe, 412
infinitesimal, 146
infinitesimal calculus, 553
inflation
  model, 405
information
  algorithmic, 407
information theory, 350, 354, 402, 414
  algorithmic, 401

epistemology, 403
information-processing machine, 734
information-theoretic teleodynamics, 353
information-theoretic teleodynamics (ITT), 350, 351
inhibitor, 191
inhibitor arc, 202
initial
   conditions, 402, 406
initial state, 194
input alphabet, 200
integer base, 705
Intel, 708
intensional, 66
intentional stance, 75, 361
interactive process, 194, 204
internal scaling dimension, 468
interpretation
   many-worlds, 404
interval arithmetic, 517
intuitionistic logic, 144
Inverse Relationship Principle (IRP), 358
Inverse Relationship Principle (IRP), 358
Inverse Relationship Principle (IRP), 354, 357, 359
invertible cellular automata, 683
it from bit, 15
Itai, A., 251

Jacquard loom, 569
Java, 164
JavaScript, 164
Johnson G., 673
joint complexity, 299
Joosten, J.J., xliv
Jozsa, R., 615
jump to universality, 440

Kahn, J. M., 246
Kant. E., 559
Kari, 683
Katz, R. H., 246
Kelley, J.L., 83

Kelly, K., 522
Klaus, S., xli
Kleene tree, 155
Kleene, S., xvii, xxxviii, 48, 79, 151, 155, 159, 257, 614, 787
Knight, T. F., Jr., 244, 246
knowledge
   objective, 415
   subjective, 415
   theory, 403, 412
Knuth, D.E., 80, 83
Kochen and Specker theorem, 15
Kolmogorov
   complexity, 407
Kolmogorov complexity, 6
Kolmogorov, A., 144, 724
Kolmogorov-Chaitin complexity, 6
Kreisel's criterion, 517
Kreisel, G., 480
Kronecker, 713
Kronheimer, E., xxiv
Kurzweil, R., 246

lambda calculus, 720
   typed, 720
Lancaster, D., 705
Landauer's principle, 18
Landauer, R., 18, 438, 709
Lange, T., 352
language
   program, 409
Lardner, D., 27
Last, M., 244, 246
Lattimore, T., 415
law
   of nature, 401
law of cancellation, 148
Law of excluded middle, 144
lawfulness hypothesis, 529
lawless, Universe, 526
lawlessness hypothesis, 530, 535
Laws for Actors, 165
laws of computation, 553
laws of the Universe, 528
Lawvere, W.F., 147
LCAU21, 264

learning
  machine, 401
Leggett, A., xxxvii, 695, 716
Leibnitz, G., 24
Leibniz, G., 1, 146
length
  minimize, 411
Lerman, M., 441
Levin's universal distribution, 5
Levin, L., 2
Levin, L.A., 458
Li, M., 415
Liebowitz, B., 244, 246
limited
  observer, 410
Link, J. R., 246
Lloyd, S., xviii, xxi, xli, 3, 452, 577, 687, 694, 718
locality, 683
localization
  observer, 403, 405, 406, 408, 413, 414
  within universe, 409
localizer, 244
Loewenheim-Skolem
  theorem, 412
Logic Programming, 167
Loll, R., xxxvii
Lorentz contraction hypothesis, 740
Lorentz transformations, 768
Lovász number, 603
Lovelace, A., xvii, 36, 569
Lucas, J.R., xl

machine
  learning, 401
  Turing, 409
macroscopic quantum coherence, 699
many-worlds
  interpretation, 404
many-worlds interpretation, 681
Marconi, 701
Margenstern, M., xvi, xxiii, xxxiii, xli
marking, 203
Markov, A.A. (the younger), 79
Marshall, J., xliv

Martínez, G.J., xvi
Martin, A.M., 673
Martin-Löf random sequence, 530
Martinez, G.J., xxxix
materialism, 712
Mathematica, 685, 705–707, 721, 722
mathematical
  consistency, 407
Mathematical Structure Hypothesis, xxxix
Mauchly, J., 573
Maximally incompatible experiments, 597
Maxwell equations, 740
McAllister, J., 8
McCarthy, J., 677
McClelland, J., 352
MDL
  principle, 401, 414
meaningful
  theory, 414
measurement from inside, 610, 612
measurement function, 207
measurement problem, 5
mechanical
  universe, 400
mechanics
  Newton, I., 402
mechanistic theories, 480
Meinberg, F., xliii
memory, 352, 356
Menabrea, L., 39
Mendel, G., 6
Mermin, N. D., 590
metabiology, 278
Meyer, D., 452
Microprogramming, 30
Milner, R., 171
minimize
  length, 411
minimum-size program, 282
Minkowski, H.
  space, 412
Minsky, M., 53, 55, 675, 677
mitochondria, 719
MML

principle, 401, 414
model, 401
  countable, 412
  deterministic, 402
  egocentric, 403
  geocentric, 403
  heliocentric, 403, 404
  inflation, 405
  standard, 407
model selection
  rational criterion, 403
models
  all-a-carte, 404
  bogus, 403
  cosmological, 404, 405
  for our world, 403
  perfect, 404, 406
  physics, 402
  speculative, 403
module, 207
molecular communication, 253
molecular computation, 718
molecular computing, 718
Monte Carlo simulations, 706
Monte-Carlo method, 781
Moore's law, 714
morphogenesis, 277
Mosconi, J., xliv
MSDOS, 264
multiple
  theories, 412
multiverse
  theories, 404
  universal, 404
  universe, 415
mutation, 281
mutation distance, 282
Myrheim-Meyer dimension, 456

Nagpal, R., 244, 246
nano-bots, 246
nano-machines, 253
Napier's Bones, 24
Napier, J., 569
natural computing, 189
Natural Deduction, 183

natural selection, 7
nature, 347–350, 356, 361
  law, 401
Newman, M., 38
Newton, I, 146
Newton, I., 1, 527, 693
  mechanics, 402
Newtonian mechanics, 441
NextStep, 264
Nielsen, M.A., 605
Nikoletseas, S., 244
NKS, 675, 686, 701, 704
noise
  random, 406
nomological, 348, 349
non-commuting operators, 13
Non-Computable Universe
  Hypothesis, xl
Non-contextual computer, 597
non-materialistic view, 712
non-standard analysis, 147, 697
non-standard arithmetics, 711
non-Turing computable functions, 556
Non-Turing Computable Universe
  Hypothesis, xxxix
non-uniform models, 254
noncontinuous processes, 735
Nondeterministic Turing Machine, 162
normal
  number, 404
normal form, 52
normal form theorem, 53
normal proof, 720
NP complete, 674
NP completeness, 674
number
  Champernowne, D. G., 404
  normal, 404
number of planar triangulations, 475
Nygaard, K., 165

objective
  knowledge, 415
  theory of everything, 406, 409
  world, 412, 415

Objective C, 164
observation, 401
  direct, 410
  process, 410, 414
  string, 402
  time-series, 402
observation language, 445
observations
  future, 411
observer, 445
  limited, 410
  localization, 403, 405, 406, 408, 413, 414
  perception ability, 410
observer process
  computable, 413
Ockham's razor
  justification, 414
  principle, 407, 413
  science, 407, 413
  validity, 414
ontology
  bit-string, 402, 412
OpenStep, 264
oracle, 300
oracle machines, xxxiv
oracles, 80, 84–86, 89–92, 278
oracles for real numbers, 486, 507
  complete, 507
  nested, 507
  standard decimal, 486
ordering fraction, 456
Ortiz, G., xxxvii, 673, 700
oscillating
  universe, 404
Overflow, 30
overlap
  domain, 412

P vs. NP problem, 714
Pachner move, 464
Packard, N., 676
Page, S., 355
paradigm shift, 568
parameter
  complexity, 412

parametric
  theory, 412
Park, M., 246
Parkinson's law, 724
partial
  theory, 411
particle physics, 681
Pascaline, 24
Peano arithmetic, xxxi, 711
Peano, G., 722
Penrose, R., xxxviii, xl, xliii, 529, 559
Pentium chip, 707, 717
perception ability, 406
  observer, 410
perfect
  models, 404, 406
perfect randomness, 707
Petrů, L., 248–253
Petri net, 202
Phdrus, 709
phasespace, 443
philosophy of computation, 348
philosophy of mind, 618
physical computation, 64
physical laws, 528
physical machines, 63
physical models, 479
  basic representations of, 514
  computable, 482
  determined by partial recursive functions, 504
  epimorphic, 505
  faithful, 480, 506, 514
  induced by a function, 516
  isomorphic, 503
  non-negative integer, 504
  normal forms of, 513
  observationally equivalent, 505
  reduced, 505
physical reality, 11
physics
  models, 402
Piccinini, G., 66
pinball machines, 441
Pipelining, 30
Pirsig, R., 709

Pister, K. S. J., 244, 246
place, 202
Planck area, 18
Plank, M., 14
plankalkül, 791
Plato, 529, 709
pluralism, 412
Podolsky, B., 11
Poincaré, H., 531
Polling, 30
polynomial primitives, 674
polynomial time, 674
Popper, K.
  falsifiability, 407
population protocols, 248
Post production systems, 52
Post tag systems, 787
Post's formulation 1, 54, 56
Post's problem, 439
Post's thesis I, 49, 50
Post's thesis II, 49
Post, E.L., 45, 79, 84, 159, 257, 571, 787
postdictive
  success, 407
power
  predictive, 403, 405, 406
PowerPC, 695
pre-Socratics, 688
precision amplitudes, 674
predicate logic, 436
prediction, 401
predictive
  indistinguishable, 408
  power, 403, 405, 406
Presburger arithmetic, 437
Prevost, H., 36
Principia Mathematica, 53, 722
principle
  MDL, 401, 414
  MML, 401, 414
  Ockham's razor, 407, 413
  simplicity, 407
Principle of Comput. Equivalence, 440
principle of micro-affinity, 147

principle of sufficient reason, 698
priority argument, 439
probabilistic
  theory, 411
probability
  subjectivist interpretation of, 492
process, 437
  complete, 446
  intermediate, 446
  observation, 410, 414
  undecidable, 446
Process calculi, 171
program
  complexity, 407
  language, 409
program extraction from proofs, 151
program-size complexity, 299
Projective measurement, 596
proof normalization, 720
proof systems, 722
proof theory, 435, 437
protein folding, 7
Pulse-shaping, 30
punched card, 788
Putnam, H, 69
Pythagoras, 568
Pythagoras syndrome, 568

quantum
  universes, 404
quantum algorithms, 682
quantum cellular automata, 559
quantum computation, 555, 568, 718
quantum computer, 4, 577, 674, 675, 687
quantum computing, 679, 707
quantum decoherence, 19
quantum entanglement, 613
Quantum Field Theory, 618
quantum gravity, 17
Quantum Information, 617
quantum information, 583, 679, 682
quantum measurement, 607
quantum measurement paradox, 681
quantum mechanics, 1, 576, 623, 674, 677, 699, 707

ensemble interpretation of, 496
many-worlds interpretation of, 490
Quantum Meta-Language, 617
Quantum Object-Language, 617
quantum randomness, 706
quantum reality, 11
quantum system, 675
quantum teleportation, 615
quantum theory, 623
qubit, 583, 597
quit, 627
quorum sensing, 253

radioactive decay, 707
RAND corporation, 707
random
    noise, 406
random number, 691
random number generator, 691
random numbers, 706
Rathmanner, S., 415
rationalism, 702
Rauch, E., 244, 246
RCA, 708
reachability, 443
reachability problem, 52
reactant, 191
reaction, 191
    enabled, 192
    result, 192
reaction system, 194
real algebra, 722
real-world realizer, 151
realism, 412
reality, 712
realizability, 150
realizability relation, 150
realizer, 150
reasoning
    deductive, 401
    inductive, 401
Rechnender Raum, 2
recursion, 356, 359
Recursive functions, 78–80, 787
    Oracular, 90, 92
    Partial, 79, 90–92

reductionism, 412
regular expressions, 260
Reisig, W., 86
Repeatable experiment, 596
representation
    compact, 402
result of reaction, 192
result sequence, 194
reversibility, 683
Riemann hypothesis, 699
RNA, 701
Robinson, A., 147
Rojas, R., 787
Rosen, N., 11
Rosen, R., 481, 503
Rosser, J.B., 48, 614, 615
Rowland, T., xxxvii, xliv
Rozenberg, G., xxxiii, xxxix
rule, 401
Rumelhart, D., 352
Ruppert, E., 248
Russell's system, 179
Russell, B., 570, 712, 720, 722, 724

Sailor, M. J., 246
Salomaa, A., xxxviii, xl
Schönfinkel, M., 787
Schickard, W., 24
Schmidhuber, J., xxi, xxiv, xxxvii, 3, 415, 452, 787
Schrödinger's cat, 597, 716
Schrödinger, E., 493
science
    foundation, 414
    Ockham's razor, 407, 413
scientific
    discovery, 401
Searle, J., xl, 444
second law of thermodynamics, 685
selection principle
    theory of everything, 408, 410, 415
self-assembling, 7
self-sampling
    anthropic, 414
    assumption, 414
    universal, 414

Seligman, J., 354
semantic, 354, 360
semantic information, 351, 354
semantics, 159
semidecidable, 436
separable
  space, 412
sequence
  binary, 402, 412
Sequential postulates, 81, 89
SETI, 701, 703, 716
Shah, M. J., 246
Shannon, C., 350, 573
Sharp measurement, 596
Shoenfield, J.R., 79
Shor's algorithm, 717
Shor, P., 577
Shore, R.A., 80
Sieg, W., xvii, xxx, xli, xliv, 48, 159
simple mapping account (SMA), 69
simplicity
  principle, 407
simulation, 91, 92, 675
singularity, 246
SKI combinators, 787
smart dust, 246
Smolin's baby universe theory, 404
Smolin, L., 404, 415
SMP, 685
Soare, R.I., 47
solipsism, 412
Sorkin, R.D., 452
space
  Minkowski, H., 412
  separable, 412
space-time, 149
speckled computing, 246
Sprevak, M., 72
sprinkling, 454
SSEM, 38
standard
  model, 407
Standish, R., 415
state, 194
  evolution, 406
  initial, 194

transition, 192
State machine, 165
state sequence, 194
statistical mechanics, 680
Statman, R., 521
steampunk, 570
step, 203
step sequence, 203
Sterling, B., 570
Stove, D. C., 415
string
  observation, 402
  theory, 404, 407
Structures, first order, 81–83
subjective
  knowledge, 415
  theory of everything, 406, 410
  universe, 400
success
  postdictive, 407
super-Turing computing power, 254
Susskind, L., 475
Sussman, G. J., 244, 246
Sutner, K., xvi
Svozil, K., 631, 673
Swade, D., xvii, xli
switching circuit, 197
synchronous amorphous computing
  system, 249
synthetic differential geometry, 146
Szilárd, L., 18
Szudzik, M., xxii, xliii, 8, 479

T-cells, 720
tabulator, 24
tag system, 51
Tarski's result, 674
Tarski, A., 157, 624, 674
Tegmark, M., 415, 452
teleodynamic, 348, 349, 351, 356, 360, 361
teleodynamic game page, 360
teleological, 348, 350, 360
teleonomic, 349, 351, 360, 361
temporal
  binary sequence, 402, 412

Tesla, 702
Teuscher, C, xli
Teuscher, C., xvii, xxiii
The Algorithmic Information
  Hypothesis, xl
the Bombe, 38
The Cellular Automaton Hypothesis,
  xxxix
The Computable Universe
  Hypothesis, xxxviii
The Computational Pragmatic
  Hypothesis, xxxix
The Hyper-computable Hypothesis, xl
The Informational Universe
  Hypothesis, xxxix
The Non-Turing Computable
  Universe Hypothesis, xl
The Random Universe Hypothesis, xl
The Standard Quantum Universe
  Hypothesis, xli
theorem
  Loewenheim-Skolem, 412
theories
  multiple, 412
  multiverse, 404
theory, 72
  approximate, 411, 412
  correspondence, 410
  deterministic, 402
  fundamental, 412
  knowledge, 403, 412
  meaningful, 414
  of nothing, 404, 414
  parametric, 412
  partial, 411
  probabilistic, 411
  string, 404, 407
theory of automatons, 732
theory of everything, xxxix, 4, 401,
  414
  assumptions, 412
  best complete, 411
  complete, 406, 409, 414
  complexity, 407
  conventional, 414
  extensions, 411

formal quest, 414
objective, 406, 409
selection principle, 408, 410, 415
subjective, 406, 410
universal, 404, 408
thermodynamic, 349, 353
thermodynamical equilibrium, 700
thermodynamics, 623, 681, 701
time-series
  observation, 402
timed probabilistic automaton, 253
Toffoli gate, 683
Toffoli, T., xxxvii, 266, 673, 676, 700
token, 347, 348
topologies
  computable, 507
  effective, 507
touring ant, 462
transducer, 442
transition, 202
transition function, 200
transition system, 200
transitive percolation, 456
trinet, 464
trinet mobile automata, 464
true randomness, 13
Turing barrier, 438, 682
Turing computable functions, 559
Turing machine, 5, 47–49, 54, 56, 59,
  60, 542, 571, 674
  universal, 409
Turing machine, elementary, 461
Turing machines, 79, 80, 88, 92
  Oracular, 85, 86, 92
Turing universality, xxxviii
Turing's Model, 159
Turing's thesis, 47, 48, 79
Turing, A., 719, 723
Turing, A.M., 23, 47, 48, 78, 79, 85,
  90, 92, 112, 159, 257, 277, 347, 352,
  353, 436, 480, 558, 568, 614, 615
Turing, Alan, 583
Turing-complete, 352
turmite, 470
Turner, R, xxii
Turner, R., xli

TX2, 706
type-two effectivity, 507
typing monkeys, 579

Ulam, S., 575
Unbounded Nondeterminism, 167
unconventional computation, 718
unconventional computing, 257
undecidability, 172
uniformity, 683
universal
   induction, 414
   multiverse, 404
   self-sampling, 414
   theory of everything, 404, 408
   Turing machine, 409
universal computer, 437
universal programming language, 297
universal Turing computer, 3
universal Turing machine, 674
universal-computing CA, 266
universe
   coding, 410
   computable, 400, 404, 413
   continuous, 412
   countable, 412
   infinite, 412
   mechanical, 400
   multiverse, 415
   oscillating, 404
   subjective, 400
   visible, 405
universes
   baby, 404
   quantum, 404

validity
   Ockham's razor, 414
van Leeuwen, J., 254
variables
   inter-universes, 404
Velupillai, K., xiv
Velupillai, V., xxxix, xli
Vinge, V., 245
visible
   universe, 405

Vitányi, P. M. B., 415
von Neumann's normalisation, 546
von Neumann, J., xxvi, 15, 33, 436, 570, 623, 678, 684, 724, 790

Wallace, C. S., 415
Wang, H., 439
Warneke, B., 244, 246
Watson, J., 6, 277, 719
wave-function
   collapse, 402, 404
Weierstrass, K., 146, 154
Weihrauch, K., 522
Weihs, H., 12
Weinfurter, Harald, 590
Weiss, R., 244, 246
Weyl, H., 49
Wheeler, J.A., 1, 404
Whitehead, 722, 724
Wiedermann, J., xvi, xxiii, xli, 248, 249, 251–254
Wiener, N., 354
wireless sensory network, 248
Wittgenstein, L., 178
Wolfram
   Principle of computational equivalence, 674, 701, 716
Wolfram rule 30, 9
Wolfram's rule 110, 266, 440
Wolfram, S., xvi, xxiv, 3, 259, 415, 440, 442, 451, 458, 482, 483, 522, 575, 631, 673, 676
Wong, K. J., 246
world
   models, 403
   objective, 412, 415
worlds
   many, 404

XOR, 678

Z-series machines, 38
Zambelli, S., xxxix, xli
Zeilinger, A., 12
Zenil, H., xvi, xxiv, xxxiii, xl, 522, 534, 631, 673, 700, 729, 790

Zeno, 555
Zeno machine, 542
Zeno of Elea, 554
Zermelo-Fraenkel set theory, 723
Zizzi, P., xxix, xxxiii, xxxviii
Zoller, P., 577

Zukowski, M., xxi
Zuse, H., 790
Zuse, K., xvi, xxxix, 38, 415, 442, 452, 482, 483, 572, 787
   Calculating Space, 729
   Rechnender Raum, 729

Lightning Source UK Ltd.
Milton Keynes UK
UKHW021951030621
384893UK00002B/158